PROCESS/ INDUSTRIAL INSTRUMENTS AND CONTROLS HANDBOOK

Other McGraw-Hill Books of Interest

PROCESS/ INDUSTRIAL INSTRUMENTS AND CONTROLS HANDBOOK

Douglas M. Considine, P.E.   Editor-in-Chief

Registered Professional Engineer (California)
in Control System Engineering

Fourth Edition

McGRAW-HILL, INC.

New York San Francisco Washington, D.C. Auckland Bogotá
Caracas Lisbon London Madrid Mexico City Milan
Montreal New Delhi San Juan Singapore
Sydney Tokyo Toronto

Library of Congress Cataloging-in-Publication Data

Process/industrial instruments and controls handbook / Douglas M. Considine. —
4th ed.
    p.   cm.
   Includes index.
   ISBN 0-07-012445-0 (alk. paper)
    1. Process control—Handbooks, manuals, etc.   2. Automatic
control—Handbooks, manuals, etc.   3. Engineering instruments—
Handbooks, manuals, etc.   I. Considine, Douglas M.
   TS156.8.P764   1993
   629.8—dc20   93-20567
                                              CIP

  4 5 6 7 8 9 0 DOC/DOC 9 9 8 7 6 5

ISBN 0-07-012445-0

*The sponsoring editor for this book was Harold B. Crawford, the editing
supervisor was Kimberly A. Goff, and the production supervisor was
Donald F. Schmidt. This book was set in Times Roman by North Market
Street Graphics.*

Printed and bound by R. R. Donnelley & Sons Company.

CONTENTS

SECTION 1
INTRODUCTORY REVIEW

Industrial instrumentation is becoming increasingly interdisciplinarian in nature. Indeed, it is difficult to define what really is the hard core of instrumentation and control technology. Prior to the introduction of solid-state electronics nearly a half-century ago, the topic was relatively easily organized in terms of (1) the variables that require measurement and (2) the means used to control those variables automatically. Temperature, pressure, flow, and liquid level were the major variables, followed by perhaps a dozen less frequently encountered variables such as chemical composition, viscosity, density, humidity, moisture content, and so forth. In the area of discrete-piece manufacturing, certain geometric and motion sensors predominated. The means used for connecting measurement sensors with the final controlling devices were at one time regarded as adjuncts. These matters required a practical knowledge of electricity, pneumatics, and the fundamental principles of measurement and control. Essentially, this comprised the grouping of knowledge that could be called instrumentation and control engineering, technology, or, maybe, science.

With the addition of electronics, which fully transformed former electrical circuitry and made better, smaller, more convenient devices possible, a new area of knowledge had to be mastered by the control engineer. An earlier edition of this handbook included long discussions of basic electronics, but, progressively, electronics became part of the "woodwork" so to speak, such that, in response to severe space limitations, such a discussion would be out of place in the present edition.

Only a few decades ago, computer technology became an intimate part of control system engineering. Again, to accommodate this new body of knowledge, material of a tutorial nature was included in a preceding edition of this handbook. Today a vertical handbook on instrumentation and control cannot fill such a knowledge gap adequately. The control engineer requires the support of more specifically oriented references on computer technology.

With the introduction of the concept of computer integrated manufacturing (CIM) in the 1970s, the thirst for information had no limits. This required much greater emphasis on many ancillary values of data acquisition (those beyond the data needed strictly for immediate control purposes). This definitely falls within the hard core of instrumentation, and it is hoped that this subject is covered adequately in the present edition. This is simply an extension of what one may interpret as "field instrumentation." But a CIM environment (token, partial, or nearly complete) encompasses areas of data usage that involve statistical quality control (SQC) or statistical process control (SPC), in the past not necessarily within the purveyance of the typical control engineer. In this handbook edition these topics are defined and described succinctly. However, beyond a certain degree of expertise, the control engineer again requires the support of more specifically oriented references on quality control technology, a topic that dates back many decades, particularly in the discrete-piece manufacturing industries.

Concurrently with the introduction of CIM environments came the need to transmit (communicate) instrumentation data and management information to users whose needs go well beyond the information necessary to control a steady operating state of processes or machines. The science of communications (networking) predates the CIM concept considerably. The translation of networking techniques, previously proven for data transmission in other fields, to the communication of instrument and control system data has not been easy. Millions of skill hours have gone into a study of this problem, and with considerable success, but most experts currently agree that they are far from reaching the ultimate solution. Networking thus becomes one more body of knowledge that must be added to the control engineer's expertise. Networking principles and examples are given in several portions of this handbook as they relate to specific situations. But again, the control communications engineering specialist also requires the support of references on networking technology.

CONCEPT OF DISTRIBUTED KNOWLEDGE

It has been mentioned that certain special topics, such as SPC and SQC, are not described by separate articles, but rather are "distributed" among articles on related topics. This also holds true for a number of other important topics, including some of the major trends toward the future. These would include such topics as open-system architecture, self-diagnostics, smart transducers and transmitters, process modeling and simulation, artificial intelligence, and fiber optics in data networks and sensors.

This handbook was designed with the control engineers in the process industries and the discrete-piece manufacturing industries equally in mind. Following is a "walk-through" of the 50 technical articles, supported by 775 illustrations and 70 tables, contained in this handbook.

SECTION 2: CONTROL SYSTEM FUNDAMENTALS

Control Principles

As was observed by readers of earlier editions, this has been one of the most widely used articles in this handbook. For the fourth edition it has been carefully updated to reflect the technological changes of the past few years. This article is intended not only for individual study, but also for use by groups of scholars in college, technical school, and in-plant training programs. The article commences with the nontheoretical analysis of a typical process or machine control system. Discussed are process reaction curves, transfer functions, control modes, and single- and multicapacity processes—relating control characteristics with controller selection.

Techniques for Process Control

This article reviews from both the practical and the theoretical viewpoints the numerous advancements achieved in solving difficult control problems and in improving the performance of control systems where fractional gains in response and accuracy can be translated into major gains in yield and productivity. This article is the logical next step for the instrumentation and control engineer who understands the fundamentals of control, but who desires to approach this complex subject in a well-organized mathematical and theoretical manner. When astutely applied, this advanced knowledge translates into very practical solutions. The author proceeds in an orderly manner to describe state-space representation, transfer-operator representation, the mathematics of open-loop, feedback, feedforward, and multiple-loop control, followed by disturbance representation, modeling, the algebraics of

PID (proportional-integral-derivative) design, adaptive control, pattern recognition, and expert systems. The techniques of least squares, batch parameters, the Kalman filter, recursive parameter identification, and projection also are described.

Basic Control Algorithms

Continuous process control and its counterpart in discrete-piece manufacturing control systems traditionally were developed on an analog base. Experience over the past few decades has shown that digital control provides many advantages over analog systems, including greater flexibility to create and change designs on line, a wider range of control functions, and newer functions, such as adaptation. But digital computation is not naturally continuous like the analog controller. The digital approach requires sophisticated support software.

This article addresses the basic issues of carrying out continuous control in the digital environment, emphasizing the characteristics which must be addressed in the design of operationally natural control algorithms. The author describes number systems and basic arithmetic approaches to algorithm design, including fixed-point and floating-point formats. Lag, lead/lag, and dead-time calculations required in the development of a basic control algorithm are presented. Also included are descriptions of quantization and saturation effects, the identification and matrix-oriented issues, and software and application issues. A closing appendix details the generalized floating-point normalization function.

Safety in Instrumentation and Control Systems

Never to be taken lightly are those features that must be engineered into control systems on behalf of protecting plant personnel and plant investment, and to meet legal and insurance standards. This is a major factor of concern to users and suppliers alike. Even with efforts made toward safety design perfection, accidents can happen.

The author of this article carefully defines the past actions and standards that have been set up by such organizations as the International Electrotechnical Commission (IEC). He gives descriptions of numerous techniques used to reduce explosion hazards, including design for intrinsic safety, the use of explosionproof housings, encapsulation, sealing, and pressurization systems. Obtaining certification approval by suppliers and users of intrinsically safe designs is discussed in some detail, along with factors pertaining to the installation of such equipment.

SECTION 3: CONTROLLERS

Distributed Control Systems

To say that millions of words pertaining to distributed control systems (DCS) have been written since the concept originated in the early 1970s would indeed be an understatement. In this article the author condenses the essence of the topic, commencing with a description of how and why the DCS concept emerged, followed by a status quo delineation and a penetrating probe of the DCS architecture of the future. The result is an encapsulation of past verbiage and an insight into the future of DCS. Topics covered in considerable detail include DCS architecture present and future, including open systems, operations workstations, controller subsystems, data collection subsystems, process computing subsystems, and communications subsystems, stressing networking.

Programmable Controllers

As a backdrop to this article, the author redefines the characteristic functions of a programmable logic controller (PLC) as used in contemporary manufacturing and processing. Present software and programming needs are delineated, as well as input/output (I/O) systems and modules. Packaging of the PLC for demanding factory environments is addressed and, very importantly, designing and applying the PLC from a communications viewpoint—point-to-point connections and networking—are presented. The required designed-in features to ensure reliability, noise immunity, and availability are discussed.

Stand-Alone Controllers

The continuing impressive role of these controllers, particularly in non-CIM environments, is emphasized. Descriptions include revamped and modernized versions of these decades-old workhorses. A potpourri of currently available stand-alone controllers is included, with emphasis on new features, such as self-tuning and diagnosis, in addition to design conformation with European DIN (*Deutsche Industrie Norm*) standards.

Timers and Counters

Time-sequencing applications are found in numerous industrial control situations. Like other control elements, these devices have undergone continuous development and improvements during recent years. Electronic methodologies have replaced prior mechanical means to a large extent. Basic timer formats, such as switches, delay-on-make, interval, percentage, repeat-cycle, reset, and programmable timers, are discussed concisely. Contemporary time-schedule controllers and counters and totalizers are described.

Hydraulic Controllers

The important niche for powerful hydraulic methods continues to exist in the industrial control field. The principles, which were established decades ago, are described, including jet pipe, flapper, spool, and two-stage valves. Contemporary servo valves are discussed. Hydraulic fluids, power considerations, and the selection criteria for servo or proportional valves are outlined. A tabular summary of the relative advantages and limitations of various hydraulic fluids, including the newer polyol esters, is included.

Pneumatic Devices

The well-established principles of pneumatic devices and controllers are described. Although no longer the most popular control methodology, pneumatics have not disappeared into oblivion but continue to be specified for certain environments. Further, pneumatic methodologies persist for control valve actuation. The author describes baffle nozzles, pilot relays, pressure divider circuits, and force- and motion-balance feedback amplifiers. The accomplishment of the various modes of control action is described.

Batch Process Control

During the last few years much attention has been directed toward a better understanding of the dynamics of batch processes in an effort to achieve greater automation by applying advanced control knowledge gained from experience with continuous process controls and computers. This has proved to be more difficult and to require more time than had been anticipated. Standards organizations, such as the Instrument Society of America, continue to work up standards for a batch control model. In this article an attempt has been made to cut through some of the complexities and to concentrate on the basics rather than on the most complex model one can envision. Batching nomenclature is detailed, and definitions of the batch process are given in simplified, understandable terms. To distinguish among the many methods available for accomplishing batch control, a tabular summary of process types versus such factors as duration of process, size of lot or run, labor content, process efficiency, and the input/output system is given. Interfacing with distributed control systems and overall networking are considered.

Automatic Blending Systems

Although the need to blend various ingredients in pots and crocks dates back to antiquity, contemporary blending systems are indeed quite sophisticated. The author contrasts the control needs for batch versus continuous blending. A typical blend configuration is diagrammed in detail. Some of the detailed topical elements presented include liquid or powder blending, blending system sizing, blend controllers, stations, and master blend control systems. The application of automatic rate control, time control, and temperature compensation is delineated.

Distributed Numerical Control and Networking

An expert in the field redefines numerical control (NC) in the contemporary terms of distributed numerical control (DNC), tracing the developments which have occurred since the days of paper-tape controlled machines. The elements of the basic DNC configuration are detailed in terms of application and functionality. Much stress is given to behind-the-tape readers (BTRs). The numerous additional features which have been brought to NC by sophisticated electronic and computer technology are described. The tactical advantages of the *new* NC are delineated. The manner in which numerical control can operate in a distributed personal computer (PC) network environment is outlined. UNIX-based networks, open architectures, and the Novell networks, for example, are described.

Computers and Controls

This article, a compilation by several experts, commences by tracing the early developments of the main-frame computer, the 1960–1970 era of direct digital control (DDC), up to the contemporary period of personal computers (PCs) and distributed control systems (DCSs). Inasmuch as there is another article in this handbook on DCSs, primary attention in this article is on PCs. The basic PC is described in considerable detail, including its early acceptance, its major components (microprocessor, memory, power supply, keyboard, and I/O). The advantages and limitations of the PC's "connectability" in all directions, including networks, are discussed. Internal and external bus products are compared. PC software is discussed, with examples of specific languages and approaches. Software control techniques are presented in some detail. Progressive enhancement of the PC toward making it more applicable to process and factory floor needs is reviewed. In consideration of the fact that minicomputers and main-frame computers enter into some control situations, a few basic computer definitions are included in the form of an alphabetical glossary. This is not intended as a substitute for a basic text on computers, but included as a convenient tutorial.

SECTION 4: PROCESS VARIABLES—FIELD INSTRUMENTATION

Temperature Systems

Commencing with definitions of temperature and temperature scales and a very convenient chart of temperature equivalents, the article proceeds to review the important temperature measurement methodologies, such as thermocouples and resistance temperature detectors (RTDs), with a convenient tabular summary of each for selection purposes. Smart temperature transducers are illustrated and described. Other temperature measurement methods described include thermistors, solid-state temperature sensors, radiation thermometers, fiber-optic temperature sensors, acoustic pyrometers, and filled-system thermometers.

Fluid Pressure Systems

This article commences with a terse review of the ubiquitous elastic-element mechanical pressure gages, so commonly used today even in an environment of sophisticated devices. Important contemporary pressure sensors, transducers, and transmitters of more recent vintage are presented. These include strain-gage, capacitive, piezoresistive, piezoelectric, resonant-wire, linear variable differential transformer, carbon resistive, reluctive, and optical pressure transducers. Also included are devices for vacuum measurement, such as Pirani or thermocouple, hot-filament ionization, cold-cathode ionization, and spinning-rotor friction vacuum gages as well as partial pressure analyzers. The article ends with a detailed discussion of smart (intelligent) pressure transmitters and features pressure sensor selection guidelines for equipment procurement.

Flow Systems

There is a very wide range of flow sensors, and selection decisions can be difficult. Traditional differential pressure flowmeters, which remain in wide usage, are described. These include differential producers, venturis, Pitot tubes, and target flowmeters. A tabular summary of differential pressure flowmeters is featured. Then follow descriptions of magnetic, turbine, and oscillatory flowmeters. Special attention is given to more recent designs, such as mass and ultrasonic flowmeters. The article ends with descriptions of positive-displacement meters and open-channel and bulk-solids flow measurements.

Fluid Level Systems

Measuring the level of fluids can present the control engineer with the problem of finding the best choice of a wide variety of methodologies. There is usually a best method for even the most difficult to handle fluids. The article commences with a discussion of the importance of level measurement and control, frequently not as well understood as the reasons for measuring temperature, for example. The mathematics of level measurement is described because there are numerous variations of tank and vessel design. The remainder of the article is devoted to specific level-measurement means, including visual, buoyancy, float, hydrostatic pressure, radio-frequency, ultrasonic, microwave, nuclear, resistance tape, magnetostriction, and thermal-type level systems.

Industrial Weighing and Density Systems

Strain-gage and pneumatic load cells for weighing various hopper and tank vessels as may be used in batching systems are described, as well as a microprocessor-based automatic drum-filling scale. Numerous fluid-density measuring systems are reviewed, including the photoelectric hydrometer and the inductance bridge hydrometer. Specific-gravity sensors described include the balanced-flow vessel, the displacement meter, and the chain-balanced float gage. Several density and specific-gravity scales are defined.

Humidity and Moisture Systems

This article opens with a discussion of the physical parameters of humidity, including very convenient psychometric charts graduated in English and metric units. Wet- and dry-bulb measurements and percent relative humidity are described. Contemporary electronic humidity sensors, including the Dunmore and Pope cells, are presented. The principles of dew-point hygrometers are delineated. Other hygrometers described include the condensation type, the electrolytic type, and the aluminum oxide sensor. The need for moisture measurement is outlined, followed by laboratory moisture determinations and infrared absorption, electrical capacitance, and electrical conductivity methods. Other moisture-measurement instruments described include the aluminum oxide impedance and vibrating quartz crystal systems. Humidity parameters and representative applications are given in a tabular summary.

SECTION 5: GEOMETRIC AND MOTION SENSORS

Basic Metrology

Of fundamental interest to the discrete-piece manufacturing industries, this article includes the very basic instrumental tools used for the determination of dimension, such as the interferometer, optical gratings, clinometer, sine bar, optical comparator, and positioning tables.

Metrology, Position, Displacement, and Thickness Transducers

Described are the fundamentals of metrology and rotary and linear motion and the instrumental means used to measure and control it, such as various kinds of encoders, resolvers, linear variable differential transformers, linear potentiometric, and the new magnetostrictive linear displacement transducers. Noncontacting thickness gages, including the nuclear, x-ray, and ultrasonic types, are described. The importance and measurement of surface texture are described. The fundamentals of production gaging and statistical quality control (SQC) are outlined.

Speed, Velocity, Acceleration, and Vibration Instrumentation

Following definitions of terms, the many kinds of tachometers available are presented, including dc, ac, voltage-responsive, variable-reluctance, photoelectric, and eddy-current. The tachometerless regulation of servo speed is described as are governors. Air and gas velocity measurements, including air-speed indicators and anemometers, are delineated. Vibration measurement and numerous kinds of accelerometers, including piezoelectric, piezoresistive, and servo accelerometers, are described. Velocity transducers for sensing relative motion are discussed.

SECTION 6: PHYSICOCHEMICAL AND ANALYTICAL SYSTEMS

Classification of Analysis Instruments

The classification of analytical instruments in accordance with their energy-material reactions provides for a more penetrating understanding of this complex and widely varied field. A classification in tabular form is also presented. A very helpful table pertaining to the interconversion of concentration units of gases and vapors is included.

Sampling for On-Line Analyzers

The author describes some new approaches to the problem of sampling, gives basic instructions pertaining to sample preparation, and breaks this knotty problem down into four fundamental dimensions—separation, temperature, flow rate, and carrier. The advantages of the trap and transfer technique are delineated.

pH and Redox Potential Measurements

As one of the principal analytical determinations made today, two experts describe the fundamentals of acid-base theory, the details of pH measurement, including the types of electrodes used, temperature compensation, relatively recent improvements resulting from microprocessor technology, and exemplary applications of pH instruments. Redox potential measurements, including specific ions, are presented.

Electrical Conductivity Measurements

Although used for decades, electrical conductivity sensors have been greatly improved in recent years. The author describes the basic measurement units used, temperature compensation of cells, and contacting-type and electrodeless conductivity measurements. Typical applications are tabulated and a convenient table relating total dissolved solids content to electrical conductivity is included.

Thermal Conductivity Gas Analyzers

In principle these analyzers are decades old, but they have been grossly improved through the incorporation of modern electronic techniques. The authors outline applications and limitations of the method and provide very helpful selection guidelines. Convenient tables that relate the thermal conductivity of gases and gas mixtures and of comparison gases used in thermal conductivity determinations are included.

Process Chromatography

Two veterans in this very important area of analysis review the fundamentals of chromatography, trace the progress made in recent years, and zero in on both gas and liquid column chromatography. Chromatography data reduction is addressed in detail. Applications are described.

Oxygen Determination

The principal methods of oxygen determination are reviewed. These include magnetic susceptibility, zirconium oxide sensors, polarographic methods, electron capture analyzers, and catalytic combustion. Several illustrations explain the various methodologies.

Gas and Process Analyzers

Developments in these instruments have centered on improving older approaches through the application of microprocessors and other electronic circuitry, thus improving performance. However, some entirely new techniques have developed, including tin dioxide and other semiconducting metal oxide sensors. Ultraviolet, infrared, near infrared (a more recent development), combustibles, total hydrocarbons, carbon monoxide, reaction product, and mass spectrometric analyzers are described.

Spectroterminology

This abbreviated article is designed to explain the often complex and hyphenated terms used in describing instruments, all of which have emanated from the original simple spectroscope.

Refractometers

One of the very early analytical methods, refractometry has recaptured its former popularity through the inclusion of improved optics and fiber optics. The author reviews the fundamentals of refractometry and presents categories of applications, notably in the field of food processing.

Turbidity Measurement

In recent years, because of the emphasis on water purity and pollution control, the measurement of turbidity has taken on many new and extensive applications. Often poorly understood, the author explains the effects of particle size, particle shape, and particle color on the index of refraction measured. Optical design limits are described. Basic types of instruments incorporating the dual-beam technique or the ratio four-beam technique are described. Standard turbidity specifications are delineated and a table that compares turbidity measurement standards is included.

Rheological Systems

Several experts were involved in the preparation of this article, which is broken down into two main parts: (1) viscosity and (2) consistency. The viscous behavior of different classes of fluids is described. This is followed by descriptions and illustrations of contemporary viscometer designs and a listing of ASTM viscosity-related test specifications. Consistency, often poorly understood, is explained by considering those physical factors that influence it. The article closes with descriptions of the principal continuous consistency measurement schemes— blade-type transmitters, rotating consistency sensor, and web consistency measurement.

SECTION 7: CONTROL COMMUNICATIONS

Data Signal Handling in Computerized Systems

Networking, whether simple or complex, cannot succeed unless the raw data fed to the network are reliable, accurate, and free from competing signals. The author defines signal types, termination panels, field signals and transducers, sampled data systems, analog input systems, analog outputs, and digital inputs and outputs. Stressed are signal conditioning of common inputs, such as from thermocouples, solid-state temperature sensors, and resistance temperature detectors (RTDs). Amplifiers, common-mode rejection, multiplexers, filtering, analog signal scaling, and analog-to-digital and digital-to-analog converters are among the numerous topics covered and profusely illustrated.

Noise and Wiring in Data Signal Handling

The basic problems which a control engineer must seek to correct or avoid in the first place, including grounding and shielding, are delineated. Troubleshooting for noise is highlighted. A tabular troubleshooting guide is included.

Industrial Control Networks

Early networking and data highway concepts are described as a basis for understanding the many more recent concepts. Network protocols, including CSMA/CD, token bus, and token ring, are defined. Communication models and layers are defined as well as open systems and

Fieldbus. The important more recent roles of fiber-optic cables and networks are described, including the characteristics of optical fibers or cables and light sources and detectors. Note that this topic appears also in several other articles of the handbook.

SECTION 8: OPERATOR INTERFACE

Operator Interface—Design Rationale

The basics of good design are brought to the process and machine operator interface. There are discussions of the fundamental factors that determine good interface design, including human, environmental, and aesthetic considerations. Graphics used in panels are described as well as visual displays. The role of color is included. The article ends with a discussion of interface standards and regulations, maintainability, and miniaturization.

Process Operator Task Analysis and Training

The author reports on special studies of the operator interface from an industrial engineering standpoint and explores in particular the cognitive skills required of an operator.

Distributed Display Architecture

This article essentially zeros in on the CRT and equivalent interfaces which do not enjoy the attributes of larger panels. Interactive graphics is described in some detail.

Indicators, Recorders, and Loggers

Even in CIM environments there is a need for simple indicators, recorders, and sometimes stand-alone loggers. The demand for stand-alone hardware of this type remains healthy. The various indicator and recorder geometries used are delineated. The roles for process recorders as well as other industrial needs for recorders are discussed. Strip and round chart recorders as well as X-Y and other functionally complex recorders are described. Some emphasis is given to chart marking means, including advanced pens and inks, impact printing, thermal writing, and thermal array, electrostatic, ultraviolet, and fiber-optic recording. Hybrid recorders and their relationship to stand-alone data loggers are discussed.

Annunciators and Hazard Management

The role of stand-alone annunciators and alarm systems is described. The standardized sequence of alarm operations is discussed and role of computer alarms weighed. The article ends with a tabular comparison of annunciator, CRT, and combined alarm systems.

SECTION 9: VALVES, SERVOS, MOTORS, AND ROBOTS

Process Control Valves

The author describes and illustrates very methodically, with unusually clear diagrams of complex hardware, major control valve bodies and how to select valve styles and materials for given applications. Flow characteristics, valve sizing, actuators, and control valve accessories are addressed separately. The article includes five tables designed to assist in control valve selection.

Control Valve Cavitation

The author provides knowledge from years of study of the fundamentals of cavitation, emphasizing cavity behavior and its negative effects on valve and system performance. The importance of valve sizing and selection toward the avoidance of cavitation problems is stressed.

Control Valve Noise

This research specialist addresses the serious problem of valve noise. Noise terminology is defined. The kinds of noise encountered—mechanical, hydrodynamic, and aerodynamic—are delineated. Suggestions for reducing noise are given.

Servomotor Technology in Motion Control Systems

This rather exhaustive article, directed mainly to engineers in the discrete-piece manufacturing industries, also finds generous application in the process industries. It embraces factors in selecting a servomotor, describing the basic kinds of dc motors, hybrid servos, stepper motors, linear steppers, power transmission drives, stepper motor drives, emergency stop measures, machine motion control systems, and a potpourri of motion control systems.

Solid-State Variable-Speed Drives

As of the early 1990s there has been a profusion of solid-state variable-speed motor drives, ranging from subfractional to the multithousand-horsepower rating. Thyristors and insulated-gate bipolar transistors are described. The discussion is divided between dc and ac drives. Induction motor variable-speed drives, including the cycloconverter, voltage-source inverter, and pulse-width modulated (PWM) drives, are presented.

Robots

The technology of robotics, after an amazing surge of activity, now has reached a reasonable stage of maturity and acceptance. In this article the basic format of the robot is described, that is, its characteristics, including axes of motion, degrees of freedom, load capacity, and power requirements, as well as its dynamic properties, including stability, resolution and repeatability, and compliance among other characteristics. End effectors or grippers are presented. Workplace configurations are analyzed. Robot programming and control are explained and numerous styles of robots are illustrated.

Current-to-Pressure Transducers for Control Valve Actuation

Diaphragm-motor valves (pneumatically operated) remain the principal choice as final controlling elements for fluid flow. Although the demand for pneumatic control generally has diminished over the past few decades, the process control valve is operated by pneumatic force. Thus modern electronic controllers with digital outputs must utilize some form of current-to-pressure (air) transducer at the valve site. Several forms are available, including the older flapper-nozzle designs and the more recent technologies involving piezoceramic bender-nozzle principles. This article also describes the combination of the newer pressure sensors with electronic feedback control.

SECTION 2
CONTROL SYSTEM FUNDAMENTALS*

E. H. Bristol
Bristol Fellow, Research Department, The Foxboro Company (A Siebe Company), Foxboro, Massachusetts. (Basic Control Algorithms)

G. A. Hall, Jr.
Westinghouse Electric Corporation, Pittsburgh, Pennsylvania. (Control Principles—prior edition)

Peter D. Hansen
Bristol Fellow, Systems Development and Engineering, The Foxboro Company (A Siebe Company), Foxboro, Massachusetts. (Techniques for Process Control)

Stephen P. Higgins, Jr.
Retired Senior Principal Software Engineer, Honeywell Inc., Phoenix, Arizona. (Techniques for Process Control—prior edition)

Richard H. Kennedy
The Foxboro Company (A Siebe Company), Foxboro, Massachusetts. (Techniques for Process Control—prior edition)

E. C. Magison
Consultant, Ambler, Pennsylvania. (Safety in Instrumentation and Control Systems)

Joe M. Nelson
Applications Systems, Honeywell Inc., Billerica, Massachusetts. (Techniques for Process Control—prior edition)

Robert L. Osborne
Manager, Diagnostics and Control, Westinghouse Electric Corporation, Orlando, Florida. (Techniques for Process Control—prior edition)

John Stevenson
Engineering Manager, West Instruments, East Greenwich, Rhode Island. (Control Principles)

---

* Persons who authored complete articles or subsections of articles, or who otherwise cooperated in an outstanding manner in furnishing information and helpful counsel to the editorial staff.

CONTROL PRINCIPLES

by John Stevenson*

In this article the commonly measured process variable temperature is used for the basis of discussion. The principles being discussed apply also to other process variables, although some may require additional sophisticated attention, as discussed in the subsequent article which deals with techniques for process control.

In contrast with manual control, where an operator may periodically read the process temperature and adjust the heating or cooling input up or down in such a direction as to drive the temperature to its desired value, in automatic control, measurement and adjustment are made automatically on a continuous basis. Manual control may be used in noncritical applications, where major process upsets are unlikely to occur, where any process conditions occur slowly and in small increments, and where a minimum of operator attention is required. However, with the availability of reliable low-cost controllers, most users opt for the automatic mode. A manual control system is shown in Fig. 1.

In the more typical situation, changes may be too rapid for operator reaction, making automatic control mandatory (Fig. 2). The controlled variable (temperature) is measured by a suitable sensor, such as a thermocouple, a resistance temperature detector (RTD), a thermistor, or an infrared pyrometer. The measurement signal is converted to a signal that is compatible with the controller. The controller compares the temperature signal to the desired temperature (set point) and actuates the final control device. The latter alters the quantity of heat added to or removed from the process. Final control devices, or elements, may take the form of contactors, blowers, electric-motor or pneumatically operated valves, motor-operated variacs, time-proportioning or phase-fired silicon-controlled rectifiers (SCRs), or saturable core reactors. In the case of automatic temperature controllers, several types can be used for a given process. Achieving satisfactory temperature control, however, depends on (1) the process characteristics, (2) how much temperature variation from the set point is acceptable and under what conditions (such as start-up, running, idling), and (3) selecting the optimum controller type and tuning it properly.

PROCESS (LOAD) CHARACTERISTICS

In matching a controller with a process, the engineer will be concerned with process reaction curves and the process transfer function.

Process Reaction Curve

An indication of the ease with which a process may be controlled can be obtained by plotting the process reaction curve. This curve is constructed after having first stabilized the process temperature under manual control and then making a nominal change in heat input to the process, such as 10 percent. A temperature recorder then can be used to plot the temperature versus time curve of this change. A curve similar to one of those shown in Fig. 3 will result.

Two characteristics of these curves affect the process controllability, (a) the time interval before the temperature reaches the maximum rate of change, *A,* and (2) the slope of the max-

* Engineering Manager, West Instruments, East Greenwich, Rhode Island.

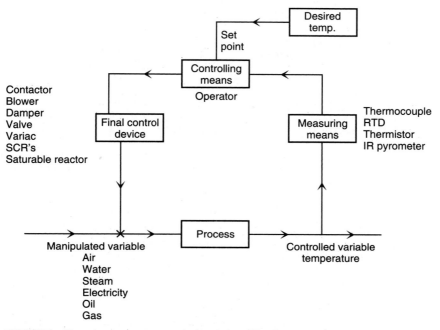

FIGURE 1   Manual temperature control of a process. (*West Instruments.*)

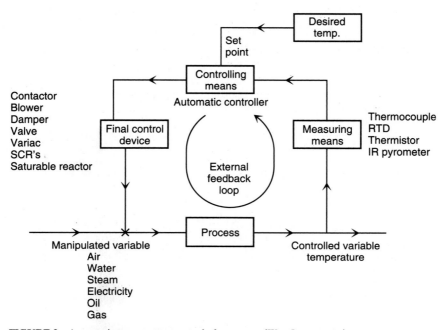

FIGURE 2   Automatic temperature control of a process. (*West Instruments.*)

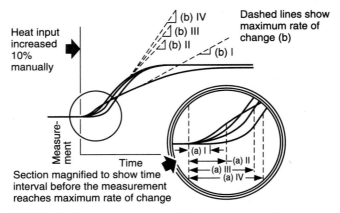

FIGURE 3 Process reaction curves. The maximum rate of temperature rise is shown by the dashed lines which are tangent to the curves. The tangents become progressively steeper from I to IV. The time interval before the temperature reaches the maximum rate of rise also becomes progressively greater from I to IV. As the S curve becomes steeper, the controllability of the process becomes increasingly more difficult. As the product of the two values of time interval A and maximum rate B increases, the process controllability goes from easy (I) to very difficult (IV). Response curve IV, the most difficult process to control, has the most pronounced S shape. Similar curves with decreasing temperature may be generated by decreasing the heat input by a nominal amount. This may result in different A and B values. (*West Instruments.*)

imum rate of change of the temperature after the change in heat input has occurred, B. The process controllability decreases as the product of A and B increases. Such increases in the product AB appear as an increasingly pronounced S-shaped curve on the graph. Four representative curves are shown in Fig. 3.

The time interval A is caused by dead time, which is defined as the time between changes in heat input and the measurement of a perceptible temperature increase. The dead time includes two components, (1) propagation delay (material flow velocity delay) and (2) exponential lag (process thermal time constants). The curves of Fig. 3 can be related to various process time constants. A single time-constant process is referred to as a first-order lag condition, as illustrated in Fig. 4.

This application depicts a water heater with constant flow, whereby the incoming water is at a constant temperature. A motor-driven stirrer circulates the water within the tank in order to maintain a uniform temperature throughout the tank. When the heat input is increased, the temperature within the entire tank starts to increase immediately. With this technique there is no perceptible dead time because the water is being well mixed. Ideally, the temperature should increase until the heat input just balances the heat taken out by the flowing water. The process reaction curve for this system is shown by Fig. 5.

The system is referred to as a single-capacity system. In effect, there is one quantity of thermal resistance R_1 from the heater to the water and one quantity of thermal capacity C_1, which is the quantity of water in the tank. This process can be represented by an electrical analog with two resistors and one capacitor, as shown in Fig. 6. R_{Loss} represents the thermal loss by the flowing water plus other conduction, convection, and radiation losses.

It should be noted that since the dead time is zero, the product of dead time and maximum rate of rise is also zero, which indicates that the application would be an easy process to control. The same process would be somewhat more difficult to control if some dead time were introduced by placing the temperature sensor (thermocouple) some distance from the exit pipe, as illustrated in Fig. 7. This propagation time delay introduced into the system would be

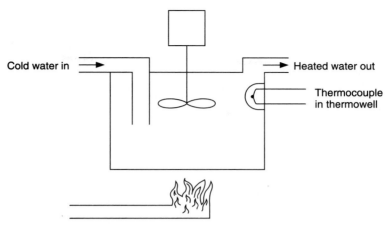

FIGURE 4   Single-capacity process. (*West Instruments.*)

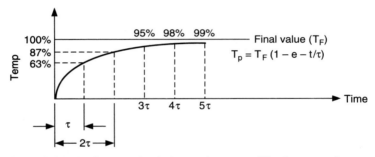

FIGURE 5   Reaction curve for single-capacity process. (*West Instruments.*)

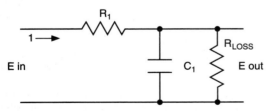

FIGURE 6   Electrical analog for single-capacity process. (*West Instruments.*)

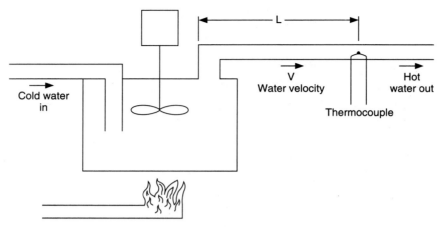

FIGURE 7   Single-capacity process with dead time. (*West Instruments.*)

equal to the distance from the outlet of the tank to the thermocouple divided by the velocity of the exiting water. In this case the reaction curve would be as shown in Fig. 8. The product AB no longer is zero. Hence the process becomes increasingly more difficult to control since the thermocouple no longer is located in the tank.

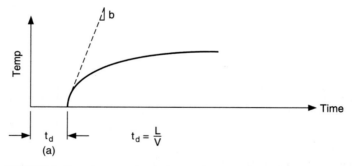

FIGURE 8   Reaction curve for single-capacity process with dead time. (*West Instruments.*)

A slightly different set of circumstances would exist if the water heater were modified by the addition of a large, thick metal plate or firebrick on the underside of the tank, between the heater and the tank bottom, but in contact with the bottom. This condition would introduce a second-order lag, which then represents a two-capacity system. The first time constant is generated by the thermal resistance from the heater to the plate and the plate heat capacity. The second time constant comes from the thermal resistance of the plate to the water and the heat capacity of the water. The system is shown in Fig. 9. The reaction curve for the system is given in Fig. 10. There is now a measurable time interval before the maximum rate of temperature rise, as shown in Fig. 10 by the intersection of the dashed vertical tangent line with the time axis. The electrical analog equivalent of this system is shown in Fig. 11. In the diagram the resistors and capacitors represent the appropriate thermal resistances and capacities of the two time constants. This system is more difficult to control than the single-capacity system since the product of time interval and maximum rate is greater.

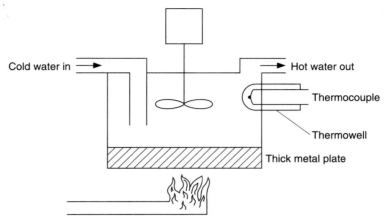

FIGURE 9   Two-capacity process. (*West Instruments.*)

The system shown in Fig. 9 could easily become a third-order lag or three-capacity system if there were an appreciable thermal resistance between the thermocouple and the thermowell. This could occur if the thermocouple were not properly seated against the inside tip of the well. Heat transfer from the thermowell to the thermocouple would, in this case, be through air, which is a relatively poor conductor. The temperature reaction curve for such a system is given in Fig. 12, and the electric analog for the system is shown in Fig. 13. This necessitates the addition of the R_3, C_3 time constant network.

Process Transfer Function

Another phenomenon associated with a proces or system is identified as the steady-state transfer-function characteristic. Since many processes are nonlinear, equal increments of heat input do not necessarily produce equal increments in temperature rise. The characteristic transfer-function curve for a process is generated by plotting temperature against heat input under constant heat input conditions. Each point on the curve represents the temperature under stabilized conditions, as opposed to the reaction curve, which represents the temperature under dynamic conditions. For most processes this will not be a straight-line, or linear, function. The transfer-function curve for a typical endothermic process is shown in Fig. 14, that for an exothermic process in Fig. 15.

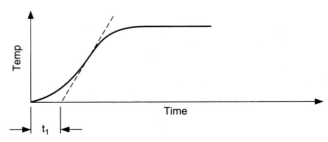

FIGURE 10   Reaction curve for two-capacity process. (*West Instruments.*)

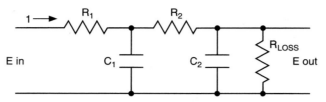

FIGURE 11   Electrical analog for two-capacity process. (*West Instruments.*)

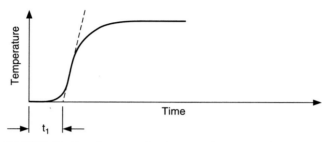

FIGURE 12   Reaction curve for three-capacity process. (*West Instruments.*)

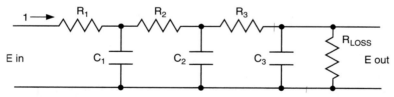

FIGURE 13   Electrical analog for three-capacity process. (*West Instruments.*)

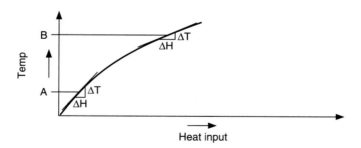

FIGURE 14   Transfer curve for endothermic process. As the temperature increases, the slope of the tangent line to the curve has a tendency to decrease. This usually occurs due to increased losses through convection and radiation as the temperature increases. This process *gain* at any temperature is the slope of the transfer function at that temperature. A steep slope (high $\Delta T/\Delta H$) is a high gain; a low slope (low $\Delta T/\Delta H$) is a low gain. (*West Instruments.*)

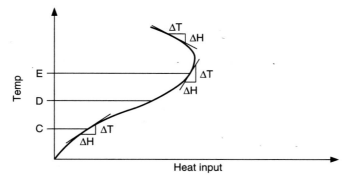

FIGURE 15   Transfer curve for exothermic process. This curve follows the endothermic curve up to the temperature level *D*. At this point the process has the ability to begin generating some heat of its own. The slope of the curve from this point on increases rapidly and may even reverse if the process has the ability to generate more heat than it loses. This is a negative gain since the slope $\Delta T/\Delta H$ is negative. This situation would actually require a negative heat input, or cooling action. This type of application is typical in a catalytic reaction process. If enough cooling is not supplied, the process could run away and result in an explosion. Production of plastics from the monomer is an example. Another application of this type is in plastics extrusion, where heat is required to melt the plastic material, after which the frictional forces of the screw action may provide more than enough process heat. Cooling is actually required to avoid overheating and destruction of the melt material. (*West Instruments.*)

CONTROL MODES

Modern industrial controllers are usually made to produce one, or a combination of, control actions (modes of control). These include (1) on-off or two-position control, (2) proportional control, (3) proportional plus integral control, (4) proportional plus derivative (rate action) control, and (5) proportional plus integral plus derivative (PID) control.

On-Off Control Action

An on-off controller operates on the manipulated variable only when the temperature crosses the set point. The output has only two states, usually fully on and fully off. One state is used when the temperature is anywhere above the desired value (set point), and the other state is used when the temperature is anywhere below the set point.

Since the temperature must cross the set point to change the output state, the process temperature will be continually cycling. The peak-to-peak variation and the period of the cycling are mainly dependent on the process response and characteristics. The time-temperature response of an on-off controller in a heating application is shown in Fig. 16, the ideal transfer-function curve for an on-off controller in Fig. 17.

The ideal on-off controller is not practical because it is subject to process disturbances and electrical interference, which could cause the output to cycle rapidly as the temperature crosses the set point. This condition would be detrimental to most final control devices, such as contactors and valves. To prevent this, an on-off differential or "hysteresis" is added to the controller function. This function requires that the temperature exceed the set point by a certain amount (half the differential) before the output will turn off again. Hysteresis will pre-

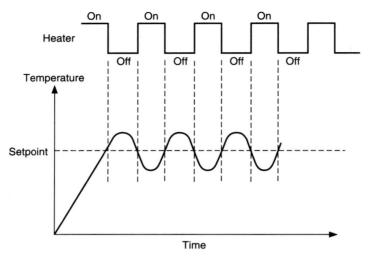

FIGURE 16   On-off temperature control action. (*West Instruments.*)

vent the output from chattering if the peak-to-peak noise is less than the hysteresis. The amount of hysteresis determines the minimum temperature variation possible. However, process characteristics will usually add to the differential. The time-temperature diagram for an on-off controller with hysteresis is shown in Fig. 18. A different representation of the hysteresis curve is given in the transfer function of Fig. 19.

Proportional Control

A proportional controller continuously adjusts the manipulated variable so that the heat input to the process is approximately in balance with the process heat demand. In a process using electric heaters, the proportional controller adjusts the heater power to be approximately equal to the process heat requirements to maintain a stable temperature. The range of

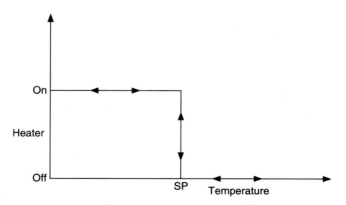

FIGURE 17   Ideal transfer curve for on-off control. (*West Instruments.*)

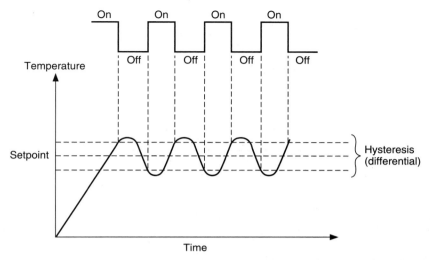

FIGURE 18 Time-temperature diagram for on-off controller with hysteresis. Note how the output changes state as the temperature crosses the hysteresis limits. The magnitude, period, and shape of the temperature curve are largely process-dependent. (*West Instruments.*)

temperature over which power is adjusted from 0 to 100 percent is called the proportional band. This band is usually expressed as a percentage of the instrument span and is centered about the set point. Thus in a controller with a 1000°C span, a 5 percent proportional band would be 50°C wide and extend 25°C below the set point to 25°C above the set point. A graphic illustration of the transfer function for a reverse-acting controller is given in Fig. 20.

The proportional band in general-purpose controllers is usually adjustable to obtain stable control under differing process conditions. The transfer curve of a wide-band proportional controller is shown in Fig. 21. Under these conditions a large change in temperature is

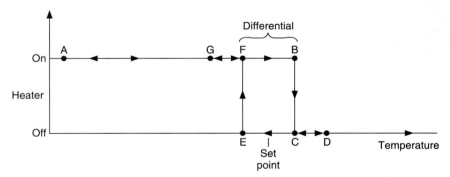

FIGURE 19 Another representation of the hysteresis curve—transfer function of on-off controller with hysteresis. Assuming that the process temperature is well below the set point at start-up, the system will be at *A*, the heat will be on. The heat will remain on as the temperature goes from *A* through *F* to *B*, the output turns off, dropping to point *C*. The temperature may continue to rise slightly to point *D* before decreasing to point *E*. At *E* the output once again turns on. The temperature may continue to drop slightly to point *G* before rising to *B* and repeating the cycle. (*West Instruments.*)

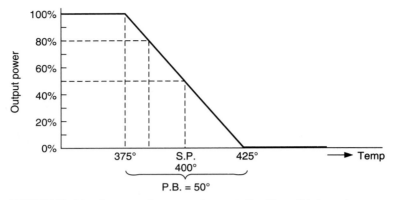

FIGURE 20 Transfer curve of reverse-acting controller. The unit is termed reverse-acting because the output decreases with increasing temperature. In this example, below 375°C, the lower edge of the proportional band, the output power is on 100 percent. Above 425°C the output power is off. Between these band edges the output power for any process temperature can be found by drawing a line vertically from the temperature axis until it intersects the transfer curve, then horizontally to the power axis. Note that 50 percent power occurs when the temperature is at the set point. The width of the proportional band changes the relationship between temperature deviation from set point and power output. (*West Instruments.*)

required to produce a small change in output. The transfer curve of a narrow-band proportional controller is shown in Fig. 22. Here a small change in temperature produces a large change in output. If the proportional band were reduced to zero, the result would be an on-off controller.

In some applications the proportional band is usually expressed as a percent of span, but it may also be expressed as controller gain in others. Proportional band and controller gain are related inversely by the equation

$$\text{Gain} = \frac{100 \text{ percent}}{\text{proportional band (percent)}}$$

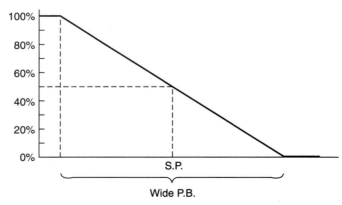

FIGURE 21 Transfer function for wide-band porportional controller. (*West Instruments.*)

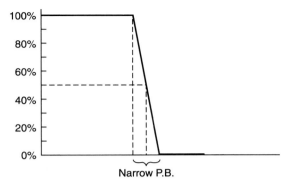

FIGURE 22   Transfer function for narrow-band proportional controller. (*West Instruments.*)

Thus narrowing the proportional band increases the gain. For example, for a gain of 20 the proportional band is 5 percent. The block diagram of a proportional controller is given in Fig. 23. The temperature signal from the sensor is amplified and may be used to drive a full-scale indicator, either an analog meter or a digital display. If the sensor is a thermocouple, cold junction compensation circuitry is incorporated in the amplifier. The difference between the process measurement signal and the set point is taken in a summing cicuit to produce the error or deviation signal. This signal is positive when the process is below the set point, zero when the process is at the set point, and negative when the process is above the set point. The error signal is applied to the proportioning circuit through a potentiometer gain control. The proportional output is 50 percent when the error signal is zero, that is, the process is at the set point.

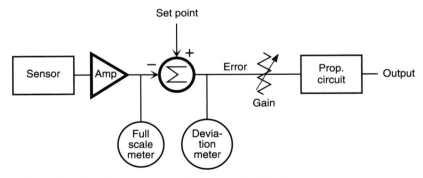

FIGURE 23   Block diagram of proportional controller. (*West Instruments.*)

Offset

It is rare in any process that the heat input to maintain the set-point temperature will be 50 percent of the maximum available. Therefore the temperature will increase or decrease from the set point, varying the output power until an equilibrium condition exists. The temperature difference between the stabilized temperature and the set point is called offset. Since the stabilized temperature must always be within the proportional band if the process is under control, the amount of offset can be reduced by narrowing the proportional band. However, the proportional band can be narrowed only so far before instability occurs. An illustration of a process coming up to temperature with an offset is shown in Fig. 24. The mechanism by which offset occurs with a proportional controller can be illustrated by superimposing the temperature controller transfer curve on the process transfer curve, as shown in Fig. 25.

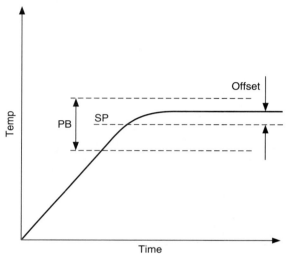

FIGURE 24   Process of coming up to temperature with an offset. (*West Instruments.*)

Manual and Automatic Reset

Offset can be removed either manually or automatically. In traditional instrumentation, manual reset uses a potentiometer to offset the proportional band electrically. The amount of proportional band shifting must be done by the operator in small increments over a period of time until the controller power output just matches the process heat demand at the setpoint temperature (Fig. 26). A controller with manual reset is shown in the block diagram of Fig. 27.

Automatic reset uses an elecctronic integrator to perform the reset function. The deviation (error) signal is integrated with respect to time and the integral is summed with the deviation signal to move the proportional band. The output power is thus automatically increased or decreased to bring the process temperature back to the set point. The integrator keeps changing the output power, and thus the process temperature, until the deviation is zero. When the deviation is zero, the input to the integrator is zero and its output stops changing. The integrator has now stored the proper value of reset to hold the process at the set point. Once this condition is achieved, the correct amount of reset value is held by the integrator. Should process heat requirements change, there would once again be a deviation, which the integrator would integrate and apply corrective action to the output. The integral term of the controller acts continuously in an attempt to make the deviation zero. This corrective action has to be applied rather slowly, more slowly than the speed of response of the load. Otherwise oscillations will occur.

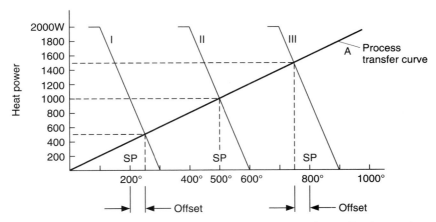

FIGURE 25   Mechanism by which offset occurs with a proportional controller. Assume that a process is heated with a 2000-watt heater. The relationship between heat input and process temperature, shown by curve *A*, is assumed to be linear for illustrative purposes. The transfer function for a controller with a 200°C proportional band is shown for three different set points in curves I, II, and III. Curve I with a set point of 200°C intersects the process curve at a power level of 500 watts, which corresponds to a process temperature of 250°C. The offset under these conditions is 250 to 200°C, or 50°C high. Curve II with a set point of 500°C intersects the process curve at 1000 watts, which corresponds to a process temperature of 500°C. There is no offset case since the temperature corresponds to the 50 percent power point. Curve III with a set point of 800°C intersects the process curve at 1500 watts, which corresponds to a temperature of 750°C. The offset under these conditions is 750 to 800°C, or 50°C low. These examples show that the offset is dependent on the process transfer function, the proportional band (gain), and the set point. (*West Instruments.*)

Automatic Reset—Proportional plus Integral Controllers

Automatic reset action is expressed as the integral time constant. Precisely defined, the reset time constant is the time interval in which the part of the output signal due to the integral action increases by an amount equal to the part of the output signal due to the proportional action, when the deviation is unchanging. A controller with automatic reset is shown in the block diagram of Fig. 28.

If a step change is made in the set point, the output will immediately increase, as shown in Fig. 29. This causes a deviation error, which is integrated and thus produces an increasing change in controller output. The time required for the output to increase by another 10 percent is the reset time—5 minutes in the example of Fig. 29.

Automatic reset action also may be expressed in repeats per minute and is related to the time constant by the inverse relationship

$$\text{Repeats per minute} = \frac{1}{\text{integral time constant (minutes)}}$$

Integral Saturation

A phenomenon called integral saturation is associated with automatic reset. Integral saturation refers to the case where the integrator has acted on the error signal when the temperature is outside the proportional band. The resulting large output of the integrator causes the proportional band to move so far that the set point is outside the band. The temperature must

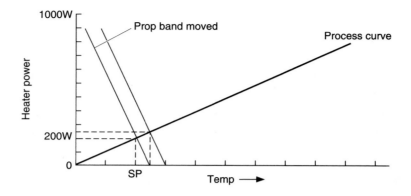

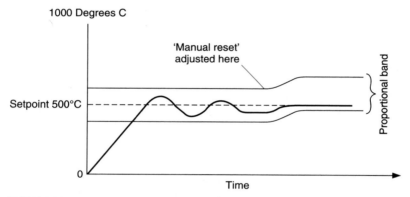

FIGURE 26   Manual reset of proportional controller. (*West Instruments.*)

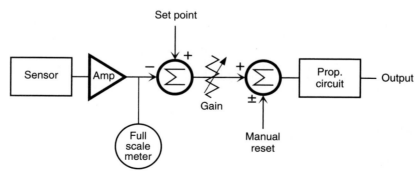

FIGURE 27   Block diagram of proportional controller with manual reset. (*West Instruments.*)

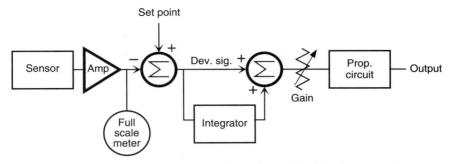

FIGURE 28   Block diagram of proportional plus integral controller. (*West Instruments.*)

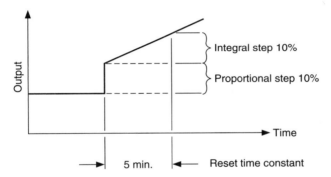

FIGURE 29   Reset time definition. (*West Instruments.*)

pass the set point before the controller output will change. As the temperature crosses the set point, the deviation signal polarity changes and the integrator output starts to decrease or desaturate. The result is a large temperature overshoot. This can be prevented by stopping the integrator from acting if the temperature is outside the proportional band. This function is called integral lockout or integral desaturation.

One characteristic of all proportional plus integral controllers is that the temperature often overshoots the set point on start-up. This occurs because the integrator begins acting when the temperature reaches the lower edge of the proportional band. As the temperature approaches the set point, the reset action already has moved the proportional band higher, causing excess heat output. As the temperature exceeds the set point, the sign of the deviation signal reverses and the integrator brings the proportional band back to the position required to eliminate the offset (Fig. 30).

Derivative Action (Rate Action)

The derivative function in a proportional plus derivative controller provides the controller with the ability to shift the proportional band either up or down to compensate for rapidly changing temperature. The amount of shift is proportional to the rate of temperature change. In modern instruments this is accomplished electronically by taking the derivative of the temperature signal and summing it with the deviation signal (Fig. 31*a*). [Some controllers take the derivative of the deviation signal, which has the side effect of producing upsets whenever the set point is changed (Fig. 31*b*).]

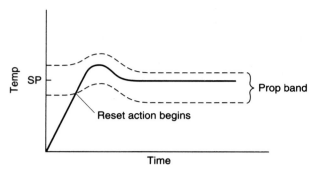

FIGURE 30 Proportional plus integral action. (*West Instruments.*)

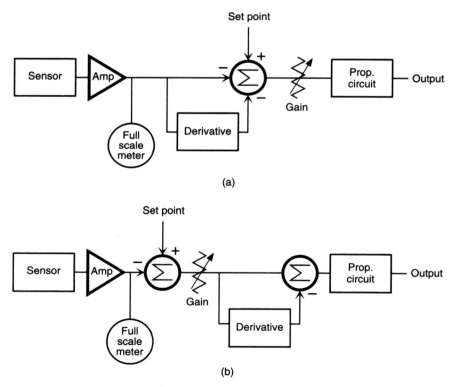

FIGURE 31 Block diagrams or proportional plus rate controller. (*a*) The derivative of the sensor (temperature) signal is taken and summed with the deviation signal. (*b*) The derivative of the deviation signal is taken. (*West Instruments.*)

The amount of shift is also proportional to the derivative time constant. The derivative time constant may be defined as the time interval in which the part of the output signal due to proportional action increases by an amount equal to that part of the output signal due to derivative action when the deviation is changing at a constant rate (Fig. 32).

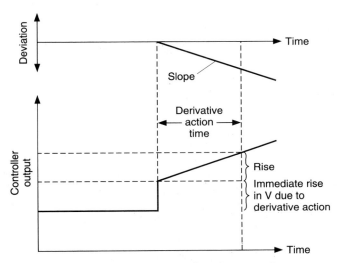

FIGURE 32   Derivative time definition. (*West Instruments.*)

Derivative action functions to increase controller gain during temperature changes. This compensates for some of the lag in a process and allows the use of a narrower proportional band with its lesser offset. The derivative action can occur at any temperature, even outside the proportional band, and is not limited as is the integral action. Derivative action also can help to reduce overshoot on start-up.

Proportional plus Integral plus Derivative Controllers

A three-mode controller combines the proportional, integral, and derivative actions and is usually required to control difficult processes. The block diagram of a three-mode controller is given in Fig. 33. This system has a major advantage. In a properly tuned controller, the temperature will approach the set point smoothly without overshoot because the derivative plus deviation signal in the integrator input will be just sufficient for the integrator to store the required integral value by the time the temperature reaches the set point.

Time- and Current-Proportioning Controllers

In these controllers the controller proportional output may take one of several forms. The more common forms are time-proportioning and current-proportioning. In a time-proportioning output, power is applied to the load for a percentage of a fixed cycle time. Figure 34 shows the controller output at a 75 percent output level for a cycle time of 12 seconds.

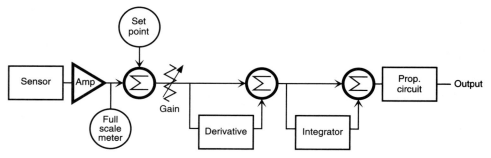

FIGURE 33   Proportional plus integral plus derivative controller. (*West Instruments.*)

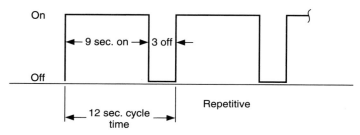

FIGURE 34   Time-proportioning controller at 75 percent level. (*West Instruments.*)

This type of output is common with contractors and solid-state devices. An advantage of solid-state devices is that the cycle time may be reduced to 1 second or less. If the cycle time is reduced to one-half the line period (10 ms for 50 Hz), then the proportioning action is sometimes referred to as a stepless control, or phase-angle control. A phase-angle fired output is shown in Fig. 35.

The current output, commonly 4 to 20 mA, is used to control a solid-state power device, a motor-operated valve positioner, a motor-operated damper, or a saturable core reactor. The relationship between controller current output and heat output is shown in Fig. 36.

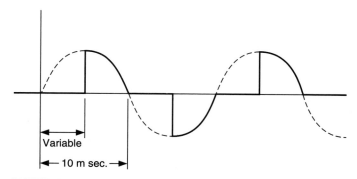

FIGURE 35   Phase-angle-fired stepless control output. (*West Instruments.*)

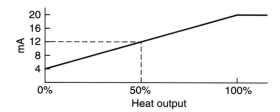

FIGURE 36 Current-proportioning controller. (*West Instruments.*)

Heat-Cool PID Control

Certain applications which are partially exothermic demand the application of cooling as well as heating. To achieve this, the controller output is organized as shown in Fig. 37. The controller has two proportional outputs, one for heating and one for cooling.

The transfer function for this type of controller is shown in Fig. 38. Below the proportional band, full heating is applied; above the proportional band, full cooling is applied. Within the proportional band (X_{p1}) there is a linear reduction of heating to zero, followed by a linear increase in cooling with increasing temperature. Heating and cooling can be overlapped (X_{sh}) to ensure a smooth transition between heating and cooling. In addition, to optimize the gain between heating and cooling action, the cooling gain is made variable (X_{p2}).

PROCESS CONTROL CHARACTERISTICS AND CONTROLLER SELECTION

The selection of the most appropriate controller for a given application depends on several factors, as described in the introduction to this article. The process control characteristics are very important criteria and are given further attention here. Experience shows that for easier controller tuning and lowest initial cost, the simplest controller that will meet requirements is usually the best choice. In selecting a controller, the user should consider priorities. In some cases precise adherence to the control point is paramount. In other cases maintaining the temperature within a comparatively wide range is adequate.

In some difficult cases the required response cannot be obtained even with a sophisticated controller. This type of situation indicates that there is an inherent process thermal design

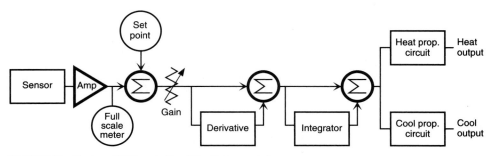

FIGURE 37 Heat-cool PID controller. (*West Instruments.*)

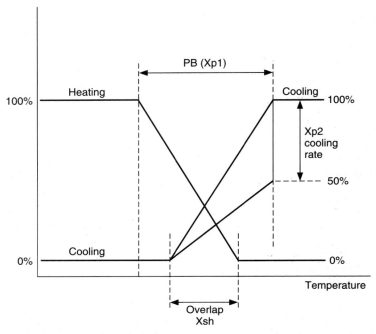

FIGURE 38   Transfer function for heat-cool PID controller. The controller has two proportional outputs, one for heating and one for cooling. (*West Instruments.*)

problem. Thermal design should be analyzed and corrected before proceeding with controller selection. A good thermal design will provide more stable control and allow the use of a less complicated and usually less expensive controller.

Controller Selection

Selection of the controller type may be approached from several directions:

1. Process reaction curve
2. Physical thermal system analysis
3. Previous experience
4. Experimental testing

The process reaction curve may be generated and observed to classify the process as easy or difficult to control, single- or multicapacity. This knowledge should be compared to the process temperature stability requirements to indicate which type of controller to use.

The process controllability may be estimated by observing and analyzing the process thermal system. What is the relative heater power to load heat requirements? Are the heaters oversized or undersized? Oversized heaters lead to control stability problems. Undersized heaters produce slow response. Is the thermal mass large or small? What are the distance and the thermal resistance from the heaters to the sensor? Large distances and resistances cause lag and a less stable system. Comparing the controllability with the process temperature stability requirements will indicate which type of controller to use. This same system of analysis can be applied to process variables other than temperature.

Prior experience often is an important guideline, but because of process design changes, a new situation may require tighter or less stringent control. A method often used is to try a simple controller, such as proportional plus manual reset, and to note the results compared to the desired system response. This will suggest additional features or features that may be deleted.

Single-Capacity Processes

If the process reaction curve or system examination reveals that the process can be classified as single-capacity, it may be controlled by an on-off controller. However, two conditions must be met: (1) a cyclical peak-to-peak temperature variation equal to the controller hysteresis is acceptable, and (2) the process heating and cooling rates are long enough to prevent too rapid cycling of the final control devices. Controller hysteresis also has an effect on the period of temperature cycling. Wider hysteresis causes a longer period and greater temperature variation. A narrow hysteresis may be used with final control devices, such as solid-state relays, triacs, and SCRs, which can cycle rapidly without shortening their life. Typical system responses for over- and undersized heater capacity are shown in Figs. 39 and 40, respectively.

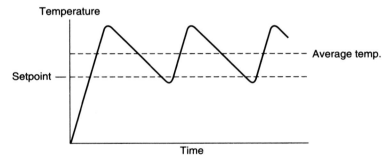

FIGURE 39   Typical system response for oversized heater condition. (*West Instruments.*)

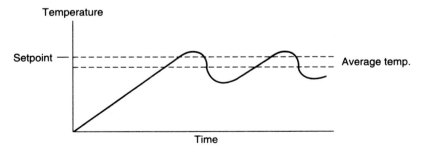

FIGURE 40   Typical system response for undersized heater condition. (*West Instruments.*)

If the previously mentioned two conditions are not acceptable, then the use of a proportional controller is indicated. A proportional controller would eliminate the temperature cycling. In a controller with adjustable proportional band, the band usually may be adjusted quite narrow and still maintain stability so that offset will not be a problem. If the controller has a fixed proportional band, at a value much larger than optimum, the resulting offset may be undesirable. Manual reset may be added to reduce the offset. A narrow proportional band will make the offset variations minimal with changes in process heat requirements so that automatic reset usually will not be required.

A single-capacity process usually will not require derivative action. However, control action during process upsets may be improved by the addition of some derivative action. Adding too much derivative action (too long a derivative time constant) can cause instability with some controllers.

Multicapacity Processes

A multicapacity process or a single-capacity process with transport delay is generally not suited to on-off control because of the wide temperature cycling. These processes require proportional control. Depending on the process difficulty, as evidenced by the process reaction curve and the control precision requirements, a proportional controller or one with the addition of derivative and integral action will be required.

Proportional controllers must be "tuned" to the process for good temperature response. The question is—what is good response? Three possible temperature responses under cold start-up conditions are shown in Fig. 41.

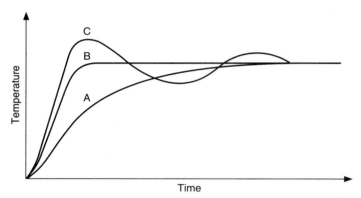

FIGURE 41   Three possible temperature responses of proportional controllers under cold start-up conditions. Curve *A* could be considered good response if a slow controlled heat-up is required. Curve *B* would be considered good response if the fastest heat-up without overshoot is required. Curve *C* could be good response if the fastest heat-up is required. The definition of "good response" varies with the process and operational requirements. (*West Instruments.*)

Control systems usually are tuned under operating conditions rather than for start-up conditions. Tuning a controller requires first that the process temperature be stable near the operating point with the system in operation. Then a known process disturbance is caused and the resulting temperature response observed. The response is best observed on a recorder. The proper disturbance for tuning is one which is likely to occur during actual operation, such as product flow change or speed change. However, this may be impractical and thus a small set-point change usually is used as the disturbance. The optimum tuning for set-point changes may not produce optimum response for various process disturbances.

Process characteristics for a proportional-only controller are given in Fig. 42. The curves show the resulting temperature change after decreased process heat demand. Similar curves would result if the set point were decreased several degrees.

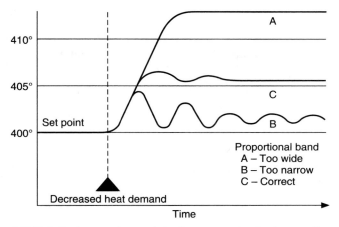

FIGURE 42 Process characteristics for a proportional-only controller. Curve *A* results when the proportional band is too wide. Note the large offset. The offset can be reduced by narrowing the proportional band. Instability results if the proportional band is too narrow, as shown by curve *B*. Optimum control, as shown by curve *C,* is achieved at a proportional band setting slightly wider than that which causes oscillation. If process parameters change with time or if operating conditions change, it will be necessary to retune the controller or avoid this by using a proportional band wider than optimum to prevent future instability. (*West Instruments.*)

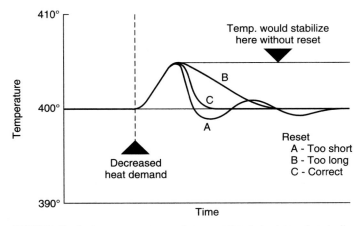

FIGURE 43 System response curves for proportional plus integral controller for an application such as a heat exchanger when there is a decrease in heat demand. An integral time constant which is too long for the process will take a long time to return the temperature to set point, as shown in curve *B*. An integral time constant that is too short will allow integral to outrun the process, causing the temperature to cross the set point with damped oscillation, as shown by curve *A*. If the integral time is much too long, continuous oscillation results. The integral time constant usually considered optimum is that which returns the temperature to set point as rapidly as possible without overshooting it, as shown by curve *C*. However, a damped oscillation (curve *A*) may be more desirable if the temperature must return to set point faster and some overshoot can be allowed. (*West Instruments.*)

Processes with long time lags and large maximum rates of rise, such as a heat exchanger, require a wide proportional band to eliminate oscillation. A wide band means that large offsets can occur with changes in load. These offsets can be eliminated by the automatic reset function in a proportional plus integral controller. The system response curve will be similar to those shown in Fig. 43 for a decrease in heat demand.

Derivative (rate) action may be used to advantage on processes with long time delays, speeding recovery after a process disturbance. The derivative provides a phase lead function, which cancels some of the process lag and allows the use of a narrower proportional band without creating instability. A narrower proportional band results in less offset. The response of a proportional plus derivative controller in a system is dependent not only on the proportional band and the derivative time constant, but also on the method used to obtain the derivative signal. The curves of Fig. 44 show some typical results of the proportional plus derivative control algorithm as used in the West MC30 and previous controllers as well as most other brands of proportional plus derivative controllers. Figure 45 shows the response to a decrease in heat demand for the controllers described previously (Fig. 44).

As noted from the aforementioned illustrations, the problem of superimposed damping or continuous oscillation remains. This condition may be corrected by either decreasing the derivative time constant or widening the proportional band. The oscillations result from a loop gain that is too great at the frequency of oscillation. The total loop gain is the process

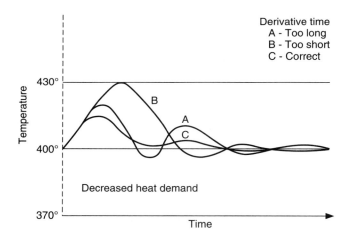

FIGURE 44 Response curves for controller using proportional plus derivative algorithm. A derivative time constant which is too long causes the temperature to change too rapidly and overshoot the set point with damped oscillation (curve *A*). A derivative time constant which is too short allows the temperature to remain away from the set point too long (curve *B*). The optimum derivative time returns the temperature to set point with a minimum of ringing (curve *C*). The damped oscillation about the final value in curve *A* can be due to excessive derivative gain at frequencies above the useful control range. Some controllers have an active derivative circuit which decreases the gain above the useful frequency range of the system and provides full phase lead in the useful range. The results of too short a derivative time constant remain the same as shown, not enough compensation for process lags. However, this method improves response at the optimum derivative time constant. It also produces two more possible responses if the time constart is longer than optimum. (*West Instruments.*)

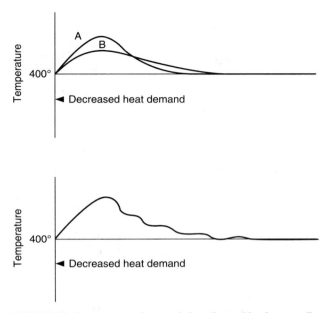

FIGURE 45   Response to a decrease in heat demand for the controller of Fig. 44. Curve *A* (top digram) is for the optimum derivative time constant. The temperature returns smoothly to set point. Curve *B* shows one possibility for a derivative time constant which is too long. The temperature deviation is less, but the temperature returns to set point on the derivative time constant curve. The curve in the bottom diagram shows another possibility for a derivative time constant which is too long. The temperature returns to set point on the derivative time constant curve, but has either damped or continuous oscillation superimposed. (*West Instruments.*)

gain times the proportional gain times the derivative gain. Decreasing any one of these gains will decrease the total loop gain and return stability.

The proportional plus derivative controller may be used to advantage on discontinuous processes such as batching operations involving periodic shutdown, emptying, and refiling. Here the proportional plus integral controller would not perform well because of the long time lags and intermittent operation. Derivative action also reduces the amount of overshoot on start-up of a batch operation.

The most difficult processes to control, those with long time lags and large maximum rates of rise, require three-mode or proportional plus integral plus derivative (PID) controllers. The fully adjustable PID controller can be adjusted to produce a wide variety of system temperature responses from very underdamped through critically damped to very overdamped.

The tuning of a PID control system will depend on the response required and also on the process disturbance to which it applies. Set-point changes will produce a different response from process disturbances. The type of process disturbance will vary the type of response. For example, a product flow rate change may produce an underdamped response while a change in power line voltage may produce an overdamped response.

Some of the response criteria include rise time, time to first peak, percent overshoot, settling time, decay ratio, damping factor, integral of square error (ISE), integral of absolute error (IAE), and integral of time and absolute error (ITAE). Figure 46 illustrates some of these criteria with the response to an increase in set point.

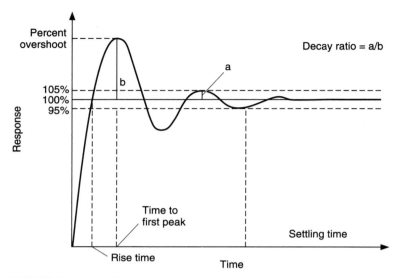

FIGURE 46   Tuning of PID control system depends on response required as well as on process disturbance to which it applies. (*West Instruments.*)

In many process applications a controller tuning which produces a decay ratio of ¼ is considered good control. However, this will vary with the application. Also the tuning parameters which produce a decay ratio of ¼ are not unique and neither are the responses, as illustrated in Fig. 47. In some applications the deviation from set point and the time away from set point are very important. This leads to imposing one of the integral criteria, as shown in Fig. 48.

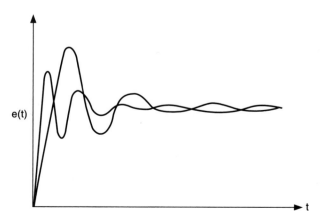

FIGURE 47   Nonunique nature of quarter decay ratio. (*West Instruments.*)

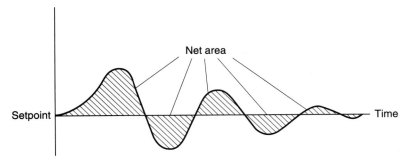

FIGURE 48   Application of integral criteria. (*West Instruments.*)

TECHNIQUES FOR PROCESS CONTROL

by Peter D. Hansen*

This article presents a number of techniques useful in the analysis and design of modern process control systems. Particular emphasis is given to transfer-function and adaptive methods, which lead to designs that cope with process delay (dead time), loop interaction, nonlinearity, and unmeasured disturbances.

The mathematical approaches described here can be used (1) by manufacturers of industrial controllers to achieve an improved and more versatile design, (2) by control engineers seeking a solution to difficult control applications, and (3) by students and researchers to achieve a more thorough understanding of control system dynamics.

An effort has been made to use consistent notation throughout this article. Uppercase letters represent transfer operators, functions of the differential operator s or the backward shift operator z^{-1}. Matrices are in boldface type. Scalar or vector variables and parameters are represented by lowercase letters. Key equations referenced throughout are the process equation [Eq. (5)], the target performance equation [Eq. (8)], the design equation [Eq. (10)], the open-loop controller equation [Eq. (11)], and the feedback controller equation [Eq. (13)].

DIGITAL CONTROL

A digital controller is generally considered to be superior to an analog controller. However, if it is used to emulate an analog controller, the digital device may be less effective because of

* Bristol Fellow, Systems Development and Engineering, The Foxboro Company (A Siebe Company), Foxboro, Massachusetts.

phase (or delay) and resolution errors introduced by sampling and converting. The digital controller's advantage is its algorithmic flexibility and precision with respect to both calculations and logic, thereby facilitating on-line restructuring and parameter adaptation.

A digital control algorithm utilizes samples of its input signals which are discrete in both magnitude and time. Usually, continuous signals are sampled at a constant rate. Sampling the controlled variable introduces phase lag (effective delay) into the feedback loop because of:

1. Low-pass filtering
2. Computation and transmission
3. Output holding between updates

Effective delay or parasitic lag, whether in the digital or the analog portion of a feedback loop, has an adverse effect on performance.

Loop delay, measurement noise, and output saturation determine the performance achievable with feedback control. Minimum integrated absolute error in response to an unmeasured load increases in proportion to the delay time for a dominant-delay process and in proportion to the square of delay for a dominant-lag process. Consequently the sampling-related delays should be made a small fraction of the total loop delay by using a small sampling interval.

State-Space Representation

Linear dynamic systems can be represented in terms of a state-variable vector x as a set of simultaneous first-order difference equations,

$$x\{t + h\} = \mathbf{M}x\{t\} + \mathbf{N}u\{t\}$$

$$y\{t\} = \mathbf{C}x\{t\} \tag{1}$$

where h is the computing interval, or as differential equations,

$$\frac{dx}{dt} = \mathbf{A}x + \mathbf{B}u$$

$$y = \mathbf{C}x \tag{2}$$

In these equations u is a vector of inputs, and y is a vector of measured variables. Matrices, but not vectors, are capitalized and in boldface type. $\mathbf{M}$ and $\mathbf{A}$ are square matrices and $\mathbf{N}$, $\mathbf{C}$, and $\mathbf{B}$ are rectangular (noninvertible). When a zero-order hold drives the continuous process, the representations are related at sampling instants by

$$\mathbf{M} = e^{\mathbf{A}h} = \frac{\sum_{0}^{\infty}(\mathbf{A}h)^{n}}{n!}$$

$$\mathbf{N} = \mathbf{A}^{-1}(\mathbf{M} - \mathbf{I})\mathbf{B} = \left[\frac{\sum_{0}^{\infty}(\mathbf{A}h)^{n}}{(n + 1)!}\right]\mathbf{B}h \tag{3}$$

The inversion of $\mathbf{A}$ can be avoided by replacing $\mathbf{M}$ with its Taylor series, which converges (possibly slowly) for all $\mathbf{A}h$.

The state-space approach may be used to model multivariable systems whose characteristics are time-varying and whose controlled variables are not measured directly. However, the representation may be inefficient because the matrices are often sparse. The approach can be generalized to characterize nonlinear systems by considering the right-hand sides of Eqs. (1)

or (2) to be vector functions of the state variables and inputs. However, a process with a time delay cannot be represented directly with a finite differential-equation form. The difference-equation form introduces an extra state variable for each time step of delay.

Methods of analyzing observability, controllability, and stability of state-space representations are discussed in many control texts [1]–[3], as are design methods for predictors and controllers. The state-space feedback-controller design procedures lead to inflexible global control structures, which are usually linear. All manipulated variables are used to control each controllable state variable, and all measured variables are used to calculate each observable state variable. Consequently an on-line redesign (adaptive) capability may be needed to retune for process nonlinearity and to restructure following either an override condition or a loss of a measurement or manipulator.

Transfer-Operator Representation

The state-space equations can be expressed in transfer-function form, using algebraic operators to represent forward shift z and differentiation s,

$$y = \mathbf{C}(z\mathbf{I} - \mathbf{M})^{-1}\mathbf{N}u$$
$$y = \mathbf{C}(s\mathbf{I} - \mathbf{A})^{-1}\mathbf{B}u \tag{4}$$

For a single-input, single-output time-invariant system, these equations can be expressed as

$$Ay = BDu + Ce \tag{5}$$

where y is the controlled or measured variable, u is the manipulated variable, and e is a load (or disturbance) variable. A, B, C, and D are polynomial functions of s or the backward shift $z^{-1} = e^{-hs}$. B contains stable zeros and is therefore cancelable. D may be noncancelable and has unity steady-state gain. A zero of a polynomial is a root, a value of its argument (s or z^{-1}) that causes the polynomial to be zero. Unstable (noncancelable, nonminimum phase) zeros are in the right half of the complex s plane or inside the unit circle in the complex z^{-1} plane. The delay operator, whose zeros are at the origin of the z^{-1} plane, is noncancelable and nonminimum phase. Its inverse, a time advance, is physically unrealizable.

For sinusoidal signals the differentiation operator becomes $s = j\omega$, and the backward shift becomes $z^{-1} = e^{-j\omega h}$. In steady state the radian frequency ω is zero, allowing s to be replaced by 0 and z^{-1} by 1 in the polynomial operators. The role of the C polynomial is played by an "observer" in state-space design. When e is dominated by measurement noise, e appears unfiltered at y; hence C is (almost) equal to A. When e is a load upset, e appears at y filtered by the process dynamics $1/A$; hence C is (nearly) 1. When e is considered an impulse, a process that would have a nonzero steady-state response to a steady-state e input has an additional zero at $s = 0$ or $z^{-1} = 1$ in its A and B polynomials.

An example of the conversion from the s domain to the z^{-1} domain is shown in [1]. A sampled lag $\{\tau_L\}$–delay $\{\tau_D\}$ process with gain $\{k\}$, whose input u is constant between sampling instants (because of a zero-order hold), is represented as

$$(1 - z^{-1}e^{-b})y = kz^{-n}[1 - e^{-a} + z^{-1}(e^{-a} - e^{-b})]u \tag{6}$$

where the delay is between n and $n + 1$ sampling intervals, $nh < \tau_D < (n + 1)h$, and

$$a = \frac{(n+1)h - \tau_D}{\tau_L}$$

$$b = \frac{h}{\tau_L}$$

When $e^{-a} - e^{-b} < 1 - e^{-a}$, the first-order factor in parentheses on the right of Eq. (6) is cancelable and can be part of B. Otherwise it must be part of D.

When b is very small, because the sampling interval is very small, Eq. (6) becomes

$$[1 - z^{-1}(1-b)]y = \left(\frac{k}{\tau_L}\right)z^{-n}[h(n+1) - \tau_D + z^{-1}(\tau_D - nh)]u \tag{7}$$

Except for the b term on the left, this is indistinguishable from an integral$\{\tau_L/k\}$–delay$\{\tau_D\}$ process, signaling the likelihood of numerical difficulty in applications such as parameter (k and τ_L) identification.

Presuming that the desired behavior is a function of the measured variable y, the target closed-loop performance can be expressed as

$$Hy = Dr + Fe \tag{8}$$

where H is a (minimum-phase) polynomial with unity steady-state gain and stable zeros. H and F may be totally or partially specified polynomials. If e may have an arbitrary value in steady state, the steady-state value of F must be zero for y to converge to the set point (or reference input) r. Eliminating y from Eqs. (5) and (8) results in

$$ADr + AFe = HBDu + HCe \tag{9}$$

Because D is not cancelable, this equation cannot be solved directly for u. However, the product HC may be separated into two parts, the term on the left-hand side AF and a remainder expressed as the product DG, so that D becomes a common factor of Eq. (9),

$$HC = AF + DG \tag{10}$$

This equation is key to the controller design: selecting some and solving for other coefficients of H, F, and G when those of A, B, C, and D are known or estimated.

OPEN-LOOP CONTROL

The open-loop controller design results when Eq. (10) is used to eliminate AFe from Eq. (9), since D is not zero,

$$u = \frac{Ar - Ge}{BH} \tag{11}$$

Of course, if e is unmeasured, open-loop control will not reduce e's effect on y. This causes F to be an infinite-degree polynomial HC/A and G to be zero. If e is measured, G is a feedforward operator. Substituting u from Eq. (11) back into the process equation, Eq. (5), and not canceling terms common to both numerator and denominator, results in the open-loop performance equation

$$y = \frac{BA\,(Dr + Fe)}{HBA} \tag{12}$$

To avoid (imperfect) canceling of unstable roots, A as well as H and B must contain only stable zeros.

High-performance ($H \approx 1$) open-loop control applies the inverse of the process characteristic A/B to a set-point change. Because a dominant-lag process has low gain at high frequencies, its controller has high gain there. A rapid set-point change is likely to saturate the manipulated variable, but otherwise leaves its trajectory unchanged. The early return of this variable from its limit causes slower than optimal controlled variable response. This can be avoided by using nonlinear optimization (such as quadratic programming suggested in [4]) to compute the optimal controller-output trajectory, taking into account output limits, load level, and other process equality and inequality constraints.

The performance of an open-loop controller may be degraded by an unmeasured load or by mismatch between the process and the inverse controller at low frequencies. Mismatch at high frequency will not cause significant difficulty, however.

FEEDBACK CONTROL

Combining Eqs. (10) and (11) with the target equation, Eq. (8), to eliminate e results in the closed-loop control law

$$u = \frac{Cr - Gy}{BF} \tag{13}$$

This equation also results when e is eliminated from the process and target equations, Eqs. (5) and (8), and D is made a common factor with the design equation, Eq. (10). From Eq. (13) it is clear that the disturbance polynomial C and its Eq. (10) decomposition terms F and G play a key role in the feedback-controller design. Various methods for determining these polynomials will be discussed. Except in the special case where $C = G$, the control output u does not depend exclusively on the control error $r - y$. However, the controller will provide integral action, eliminating steady-state error, if the steady-state values of G and C are equal and not zero, and either those of $e\{t\}$, A, and B are zero or that of F is zero.

Substituting u from Eq. (13) back into the process equation and not canceling terms common to both numerator and denominator results in the closed-loop performance equation

$$y = BC \, \frac{Dr + Fe}{HBC} \tag{14}$$

To avoid (imperfect) canceling of unstable roots, C as well as H and B must contain only stable zeros. However, it is not necessary that all of the zeros of A be stable when the control loop is closed. Zeros of C not common to A correspond to unobservable modes of the disturbance variable e. Zeros of B or D not common to A correspond to uncontrollable modes of y.

When the manipulated variable saturates, it is necessary to stop (or modify) the controller integral action to prevent "windup." If the integration were allowed to continue, the prelimited controller output would continue to rise (wind up) above the limit value, requiring a comparable period after the control error reverses sign before the manipulated variable could recover from saturation. This would cause a significant (avoidable) controlled-variable overshoot of the set point. A controller of a dominant-lag process, designed for good unmeasured-load rejection, employs significant proportional (and derivative) feedback G. When responding to a large set-point change, this feedback keeps the output saturated and the controlled variable rate limited longer than would a linear open-loop controller, resulting in a faster response [5]. Halting the integral action, while the manipulated variable remains limited, prevents appreciable overshoot.

The performance of the feedback loop is most sensitive to the process behavior in the frequency range where the absolute loop gain is near 1. Performance at significantly lower frequencies is often quite insensitive to the process characteristics, load, and controller tuning.

Robustness

The ability of a feedback loop to maintain stability, when the process parameters differ from their nominal values, is indicated with robustness measures. Denoting the locus of combined shifts of the process gain by the factor b and the process delay by the factor d, which make the loop marginally stable, is a useful indicator of robustness, providing, in a more physical form, the information contained in gain and phase margins. The use of the two parameters b and d is based on the idea that the process behavior, in the frequency range critical for stability, can be approximated with an n-integral-delay two-parameter model. The integrals, whose number n may range from 0 to 1 plus the number of measurement derivatives used in the controller, contribute to the gain shift b. The phase, in excess of the fixed contribution of the integrals, and the shift in phase $d\omega$ can be considered to be contributed by "effective" delay.

At marginal stability the return difference (1 plus the open-loop gain) is zero,

$$1 + \frac{G}{FB}\,\frac{bBD^d}{A} = 1 + \frac{bGD^d}{AF} = 0 \tag{15}$$

Here it is assumed that D is effective delay, which may include small lags not included in A and a $(1 - \tau s)/(1 + \tau s)$ factor for each nonminimum-phase zero.

Figure 1 is a plot of d versus b, using logarithmic scales, for PID control of pure-delay, integral-delay, and double-integral-delay processes. The proportional band (PB) and the integral time (IT) were determined for minimum overshoot using the algebraic PID design method described in a later section. In [6] a single number, characterizing robustness, is derived from a robustness plot. This robustness index is -1 plus the antilog of the length of the half-diagonal of the diamond-shaped box centered at the nominal design point $(d = b = 1)$ that

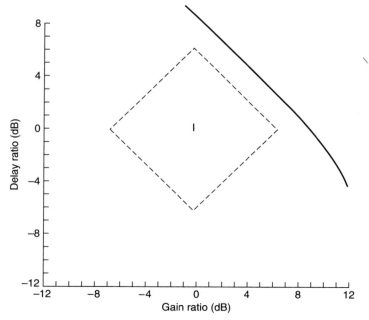

FIGURE 1a Robustness plots. The diamond corresponds to a factor of 2 changes in the product or quotient of delay and gain from their nominal values. PI control of pure delay process. (*Foxboro.*)

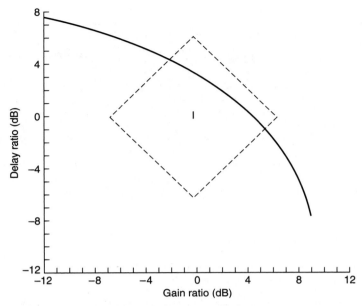

FIGURE 1b   PID control of integral-delay process. (*Foxboro.*)

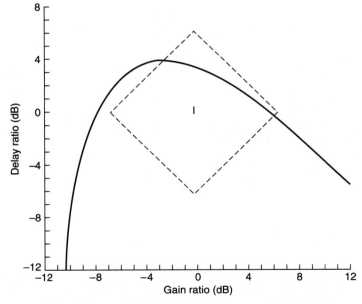

FIGURE 1c   PID control of double-integrator-delay process. (*Foxboro.*)

touches the d versus b curve at its closest point. A value of 1 indicates that the product or ratio of d and b can be as large as 2 or as small as 0.5 without instability. For the three cases of Fig. 1 the robustness index is 1.67, 0.47, and 0.30, each determined by sensitivity to delay shift. The diamond-shaped box in the figure would correspond to a robustness index of 1.

Most control schemes capable of providing high performance provide poor robustness (a robustness index near 0). Adaptive tuning may be required to keep a controller that is capable of high performance current with changing process conditions.

Digital simulation provides a useful means for exploring robustness experimentally. Nonlinearities can be included naturally in the time domain. It may not be necessary to use an exotic integration algorithm if the process can be modeled in real-factored form. The factored form can be much less sensitive to roundoff error than unfactored polynomial and state-space forms.

The simulation equations should be solved in causal sequence. Each equation's dependent variable should be updated based on the most current computed value of its independent variables (as is done in the Gauss-Seidel iterative solution of algebraic equations). A useful such model for a first-order factor, $y/x = 1/(1 + \tau s)$, is

$$y\{t\} = y\{t - h\} + \frac{h}{(\tau + h)}\ (x\{t\} - y\{t - h\}) \tag{16}$$

and for a damped second-order factor, $y/x = 1/(1 + Ts + T\tau s^2)$, is

$$v\{t\} = v\{t - h\} + \frac{h}{\tau + h}\ (x\{t\} - v\{t - h\} - y\{t - h\})$$

$$y\{t\} = y\{t - h\} + \frac{h}{T + h}\ v\{t\} \tag{17}$$

The internal variable v is a measure of the derivative of the output y,

$$\frac{dy}{dt} \approx \frac{v\{t\}}{T + h} \tag{18}$$

These models both give the correct result when $T = \tau = 0$: $y\{t\} = x\{t\}$. When the sampling interval h is very small compared with T and τ, it may be necessary to compute with double precision to avoid truncation, because the second term on the right of Eqs. (17) and (18) may become much smaller than the first before the true steady state is reached.

A fixed time delay may be modeled as an integer number of computing intervals, typically 20 to 40. At each time step an old data value is discarded and a new value added to a storage array. Incremented pointers can be used to keep track of the position of the delay input and output, as in a ring structure. This avoids shifting all the stored data each time step, as in a line structure.

FEEDFORWARD CONTROL

Feedforward control, to counteract the anticipated effect of a measured load e_M, combined with feedback control to mitigate the effect of an unmeasured load e_U, makes use of two design equations like Eq. (10), one for each load type,

$$HC_U = AF_U + DG_U$$

$$HC_M = AF_M + DG_M \tag{19}$$

C_M need not be cancelable and may include a delay factor. Combining the process equation, like Eq. (5), with the target equation, like Eq. (8), with the design equations (19), like Eq. (10), to eliminate e_U and D results in the combined feedback and feedforward control law, like Eqs. (11) and (13),

$$u = \frac{C_U r - G_U y}{B F_U} - \left(G_M - \frac{F_M\, G_U}{F_U} \right) \frac{e_M}{BH} \tag{20}$$

The second (e_M) term is an additive feedforward correction. If $F_M G_U / F_U = G_M$, feedforward control is not capable of improving upon feedback performance. The $F_M G_U / F_U$ term represents the reduction, from the open-loop feedforward correction G_M, needed to prevent redundant (overcorrecting) contributions. F_M can be made (nearly) zero, at least at low frequencies, when there is no more effective delay in the manipulated-variable path D to the controlled variable y than in the measured disturbance path C_M. Then from Eqs. (19), $G_M = HC_M/D$ because D is a factor of C_M. The measured disturbance e_M is (almost) perfectly rejected with the feedforward correction $u_{FF} = -(C_M/BD)e_M$, provided the controller output does not limit.

Feedforward provides a means for this single-output transfer function approach to be applied to a process with interacting loops. Unlike the state-space approach, it is necessary to associate each controlled variable with a particular manipulated variable. Then the effect of other manipulated variables on that controlled variable can be removed or reduced with feedforward corrections. This approach has the advantage that the appropriate compensation can be applied to active loops even when other loops are saturated or under manual control.

Furthermore, feedforward compensation may be structured to multiply the feedback correction. Multiplicative compensation is particularly effective for a temperature or composition loop manipulated with a flow. This configuration is intended to make the process appear more linear, as seen from the feedback controller. Thus the feedforward can be considered to provide gain scheduling for the feedback controller. Alternatively, the feedback controller can be viewed as adaptively tuning the gain of the feedforward compensator.

Other nonlinearities, even though they may involve the feedback variable, may be removable with additive or multiplicative feedforwardlike corrections [7]. For example, consider a nonlinear dominantly second-order process, such as a robot arm with negligible actuator delay and linkage flexibility. The process

$$g\{y\}\, \frac{d^2 y}{dt^2} + f\left\{ y,\, \frac{dy}{dt} \right\} = u \tag{21}$$

is controllable with u when the functions f and g are known,

$$u = f\left\{ y,\, \frac{dy}{dt} \right\} + g\{y\} \left[K(r - y) - D_M\, \frac{dy}{dt} \right] \tag{22}$$

This control, shown in Fig. 2, achieves linear closed-loop response to set point r,

$$y = \frac{Kr}{K + D_M s + s^2} \tag{23}$$

Here the proportional (K) and derivative (D_M) feedback terms should be chosen to achieve desired closed-loop performance, taking the neglected effective delay and high-frequency resonances into account, perhaps adaptively. If the K and D_M terms can be made large compared with f/g, the closed-loop performance may be quite insensitive to imperfect compensation for f. An integral term added to the proportional plus derivative controller would help to adapt the effective gain of the multiplicative g term. Dominantly first- or zero-order processes can be linearized similarly.

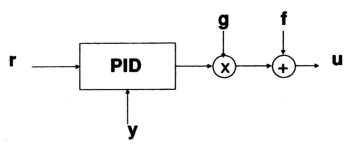

FIGURE 2   Feedback controller with multiplicative and additive compensation. (*Foxboro.*)

MULTIPLE-LOOP CONTROL

A cascade of control loops, where the output of a primary (outer-loop) controller is the set point of the secondary (inner-loop) controller, may improve performance of the outer loop, particularly when the primary measurement responds relatively slowly. Nonlinearity, such as results from a sticking valve, and disturbances within the fast inner loop can usually be made to have little effect on the slow outer loop. Limits on the primary output constrain the set point of the secondary loop. Typical secondary controlled variables are valve position and flow. Jacket temperature may be the secondary variable for a batch reactor. The design of the primary controller should provide means of preventing integrator windup when the secondary controller limits or is in manual [8].

Controllers also may be structured in parallel to provide a safety override of a normal control function. For example, the normal controlled variable may be a composition indicative of product quality. In an emergency, a pressure controller may take over its manipulated variable. This may be done by selecting the controller with the smaller (or larger) output or error to drive the manipulated variable. Means for preventing integrator windup of the unselected controller should be provided.

When there are multiple interacting controlled and manipulated variables, every controlled variable should be paired with a controller output in order to make each control loop as insensitive as possible to the status of the others. Bristol's relative gain array (RGA) [9] can help in the evaluation of potential pairs. A good pair choice may result by considering some controller outputs to be the sum (or ratio) of measurable variables. Control is then implemented with a cascade structure. One of the summed (or ratioed) variables, acting as a feedforward, subtracts from (or multiplies) the primary output to get the secondary set point for the other variable. Again, means to prevent integrator windup, when the secondary saturates, should be provided.

For example, in distillation column control (Fig. 3), the distillate (DF) and reflux (LF) flows may be manipulated to control the distillate (impurity) composition and condenser level. Normally LF is much larger than DF. If LF and DF were the controller outputs, an RGA would show that the larger flow LF could be paired with level and DF with composition [8]. However, if the composition controller adjusts [DF/(LF + DF)] and the level controller adjusts (LF + DF), an RGA would indicate minimal interaction since the ratio has no effect on level when the sum is constant. In this case the distillate set point is found by multiplying the composition controller output by the measured LF + DF, and the reflux set point is found by subtracting the measured DF from the level controller output. Dynamic compensation of the feedforward terms will not improve performance, since LF affects the composition with no more effective delay than DF.

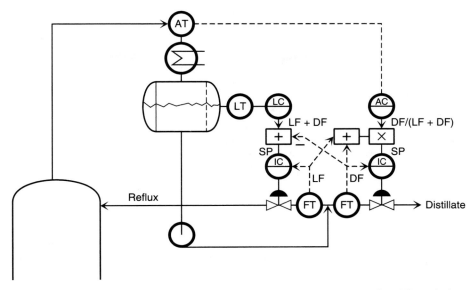

FIGURE 3 Level controller LC manipulates total flow and composition controller AC manipulates reflux ratio. (*Foxboro.*)

The RGA is an array where each element Γ_{ij} is a ratio of the sensitivities of a measurement y_i to an output u_j, the numerator having all other outputs fixed and the denominator having all other measurements fixed. For the process

$$A_1y_1 = b_{11}D_1u_1 + b_{12}D_2u_2$$
$$A_2y_2 = b_{21}D_1u_1 + b_{22}D_2u_2$$

(24)

where A_i and D_j are dynamic operators and b_{ij} are constants, the RGA elements are also constants,

$$\text{RGA} = \begin{bmatrix} \Gamma & 1-\Gamma \\ 1-\Gamma & \Gamma \end{bmatrix}$$

(25)

with only one interaction parameter Γ,

$$\Gamma = \frac{b_{11}b_{22}}{b_{11}b_{22} - b_{12}b_{21}}$$

(26)

[The number of interaction parameters is $(n-1)^2$, where n is the number of interacting loops, because each RGA row and column sums to 1.]

When Γ is between 0.5 and 2, (u_1, y_1) and (u_2, y_2) could be pairs. When Γ is between -1 and 0.5, the opposite pairs could be used. Least interaction occurs when Γ is 1 or 0, which happens when one of the b_{ij} terms is zero. Values of Γ smaller than -1 or larger than 2 indicate that neither set of pairs should be used because the interaction is too severe. Saturating one of the loops, or placing it in manual, would change the gain in the other by more than a factor of 2.

When there is effective delay associated with each of the b_{ij} terms (here b_{ij} is not entirely cancelable), it is useful to compare the sum of the b_{11} and b_{22} delays with the sum of the b_{12}

and b_{21} delays. If the interaction is significant (Γ not within 0.2 of either 0 or 1) and the combination with the smaller delay sum does not confirm the pairing based on Γ, a different choice of controller output variables (using a decoupling feedforward compensation) may be indicated.

DISTURBANCE REPRESENTATION

If the disturbance is a gaussian random variable, e is assumed to be a zero-mean white gaussian noise source. Colored noise is assumed to result from stably filtering the white noise e. The filter moving-average (numerator) characteristic is included in the C polynomial, and its autoregressive (denominator) characteristic is included in the A and B (or D) polynomials. The cross-correlation function of a linear filter's output with its input is equal to its impulse response convolved with the autocorrelation function of its input. When the input is white noise, its autocorrelation is an impulse function (the derivative of a step function). The cross correlation is then equal to the filter's impulse response [1]–[3].

Similarly, a deterministic disturbance may be considered to be the step or impulse response of a stable linear filter. A more complicated disturbance may be represented as the filter response to a sequence of steps (or integrated impulses). Any delay associated with an unmeasured disturbance is considered to determine the timing of the step in order that C_U be cancelable (stable zeros).

Pole-Placement Design

Pole-placement design [1] requires that A, B, C, and D be known and that a suitable H be selected. There may be difficulty in separating the BD product. As an expedient, B may be considered a constant and D allowed to include both stable and unstable zeros. If the degree of all of the polynomials is no greater than n, Eq. (10) provides $2n + 1$ equations, one for each power of s or z^{-1}, to solve for the coefficients of F and G. If the unmeasured disturbance e is considered an impulse, the degree of the G polynomial should be less than that of A, otherwise the degrees may be equal.

A design resulting in a B or F polynomial with a nearly unstable zero, other than one contributing integral action, should probably be rejected on the grounds that its robustness is likely to be poor. It may be necessary to augment H with additional stable factors, particularly if the degree of F or G is limited. The set point can be prefiltered to shift any undesired poles in the set-point function D/H to a higher frequency. However, the resulting uncompensatable unmeasured-load rejection function F/H may be far from optimal.

Linear-Quadratic Design

Linear-quadratic (LQ) design [1] provides a basis for calculating the H and F polynomials, but is otherwise like the pole-placement approach. In this case D contains both the stable and the unstable zeros and B is a constant. C/A is an impulse response function since e is specified to be an impulse. The H polynomial, which contains only stable zeros, is found from a spectral factorization of the steady-state Riccati equation,

$$\sigma H\{z\}H\{z^{-1}\} = \mu A\{z\}A\{z^{-1}\} + D\{z\}D\{z^{-1}\}$$

or

$$\sigma H\{s\}H\{-s\} = \mu A\{s\}A\{-s\} + D\{s\}D\{-s\} \tag{27}$$

The parameter σ is chosen to make the steady-state value of H unity and μ is an arbitrary parameter in the criterion function J,

$$J = E\{(r - y)^2 + \mu u^2\} \tag{28}$$

E is the expectation operator. The u term imposes a soft constraint on the manipulated variable with a penalty factor μ. A polynomial X, satisfying

$$X\{z\}A\{z^{-1}\} + \sigma H\{z\}G\{z^{-1}\} = D\{z\}C\{z^{-1}\}$$
$$X\{z\}H\{z^{-1}\} + \mu A\{z\}G\{z^{-1}\} = D\{z\}F\{z^{-1}\} \tag{29}$$

and Eq. (27) also satisfies Eq. (10) and minimizes Eq. (28) for an impulse disturbance e. The equations in s are similar. When the degree of A is n, the first equation of Eqs. (29) provides $2n$ equations, one for each power of z (or s). These can be solved for the $2n$ unknown coefficients of X and G, after H and σ are found from Eq. (27). G has no nth-degree coefficient and $X\{z\}$ and $D\{z\}$ have no zero-degree coefficient. [$X\{-s\}$ and $D\{-s\}$ have no nth-degree coefficient.] None of the polynomials is more than nth degree. F can then be found by polynomial division from Eq. (10) or the second of Eqs. (29).

For the optimization to be valid, the penalty factor μ must be large enough to prevent the manipulated variable u from exceeding its limits in responding to any input. However, if μ were chosen to be too large, the closed-loop response could be as sluggish as the (stabilized) open-loop response. The LQ problem may be solved leaving μ as a tuning parameter. Either experiments or simulations could be used to evaluate its effect on performance and robustness.

Despite the LQ controller being optimal with respect to J for a disturbance input, a switching (nonlinear) controller that takes into account the actual manipulated variable limits and the load can respond to a step change in set point r in less time and with less integrated absolute (or squared) error.

Minimum-Time Switching Control

The objective for the switching controller is to drive the controlled variable y of a dominant-lag process from an arbitrary initial value so that it settles at a distant target value in the shortest possible time. The optimal strategy is to maintain the manipulated variable u at its appropriate limit until y nears the target value r. If the process has a secondary lag, driving u to its opposite limit for a short time will optimally slow the approach of y to r, where it will settle after u is stepped to the intermediate value q needed to balance the load. Until the last output step, switching control is the same as "bang-bang" control. Determination of the output switching times is sufficient for open-loop control. The switching criteria must be related to y and its derivatives (or the state variables) in a feedback controller. Either requires solving a two-point boundary-value problem.

As an example, consider a linear integral $\{T\}$–lag $\{\tau_L\}$–delay $\{\tau_D\}$ process with constant manipulated variable (controller output) u. The general time-domain solution has the form

$$x\{t\} = at + b \exp\left\{-\frac{t}{\tau_L}\right\} + c$$
$$y\{t\} = x\{t - \tau_D\} \tag{30}$$

where x is an unmeasured internal variable and a, b, and c are constants that may have different values in each of the regimes. At time zero the controlled variable y is assumed to be approaching the target value r from below at maximum rate,

$$\frac{dy\{0\}}{dt} = \frac{dx\{0\}}{dt} = \frac{u_M - q}{T} = a - \frac{b}{\tau_L}$$

$$x\{0\} = b + c \tag{31}$$

$$y\{0\} = x\{0\} - \frac{(u_M - q)\tau_D}{T}$$

where u_M is the maximum output limit and q is the load. At that instant the output is switched to the minimum limit, assumed to be zero. If the next switching were suppressed, x and y would eventually achieve their negative rate limit,

$$\frac{dy\{\infty\}}{dt} = \frac{dx\{\infty\}}{dt} = -\frac{q}{T} = a$$

Combining with Eqs. (31) to eliminate a,

$$b = -\frac{u_M \tau_L}{T}$$

$$\tag{32}$$

$$c = x\{0\} - b$$

At time t_1, x reaches the target value r with zero derivative,

$$\frac{dx\{t_1\}}{dt} = 0 = -\frac{q}{T} + \frac{u_M}{T} \exp\left\{-\frac{t_1}{\tau_L}\right\}$$

$$\tag{33}$$

$$x\{t_1\} = r = x\{0\} - \frac{q(t_1 + \tau_L)}{T} + \frac{u_M \tau_L}{T}$$

Consequently,

$$t_1 = \tau_L \ln\left\{\frac{u_M}{q}\right\} \tag{34}$$

and switching from maximum to zero output occurred when

$$r - y\{0\} - \left(\tau_D + \tau_L - t_1 \frac{q}{u_M - q}\right) \frac{dy\{0\}}{dt} = 0 \tag{35}$$

If q is zero, as it may be when charging a batch reactor, the qt_1 product is zero, even though t_1 is infinite.

Optimal switching, from zero output to that required to match the load q, occurs at time t_1. If t_1 is less than the delay time τ_D,

$$r - y\{t_1\} - \left(\tau_D + \tau_L - t_1 \frac{u_M}{u_M - q}\right) \frac{dy\{t_1\}}{dt} = 0 \tag{36}$$

Otherwise

$$r - y\{t_1\} - \left(\tau_L - \frac{\tau_D}{\exp\left\{\dfrac{\tau_D}{\tau_L}\right\} - 1}\right) \frac{dy\{t_1\}}{dt} = 0 \tag{37}$$

The controlled variable y will settle at the target value r at time $t_1 + \tau_D$. At this time the switching controller could be replaced by a linear feedback controller designed to reject unmeasured load disturbances and correct for modeling error. This combination of a switching controller with a linear controller, called dual mode, may be used to optimally start up and regulate a batch process [9].

Minimum-Variance Design

The minimum-variance design is a special case of the LQ approach where μ is zero. This may result in excessive controller output action with marginal improvement in performance and poor robustness, particularly when a small sampling interval h is used. Performance is assumed limited only by the nonminimum-phase (unstable) zeros and delay included in D. The H polynomial contains the stable zeros and the reflected (stable) versions of unstable zeros of D.

When D is a k time-step delay and e is considered an impulse, the minimum-variance solution for $F\{z^{-1}\}$ and $G\{z^{-1}\}$ can be found from Eq. (10) by polynomial division of HC by A. $F\{z^{-1}\}$ consists of the first $k-1$ quotient terms. The remainder $(HC - AF)$ is $DG\{z^{-1}\}$. From Eq. (27) H is unity. However, if H is arbitrarily assigned, this design becomes pole placement.

As a minimum-variance example, consider D to be a delay. Again, H is unity. The disturbance e is assumed to be a step applied downstream of the delay. C is unity. B is a gain b. The A polynomial represents a kind of lag,

$$A = 1 - aD \tag{38}$$

In one delay time the controlled variable can be returned to the set point r, hence $F = 1 - D$. Equation (10) becomes

$$1 = (1 - aD)(1 - D) + DG \tag{39}$$

Solving for G,

$$G = 1 + a - aD \tag{40}$$

and the controller from Eq. (13) becomes

$$u = \frac{(r - y)/(1 - D) - ay}{b} \tag{41}$$

When the delay is one computing step h, the process can be considered a sampled first-order lag with time constant $h/\ln\{1/a\}$ and a zero-order hold. Equation (41) has the form of a digital proportional plus integral controller with proportional band PB $= b/a$ and integral time IT $= ah$. When $a = 1$, the process can be considered a sampled integral with time constant h/b and a zero-order hold. When a is zero, the process is a pure delay and the controller is "floating" (pure integral, PB IT $= bh$). The controlled variable y has dead-beat response in one time step for either a set point or a load step,

$$y = Dr + (1 - D)e \tag{42}$$

The response is also optimum with respect to the minimum largest absolute error and the minimum integrated absolute error (IAE) criteria.

When the delay has k time steps, the $F = 1 - D$ factor in the controller equation has k roots equally spaced on the unit circle. One provides infinite gain at zero frequency (integral action). The others make the controller gain infinite at frequencies that are integer multiples of $1/kh$, not exceeding the Nyquist frequency $1/2h$. These regions of high gain cause the loop stability to be very sensitive to mismatch between the actual process delay and kh used in the controller. As k approaches infinity, the robustness index approaches zero, indicating no tolerance of delay mismatch. A low-pass filter that improves the robustness by attenuating in these regions of high gain also degrades the nominal performance.

The control loop may be structured as two loops, the outer-loop integral controller providing the set point r_I to the inner-loop proportional controller,

$$r_I = \frac{r - y}{1 - D} = r - y + Dr_I \tag{43}$$

and

$$u = \frac{r_I - ay}{b} \tag{44}$$

The effect of closing only the inner loop is to create a delay process as seen by the outer-loop controller,

$$y = Dr_I + e \tag{45}$$

The outer-loop controller can be considered model feedback in relation to the closed inner loop, since the difference between the controlled variable's measured (y) and predicted (Dr_I) values is fed back as a correction to an open-loop (unity-gain) controller.

Model-Feedback Control

Model-feedback control, whose variations include, among others, Smith predictor, Dahlin, dynamic matrix, and model predictive (unless a special unmeasured disturbance model is used [10]), consists of an open-loop controller with a feedback correction equal to the model prediction error. If model-feedback control is applied without first closing a proportional inner loop, there results

$$u = \frac{A}{BH} \left[r - \left(y - \left(\frac{BD}{A} \right) u \right) \right]$$

$$y = \frac{D}{H} r + (1 - D) \left(\frac{C}{AH} \right) e \tag{46}$$

assuming no process-model mismatch and no unstable zeros of A. H is a filter chosen to improve robustness. For the above process, assuming H to be unity, the response to a step disturbance can be easily calculated by polynomial division,

$$F = (1 - D) \frac{C}{A} = \frac{1 - D}{1 - aD}$$

$$= 1 - (1 - a)D \left[1 + (aD) + (aD)^2 + (aD)^3 + \cdots \right] \tag{47}$$

The maximum error occurs during the first delay interval when $F\{0+\} = 1$. After n delay steps $F\{n+\}$ is reduced to a^n. Even though this result is optimum with respect to the minimum largest absolute error criterion, the recovery from a load upset can be very slow when a is close to 1 (or divergent, when a is greater than 1). The ratio of the IAE to the optimum is $1/(1 - a)$. Consequently model-feedback control may not adequately reject an unmeasured load disturbance, when the process has a dominant lag (a near 1), unless well-chosen inner-loop feedback is applied or the model deliberately mismatches the process, as recommended in [6]. However, design of the inner-loop controller or the model mismatch, for near optimal load rejection, may require more detailed high-frequency knowledge (for example, a spectral factorization of process polynomials) than is necessary for selecting an output trajectory (open-loop controller) to achieve good set-point tracking.

The stability of a tightly tuned matched-model feedback loop is very sensitive to mismatch between the model and process delays. To achieve adequate robustness, it may be necessary to detune the controller with H, further sacrificing unmeasured-load rejection capability.

Without the inner loop or deliberate model mismatch, early return of the output from saturation may cause excessively slow controlled-variable response of a dominant-lag process to a large set-point step. As with open-loop control, an on-line nonlinear optimizer may be used to avoid this suboptimal behavior.

Algebraic Proportional plus Integral plus Derivative Design

In this section a two-phase method for applying Eq. (10) to the design of analog (or fast-sampling digital) proportional plus integral plus derivative (PID) controllers is described. Unlike Bode and root-locus design methods, this method allows all of the controller parameters as well as the closed-loop performance parameters (time scale and load sensitivity) to be found directly, without trial and error.

The process is represented with an A polynomial in s. B, D, and C are assumed 1. The inverse of a delay or small numerator zero factor (if representable as a convergent Taylor series up to the frequency range critical for stability) is included in the A polynomial,

$$(a_0 + a_1 s + a_2 s^2 + a_3 s^3 + \cdots)y = u + e \tag{48}$$

A large stable zero is unusual and requires special consideration. It should be approximately canceled by a controller or process pole. Such a pole is not available from a PID controller. However, an effective process cancellation, without a factorization, may result by disregarding the zero-order terms of both the process numerator and the denominator before determining A by polynomial division. When the process zero and pole are sufficiently dominant, mismatch in zero-order terms, which affects the very low-frequency behavior, will be corrected by high controller gain in that frequency range.

This two-phase design process implicitly imposes the performance limitation that would normally be imposed by including delay and nonminimum-phase zeros in D. The first design phase prevents inner-loop feedback when the open-loop process already approximates a pure delay. This is done by selecting the inner-loop gain and derivative terms,

$$G_I = K_M + D_M s \tag{49}$$

to make the closed inner loop H_I^{-1} approximate a delay at low and moderate frequencies. As many low-order terms of Eq. (10) are matched as are needed to determine the unknown controller and performance parameters.

The H_I polynomial is chosen to be the Taylor-series expansion of an inverse delay whose time τ_I and gain h_0 is to be determined,

$$H_I = h_0 \left[1 + \tau_I s + \frac{(\tau_I s)^2}{2} + \frac{(\tau_I s)^3}{6} + \cdots \right] \tag{50}$$

When F_I is chosen as 1 (instead of choosing h_0 as 1), Eq. (10) gives

$$H_I = A + G_I \tag{51}$$

The limited-complexity inner-loop proportional plus derivative controller can significantly influence only the low-order closed-loop terms and hence can shape only the low-frequency closed-loop behavior. Only the most dominant two poles (lowest in frequency) of the open-loop process may be unstable, since only they can be stabilized with proportional and derivative feedback. As a result the limiting closed inner-loop performance measures, the values of τ_I and h_0, are determined by a_2 and a_3, provided the latter have the same sign. Equating term by term and rearranging gives

$$\tau_I = \frac{3a_3}{a_2}$$

$$h_0 = \frac{2a_2}{\tau_I^2} \tag{52}$$

$$D_M = h_0\tau_I - a_1$$

$$K_M = h_0 - a_0$$

When the sign of D_M is different from that of K_M, or derivative action is not desired, the parameters should be calculated with

$$\tau_I = \frac{2a_2}{a_1}$$

$$h_0 =$$

$$\frac{a_1}{\tau_I} \tag{53}$$

$$D_M = 0$$

$$K_M = h_0 - a_0$$

For a pure delay process both K_M and D_M are zero. The closed inner loop becomes

$$y = \frac{r_I + e}{H_I} \tag{54}$$

The outer loop uses gain and integral terms applied to the error,

$$r_I = \left(\frac{1}{I_E s} + K_E \right)(r - y) \tag{55}$$

Using this equation to eliminate r_I from the previous gives

$$[1 + I_E s(K_E + H_I)]y = (1 + K_E I_E s)r + I_E s e \tag{56}$$

The target closed-loop set-point behavior is chosen to approximate a nonovershooting delay-like model, n equal lags. The shape parameter n and the time constant τ_0 are to be determined, as are the controller parameters K_E and I_E,

$$\left(\frac{1 + \tau_0 s}{n} \right)^n y = r + \frac{I_E s e}{1 + K_E I_E s} \tag{57}$$

Equating term by term and solving the four simultaneous equations gives

$$n = 10.4$$

$$\tau_0 = 1.54\tau_I$$

$$K_E = 0.198h_0 \tag{58}$$

$$I_E = \tau_0/h_0$$

A small value of I_E is desirable because its product with the output change is equal to the integrated error response to a load change. Since the controller is designed to achieve very small overshoot, the product is also nearly equal to the IAE for a step-load change. If the response shape parameter n were made infinite (corresponding to a pure delay target) instead of matching the fourth-degree terms, a faster but more oscillatory and less robust response would result (with near minimum IAE for a load step). Then the closed-loop time constant τ_0 would become $1.27\tau_I$ and K_E would equal $0.289h_0$. Equations (58) still would apply for I_E.

The resulting controller is a four-term noninteracting (sum of terms) type. The following equations may be used to convert these tuning values to those for a conventional three-term PID interacting (product of factors) type, adjusted for good load rejection,

$$K = K_E + K_M \tag{59}$$

If K^2 is greater than four times the ratio of D_M to I_E,

$$\frac{1}{PB} = 0.5\left[K + \left(K^2 - \frac{4D_M}{I_E}\right)^{0.5}\right]$$

$$IT = I_E/PB \tag{60}$$

$$DT = D_M\,PB$$

Otherwise,

$$PB = \frac{2}{K}$$

$$IT = DT = (I_E D_M)^{0.5} \tag{61}$$

To achieve the designed set-point response also, the set-point input should be applied to the controller through a lead-lag filter. The lag time is matched to the integral time IT. The ratio of lead-to-lag time α is made equal to $K_E\,PB$. The resulting controller equation is

$$u = \frac{(1 + \alpha ITs)r - (1 + ITs)(1 + DTs)y}{PB\,ITs} \tag{62}$$

Derivative action is applied only to the controlled measurement, not to the set point. To prevent excessive valve activity at frequencies beyond the closed-loop bandwidth, it is customary to condition the controlled measurement y with a low-pass filter whose time constant is a small fraction (≈ 0.1) of DT. In order that the sampling process of a digital controller not further diminish the effectiveness of the derivative term, the sampling interval should be less than the effective filter time.

For example, consider a thermal process with a 300-second lag and a 10-second effective delay. The algebraic design gives an integral time IT = 24 seconds, derivative time DT = 3.6 seconds, and the closed-loop time constant $\tau_0 = 23.1$ seconds, all more sensitive to the delay than the lag. The sampling interval should not exceed 0.4 second in order not to compromise closed-loop performance. This is a surprisingly small interval, considering the relatively slow open-loop response.

Antialias Filtering

When an analog signal contains a component with frequency higher than the Nyquist frequency (half the sampling frequency f_S), the sampled signal component appears to have a frequency less than the Nyquist frequency, as shown in Fig. 4. If the analog signal component frequency f lies between odd integers $(2n - 1)$ and $(2n + 1)$ times the Nyquist frequency,

$$(n - 0.5)f_S \leq f < (n + 0.5)f_S \tag{63}$$

the sampled signal component has the same amplitude but appears to be shifted to the frequency $|f - nf_S|$. This frequency shifting is called aliasing [1].

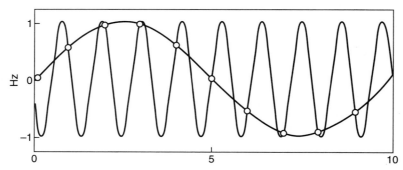

FIGURE 4   Two signals with different frequencies (0.1 and 0.9 Hz) have same values at 1-Hz sampling instants. The 0.9-Hz signal is aliased as 0.1 Hz after sampling. (*Foxboro.*)

To achieve good performance with digital control, it is necessary to sample the controlled variable at a rate faster than twice the highest frequency significant for control and to attenuate, before sampling, components with higher than Nyquist frequency. This should be done with minimum attenuation or phase shifting of the lower frequency components that are important for control.

It is particularly important to remove, before sampling, a signal component that is an integer multiple of the sampling frequency, because this component would shift to zero frequency, causing a steady-state offset error. An analog filter that averages between sampling instants removes such components completely. This filter can be realized with an analog-to-frequency converter and a sampled digital counter. The longer the sampling interval, the higher is the count and the greater is the digital signal resolution.

However, the analog averaging filter does not attenuate sufficiently near the Nyquist frequency. A stage of digital filtering, a two-sample average, completely removes a Nyquist-frequency component in the sampled signal. Its passband is flatter and its cutoff is sharper than those of a digital Butterworth (autoregressive) filter having the same low-frequency phase (effective delay).

This measurement antialias filter together with the manipulated variable zero-order hold adds an effective delay of $1.5h$ to the analog process. Calculation delay, which may be as large as the sampling interval h, is additional. Consequently the sampling interval should be small compared with the effective delay of the process, in order that feedback loop performance (unmeasured-load rejection) not be compromised. On the other hand, resolution is improved by averaging over a longer sampling interval.

An antialiasing filter removes or diminishes the effects of higher than Nyquist frequency process signals on the digital measurements. However, digital outputs apply a sequence of steps to the process that may excite a higher than Nyquist-frequency open-loop resonance (such as water hammer). This effect will be accentuated in high-performance loops with large output steps and in loops where the resonance is synchronized with a multiple (harmonic) of the output update frequency.

ADAPTIVE CONTROL

Time-varying or nonlinear process dynamics, variable operating conditions, slow response to upsets, dominant unmeasured disturbances, performance degradation resulting from deliberate upsets, and lack of tuning expertise are all reasons to consider adaptive (self-tuning) control. Economic incentives may result from improved control of product quality and yield, a higher production rate, less energy usage, improved plant safety, and less pollution.

Adaptive control schemes may be classified according to several design alternatives. Associated with each of these are concerns that should be addressed in order to assure robust adaptation.

First, the adaptor may be open- or closed-loop. An open-loop adaptor programs the controller tuning based on a model of the process. The model may be fixed and nonlinear (such as a neural net) or time-varying and linear, with parameters updated to reconcile measured process inputs and outputs. Mismatch between the model and the process may be caused by misidentification as a result of large nonstationary unmeasured disturbances, insufficiently rich inputs, or by model structural deficiencies. Mismatch may lead to inappropriate controller tuning that will not be corrected until (or unless) the model is revised. For example, even though the effective process delay limits how tightly the controller can be tuned, the model delay may be assigned arbitrarily (it is not easily identified) or the process delay masked by a larger sampling interval. On the other hand, a closed-loop adaptor monitors one or more performance measures for the closed control loop and adjusts controller parameters to drive these measures to target values. Desired performance is assured when the feedback adaptor converges. Its issues include the rate of convergence and the robustness of the adaptive loop.

Second, the adaptation may be based on observations of responses to deliberate or natural disturbances. For example, a performance-optimizing adaptor, using a hill-climbing strategy, requires the comparison of process performance measures for responses to identical, hence deliberate, disturbances. However, a deliberate disturbance degrades a well-tuned loop's performance. Therefore deliberate upsets should be applied infrequently and only when there is high likelihood that the controller is mistuned and is not likely to recover soon. A deliberate disturbance has the advantage that it is known and can be made large enough to dominate unmeasured disturbances and rich enough to excite important process modes. On the other hand, natural disturbances may be unmeasured and nonstationary, often consisting of significant isolated filtered steplike events and low-level noise. However, a response to a natural disturbance may contain incomplete information to observe all process modes and make an unambiguous adaptation.

Third, the target control-loop time scale may be fixed or optimal. One may choose to drive the closed-loop performance to that of a fixed model, as is done with pole-placement, fixed model-delay minimum-variance, or model-reference adaptors. A fixed time scale for all operating conditions must be user-selected to be as large as the largest minimum, and hence suboptimal for other operating conditions. Extensive knowledge of the process dynamics helps in choosing the time scale. Alternatively the control-loop time scale may be adapted in an open-loop minimum-variance adaptor by updating the effective delay time, or in a performance-feedback adaptor by choosing response shape parameters that are sensitive to the relative positions of the three (or more) most dominant closed-loop poles. On the other hand, a tightly tuned controller may not have an adequate stability margin, particularly if the (nonlinear) process may experience sudden large unmeasured load changes.

Fourth, the adaptations may be performed continuously, with updates each sampling interval, or aperiodically, following the response to each significant disturbance. When the controller tuning is updated each sampling instant, the update must recursively take into account a portion of past history. Over this time interval both the process parameters and the statistical measures of unmeasured disturbances are assumed to be stationary. However, real processes may be subjected to minimal load changes for extended intervals and to large unmeasured nonstationary load changes at other times. This makes choosing an adequate time interval difficult, particularly when effective adaptation requires that a time-varying linear model track a nonlinear process. On the other hand, an adaptive scheme designed to update controller parameters following significant isolated disturbance responses must also cope with cyclical and overlapping responses. It should distinguish between a loop instability, a stable limit cycle, and a cyclical disturbance.

Cycling can also be a problem for an adaptor based on model identification, because the signals may provide incomplete information for identifying more than two process parameters. An incorrect identification may lead to worse controller tuning. Furthermore, if the cycle amplitude is large, it may be impractical to make the identification unique by superimposing sufficiently significant deliberate disturbances.

An event-triggered adaptor provides an additional opportunity: the state and measured inputs, existing at the moment a new disturbance is sensed, can be used to select among several sets of stored tunings to determine the set to be used during, and updated following, the response interval. This capability (such as gain scheduling and multiplicative feedforward compensation) enables the controller to anticipate and compensate for the effect of process nonlinearity.

Three types of adaptive controllers will be discussed. First is the performance-feedback type, using expert system rules to cope with incomplete information, and using nonlinear dead-beat adaptation when information is complete. The second is an open-loop type that uses a recursive parameter identifier to update the parameters of a difference equation model. The controller is tuned as if the model were the process, thus invoking the "certainty equivalence" principle. Third is another open-loop type. This one identifies the low-order parameters of a differential equation model in order to update the coefficients in a feedforward compensator. It uses the moment-projection method on each isolated response.

Other types of adaptors, including the model reference type, may be more suitable for set-point tracking than for unmeasured load rejection. For example, the extended-horizon type uses a nonlinear optimizer to determine an open-loop controller's constrained output trajectory. The controlled-variable trajectory must be optimized over a prediction interval that exceeds that of the output by the process delay time. An extended-horizon optimizer can be used in conjunction with predictive-model feedback, useful (as explained earlier) for load rejection with a dominant-delay process and for model mismatch correction with set-point tracking. The nonlinear optimization calculations may require a large time step that compromises control loop performance with additional effective delay. Also, a linear moving-average process model may have so many degrees of freedom that on-line identification is impractical, both because the computational load is excessive and because a reasonable interval of natural-signal history does not persistently excite all of the model and process modes. A fixed nonlinear neural-net model may be more practical for a time-invariant process, even though its programming requires training with extensive signal records spanning the range of expected operating conditions.

PATTERN RECOGNITION AND EXPERT SYSTEMS, PERFORMANCE FEEDBACK ADAPTOR

A performance feedback adaptor monitors a single-loop variable control error [11]. Since the set point and process measurement are known, the controller output does not contain independent information, because it is determined by the controller equation. Pattern features are measured of the error response to a significant (typically unmeasured) disturbance. When enough of the response has been observed, the controller parameters are adjusted in order to make the feature values of the next response approach target values.

A significant error event is detected when the absolute control error exceeds a set threshold. The threshold value is chosen large enough so that an error event will not be triggered by low-level process or measurement noise. Since error peaks are the most prominent features of an oscillatory response, peak amplitudes and times are sought. Expert system (heuristic) rules may be used to distinguish response peaks from noise peaks.

Zero-to-peak $(-E_2/E_1)$ and peak-to-peak $[(E_3 - E_2)/(E_1 - E_2)]$ error amplitude ratios may be chosen as shape features. These are independent of the response amplitude and time scales and are called overshoot and decay, respectively. Target values for these ratios may be chosen to minimize a criterion function, such as minimum IAE, for a specific process and disturbance shape. The ratio of times between peaks is not recommended as a controlled-shape feature, because it is relatively sensitive to nonlinearity and noise and insensitive to the relative location of closed-loop poles.

The second and third peaks, E_2 and E_3, do not exist for an overdamped response. In this case a "knee" (or quasipeak, defined by "expert" criteria) is sought to determine a response time scale used to terminate the peak search and to estimate the (negative) effective overshoot. Decay is zero in this case.

If the error response were a damped quadratic function, the first three error peaks E_i are related by

$$E_1E_3 = E_2^2 \qquad \text{decay = overshoot} \tag{64}$$

This response, like the overdamped response, does not contain complete information for controller tuning [12]. A response containing both lag and underdamped quadratic terms provides complete information for tuning a PID controller. For the response

$$E\{t\} = \alpha e^{-at} + \beta e^{-bt} \cos\{\omega t\} \tag{65}$$

three shape ratios β/α, b/a, and b/ω provide sufficient information to update three controller parameters PB, IT, and DT. However, if either α or β were 0, values of a, or b and ω, would be unmeasurable and the information incomplete. It is desirable that the features used for adaptation reflect the relative pole positions indicated by the last two ratios, because the dominant error poles are usually closed-loop poles. Furthermore, the features should be insensitive to the first ratio, β/α, because this ratio is sensitive to the unmeasured disturbance shape and point of application, as indicated by the relative location of the error signal zeros.

When information is complete from the error response to a load step applied upstream of a dominant lag or delay,

$$E_1E_3 > (E_2)^2 \qquad \text{decay > overshoot} \tag{66}$$

This corresponds to the lag (α) and quadratic (β) terms making contributions of the same sign to the first peak.

These inequalities can be reversed if the disturbance has a different shape or is applied at a different location. Reversal would result if the disturbance were a narrow pulse or if it were a step applied downstream of a dominant lag. Reversal also would result if the disturbance were statically compensated by a feedforward controller and could result if the disturbance were to affect interacting loops. The final part of this type of error response has a shape similar to the "usual" response so that, when the first peak, or peaks, is discarded and the remaining peaks are renumbered, decay becomes greater than overshoot.

Peak shifting desensitizes the pattern features to open-loop and disturbance-signal zeros, while maintaining sensitivity to the relative positions of the three most dominant error-signal poles. These poles usually are the closed-loop poles, but may include the poles of the disturbance signal if the disturbance is not applied suddenly.

The changes in the controller-parameter vector P are computed from the deviation of the measured-feature vector F from its target-value vector F_t, according to the adaptor's nonlinear gain function matrix $\mathbf{G}$,

$$P\{i + 1\} = P\{i\} + \mathbf{G}(F_t - F\{i\}) \tag{67}$$

The response feature vector $F\{i\}$, measured after the ith response, is a nonlinear function of the controller parameter vector $P\{i\}$ (existing during the ith response), the process type, and the disturbance shape. For a given process and disturbance shape, simulation can be used to map feature deviations as a function of control parameter deviations $\delta \mathbf{F}/\delta \mathbf{P}$, allowing feature deviations to be predicted with

$$F\{i+1\} = F\{i\} + \frac{\delta \mathbf{F}}{\delta \mathbf{P}} \ (P\{i+1\} - P\{i\}) \qquad (68)$$

If the unique inverse of the function matrix $\delta \mathbf{F}/\delta \mathbf{P}$ exists, the latest response contains complete information. Then a dead-beat adaptation $(F\{i+1\} = F_t)$ is possible with

$$\mathbf{G} = \left(\frac{\delta \mathbf{F}}{\delta \mathbf{P}} \right)^{-1} \qquad (69)$$

This multivariable adaptive loop is quite robust, being particularly tolerant of smaller than optimum $\mathbf{G}$. For example, if the optimum $\mathbf{G}$ is multiplied by a factor ranging between 0 and 2, the adaptive loop will remain stable. The eigenvalues of the $I - (\delta \mathbf{F}/\delta \mathbf{P})\mathbf{G}$ matrix, nominally located at the origin of the complex z plane, must stay within the unit circle.

If the process were a lag delay, a process-type variable, sensitive to the ratio of the lag-to-delay times, could be used to interpolate between the two extremes. The ratio of the controller IT to the response half-period is an indicator of the process type when decay and overshoot are fixed. This property may be used to identify the process type. The optimal IT-to-half-period ratio IT/T is smaller for the pure delay than for the integral delay.

When only two features, such as overshoot and decay, reliably characterize a response shape, only two controller parameters (PB and IT of a PID controller) are determined through performance feedback. However, because the optimal derivative-to-integral ratio is also a function of the process type, DT can be calculated after IT and the process type have been determined.

The process type, that is, the proportional-band ratio PB/PB_t and the integral-time ratio IT/IT_t, is determined by interpolating stored data from performance maps for the process-type extremes, given overshoot, decay, IT/T, and DT/IT. The half-period T and the controller parameters PB, IT, and DT are values for the latest response. PB_t and IT_t are the newly interpolated values of the controller parameters predicted to produce the target features on the next response.

When the error-shape information is incomplete, as when the response is overdamped, quadratic, nonisolated, or nonlinear (because the measurement or controller output has exceeded its range), expert system rules are used to improve the controller tuning. These rules, of the if-then-else type, invoke a special strategy for each of these contingencies. Several retunings may be needed before a response shape contains sufficient information to achieve the desired performance on the next response. Even when the information is incomplete, robust tuning rules are possible, provided derivative action is not essential for stabilizing the control loop.

A nonisolated response is recognized if its start is detected while waiting for the last response to settle. A nonisolated response may be caused by the failure of the preceding response to damp quickly enough. If the decay of a nonisolated response is sufficiently small, even though it may be bigger than the target, the existing tuning is retained. Typically a continuing oscillation will be dominated by a quadratic factor, giving rise to incomplete information for retuning.

A nonisolated response may also be caused by a rapid sequence of load changes, the next occurring before the response to the last has settled. Peak shifting tends to desensitize the adaptor to a strange sequence of peaks, allowing detuning only when a conservative measure of decay is excessive.

A marginally stable loop is distinguished from a limit cycle or response to a cyclical load by observing the improvement in decay caused by an adaptive retuning. If retuning fails to reduce the decay measure, the last successful tunings may be restored and adaptation suspended until settling or an operator intervention occurs.

DISCRETE-MODEL IDENTIFICATION, OPEN-LOOP ADAPTATION

Adaptation of a feedback controller based on an identification of an input-output process model is most effective when the important process inputs are measured and (at least partially) uncorrelated with one another. An unmeasured disturbance is assumed to come from a stationary filtered white gaussian-noise source uncorrelated with the measured inputs. When a large unmeasured disturbance violates this assumption, the identified model may be a poor match for the process and poor controller tuning may result. Process-model mismatch may also result when the disturbance fails to independently excite process or model modes, a condition called nonpersistent excitation. A poor model structure, such as one having an incorrect unadapted delay or insufficient model degrees of freedom, may also cause mismatch that leads to poor control. Coefficients for a model having both linear and nonlinear terms may not be uniquely identifiable if the process input and output changes are small.

Two types of models may be identified, called explicit and implicit. An explicit model relates the process inputs and output with parameters natural to the process, such as Eq. (5). The explicit model is most useful for the design of an open-loop or model-feedback controller. A complicated design process involving Eq. (10) would be needed to compute the feedback controller parameters of Eq. (13). An implicit model combines the target equation, Eq. (8), and the feedback control equation, Eq. (13), so that the parameters needed for control are identified directly. In either case the identification model may be put in the prediction form

$$\Omega\{t + k\} = \Phi\{t\}^T \Theta + \varepsilon\{t + k\} \tag{70}$$

which predicts the value of Ω, k time steps ahead, given present and past values of the process inputs and outputs concatenated in the vector Φ. Θ is a corresponding vector of parameters determined by the identifier; ε is the identification error.

For the explicit model,

$$Ay = DBu + Ce + e_0 \tag{71}$$

Here e_0 is the steady-state offset and

$$A = 1 + a_1 z^{-1} + \cdots$$
$$C = 1 + c_1 z^{-1} + \cdots \tag{72}$$
$$B = b_0 + b_1 z^{-1} + \cdots$$

The time step h is assumed to be one time unit. The time step should be chosen small enough that the antialias filter, the digital computation, and the output hold do not dominate the effective delay, but large enough that roundoff or data storage do not cause difficulty. Here D is assumed to be a known and fixed k-time-step delay. The value of k must be large enough that B is stably cancelable. Of course, other choices for D are possible. The prediction model variables become

$$\Omega\{t\} = y\{t\}$$
$$\Phi\{t - 1\}^T = [u\{t - k\}, \ldots, -y\{t - 1\}, \ldots, \varepsilon\{t - 1\}, \ldots, 1] \tag{73}$$
$$\Theta^T = [b_0, \ldots, a_1, \ldots, c_1, \ldots, e_0]$$

If the model matched the process exactly, the prediction error $\varepsilon\{t\}$, which is uncorrelated with the variables in $\Phi\{t - 1\}$, would equal the white-noise disturbance $e\{t\}$. In order to identify C, past values of the prediction error are needed, but these cannot be found until Θ is identified. This difficulty can be overcome by solving for Θ recursively. When Θ is updated each time step, $\varepsilon\{t\}$ can be calculated using the most recent Θ. C must be constrained to have stable zeros. An algorithm that identifies C is said to be "extended." When Φ and Ω are prefiltered by C^{-1}, the algorithm is "maximum likelihood." If the identifier inputs are prefiltered by E^{-1}, $(E/C) - 0.5$ must be positive real in order to ensure that Θ can converge to its true value [13]. Convergence also requires no structural mismatch.

The model form for implicit identification uses the target equation, Eq. (8), to eliminate $r\{t - k\}$ from the controller equation, Eq. (13),

$$Hy\{t\} = BFu\{t - k\} + Gy\{t - k\} - \acute{C}r\{t - k - 1\} + e_0 + Fe\{t\} \tag{74}$$

where $\acute{C} = C - 1$ and D is a k-step delay ($h = 1$). The prediction model variables become

$$\Omega\{t\} = Hy\{t\} = h_0 y\{t\} + h_1 y\{t - 1\} + \cdots \tag{75}$$

where H is specified.

$$
\begin{aligned}
BF &= \beta_0 + \beta_1 z^{-1} + \cdots \\
G &= \alpha_0 + \alpha_1 z^{-1} + \cdots \\
\Phi\{t - k\}^T &= [u\{t - k\}, \ldots, y\{t - k\}, \ldots, -r\{t - k - 1\}, \ldots, 1] \\
\Theta^T &= [\beta_0, \ldots, \alpha_0, \ldots, c_1, \ldots, e_0]
\end{aligned}
\tag{76}
$$

To identify the C polynomial, the set point r must be active, the updated control law implemented, and a recursive algorithm used. C must be constrained to have stable zeros. Also, k must be large enough that the zeros of BF are stable. If the model matched the controlled process exactly, the modeling error $\varepsilon\{t\}$, which is uncorrelated with any of the variables in $\Phi\{t - k\}$, would equal the closed-loop noise response $Fe\{t\}$. If $H = 1$, this implies a minimum-variance design, otherwise pole placement.

The same positive real and structural consistency requirements apply for convergence of an implicit parameter set as apply for the explicit set. The positive real condition assures that the component of control error Fe, in phase with the model error ε, is positive and at least half as big.

Control based on either model will not have integral action when the identifier is turned off. Integral action depends on updating the offset e_0. The effective integral time constant depends on the quantity of past history taken into account in calculating Θ. Consequently it is likely to be significantly larger than optimal. On the other hand, integral action can be implemented explicitly in an outer-loop controller, such as

$$\delta r_I = \frac{r - y}{2k - 1} \tag{77}$$

without identifying e_0, if an incremental identification model is used to design the inner-loop controller,

$$\delta u = \frac{(C \, \delta r_I - G \, \delta y)}{BF} \tag{78}$$

For an incremental model, the values of variables in Ω and Φ are the changes δu and δy from one time step to the next. If C were 1 in such a model, the autocorrelation function of the unmeasured disturbance noise would be a step instead of an impulse. The inner-loop set point δr_I will be active, providing excitation to a mode (allowing identification of β_0 as well as α_0) that would not be excited in a single loop structure when the set point r is fixed.

A restricted complexity model has fewer modes than the process. Consequently its modeling error will have components resulting from structural mismatch as well as unmeasured disturbances. It may have substantially fewer parameters than an "exact" model. For example, the C and β polynomials may be restricted to one term ($C = 1$, $BF = \beta_0$) to be certain that they have no unstable zeros. Less past history is needed to reliably identify a small number of parameters since fewer equations are needed to solve for fewer unknowns. Therefore a restricted complexity model can be updated more quickly, allowing it to better track a changing or nonlinear process. The identifier inputs Φ and Ω should be filtered in order to make the process-model match best in the critical frequency range, where the open-loop absolute gain is near 1 for feedback control or near steady state for open-loop, or feedforward, control.

The implicit model form can be used to identify the delay time. The same Φ vector can be used for a number of predictor models, each predicting Ω a different number of time steps k into the future. If d is the largest possible value of k, the identifier equations can be time-shifted to yield a common Φ [14]. At time step t,

$$\Omega_k\{t - d + k\} = \Phi\{t - d\}^T \Theta_k \tag{79}$$

The prediction model with the largest β_0 coefficient can be chosen for the controller design, since this model indicates the greatest sensitivity of the controlled (predicted) variable to the present manipulated variable. Hence it will result in the smallest controller gain. Furthermore, if more than one β coefficient is identified, the model with the largest β_0 is most likely to have stable zeros. The model with the smallest prediction error is most likely the one with the smallest k because of the autoregressive α terms, but this model would not necessarily be best for control. The identification filter time constants can be made proportional to the identified k, since k determines the closed-loop performance and the critical frequency range.

CONTINUOUS-MODEL IDENTIFICATION, OPEN-LOOP ADAPTATION

A continuous (differential equation) model, in contrast to a difference equation model, is insensitive to the computing interval h, provided h is small. A restricted complexity identifier, for a process that includes delay, can be based on the method of moments [15].

The Laplace transform $X\{s\}$ of each signal's derivative $x\{t\}$ can be expanded into an infinite series of moments,

$$X\{s\} = \int_0^\infty e^{-st} x\{t\} \, dt = M_0\{x\} - s M_1\{x\} + \cdots \tag{80}$$

Signal derivatives are used so that the moment integrals, for an isolated response, converge to near final values in the finite time τ from the disturbance start,

$$M_n\{x\} \approx \int_0^\tau t^n x\{t\} \, dt \approx \sum_{k=1}^{\tau/h} (kh)^n \times \{k\} h \tag{81}$$

The signal transforms are related to the model polynomials on a term-by-term basis. Choosing the B and D polynomials to be 1 and

$$A\{s\} = a_0 - s a_1 + \cdots$$
$$C\{s\} = c_0 - s c_1 + \cdots \tag{82}$$

in the process equation

$$u\{s\} = A\{s\} y\{s\} - C\{s\} e\{s\} \tag{83}$$

gives

$$M_0\{u\} = a_0 M_0\{y\} - c_0 M_0\{e\}$$

$$M_1\{u\} = a_1 M_0\{y\} + a_0 M_1\{y\} - c_1 M_0\{e\} - c_0 M_1\{e\}$$

$$\tag{84}$$

.

.

.

For each additional equation there are one plus the number of measured disturbances e of additional unknown parameters. The projection algorithm can be used to find the smallest sum of weighted squared parameter changes that will satisfy the equations. Using projection, only those parameters weighting signals that are significantly active are updated. Eq. 84 expressed in vector and matrix form, for use in the projection algorithm, is

$$\Omega = \Phi^T \Theta$$

$$\Omega^T = [M_0\{u\}, M_1\{u\}, \dots]$$

$$\Theta^T = [a_0, a_1, \dots, c_0, c_1, \dots]$$

$$\Phi^T = \left| \begin{array}{ccccc} M_0\{y\}, & 0, & \dots, & -M_0\{e\}, & 0, \dots \\ M_1\{y\}, & M_0\{y\}, & \dots, & -M_1\{e\}, & -M_0\{e\}, \dots \end{array} \right|$$

$$\tag{85}$$

The moment-projection approach is particularly suited for adapting feedforward gain and delay compensators, because the inputs need not be persistently excited. Only two moments need be computed for each signal and two moment equations solved by projection. However, when signals are cycling or responses overlap, the moment integrals do not converge and the adaptation must be frozen. Since an adaptive feedback controller should be capable of stabilizing a stabilizable unstable loop, the moment-projection method is not suited for adaptation of a feedback controller.

LEAST-SQUARES METHOD, BATCH PARAMETER IDENTIFICATION

A batch identifier calculates the model parameters that best fit a block of measured data. The start may be triggered when a significant disturbance is sensed and the end may follow the settling of an isolated response or the detection of a preset number of peaks for a cycling response. A least-squares identifier finds the parameter vector Θ that minimizes the sum of squared prediction errors,

$$\varepsilon\{t\} = \Omega\{t\} - \Phi\{t - k\}^T \Theta \tag{86}$$

When the inverse of the matrix $\mathbf{P}$ exists,

$$\mathbf{P}^{-1} = \sum_i \Phi\{i - k\}\Phi\{i - k\}^T \tag{87}$$

the result is given by

$$\Theta = \mathbf{P} \sum_i \Phi\{i - k\}\Omega\{i\} \tag{88}$$

In order that $\mathbf{P}^{-1}$ not be dominated by steady-state components of Φ, it is customary to choose Φ and Ω to have nearly zero mean. If the means were completely removed, $\mathbf{P}^{-1}$ would be the covariance of Φ and $\mathbf{P}$ would be the covariance of Θ. $\mathbf{P}$ also appears in the recursive algorithm of the next section.

For Θ to be calculable, $\mathbf{P}^{-1}$ must not be singular. Nonsingularity is difficult to guarantee. If any of the process inputs were quiescent over the identification period, $\mathbf{P}^{-1}$ would be singular. This could happen if the controller output were limited or if the controller were in the manual mode. When $\mathbf{P}$ exists, the process is said to be persistently excited. It may be necessary to add otherwise undesirable probing signals to the normal process inputs to achieve persistent excitation.

KALMAN FILTER, RECURSIVE PARAMETER IDENTIFICATION

The Kalman filter [16] provides a one-step-ahead prediction of the parameter vector Θ (treated as a state variable) in a model equation,

$$\Theta\{t\} = \Theta\{t - h\} + v\{t\} \tag{89}$$

modified by observations of a related measured variable scalar (or vector) Ω,

$$\Omega\{t\} = \Phi\{t - k\}^T\Theta\{t\} + \varepsilon\{t\} \tag{90}$$

The model equation, Eq. (89), has a zero-mean white gaussian noise source vector v with covariance matrix $\mathbf{Q}$, which causes Θ to change randomly. The scalar (or vector) observation equation has a zero-mean white gaussian noise source scalar (or vector) ε, with covariance value (or matrix) R, in this case uncorrelated with v. Depending on which noise source dominates, the Kalman filter weights the other equation more heavily,

$$\Theta\{t\} = \Theta\{t - 1\} + K\{t\}(\Omega\{t\} - \Phi\{t - k\}^T\,\Theta\{t - 1\}) \tag{91}$$

Θ is the predicted parameter vector and K is the time-varying Kalman gain vector (or matrix) which can be precalculated using

$$K\{t\} = \mathbf{P}\{t - 1\}\Phi\{t - k\}\,(R + \Phi\{t - k\}^T\,\mathbf{P}\{t - 1\}\Phi\{t - k\})^{-1} \tag{92}$$

$$\mathbf{P}\{t\} = (\mathbf{I} - K\{t\}\Phi\{t - k\}^T)\,\mathbf{P}\{t - 1\} + \mathbf{Q} \tag{93}$$

When $\mathbf{Q}$ is a null matrix, this algorithm is the recursive least-squares algorithm, finding the parameter set Θ that minimizes the sum of squared model errors, equally weighted over all past observations. As time progresses the $\mathbf{P}$ matrix and gain K approach zero, so that eventually each new observation has almost no effect on the identified Θ. Therefore an unmodified recursive least-squares solver does not allow a model to adapt to a time-varying or nonlinear process.

Whereas R tends to reduce K and $\mathbf{P}$ by a factor each iteration, $\mathbf{Q}$ increases $\mathbf{P}$ by a fixed increment. Therefore $\mathbf{Q}$ has a relatively greater influence when $\mathbf{P}$ is small and vice versa when $\mathbf{P}$ is large. Thus when neither R nor $\mathbf{Q}$ is zero, $\mathbf{P}$ tends toward a midrange value. As a result, the Kalman gain K remains finite, so that the most recent observations affect the identified Θ, allowing the model to adapt to a time-varying or nonlinear process. In effect this method, in contrast to the variable forgetting factor approach, weights a different quantity of past history for each variable Φ_i, depending on its activity and the ratio of $\mathbf{Q}_{ii}$ to R.

When both R and $\mathbf{Q}$ are zero, the predictor becomes an orthogonal projection algorithm. Θ converges in a determinate number of iterations to a fixed vector, provided each of these observations contains some independent information. The number of iterations is the number of parameters in Θ divided by the number of components in Ω.

PROJECTION

If $\mathbf{P}$ is not updated and R is zero, the Kalman filter algorithm performs a projection each iteration, finding the smallest set of weighted squared parameter (state-variable) changes that satisfy the model equations exactly. The weighting matrix $\mathbf{P}$ is the a priori covariance of Θ.

REFERENCES

1. Astrom, K. J., and B. Wittenmark, *Computer Controlled Systems,* Prentice-Hall, Englewood Cliffs, New Jersey, 1984.
2. Takahshi, Y., M. J. Rabins, and D. M. Auslander, *Control Systems,* Addison-Wesley, Reading, Massachusetts, 1970.
3. Franklin, G. F., and J. D. Powell, *Digital Control,* Addison-Wesley, Reading, Massachusetts, 1980.
4. *Chemical Process Control Series,* American Institute of Chemical Engineers, New York, 1991.
5. Hansen, P. D., "Recovery from Controller Bounding," presented at the ASME Winter Annual Meeting, 1987.
6. Shinskey, F. G., "Putting Controllers to the Test," *Chemical Engineering,* pp. 96–106, Dec. 1990.
7. Slotine, J. J. E., and W. Li, *Applied Nonlinear Control,* Prentice-Hall, Englewood Cliffs, New Jersey, 1991.
8. Shinskey, F. G., *Process Control Systems,* 3d ed., McGraw-Hill, New York, 1988.
9. Bristol, E. H., "On a New Measure of Interaction for Multivariable Process Control," *IEEE Trans. Automatic Control,* Jan. 1966.
10. Ricker, N. L., "Model Predictive Control: State of the Art," *Chemical Process Control Series,* American Institute of Chemical Engineers, New York, 1991.
11. Kraus, T. W., and T. J. Myron, "Self-Tuning PID Controller Uses Pattern Recognition Approach," *Control Engineering,* June 1984.
12. Hansen, P. D., "Recent Advances in Adaptive Control," presented at the 40th Chemical Engineering Conf., Halifax, Nova Scotia, Canada, July 1990.
13. Ljung, L., and T. Soderstrom, *Theory and Practice of Recursive Identification,* M.I.T. Press, Cambridge, Massachusetts, 1982.
14. Hansen, P. D., and T. W. Kraus, "Expert System and Model-Based Self-Tuning Controllers," in *Standard Handbook of Industrial Automation,* D. M. Considine, Ed., Chapman and Hall, New York, 1986, p. 216.
15. Bristol, E. H., and P. D. Hansen, "Moment Projection Feedforward Control Adaptation," in *Proc. 1987 American Control Conf.,* p. 1755.
16. Goodwin, G. C., and K. S. Sin, *Adaptive Filtering, Prediction and Control,* Prentice-Hall, Englewood Cliffs, New Jersey, 1983.

BASIC CONTROL ALGORITHMS

by E. H. Bristol*

Digial control provides many advantages over traditional analog controllers. These include greater flexibility to create and change designs on line, a wider range of traditional control functions, and newer functions, such as adaptation. But the digital computation is not naturally continuous like the analog controller, and it requires sophisticated support software. This article addresses the basic issues of carrying out continuous control in the digital envi-

* Bristol Fellow, Research Department, The Foxboro Company (A Siebe Company), Foxboro, Massachusetts.

ronment, emphasizing the characteristics which must be addressed in the design of operationally natural control algorithms.

CHRONOLOGY

Continuous process control traditionally used analog control devices, naturally related to the process. The modern era has largely replaced these with microprocessors and digital computations to reduce cost, but also for greater flexibility.[1] The initial approach to digital control has been to convert each analog control function (perhaps as mathematically generalized) into some digital equivalent [1]–[4].

The conversion requires an appropriate software architecture and the matching of the digital algorithms to analog behavior. Whereas the analog controller sensed the process state and manipulated the actuators continuously, the digital controller must sample that state repeatedly, convert it to a quantized number, use that number to compute control actions, and output those actions. Each of these steps involves its own problems and errors.

Standard digital control texts address the sampling problem thoroughly [5], but treat the broader control design in terms of neutral, computed parameters not clearly related to the process gains and time constants. Process control depends on standard control algorithms whose parameters may be set or tuned in terms of known process properties. The approach defined herein emphasizes control with traditional parameters, naturally related to the process and practice. This includes compensating sampled algorithms and their tuning properties for the sampling time, or even for irregular sampling. The consideration of alternatives to the traditional practice is a challenge to be resolved elsewhere.

One aspect of sampling should be understood, namely, aliasing. This is a bizarre effect where sampled higher-frequency data appear to be at a much lower, more significant frequency (Fig. 1). The same effect limits the minimum apparent width in time of any disturbance pulse to one sampling period.

Aliasing is not normally a problem because most commercial controllers sample frequently and filter out the still higher frequencies. But users should be aware of aliasing if they construct home-brewed control algorithms of low sampling frequency.[2]

Except for aliasing, the choice of sampling time is not an issue today. Increasing the sample times faster than the dominant closed-loop time constants of the economically important process variables gives rapidly diminishing performance returns.[3]

For example, with end-product quality the sole criterion on a process which takes an hour to respond, even fast-flow loops can be sampled once every 5 minutes. Of course, many "housekeeping functions," such as flow control, may have constraining side effects the violation of which would involve real costs if this logic were actually implemented. Sampling times faster than 1 second, well filtered, are rarely needed in continuous fluid process control, even when the local process dynamics are faster.

As with other hardware considerations, the inexpensive microprocessor has caused a conservative design tradeoff to favor faster sampling times, eliminating the issues where the cost is so small. Nevertheless, a 10:1 reduction in sampling time does correspond to a 10:1 reduction in computing resources if one has the tools and imagination to use the capability. Faster sampling times also require that internal dynamic calculations be carried out to a greater precision.

---

[1] Digital control also permits the implementation of control forms, such as Smith predictor controllers, that would otherwise be impractical.

[2] Model-based control techniques, such as internal model control and dynamic matrix control, are often constrained to operate at low sample times for reasons of modeling sensitivity.

[3] Slow sampling frequency also causes tuning sensitivity, even for frequencies fast enough to give good control. Further doubling or tripling the frequency beyond this point should eliminate even this problem.

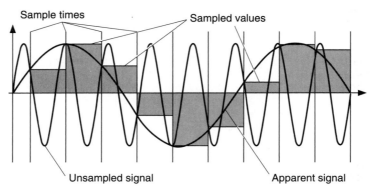

FIGURE 1 Aliasing of sampled to apparent signal. (*Foxboro.*)

Process control algorithm design also differs from academic treatments in respect to accuracy considerations. Practice designs for control and human operation, not for simulation or computation. When accuracy is important, a common thread through the discussion is the effect of differences between large numbers in exaggerating errors and the role of multiplication in creating these large numbers.

Rarely is the parameter or performance precision (say, to better than 10 percent) important to control, even under adaptation. While there are sensitive high-performance control situations, a 2:1 error in tuning is often unimportant. In contrast, there are situations *within* an algorithm where a small error (say, a quantization error of 1 part in 100,000) will cause the control to fail. Computational perfection is unimportant; control efficacy is crucial.

A casual experimenter or designer of a one-use algorithm should be able to tailor a simple controller in Fortran floating point without any special considerations. The normal debugging and tuning should discover any serious deficiencies. A multiuse design requires a deeper understanding of algorithmic and fixed- versus floating-point issues and tradeoffs. The difficulties of programming in fixed point and machine language are overstated. The advantages in speed and exactitude are worth consideration, particularly on small control computing platforms, where they may be essential. The problems are simply those of understanding, and of effective specification and testing of the algorithm.

The discussion thus addresses the refined design of process-control-oriented continuous control algorithms and their software support. It emphasizes high-quality linear dynamic algorithms. This still includes representative examples of the effects of discrete computations on modeled continuous control activities:

- Imprecision, as it limits control performance
- Quantization, as it artificially disturbs the process
- Sampling, as it affects time-continuous computations such as dead time

The user interested in the computation of nonlinear compensating functions should consult the general computing literature [6], [7]. This discussion briefly touches on issues relevant to advanced forms such as adaptive control, but their details are outside the intended scope.

NUMBER SYSTEMS AND BASIC ARITHMETIC

Operation on analog data involves not simply converting to decimal numbers, but also converting to the format that those numbers take in the digital computer. There are two such formats—fixed-point for representing integers and floating-point for representing real numbers. Even though analog data are usually conceived of in terms of real numbers, the most efficient processing of continuous control data, particularly for small microprocessors, is in fixed-point arithmetic.

In this case the data must be (painfully) scaled in the same way that analog control systems and simulations were scaled. Scaling and multiple-precision computations and conversions are the explicit responsibility of the programmer. Apart from computational necessity, scaling is an art which is inherently significant to proper control design, since the process is itself fixed-point. Meaningful control actions can only be guaranteed if the user is aware that the scale of the calculations is in fact appropriate to the process.

Floating-point data are automatically scaled in that they have a fractional part f, or mantissa, corresponding to the integer part of fixed-point data, and an exponent part e, which defines an automatically adjusted scaling factor equal to a base value b (usually 2) raised to that exponent. Although the fractional part is usually viewed as a fraction, the discussion will be less confusing if e is chosen so that f is an appropriate integer. Any such value is then expressible entirely in terms of integers, as $f \times b^e$.

Until recently floating-point formats were not standardized [8]. Moreover, the floating-point format inherently involves arbitrary truncations and uncertain relationships between single- and double-precision computation. For this reason, fixed-point algorithms still permit the most precise designs.

FIXED-POINT FORMAT

A fixed-point format represents an integer as a binary number stored in a word containing a fixed number (here n) of bits, usually in what is called two's complement format. This format represents positive numbers in a range of 0 to $2^n - 1$ and signed numbers in a range of -2^{n-1} to $2^{n-1} - 1$. In two's complement arithmetic, negative numbers are represented in binary notation as if they were larger positive integers. As a result, when the numbers are added, the natural additive truncation achieves the effect of signed addition.

Thus in Fig. 2 (with $n = 3$) negative 2 is represented by a binary number corresponding to a positive 6. And 6 added to 2 becomes an 8 which, with truncation of the carry, corresponds to 0. Thus the 6 is a perfectly good negative 2.

Sign and Value	Two's Complement	
Decimal	Binary	Positive Decimal Equivalent
0	000	0
1	001	1
2	010	2
3	011	3
−4	100	4
−3	101	5
−2	110	6
−1	Carry bit 111	7
(2)+(−2)	1000	(3)+(5)=8

FIGURE 2 Two's complement arithmetic. (*Foxboro.*)

Two's complement arithmetic is also related to modular, or "around the clock," arithmetic. Because of this behavior, results too large to fit into a word must be taken care of in one of three ways:

- Data values can be distributed in a sufficiently large set of words to represent them. The arithmetic hardware or software then utilizes carry bits to pass the data between words when a carry is required.
- Results too large to fit a word may be saturated.[4]
- One may ignore the problem in parts of the calculation where it is clear that results will never overflow.[5]

The remaining discussion will be framed in decimal arithmetic, rather than in the unfamiliar binary arithmetic, for clarity's sake. Suffice it to say, actual designs are implemented by good fixed-point hardware, including the basic binary operations and carry/overload bits to support effective single- and multiple-precision arithmetic. For this reason, multiplication and division of powers of 2 will be particularly efficient, to be preferred when there is a choice.

FIXED-POINT SCALING

In the discussion of multiple precision and scaling it is convenient to distinguish different tag ends to the variable name to represent different aspects of the variables under consideration. Thus if V is the name of a variable, then:

- $V.S$ will become the scaled representation of the variable, taken as a whole.
- $V.0$, $V.1$, $V.2$, ..., $V.m$ (or $V.S.0$, $V.S.1$, $V.S.2$, ..., $V.S.m$) will be different storage words making up the multiple-precision representation of V (or $V.S$), with $V.0$ being the rightmost word and $V.1$ being the next rightmost word, and so on, as shown in Fig. 3.

$$\frac{V}{123456} \quad \Bigg| \quad \frac{V.0}{12} \quad \frac{V.1}{34} \quad \frac{V.2}{56}$$

FIGURE 3   Decimal multiple precision. (*Foxboro.*)

- When V is scaled by a fixed-point fraction, the numerator will be $V.N$ and the denominator will be $V.D$ [either of which may be itself in multiple-precision form (with $V.N.0$, $V.N.1$, ..., or $V.D.0$, $V.D.1$, ...)]:

$$V.S = \frac{V. \times V.N}{V.D}$$

Normally the actual scaling computations will take place only when the data are processed for input/output (I/O) or display. Control calculations will generally take place with respect to $V.S$. For this reason, when no confusion arises, the discussion will refer to V, substituting for $V.S$. Scaling conversions, like binary and decimal arithmetic and conversions, are straightforward, though tedious; they need not be addressed further.

The special power of this notation is that it allows all scaling and multiple-precision computations to be expressed entirely in conventional algebra. In particular, for analytical purposes a multiple-precision value can be expressed and operated on simply as a sum of the normal values $V.0$, $V.1$, $V.2$, ..., $V.m$, each with its own scaling factor

$$V = V.0 \times B^m + V.1 \times B^{m-1} + V.2 \times B^{m-2} + \cdots + V.m$$

---

[4] That is, a value too large to fit the word is replaced by the largest number of the right sign that will fit.
[5] Where they will never become too large or too small to fit the word. This is a dangerous approach.

The value of B is one greater than the largest positive number represented in a single-precision word. ($B = 2^n$ for unsigned binary data, $B = 2^{n-1}$ for signed data.)

Control algorithm parameter scaling is chosen according to need, computational convenience, and best use of the available data range, working with single-precision representation as much as possible. For a controller measurement or value on a 16-bit word machine, ½ percent precision is normally minimally adequate. For positive data a byte represents a data range of 0 to 255 ($2^8 - 1$), which is better than ½ percent precision.

On the other hand, with a 16-bit word there are eight more bits in the word, which could be used to give a smoother valve action and simplify calculations. Working only with positive values, 16 bits corresponds to the normally unnecessary precision of 1 part in 65,535; with signed data, 1 part in 32,767; or with a 1-bit safety margin, 1 part in 16,384. This scaling is well above the minimum process data precision, while still not forcing multiple-precision arithmetic in most cases. The corresponding 14-bit analog-to-digital (A/D) and digital-to-analog (D/A) converters for the process data are still reasonable.

Control parameters may have different natural data ranges or scalings. A controller gain might be scaled so that the minimum gain is ½₅₆ and the maximum gain is 256. In this way the range of values equally spans high-gain and low-gain processes about a nominal unity gain. Time constants may require a certain resolution. When a minimum reset time of 1 second is adequate, 1 hour corresponds to 3600 seconds and 2^{16} corresponds to more than 18 hours.

On the other hand, using the full range of data storage for the control parameters may require arithmetic routines which mix signed and unsigned arguments. At this level there is a tradeoff between the increased efficiency and the added complexity. Certainly it is more convenient to program uniformly in signed fixed-point (or even floating-point) arithmetic. But the costs of this convenience are also significant.

RANGE AND ERROR IN FIXED-POINT ARITHMETIC

Good design for quantization and multiple precision avoids poor control arising from inaccuracies not inherent in the real process. Normal control algorithms involve the standard combinations of additions, subtractions, multiplications, and divisions, and rational functions of their data. Often the basic calculation allows many ways to order these calculations, which are theoretically identical in their results as long as precision is indefinite.

Practical fixed-point programming requires a more careful understanding of the basic operations and the effect of ordering on the calculation. First, one should examine how each operation affects the worst-case range[6] and the error accumulation of the result. As will be seen, range effects are typically more important.

Error can be considered in terms of absolute error, that is, the actual worst-case error, or relative error, namely, the worst-case error as a percentage of the nominal value, or scale. Usually the relative error is important in final results and products, or quotients, whereas the absolute error is important in determining how the error accumulates in a sum.

When two numbers are added or subtracted, their worst-case range doubles, requiring either a carry (into a multiple-precision result), or a sign bit[7] (Fig. 4a). Additions and subtractions also cause the absolute error to increase. Addition of numbers of the same sign can never increase the relative error over the worst error of the added numbers. But their subtraction increases it, as does the addition of numbers whose sign is uncertain. In Fig. 4b[8] two 10 percent relative error numbers, when added, result in a 10 percent relative error. But when subtracted, the relative error can be arbitrarily large (30 percent in this case). In this way, differences between large numbers explain most computing accuracy problems.

---

[6] The range of the result for all possible combinations of the input data.
[7] When positive numbers are subtracted.
[8] In Fig. 4b the underlined numbers represent an error term being added to or subtracted from the "ideal" value of a computational input or output.

$$(20 \pm \underline{2}) + (10 \pm \underline{1}) = (30 \pm \underline{3})$$

$$99 + 99 = 198$$
$$99 + (-99) = 0$$

$$\frac{(20 \pm \underline{2}) + (10 \pm \underline{1})}{20 + 10} = \left(\frac{30}{30} \pm \frac{3}{30} \right)$$

$$99 - 0 = 99$$
$$0 - 99 = -99$$

$$\frac{(20 \pm \underline{2}) - (10 \pm \underline{1})}{20 - 10} = \left(\frac{10}{10} \pm \frac{3}{10} \right)$$

a. Range effects   b. Error effects

FIGURE 4 Addition and subtraction. (*Foxboro.*)

Fixed-point multiplication does more than just increase the range of the result; it doubles the required storage of the result (Fig. 5a). For this reason most hardware implements single-precision multipliers to return a double-word result. By analogy, most hardware division divides a double-precision dividend by a single-precision divisor to obtain a single-precision quotient with a single-precision remainder. But as Fig. 5a shows, such a division can still give rise to a double-precision quotient. The hardware expresses this as an overload (similar to division by zero).

Fixed-point hardware is designed to permit the effective programming of multiple-precision arithmetic. One of the strengths of fixed-point arithmetic, with its remainders, overflow, and carry bits, and multiple-precision results is that no information is lost except by choice. The precision, at any point in the calculation, is entirely under the control of the programmer. This is not true of floating-point arithmetic. As shown in Fig. 5b, multiplication and division always increase (that is, double) the relative error.

$$\overline{99} \times \overline{99} = \overline{98} \ \overline{01}$$

$$\overline{98} \ \overline{01} / \overline{99} = \overline{99}$$

$$\overline{98} \ \overline{00} / \overline{99} = \overline{98} \ \text{Rem.:} \ \overline{98}$$

$$\text{but:} \ \overline{98} \ \overline{00} / \ \overline{1} = \overline{98} \ \overline{00}$$

$$(10 \pm 1) \times (10 \pm 1) = (100 \pm 21)$$

$$\frac{(100 \pm 8)}{(10 \pm 1)} = 10 \pm 2$$

a. Range effects   b. Error effects

FIGURE 5 Multiplication and division. (*Foxboro.*)

FIXED-POINT MULTIPLICATION AND DIVISION

Multiplication's range explosion is its most problematic aspect. Among other consequences, it generates large numbers whose differences may cause large errors. It makes smaller numbers still smaller, compared to errors caused by other large numbers. The important issue is the ratio of the large numbers (which make the errors) to the small numbers (which end up as the result). Multiplication's range expansion forces a choice: precision can be preserved, or intermediate values can be rescaled and truncated back to their original data size.

Thus it is usually desirable to avoid repeated multiplications. Often multiplications and divisions occur together in a manner in which they can be alternated, the multiplication generating a double-precision value and the division returning that value to a single-precision quotient. The basic proportional controller calculation is a good example,

$$\text{Output} = \frac{100 \times \text{error}}{\text{proportional band}} + \text{bias}$$

Common combined operations such as this may usefully call for specially designed routines. In general one should try to maintain a constant data range and size throughout the computation. Often a product is naturally intended to return a value in the same range as a process variable (output). There are two natural cases—a value is multiplied either by a gain or by a proper fraction (Fig. 6). In the controller calculation the 100/proportional band gain might be limited to the range of $\frac{1}{256}$ to 256. If the gain is expressed as a single-precision integer (with decimal point effectively in the middle of its digits) rather than as a fraction, one can still generate an appropriate single-precision result by taking the middle single-precision set of digits out of the double-precision scaled result (Fig. 6), saturating and truncating the extraneous data.

$$\text{(a):} \quad \overbrace{\text{XX.XX}}^{\text{Gain}} \times \overbrace{\text{XXXX}}^{\text{Natural Range}} = \overbrace{\text{XXXXXX.XX}}^{\text{Natural Range}}$$

$$\text{(b):} \quad \overbrace{\text{.XXXX}}^{\text{Fraction}} \times \overbrace{\text{XXXX}}^{\text{Natural Range}} = \overbrace{\text{XXXX.XXXX}}^{\text{Natural Range}}$$

FIGURE 6 Scaled multiplication. (*a*) Gain. (*b*) Fraction. (*Foxboro.*)

As a common case of proper fractional multiplication, a lag calculation (time constant T) can calculate its output X as a weighted average of the past output Y' and the current input X,

$$Y = \frac{1}{T+1} X + \frac{T}{T+1} Y' = Y' + \frac{1}{T+1}(X - Y')$$

In this case, when the proper fractions are multiplied (scaled as integers), the proper result can nevertheless be returned as the leftmost part of the double-precision result (Fig. 6). It is often appropriate to combine constants or parameters together into a common effective parameter which is the above kind of gain or proper fraction. This ensures that the linear operation with the process data is truncated by only the one final multiplication.

The combined parameters can be made to act consistently on the data, even if in error as calculated. The errors can be reinterpreted as (presumably insignificant) errors in the original parameters. Note that the rightmost expression within the preceding equation involves a feedback between Y and the difference between X and Y'. Often such a feedback can be included in the calculation to improve an otherwise error-prone algorithm, making it self-corrective, like any other feedback system.

There is a parallel between the preceding range discussion and traditional dimensional analysis [9]. Sums must be of data elements with identical units and similar ranges. Multiplications change the units and also change the range. The ultimate purpose of any of our calculations is to convert data of limited range from the process to data with limited range to drive the process. Thus whatever the internal gyrations, the process constrains the results to be reasonable. As an illustration consider two sets of equations between limited-range process data elements X and Y,

$$AX = Y \qquad A_{11}X_1 + A_{12}X_2 = Y_1$$
$$A_{21}X_1 + A_{22}X_2 = Y_2$$

The corresponding solutions are

$$X = \frac{Y}{A} \qquad X_1 = \frac{A_{22}Y_1 - A_{12}Y_2}{A_{11}A_{22} - A_{12}A_{21}}$$
$$X_2 = \cdots$$

The challenge is to carry out the calculations so that the inherently limited practical range prevails throughout the calculation.

DIGITAL INTEGRATION FOR CONTROL

Figure 7 shows a different kind of division problem—process control integration. The object is to integrate the process error in a control (reset) calculation. This can be done by computing a shared fractional multiplier $\Delta t/T$, which has been scaled to give a good range [which might also include the proportional band action PB, as in $100 \times \Delta t/(PB \times T)$]. This result would be multiplied by the error and added in double precision to the previous "integrated" (that is, summed) value to get the current control value.

$$\sum_{i=0}^{n} \frac{Error\ (i)\ \Delta t}{T} =$$

$$\frac{\sum_{i=0}^{n} Error\ (i)\ \Delta t}{T} = \frac{\sum_{i=0}^{n-1} Error\ (i)\ \Delta t}{T} + Error(n)\ \frac{\Delta t}{T}$$

$$= \sum_{i=0}^{n-1} \frac{Error\ (i)\ \Delta t}{T} + Quotient \left(\frac{Error(n)\ \Delta t\ +\ Remainder}{T} \right)$$

Note:

$$Error\ (n)\ \Delta t \ = \ T \cdot Quotient \left(\frac{Error(n)\ \Delta t\ +\ Remainder}{T} \right)$$

FIGURE 7   Integration trick. (*Foxboro.*)

Double precision is essential here, since a large value of T corresponds to a small $\Delta t/T$ and a truncation loss of significant process errors. For example, suppose that the error and sum are both scaled as signed single-precision integers, ranging from $-10,000$ to $+10,000$. With $\Delta t = 1$ and the minimum $T = 1$ (corresponding to 1 second), the corresponding maximum $\Delta t/T$, which equals 1, must be scaled to a value of 10,000. The natural scaling of the controller output can be achieved by dividing by 10,000 (see earlier scaling discussion), that is, the scaled equations should be

$$Sum.S = old\ sum.S + \left[\frac{(\Delta t/T).S \times error.S}{10,000} \right]$$

In normal single-precision division only the integer quotient is considered. Thus a product $(\Delta t/T).S \times error.S$ less than 10,000 is truncated as zero—the error is ignored and the integration stops, causing a permanent offset. For T large (equal to 10,000, which corresponds to about 3 hours) and $(\Delta t/T).S$ small (equal to 1.0 in this case), any error less than 100 percent (scaled to 10,000) will be lost. Under control the result is a 100 percent offset. Double precision alleviates the offset, but still loses some information to truncation.[9]

---

[9] One alternative to multiple-precision integration uses random numbers (or dither in mechanical engineering terms). The computation is carried out in normal double precision. However, the higher-precision (lower-order) data word is ignored and not saved. It is replaced by a random number in the next sample time's calculation when it is needed again. This method works well practically and theoretically. A similar method was incorporated into the Digital Equipment Corp. PDP-1 floating-point package. But who would have the courage to use it on a real process?

The better method, shown in Fig. 7, preserves the division and achieves an exact result with the same storage as double precision. In this case the product of the error and the sample time (the sample time may equal 1) is formed as a double-precision value, then added to the remainder from the previous sample calculation's division. This net value is divided by T to get back a single-precision quotient to be added to the old sum, and a remainder to be saved for the next sample time. Data results truncated from any quotient are never lost, but preserved and accumulated in the remainder, to show up in some later quotient.[10]

FLOATING-POINT FORMAT

Modern microprocessors are often supported with high-speed floating-point processors. Without these, floating-point calculations run an order of magnitude slower than fixed-point calculations. With the floating-point format, all remainders, carries, and multiple-precision results of single-precision computations disappear. Instead, the user chooses a fixed level of precision whose truncations show up in guard bits [8].

Floating-point errors introduced in small differences of large numbers become more severe and at the same time less obvious because their processing is automated and invisible. For example, when a large number L (such as 10,000, stored to three-digit accuracy) is added to a small number S (such as 1), the smaller number totally disappears. The sum $L - L + S$ should equal S. If the calculation is carried out in the natural ordering, $(L - L) + S$, then S will result. But the nominally equivalent $(L + S) - L$ will generate zero.

There are a number of intricate ways of avoiding this problem:

- Convert all floating-point numbers to ratios of integers and operate in the fixed-point format. This is one way to study the properties of the algorithm.
- Use adequate precision. Practically this is often unpredictable, and theoretically it is impossible because repeating decimals require infinite data. This suggests also deferring division to the last operation.
- Reorder the calculation for the most favorable computation, either statically based on algorithm properties or by using an arithmetic package which continually reorders the operands.

It is useful, in the following discussion, to consider every list of terms to be added (subtraction being taken as addition of a negated number) to be ordered by magnitude. The individual operations are carried out according to the following rules. Under addition or subtraction:

- Combine small numbers together before large numbers (to let them accumulate before being lost).
- Take differences (subtractions or additions of numbers of opposite sign) before sums (to achieve all possible subtractive cancellation between larger numbers before these can lose the small numbers).
- From a list of numbers to be multiplied and divided, cancel or divide out most nearly equal numbers first (to minimize floating-point overload[11]).
- When all denominator terms are gone, multiply largest with smallest numbers (to minimize floating-point overflow).

---

[10] As pointed out in the next section, floating-point arithmetic lacks the support for such refined tricks. It is particularly subject to integration problems because a large sum may be big enough to prevent the addition of a small term even though that term is rescaled to avoid its being truncated to zero.

[11] The process of the floating-point exponent getting too large for the data space provided.

One other advantage of following such a set of ordering rules is that it will give identical results to identical data, even when they originally occurred in a different programmed order.

GENERALIZED MULTIPLE-PRECISION FLOATING POINT

Normally the multiple-precision floating-point format is the same as the single-precision format, but with larger fraction and exponent data fields. The author has experimented with a more open-ended multiple-precision floating point as illustrated in Fig. 8. The multiple-precision floating-point number is represented by a summed, ordered set of signed single-precision floating-point numbers, each mantissa having M digits.[12]

$$12340.36789 \Rightarrow 1234 \times 10^1 + 3679 \times 10^{-4} - 1000 \times 10^{-8}$$

Addition:
$$A = 3057 \times 10^6 \quad B = 4263 \times 10^4$$
$$A + B = 309963 \times 10^4 = 3100 \times 10^6 - 3700 \times 10^2$$

FIGURE 8   Generalized multiple-precision floating point ($M = 4$). (*Foxboro.*)

The numbers in the set are chosen so that the set can be truncated at any point to give the best possible truncated approximation. In this case the value is considered to be made up of the truncated value and a signed ($\pm$) remainder, which expresses the part cut off in truncation.[13] The generalized floating-point arithmetic would be supported by the following operations:

- *Text conversion.* Converting to or from text general precision floating point to the internal format consisting of the summed set of single-precision values.
- *(Set) normalization.* Consisting of reprocessing the set of single-precision values so that they are ordered in magnitude with largest first,[14] so that the magnitude of each mantissa is between b^{M-1} and b^M, and so that consecutive values have exponents whose difference is greater than M (or, if the difference equals M,[15] then the magnitude of the second mantissa is less than or equal to $b^M/2$).[16]
- *Addition and subtraction.* Consisting of the simple merging of entries into a final value set followed by a normalization of that set.
- *Multiplication.* Consisting of the multiplication of every pair of elements, one from each of the multiplicand and multiplier, to generate a double-precision result, followed by the merging and normalization of the accumulated set of results.
- *Division.* Consisting of the division of the largest elements in the dividend by the divisor, subtracting the divisor multiplied by the quotient from the dividend to obtain the remain-

---

[12] Analogous to multiple-precision fixed-point data represented as a summed scaled set of single-precision numbers. Each element is normalized (shifted) to use the full range of M digits. Recall that the general representation of a single-precision floating-point number is $f \times b^e$, with f, e, and b being integers. As before, the format is described in terms of a general b and illustrated with $b = 10$. In practice b will equal 2. Note that the different members in the summed set may have different signs. However, the sign of the total (set) value is still the sign of its initial and largest element.

[13] The magnitude of the remainder is always less than the value of half the units of the next larger element in the set, since the best approximation requires that the remainder round to zero.

[14] In the form $f_0 \times b^{e_0}, f_1 \times b^{e_1}, f_2 \times b^{e_2}, \ldots$.

[15] Not possible after normalization if $M = 2$.

[16] The Appendix presents a set-normalization algorithm as an example of the design issues.

der. This division is designed to select that quotient which returns the remainder with the smallest magnitude (of whatever sign). The remainder can be redivided by the divisor to compute any desired level of multiple-precision quotients, with the final resulting quotient and remainder being normalized. The remainder so developed can be used in the earlier integration procedure.

This type of generalized floating point can be used to develop calculations whose precision expands indefinitely as needed. Such a system could give absolutely error-free results, without any special care.

SPECIFICATION OF FIXED-POINT ALGORITHMS

Clear fixed-point specification includes the proper statement of computation order and of scaling of intermediate and final results. This can be superposed on a conventional algebraic notation. For example, in the following control computation the parentheses define any required ordering:

$$\text{Output}_5 = \left[\left(\frac{100 \times \text{error}_1}{\text{proportional band}_2}\right)_4 + \text{bias}_3\right]_5$$

Unparenthesized addition and subtraction is assumed to be from left to right, and multiplication and division are assumed to alternate as described earlier. Any constants which can be combined would certainly be combined in a working system. This should include any scaling constants. The subscripts refer to scaling specifications in the table of Fig. 9.

Subscript	I/O	Saturation Hi	Lo	V.N	V.D	Sign	Precision
1	IN	100	−100	16384	100	±	1
2	IN	16384	0	2	1	+	1
3	IN	100	−100	16384	100	±	1
4	—	YES		16384	200	±	1
5	OUT	100	0	16384	100	+	1

FIGURE 9   Tabular listing of variable scaling-related properties. (*Foxboro.*)

Such a table would completely specify all scalings, saturations, and conversions appropriate to the values within the calculation. Remembering that the internal scaling is reflected in the relation

$$V.S = \frac{V.N \times V}{V.D}$$

the conversion from functional algebraic equations to internal computational form is then carried out by the computation

$$V = \frac{V.D \times V.S}{V.N}$$

When all of these scalings are incorporated back into the original proportional controller calculations, its internal scaled form becomes

$$\text{Output.}S \times \frac{100}{16{,}384} = \frac{100 \times \dfrac{100}{16{,}384}\ \text{error.}S}{\dfrac{1}{2} \times \text{proportional band.}S} + \frac{100}{16{,}384} \times \text{bias.}S$$

or

$$\text{Output.}S = \left[\left(\frac{200 \times \text{error.}S}{\text{proportional band.}S}\right) + \text{bias.}S\right]$$

Such a simplification is to be expected in practical calculations as part of the combination of similar scaling terms and application constants. The final equation then becomes the programmed calculation, except that each operation would be saturated as specified. Saturation can, in fact, make the combination of terms invalid, but in this case it may be worth considering whether or not the saturations might not better be left out in the interests of a more perfect result.

Definition and implementation of an algorithm then has three parts:

- Specification and programming of the necessary arithmetic and saturation routines
- Manual development of the scaled calculation, in terms of the routines
- Programming of the algorithm

Ideally, an algorithm should be tried out first in a higher-level language (such as Fortran or C). Here it can first be expressed in floating-point and then in scaled fixed-point format. If later the final form is needed in machine language, the three different forms can be run comparatively, greatly facilitating debugging.

OPERATIONAL ISSUES

The basic controls are normally expressed as linear algorithms, defined as if the process measurements and actuators were capable of perfect operation to whatever range might be needed. In fact, valves limit and sensors fail. The algorithms must be designed to accommodate temporary valve saturation or loss of sensor data. Moreover they must be designed to allow the system to be restarted smoothly after longer-term failure or shutdowns.

The most common such problem is windup—the property of the controller which continues to integrate under error even after the actuator has limited and is incapable of further change. Windup requires recovery time even after the error reverses sign, blocking effective control, for that interval. The response to this problem requires that the algorithm be provided some indication of the limiting so that it can alter its behavior. The information can take the form of flags which inform the controller of the limiting actions being taken, propagated back to any affected controller. The flag then causes the integration to stop in some appropriate way.

This approach is proposed in algorithms currently part of the ISA SP50 field bus standardization effort. A superior approach, called external feedback, senses the actual state of the manipulated variable and feeds it back, in comparison to its intended value. By working with the true state of the process, the algorithm can make much more refined accommodation strategies. Controller windup is not the only problem arising from actuator limiting; blend pacing, split range control, multiple output control, all relate to standard effects of valve limiting and its accommodation. The external feedback strategy effectively unifies the handling of all of these issues.

The problem is complicated because the effects of limiting must propagate back from the actual valve through any control functions (such as cascaded controllers, ratio units) to the controller under consideration. Thus there is not only a need to notify controllers of valve limitings or sensor failures but to propagate this information to all affected control elements. Software must be designed to accomplish the propagation.

The external feedback approach is particularly advantageous because it recognizes loss of control locally to the affected controller, and it responds only when the loss is material to it. Of course, the loss of control at any level may be a useful basis for alarming, independent of immediately recognized effects on control. The present discussion is largely limited to the algorithmic consequences of the problem, but the software consequences are just as important.

A common problem with inexpensive controller designs is that they bump the output whenever the parameters are changed for tuning. The proportional plus integral controller computation illustrates the problem:

$$O(t) = \left(\text{error} + \frac{1}{\tau} \int_0^t \text{error } dt \right) \frac{100}{\text{PB}}$$

$$= \text{error} + \int_0^t \frac{100}{\text{PB} \cdot \tau} \text{ error } dt$$

The first of two nominally equivalent expressions computes the output with the tunings acting after the integration. The practical result is that any change in settings immediately bumps the output. In the second expression, the tunings act on the error value prior to integration. Any tuning changes only affect errors occurring after the change.

Apart from output limiting and tuning bumps there are a number of similar operational modalities which the controller should support in relation to either manual or automatic operation. For example:

- *Cold start initialization.* Sometimes it is desirable as part of the process start-up to start the controller in automatic mode so that its output has no tendency to move, letting the operator move the process on his or her own schedule. This is also the natural way to initialize secondary controllers when the primary controller goes to automatic mode. Cold start can be implemented by setting the controller set point to the present measurement value and initializing the internal states of the controller so that the output matches the externally fed-back or operator-set value of the output. Automatic control then continues naturally.

- *Bumpless transfer.* The purpose of this mode is to transfer control to automatic control in such a way that the process does not receive any immediate valve change, but moves from its prior position smoothly. In this case the controller set point is left in place, but all internal states are initialized to be consistent with the current error and external feedback (and current controller output value).[17]

- *Batch preload.* In circumstances where a known set-point change is to be applied (for example, in batch production) the controller may be set up to pick up on the set-point change with a preset initial internal integration value. The purpose is to give the process an initial kick in order to get it to the set point in the fastest time. This strategy has a number of elaborations of varying sophistication.

- *Ramped set-point change.* The controller may be designed to limit the rate at which it responds to set-point changes to minimize the bumps to the process.

The variations on cold start and bumpless transfer depend on back calculation—the recomputation of the internal data[18] to be consistent with the unchanged output and external feedback. For example, a PID computation developed later computes its output O from the collective effect X of the proportional and derivative effects and on an internal bias B. The bias is computed in turn by applying a lag computation to the external feedback O_{FB},

---

[17] This mode can be applied to all controllers in a cascade, but because the secondary set points will then be matched by their primary controllers to their sensors, the result for them should be the same as if they had been initialized under a cold start. This is only true if the controller calculations are carried out in an appropriate order relative to each other. In lieu of this, it may be better to use the cold start on the secondaries anyway.

[18] Particularly integration data. This guarantees that the integration will resume as if the controller had always been operating under the current error and output conditions.

$$O(t) = X(t) + B(t)$$

$$B(t) = \frac{O_{FB}(t) \cdot \Delta t + B(t - \Delta t) \cdot \tau}{\tau + \Delta t}$$

The bumpless transfer can be accomplished by computing the new value of X and then rearranging the first of the two equations to back-calculate the bias from the new X and the old B,

$$B(t) = O(t) - X(t)$$

When the O is later calculated from this B, its value will remain initially at its old value, irrespective of manual changes in O, or measurement, set-point, or tuning changes reflected in X. The second equation might be used to back-calculate O_{FB}, but this value will be overridden by later computations anyway.

OUTPUT LIMITING—EXTERNAL FEEDBACK

As already indicated, external feedback represents a precise and smooth way of handling windup. However, implemented in a PID controller algorithm, it has the effect of including, in the integrated term, the difference between the controller output and the corresponding externally fed-back measured state. The difference is added to oppose the normal integration. Thus whenever the external feedback fails to follow the controller output, the difference builds up to stop the integration.

Feedback controllers are built about the processing of an error signal on their input. External feedback extends the principle to the output. It allows the control algorithm to be designed to alter its approach in the face of output failure to act. This same strategy can be generalized to apply to any control function:

- *Blending.* When several ingredient flows are ratioed to generate a blended product, two basic product properties are involved—product quality and product flow. Under the standard strategy of pacing, if one ingredient flow limits, the remaining flows are ratioed, not off their original target, but off the external feedback from the limited flow. Such a system extends the external feedback concept to give up the control of product flow in favor of the more important product quality.

- *Fuel/air ratio control.* With certain liquid fuels an excess accumulation of fuel in the burners constitutes a fire hazard. The fuel controller is designed to limit the fuel to be less than the combustible ratio to the measured air flow, at the same time limiting the air flow to be greater than the combustible ratio to the measured fuel flow. If either limits, the other is held to a safe value. Such a system extends the external feedback concept to safety control.

- *Multiple-output control.* In certain cases a manipulated resource may be duplicated so that several devices share a load. It is desirable that operating personnel be able to take one or more of such devices out of service in such a way that the others take up the load. In this case a multiple-output controller computes an output value such that when the value is ratioed as the set point to each device, the sum of the external feedbacks from all devices equals the net desired load. In this way, as one device is taken over in manual, the others will take up the slack.

- *Backup (also split range control).* The normal external feedback antiwindup action can be extended, using the recognized output/external feedback error to drive other actuators. Figure 10 illustrates a refined handling of this function. In this design the controller output is normally connected to actuator 1. But if the calculated difference (the subtraction element Δ marked 1) between actuator 1 and its external feedback exceeds the gap parame-

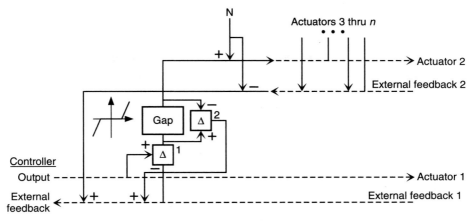

FIGURE 10 External feedback-based backup. (*Foxboro.*)

ter, the excess difference will serve to transfer the active control to actuator 2. In this way the controller can be backed up against any downstream failures or overrides. The nominal bias N reflects the preferred inactive value for actuator 2. (In more refined designs N may be established dynamically by a separate controller.) Differences can further be passed on indefinitely to actuators 3 through n. Effectively this arrangement generalizes the behavior of split range control, taking into account any downstream loss of control action. The external feedback 2 is added into the controller external feedback to allow the controller to continue its full (integrating) control action. The purpose of the gap is to ensure that actuator transfer does not give rise to a chattering between actuators, but acts only for significant loss of actuator control. However, to guarantee that the resulting temporary loss of control does not cause the controller to stop integrating, the actual amount of gapping action (the Δ marked 2) is added into the controller external feedback as well. The external feedback sees neither gap nor actuator transfer. In actuality, the different actuators might call for different control dynamics and different compensation. This could be built into the control transferring paths. However, digital implementation allows the switching of controller tunings, as a function of the active actuator, to be carried out as part of the controller computation, a more natural arrangement. Digital implementation also allows the above structure to be black-boxed flexibly, taking the confusing details out of the hands of the user.

- *Linear programs and optimization.* It has been argued that external feedback is incapable of dealing with connection to higher-level supervisory functions such as linear programs or optimizers. This position reflects higher-level functions not designed for operations, rather than any inherent problem with external feedback. The operationally correct optimizer will benefit from external feedback data like any other control computation. In this case each optimization target, with its external feedback, is associated with an implied output constraint. Whenever a difference develops between the two, the constraint limit and the violation become apparent.

   Thus the external feedback value should be fed into the optimizer, parameterizing a corresponding optimization constraint. There are three special considerations:

- The control action actually implemented must push beyond the constraint so recognized. Otherwise the constraint becomes a self-fulfilling prophecy which, once established, never gets retracted. Since the control action is presumed to be up against a real process constraint, it does not matter how much further into the constraint the target variable is set.

However, it is probably better to exceed the constraint by some small number (for example, 1 to 5 percent of scale).

- The optimization computation is likely to be run infrequently compared to the normal regulatory dynamics. For this reason some lag filtering or averaging should be built into all external feedback paths to minimize noise effects and increase their meaningful information content. The filter time constant should correspond to the optimization repetition interval.

- All of this assumes that the optimizer addresses the economic constraint dimensions only. Significant safety or quality constraint effects must always be addressed separately at the regulatory level.

Once one begins to address operational nonlinearities, many issues come up, calling for many different kinds of thinking. Of course these same differences must fit nonintrusively and naturally with the intentions and expectations of operating people. While the operational user will normally not be aware of the technology behind these techniques, he or she will become intuitively aware of any inconsistencies between the handling of similar functions in different control elements. External feedback provides a powerful strategy for addressing many of these problems, the application uniformity of which the end user will appreciate.

BASIC CONTROL ALGORITHMS

Lag Calculation

The lag calculation corresponds to the following continuous transfer function:

$$\frac{L(s)}{I(s)} = \frac{1}{\tau s + 1}$$

where L is the output, I is the input, and τ is the lag-time constant. The usual practice of going to the z transform for the corresponding sampled-data form should not be overemphasized. No exact approximation is possible. Instead, the algorithms are designed primarily to avoid operationally unnatural behavior. With this in mind, the best direct sampled-data approximation of the preceding differential equation is

$$I(t) = \frac{\tau \Delta L}{\Delta t} + L(t)$$

$$= \frac{\tau}{\Delta t}[L(t) - L(t - \Delta t)] + L(t)$$

or
$$L(t) = \frac{I(t) + \frac{\tau}{\Delta t} L(t - \Delta t)}{1 + \frac{\tau}{\Delta t}}$$

$$= \frac{I(t)\Delta t + \tau L(t - \Delta t)}{\tau + \Delta t}$$

This approximation amounts to a weighted average of the new input and the old output. From the scaling point of view, each of the products has the same range as the sum; scaling is simple. The calculation is stable for all positive τ, accurate for large τ, and qualitatively natural as τ approaches 0, with $L(t)$ equaling $I(t)$ if $\tau = 0$, as intended.

The calculation is usable even in single precision if the term $I\Delta t$ is truncated up in magnitude (and τ is not too large). In this case the product is never truncated to zero unless the product is truly zero. This guarantees that the output will always settle out at any steady-state input value. Normal truncation would leave the result below its "theoretical" steady-state value, a situation similar to the integral offset described earlier.

But a trick, similar to the one used with the integrators before, can be applied to calculating lags exactly:

$$L(t) = \text{quotient}\left[\frac{I(t)\,\Delta t + \tau L\,(t - \Delta t) + \text{remainder}}{\tau + \Delta t} \right]$$

with the remainder being saved for use in the next sampled calculation.

Lead/Lag Calculation

Filtered derivative and lead/lag calculations are most easily and reliably developed from the preceding lag calculation, by analogy with transfer-function calculations,

$$\frac{O(s)}{I(s)} = \frac{\tau \cdot s}{k \cdot \tau \cdot s + 1} = \frac{1}{k}\left[1 - \frac{1}{(k \cdot \tau)s + 1} \right]$$

Similarly,

$$\frac{O(s)}{I(s)} = \frac{\tau \cdot s + 1}{(k \cdot \tau)s + 1} = \frac{1}{k}\left[1 - \frac{1}{(k \cdot \tau)s + 1} \right] + \frac{1}{(k \cdot \tau)s + 1}$$

$$= \frac{1}{k}\left[1 + \frac{k - 1}{(k \cdot \tau)s + 1} \right]$$

or $\qquad \dfrac{O(s)}{I(s)} = \dfrac{\sigma \cdot s + 1}{\tau \cdot s + 1} = \dfrac{1}{\tau}\left(\sigma + \dfrac{\tau - \sigma}{\tau \cdot s + 1} \right)$

The translation to digital form consists in carrying out all of the algebraic steps directly and replacing the lag transfer function by the digital lag algorithmic calculation described in the preceding section. Considering the last form and proceeding in reverse order, the output of a lead/lag calculation would be determined from the output of the lag calculation (with time constant τ),

$$O(t) = \frac{\sigma I(t) + (\tau - \sigma)L(t)}{\tau}$$

A basic filtered derivative can be calculated using a lag calculation, now indicated as L_D, and assuming, typically, $k = 0.1$ (in fixed-point, k would be a power of 2: $\frac{1}{8}$ or $\frac{1}{16}$) and the lag time constant $0.1\tau_D$,

$$D(t) = \frac{1}{k}\,[I(t) - L_D(t)] = 10\,[I(t) - L_D(t)]$$

PID Controller Calculation

PID controller designs are expressed in many forms:

$$O(s) = \left(\tau_D s + 1 + \frac{1}{\tau I s} \right) \frac{100}{\text{PB}} \cdot \text{error}$$

$$O(s) = (\tau_D s + 1)(\tau_D I s + 1) \frac{100}{\text{PB} s} \cdot \text{error}$$

$$O(s) = (\tau_D s + 1) \left(1 + \frac{1}{\tau_I s} \right) \frac{100}{\text{PB}} \cdot \text{error}$$

Each of these forms is capable of the same performance, with one exception—the first form is capable of providing complex zeros. There is no generally argued requirement for complex zeros, but this is nonetheless a real distinction. .

There is also some disagreement as to whether the $1/\tau$ terms should be replaced by gains, or whether the proportional-band terms should be combined into independent proportional, integral, or derivative terms. The reason for giving all terms as gains is that this then places the most stable setting for all terms at zero (not entirely true of the derivative). This argument is pitched to operators. The reason for leaving the terms as before is that the time constants have a process-related meaning for engineers who understand the control issues. The separate proportional band then becomes a single stabilizing setting for all terms.

Different implementations also apply different parts of the algorithm differently to the set-point and measurement terms within the error. This reflects that these terms have different effects within the process. Ideally one would provide separate tunings for load and set-point changes. A practical compromise is to apply all three actions to the measurement, but only the proportional and integrating action to the set point. The controller will then be tuned for load disturbances.

The preceding forms have a particular difficulty if the integrating calculation term is interpreted literally. The integrating term is most naturally translated digitally as

$$\sum_{i=0}^{t/\Delta t} \frac{\Delta t}{\tau} \cdot \text{error}\,(i\Delta t)$$

However, the individual summed terms get unnaturally large when τ approaches zero. A practical way of bounding the value is to replace τ by $\tau + \Delta t$. When τ approaches zero, this still leaves an unnaturally large term for $i = t/\Delta t$, in competition with the proportional term. The solution is to replace $t/\Delta t$ by $t/\Delta t - 1$,

$$\sum_{i=0}^{t/\Delta t - 1} \frac{\Delta t}{\tau + \Delta t} \cdot \text{error}\,(i\Delta t)$$

[The summation (integration) can be carried out according to the earlier discussion.] The result can be justified in another way. Consider the last PID form introduced,

$$O(s) = (\tau_D s + 1) \left(1 + \frac{1}{\tau_I s} \right) \frac{100}{\text{PB}} \cdot \text{error}$$

The differentiation must be carried out lagged or filtered, as described previously. If all but the integrating calculations are calculated as a combined result X (taking into account any separation of the treatment of set point and measurement), the result is

$$O(s) = \left(1 + \frac{1}{\tau_I s} \right) X(s) = X(s) + \frac{X(s)}{\tau_I s}$$

This has an alternative formulation, which introduces a calculated bias B, particularly convenient for implementing the external feedback,

$$O(s) = X(s) + B(s)$$

$$B(s) = \frac{O_{FB}(s)}{\tau s + 1}$$

where O_{FB} is the external feedback term, nominally equal to O. When this pair of transfer functions is translated to an algorithm, they become

$$O(t) = X(t) + B(t)$$

$$B(t) = \text{quotient} \left[\frac{O_{FB}(t) \cdot \Delta t + B(t - \Delta t) \cdot \tau + \text{remainder}}{\tau + \Delta t} \right]$$

The collective effect of this calculation corresponds to the algorithmic expression of the more direct PID form with the modified integration proposed,

$$O(t) = X(t) + \sum_{i=0}^{t/\Delta t - 1} \frac{\Delta t}{\tau + \Delta t} X(i\Delta t)$$

As indicated earlier, external feedback must be added to this form by subtracting the difference between output and external feedback from the $X(i\Delta t)$ term before integration.

Dead-Time Calculation

The dead-time calculation corresponds to the following continuous transfer function:

$$e^{-Ts}$$

Dead time represents a black box whose output exactly repeats the time form of its input, but delayed by some amount in time. It represents the behavior of the state variables of the product—input into a pipe or onto a conveyor belt, and output (delayed) at the other end. It is an essential modeling element for typical processes.

Dead time is conventionally approximated in two ways—through Padé (continued fraction) approximations of the preceding transfer function, and through what is called a bucket brigade. The bucket brigade is implemented in an array of data cells, with the input entered at one end and then shifted down the array, one element per sample time, until it comes out at the end, n sample times later, as illustrated in Fig. 11.

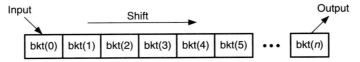

FIGURE 11 Bucket brigade. (*Foxboro.*)

On the face of it, the Padé is capable of modeling a continuous range of dead times, whereas the bucket brigade is capable of representing only integral delays, for instance, by varying n. Either mechanism can represent a fixed dead time. A further problem arises if the goal is to represent changing dead times. In this case neither the Padé nor the bucket brigade with varying n really reflects the physical or theoretical behavior of the process.

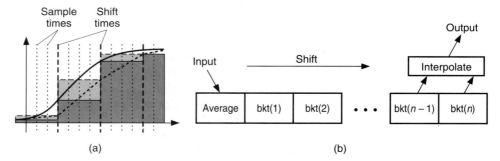

FIGURE 12 Bucket brigade for slow shifts. (*Foxboro.*)

However, the bucket brigade more closely models the state behavior of the product in a delay element; the internal bucket states represent the internal propagation. The states in the Padé are unrelated to the internal product propagation.

Thus if the bucket brigade delay can be changed by "speeding it up or slowing it down" rather than by varying n (the output bucket), it can model changing delay times resulting from changes in flow rates. There are still design questions:

- How does one smoothly achieve dead times smaller than $n\Delta t$ (the number of buckets multiplied by the sample time)?
- How does one smooth the data between stored buckets (between shifts, as shown in Fig. 12a[19])?
- How can one modify the discrete dead time (modeled by an integral number of sample time shifts) to represent a continuous range of (changing) dead times?

The first question is the easiest to answer. Shift more than once at a time.[20] The second question is almost as easy to answer. On the input side, average the sampled inputs between shifts; on the output side interpolate between the last two buckets to smooth the effect of the shift (Fig. 12a[21] and b[22]). The effect of all this is to create a dead-time approximation whose effective dead time T corresponds to $n + 0.5$ shift times.[23]

The last question is the more subtle one. It is answered by using an irregular shifting frequency whose immediate average value is equal to the current desired shifting frequency, corresponding to the desired (fractional) dead time. This solution is somewhat similar to the method used to achieve fine color variations on a CRT display which supports only a few basic colors—mix a number of different pixels irregularly for the intended average effect.

The flowchart of Fig. 13 shows the simplest way of achieving the desired irregular shift time. An accumulator variable (initially zero) is incremented by $(n + 0.5)/T$, corresponding to the desired fractional number of shifts per sample time.[24] If the accumulator variable has been incremented to 1 (indicating a net requirement for one full shift), a bucket brigade shift takes place and the accumulator is decremented by 1. Shifting and decrementing are repeated until

---

[19] In Fig. 12(a), process record is indicated by a solid black line. Thin, vertical dashed lines show process as sampled. Heavy, vertical dashed lines show process as effectively sampled by the delay under infrequent shifts. The process as sampled and held is indicated by the shaded histogram.

[20] Also averaging the output values coming from such a multiple shift.

[21] The more lightly shaded histogram shows the effect of averaging the shifted values. The dashed record reconstruction shows the effect of interpolation. Note the half shift time delay due to averaging and interpolation.

[22] Figure 12b shows the modified bucket brigade with averaging at the input and interpolation at the output.

[23] A more refined approximation would view it as a dead time of n shifts with a lag time of one shift time.

[24] If $(n + 0.5)\Delta t = T$, then $1/\Delta t = (n + 0.5)/T$.

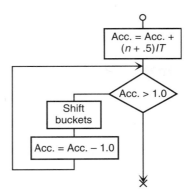

FIGURE 13 Flowchart for bucket brigade shift calculation. (*Foxboro.*)

the accumulator value drops back below 1.[25] The result is a compensator algorithm capable of modeling dead time in fully time-varying situations.

QUANTIZATION AND SATURATION EFFECTS

All of the calculations can now be carried out. The discussion has not addressed the effects of truncation on the derivative, but as developed here, they are no more serious than for proportional control. In both cases truncation will cause a very small limit cycle, on the order of the minimum quantization of the D/A converter. However, if the derivative is not carefully filtered as part of its computation (as has been shown in integrated form), severe problems arise. An approximate unfiltered derivative calculation has the form

$$D(t) = \tau_D \frac{\Delta\text{measurement}}{\Delta t} = \frac{\tau_D}{\Delta t} \cdot \Delta\text{measurement}$$

Quantization has the awkward effect of forcing a minimum nonzero measurement change for any sampled calculation.[26] This value is multiplied by the gain $\tau_D/\Delta t$, which is usually very large. The result is very large pulses on the output of the controller. For Δmeasurement quantized to one part in 1000 (0.1 percent), $\Delta t = 1$ second, and $\tau_D = 10$ minutes $= 600$ seconds, the minimum nonzero derivative equals

$$D(t) = 600 \times 0.1 = 60 \text{ percent}$$

The use of filtering smears out the pulses and limits their heights. The simple 0.1 time constant (developed earlier) lagged the derivative filter and thus limits the maximum pulse height to 10 times the quantization value. More refined algorithms can minimize the problem further by dynamically broadening the effective Δt in the calculation to get a better average derivative. The challenge is to get effective derivative action with a quantization that represents the realistic accuracy bounds of the data.

---

[25] Actually the shift test value for the accumulator is irrelevant (and would be set to zero in practice) because the continual incrementing and decrementing balances itself out to the same frequency, whatever the test point. A further simplification is to increment by $n + 0.5$ and decrement by T, further saving the need for floating point or the division.

[26] Conventional resistor ladder A/D converters may be in substantial error in the calibration of their lowest-order bits to the extent that a constant slope measurement may appear to wander up and down as converted. This makes these quantization effects several times worse in practice.

These derivative problems can be made far worse by internal saturations after integration in the PID algorithm, particularly in incremental algorithms (algorithms which output the desired change in valve position rather than the desired position). Such algorithms involve a second derivative action. The problem arises because the saturation is likely to be unsymmetrical. When reintegrated later in the algorithm or system, the result is a significant offset. In the previous case, the second differencing causes a doublet which extends 60 percent of scale in both directions. One-sided saturation, reintegrated, would create a 60 percent valve offset (bump).

IDENTIFICATION AND MATRIX-ORIENTED ISSUES

Theory-motivated control thinking emphasizes matrix-oriented formulations. These are becoming more common as engineers are trained in them. Properly understood, traditional methods are capable of equally good control, but there are aspects of normal control algorithm design where these newer methods may be more appropriate. Adaptive control often suggests such methods. Space permits only an introduction to the problem but [10]–[14] cover many important considerations.

A typical equation of this class defines least-squares data fitting of overdetermined parameters, as used in adaptation,

$$A^T A x = A^T y$$

In this equation y is a data vector, x a vector of parameters, and A an $n \times m$ matrix $(n \geq m)$ of data vectors, occurring in a set of equations of the following form:

$$a_{i1} x_1 + a_{i2} x_2 + \cdots = y$$

The problem is to find the best fit for the parameters x, given known A and y, solving the first equation:

$$x = (A^T A)^{-1} A^T y$$

The problem can be solved directly or recursively. However, work has shown that both direct and recursive techniques are unnecessarily sensitive to numerical rounding and truncation errors. The related eigenvalue problems were equally difficult until the problems were understood [10]–[13].

The problem relates to the earlier discussion on avoiding differences of large numbers, and the explosion of the digital data range under multiplication by numbers not close to 1. In matrix computations the concept of being "close to 1" is formalized in several ways. Is $|A|$ close to 1? Because matrices involve several "directions," the determinant is misleading. Another measure of a matrix is its norm,

$$\| A \| = \max_x \frac{\sqrt{x^T A^T A x}}{\sqrt{x^T x}}$$

Underlying the norm concept is the theory of orthogonal matrices and singular values [10]. An orthogonal matrix is one whose inverse equals its transpose, $Q^T Q = I$. (The letter Q will designate an orthogonal matrix.) In essence orthogonal matrices are "equal to 1" in every possible way except being the identity:

1. Their determinant is equal to 1, but for sign.
2. They do not change the length of vectors that they multiply.
3. Therefore their norm equals 1.

4. Products of orthogonal matrices are orthogonal.

5. Every element of Q has magnitude ≤ 1.

6. The elements of the RGA [15] or interaction measure of an orthogonal matrix are all between 0 and 1. Thus easy computation corresponds to easy control.

The singular values σ_i of a matrix A are the square roots of the eigenvalues of $A^T A$. More interestingly, every matrix (square or not) obeys the singular value decomposition theorem [10]. This theorem states that

$$A \equiv Q_1 \Sigma Q_2$$

for Σ, the diagonal matrix of the singular values σ_i, and some orthogonal Q_1, Q_2. Also,

7. The singular values of an orthogonal matrix are all equal to 1.

Under the singular-value decomposition, when A multiplies a vector, the immediate Q_1 or Q_2 twists the vector into an orientation where each component is multiplied by one of the singular values. Thus the vector most amplified in length by A is one oriented so that it is multiplied by the largest σ_i. For this reason the norm of A equals its largest σ_i. Also the vector most diminished in length by A is one oriented so that it is multiplied by the smallest σ_i.

Generally, a matrix is hard to compute with (is effectively "much larger than 1") if there are significant off-diagonal terms and the ratio of largest σ_i to smallest σ_i is much larger than 1. (This also corresponds to large RGA elements.) Computing $A^T A$, as in the least-squares equation, squares this ratio, making computation that much worse. Effective algorithms minimize such operations.

As a simple example, consider the usual solution of the equation $Ax = b$, with A the matrix shown in Fig. 14. The conventional gaussian solution[27] involves reducing the matrix to a triangular form, which can then be "back-solved." This is equivalent to factoring that same triangular form out, leaving a second triangular form (L and U in Fig. 14). Note that that result involves large numbers (10 and −9).

Gaussian (LU):

$$A = \begin{bmatrix} 0.1 & 1 \\ 1 & 1 \end{bmatrix} = LU = \begin{bmatrix} 1 & 0 \\ 10 & 1 \end{bmatrix} \times \begin{bmatrix} 0.1 & 1 \\ 0 & -9 \end{bmatrix}$$

Orthogonal (QR):

$$A = \begin{bmatrix} 0.1 & 1 \\ 1 & 1 \end{bmatrix} = QR$$

$$= \begin{bmatrix} -0.0995 & -0.995 \\ 0.995 & 0.0995 \end{bmatrix} \times \begin{bmatrix} -1.005 & -1.0945 \\ 0 & -0.8955 \end{bmatrix}$$

FIGURE 14 Matrix for external feedback-based backup. (*Foxboro.*)

The same matrix can also be reduced to a triangular (back-solvable) form by multiplication by an orthogonal matrix. The corresponding factorization is called QR factorization. Note that all the calculations will now involve small numbers (0.995 and −1.005). The thrust of the newer matrix methods is to avoid matrix multiplication if possible (it expands the data range) and to try to restrict any multiplications to those involving orthogonal, diagonal, or triangular matrices.

---

[27] The example cheats a little, bypassing any "pivot" operation. These are imprecise compared to the orthogonal matrix methods.

SOFTWARE AND APPLICATION ISSUES

Earlier there was a brief reference to software implications of control saturation. No modern discussion of control would be complete without observing the critical role of software [16]–[22]. Matrix-oriented control has been based on Fortran and purely mathematical approaches to control thinking. This has separated it from the major operational concerns of the industry. A control system should support not only the control computation, but the operational access to control data in a framework that is as easy to use as possible. It should include some model of sensible operator intervention.

These considerations are accommodated naturally by the software for traditional control. Standard process regulatory control is based on "blocks." These blocks represent the digital equivalents of the old analog blocks in the block diagrams. They also are data blocks whose data elements correspond to the signals and parameters of the analog block controller or are data pointers which make the connections between the blocks.

The associated control software is designed to support the implementation of control designs by the configuration of blocks. It supports the repeated on-line interpretive computation of the block data against their algorithms in an environment where these data must also be available for access from supervisory computations and operator displays.

Existing commercial systems approach these issues in many different ways, encouraging an interest in standardization. But neither the existing systems nor the proposed standards offer attractive solutions supporting the necessary flexibility and ease of use. The challenge is to provide flexible, future-oriented systems in which the goals and the structure of the control system are transparent; where new sensors, actuators, and algorithms are easily added; and which include sufficient standards so that control systems can contain elements from many vendors [21], [22].

Two issues are especially crucial—smart field devices and interoperability. Modern sensors and actuators now include significant amounts of computational power themselves. The uses of this power will expand indefinitely. But both the field devices and the central controls must be developed in a standard model which permits control and its orderly operation to be supported without special or redundant programming of either.

There is a broader challenge—to solve the control software problem with solutions that capitalize on the computer as a truly intelligent control device, beyond the rigid scientific computations envisioned by the matrix control approaches, or the programmable block diagram mimicking the dated analog controls.

SUMMARY

Digital control algorithms can be designed for experimental or single applications using the easiest tools available—Fortran, C, BASIC, and floating-point arithmetic. Here there is reasonable hope that normal commissioning debugging will weed out all the problems. But if a sense of workmanship prevails, or if the algorithm is to be used in numerous applications, then refinement and foolproofing are necessary and the following need to be considered:

1. Numerical effects of fixed- and floating-point arithmetic
2. Documentation, control, and testing of detailed scaling, precision, and saturation within the algorithm
3. Design for natural tuning and qualitative behavior, predictable from analog intuition
4. Nasty quantization surprises
5. Accommodations of windup and other control-limiting effects

6. Bumpless transfer and operational considerations

7. Software, architectural, and configuration considerations

General-purpose digital programming languages and tools do not remotely address these issues.

APPENDIX
GENERALIZED SET NORMALIZATION

This appendix details the generalized floating-point normalization[28] function to illustrate the concepts. Set normalization reprocesses the set of component elements in the generalized multiple-precision floating-point format so that the data ranges of the consecutive elements do not overlap, each element has the largest possible mantissa, and successive elements are fully rounded down relative to their predecessors.

Several basic operations and their requisite conditions of execution need to be outlined. Note that some of these basic operations, once executed on a given element, may further require the recursive execution of basic operations on preceding elements as well. Thus a carry from one element may progressively require a carry out of its preceding elements as well. The basic operations generalize traditional operations such as carry and rounding[29]:

- *(Element) normalization at nth position.* Required whenever $|f_n| < b^{M-1}$. It requires that f_n be multiplied by b, and e_n be decremented by 1 repeatedly until $|f_n| < b^{M-1}$ no longer applies. [This condition corresponds to the normal (single-precision) floating-point normalization. It recognizes that the higher-order digits are zero and thus underutilized.]

- *Clear at nth position.* Needed when a calculation results in an f_n that would equal zero. It requires the elimination of the resulting element altogether.

- *Round at nth position.* Needed after the first element ($n > 1$) when both $e_{n-1} - e_n = M$ and $|f_n| \geq 0.5 \times b^M$.[30] Whenever this condition is recognized, f_n is replaced by $f_n - \text{sgn}(f_n) \times b^M$ and f_{n-1} is replaced by $f_{n-1} + \text{sgn}(f_n)$. Any carry or round required at the $(n-1)$th position must then be serviced appropriately. Necessary element normalization resulting at the nth position is carried out. [This condition recognizes that the higher-order element will more accurately (and more efficiently) reflect the result if the lower-order result is rounded and carried into the higher-order result.]

- *Carry at nth position.* Needed when a calculation results in an overrange f_n which, if preserved, would make its magnitude equal to or larger than b^M (and require more than M digits). It requires the appropriate execution of a carry up or down as described next. [This corresponds to the normal carry within multiple-precision data elements. However, in the floating-point context, depending on the preceding and following elements, this situation is accommodated by carrying the data up into the higher-order elements or down into the lower-order elements.]

---

[28] In normal single-precision floating-point format, normalization shifts the mantissa and decrements the exponent to eliminate high-order zeros, ensuring that all later computations will include the maximum possible number of lower-order digits.

[29] Note that all divisions and multiplications of terms of the form b^e in the computations described become simple shift operations (conventionally so in the expected binary implementation).

[30] In binary arithmetic the second part of this condition consists of checking whether the highest data bit and the sign bit (of f_n) disagree. After correction, binary rounds will always require a normalization at the nth position, incorporated simply in the correction operation. The combined effect of binary rounds and simple normalization is to then rule out situations where $e_{n-1} - e_n = M$ after the generalized normalization has been completed.

- *Carry up at nth position.* Needed after the first element ($n > 1$) to correct a carry at the nth position when $e_{n-1} - e_n \leq M$. Whenever this condition is recognized, the overrange f_n is divided by b^M, the actual f_n is replaced by the remainder, and the quotient, scaled by $b^{M+e_n-e_{n-1}}$ [normally then equal to $\mathrm{sgn}(f_n) \times b^{M+e_n-e_{n-1}}$ or, simply, $\mathrm{sgn}(f_n)$], is added to f_{n-1}. The addition to f_{n-1} must in turn be checked and processed for any needed carry and round at the $(n-1)$th position. Necessary element clear or normalization at the nth position must be carried out. [This corresponds to the normal action of a carry when high order elements are available to receive its overranged bit.]

- *Carry down at nth position.* Needed to correct a carry at the nth position when $n = 1$ or $e_{n-1} - e_n > M$. Whenever this condition is recognized, f_n is divided by b and replaced by the resulting quotient, e_n is incremented by 1. The remainder, multiplied by b^{M-1} (if nonzero), is entered as the mantissa of a new element in the list, with an exponent equal to $e_n - M$.[31] The new element is then normalized. [This resolves carries when there is no appropriate higher-order term to receive the carry data by raising the exponent and passing the data into lower-order elements.]

The generalized set-normalize function, applicable to the set of single-precision component elements, is carried out, in terms of the basic operations, according to the following steps:

1. Order the elements, largest first.
2. Clear or normalize each element (at its nth position) if needed.
3. Then, scanning the set, starting with the largest element and continuing to the next to last element, carry out the following steps, recombining, where appropriate, the nth and $(n + 1)$th elements:
   a. If the difference in exponents is less than M, divide f_{n+1} by $b^{e_n - e_{n+1}}$. Reserve the remainder as the new f_{n+1} multiplied by $b^{M-(e_n - e_{n+1})}$, unless the result is zero. In that case delete this $(n + 1)$th value from the set. Set e_{n+1} to $e_n - M$ and add the quotient to f_n. Apply any necessary carry (up or down) or round operations at the nth position to the resulting sum. Normalize the element at the $(n + 1)$th position if necessary.
   b. If the difference in exponents is equal to M (whether or not step a was required), check for a round at the $(n + 1)$th position.
   c. Reorder all resulting new or altered values as required (step 1) to continue the scan.

REFERENCES

1. Bristol, E. H., "On the Design and Programming of Control Algorithms for DDC Systems," *Control Engineering,* Jan. 1977.

2. Clagget, E. H., "Keep a Notebook of Digital Control Algorithms," *Control Engineering,* pp. 81–84, Oct. 1980.

3. Fehervari, W., "Asymmetric Algorithm Tightens Compressor Surge Control," *Control Engineering* pp. 63–66, Oct. 1977.

4. Tu, F. C. Y., and J. Y. H. Tsing, "Synthesizing a Digital Algorithm for Optimized Control," *Instrum. Tech.,* pp. 52–56, May 1979.

5. Franklin, G. F., and J. D. Powell, *Digital Control of Dynamic Systems,* Addison-Wesley, Reading, Massachusetts, 1980.

6. Knuth, D. E., *The Art of Computer Programming,* vol. 2, Addison-Wesley, Reading, Massachusetts, 1981.

7. *Collected Algorithms from ACM, 1960–1976.* American Computer Manufacturers, 1978.

---

[31] Using the value of e^n just incremented. Alternatively, if the overrange f_n actually equals b^M, then f_n can equivalently be replaced by $\mathrm{sgn}(f_n) \times b^{M-1}$ and e^n by $e^n + 1$, avoiding the division and the remainder altogether.

8. ANSI/IEEE Std. 754-1985 "Binary Floating-Point Arithmetic," Inst. of Elect. and Electron. Eng., New York, 1985.

9. Perry, R. H., D. W. Green, Eds., *Chemical Engineers' Handbook,* 5th ed., McGraw-Hill, New York, 1973, pp. 2-81–2-85.

10. Strang, G., *Linear Algebra and Its Applications,* Academic Press, New York, 1976.

11. Bierman, G. J., *Factorization Methods for Discrete Sequential Estimation,* Academic Press, New York, 1977.

12. Lawson, C. L., and R. L. Hansen, *Solving Least Squares Problems,* Prentice-Hall, Englewood Cliffs, New Jersey, 1974.

13. Laub, A. J., and V. C. Klema, "The Singular Value Decomposition: Its Computation and Some Applications," *IEEE Trans. Autom. Control,* vol. AC-25, pp. 164–176, Apr. 1980.

14. MacFarlane, A. G. J., and Y. S. Hung, "A Quasi-Classical Approach to Multivariable Feedback Systems Design," in *Proc. 2d IFAC Sympery Computer Aided Design of Multivariable Technological Systems* (Purdue University, West Lafayette, Indiana, Sept. 1982), pp. 39–48.

15. Bristol, E. H., "On a New Measure of Interaction for Multivariable Control," *IEEE Trans. Autom. Control,* vol. AC-11, pp. 133–134, Jan. 1966.

16. Shinskey, F. G., "An Expert System for the Design of Distillation Controls," presented at Chemical Process Control III, Asilomar, California, Jan. 12–17, 1986.

17. Bristol, E. H., "Strategic Design: A Practical Chapter in a Textbook on Control," presented at the 1980 JACC, San Francisco, California, Aug. 1980.

18. Bristol, E. H., "After DDC—Idiomatic (Structured) Control," presented at the 88th National Meeting of the AIChE, Philadelphia, Pennsylvania, June 1980.

19. Prassinos, A., T. J. McAvoy, and E. H. Bristol, "A Method for the Analysis of Complex Control Schemes," presented at the 1982 ACC, Arlington, Virginia, June 1982.

20. Bristol, E. H., "A Methodology and Algebra for Documenting Continuous Process Control Systems," presented at the 1983 ACC, San Francisco, California, June 1983.

21. Bristol, E. H., "Super Variable Process Data Definition," Instrument Society of America, Working Paper SP50.4, Oct. 24, 1990, and updates.

22. Bristol, E. H., "An Interoperability Level Model: Super-Variable Categories," Instrument Society of America, Continuing Working Paper SP50.4.

SAFETY IN INSTRUMENTATION AND CONTROL SYSTEMS

by E. C. Magison*

Extensive use of electrical and electronic control systems, computers, sensors, and analyzers in connection with process instrumentation continues to focus attention on reducing the probability of fire or explosion due to electric instrument failure. At one time explosionproof housing was the common method of providing protection. Attention then turned to other

---

* Consultant, Ambler, Pennsylvania.

means which can provide the same or higher levels of safety, but with less weight and easier accessibility for maintenance and calibration and at equivalent or lower costs.

Because instrument manufacturers serve an international market, increased activity within national jurisdictions is being matched by recognition that standardization must be accomplished at the international level as well.

AREA AND MATERIAL CLASSIFICATION

North America

In the United States, Articles 500 to 504 of the National Electrical Code (NEC) provide basic definitions of hazardous areas and the requirements for electrical installations. Article 500 defines the classification of hazardous locations broadly in terms of kind and degree of hazard. The kind of hazard is specified by class and group. The degree of hazard is designated by division. Typical industrial locations, for example, may be classified as Class I, Group D, Division 1; or Class II, Group G, Division 2. The principal features of the NEC classification are summarized in Table 1. Many additional materials are listed in National Fire Protection Association publication NFPA 497M. Similar definitions are given in the Canadian Electrical Code.

International Electrotechnical Commission

Most industrial nations have adopted or are adopting the area and material classification definitions of the International Electrotechnical Commission (IEC). Locations where a flammable concentration may be present are designated Zone 0, Zone 1, or Zone 2.

A Zone 0 location is a location where the atmosphere may be in the explosive range such a high percentage of the time (above 10 percent) that extraordinary measures must be taken to protect against ignition by electrical apparatus.

Zone 1 locations have a probability of hazard between Zone 2 and Zone 0. A Zone 2 location is similar to North American Division 2. Taken together, Zone 1 and Zone 0 equate North American Division 1.

In Zone 2, requirements analogous to those in North America are accepted in principle in many countries, but in practice Zone 1 types of protection are often used in Zone 2 because there is no accepted standard for Zone 2 apparatus. The advantage of distinguishing between the extraordinary hazards of Zone 0 and the lesser hazards of Zone 1 is that apparatus and installation requirements can be relaxed in Zone 1. For example, intrinsically safe systems in North America are judged on the basis of two faults because of the encompassing definition of Division 1. For use in Zone 1, consideration of only one fault is required, although two faults are assessed in Zone 0.

Material classification in most countries now uses IEC terminology. A Group I hazard is due to methane (firedamp) in the underground works of a mine. The presence of combustible dusts and other environmental aspects of mining works are assumed when preparing apparatus requirements for Group I.

Group II gases and vapors are flammable materials found in industrial aboveground premises. They are divided into Groups IIA, IIB, and IIC, which are similar although not identical to North American Groups D, C, and B, respectively.

Classifying a Hazardous Location

The NEC definitions provide guidelines, but do not give a quantitative method for classifying a specific hazardous location. Factors to consider include the quantity of hazardous material

TABLE 1 National Electrical Code Area Classification System

Class I Gases and vapors	Class II Dusts	Class III Flyings
Group A—Acetylene Group B—Hydrogen or gases of similar hazardous nature, such as manufactured gas, butadiene, ethylene oxide, propylene oxide Group C—Ethyl ether, ethylene, cyclopropane, unsymmetrical dimethylhydrazine, acetaldehyde, isoprene Group D—Gasoline, hexane, naphtha, benzene, butane, propane, alcohol, acetone, benzol, lacquer solvent, natural gas, acrylonitrile, ethylene dichloride, propylene, styrene, vinyl acetate, vinyl chloride	Group E—Metal dusts Group F—Carbon black, coal, coke dusts Group G—Grain dust, flour, plastics, sugar	No group assigned. Typical materials are cotton, kapok, nylon, flax, wood chips—normally not in air suspension
Division 1*		
For heavier-than-air vapors, below-grade sumps, pits, etc., in Division 2 locations. Areas around packing glands; areas where flammable liquids are handled or transferred; areas adjacent to kettles, vats, mixers, etc. Where equipment failure releases gas or vapor and damages electrical equipment simultaneously.	Cloud of flammable concentration exists frequently, periodically, or intermittently—as near processing equipment. Any location where conducting (resistivity less than 10^5 Ω-cm) dust may accumulate.	Areas where cotton, spanish moss, hemp, etc., are manufactured or processed.
Division 2*		
Areas adjacent to a Division 1 area. Pits, sumps containing piping, etc., in nonhazardous location. Areas where flammable liquids are stored or processed in completely closed piping or containers. Division 1 areas rendered nonhazardous by forced ventilation.	Failure of processing equipment may release cloud. Deposited dust layer on equipment, floor, or other horizontal surface	Areas where materials are stored or handled.

* In a Division 1 location there is a high probability that a flammable concentration of vapor, gas, or dust is present during normal plant operation, or because of frequent maintenance. In a Division 2 location there is only low probability that the atmosphere is hazardous—for example, because of equipment failure.

Until the 1971 revision, material classification in the NEC differed from the practice in almost all other countries except Canada. Material groupings were based on consideration of three parameters: autoignition temperature (AIT) (or, spontaneous ignition temperature SIT), maximum experimental safe gap (MESG), and the maximum pressure rise in an explosion test chamber. In Europe materials long have been grouped by AIT and separately by MESG. Pressure rise is not a material classification criterion. It is now recognized in the United States that there is no correlation between MESG and AIT. Hydrogen, for example, has a very small MESG and a very high AIT. Many Group C and D materials have lower ignition temperatures but wider experimental safe gaps. Because the NEC classification was based on two uncorrelated parameters, United States experts could not use the results of experimental work on new material in other countries, or use other classification tables. The 1971 NEC revisions separate AIT from considerations of MESG. Explosionproof housings now can be designed for MESG typical of a group of materials. External surface temperatures shall not exceed the AIT of the hazardous gas or vapor of concern.

which may be released, the topography of the site, the construction of the plant or building, and the past history of fire and explosion (of a particular location or plant as well as of an entire industry). Although authorities recognize the need, there are no concise rules for deciding whether a location is Division 1 or Division 2. The best guides to area classification known to the author are American Petroleum Institute (publications API RP500A, B, and C for petroleum installations, and NFPA 497 for installations in chemical plants). These documents are applicable to any industry.

Special Cases of Area Classification

It is common practice to pressurize instrument systems to reduce the area classification inside the enclosure. The inside of an instrument enclosure provided with a simple pressurization system, located in a Division 1 area, can be considered a Division 2 location because only by accidental failure of the pressurization system can the internal atmosphere become hazardous. If the pressurization system is designed to deenergize all equipment within the enclosure when pressurization fails, the interior can be considered a nonhazardous location. Two failures are required—(1) of the pressurization system and (2) of the interlock system—before an explosion can occur.

An important limitation of this philosophy is that if any single failure can make the enclosure hazardous, the interior of the enclosure must not be classified less hazardous than Division 2, regardless of the pressurization system design. Such is the case with bourdon-tube or diaphragm-actuated instruments where process fluid is separated from the instrument interior only by a single seal, namely, the bourdon or diaphragm. Unless the pressure is high enough or the enclosure air flow is great enough to prevent a combustible concentration inside the enclosure should the measuring element fail, the interior never should be classified less hazardous than Division 2. Such systems often are referred to as singly sealed systems.

In a doubly sealed system two seals are provided between the process fluid and the area being purged, and a vent to the atmosphere is provided between the seals. Failure of both seals is required to make the enclosure interior hazardous. Even so, pressurization can prevent the hazardous material from entering the compartment because the hazardous material is at atmospheric pressure. Article 501-5(f) of the NEC mandates a double seal wherever failure of a sealing element could force flammable material into the conduit system.

TECHNIQUES USED TO REDUCE EXPLOSION HAZARDS

The predisposing factors to fire or explosion are (1) the presence of a flammable liquid, vapor, gas, dust, or fiber in an ignitable concentration, (2) the presence of a source of ignition, and (3) contact of the source with ignitable material. The most obvious way to eliminate the possibility of ignition is to remove the source to a location where there is no combustible material. This is the first method recognized in the NEC, Article 500. Another method is to apply the principle of intrinsic safety. Equipment and wiring which are intrinsically safe are incapable, under normal or abnormal conditions, of igniting a specifically hazardous atmosphere mixture. For practical purposes there is no source of ignition.

Figure 1 summarizes the techniques used to reduce explosion hazards. Methods based on allowing ignition to occur force combustion under well-controlled conditions so that no significant damage results. A continuous source of ignition, such as the continuous pilot to localize combustion in gas appliances, is commonplace. Explosionproof enclosures contain an explosion so that it does not spread into the surrounding atmosphere. Historically this has been the most common technique. In Zone 2, enclosed break devices in which the enclosed volume is so small that a weak incipient explosion cannot escape the enclosure, are permitted in some countries.

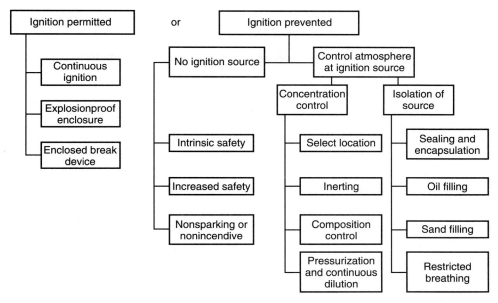

FIGURE 1   Techniques used to reduce explosion hazards.

There are several methods for reducing hazard by preventing the accumulation of combustible material in an explosive concentration or for isolating an ignition source from flammable material. Pressurization of instruments is common. Continuous dilution in which the interior of an enclosure is pressurized to exclude flammable material from entering and is also continuously purged to dilute any internal release of flammable material is applicable to analyzers and other devices in which flammable material may be released inside the enclosure. In hydrogen-annealing furnaces and hydrogen-cooled electric generators the concentration is held above the upper explosive limit. Blanketing of tanks with nitrogen or carbon dioxide (CO_2) and rock dusting of coal mine galleries and shafts are examples of using inert materials to suppress a combustible mixture.

Several techniques are used to isolate the ignition source. Oil immersion prevents contact between the atmosphere and the ignition source. In Europe sand-filled equipment sometimes is used. Sealing and encapsulation both provide a barrier to impede contact.

Increased safety is a technique used for transformers, motors, cables, and so on, which are constructed with special attention to ruggedness, insulation, reliability, and protection against overtemperatures so that an ignition-capable failure is of very low probability. Increased safety, developed first in Germany and now accepted widely in Europe, may be used in Zone 1.

Nonsparking apparatus and nonincendive apparatus are suitable for use in Division 2 and Zone 2 because they have no normal source of ignition.

Restricted breathing is the technique of using a tight, but not sealed enclosure in Zone 2 to allow only slow access of flammable vapors and gases to the source of ignition. This technique, developed in Switzerland, is slowly achieving acceptance in Europe.

Explosionproof Housings

Termed flameproof enclosures in international English, explosionproof housings remain the most practical protection method for motor starters and other heavy equipment which produces sufficient energy in normal operation to ignite a flammable atmosphere. Explosion-proof enclosures are not vaportight; it is expected that a flammable atmosphere will enter the enclosure.

A pressure rise of 100 to 150 lb/in^2 (690 to 1034 kPa) is typical for the mixture producing the highest explosion pressure. In small enclosures, loss of energy to the enclosure walls decreases the pressure rise. Because the enclosure must contain the explosion and also must cool escaping gases, cast or heavy metallic construction with wide, close-fitting flanges or threaded joints is typical. Nonmetallic construction is permitted.

For specific design criteria in the United States, reference should be made to the standards of the intended certifying agency. Although requirements of all agencies are similar, there are many differences in detail. In general, in addition to tests to ensure that an internal explosion is not transmitted to the outside, the enclosure must withstand a hydrostatic pressure of four times the maximum pressure observed during the explosion test and must not have an external case temperature high enough to ignite the surrounding atmosphere. In Canada the applicable standard is CSA C22.2, No. 30.

In Europe the applicable standards are CENELEC EN50014, "General Requirements," and CENELEC EN50018, "Flameproof Enclosure 'd'." These are available in English as British Standard BS 5501, Parts 1 and 5.

European requirements emphasize special fasteners to prevent unauthorized opening of flameproof enclosures. Requirements for flange gaps are less restrictive than North American standards. Routine testing of enclosures at lower pressures than those of the North American test is common. Type testing is achieving recognition. In North America, wider permissible flange gaps and routine testing are gaining acceptance.

Encapsulation, Sealing, and Immersion

These techniques are not likely to be applied to a complete instrument. They serve to reduce the hazard classification of the instrument by protecting sparking components or subassemblies. Oil immersion and sand filling are applied to power-handling apparatus, but neither technique has important applications in instrument systems, although oil immersion may be a convenient technique for some hydraulic control elements.

Sealing

Article 501-3(b)(2) of the NEC states that general-purpose enclosures may be used in Division 2 locations if make-and-break contacts are sealed hermetically against the entrance of gases or vapors. The NEC provides no definition of a satisfactory hermetic seal, however. Seals obtained by fusion, welding, or soldering and, in some instances, plastic encapsulation are widely accepted. In reality, the leak rate of soldered or welded seals is lower than that required for protection against explosion in a Division 2 location.

The long-time average concentration inside a sealed enclosure approaches the average concentration outside. The function of a seal is to prevent transient excursions above the lower explosive limit (LEL) outside the device from raising the concentration inside the device to the LEL. Three mechanisms can force material through a seal, (1) changes in ambient temperature, (2) changes in barometric pressure—both effects tending to make the seal breathe, and (3) wind and strong air currents. The last named mechanism usually can be ignored because a sealed device must be installed in a general-purpose enclosure to protect it from such conditions.

Encapsulation involves enclosing a component or subassembly in a plastic material, a tar, or a grease, with or without the additional support of a can. If the encapsulated assembly is robust and has mechanical strength and chemical resistance adequate for the environment in which it is used, it can be considered the equivalent of a hermetic seal. An external hazardous atmosphere must diffuse through a long path between the encapsulating material and the device leads to reach the interior. Standards for sealed devices can be found in Instrument Society of America Standard ISA S12.12.

Pressurization Systems

Lowering the hazard classification of a location by providing positive-pressure ventilation from a source of clean air has long been recognized in the NEC, and instrument users have pressurized control room and instrument housings for many years. The first detailed standard for instrument purging (pressurizing) installations was ISA SP12.4 (now withdrawn). These requirements in essentially the same form make up the first section of NFPA 496, which also covers purging of large enclosures, ventilation of control rooms, Class II hazards, and continuous dilution. NFPA 496 defines three types of pressurized installation:

> *Type Z.* Pressurization to reduce the classification within an enclosure from Division 2 to nonhazardous
>
> *Type Y.* Pressurization to reduce the classification in an enclosure from Division 1 to Division 2
>
> *Type X.* Pressurization to reduce the classification within an enclosure from Division 1 to nonhazardous

Type Z Pressurization.   This permits the installation of ignition-capable equipment inside the enclosure. For an explosion to occur, the pressurized system must fail, and also, because the surrounding area is Division 2, there must be a process equipment failure which releases flammable material. Thus there must be two independent failures, and no additional safeguards in the pressurization system are necessary.

Figure 2 shows a typical installation for Type Z pressurization from NFPA 496. Only a pressurization indicator is required. The probability that a process fault will make the location hazardous before any failure of the pressurization system is recognized and corrected is assumed to be extremely low. An electric indicator or alarm must meet the requirements of its location before pressurization. If a pressure indicator is used, no valve may be installed between the pressure indicator and the case. Any restriction between the case and the pressure device must be no smaller than the smallest restriction on the supply side of the pressure device. The case must be opened only if the area is known to be nonhazardous or if power has been removed. In normal operation the pressurization system must maintain a minimum pressure of 0.1 inch (2.5 mm) of water gage. The flow required to maintain this pressure is immaterial. The temperature of the pressurized enclosure must not exceed 80 percent of the ignition temperature of the gas or vapor involved when it is operated at 125 percent of rated voltage. A red warning nameplate must be placed on the instrument to be visible before the case is opened.

Type Y Pressurization.   In this case all requirements of Type Z pressurization must be met. Equipment inside the enclosure must be suitable for Division 2, that is, it is not an ignition source in normal operation. For an explosion to occur, the pressurization must fail, and the equipment inside must fail in a way to make it an ignition source.

Type X Pressurization.   In this system the pressurization is the only safeguard. The atmosphere surrounding the enclosure is presumed to be frequently flammable. The equipment within the enclosure is ignition-capable. The pressurization system failure must automatically deenergize internal equipment. All requirements for Type Z and Type Y pressurization must be met. The interlock switch may be pressure- or flow-actuated. The switch must be suitable for Division 1 locations, even if it is mounted within the instrument case, because it may be energized before purging has removed all flammable material. A door that can be opened

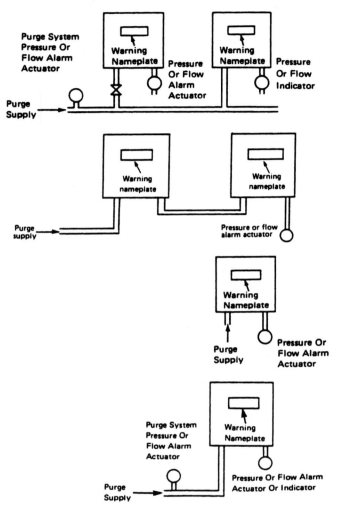

FIGURE 2   Typical installations for Type Y and Type Z pressurization. (*From NFPA 496.*)

with the use of a tool must be provided with an automatic disconnect switch suitable for Division 1. A timing device must prevent power from being applied until four enclosure volumes of purge gas can pass through the instrument case with an internal pressure of 0.1 inch (2.5 mm) of water gage. The timing device also must meet Division 1 requirements, even if inside the case (Fig. 3).

IEC and CENELEC standards for pressurization systems are similar to those of NFPA, although the requirements are not phrased in terms of reduction in area classification. The minimum pressure is 0.2 inch (5.1 mm) of water gage.

The IEC and NFPA standards also cover continuous dilution. The hardware is similar to that required for pressurization, but the rationale for selecting the level of protection needed is based on the presence of a source of flammable material within the enclosure, as in an analyzer. The objectives are to prevent entry of an external flammable atmosphere (pressuriza-

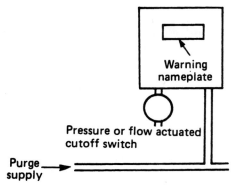

FIGURE 3   Typical installation for Type X pressurization. (*From NFPA 496.*)

tion) and also to dilute any internal release to a low percentage of the lower flammable limit
(continuous dilution).

INTRINSIC SAFETY

Experiment and theory show that a critical amount of energy must be injected into a combustible mixture to cause an explosion. If the energy provided is not greater than the critical
ignition energy, some material will burn, but the flame will not propagate. An explosion
occurs only when enough energy is injected into the mixture to ignite a critical minimum volume of material. The diameter of a sphere enclosing this minimum volume is called the
quenching distance or quenching diameter. It is related to the maximum experimental safe
gap (MESG), but is about twice as large. If the incipient flame sphere does not reach this
diameter, it will not propagate.

The energy required for ignition depends on the concentration of the combustible mixture. There is a concentration at which the ignition energy is minimum. The curve of ignition
versus concentration is asymptotic to limits of concentration commonly called the lower
explosive limit (LEL) and the upper explosive limit (UEL). Figure 4 illustrates the influence
of concentration on the critical energy required to cause ignition. A hydrogen-air mixture,
one of the most easily ignited atmospheric mixtures, supports combustion over a wide range
of concentrations. A propane-air mixture, which is typical of many common hazardous materials, is flammable only over a narrow range of concentrations. The amount of energy
required to ignite the most easily ignited concentration of a mixture under ideal conditions is
the minimum ignition energy (MIE).

Definition

The NEC defines intrinsically safe equipment and wiring as "incapable of releasing sufficient
electrical or thermal energy under normal or abnormal conditions to cause ignition of a specific hazardous atmosphere mixture."

Early Developments

The British first applied intrinsic safety in direct-current signaling circuits, the first studies
beginning about 1913. In 1936 the first certificate for intrinsically safe equipment for other

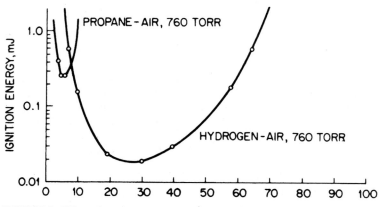

FIGURE 4 Effect of concentration on ignition energy.

than mining was issued. By the mid-1950s certification in Great Britain for industrial applications was common. At the U.S. Bureau of Mines work on intrinsically safe apparatus, although the term was not used, commenced about the same time as the British investigations. Rules for telephone and signaling devices were published in 1938.

During the 1950s increased use of electric equipment in hazardous locations stirred worldwide interest in intrinsic safety, and by the late 1960s almost every industrial country had either published a standard for intrinsically safe systems or drafted one. The major industrial countries also were active in the IEC Committee SC31G, which prepared an international standard for intrinsically safe systems.

The first standards for intrinsically safe equipment intended for use by the instrument industry were published as ISA RP12.2, issued in 1965. The NFPA used ISA RP12.2 as a basis for the 1967 edition of NFPA 493.

During the years following the publication of ISA RP12.2 and NFPA 493-1967 the certification of intrinsically safe systems by independent approval agencies, such as Factory Mutual and Underwriters Laboratories in the United States, CSA in Canada, BASEEFA in the United Kingdom, and PTB in Germany, became a legal or marketing necessity in most countries. All standards for intrinsic safety have therefore become much more detailed and definitive. Adherence to the standard is the objective, not a judgment of safety. The work of the IEC has served as the basis for later editions of NFPA 493 and for UL 913, which is now the U.S. standard for intrinsically safe equipment, as well as for Canadian Standard CSA C22.2-157 and CENELEC Standard EN50020. Any product marketed internationally must meet all these standards.

All the standards agree in principle, but differ not only in detail, but also in the way they are interpreted by the approval agencies.

CSA and U.S. standards are based on safety after two faults, because in these cases Division 1 includes Zone 1 and Zone 0. European standards provide *ia* and *ib* levels of intrinsic safety for Zone 0 and Zone 1 application, based on consideration of two faults and one fault, respectively.

Standards for intrinsic safety can be less intimidating to the user if it is appreciated that most construction details are efforts to describe what can be considered a fault or what construction can be considered so reliable that the fault will never occur. When viewed in this light, creepage and clearance tables, transformer tests, and tests of protective components make much more sense. They are guidelines for making design decisions—not mandated values for design. They apply only if safety is affected.

Design of Intrinsically Safe Systems

The objective of any intrinsically safe design, whether produced by an equipment manufacturer or by a user attempting to assemble a safe system from commercially available devices, is the same—to ensure that the portion of system in the Division 1 location is unable to release sufficient energy to cause ignition, either by thermal or by electrical means, even after failures in system components. It is not necessary that the associated apparatus, that is, the apparatus located in Division 2 or a nonhazardous location connected to the intrinsically safe circuit, be itself intrinsically safe. It is only necessary that failures, in accordance with the accepted standard for intrinsic safety, do not raise the level of energy in the Division 1 location above the safe level.

BASIC TECHNIQUES USED BY MANUFACTURERS

Techniques used by manufacturers in the design of intrinsically safe apparatus are relatively few in number, and all manufacturers use the same fundamental techniques.

Mechanical and Electrical Isolation

The most important and most useful technique is mechanical isolation to prevent intrinsically safe circuits and nonintrinsically safe circuits from coming in contact. Often mechanical isolation is achieved solely by appropriate spacing between the intrinsically safe and nonintrinsically safe circuits. In other cases, especially at field connections or in marshaling panels, partitions or wireways ensure that the nonintrinsically safe wiring and intrinsically safe wiring are separate from one another. Encapsulation is sometimes used to prevent contact between the two types of circuits.

Related to mechanical isolation is what can be called electrical isolation. Except in battery-operated systems, intrinsically safe systems have some connection to the power line, usually through a power transformer. The designer must consider the possibility of a transformer fault which connects the line voltage primary winding to the low-voltage secondary winding. In many systems, if one must consider the presence of line voltage on secondary circuits, the value and power rating of limiting elements would make the design of an intrinsically safe system both functionally and economically impractical. Therefore in modern standards for intrinsically safe construction, several varieties of transformer construction are recognized to be so reliable that one can assume that a primary-to-secondary short circuit will never occur. In one such reliable construction, a grounded shield between primary and secondary ensures that any fault is from the primary to the grounded shield, so that the secondary winding potential is not raised to an unsafe voltage. In addition to special attention to transformer construction and testing, it is also necessary that the wiring layout prevent any accidental connection between wiring on the primary side of the transformer and wires connected to the transformer secondary.

Current and Voltage Limiting

Except in some portable apparatus, almost all intrinsically safe circuits require both current and voltage limiting to ensure that the amount of energy released under fault conditions does not exceed safe values. Voltage limiting is often achieved by use of redundant zener diodes to limit the voltage, but zener-triggered silicon-controlled rectifier (SCR) crowbar circuits are also used to limit the voltage. Redundancy is provided so that, in the case of the failure of a single diode or limiting circuit, the second device continues to provide voltage limiting. Cur-

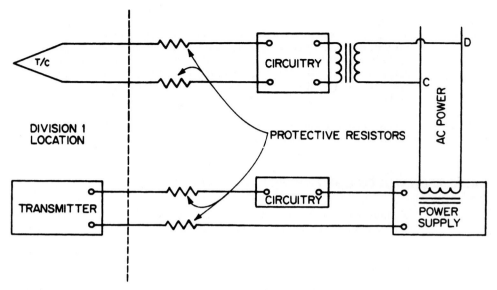

FIGURE 5   Use of resistors to limit current in hazardous location.

rent limiting in dc circuits and in most ac circuits is provided by film or wirewound resistors of high reliability (Fig. 5). Properly mounted resistors which meet the requirements for protective resistors in the applicable standard need not be redundant. They are of a level of quality that they will not fail in a way that allows current to increase to an unsafe level.

One very common use of current and voltage limiting is in the Redding zener diode barrier (Fig. 6). The unique feature of these barriers is the incorporation of a fuse in series with the zener diodes, so that when a fault causes current to flow through the zener diode, the fuse will open before the power in the zener diode reaches a level at which the diode may open the circuit. In the design shown in Fig. 6, the 20-ohm resistor does not perform a safety function. It allows testing of the barrier to determine that the diodes are still intact. The current limiting function is performed by the 300-ohm resistor.

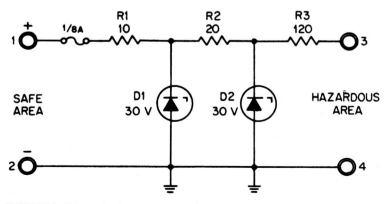

FIGURE 6   Schematic of zener diode barrier, positive type.

Devices to be connected to terminals 3 and 4 must be approved as intrinsically safe, but any device incapable of applying voltage to terminal 1 higher than the barrier rating may be connected. If the barrier is designed to limit against full power line potential, then the equipment in the nonhazardous area may be selected, connected, and intermingled without regard to safety in the field circuits if no potential above power line voltage is present.

In use, terminals 2 and 4 are both bolted to a busbar which is grounded through a very low (usually less than 1-ohm) ground resistance. The power supply also must be grounded. In operation, diodes D_1 and D_2 conduct only leakage current, which is small compared with the normal circuit flowing between terminals 1 and 3. When high voltage is applied to terminal 1, the diodes conduct and limit the voltage at terminal 3 to a safe value. R_2 limits the current into the hazardous area. Under fault conditions, the barrier looks like a low-voltage resistive source from the intrinsically safe side, and like a very low-impedance load at terminals 1 and 2.

The values in Fig. 6 are for a 28-volt 93-mA barrier. Under fault conditions, the intrinsically safe circuit will appear to be driven from a nominal 28-volt source with a source resistance of 300 ohms. Safety is provided by the diodes and resistors. The resistors can be presumed not to fail. The diodes are redundant. Should one fail, limiting would still take place. The fuse serves no purpose regarding ignition and could be replaced by a resistor. Its function is to make the diode barrier economical. Should a fault occur, the zener diodes would connect heavily and, except for the fuse, would have to be impractically large and costly. The fuse is selected to blow at a current much lower than that which would damage the diode, permitting lower power, less costly diodes to be used.

Shunt Elements

These devices are used to absorb the energy which would otherwise be released by an inductor to the arc. The function of a shunt diode is shown in Fig. 7. Although capacitors and resistors can be used to shunt inductors, in dc circuits diodes are placed so that in normal operation they are back-biased and draw no current. The shunt elements must be redundant so that if one fails, the other will continue to protect the inductor. Both must be connected close to the inductor being protected. Connection to the inductor must be especially reliable so that a fault between the inductor and the protective shunt diodes can be assumed not to occur. If such a fault occurs, ignition is possible because of the release of energy from the inductor. The purpose of the shunt diodes is to absorb energy stored in the inductor if the circuit external to the protected inductor opens.

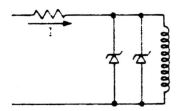

FIGURE 7   Shunt diodes reduce incendivity of inductor.

Analytical Method for Circuit Design

The outline presented here can be considered both as a method of assessing a circuit and a mechanical design which already exists, or as a means for both analyzing the circuit and designing the layout. Only a slight difference in point of view is required. The steps in the analysis are essentially the same in both cases.

The first step is to identify the portion of the circuit which is to be intrinsically safe. Only the circuit in a Division 1 location need be intrinsically safe. A fault occurring in a nonhazardous or Division 2 location is of no consequence from the standpoint of the energy that may be released at that location. The fault is of concern if it affects the amount of energy which can be released in a Division 1 location.

Second, review the circuit or the hardware for isolating constructions which will allow one to assume that certain interconnections will not occur. If the hardware exists, review the mechanical layout to determine whether the use of adequate spacings or partitions allows one to avoid considering an interconnection between some nonintrinsically safe circuits and the intrinsically safe circuit. Usually the transformer must be one of the special protective construction. Otherwise it is unlikely that the circuits can be both functional and safe if one assumes a line voltage connection to the transformer secondary. If there is a transformer of special construction, then the wiring or the printed wiring board layout must ensure that the primary leads of the transformer are separated from the secondary leads by sufficient space (as defined in the relevant standard), or by a partition, so that connection between the two can be ignored. If the hardware has not been designed, one must determine where intrinsically safe portions of the circuit must be separated by appropriate spacing or partition from the higher-energy portion of the circuits. The specific nature of the spacings must be determined from further detailed analysis.

Having reviewed for isolating construction, one should assume normal operation of the circuit. Compute the current and voltage in the circuit and compare it to the reference curves to determine whether the appropriate test factor has been observed. If not, adjust the circuit constants until the requirements are met. In this step, and in subsequent steps, it is essential that an orderly approach to record keeping be adopted. Even a relatively simple circuit may require the consideration of many steps. It is essential to record each combination of open, short, or ground of field wiring (these are not counted as faults) and component failures, so that when the analysis is complete, one can verify that the worst-case situation has been analyzed. If one is submitting the circuit for later review by certifying agencies, the availability of a well-organized fault table will ease and expedite the assessment process.

Third, if the hardware already exists, one considers the layout and spacing to determine what circuits must be assumed to be shorted together, what connections can be considered to be a fault, and what connections can be considered never to occur. After identifying these, recompute the current and voltage under fault conditions. This must be done for a single fault in combination with opening, grounding, and shorting of the external wires—and also for two faults. One cannot assume that two-fault situations will be the most hazardous, because of the difference in test factor required when only a single fault is considered. Adjust the circuit constants until the voltages and currents are suitable.

After the analysis for arc ignition, consider whether current flowing in the circuit under fault conditions may produce a high surface temperature on resistors, transistors, and so on. The temperature rise of small components is typically 50 to 100°C/watt. If the hardware exists, of course, one can measure the temperature of components which are suspect under fault conditions. In some approval laboratories, because a barrier or other intrinsically safe device powering the device in the Division 1 location is marked with its maximum voltage and maximum current output, it is assumed that the power supply has a rectangular VI characteristic, so that maximum voltage and maximum current can occur together unless a maximum power input is also specified. The temperature rise must be assessed, therefore, at the wattage of the maximum voltage and current occurring simultaneously. The limiting temperature to be considered must, of course, be determined from the standard being used as a criterion for design.

Simplifying Assumptions

If one is analyzing a circuit, the validity of the analysis is only as good as the data to which the analysis can be compared. In general, the available data are limited to simple $RL, RC,$ or resistive circuits. It may not be possible to analyze the effects of shunt elements included for some functional purpose. For example, an inductive force coil might be shunted with a variable resistor for a span adjustment. There is no reliable way to assess the additional safety provided by the shunt resistor. In general, one ignores such a parallel component in the analysis. Similarly, a capacitor might be wired in shunt with the force coil to provide damping of the electromechanical system. This, too, is ignored in the analysis.

It is assumed by almost all experts that an iron-core inductor is less efficient in releasing stored energy than an air-core inductor, because some of the energy, rather than being released to the arc, is dissipated in eddy current and hysteresis losses. If one is analyzing a circuit in which the inductor has a ferromagnetic core and the circuit will be safe with an air-core inductor of the same value, then it will certainly be safe. The converse is not true. Inductance is a measure of the slope of the B–H curve of the core material. Many small inductors, especially those with ferrite cores, have high inductance because the core material has a very high initial permeability. However, if the volume of material and the level at which it saturates are low, the amount of stored energy in the inductor may be considerably less than that calculated from a measured inductance value. Testing may verify safety.

Another simplifying assumption, which in many circuits reduces the amount of analysis considerably, is to determine the highest possible power supply voltage which may ever exist under fault conditions and with a high line voltage. This value is then used to determine the stored energy in all the capacitors or to determine the current and the resulting stored energy in all the inductors. If all the calculations are safe, one need not calculate the actual circuit currents and voltages.

Another simplification results from the need to prevent the discharge of a large capacitor in a hazardous location. Although curves giving the value of ignition voltage on a capacitor discharging through a resistor are available, one can assume that the capacitor is a battery charged to the fault voltage and select a series resistor based on the resistive circuit ignition characteristics. The resistor selected will be higher than that based on ignition tests of capacitors because the voltage on the capacitor decays when the current flows through the resistor. The connection between the resistor and the capacitor must be prevented from contacting any surrounding circuit. Therefore spacing, potting, or some other technique must be used to ensure that the capacitor cannot discharge except through the resistor.

Testing of Special Cases

It is not possible to determine the safety of all circuits by analysis alone. For example, some experts feel that one should verify the safety of diode-shunted inductors by conducting ignition tests. Inductors, especially small ones with ferromagnetic cores, may require testing to verify that they are safe despite high measured inductance.

Another common piece of apparatus which may have to be tested to determine safety is a regulated power supply. The reference curves of open-circuit voltage and short-circuit current for ignition in resistive circuits assume that the source impedance of the circuit, that is, the Thévenin equivalent impedance, is resistive. If the power supply is regulated, the voltage will remain essentially constant until a critical level of current is reached, beyond which the voltage will drop off with a further increase in current. The safety of maximum voltage and maximum current from such a supply cannot be determined from the reference curves for resistive circuits. In general, safety must be established by test, although some good special methods of analysis may be used.

Transmission lines are another special case. The common method of assessing safety, that is, multiplying capacitance per foot (meter) or inductance per foot (meter) by the number of feet (meters) of cable and comparing these values with limit values for the voltage and current in the cable, yields conservative results. It is well known that because the resistance, capacitance, and inductance are distributed, the actual cable will be safer than this lumped constant analysis suggests. However, there are not sufficiently good reference data available to allow one to analyze on a more scientific basis. If answers from lumped constant approaches are not satisfactory, then the cable must be tested.

However, it is recognized that the L/R ratio of the cable, if it is sufficiently low, may be such that no hazard exists, even though the total inductance exceeds that which would be safe if the inductance were lumped. If one assumes a cable of resistance R and inductance L_x per foot (meter) operating from a circuit of maximum voltage V_{max} and source resistance R, then the maximum energy will be stored in the cable when the cable resistance is equal to the source impedance. The current will be V_{max}/R. Therefore the maximum inductance permitted will be four times that permitted at the short-circuit current of the source. If the ratio L_x/R_x does not exceed $4L_{max}/R$, where L_{max} is the maximum permitted connected inductance, the cable will be safe regardless of length.

In summary, the design techniques used in all commercially available systems are similar. In this section we noted that the fundamental techniques are very few in number. Manufacturers may introduce variants of the basic techniques, some quite imaginative, but the fundamental design techniques are similar in all commercially available systems. Although some systems have used current limiting resistors and have put voltage limiting in the power supply, some have used active barrier isolators, and some have used Redding-type barriers, there is no difference in safety among the various approaches. Any of the techniques properly applied will yield a safe system.

CERTIFICATION OF INTRINSICALLY SAFE APPARATUS

In the early years almost all intrinsically safe apparatus was certified as part of a complete loop, now called *loop certification*. The apparatus in the Division 2 or nonhazardous location, now called *associated apparatus,* was specified either by model number or, somewhat later, in the case of intrinsic safety barriers, by the electrical characteristics V_{max} and I_{max}. The intrinsically safe apparatus was also specified by model number.

In Germany a different scheme developed. Associated apparatus is characterized by the maximum open-circuit voltage V_{max}, the maximum short-circuit current I_{max}, and the maximum permissible connected capacitance and inductance C_a and L_a.

Intrinsically safe apparatus is characterized by V_{max} and I_{max}, the maximum voltage and current that can be safely applied at the terminals, and C_i and L_i, the effective capacitance and inductance seen at the terminals. (A large capacitor discharging into the external circuit through a current limiting resistor may appear at the terminals to be equivalent to a much smaller capacitor, and an inductor shunted by diodes may appear to be a small inductor.)

IEC and CENELEC standards are written around this kind of specification, now being adopted in North America as *entity approval.*

System Design Using Commercially Available Hardware

The objective of the system design produced by an engineering consultant or a user is to ensure that the voltage and currents in a Division 1 location are limited to recognized safe values. Depending on the jurisdiction under which the plant is being built, the objective may be to make the system safe or to have the system safe and certified. This discussion assumes that the system designer has the option of at least a partially uncertified system. Even in the United States, where government regulations apply, for special combinations of apparatus there is some allowance for the use of uncertified apparatus (see 29CFR 1910.307). The most attractive option for most plants and most jurisdictions is to select a completely certified system. It is often possible to specify that all the hardware in a measuring and control system be procured from one vendor or to select hardware which, although manufactured by different vendors, has either been listed as an intrinsically safe combination or has been shown to be safe from entity parameters. The latter situation exists for many varieties of transmitters, valve positioners, and current-to-pressure transducers. The user often has a relatively wide choice of vendors because the vendors have already secured certification of various combinations of intrinsic safety interface elements, such as barriers, and their field-mounted products. For less common field-mounted devices, however, the hardware selection may be very restricted if certification is an essential criterion.

Even if the products of two manufacturers are not already certified together, it may be possible to specify these products if the procurement has a lead time of 6 to 12 months so that certification can be obtained by one of the manufacturers. It is also necessary that both products already be certified so that certification of the combination is a paper exercise rather than a complete assessment of either piece of apparatus.

Specifying a completely certified system is often impractical if the lead times are short and if a system requires products from many vendors. Certification as a primary criterion may also fail if highly specialized devices such as analyzers are needed.

The user or contractor may occasionally feel that the responsibility for obtaining listings can be assigned to the major supplier of hardware for the system. If only cross-listings, that is, certification of combinations of devices which have already been otherwise certified, are involved, this may be practical. If any of the products have not been certified as intrinsically safe, however, only the vendor of that product can obtain the basic intrinsic safety listing. If such products are involved, then the time required to obtain certification may be 12 to 18 months.

Self-Certification by the Manufacturer

Under these circumstances, the supplier of the major portion of hardware for a system may be willing to assume total responsibility for the safety of the system. A user contemplating such an arrangement should insist on written justification of safety from the supplier. The user also must allow the manufacturer veto power on equipment selection in order to ensure safety.

If all the pieces have been certified in some way, then the system may perhaps be procured from many sources. If uncertified apparatus is intended for a Division 1 location, it may be impossible for anyone but the vendor of the apparatus to self-certify unless it is very simple in construction. Certification of Division 1 apparatus is time-consuming and requires complete documentation, and it always involves an assumption that the design evaluated is indeed the design that will be subsequently produced. Third-party certification always imposes a responsibility on the manufacturer to ensure that this is indeed the case, and in some countries there is also a periodic audit to check that the products shipped are the same as those whose design was certified.

Self-Certification by the User or Contractor

This option is not often a viable one. It is necessary to ensure that the required expertise is available. However, it is advantageous because it retains responsibility for the safety and selection of apparatus in the same organization and yields maximum flexibility. Techniques for design by the user or contractor are discussed for three situations.

1. Field-Mounted Equipment Certified, but Control House Apparatus Not Certified. This situation commonly exists if the user wishes to use panel board equipment which has not been certified for intrinsic safety. The common solution is to purchase a barrier which has been certified with the field-mounted device and install it between the field-mounted device and the panel-mounted device. If the field-mounted device is a low-energy device, such as a thermocouple or an RTD sensor, most barrier vendors can supply a suitable barrier because specific designs need not be certified with a barrier.

In the case of a field-mounted transmitter, one should purchase barriers whose ratings meet the certification data of that field-mounted transmitter. In this case the user must ensure that a barrier that provides safety also provides the desired function. The user should make certain that the sum of the voltage drops through the load and barrier and the voltage drop across the transmitter do not exceed the available minimum voltage from the power supply. An obvious step, often overlooked, is to make certain that the voltage ratings of the zener diodes in the protective barrier are higher than the nominal voltage of the power supply. If not, the diodes may fire, blowing out the protective fuse the first time the system is energized. In some cases it may be necessary also to verify that the current leakage through the zener diodes does not impair the desired accuracy of the system.

After ensuring that the system will function, it is also necessary to make certain that no voltage in any of the apparatus connected behind the barrier exceeds the rating of the barrier, usually 250 volts root mean square (rms), but occasionally higher. The voltage of the resonant winding in ferroresonant power supplies, voltages in switching power supplies, and voltages in cathode ray tubes (CRTs) commonly exceed the rating of barriers. Most vendors of such apparatus who also supply barriers have had the pieces of apparatus containing high voltages especially certified by their certifying agency to be appropriate for use behind barriers. If such certification exists, it implies safety as well for any other barrier of the same input rating. However, this is the author's opinion and not necessarily the position of the approval bodies.

After determining that the barrier selected and the field-mounted apparatus both function and provide a safe system within the limits of the certification of the barrier, it is necessary to design the wiring system to ensure that the parameters of the cabling and any storage elements in the field-mounted devices do not exceed the levels permitted by the barrier certification. It is also essential to design the wiring system to maintain isolation of intrinsically safe wiring from all nonintrinsically safe wiring. ISA RP12.6, the manufacturer's instructions, and the applicable electrical code, for example, NEC Article 504, should be used as guides.

2. Field-Mounted and Control House-Mounted Equipment Both Certified, but Not as a Connected System. Apparatus which has been marked in accordance with the entity marking system is in principle easy to mix and match simply by reading the labels of the devices to be interconnected. One simply compares the marked voltage and current values on the field and associated apparatus. If the values of the field apparatus are equal to or greater than those of the associated apparatus, and if the inductance and the capacitance of the field apparatus are less than those marked on the associated apparatus, the combination is safe.

This system worked very well in Germany because of a constraint on the marking which for many years was not recognized as being one of the factors that made the system usable. Most German apparatus has been marked so that the maximum power which can be delivered by or to the apparatus is 3 watts. Therefore many field-mounted pieces of equipment were marked for a maximum of 29 volts and 150 mA, or a maximum of 30 volts and 100 mA.

Although in North America many transmitters and barriers have been entity approved and are being so marked, the great variation among the maximum current and voltage levels for which barriers have been designed in the United States makes it difficult to simply read the data and arrive at an immediate conclusion. In addition, simply comparing the parameters of associated apparatus and the intrinsically safe apparatus is applicable only when the system contains only one two-terminal associated apparatus (one barrier, for example). The rules for judging safety have not been standardized and published for the case when more than one associated apparatus is connected to the intrinsically safe apparatus. For this reason, control drawings are mandated in the United States which specify the connections between associated and intrinsically safe apparatus and the restrictions placed on the selection of apparatus.

If the apparatus is loop-certified with a barrier, as has often been the case in the United States, Canada, and the United Kingdom, one must first secure from the manufacturer of the control house apparatus the maximum current and voltage values and the permissible connected inductance and capacitance values.

The second step is to determine from the manufacturer of the field-mounted apparatus the values of current, voltage, inductance, and capacitance of the control house apparatus with which certifications are held.

One then compares the second set of data with the first. The analysis is similar to that described previously, but may be more complicated because there has been relatively little commonality of barrier design or field-mounted apparatus design in North America.

An additional complication arises because of the question whether a transmitter certified with a barrier whose maximum voltage is 24 volts is also safe relative to hot-wire ignition or hot-surface ignition if it is used with a barrier certified for the same group of gases, but with a maximum voltage of 30 volts and a lower current. If both barriers are passive so that their output *VI* characteristic is resistive, then one can be certain that safety against ignition by hot surfaces will be achieved if the maximum power output of the second barrier is less than that of the barrier with which the device was originally certified.

3. Field-Mounted Equipment Not Certified.   Unless the field-mounted equipment is very simple, it is usually impractical for anyone to certify it except the manufacturer or a certifying agency working with the manufacturer. It is necessary to perform a detailed analysis of the faults, construction, and stored energies, just as the manufacturer must do. This requires a complete set of manufacturing documentation, as well as a commitment on the part of the manufacturer not to deviate from the documentation supplied. In addition, the investment in time required to perform such an assessment is large, and a user or contractor is not likely to attempt this task.

Simple apparatus is defined as apparatus which can supply no more than 1.2 volts, 0.1 ampere, 20 μJ, or 25 mW. Simple apparatus includes RTDs, thermocouples, switches, and so on, which need not be specifically certified.

If simple apparatus is in the same enclosure as other electrical apparatus, such as a status switch in a contactor assembly, it is not necessarily exempted from certification. The author feels that if the switch and all associated wiring are at least 2 inches (50 mm) from any nonintrinsically safe wiring and live parts, the device can be used safely.

INSTALLATION OF INTRINSICALLY SAFE SYSTEMS

Because an intrinsically safe system is incapable of igniting a flammable atmosphere even under fault conditions, cables of special construction or conduit are unnecessary. It is, however, necessary to install intrinsically safe systems so that ignition-capable energies will not intrude from another circuit.

An intrinsically safe system will be safe *after* installation if the installation:

1. Conforms to the limiting parameters and installation requirements on which approval was based

2. Prevents intrusion of nonintrinsically safe energy on intrinsically safe circuits

3. Prevents power system faults or differences in ground potential from making the circuit ignition-capable

To minimize the probability of interconnection of circuits in a way not envisioned when the system was approved, NEC Article 504 mandates separate cables for systems supplied from power supplies of a different voltage or with different ground reference points. Circuits may be run in the same cable if all the conductors have a minimum insulation thickness of 0.25 mm or each circuit is within a grounded metallic shield.

To prevent the intrusion of other circuits, intrinsically safe circuits must be run in separate cables, wireways, or conduits from other circuits. Terminals of intrinsically safe circuits must be separated from other circuits by spacing (50 mm) or partitions. The grounding of intrinsically safe circuits must be separate from the grounding of power systems, except at one point.

Nonincendive Equipment and Wiring

It is not necessary to provide intrinsically safe equipment in Division 2 locations. The equipment need only be nonincendive, that is, incapable in its normal operating conditions of releasing sufficient energy to ignite a specific hazardous atmospheric mixture. Such equipment has been recognized without specific definition in the NEC. In Division 2 locations, equipment without make-or-break or sliding contacts and without hot surfaces may be housed in general-purpose enclosures.

Requirements for apparatus suitable for use in Division 2 and Zone 2 locations have been published by ISA and IEC. These documents provide more detail than is found in the NEC, partially because of a trend toward certification of nonincendive circuits which may be normally sparking, but release insufficient energy in normal operation to cause ignition. These documents also define tests for sealed devices which are needed by industry. The documents are similar, except that the ISA standard does not cover restricted breathing and enclosed-break devices.

ACKNOWLEDGMENT

The cooperation of the Instrument Society of America in permitting extractions and adaptations of text and illustrations from *Electrical Instruments in Hazardous Locations* is greatly appreciated.

BIBLIOGRAPHY

ANSI/NFPA 70, "National Electrical Code (NEC)," National Fire Protection Association, Quincy, Massachusetts, Art. 500–504.

API RP 500A, "Classification of Areas for Electrical Installations in Petroleum Refineries," American Petroleum Institute, New York.

BS 5501, "Electrical Apparatus for Potentially Explosive Atmospheres," British Standards Institution, London, 1978.
  Part 1: "General Requirements" (EN 50014)
  Part 2: "Oil Immersion 'o' " (EN 50015)
  Part 3: "Pressurized Apparatus 'p' " (EN 50016)
  Part 4: "Powder Filling 'q' " (EN 50011)
  Part 5: "Flameproof Enclosures 'd' " (EN 50018)
  Part 6: "Increased Safety 'e' " (EN 50019)
  Part 7: "Intrinsic Safety 'i' " (EN 50020)

ISA RP12.1, "Electrical Instruments in Hazardous Atmospheres," Instrument Society of America, Research Triangle Park, North Carolina.

ISA Monographs 110-113, "Electrical Safety Practices," Instrument Society of America, Research Triangle Park, North Carolina.

ISA RP12.6, "Installation of Intrinsically Safe Instrument Systems in Class 1 Hazardous Locations," Instrument Society of America, Research Triangle Park, North Carolina.

ISA S12.12, "Electrical Equipment for Use in Class 1, Division 2 Locations," Instrument Society of America, Research Triangle Park, North Carolina.

Magison, E. C., *Electrical Instruments in Hazardous Locations,* 3d ed., Instrument Society of America, Research Triangle Park, North Carolina, 1978.

Magison, E. C., *Intrinsic Safety, an Independent Learning Module*, Instrument Society of America, Research Triangle Park, North Carolina, 1984.

NFPA 496, "Purged and Pressurized Enclosures for Electrical Equipment," National Fire Protection Association, Quincy, Massachusetts.

NFPA 497M, "Classification of Gases and Vapors and Dusts for Electrical Equipment," National Fire Protection Association, Quincy, Massachusetts.

REC Pub. 79–15 "Electrical Apparatus with Type of Protection 'n,' " International Electrotechnical Commission, Geneva, Switzerland.

UL 913, "Intrinsically Safe Apparatus and Associated Apparatus for Use in Class I, II, and III, Division 1 Hazardous (Classified) Locations," Underwriters Laboratories, Northbrook, Illinois.

SECTION 3
CONTROLLERS*

L. Arnold
Johnson Yokogawa Corporation, Newnan, Georgia (Stand-Alone Controllers)

Douglas M. Considine
Consultant, Columbus, Georgia (Computers and Controls)

Terrence G. Cox
Vice President, CAD/CAM Integration, Inc., Woburn, Massachusetts (Distributed Numerical Control and Networking)

G. L. Dyke
Bailey Controls Company, Wickliffe, Ohio (Stand-Alone Controllers)

Thomas J. Harrison
IBM Corporation, Boca Raton, Florida (Computers and Controls)

John P. King
The Foxboro Company (A Siebe Company), Rahway, New Jersey (Distributed Control Systems)

J. Kortwright
Leeds & Northrup (A Unit of General Signal), North Wales, Pennsylvania (Stand-Alone Controllers)

Ralph Mackiewicz
Vice President, SISCO, Sterling Heights, Michigan (Programmable Controllers)

C. L. Mamzic
Manager, Systems and Application Engineering, Moore Products Company, Spring House, Pennsylvania (Stand-Alone Controllers)

Donald McArthur
Hughes Aircraft Company, Culver City, California (Computers and Controls—prior edition)

---

* Persons who authored complete articles or subsections of articles, or who otherwise cooperated in an outstanding manner in furnishing information and helpful counsel to the editorial staff.

T. A. Morrow

Assistant Division Manager, Omron Electronics, Inc., Schaumburg, Illinois (Timers and Counters)

J. W. Pogmore

President, Cramer Company, Old Saybrook, Connecticut (Timers and Counters)

Raymond G. Reip

Consulting Mechanical Engineer, Sawyer, Michigan (Hydraulic Controllers)

B. R. Rusch

Gould, Inc., Andover, Massachusetts (Programmable Controllers—prior edition)

H. L. Skolnik

Marketing Manager, Intelligent Instrumentation (A Burr-Brown Company), Tucson, Arizona (Computers and Controls)

D. N. Snyder

LFE Instruments, Clinton, Massachusetts (Stand-Alone Controllers and Timers and Counters)

J. Stevenson

Engineering Manager, West Instruments, East Greenwich, Rhode Island (Stand-Alone Controllers)

Stan Stoddard

Marketing Manager, Waugh Controls Corporation, Chatsworth, California (Automatic Blending Systems)

W. C. Thompson

Fisher Controls International, Inc., Austin, Texas (Batch Process Control)

J. D. Warnock

(Deceased) formerly Manager, Systems Engineering, Moore Products Company, Spring House, Pennsylvania (Pneumatic Controllers)

T. J. Williams

Purdue University, West Lafayette, Indiana (Computers and Controls—prior edition)

Douglas Wilson

Hughes Aircraft Company, Canoga Park, California (Computers and Controls—prior edition)

3

DISTRIBUTED CONTROL SYSTEMS

by John P. King*

Distributed control systems (DCSs) have been evolving rapidly since the mid-1980s from being essentially panelboard replacements at their inception to become comprehensive plant information, computing, and control networks fully integrated into the mainstream of plant operations. This progress has been fueled, in part, by the technological revolution in microprocessor and software technology and, in part, by economic necessity.

Prior to the early 1970s, when fuel feedstock was plentiful, energy was cheap, markets were stable, and competition was predictable, the role of instrumentation essentially was to ensure stable process operation within the equipment design constraints. Feedforward and

* The Foxboro Company (A Siebe Company), Rahway, New Jersey.

other optimizing techniques were only being applied in a general way to a few well-studied unit operations. These included compressor surge control, multiple-effect evaporators, and certain classes of light ends distillation columns.

Calibrating, commissioning, and maintaining these advanced control strategies using analog panelboard instrumentation was a formidable task. Modifying these strategies to reflect changing process or economic conditions required considerable expertise and excellent documentation on how the original system was installed and subsequently modified. Instruments usually had to be rewired, and the addition of new components was often necessary.

A few progressive companies were implementing advanced control across a wide spectrum of unit operations, but these were the exception and, in most cases, were utilizing computers hard-wired to panelboard instruments for much of the computing requirements.

The Arab oil embargo of the early 1970s heightened the growing realization of the need to utilize the available resources more efficiently and intelligently in order to cope with scarcity, escalating costs, and market instability. Efficiency, flexibility, and product quality joined production throughput as primary driving forces. Working "smarter" by taking advantage of technology became the strategy by which the United States and the industrial world would compete to secure their standard of living. One very cost-effective solution was to harness the power and speed of the microprocessor, including its calculation and logic capabilities, its information storage and retrieval, and its complex instruction capabilities to improve competitiveness.

The DCS, long in the development laboratories of many of the progressive control system manufacturing companies, became commercially viable in fulfilling its potential to improve the accuracy, operability, computational and logical ability, calibration stability, and ease of control strategy modification. DCS boosted the confidence of engineers to implement advanced control strategies and thereby enable process plants to:

- Reduce operating costs, particularly fuel savings, via optimizing control strategies
- Reduce product giveaway through more uniform operation and tighter control enforcement
- Reduce capital investment through improved scheduling of capital equipment and increased production from existing equipment
- Improve responsiveness to changes in processes, production rates, product mix, product specifications, additions of new products, and external demands, such as government regulations
- Incorporate contingency strategies to minimize production upsets resulting from plant equipment failures or unanticipated process conditions and to improve plant safety
- Improve more timely information to plant operations and maintenance managers to enable them to keep a plant running longer and more efficiently and to plant managers to enable them to run more profitably
- Improve integration of plant operations through coordinated control strategies

EVOLUTION OF DISTRIBUTED CONTROL SYSTEMS

Microprocessor-based DCSs made their debut in the mid-1970s. Initially they were conceived as functional replacements for electronic panelboard instrumentation and were packaged accordingly. The initial systems utilized discrete panelboard displays similar to their electronic instrument counterparts.

These systems evolved quickly, adding video-based workstations and shared controllers capable of expressing complex unit-operations-oriented regulatory and sequence control strategies containing scores of functional elements, such as PID (proportional-integral-derivative), lead/lag totalizers, dead-time elements, elapsed timers, logic circuits, and general-purpose calculators (Figs. 1 and 2).

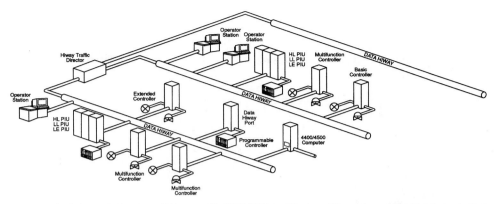

FIGURE 1 Schematic drawing of Honeywell's TDC 2000 architecture. The system utilized basic controllers which were the digital equivalents of conventional panelboard instruments and offered the choice of video workstations or discrete panelboard display modules.

Emergence of the Open System

By the early to mid-1980s the personal computer industry matured into a multibillion dollar-per-year market with the IBM PC disk operating system (DOS) as the standard. This gave birth to a new industry—"shrink-wrap" software. Feature-laden high-quality inexpensive software packages opened the minds of the professional community. Of course this was only possible because the manufacturer was able to amortize the investment in the intellectual property over tens of thousands of sales.

The opportunity for system integrators and value-added resellers was clear. One could develop a relatively inexpensive scan, control, alarm, and data acquisition (SCADA) package for a personal computer platform and integrate it with these general-purpose shrink-wrap software packages, such as spreadsheet, desktop publishing, or database management, and one could have a very cost-effective alternative to DCS.

Because of performance and the general suitability limitations of these PC offerings, this approach had appeal mostly in cost-sensitive noncritical applications and where there existed a low safety or hazard risk. This concept, however, created an expectation and vision of the future, that is, open architectures.

DCS vendors felt a compulsion to enrich their arsenal of tools to address real-time process control applications by incorporating the low-cost shrink-wrap packages into their systems. Such packages include:

- Relational database management
- Spreadsheet packages
- Statistical process control capabilities
- Expert systems
- Computer-based process simulation
- Computer-aided design and drafting
- Desktop publishing
- Object-oriented display management
- Windows-oriented workstations
- Information exchange with other plant systems

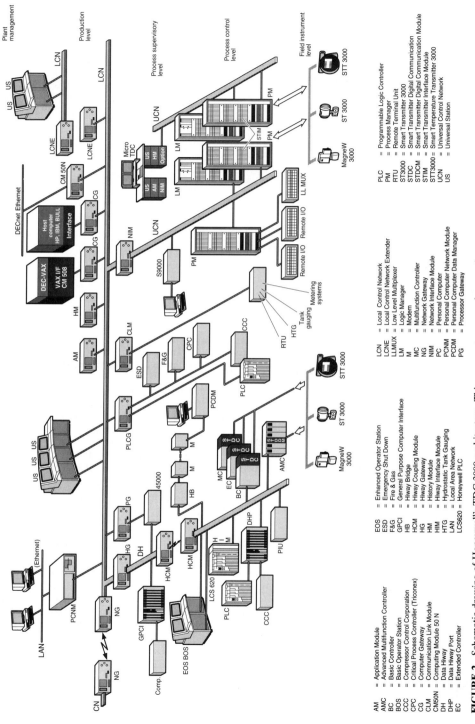

FIGURE 2 Schematic drawing of Honeywell's TDC 3000 architecture. This represents a second-generation DCS, offering a network file server for collection of historical information, multifunction controller for executing advanced control and batch using procedural statements, as well as application modules for optimization and other advanced functions.

AM   = Application Module
AMC   = Advanced Multifunction Controller
BC   = Basic Controller
BOS   = Basic Operator Station
CCC   = Compressor Control Corporation
CPC   = Critical Process Controller (Triconex)
CG   = Computer Gateway
CLM   = Communication Link Module
CM50N = Computing Module 50 N
DH   = Data Hiway
DHP   = Data Hiway Port
EC   = Extended Controller

EOS   = Enhanced Operator Station
ESD   = Emergency Shut Down
F&G   = Fire & Gas
GPCI = General Purpose Computer Interface
HB   = Hiway Bridge
HCM   = Hiway Coupling Module
HG   = Hiway Gateway
HM   = History Module
HIM   = Hiway Interface Module
HTG   = Hydrostatic Tank Gauging
LAN   = Local Area Network
LCS620 = Honeywell PLC

LCN   = Local Control Network
LCNE   = Local Control Network Extender
LLMUX = Low Level Multiplexer
LM   = Logic Manager
M   = Modem
MC   = Multifunction Controller
NG   = Network Gateway
NIM   = Network Interface Module
PC   = Personal Computer
PCNM   = Personal Computer Network Module
PCDM   = Personal Computer Data Manager
PG   = Processor Gateway

PLC   = Programmable Logic Controller
PM   = Process Manager
RTU   = Remote Terminal Unit
ST3000 = Smart Transmitter 3000
STDC   = Smart Transmitter Digital Communication
STDCM   = Smart Transmitter Digital Communication Module
STIM   = Smart Transmitter Interface Module
STT3000 = Smart Temperature Transmitter 3000
UCN   = Universal Control Network
US   = Universal Station

3.8

NETWORKED COMPUTING SYSTEMS

During the late 1980s and early 1990s the computer industry continued its transformation. Networking of systems into a cohesive whole promised to (again) revolutionize an industry which had barely absorbed the impact of the PC revolution. Software and communications standards began to take hold, making interoperability among disparate computing platforms and application software a near-term reality. The business enterprise, including the factory floor, could be molded into a cohesive whole by making the various departmental systems work cooperatively at an acceptable integration cost. These standards include:

- Open operating system standards, such as UNIX or POSIX
- Open-system interconnect (OSI) communications model
- Client-server cooperative computing model
- X-window protocols for workstation communications
- Distributed relational database management systems
- SQL access to distributed relational databases
- Object-oriented programming and platform-independent languages
- Computer-aided software engineering

ARCHITECTURE

DCSs are becoming distributed computing platforms with sufficient performance to support large-scale real-time process applications and scalable to address small unit applications.

Open-system standards are enabling DCSs to receive information from a diverse set of similarly compatible computing platforms, including business, laboratory information, maintenance, and other plant systems as well as to provide information to these systems in support of applications, such as:

- Automated warehousing and packaging line systems so that a complete order can be coordinated from the receipt of raw materials to the shipment of the final product.
- Laboratory information management systems (LIMS), which perform in-process analysis as well as quality assurance inspections. Coordination of the information between these systems can result in more timely decision making and reduced waste, rework, and loss of efficiency.
- Automated production scheduling for a plant accessing the business system and tying into MRP II systems and finite-capacity scheduling packages.
- Automated troubleshooting by using process alarms in the DCS to identify and isolate a problem and develop a corrective action strategy; by merging information from the technical services area, such as drawings and equipment performance specifications; from the maintenance shop, parts inventory, and despatching schedule; and the DCS information regarding current process values, historical trends, and the status of various devices.

DCSs traditionally are organized into five major subsystems, namely, (1) operations workstations, (2) controller subsystems, (3) data collection subsystems, (4) process computing subsystems, and (5) communications networks. A typical third-generation architecture is depicted in Fig. 3.

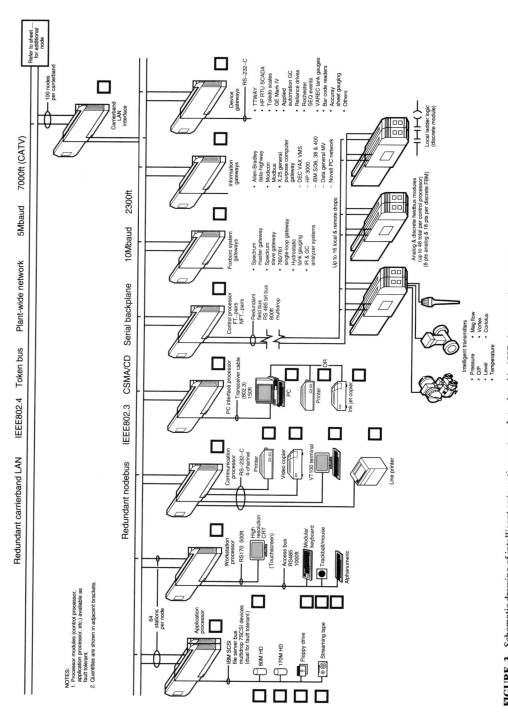

FIGURE 3 Schematic drawing of intelligent automation system. Introduced in 1987, the system incorporates graphic user interface technology; symbolically referenced data objects across the network; remote file access across the network; distributed computing; conformance to open-system standards framework communication. It uses the UNIX operating system for workstation and application processor functionality. This is an example of a third-generation system. (*Foxboro.*)

OPERATIONS WORKSTATIONS

The primary criterion for the early workstations was operator acceptance. Mimicking a conventional panel in a manner which could be intuitively understood by the process operator was the primary approach. This was done by emulating the panel front arrangements by displaying video representations of conventional panelboard instrumentation (controllers, indicators, totalizers, lamps, switches, trend recorders, alarm annunciators, and so on). These video faceplates were organized into different views, which were intended to enable operators to interact with the video displays in a manner similar to panelboard operations. The advantages of this approach over their conventional equivalents were both operational (greater flexibility in the arrangement of displays) and economic (reduced control room size, fewer operators, and lower design and installation costs) (Fig. 4).

Displays were organized to provide the same information as hundreds of feet (tens of meters) of panel on the limited "real estate" provided. This was accomplished by organizing the displays into hierarchies where the entire screen at one level would be represented by only a small portion of the screen at the next higher level. A loop display containing all of the information about a single faceplate might also be displayed along with several other loops at the group display level. The group would be only a portion of the area display, and the area would be only a portion of the plant. If the ratio were 8:1, as it was for many of these systems, at the plant level over 1000 loops would be represented. Obviously the information regarding any single loop at the plant level was quite limited. It typically presented only a normalized deviation from target and alarm status.

Process graphics were added later to the display repertoire. These offered pictorial representations of a process unit or a section of a process area with process information updated onto it. This was an alternative to the area-level display and eventually replaced it (Fig. 5). Keyboards were used to manipulate these displays. Their design was quite innovative, with many utilizing display-relative function keys. One such approach adopted by Foxboro assigned keys for traveling up, down, or across the hierarchy. This provided an intuitive manner for an operator to manage the electronic panel, such as to move sideways at a particular distance from the electronic panelboard (as depicted by the size and resolution of information on the screen) or to move toward or away from the electronic panel. This was accomplished by utilizing relatively few keys on a custom-designed keyboard. In the case of the Videospec system, eight keys would be used to move toward one of eight selected areas on the screen, one key would be required to move away from the electronic panel, and two to traverse the board (one for left and one for right). Only 11 keys were needed to manage travel around the electronic panel containing over 1000 faceplates. Also to travel from the top-level display to a particular loop required only three keystrokes (Fig. 6).

Alarm management was an area of concern to operations personnel. A conventional panelboard enabled process operators to view all instruments simultaneously. With a video-based workstation the limited real estate available on the screen made this impossible from all but the highest-level display. To compensate for this, auxiliary alarm lamps were offered. They represented specific plant areas and would alert the operator of upset conditions regardless of the display on the screen. An audible alarm typically would also sound if the alarm was specified as critical. These physical annunciators would later be replaced by dedicated areas on the display screen itself, which flashed and changed color to indicate an alarm in a section of the plant not currently being displayed.

Once alerted to an alarm condition, the operator must find the alarm and take action. Several innovative techniques were offered to address this. One was to offer a push button for each annunciator. A second was to have the system search through the display hierarchy and present the display associated with an alarm condition whenever an alarm button was pressed. Successive depressions of the alarm key would present other displays containing alarms until they were all displayed and the first would again appear.

Often alarms cascade. A low flow condition may cause a low-level condition which may result in a high temperature, and so on. Managing these situations with a limited view of the

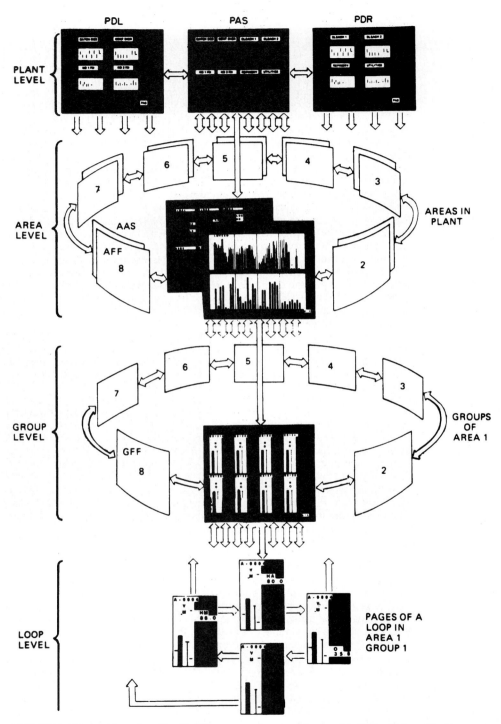

FIGURE 4 Display hierarchy utilized in Videospec workstation introduced in 1976, which enables the user to view different panelboard representations of the plant by relatively few keystrokes. (*Foxboro.*)

FIGURE 5 Multistation, the operator display module introduced in the spectrum-based DCS in 1980. (*Foxboro.*)

operations can result in critical information being ignored or improper conclusions being drawn and wrong actions taken. To deal with this, many systems enable the user to prioritize alarms so that the most important are viewed first. Individual variables may change their priority based on the nature of the alarm condition (such as high deviation or high absolute) or based on the time the variable has been in alarm.

Providing operations personnel information about the current trend of process conditions is an area where DCSs clearly excelled. The magnetic media available to these computerized systems were considerably more cost-efficient and convenient than their instrumentation counterparts (multipoint trend recorders). They could literally store information on thousands of variables. Recent history could be played back to enable operations or maintenance personnel to determine the current stability of process loops, units, or even plant areas. Longer-term information might be used for analysis to improve plant operations or for business purposes such as inventory management.

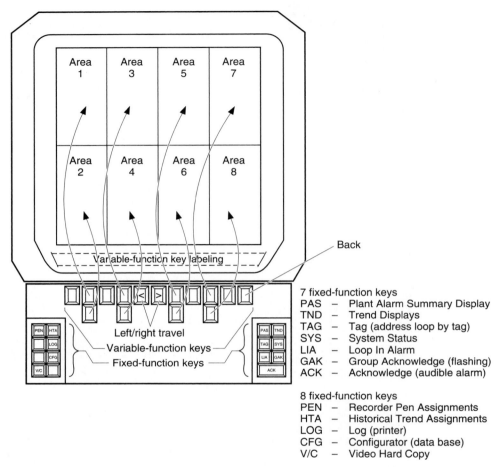

FIGURE 6   Variable-function keyboard. Videospec workstation shown schematically. (*Foxboro.*)

As operations workstations evolved, the concept of a universal workstation began to take root. This is an operations console which provides all of the information needed by operators for process regulation as well as access to selected information in other operations so as to enable the operator to make better tactical decisions such as when to start up a unit or adjust process throughput to compensate for a downstream bottleneck. As the concept of the universal workstation matured, it became apparent that the potential uses and applications could not be anticipated completely by the original designers. A methodology enabling access to applications not yet specified had to be provided in a general set of utilities. Furthermore the management of the console had to be consistent and intuitive with regard to accessing information and interacting with displays regardless of the origin of the application. The industry began adopting the technologies emerging in the network computing arena, particularly X-windows and distributed computing.

Windows-oriented workstations essentially enable a workstation screen to be divided into scalable sections, with each section acting as though it were an independent display subsystem. Supervisory and operations personnel can concurrently view information from dif-

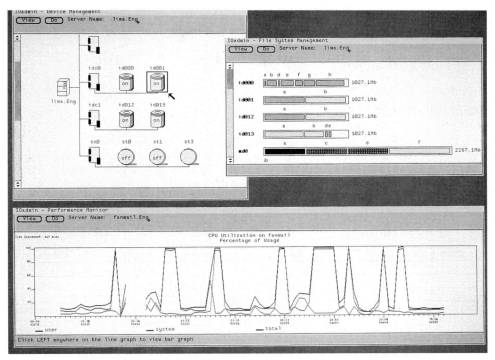

FIGURE 7   Use of multiple-window display in highly different applications. (*SUN SPARC Station.*)

ferent programs by simply directing the display output for that program to the partitioned window (Fig. 7).

No advanced knowledge of the intended applications and no advanced planning or application-specific programming are required to support the set of displays other than to conform to the standards imposed by the X-windows protocols. An operator could request a process graphic for a process unit to be directed to one window; a statistical analysis of the past 25 batches made on that unit directed to a second; an inventory report of available raw materials directed to a third; a real-time trend of specific variables directed to a fourth; a product formulation for the current batch directed to a fifth; and a production schedule summary for this week directed to a sixth—all without the need for the designer of the application display hierarchy to have anticipated that composite display. The information would appear in separate independent windows on the same screen.

More advanced uses (combined with seamless integration to other plant computers) would combine information from different systems to offer a mosaic on a single screen regarding all aspects of the process operation. This might include:

- A quality assay of the ingredients charged to a designated reactor derived from the batch lot id and obtained from quality assurance's laboratory information management system (LIMS).
- The accumulated time over the past 30 days a worker had been in an area where there was potential to exposure to specific chemicals. This information might be accessed from the individual's personnel records.
- The production quota for the product currently in production and current actual production against planned production derived from the MRP II system.

Windowing effectively decouples the display from the hardware platform, that is, the display and the information are sent to whichever workstation makes the request. If designed properly, multiple workstations at various locations can concurrently view the same display. In effect all workstations could be made to have independent and total access to all information and applications in the extended network.

CONTROLLER SUBSYSTEMS

The first multiloop controller was introduced by Honeywell in 1976. It offered eight loops per controller, with some of the signal conditioning and computing external to the microprocessor. The algorithm set offered a variety of functions, such as PID, ratio stations, totalizers, lead/lag, summers, and selector circuits. The basic controller offered a great deal of flexibility regarding the configuration of each of the control elements or blocks, but was limited in its ability to connect the various blocks together into complex control schemes without external wiring.

By the late 1970s these controller subsystems had expanded in size and function to offer a relatively large number of loop elements or blocks which could be "soft-wired" together to implement complex control strategies. The advanced control capabilities of these shared controllers was a primary reason for selecting DCSs over conventional instrumentation. The tool set to accomplish this is essentially an emulation of electronic instruments as function blocks, which can be linked together to form control strategies or loops (Fig. 8). By 1980 interlocking and sequencing functions were added, enabling the systems to include pumps, on-off valves,

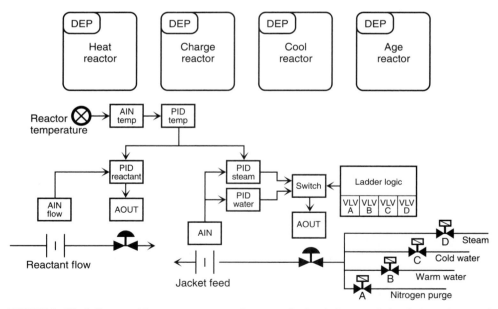

FIGURE 8   Block diagram of the temperature control strategy of a chemical reactor during the heat, charge and cool, and age phases. Heating is accomplished by flowing steam through the reactor jacket. Temperature is controlled by using a reactor temperature over jacket flow-rate cascade. As reactant is charged, an exothermic reaction begins and cold water replaces the steam in the jacket to remove the excess heat. Temperature is controlled by manipulating the reactant flow in a reactor temperature over reactant flow cascade with the jacket flow valve set at 90% open. As the reaction subsides, warm water replaces the cold water in the jacket.

and other motor-operated devices into the advanced control strategies. This approach has proved quite effective at implementing regulatory and feedforward control strategies. Typical function blocks include:

AIN	Analog input
TOT	Totalizer
CHAR	Characterizer
ADAPT	Adaptive tuning
DT	Dead-time compensator
L/L	Lead/lag dynamic compensator
RTIO	Ratio control
PID	Proportional plus integral plus derivative controller
SEQ	Sequencer
ALM	Alarm block
CALC	General-purpose calculator
AOUT	Analog output
CIN	Contact input
MTR	Motor controller
VLV	Throttling control valve
SV	Solenoid valve
LL	Ladder logic (boolean) block

Although relatively simple to apply, the design of each of these function blocks is quite complex. A PID algorithm is depicted here to illustrate the point. The classical formula for the PID algorithm is

$$m = \frac{100}{P} \left(e + \frac{1}{I} \int e \, dt - D \, \frac{dc}{dt} \right)$$

or, expressed as a difference equation,

$$m_n = \frac{100}{P} \left[e_n + \frac{t}{I} \sum_{i=0}^{n} e_i - \frac{D}{t(e_n - e_{n-1})} \right]$$

where m = manipulated variable or output
P = proportional band
I = integral term
e = error
D = derivative term
c = controlled variable or measurement
n = current sampled-data interval
t = time interval

This would require m to continue to increase on a sustained positive error. The "windup" would have to be undone by a sustained negative error before the algorithm would again function within the operating limits of the control system. Many systems use clamping to limit the adverse effects of windup. A better solution is to substitute external reset feedback through a first-order filter in place of the integral term. This eliminates adverse integral windup without compromising the intent of the algorithm. The form of the equation would then be

$$m_n = \frac{100}{\text{PB}} \left[e_n - K_D(c_n - y_n) \right] + b_n$$

where

$$y_n = y_{n-1} + \frac{t(c_n - y_{n-1})}{t + D/K_D}$$

$$b_n = b_{n-1} + \frac{t(f_n - b_{n-1})}{t + I}$$

and K_D is the derivative gain limit and f the external reset feedback. A more detailed explanation of this difference equation is given in Shinsky (1978).

The function block must also recognize the various states in which it can be placed, such as auto, manual, or off, and provide state transition logic to move from one state to another. When applying these function blocks it is important to understand that fundamental differences exist between their implementation using electronic circuitry versus digital technology. These are explained in paragraphs that follow.

Sampled-Data Control

The digital algorithms are executed at fixed intervals, not continuously. The sampled-data interval introduces dead time equal to one-half the sampled-data interval into most control loops. This can affect the loop control if the sampled-data interval is not a negligible portion of the loop time constant. Some control strategies simply cannot be copied from electronic instrumentation to digital control.

A particularly pernicious example of this is the multiple-output control system (MOCS). This is a cascade control system where the inner loop is intended to isolate the outer loop from abrupt and significant changes in controller gain resulting from switching one of several of the outputs from auto to manual or from manual to auto. The concept is to send the output of the outer loop to the set point of a high-gain proportional controller. The measurement of the controller is derived from a weighted summer, which receives its signals from those being sent to the final actuators. The weighting constants represent the fractional contribution of each actuator to the total controller output. In the electronic circuit implementation any discrepancy between the set point and measurement would force the high-gain controller output, and therefore the high-gain controller measurement, to be driven to saturation. However, as the measurement crossed the controller set point, the controller error would be zero and the output would hold at that value.

The digital implementation of this control strategy computes discrete values for the outputs of each of the blocks at prescribed sample intervals. The output, and thus the high-gain controller measurement, will not cross the set point. It will simply arrive at its new value which, because of the high gain, will be the saturation point. In the next sampled-data interval the measurement will be on the other side of the set point, forcing the output to drive in the opposite direction. The net result will be output limit cycling at twice the period of the inner loop (Figs. 9 and 10).

Synchronization

The digital algorithms are executed in the sequential order specified by the user. Strategies which depend on information from downstream blocks will be furnished old (last execution cycle) information. This introduces a delay in the current value, which is equal to the block period, and may in some instances introduce a logical inconsistency. This is particularly true in loops containing discrete elements or logic.

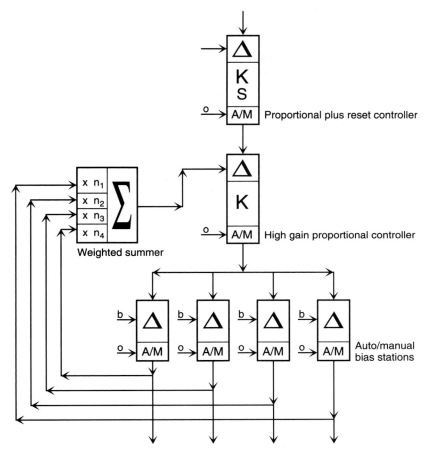

FIGURE 9 Multiple-output control system (MOCS) where a continuous feedback system, the high-gain proportional controller, drives the output of the inner loop until the set point equals the measurement and thereby isolates the outer loop from any gain changes within the inner loop as a result of changing the state of the inner loop. (*SAMA.*)

A problem may also occur when individual blocks in the control loop execute at different periods. One example is a loop which requires sampling measurement devices having different time constants, such as temperature and flow. This is commonly referred to as phasing problems.

Initialization

Another issue is how the blocks initialize either individually, when they are transferred from a manual state to an automatic state, or as a group, when the system is first commissioned or powered up after an extended power failure or similar circumstance. Many of the blocks may require data stored from recent history to function properly, while others need to return to a fail-safe state as the state of the process is indeterminate. When a chain of blocks requires initialization, they are typically processed in reverse order of their normal execution.

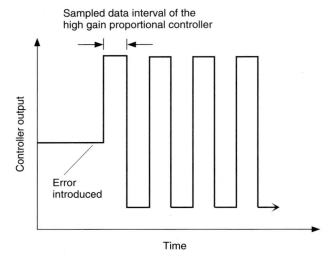

FIGURE 10   Graphic trend of the limit-cycle effect of the control loop resulting from an error introduced into the outer loop. The high-gain proportional controller, because it does not receive immediate feedback from the downstream block, drives the output to 100%. On the next cycle the measurement at 100% introduces an error in the other direction and drives the output to 0%. This limit cycling continues indefinitely. Utilizing an integral-only controller in place of the high-gain proportional controller with a reset time equal to some multiple (say 6) of the inner-loop sampled-data interval will approximate the action of the multiple-output control system (MOCS), but with not nearly as good a response time.

Exception Handling

Another factor is how the blocks respond to abnormal conditions. The objective is to manage the exception with logic closest to the condition. If a solenoid valve, with limit switches on its valve stem to confirm the proper operation of the valve, were to fail to open when directed, the problem would be detected by the function block (SV), which would send an alarm message and provide information (typically change the state of a status flag) to the control subsystem to take corrective action or inhibit further operation. If the regulatory control system can verify the nature of the problem (no flow in the line) and is capable of proceeding with the operation using an alternate route, the problem is contained and the process operation continues; if not, the exception is propagated to the next level and the exception evaluation continues.

Procedural Languages

Introduced into the controller subsystems in the early 1980s, procedural languages have evolved to a point where they have replaced much of the supervisory control accomplished by the process computer in the 1970s. They are now capable of performing applications ranging from supervisory and optimizing control strategies to message processing in support of operations and data collection activities. They have also become the basis for expanding the DCS controller into semicontinuous and batch domains.

Specialized real-time-oriented languages have evolved to express process control activities, such as OPEN_VALVE SV-104-A, as single statements or steps. This statement would operate on an instance of the VALVE block named SV-104-A and try to open it. If for some reason the valve would fail to open and the exception could not be managed within the regulatory domain, an exception message (status flag) would be detected by the procedural program, either by directly receiving the error return portion of the OPEN statement or by utilizing an independent external monitor checking for abnormal conditions and taking action. In either case normal processing of the program would be suspended and exception logic would be invoked to handle the situation. Circumstances governing transitions to and from exception logic would be specified by the control engineer during design. Examples of procedural statements or steps are

OPEN_VALVE tag

CLOSE_VALVE tag

START_MOTOR tag

STOP_MOTOR tag

RAMP_SETPOINT tag TO value AT rate

CHANGE_SETPOINT tag TO value

These statements can be combined with more generic procedural language constructs to form very expressive problem statements, such as

WAIT_UNTIL (LI-104 > 60)

OPEN_VALVE SV-104-A

WAIT_UNTIL (SV-104.FBK = ON) WHEN ERROR GO TO LABEL

START_MOTOR PMP-104-A;

or

IF (LI-104 > 70) CLOSE_VALVE SV-104-B;

Groups of expressions can be organized into subprocedures or phases which typically represent a single operation that can be performed on a process unit. An example might be HEAT_REACTOR, which might involve arranging the valves around the reactor's jacket to select steam at the inlet and the hot water condensate return line at the outlet. A cascade control strategy may also be incorporated where the reactor temperature controller output governs the set point of the jacket steam flow controller.

These phrases are capable of accepting arguments and directives, such as

HEAT_REACTOR TO 85 DEGC AT 5 DEGC/MIN

This allows them to be utilized as a higher-level language for plant chemists and process engineers to specify product recipes.

This modular building block approach to constructing batch unit processing has been incorporated into the emerging Instrument Society of America Standard SP-88 for batch control.

DATA COLLECTION SUBSYSTEMS

These subsystems provide a means for collecting information from auxiliary process variables for indicating, recording, totalizing, and alarming. They functionally replaced panelboard indicators, recorders, and so on and were initially packaged as large racks with terminations to handle clusters of inputs ranging from about 64 to over 1000. They accept a variety of signals, including millivolt, volt, 4–20 mA, thermocouple, RTD, pulse count, pulse rate, and an array of discrete signals such as dry contacts and line-voltage sense.

The information was communicated over the DCS network and reported to various substations, including workstations. This offered several advantages over panel-mounted instruments, including the ability to display the information in several formats such as trends and faceplates. Once in the system, the information can be dispatched to as many locations as needed. Geographically separated workstations could display the same information simultaneously, thereby facilitating communication. The information could also be utilized by process management subsystems to compute yields, energy consumption, production rates, efficiencies, and so on.

Device gateways were offered as extensions to the data collection systems to bring in information other than from traditional transmitters. Over time gateway functionality included programmable controllers, weigh scales, chromatographs, and other specialized devices.

During the 1980s the large data collection subsystems gradually were displaced by distributed input-output (I/O) subsystems. The I/O subsystems differed from their predecessors in both packaging and functionality. The trend in packaging is toward creating modules handling between one and 16 signals and whose size is no bigger than is necessary to terminate the signals to the modules. A variety of environmentally hardened packages are available to provide the option of locating them close to the process. Some are designed to be protected from the weather, harsh chemicals, and dust. Some conform to NFPA standards, making them suitable for use in hazardous environments. These modules communicate to the DCS via a digital communications network. A standard specifying a communications protocol for these field devices is expected to be completed by the International Standards Organization SP-50 Subcommittee soon.

These modules are more intelligent than their predecessors, with failure detection and autodiagnostic circuitry built in. Some modules include programmable features such as conversion of the process signal to engineering units, process-oriented alarming, and the inclusion of fail-safe logic. Some control functionality, particularly routine calculations which must be executed very frequently or require a very quick response to a process condition, is migrating from the control subsystem onto the I/O subsystem. Examples include the incorporation of PID and small-scale ladder logic within the I/O module.

A trend developing in the evolution of these modules is to give them a personality, particularly as they are used to interface with increasingly more complex and specialized equipments such as chromatographs, motor-control centers, large pumps, bar-code readers, and possibly specialized display stations. This approach would enable the control subsystem to interact with field devices as entities or "objects" sending high-level commands such as START to a pump. The details for starting the pump would be embedded in the I/O subsystem. This would reduce the engineering costs for implementing a project dramatically and would eventually lead to open I/O architectures. Under these conditions many of the I/O modules would be incorporated in the device manufacturer's offering and all that would be required would be a communications tie-in.

PROCESS COMPUTING SUBSYSTEMS

Initially these were general-purpose process computers which could access the controller or the data collection subsystem's database. Like the process computers attached to conventional instruments, they were used primarily for advanced alarming and reporting as well as for supervisory control optimization and batch control applications.

The primary difference between the DCS version and the system interfaced to a conventional panelboard was the interface method. Many panelboard instruments of that era required from 12 to 20 wires per controller to provide all of the relevant information, such as set point, measurement, output, or control mode. Using a DCS, all of the information within the controller or the data collection subsystem was available in digitized form via a single high-speed communications cable connection. As the DCS approach connected the process computer to all other DCS modules via a high-speed communications network, the incremental costs to send large amounts of information to the process computer were negligible since no point-to-point wiring was required. This made a significant difference in reducing the installed cost, thereby improving the general acceptance of process computers for mainstream applications. Process management could now be developed to its full potential.

Gateways were also offered, which enabled other computers to be connected to the DCS. The gateway was the translator of information between the computer and the DCS system. As the control subsystems became more powerful and increased in functionality, the role of the computing subsystem evolved to perform higher levels of process supervision and information processing and analysis. Its supervisory role typically involves a considerable amount of process or interunit coordination, as in the case of recipe management in multiproduct batch processes, or where the analysis or information processing needed is substantial, as in the case of the deployment of expert systems.

Information-processing tasks include the following:

- Production and maintenance scheduling
- Model-based optimization strategies
- Expert system applications
- Statistical analysis of product and process quality
- Electronic records management, including the archiving of product audit trails (batch lot tracking), selected process variables and events, process alarms and exceptional conditions, operator and supervisory actions, deviations from procedures, operator observations and notes, exposure of personnel to adverse environments

Relational Database Management

Relational database management (RDBMS) packages have become the hub of modern information management systems. They organize information into tables, each containing lists of information, called records. Each record contains a data set referred to as fields. This relatively simplistic structure is accessed and manipulated by specialized languages such as structured query language (SQL). These languages enable the information within the tables and across the tables to be combined, sorted, and selectively chosen to support virtually any application. The simplicity of the structure facilitates its ability to be easily modified as application requirements change. This feature is probably the reason it has become the standard for flexible information storage.

RDBMS packages and, more recently, networked or distributed versions have found application in two major areas within the DCS environment. The first is to store information on the construction of configurable packages such as control strategies, displays, and data collection routines. Used in this environment the RDBMS becomes the engine for more visible packages such as computer-aided design and drafting (CADD). Its primary value in these types of applications is that all references to specific information are to a common place, and therefore a change made on one drawing is reflected on all other drawings using the same information—significantly reducing the rework time and costs. Taken to its logical conclusion, a change in engineering units within the database would ripple through all of the relevant configurables, such as all displays, the control database, and record-keeping information. Considering the number of changes from initial design to commissioning and during the operating life of a process, RDBMS represents an enormous productivity tool.

The second is to record historically any process-related information such as process variables, alarms, operator actions and messages, and significant events. Because the various groups have different collection requirements, a different table would be used for each group. The advantage of RDBMS in this class of application is its ability to select appropriate records from different tables based upon a user-defined set of criteria and present the resultant information for use in a report, screen display, or other application, such as a spreadsheet.

It is in the real-time data collection, archiving, and reporting area where the distributed version is most useful as the information to be collected is considerable. Sharing the load among several computing subsystems improves performance dramatically, while being able to operate on the data using a networked version of the SQL makes the partitioning problem transparent to the user.

An example is its use in chemical batch processing where a product is manufactured in several units and combined at various stages of the process. In order to relate the quality of the product to the various processing steps, information concerning its processing would have to be recorded. Each process unit, however, will make several batches of product during the course of a shift. The information regarding a particular batch would therefore be embedded within several tables, possibly on several computing modules. SQL (or another database language) provides the tool sets to capture the appropriate information from among the various tables using some keyword field (such as current batch id) appended to each record.

These same applications could be (and have been) made by a vendor or user writing programs to manage files which contain essentially the same information. This, however, would be a custom application and would suffer from the same class of problems (high cost, long delivery, considerable resource commitment, difficult ongoing maintenance) which discouraged the use of advanced control prior to the introduction of DCS.

The primary drawback to using RDBMS for historical data collection has been performance, as it consumes considerable processing transactions to add information to its database. Large applications may require thousands of collections per minute. The emerging platforms for these database engines are demonstrating dramatic improvements in performance. The software technology for the management and coordination of distributed relational databases is maturing as well. These coupled with standardized approaches for entering large amounts of similar information into the database tables promise to overcome current shortcomings in performance.

Computer-Aided Design and Drafting

CADD packages have been used for over a decade for mechanical and electrical design applications. They are an electronic alternative to traditional drafting boards and offer substantial features which are impossible to accomplish using their paper analogs. Among them are the opportunity to enlarge, compress, or rotate the objects. Another is to use them in combination with other engineering programs and apply various conditions to the model to determine graphically how the created object would respond to real-world conditions. An example would be how a bridge would deform under various loads or stresses.

CADD combined with database management software provides an effective platform for designing and configuring control strategies used in DCSs. Database templates of commonly used control loops and display objects can be created once and copied as needed, substantially reducing the time necessary to configure the entire database. The effort required to produce and revise documents, particularly drawings, is greatly reduced as information entered can be propagated to all appropriate appearances, such as block connection drawings, display connections, wiring termination drawings, and process and instrument diagrams (P&ID). This is particularly useful as changes are made and all of the documents and drawings need to be updated to reflect the current configuration.

Desktop Publishing

The use of desktop publishing (DTP) tools, including word processors, graphic presentation tools, document formatting tools, and so on, has been a standard practice within the office automation environment. Because of the specialized requirements (such as shift reports) of the production environment, however, most DCSs have not been able to use office automation platform software. Allowing relational databases to act as the intermediary between the DCS and DTP environments enables the real-time and historical information to be imparted to these packages utilizing query languages such as SQL.

Computer-Aided Software Engineering

As the mission of the DCS expanded from process control to production management and it becomes an essential member in the plant information network, the scope and diversity of its applications coupled with requirements for flexibility and user friendliness have greatly compounded the complexity of its software. Computer-aided software engineering (CASE) tools provide an environment for users to develop, design, test, and modify their applications. This environment typically includes

- An integrated development environment, complete with a source editor and program testing tools such as trace and trap
- A library of reusable routines which have been tested and certified for use
- A make facility which builds executable programs from modules available in the library, keeping track of module dependence
- Document management tools to track revision histories of programs as well as their supporting documents such as functional specifications, test documents, design documents, and installation and user manuals to ensure user-developed software compatibility

Object-oriented case tools promise even greater productivity gains. They are just beginning to gain market acceptance and none are available for DCSs. Object-oriented programming has been applied predominantly in the development of graphic draw packages where complex "objects" are created via a construction process from simpler objects and, once created, are manipulated as a single entity (dragged across a screen, rotated, or made to change color).

Objects were first incorporated into DCSs as static graphics building tools, but they quickly evolved into supporting dynamic objects whose attributes were linked to process variables. Changes to the process value would create the corresponding changes to attributes of the display object (such as color, length, position).

The encapsulation of real-time information and procedural methods into objects is also being extended to include trend data, device representations (such as pumps, motors, controllers), analysis routines, and batch operations (fill, clean, and so on). The use of object-oriented techniques has greatly simplified the effort required to design sophisticated workstation tools and thereby improved the operational performance of DCSs as a whole.

Spreadsheets

Spreadsheet packages are commonly used by engineering personnel and business people alike to solve arithmetic or logical relationships in a table-driven manner. Data can be entered into elements of the table, called cells. Formulas referencing data can also be entered into these cells. The result of the calculations of that formula is displayed in the corresponding cell. Once entered, the information can be moved, copied, or operated upon by various commands built into the spreadsheet. Macros, collections of commands executed by invoking a single keystroke, can be entered into the cells of a spreadsheet to simplify their use in complex applications. The net result is a very easy to use and powerful computational tool.

DCSs utilize spreadsheets by importing real-time and historical information into their cells and then operating upon them via stored formulas or macros to perform current energy or material balances, compute inventories or yields, and perform a variety of process analysis applications. Clever use of the command set can even enable the user to perform a reduced-scope set of applications which might otherwise be accomplished by the RDBMS.

Because of its relative ease of use and its flexibility in expressing relatively complex problems, it is replacing scientific programming languages such as Fortran for many applications that do not require extensive branching or looping. Many spreadsheet packages are being designed to work cooperatively with RDBMS to extend the capabilities of both. They are incorporating or providing convenient interfaces to DTP, including word-processing packages to further facilitate their ability to present information in the most useful form.

Statistical Process Control

Statistical process control (SPC) packages are typically used to guide operations and engineering personnel in continuously improving product quality through successive refinements in the production process. The tool sets typically provide a graphic representation of selected measurable quality variables and attributes trended over time against their respective targets. Attribute data are derived from counting the amount of defective or rejected products in a set. Plots typically include a timeline graph of an X-bar (mean) and R-bar (range) chart. These charts are analyzed using a set of rules, sometimes referred to as the Western Electric rules, which are designed to detect patterns that reflect processes outside of statistical process control. These rules are as follows:

Rule 1: One point 3 σ outside the centerline (CL)

Rule 2: Three consecutive points jumping 3 σ

Rule 3: Two of three points 2 σ above or below CL

Rule 4: Four of five points 1 σ above or below CL

Rule 5: Eight consecutive points above CL

Rule 6: Eight consecutive points below CL

Rule 7: Five consecutive points increasing in value

Rule 8: Five consecutive points decreasing in value

Rule 9: Fifteen consecutive points within 1 σ of CL

Rule 10: Eight consecutive points outside 1 σ of CL

Rule 11: After a 3 σ change, three consecutive points within ±0.75 σ of the change point

SPC packages are used to measure the statistical stability of a process, and if the fluctuations are determined to be "unnatural" (not random), these tools are helpful in determining assignable causes. Once the "capability" of the process is determined, improvements can be made and the process repeated (Fig. 11). SPC and SQC are discussed elsewhere in this handbook.

Expert Systems Packages

Expert systems enable the user to impart a knowledge base regarding the process by specifying hundreds to thousands of rules and relationships. These rules interact with each other and with process information using techniques such as forward and backward chaining so as to infer conclusions regarding the likelihood of an event being caused by certain conditions. Because they are capable of assimilating and considering so many factors (some of which may be conflicting) and draw inferences or "expert" conclusions, they are referred to as artificial intelligence engines.

Typical uses include display screen guides to operations or maintenance personnel and open-loop control by using these decisions as commands to regulatory or supervisory control strategies. Applications where expert systems have been applied in a process control environment include:

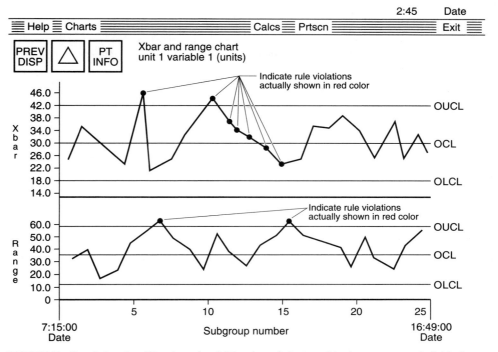

FIGURE 11 Trend plot of an *X*-bar (mean) and *R*-bar (range) chart combined on one screen. Individual samples (depicted as points) represent independent subgroups which, when combined, represent the condition of the process at that time interval. Each set of samples is tested against the Western Electric rules to determine an out-of-normal condition. (*Foxboro.*)

- Diagnosis of the causes of abnormal events such as multiple alarms
- Selection of the best among many "correct" choices such as scheduling process units in a multiproduct plant
- Interpretation of loosely structured instructions such as accepting voice commands

Simulation Packages

Simulation has been an integral part of process engineering, control engineering, and process operations for decades. Highly precise mathematical models used in applications such as chemical reaction kinetics and the design of process equipment required the resources of mainframe computer systems. Highly specialized simulators were also constructed for simulated operations of nuclear power plants and to a lesser extent for large well-understood and typically hazardous process units in petroleum refining, steel manufacturing, fossil fuel power plants, and other large-scale industries. With the advent of high-performance computing power available on-line within the control systems themselves, sophisticated process analysis and simulation tools are becoming available as general-purpose tools and incorporated into DCSs for use in applications ranging from initial testing of control system designs to training process operators. When incorporated into the DCS, these models are capable of being refined continually by the constant stream of process data available from the actual units.

COMMUNICATIONS NETWORKS

The communications network has no counterpart in conventional instrumentation but might be thought of as the wiring of the panelboard itself. It integrates the other four subsystems and facilitates the ability to tie together various plant areas within a site into a common information base to make better informed and more timely decisions. Examples include providing information regarding energy availability (steam, refrigeration, and so on) from the utilities area, determining raw-material availability and storage capacity from a tank farm area, coordinating the transfer of product between departments, or more advanced applications such as production scheduling or product quality audit trails.

Key factors in designing and evaluating an effective communications network include communications throughput, integrity of the information as it moves from source to destination, suitability of the communications network for the environment (such as noise immunity), recovery methods for retransmitting unsuccessfully communicated information, responsiveness in dispatching new information such as alarms, transmission distance, number of stations capable of sending and receiving information, types of information capable of being communicated (process variables, messages, application programs, display files, software images), and ability to modify the communications subsystem topology (add new stations, for instance) while the system is in operation.

Several communications standards have been adopted for use in DCSs during the late 1980s. These include the transmission control protocol/Internet protocol (TCP/IP), which was initially developed in 1969 in cooperation with the Department of Defense for use in the Defense Advanced Research Projects Agency. This is the most widely used protocol. It interconnected thousands of government facilities, universities, and private agencies throughout the world. Because of its wide use, reputation for reliable service, and integration into systems and applications software (such as UNIX), it is a very popular choice within the user community.

In the mid-1980s the International Standards Organization introduced a methodology for implementing network communications called the open-systems interconnect (OSI). It was developed to promote interoperability among disparate systems and is being adopted by virtually all of the systems and network communications suppliers.

OSI is a seven-layer model, consisting of the following:

1. *Physical layer.* This layer is primarily concerned with transmitting raw bits of data over the communications channel. Issues at this layer include what is the voltage level for a logical 0 or 1, what is the time duration of a bit of information, and what are the pin assignments of the communications channel connectors.

2. *Data link layer.* This layer is primarily concerned with removing the raw data transmission errors. It breaks the raw information into frames, each consisting of a group of up to several hundred bytes (8 bits). This enables the raw stream of bits to be broken into identifiable chunks such as data frames and acknowledgement frames and to thereby better define the information flow. The data link layer surrounds an information packet with a unique set of bits which clearly define the frame boundaries. These are typically referred to as headers and trailers. If an information frame is lost or garbled, the data link layer would facilitate the reconstruction and retransmission of the information.

3. *Network layer.* This layer controls the operation of the communications network. It includes packet traffic control and debottlenecking as well as the routing of information packets across communications nodes such as bridges. If a connection is established between a sender and a receiver, the network layer routes all subsequent information packets in accordance with the rules of the connection. This would typically improve communications efficiency and integrity. The role of the network layer is expanded as the information has to cross from one network subsystem to another, often running under different rules.

4. *Transport layer.* This layer manages the communication from the sending system to the receiving system, that is, its purview includes the sending and receiving devices. This differs from the network layer which limits its purview to the communications subsystem. The basic function of the transport layer is to receive information from a higher layer (session) and break it down into packets. It then establishes a communications connection with the receiving system by establishing a network layer connection.

5. *Session layer.* This layer enables users to communicate with each other and with applications across the network. Examples of the services supported at the session layer are the ability to log on to a system over the network or to transfer files across the network.

6. *Presentation layer.* This layer addresses the representation of data within the information packets themselves. It will perform the format conversions and translations necessary to transparently move information from one system to another. A simple example is the use of 7-bit versus 8-bit ASCII.

7. *Application layer.* This layer addresses the reconciliation of various disparate but common applications between systems. One such service is the network virtual terminal service, which determines a communications protocol to enable otherwise incompatible terminals to access and use applications resident on the host. An example would be the interpretation of an escape sequence issued by a terminal to a full-screen editor running on a host.

These seven layers are an abstract model of how information should be packaged to be moved from an application residing in one system to an application in an independent machine via a communications network. Instead of these being unique specifications for each layer, they incorporate sets of standards which are compatible with the model for each layer. As a result there is a great deal of freedom in implementing the OSI standard. For example, both IEEE Std. 802.3, which utilizes a technique called carrier sense multiple access with collision detection (CSMA/CD) for a packet to gain access to the network, and IEEE Std. 802.4, which utilizes a token passing scheme to gain access, are compatible with the OSI model but are incompatible with each other and can only be interconnected via a specialized device called a bridge. The manufacturing automation protocol (MAP) sponsored by General Motors is an implementation of the OSI model. This received wide attention in the late 1980s, but was deemphasized in the early 1990s as being too cumbersome for high-throughput quick-response-time communications.

SYSTEMS ISSUES

Utilizing DCSs to replace instrumentation created a new set of issues which instrumentation and control engineers did not have to face with discrete components. The most significant was the high exposure to system failures. This included the loss of a multiloop controller, which would affect as many as 30 loops, the loss of a video workstation, which would essentially blind the operator to events occurring in the process, the loss of the communications network, which could cripple the entire system except the regulatory control occurring within the multiloop controller, and the loss of bulk storage devices needed for capturing historical information, akin to all trend recorders failing simultaneously.

Improving the overall system availability can be addressed in at least five different ways:

1. Improving the inherent reliability of the system through robust circuit design and the selection of quality components which may be overrated for the task.

2. Providing redundant circuits, modules, or subsystems for critical activities or those with a propensity for failure. The difficulty with this approach is that it typically adds complexity associated with monitoring for failure and transparently switching from one hardware set to the other.

3. Offering backup systems which depend on a much smaller set of equipment to function properly and usually reduce the exposure to failure. The classic example is using panelboard displays to back up the video workstation. The primary disadvantage, aside from cost, is the degraded functionality.

4. Reducing the mean time to repair a failed component by building in comprehensive detection and diagnostic features. Modular designs facilitating repair by replacement and having the ability to replace the module while the system is operational also add to the maintainability approach. The disadvantage is that because it is not instantaneous, the failed module will adversely affect the process for some period of time.

5. Reducing the exposure to failure by limiting the number of independent activities a module can perform. Examples include the number of loops or independent units per controller or collection points per historian in a distributed RDBMS.

External failures such as the loss of power can also create serious problems for DCSs. This is because these are digital systems which perform their operations in a sequential manner and depend on their stored memory to execute procedures properly and to keep current with external events. As most working memory in systems is dynamic RAM, its contents are lost upon loss of power. To prevent the system from taking arbitrary and potentially dangerous action upon the return of power, a predetermined fail-safe condition (such as motors turned off) is placed in nonvolatile memory that governs control action until proper control action can be restored.

When power is restored, the system must undergo a restart procedure. Programs which were in dynamic RAM must be replaced using disk images. These copies do not contain the information about the recent process history. As the process may have undergone substantial perturbations resulting from the loss of power, the strategies in effect at the time of power loss may no longer be viable. Control must be regained in stages. This is typically accomplished through some level of manual intervention. Prior to going on line the control strategies must be initialized to update them on the current condition of the process. Historian files must compensate for the time when no data were collected; work in process must be reconciled. Many systems offer an elapsed timer to enable the user to choose among initialization sequences depending on how long the system was without power. Many processes can ride through a power outage of up to several seconds without affecting control while others may require intervention if the outage persists for only 100 ms. An example of the latter might be a process utilizing a gas heater with a flameout protection circuit. Loss of power would result in the heater being shut down, thereby disrupting the remainder of the process.

ARCHITECTURE OF THE FUTURE

Technological innovations are continuing to facilitate the push of intelligence and control closer to the process. The most significant factors include:

- Availability of shrink-wrap software for increasingly more sophisticated applications and the transition of these packages from single platform to networked interoperable modules utilizing techniques such as object-oriented programming

- Miniaturization, environmental hardening, reduced power requirements, and increased performance of microprocessors, power supply memories, bulk storage, and peripherals

- Increasing the sophistication of the distributed computing environment as standards begin to take hold

Intelligent transmitters and actuators of today are the signposts toward future directions. The SP-50 fieldbus standard will provide the communications vehicle whereby intelligent modules embedded in or attached to transmitters and actuators will intelligently cooperate to accomplish routine process regulation without the need for separate control devices. The types of SP-50 compliant devices supported will include measuring devices for physical properties such as temperature, flow, level, pressure, mass, volume, pH, specific ions, as well as bar-code readers, weigh scales, process analytical devices such as chromatographs, laboratory instruments such as mass spectrophotometers, specialized control devices such as blending systems, and operator display and data entry modules.

These devices will be "self-aware" and will maintain all relevant operational and diagnostic information within them, including the list of cooperating devices which they must notify of routine changes. One example would be a transmitter notifying a valve of changes to its measurement (or to change its output, depending on where the PID resides) or if its signal were suspect, notification to invoke fail-safe logic.

Process operators will be able to supervise and otherwise interact with the process via hardened display terminals directly connected to the fieldbus. Operator terminals will include remote devices communicating via radio transmissions. Configurations for these devices range from hand-held terminals to visored helmets with the information projected onto the face of the visor. Audible command processing would allow the user to communicate with the system while performing some other task.

Distributed controllers as we currently know them will evolve into supervisory controllers and will act as "gateways" to the plant real-time information network. Much of the functionality of present-day controllers which enables them to coordinate the regulatory control, invoke contingency logic, and execute batch or other unit procedures will be retained.

Application software for these devices will include distributed computing modules, whereby the controllers will be treated as satellite processors or execution modules. Product strategies will be formulated (and verified) on application servers, and those portions of the programs associated with procedural supervisory control or unit operations will be dispatched to the controllers.

As the bulk of the routine processing (such as PID) will be performed by the fieldbus modules themselves, the controller will have the capacity to perform a host of preprocessing functions for clients on the real-time network. One example would be the management of a short-term historical database which would at once service the trending needs of the fieldbus operator terminals and also reduce the transactional requirements needed for an archiving device on the real-time network to gather information.

Other functional enhancements would include advanced messaging capabilities to enable configuration data and text as well as graphic information to be routed through the controller to specialized devices on the fieldbus or to the operator workstation. Current information regarding process variables, alarms, actuator positions, set points, and relevant status information could also be routed from the devices on the fieldbus to workstations located on the real-time information network.

The workstations attached to the real-time information network will be advanced versions of the distributed computing workstations coming to the market today. They will be compliant with the graphic user interface (GUI) standards emerging today and will be capable of displaying information equally well from process and business applications.

Distributed computing modules will utilize the technology available at that time from the distributed computing companies. Application-specific software and DCS enablers will be configured into them. The major emphasis in this area will be the development of application and enabling software tool sets which utilize third-party software such as database management engines for implementation. These will include the following:

- GUI-compliant process display and trending software, including access tools to interrogate the "self-aware" fieldbus and controller modules
- Production management software capable of authoring, verifying, and executing strategies intended to manufacture products and manage production operations
- High-performance real-time historians who track production runs and reconcile them against specific products and batch lots as well as profile the performance and efficiency of plant operations by capturing transactional information from a multitude of sources including "self-aware" fieldbus devices, operator consoles (observations and actions), event noti-

fications from distributed controllers, and product strategy information from production management packages

- General-purpose interface utilities which facilitate the integration of third-party application packages such as spreadsheets, RDBMSs, CADD packages, word processors, programming language environments, SPC and other math analysis packages, and so on

The real-time information network will be a distributed computing communications standard. The real-time and office plant information network could conceivably be merged into a single entity. However, this could adversely affect the availability and integrity of real-time operations as the diverse needs of the office environment require a substantial amount of system administration. The more likely scenario would be a high-performance gateway, router, or bridge which would afford some level of isolation and possibly restrict, for security purposes, the crossing of certain types of commands or information.

Implementation of this "self-aware" approach will depend on the establishment of strict conformance to high-level standards for these various fieldbus devices to interoperate. This in turn will require substantial participation and cooperation among the vendor community. The adaptation of open-systems philosophies has in the past proven to be a slow and arduous process, full of false starts and dashed expectations. However, progress in every industry, ranging from electric power to automotive to electronics and computers, has depended upon it.

BIBLIOGRAPHY

Considine, D. M.: *Process Instruments and Controls Handbook,* 3d ed., McGraw-Hill, New York, 1985.
"Distributed Computing Roadmap: The Future of Distributed Computing," Publ. FE277-0, Sun Microsystems.
Gensym Real Times, vol. 1, Gensym Corp., Cambridge, Massachusetts, Winter 1991.
Shinsky, F. G.: *Process Control Systems,* 2d ed., McGraw-Hill, New York, 1978.
Tanenbaum, A. S.: *Computer Networks,* 2d ed., Prentice-Hall, Englewood Cliffs, New Jersey, 1988.

PROGRAMMABLE CONTROLLERS

by Ralph Mackiewicz*

In the early 1960s industrial control systems were constructed from traditional electromechanical devices such as relays, drum switches, and paper-tape readers.[1] The control relay was the most widely used device for controlling discrete manufacturing processes. Although these earlier devices are still being used today, and many of the problems associated with using them have been eliminated due to technological advances in their design, such approaches continue to suffer from some inherent problems. Relays were susceptible to mechanical failure, they required large amounts of energy to operate, and they generated

* Vice President, SISCO, Sterling Heights, Michigan.
[1] This general observation applies mainly to the discrete-piece manufacturing industries.

large amounts of electrical noise. Extreme care had to be taken in the design of relay-based control systems because it was not uncommon for the outputs to "chatter," that is, turn on and off rapidly when they changed states. The logic of the circuit was dictated by the wiring of contacts and coils, and in order to make changes, more time was required to rewire the logic than had been needed when installing it the first time.

In the late 1960s the need to design more reliable and more flexible control systems became apparent. For example, the automotive industry was spending many millions of dollars for rewiring control panels in order to make relatively minor changes to the control systems at the time of the yearly model changeovers. In 1968 a team of control engineers[2] wrote a specification for what they called a programmable logic controller. What they specified was a solid-state replacement for relay logic. Instead of wires there would be bits inside of a memory circuit that would dictate the logic. The machine would use solid-state outputs and inputs instead of control relays to control the motor starters and sense push buttons and limit switches. The first commercially successful programmable controller or, more accurately, programmable logic controller (PLC), was introduced in 1969.[3] By contemporary standards it was a massive machine containing thousands of electronic parts. It should be stressed that the machine was designed at a time prior to the availability of microprocessors. The early PLC used a magnetic core memory, which allowed retention of the program during power outages, to store a program that was written in a graphic language (relay ladder), a scheme long established in connection with relay logic (Figs. 1 and 2).

In the late 1970s the microprocessor became a reality and greatly enhanced the role of the PLC, permitting it to evolve from simply replacing relays to the sophisticated control system that it has become today. PLCs now have the ability to manipulate large amounts of data, perform mathematical calculations, and communicate with other intelligent devices such as robots and computers. Concurrent with the increased capability and flexibility of the PLC was the expansion into many other industrial applications, including the control of machine tools, material-handling systems, food-processing operations, and the continuous process control field. In the 1980s the typical PLC was further enhanced with higher speeds, more flexible communications, more flexible input and output choices, and lower prices. Because of this, the PLC in the early 1990s is used ubiquitously throughout almost every manufacturing industry. In particular, the increase in the flexibility and performance of the communications capability of the PLC has allowed the controller to become an important communications tool for collecting real-time manufacturing process data in computer-integrated manufacturing (CIM) environments.

CHARACTERISTIC FUNCTIONS OF A PLC

A programmable controller is currently defined by the National Electrical Manufacturers Association (NEMA) as a "digital electronic device that uses a programmable memory to store instructions and to implement specific functions such as logic, sequence, timing, counting, and arithmetic to control machines and processes." However, this definition is so broad as to encompass nearly every solid-state device used in manufacturing, from a simple timer to a mainframe computer. Instead it is more useful to examine the essential characteristics of a PLC that portray its unique aspects. Seven of the most important characteristics of a PLC include the following:

1. It is field programmable by the user. This characteristic allows the user to write and change programs in the field without rewiring or sending the unit back to the manufacturer for this purpose.

[2] General Motors Corporation, Hydra-Matic Division (now Powertrain Division).
[3] Bedford Associates Model 084. Bedford Associates was the predecessor of Modicon, Inc., which is now owned by AEG North America.

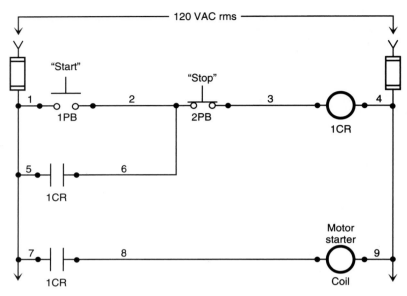

FIGURE 1   Typical motor control circuit. When the push button labeled START (1PB) is pressed, the control relay (1CR) is energized. A contact from 1CR is then closed and is used to "seal" 1CR ON after 1PB is released. Another contact from 1CR is used to energize the motor starter coil, turning the motor on. When the STOP push button (2PB) is pressed, it deenergizes 1CR, which "unseals" 1CR and deenergizes the motor starter coil to stop the motor. Implementing this motor control circuit requires nine wires, not counting the power supply wiring.

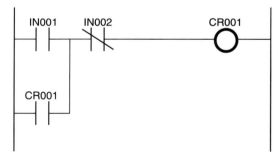

FIGURE 2   Ladder diagram used to control the motor circuit of Fig. 1. In this case all the inputs and outputs are assigned variable names, such as IN001 for the START input and CR001 in place of the control relay 1CR. This diagram is then drawn on a program loader and entered into the PLC's user memory. The PLC's processor then solves the logic that is stored in memory. Only six wires are needed between the PLC's output and the motor starter coil and between the PLC's inputs and the push buttons, not counting the power supply wiring.

2. It contains preprogrammed functions. PLCs contain at least logic, timing, counting, and memory functions that the user can access through some type of control-oriented programming language.

3. It scans memory and inputs and outputs (I/O) in a deterministic manner. This critical feature allows the control engineer to determine precisely how the machine or process will respond to the program.

4. It provides error checking and diagnostics. A PLC will periodically run internal tests of its memory, processor, and I/O systems to ensure that what it is doing to the machine or process is what it was programmed to do.

5. It can be monitored. A PLC will provide some form of monitoring capability, either through indicating lights that show the status of inputs and outputs, or by an external device that can display program execution status.

6. It is packaged appropriately. PLCs are designed to withstand the temperature, humidity, vibration, and noise found in most factory environments.

7. It has general-purpose suitability. Generally a PLC is not designed for a specific application, but it can handle a wide variety of control tasks effectively.

Block Diagram of a PLC

A simplified model of a PLC is shown in Fig. 3. The input converters convert the high-level signals that come from the field devices to logic-level signals that the PLC's processor can read directly. The logic solver reads these inputs and decides what the output states *should be*—based on the user's program logic. The output modules convert the logic-level output sig-

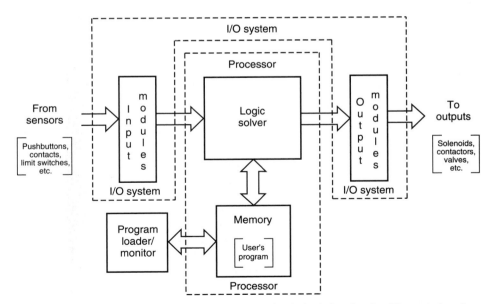

FIGURE 3 Simplified block diagram of a PLC, illustrating its basic functionality. The control engineer (user) enters the control program on the program loader. The program loader writes this program into the memory of the processor. The logic solver reads the states of the sensors through the input modules, then uses this information to solve the logic stored in the user memory (program) and also writes the resulting output states to the output devices through the output modules.

nals from the logic solver into the high-level signals that are needed by the various field devices. The program loader is used to enter the user's program into the memory or change it and to monitor the execution of the program.

PROCESSOR SECTION

A detailed block diagram of the processor section of a PLC is shown in Fig. 4. This section consists of four major elements: (1) power supply, (2) memory, (3) central processing unit (CPU), and (4) I/O interface (to be described later).

Power Supply

The basic function of the power supply is to convert the field power into a form more suitable for the electronic devices that comprise the PLC (typically +5 volts dc or ±12 volts dc). The power supply is one of the most critical components of a PLC for two reasons: (1) It is typically nonredundant. Hence a failure of the PLC power supply can cause the entire control system to fail. (2) It will typically contain high-voltage components. Hence an isolation failure

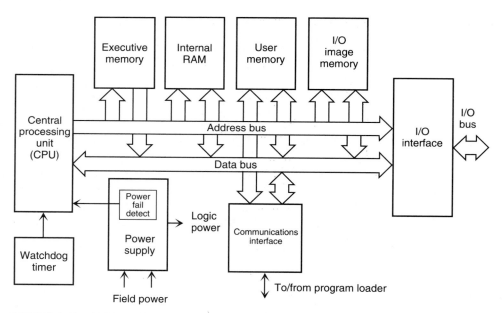

FIGURE 4   Detailed block diagram of the processor section of a PLC. The CPU, typically a microprocessor, executes a program written by the manufacturer of the PLC that is stored in the executive memory. This executive program, which the CPU executes, gives the CPU the ability to interpret the user's program. The CPU does not operate on the I/O directly. Rather, it works with an image of the I/O that is stored in the I/O image memory. The I/O interface is responsible for transferring the image outputs to the I/O system, reading the inputs from the I/O system, and writing them into the image memory. A "watchdog" timer is provided to time how long it takes the CPU to execute the user's program. If this time exceeds a predetermined value, the timer causes the processor to fault. If the CPU fails and does not execute the user's program, the timer will ensure that at least a fault will be indicated and that the processor will shut down in an orderly manner.

can create the potential for serious injury or fire. Useful guidelines when considering the power supply of a PLC include the following:

1. The power supply should be packaged properly so that the heat generated by the power supply can be removed in order to prevent overheating. This will increase reliability. If the power supply cabinet is hot to the touch at room temperature in an office environment, it will be hotter still when locked in a control panel or located on the factory floor. Care, of course, must always be taken to avoid touching any exposed power terminals. (PLC packaging is discussed further in a later part of this article.)

2. The power supply should be tested by a certification agency, such as Underwriters Laboratories (UL) or the Canadian Standards Association (CSA). These agencies perform temperature testing and electrical isolation testing on power supply components. A UL or CSA mark on the PLC power supply will indicate that the power supply was tested to comply with some basic minimum standards.

3. The power supply should meet at least one reputable standard for noise immunity. Two of these standards are NEMA ICS 2-230 (a showering arc noise test) and IEEE Std. 472 (a high-voltage impulse test). Some noise testing may also be performed by certification agencies, such as UL and CSA. The power supply should also be capable of withstanding line-voltage variations such as dropouts, brownouts, and surges, which are common to industrial facilities.

Memory

The memory of a PLC can be of two different types, volatile or nonvolatile. Volatile memory loses its contents when power is removed, whereas nonvolatile memory does not. PLCs will use nonvolatile memory for a majority of the user's memory because the program must be retained during a power-down cycle, meaning that the user will not have to reload the program each time power is lost.

It is important that all nonvolatile memory in a PLC use some form of error checking to ensure that the memory has not changed. This error checking also should be done on line, or while the PLC is controlling the machine or process, in order to ensure safe execution of the user's program.

Battery-Backed-Up CMOS RAM.   This is probably the most widely used type of memory. Although most random-access memories (RAMs) are inherently volatile, the complementary metal-oxide semiconductor (CMOS) variety consumes so little power that a small battery will retain the memory during power losses. The batteries used vary from short-life primary cells (alkaline and mercury batteries), which require periodic replacement every 6 months to a year, to long-life cells (such as lithium), which may last up to 10 years, to rechargeable secondary cells (such as NiCad and lead acid). The latter also may last several years.

EPROM Memory.   An electrically programmable read-only memory (EPROM) is programmed using electrical pulses and can only be erased by exposing the circuit to ultraviolet light, also called UV-EPROM.

EEPROM Memory.   This memory is like EPROM, but with the exception that it also can be erased by using electrical pulses. In some PLCs the use of EEPROM only alleviates the need for an ultraviolet light source, while in other PLCs EEPROM is the only type of memory used. This allows for flexibility of reprogramming, like the CMOS RAM, but without the disadvantage of having to provide battery maintenance.

For applications where the end user cannot reload programs or provide battery maintenance, it is considered better practice to use EPROM or EEPROM.

Central Processing Unit

How the CPU is constructed will determine the flexibility of the PLC (whether or not the PLC can be expanded and modified for future enhancement) as well as the overall speed of the PLC. The speed is expressed in terms of how fast the PLC will scan a given amount of memory. This measure, called the scan rate, typically is expressed in milliseconds per thousand words of memory. Faster PLCs cost more than slower PLCs. Thus it is important to choose a PLC with a scan time appropriate for the application.

It is important to note that many of the commercially available PLCs specify their scan time using contacts and coils only. A real program that uses other functions, such as timers, counters, and mathematical functions, may take considerably longer to execute. Also, in procuring a PLC one should not forget to include the scan time of the I/O, the scan time of the memory, and any additional time overheads the processor requires when making a prediction of the overall scan time for a given application.

PROCESSOR SOFTWARE

The hardware of the PLC does not differ significantly from that of a lot of computers. What makes the PLC special is the software. The executive software is the program that the PLC manufacturer provides internal to the PLC, which executes the user's program. The executive software determines what functions are available to the user's program, how the program is solved, how the I/O is serviced, and what the PLC does during power up or down and fault conditions.

Executive Software.   A simplified model of what the executive software does is shown in Fig. 5. A specific PLC may perform the basic functions shown somewhat differently. This can make a big difference in program execution. For example, some PLCs may perform diagnostics only at a single point in the executive program, while others may perform diagnostics on line, that is, while the user's program is being solved.

Close attention must be given to how the PLC runs diagnostic tests and what it does during failures. Ignoring this aspect of the PLC can result in an unsafe system.

Multitasking.   Some PLCs are capable of executing multiple tasks with a single processor. In particular, PLCs that support sequence function charts (SFCs) tend to support some limited multitasking capability. Multitasking takes several forms, two of which are (1) time-driven and (2) event-driven. In some time-driven multitasking systems the user writes programs and assigns I/O for each task separately. The user then may be allowed to configure the processor to run each task on periodic time intervals. This type of system is shown graphically in Fig. 6. This feature allows the time-critical portion of the control system, such as the portion that controls high-speed motions or machine fault detection, to run many times per second while allowing the noncritical portions, such as servicing indicator lights, to run much more slowly. Because only the time-critical logic and I/O need quick solutions, versus the entire user's program, faster throughput can be achieved.

Event-driven multitasking (also called interrupt-driven) is similar. In this case the user defines a particular event, such as an input changing state or an output turning off, that causes each task to run. Some multitasking systems allow any task to access any variable, such as an I/O point. Thus caution must be used when programming multiple tasks that access the same variables. It may be difficult to determine which task is writing which variable while trying to debug a program.

USER SOFTWARE

This is the software that the control engineer writes and stores in user memory in order to perform the required control over the machine or process. User software can contain both

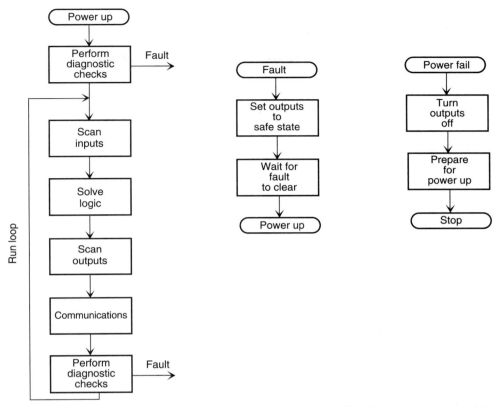

FIGURE 5 Simplified executive program controls the functionality of the PLC. It controls the actions of the CPU to perform as indicated. Diagnostic checks must be run at power-up as well as during the run loop, which is executed while the PLC is controlling the process. When faults are detected, the outputs must be set to a predetermined "safe state." The user usually has the choice of turning all the outputs off or leaving them in their last state. Advanced warning of power failure is typically provided by the power supply, which causes the executive program to shut down the CPU in an orderly manner. During the run loop, inputs are only scanned once. This allows the entire user program to operate from a consistent set of inputs because they are only determined prior to executing the user's program and do not change state in the middle of the run loop. The stability of inputs during the logic solving process is a key element of the PLC's ability to solve logic in a very deterministic manner. This method avoids many instability ("chattering"), race, and other problems associated with implementing control systems utilizing control relays.

configuration data and language programs. The configuration data contain information that tells the processor what its environment is and how it should execute the language program.

Configuration

The configuration process typically consists of assigning I/O points to particular I/O racks, telling the processor how much memory and I/O it has, assigning specific memory for tasks, determining fatal versus nonfatal faults, and assigning many other items interactively on a program loader. (This is covered in more detail later in this article.) Not all PLCs require that they be configured, but being able to configure the processor can enhance the efficiency of the PLC. For example, a PLC that scans only memory and I/O that is configured will run faster than one that scans all memory and I/O whether or not it is used. The user should inves-

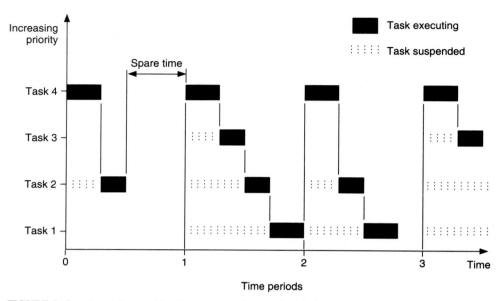

FIGURE 6   In a time-driven multitasking system, tasks are scheduled to run on predetermined time intervals. In the example shown here, tasks 4 and 2 are scheduled to run every period while tasks 3 and 1 are scheduled to run every other time period. The higher-priority tasks always execute before the low-priority tasks. During the second time period, all four tasks are scheduled to run. However, there is not sufficient time for task 1 to finish executing. Its execution is suspended until the spare time in subsequent time periods. Care must be taken to ensure that there is sufficient spare time for all the tasks to execute. Some multitasking systems will provide an indication that not enough time exists to execute all the programmed tasks, thus making it easier to debug the program.

tigate thoroughly how the configuration is done on a given PLC. Some PLCs require reprogramming if the configuration changes.

Languages

Inasmuch as the modern PLC is required to do more and more in terms of operator interfacing, communications, data acquisition, and supervisory control, greater demand is required of the language that implements these functions. Therefore it is crucial that the various aspects of the language, as described in the following several paragraphs, be considered when making decisions.

Variables.   These are the way the language allows the user to access the I/O and internal data. Variables typically take the following form:

X YYYY

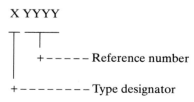

+ − − − − − Reference number

+ − − − − − − − − Type designator

The type designator will indicate the type of variable, such as input bit, output word, internal bit, or internal word. The reference number indicates which specific variable of the specified type is being accessed. Some PLCs allow the user to represent the same data in many different forms, including bits, words, tables of bits, tables of words, files, and bytes.

Languages that allow the same data to be represented in many different ways are more flexible to use and allow the user to write programs more readily and efficiently. Some languages allow the user to use variable names instead of reference-number-based variables. This feature can further enhance the readability of the program, resulting in lower software maintenance costs in the future.

On-Line versus Off-Line.   An on-line language is one that can be programmed interactively with the PLC. An off-line program requires that the programs be generated separately and then down-loaded to the PLC. The user must be careful when using an on-line language while the PLC is controlling the process. Inadvertently changing the wrong item in an active program can wreak havoc on the process.

Flow Control.   This feature dictates how easy it is to make decisions and control the program's execution. Some languages only give conditional jump instructions, while others may allow subroutines and loop functions. Care must be taken to avoid excessive use of jump instructions because it may result in programs that are very difficult to understand and debug. (In some languages the jump instruction is purposely avoided.)

Functions.   The functions provided by the language vary widely from one PLC to the next. Some languages only provide the minimum set of logic, timing, counting, and memory functions, while others may provide additional instructions, such as drum controllers, matrix operations, shift registers, mathematical functions, and many others.

Maintainability.   This refers to how easy the language is to debug, modify, and teach to others. Some computer languages, which offer the ultimate in flexibility, are not suited to applications where the people who have to maintain the machine cannot easily be trained to understand the language.

Speed.   Some languages take longer to execute than others. Choosing a slow but powerful language in time-critical applications can make the programming much more difficult than using a fast but simple language.

Efficiency.   This is a measure of how much memory is required for a language to implement a given function. Efficiency typically is measured in words of memory per function. Since PLCs are priced according to memory size, a more efficient PLC can result in significant cost savings in terms of the hardware required for a given application.

Examples of Languages Used

Ladder Language.   This is still the premier language of the PLC. It has many advantages:

1. It is readily understood and maintained by skilled workers familiar with relay logic. It simplifies training. Ladder language, however, generally lacks good flow control instructions.
2. It provides graphic display of program execution by showing power flow through the ladder diagram, thereby making it easier to debug.
3. Program is fast.
4. It generates more readable programs for sequence control.

Boolean Language. This language generally is used in very small PLCs. Boolean uses AND, OR, NOT, STORE, and RECALL instructions in order to describe the program logic. This language is not easy to debug unless a single-stepping feature is provided that slows down program execution so that the effects on the I/O can be observed visually. Although fast, boolean languages typically lack good functionality.

High-Level Languages. Examples include BASIC, Fortran, and C. These languages can be very powerful and are identical to the languages used by computers. They exhibit excellent flow control and functionality and provide for the accessing of many different types of variables. They also offer reasonable speed. Most plant-floor maintenance personnel, however, do not understand these languages, and it is difficult to monitor program execution in real time. Although some PLCs allow the user programs to be written in these high-level languages, most PLCs offer this kind of language support through an intelligent coprocessor module that contains a separate CPU for executing the language separate from the main PLC CPU. (Coprocessors are described later in this article.)

State Languages. As an example, sequence function charts (SFCs) allow the control program to be expressed in terms of the machine or process states and the I/O conditions needed to transition from one state to the next. State languages can be a valuable aid in designing large and complex control programs by allowing the engineer to describe the process graphically before using another language, such as a ladder language, to implement the control actions. A disadvantage is that all machine states and their transitions must be defined and programmed exhaustively. Care must be taken to ensure an orderly exit from any state under fault and other nonordinary circumstances. Figure 7 gives an example of SFC.

PROGRAMMING

How the user's program is entered is dictated by the programming equipment provided. How this programming equipment operates can determine how long it will take to develop a program. A majority of the time spent programming and debugging a PLC is spent interacting with a program loader, not with the PLC. A program loader can either help or hinder the user in this programming effort. Thus it is important that the user be familiar with the various functions of the program loader and how they operate. These functions include the following.

Programming. Of primary importance to the program loader, it provides an environment for entering programs. This function must be studied closely because it can determine the amount of time required to write a program. Although all PLCs boast of a user-friendly programming environment, the functionality and ease of use of competitive programming software available from the PLC manufacturer and third-party suppliers vary widely. Also, some program loaders contain sophisticated program debugging tools that will reduce the time it takes to eliminate errors from the program.

Monitoring. The program loader also provides an environment for monitoring program execution in real time (or at least near real time). Most loaders provide displays that show "power flow" through a relay ladder diagram, although some others only allow the user to see the I/O points turn on and off. The spread at which a program loader can monitor varies widely from one loader to the next.

Program Storage. This function allows the program to be stored in some machine-readable format that is separate from the PLC itself. Typically this is done on some form of magnetic medium, such as floppy disks or cassette tapes. There are also products available that will perform configuration control of user PLC programs by keeping a revision history of changes

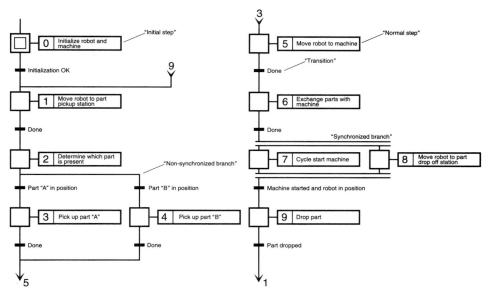

FIGURE 7   Sequence function charts (SFCs) allow the control of the process to be described graphically. An SFC consists of steps and transitions. A step represents a state and contains a program that implements an action, such as "move robot to part pickup station" (step 1). Note that the corresponding state description is "robot moving to part pickup station." A transition represents the conditions necessary for leaving the current step, or state, before proceeding to the next, such as "part in position" (transition between steps 2 and 3). Transitions typically contain a program (such as a relay ladder logic) that evaluates to a true/on or false/off state. An initial step is the step that the processor executes first upon power-up. A nonsynchronized branch (between step 2 and steps 3 and 4) is used to decide between a choice of actions. A synchronized branch (between step 6 and steps 7 and 8) is used when multiple actions are to be executed simultaneously. SFC offers an easily readable method of describing a control program's actions.

made to the program and providing security functions to prevent unauthorized tampering of master control programs.

Documentation.   Most program loaders allow the user to obtain a hard copy or printout of the program. Some program loaders allow the user to get cross references of variable usage, enter comments in the program, and define names for all the variables. These features can greatly improve the readability and maintainability of the program. Again, the functionality of the documentation features of products varies widely and there are third-party products available.

The user is well advised to pay close attention to the operating characteristics of the program loader. Although there has been an increase in the ease of use of program loaders in recent years, there remain significant differences in program loaders. If the user cannot understand how a loader works within a short period of time, this indicates a poor ergonomic design and may be indicative of future operational problems.

Types of Program Loaders

Forms of program loaders are described.

Hand-Held Loaders.   These are small but low-cost loaders that typically use liquid-crystal displays (LCD) or light-emitting diode (LED) displays. These units are primarily for applications where cost is of the utmost importance or for small, simple applications, because such loaders typically are limited in functionality and the amount of information that they can display.

Computer-Based Loaders.   Although some PLC manufacturers make proprietary computer-based program loaders, virtually all PLC manufacturers now offer program loader software for an IBM-PC[4] compatible (personal) computer (PC) running the MS-DOS[5] or Windows[6] operating systems. Some PLCs require that a special interface board for the PC be purchased for communications with the PLC while others use only standard PC hardware. With the recent advent of lap-top PCs, the computer-based program loader now can be very small and portable as well. Some program loader software uses color very intelligently to make debugging and troubleshooting more productive. Also, one must make certain that the PC hardware purchased is sufficiently fast so that the program loader software runs without unreasonable delays between keystrokes. In general, the more feature-rich and sophisticated the program loader software is, the greater will be the processing power required from the PC used in order to run without unreasonable delays.

INPUT-OUTPUT SYSTEMS

Direct I/O.   As the name implies, this is a brute-force way of getting I/O to and from the PLC's processor. There is one input signal and one output signal corresponding to the number of inputs and outputs the processor supports. This approach typically is used in the very small PLCs that have all the I/O circuits in the same package as the processor (sometimes called internal I/O). Cost is the principal advantage of internal I/O. Some flexibility, however, is lost because the processor must be changed in order to change the I/O. Direct I/O systems typically are more cost-effective when the number of I/O points is low (fewer than 64 I/O points). When the number of I/O points is larger, bus-oriented systems (parallel or serial) offer a better cost-performance ratio.

Parallel I/O Systems.   In a parallel system a parallel I/O bus emanates from the processor's I/O interface, and individual I/O modules are plugged into this bus. The I/O module contains the necessary circuitry to decode the bus signals and convert these signals into voltage levels that can drive the necessary field loads. Each module typically will have a number of input or output points on it. This multiplicity of I/O points is called the modularity of the I/O system. Most commercially available I/O systems have modularities of 4, 8, 16, or 32. Adding more I/O points on a module commonly will reduce the cost per I/O point and reduce the amount of space required to install a given number of I/O points (Fig. 8). Also, the more output points there are per module, the smaller are the loads that can be driven.

Serial I/O Systems.   Parallel systems are limited in the distance over which one can extend the I/O bus—typically less than 50 ft (15 m). If the machine should be 100 ft (30 m) long, one would have to use two PLCs. Serial I/O systems solve this problem by transmitting the I/O information over a serial data link capable of being extended over longer distances [1000–10,000 ft (300–3000 m)]. A serial bus emanates from the processor and typically is connected to a parallel bus through a serial-to-parallel converter. Since the single serial bus contains fewer wires than does the wiring to the loads, large wiring cost savings can be realized by using serial systems (Fig. 9). Some of the more recent PLC models use a completely serial system where each I/O module connects directly to the serial link. This can result in even more significant wiring cost reductions.

---

[4] IBM-PC is a registered trademark of the IBM Corporation.
[5] MS-DOS is a registered trademark of Microsoft Corporation.
[6] Windows is a registered trademark of Microsoft Corporation.

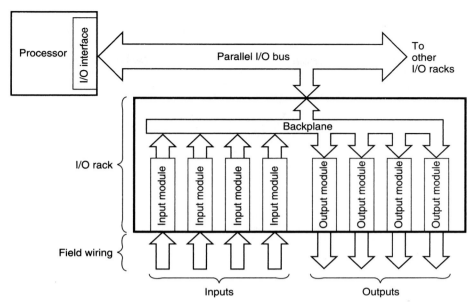

FIGURE 8   Block diagram of a parallel I/O system.

Care must be taken when using serial I/O systems in time-critical applications because two I/O buses have to be scanned, both the serial and the parallel bus, instead of one. Potentially this may make the serial system slower than the straight parallel system. Some, but not all, serial systems may "desynchronize" themselves from the logic scanning, thus making it more difficult to predict I/O responses to fast-changing signals.

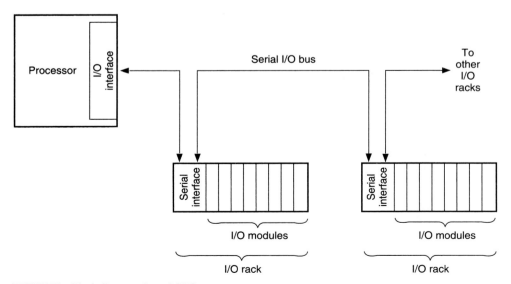

FIGURE 9   Block diagram of a serial I/O system.

INPUT-OUTPUT MODULES

An I/O module performs signal conversion and isolation between the internal logic-level signals inside the PLC and the field's high-level signals. There are many different types of I/O circuits available that are capable of driving almost any conceivable load and sensing the status of a wide variety of sensors. Most of these I/O circuits fall into one of five categories (Fig. 10). The electrical isolation of an I/O module is a critical safety feature. If the isolation fails, the PLC could become "hot" and thus an electric shock hazard. The isolation on an I/O module should withstand a dielectric test voltage of 1000 volts ac rms plus the working voltage of the field circuit (minimum) for at least 1 minute.

I/O modules include the following circuits.

Pilot Duty Outputs.   Outputs of this type typically are used to drive high-current electromagnetic loads such as solenoids, relays, valves, and motor starters. These loads are highly inductive and exhibit a large inrush current. Pilot duty outputs should be capable of withstanding an inrush current of 10 times the rated load for a short period of time without failure. They also should include some form of noise suppression because of the electrical noise that many pilot duty loads generate.

General-Purpose Outputs.   These are usually low-voltage and low-current and are used to drive indicating lights and other noninductive loads. Noise suppression may or may not be included on these types of modules.

Discrete Inputs.   Circuits of this type are used to sense the status of limit switches, push buttons, and other discrete sensors. Noise suppression is of great importance in preventing false indication of inputs turning on or off because of noise. The more noise immune an input is,

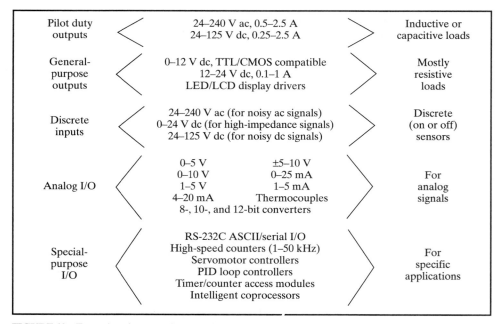

FIGURE 10   Examples of commonly available I/O circuits.

the slower it will be because of the filtering required to reject noise signals. The user should examine an application closely to make certain that an I/O module is selected with the appropriate amount of noise rejection.

Analog I/O. Circuits of this type sense or drive analog signals. Analog inputs come from devices, such as thermocouples, strain gages, or pressure sensors, that provide a signal voltage or current that is derived from the process variable. Analog outputs can be used to drive devices such as voltmeters, *X–Y* recorders, servomotor drives, and valves. Analog I/O circuits consist of either digital-to-analog (D/A) converters for outputs or analog-to-digital (A/D) converters for inputs. These analog converters transform the digital signals of the I/O bus to or from the analog signals of the devices. (More on this topic will be found elsewhere in this handbook.)

Four characteristics are of interest concerning analog I/O circuits in addition to the voltage and current levels. These include (1) conversion time, (2) accuracy, (3) drift, and (4) resolution. The faster the signal from the end device can change, the faster the converter must be to sense or control this change. The accuracy is a measure of how precisely the analog converter will sense, or drive, the analog signal.

Drift is a measure of how the accuracy of the converter changes with time or temperature. This is critical in control systems with long lifetimes. The resolution is a measure of how many different voltages the converter is capable of representing and is specified in terms of bits. An 8-bit converter can represent an analog in one of 256 (2^8) ways; a 12-bit converter can represent 4096 (2^{12}) different voltages.

Although not obvious, accuracy and resolution are independent of each other. It is possible to construct analog circuits with very high accuracy, but with very low resolution. For example, a 1-bit high-precision A/D converter is called a threshold detector. A threshold detector will turn its single bit on or off when the voltage reaches a specified value.

Special-Purpose I/O. Circuits of this type are used to interface PLCs to very specific types of circuits such as servomotors, stepping motors, PID (proportional plus integral plus derivative) loops, high-speed pulse counting, resolver and decoder inputs, multiplexed displays, and keyboards. Another popular application is the register access module. This module allows for limited access to timer and counter presets and other PLC variables without requiring a program loader. Using these special-purpose modules makes for very efficient control systems because they relieve some of the burden of control from the PLC. This has several advantages. (1) The speed of the PLC increases because it has less logic to solve for controlling the device. (2) The PLC can fail, and it still may be possible to maintain control of the device because the I/O module has not failed. (3) A wider variety of devices can be interfaced with the PLC.

Intelligent coprocessor modules also are available that contain a user-programmable computer on them. For example, most ASCII or serial I/O modules have some form of user programmability built into them. Of more recent vintage, coprocessors that have IBM-PC compatible computers and VAX computers on them have become available. These modules more typically plug into the PLC processor chassis rather than an I/O slot. The coprocessor can offer the control engineer two major advantages: (1) the ease of maintenance and understandability of a ladder-diagram-based PLC with (2) the power and flexibility of a general-purpose computer. Care should be taken to examine the software utilities provided by the manufacturer for exchanging information between the PLC processor and the coprocessor, because these capabilities vary from nonexistent to full-featured, depending on the particular product. If they are nonexistent, very specialized (read high-cost) programming skills may be needed to develop the required functionality. Another interesting point concerning the intelligent coprocessor is the availability of PLC processors that will plug into the bus of a general-purpose computer. It should be reemphasized that one should make certain that the required software to exchange information between the PLC processor and the computer will be sufficient for the intended application.

PACKAGING THE PLC

The manner in which a PLC and its I/O system are packaged is critical in determining whether or not a particular PLC is feasible for a given application. If the PLC must be mounted on the factory floor directly next to a machine or process, it must be packaged differently than, for example, a personal computer for desktop use. The constant vibration, electrical noise, and dirt plus wide ambient temperature variations found in most industrial environments will adversely affect a PLC that is not especially packaged for such usage.

The principal packaging factors of the PLC include (1) heat removal, (2) mounting, and (3) wiring.

Heat Removal

As is common with electronic equipment, means must be provided to the PLC for heat removal. Maintaining the internal temperature as low as possible adds to performance reliability and failure prevention.

Venting.   A commonly used method to transfer heat from the PLC to the environment is via convection air currents. Obviously the vents should be located near the top of the instrument package. Additional venting at the bottom increases convection airflow. In vent design, care should be taken to avoid the possibility of parts, such as screws and nuts, falling inside the equipment. Where vented, the PLC should be mounted in an upright position.

Forced-Air Cooling.   A PLC may be furnished with forced-air cooling to augment slowly moving convection air currents. This requires fans or blowers that, after a long period of operation, may fail. Thus natural convection cooling vents should be present in all designs.

Heat Sinking.   Heat sinking is used primarily to remove heat from specific devices, such as power supply transistors. In some power supplies these transistors may consume 40 percent of all the power consumed by the PLC and thus should be physically attached to a heat sink. The latter is a material of high thermal conductivity such as aluminum. A heat sink will work better if it has a radiating surface outside the PLC.

Mounting

Almost all PLCs are intended (designed) to be mounted inside some other enclosure that carries a NEMA rating, such as oiltight, dustproof, or RFI shielded. Top-of-the-line enclosures are quite expensive, but on the other hand, lower-cost enclosures may not provide the required protection. These enclosures should *not* be vented.

PLCs may be rack- or panel-mounted. In panel mounting the PLC is mounted to a flat piece of metal (panel), which then is mounted inside an enclosure. Inside the enclosure there is a large space that is only partially taken up by the PLC. The remaining space absorbs heat rejected by the PLC, as mentioned previously. Currently this is the most popular method of mounting PLCs in the United States. Rack mounting allows a much higher instrument density, but with less space for heat rejection. Frequently rack-mounted instruments will require forced-air cooling. Rack mounting is much more popular in lower-temperature environments and particularly when density is of great importance.

Wiring

The impact of wiring on installed cost cannot be minimized. These costs can easily exceed the cost of the hardware. Also, large maintenance problems can occur if the wiring is not implemented properly. Two types of wiring systems are commonly used.

1. *Fixed Wiring.* This is typically used on the PLC's power supply and on small PLCs with direct I/O. Unwiring such systems is usually quite time-consuming and costly.

2. *Removable Wiring.* This is commonly used for I/O modules. In this case the field wiring is done to a removable terminal block. In order to remove an I/O module for repair or fuse replacement, one simply detaches the terminal block and unplugs the module from the I/O rack. In some systems the terminals are fixed to the I/O rack, but the I/O module is removable. It should be mentioned that not all PLCs use removable wiring in their I/O systems.

Because wiring is so important to safety, one must make certain that the terminals are suitable for a given application. 120-volt ac pilot duty outputs and 120-volt ac inputs should use 300-volt 10-ampere terminals. The wiring system should not be "too removable" and thus unsafe where vibration is present or where a slight pull on the wiring will cause a disconnection, particularly at a time when the PLC is controlling a machine or process.

COMMUNICATIONS AND PLCS

The capability of a PLC to collect data from the manufacturing process is inherently provided through its I/O system. How easily the data can be conveyed to other PLCs and computers is a function of the communications provided by the PLC. As manufacturers attempt to integrate manufacturing process information into the business information system, a primary goal of many CIM environments, the communications capability of the PLC obviously becomes critical in determining the applicability of these devices.

Point-to-Point Communications

Most PLCs have at least one communication port built in, namely, the program loader interface. However, only a few manufacturers release the information needed in order to communicate over this interface. Even so, these ports typically use unusual protocols that can require considerable effort to implement. Some manufacturers of peripheral equipment, such as operator interfaces, have developed drivers to communicate to some PLCs directly, thereby saving the expense of writing specific communications software.

Almost all PLCs offer some form of point-to-point communications via an RS-2332C ASCII-based link that utilizes a separate intelligent I/O module for this purpose. These modules may support multidrop RS-422 and RS-485 links as well. Although these modules typically will provide the necessary software functions to allow the module to communicate with the PLC processor via the I/O system, in some cases these modules are completely user-programmable for the external ASCII communications port. This means that all the external communications software would need to be written by the user. Alternatively, some modules come preprogrammed with a fixed protocol that allows I/O and internal memory to be accessed by an external RS-232 device. With these ASCII communication modules it is possible to communicate with a wide variety of devices, such as color graphics terminals, intelligent push-button stations, bar-code readers, servomotor controllers, and ASCII terminals.

The instruction set of the PLC can have an impact on how the ASCII information is processed. PLCs with few instructions can be difficult to program for ASCII information processing. Functions, such as search, compare, ASCII-to-binary, binary-to-ASCII, block transfer functions (to transfer data to or from the ASCII communications module), and flexible conditional execution statements are crucial for the efficient handling of ASCII information. If the PLC instruction set is insufficient, then most of the processing will need to be done either in the external device or in the communications module itself.

Network Communications

Most PLC manufacturers also provide some type of network allowing for communications between their own PLCs. With these networks it is possible to distribute PLCs physically, yet have them work in unison by using the network's communications functions. Furthermore, most PLC manufacturers offer computer interfaces for these networks so that computers and other devices can also communicate with the connected PLCs. Most of these networks provide two basic functions, (1) reading and writing variables and (2) uploading and downloading programs. However, because these networks are designed to provide communications functions, not necessarily control functions, using a network inside of a control loop requires careful planning and evaluation. Some networks have difficulty transporting information from two points on the network in such a manner as to allow the use of this information in a time-critical control application. The following points should be considered when evaluating a network for a PLC application.

Response Time.   The length of time required for an input changing state on one node of a network to a second node receiving notice that the input has changed is a critical parameter when trying to implement control of a process over a network. Some networks give only a probabilistic response time based on some hypothetical installation. However, one always should know the precise response time limits before putting control information on a network.

Throughput.   Applications that require a great deal of data to be communicated will find the throughput characteristics, typically measured in bytes per second, of the network more important. Although some networks that exhibit a fast response time may offer high throughput, this is not necessarily true. It is possible for a network with long response times to exhibit a high throughput rate. Also, a high bit-signaling rate on the wire (sometimes referred to as the baud rate) does not always relate to a high throughput. Generally networks that allow large packets of data to be sent over the network will exhibit a higher throughput than networks that only support small packet sizes.

Error Checking.   Any network that is used for transferring control information should utilize extensive error checking on the information sent over it. Both ends of the network, the sender and the receiver, should be capable of detecting errors and should also perform specific and known error recovery mechanisms, such as retransmission. At a minimum both the sender and the receiver should be notified that there was an error so that the control engineers can program their own recovery scheme.

Access Mechanisms.   Because a network consists of multiple devices, some method for determining who has access to the network at any given time must be used. Two of the more popular access mechanisms are master-slave and peer-to-peer (Fig. 11).

On a master-slave system there is only one master PLC. The master sends commands out to the slave PLCs and they respond appropriately. The slaves on the network never initiate their own commands. They only respond to commands from the master.

The peer-to-peer mechanism allows any device on the network to initiate messages. However, as in the case of people talking, if everybody talks at once, nothing intelligible can be heard. Peer-to-peer networks need some mechanism for determining access between all the devices, *not* just between the master and the slave. Various mechanisms for determining access have been implemented, such as token passing and carrier-sense multiple access with collision detection (CSMA/CD).

If control information is to be sent over the network, the PLC user should make certain that either the access mechanism is deterministic, that is, the access time is known and is not probabilistic, or that the network characteristics are such that the time requirements of the control information can be met under all conditions. Control signals, such as end of travel and

Bus Access
- Peer-to-Peer
  - Token Passing (IEEE 802.4)
    (Deterministic - MAP)
  - CSMA/CD (IEEE 802.3)
    (Probabilistic - Ethernet™)
- Master/Slave

Media
- Coaxial Cable
- Twisted Pairs
- Twinaxial Cable

Signal Type
- Broadband
  (Multiple Channels)
- Carrier Band
  (Freq. Shift Keyed)
- Baseband
  (No Modulation)

Network

| Drop #1 | Drop #2 | Drop #3 | Drop #4 | Drop #5 |
| PLC | PLC | Computer | PLC | Computer |

FIGURE 11 Simplified network block diagram with typical features. In master-slave networks only the master can initiate communications. In peer-to-peer networks any drop on the network can initiate communications. A network that is used for control should have guaranteed response times and known error recovery methods. A master-slave system can be easier to maintain if variables in the drop change, because only the master and the drop in which the variable changes need be updated. Some networks alleviate this problem by allowing variables to be accessed by names instead of addresses. The network protocols and the software tools that support network communications can be the most important factors to consider. For instance, MAP/OSI networks utilizing the manufacturing message specification (MMS) offer a broad set of very high-level protocol functions with high-level software tools available to ease integration. The media and signal type affect the noise immunity of the network. Coaxial and twinaxial based networks offer good noise immunity, but may be more expensive than some twisted-wire networks. Broad-band allows for multiple communication channels on the same network (much like cable television carries multiple channels on a single cable), but is expensive for low-end applications. Baseband is the lowest cost, but does not offer the noise immunity that modulated signals provide.

emergency stop, should be hard-wired. Depending on a network to send safety interlocks and related information without hard-wired backups can result in a hazardous system should the network fail or if access times degrade.

Network Protocols.   The protocols used by the network will dictate the message formats and the functionality of the network. Because there is considerable variation between the protocols that each PLC manufacturer uses on its own networks, care must be taken to ensure that the network is capable of performing the functions that are required for the application. Most PLC manufacturers offer their networks with a set of protocols that are unique to that manufacturer. These networks typically are referred to as proprietary networks. In recent years networks based on internationally recognized standards, such as manufacturing automation protocol (MAP) networks, which are based on the International Organization for Standardization's (ISO) open-systems interconnect (OSI) networking model, have been introduced by a number of leading PLC manufacturers. These MAP/OSI networks make use of a standard application protocol called the manufacturing message specification (MMS). This is a common language for communicating using high-level commands (read, write, upload, download, start, stop, and so on), which alleviates the need to write a separate communications driver for all the different brands and types of devices in the system. Although these networks may not be completely suitable for all types of control applications, they can be useful in many applications.

In CIM environments concerned with integrating manufacturing process information from a wide variety of devices, consisting of multiple brands of PLCs and multiple brands of computers, a standards-based network, such as MAP/OSI, may have a significant advantage over proprietary networks because MAP/OSI networks can be procured from many different computer and PLC vendors. Proprietary networks in these environments may require extensive use of gateways and custom software drivers, which increases the cost for system implementation. This makes proprietary networks inappropriate for enterprisewide CIM.

RELIABILITY

Of utmost concern in any control system is the reliability of the components in that system. PLCs have gained a reputation for being very reliable, and this partly accounts for their widespread use. However, not all aspects of reliability are strictly a function of how well the PLC manufacturer designs and builds the equipment. Significant improvements in reliability can be gained by proper installation and maintenance of the equipment. Improved noise immunity and better availability (percentage of time in use) can be achieved through proper installation and maintenance of the equipment. Noise and availability are two key factors.

Noise Immunity

Many failures of control systems can be traced to noise conditions that can cause intermittent faults, resulting in lower reliability. Fortunately noise problems usually can be prevented by proper application of the equipment. Although most PLCs are designed to withstand certain levels of noise, the user will find it an advantage to go beyond the inherent protection provided by the PLC manufacturer and install the equipment in a manner that reduces the likelihood of noise problems. Among the techniques to consider are the following.

Grounding.   This is one of the least costly and most effective means for improving noise immunity. All electronic equipment requires good grounding to operate reliably and safely. The manufacturer's recommended grounding procedures should be followed carefully so that the inherent noise immunity of the equipment can be utilized to the fullest. In addition, it is good practice to ground all metal chassis, use large wires for grounding to minimize impedance for high-frequency noise, provide solid earth grounds for all electronic equipment, and avoid putting noisy devices, such as arc welders, on the same grounding system as the electronic equipment.

Isolation.   Noise immunity can be improved by separating the noise-generating devices from the noise-susceptible devices. Isolation transformers can be used on all power supplies, and field wiring should be kept separate from logic wiring (such as I/O bus and communication cables).

Suppression.   An effective way of improving noise immunity is by reducing the noise itself. Zero-crossing ac outputs should be used instead of phase-controlled ac outputs. Zero-crossing ac output generates less noise because it turns on only when the voltage across it is zero. Phase-controlled ac outputs can turn on at any point in the ac waveform, resulting in signals with fast rise times that generate high-frequency noise. Another method involves the use of noise suppressors on the noise-generating devices, such as dc solenoids and relays, or by putting noise suppressors on the input to noise-susceptible devices, such as power supplies (Fig. 12).

Availability

Even though reliability gets the most attention, it is availability (the percentage of total time that the system operates reliably and satisfactorily) that is of larger concern. For instance, suppose that system *A* fails twice a year, while system *B* fails only once every 5 years. Obvi-

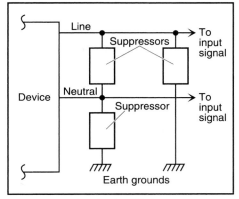

Input noise filtering

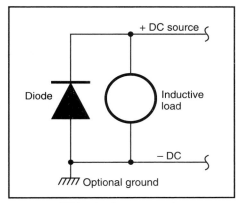

DC load noise suppression

FIGURE 12   Examples of common noise-suppression techniques. Although most PLCs are designed to be relatively noise immune, some applications require additional precautions. Input noise filtering will increase the immunity of any input circuit or power supply. The suppressors can be capacitors, resistor-capacitor networks, or metal-oxide varistors. The suppressor across the line (or +dc) and the neutral (or −dc) will improve the differential or normal-mode noise immunity, while the other suppressors, from each line to ground, will improve the common-mode noise immunity. In many cases it is more effective to eliminate the noise at the source, particularly with inductive dc loads. In this case a diode across the load, as shown, will limit the voltage surges that occur when the load is turned off. Some output circuits will not withstand the current pulses that will occur with the diode in the circuit or cannot have their negative terminal connected to the ground. In these cases the diode alone should be connected to a good earth ground.

ously system *B* is more reliable. However, further suppose that it takes 5 minutes to repair system *A,* while it takes 1 day to repair system *B*. Over the course of 5 years, system *A* will fail 10 times, thus resulting in a total of 50 minutes of downtime, while over the same time period, system *B* will experience a full day of downtime. Availability typically is expressed as a percentage of uptime and calculated by dividing the mean time between failures (MTBF), a measure of reliability, by the sum of the MTBF and the mean time to repair (MTTR) and multiplying by 100. Thus,

$$\text{Availability} = \frac{\text{MTBF}}{\text{MTBF} + \text{MTTR}} \times 100\%$$

A number of schemes can improve the availability of a system. Some of the most popular ways are (1) improving service, (2) fault tolerance, and (3) redundancy.

Serviceability.   By improving the serviceability of a system, dramatic improvements in availability can be achieved. Choosing equipment with removable wiring and modular design can significantly decrease MTTR because a failed module can be replaced more easily. Keeping spares of processor modules, I/O modules, and power supplies close at hand avoids lengthy repair cycles, particularly if the equipment must be returned to the manufacturer for repair.

Fault Tolerance.   Although they are expensive, fault tolerance techniques, such as graceful degradation and error detection and correction (EDC), can improve availability significantly.
    Graceful degradation means that the equipment still can function partially in the presence of faults. For example, most PLCs will continue to update I/O for good modules, even though some failed ones may be present. Distributed control also may be regarded as a form of graceful degradation. In a DCS the control is split up among many PLCs that are distributed over a network. In this case if any one PLC fails, control to only a small part of the process will be lost.

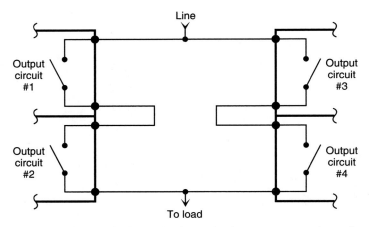

FIGURE 13 Redundant I/O circuits. Redundant output control can be achieved by wiring together four output circuits, as shown. In this case control over the output can be maintained in spite of the failure of any one output circuit. In addition, control can be maintained with two output failures, depending on how the output circuits fail (that is, whether they fail to energize or to deenergize). The major disadvantage of using redundant I/O circuits is cost. Four output circuits are required for every load.

Error detection and correction typically is applied to memories. By inserting certain codes (Hamming codes) as well as the user data into the user memory, not only does it become possible to detect errors, but the codes allow for the correction of some errors as well. Even though a memory fault occurs, it can be corrected before it causes a system failure.

Redundancy.   Some manufacturers build specialized PLCs that offer redundancy features. In these cases two or more processors are tied into one I/O system, and some means is provided that decides which of the processors is to have control over the I/O. These highly specialized high-availability PLCs typically will utilize fault-tolerance techniques as well. In addition, some PLCs can use redundant cabling for the I/O system. Many I/O circuits also can be made redundant so that the failure of any one output will not prevent the PLC from controlling the output (Fig. 13). Because of its cost, redundancy typically is limited to special high-risk applications. The inherent reliability of most PLC equipment is sufficient for many critical control applications.

The user should not disregard what happens when a PLC fails. Lack of attention to what happens during a failure in a control system design can result in an unsafe system.

STAND-ALONE CONTROLLERS

A single-loop controller (SLC) may be defined as a controlling device which solves control algorithms to produce a single controlled output. SLCs often are microprocessor-based and may be programmable or have fixed functionality.

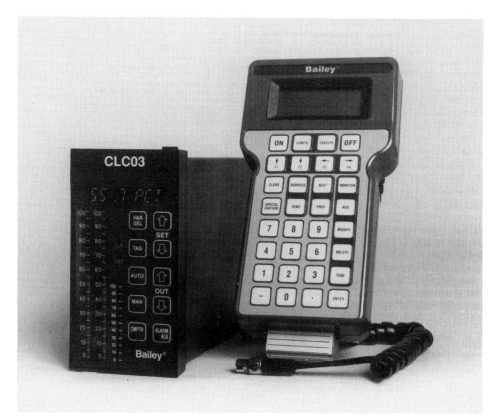

FIGURE 1   Stand-alone controller with dual-loop capability for controlling a variety of process variables. The instrument incorporates (*a*) gas plasma display for set point and control output, (*b*) on-board storage of over 75 proprietary function codes, (*c*) flexible input-output (4 analog, 3 digital inputs; 2 analog, 4 digital outputs), (*d*) optional loop bypass station permitting direct manual control of process outputs during maintenance periods, and (*e*) self-tuning. Configuration is obtained by way of a hand-held configuration and tuning terminal which uses menu-driven prompts to "walk" the operator through "fill-in-the-blanks" configuration procedures. Full monitoring, control, and configuration capability for up to 1500 control points via a personal computer (PC) platform is available. Also, in the same product family there is a sequence controller that provides additional digital I/O for controlling up to three sequences, an RS-232C serial link for connection to external devices, such as printer or data logger. Common applications include flow, temperature, and pressure control of three-element boiler feedwater control and compressor surge control, as well as motor control, burner management, or other start-up/shut-down applications. (*Bailey Controls.*)

For many years, prior to the appearance of computer and advanced electronic technology, the SLC was the mainstay of industrial process control. It also found wide usage in the discrete-piece manufacturing industries, often where temperature control was of importance in connection with such operations as forging, heat treating, or drying ovens. By today's standards, huge centralized control panelboards, say, in a chemical plant or a petroleum refinery, would contain scores of analog SLCs for controlling temperature, pressure, flow, liquid level, and so on, in similar scores of places throughout the facility. The SLCs operated electric or pneumatic control elements such as valves, dampers, and conveyors, requiring large reels of wire or pneumatic tubing to connect them to the process. Frequently the SLCs of that day incorporated large indicating pointers, scales, and circular or strip-chart recorders.

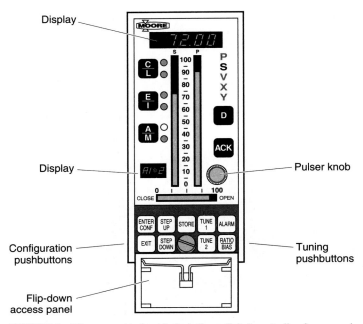

FIGURE 2 Microprocessor-based single-loop digital controller for general-purpose applications. Displays include analog liquid-crystal display (LCD) bar graphs for process, set point, and valve values; digital display for engineering unit values; alphanumeric display for status and alarm indication. Functions and operations, such as inputs, outputs, controls, and computations, are stored within the memory of the model as modular, easy to select function blocks (FBs). Typically the FBs have user-selected parameter values, calibration limits, and information specifying how an FB is linked to other FBs. The standard model includes FBs for single-loop, ratio-set, or external-set operation. An expanded version includes additional FBs for implementing advanced control strategies, namely, pressure-temperature compensation of a flow signal, dead-time compensation for transport lag, feedforward control, single-station cascade control, and override control. An additional third input option can be used to accommodate thermocouple, frequency, millivolt, RTD inputs, or computer pulse input. A serial data communications interface also is available for linking to communications networks (other controllers and computers). (*Moore Products Company.*)

Typically the contemporary SLC is digital. It is very much smaller (down to DIN sizes) and has electrical outputs (error signals) without recording or logging capability. It is built with standards in mind for those situations where a user may desire to make it part of a network, such as a distributed control system (DCS). SLCs continue to have advantages and to appeal to small- and medium-size plants that may not, within the foreseeable future, operate in a computer-integrated manufacturing (CIM) environment. There are tens of thousands of such situations. SLCs continue to enjoy a large share of the original equipment manufacturer (OEM) market for packaging with all types of equipment requiring some variable to be controlled (often temperature), which essentially is regarded by the user as isolated from other concerns with regard to mainline equipment. Even in some medium to large process and manufacturing situations, SLCs are considered as a control option, along with DCSs. Although CIM-type networking has numerous evident long-term cost-effective values (but which are sometimes difficult to justify), the cost of SLCs per loop of control needed is a handy criterion to use in making decisions on which way to go. With scores of SLC suppliers worldwide, it appears that the SLC is far from becoming obsolete. A potpourri of contemporary stand-alone controllers is presented in Figs. 1–6.

SINGLE- AND DUAL-LOOP CONTROLLERS

As is true of other instrumentation and control, stand-alone controllers have derived numerous technical advancements from digital electric circuitry, microprocessors, modern displays,

FIGURE 3   Stand-alone programmer controller in a ⅛ DIN case [96 by 48 mm (3.8 by 1.9 inches)] features a dual light-emitting diode (LED) display, and seven dedicated LEDs are used to show prompt legends during setup and the instrument status when a program is running. Full three-term control output can be offered on output 1 (heat) and output 2 (cool) with the addition of one alarm output. RS 485 serial communications option allows master-slave capabilities to profile up to 32 other similar controllers. Other features include soak hysteresis facility and dual-time-base capability to allow hour-minute or minute-second program rates. Program parameters can be revised without interruption of program. Auto-manual control allows the control of the process to be switched from automatic or closed-loop control to manual or open-loop control. Pretune and self-tune may be selected or deselected. Common applications include pottery kilns, heat-treating processes, food preparation, sterilization, and environmental chambers. (*West Instruments.*)

FIGURE 4 Microprocessor-based ¼ DIN [96- by 96-mm (3.8- by 3.8-inch)] configurable controller. Instrument features blue vacuum fluorescent dot-matrix display (four lines of 10 characters per line), and self-tuning based on one-shot calculation of the optimum PID values based on a "cold" start. Calculated PID values are stored and can be displayed or changed on demand. Both direct- and reverse-acting outputs can be self-tuned. Four PID settings can be stored for each control output. PID set selection may be programmed as part of a profile, initiated by process variable value, internal trigger, keystroke, or logic input signal. Controller uses a dynamic memory rather than fixed allocation of memory. User can assign memory as needed during configuration for best memory utilization. Security by user ID number can be selected at any program level to prevent unauthorized access and program or data changes. Self-diagnostics are used on start-up and during normal operation by monitoring internal operations and power levels. Upon detection of an out-of-tolerance condition, controller shuts off output(s), activates alarm(s), and displays appropriate message. Controller can perform custom calculations using math programs. Device maintains a history file of event triggers as a diagnostic tool. Two process inputs and four outputs can be assigned to two separate control loops creating a dual-loop controller. Optional features include parallel printer output, logic I/O, digital communications, and PC interface. (*LFE Instruments.*)

and very creative software. Contemporary stand-alone controllers differ markedly (perhaps even radically) from the devices available as recently as a decade or so ago.

Size. Not necessarily the most important, but one of the most noticeable features of the present generation of SLCs is their smaller physical size. An impressive percentage of contemporary SLCs follow the European DIN (*Deutsche Industrie Norm*) dimensions for panelboard openings. These are

¼ DIN [96 by 96 mm (3.8 by 3.8 inches)]
⅛ DIN [96 by 48 mm (3.8 by 1.9 inches)]
1⁄16 DIN [48 by 48 mm (1.9 by 1.9 inches)]

FIGURE 5   Family of single-loop process controllers. These programmable controllers incorporate computational and control functions which can be combined in the same manner as programming a pocket calculator. The self-tuning function for optimizing control is particularly useful in multiproduct batch applications where process characteristics can vary from product to product. It features a "one-shot" algorithm, triggered by set-point changes or on demand, and provides rapid response to process changes. Other features include feedforward control (with gain and bias computations), signal processing, analog inputs (4 points, 1–5 volts dc), analog outputs (3 points, 1–5 volts dc; 1 point, 4–20 mA dc). A companion programmable computing station provides data display, signal processing, and sequencing functions. The instrument features 10 status I/O points, each user-definable as input or output. Four programmable function keys on front of panel are used to start control sequences. Four associated lamps indicate sequence progress or serve as prompts. Control period is 0.1, 0.2, or 0.4 second; 500 program steps execute in 0.4 second. There are over 43 computational functions. A device of this kind can control several compressors, for example, as shown in Fig. 6. (*Johnson Yokogawa Corporation.*)

By a substantial margin, the ¹⁄₁₆ DIN is the most popular size. The adoption of DIN sizes is not surprising since an estimated (1992) survey indicates that 30 percent of SLCs sold on the U.S. market are imported from Europe and Asia.

Control Functions.   It should be emphasized that numerous controllers are also capable of handling two control loops. By way of on-board microprocessors, most suppliers offer the fundamental on-off and PID (proportional-integral-derivative) formats. Some suppliers go much further in their design sophistication, offering many mathematical functions, either built into the SLC unit per se, or easily available in companion function modules. These functions, depending on brand and model, may include those listed in Table 1.

In addition to handling a variety of process variables, SLCs find application in the discrete-piece manufacturing industries.

Self-Tuning.   This feature is available on nearly all the SLCs offered, except those in a low-cost category. The methodologies used in general are similar (Fig. 7).

Time Scheduling and Sequencing.   Many available SLCs incorporate the ability to provide sequencing as well as time versus process variable scheduling. A number of processing oper-

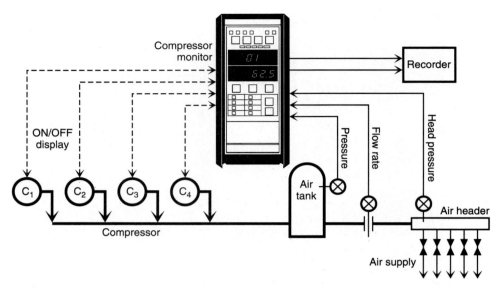

FIGURE 6   Control system for several compressors by using a stand-alone programmable computing station. (*Johnson Yokogawa Corporation.*)

TABLE 1   Mathematical Functions Offered in Stand-Alone Controllers (or Available in Computational Modules)*

Adder subtractor	$I_0 = \dfrac{G_0}{0.250}\left[G_A(V_A - b_A) \pm G_B(V_B - b_B) \pm G_C(V_C - b_C)\right] + 4\text{ mA}$
Adjustable gain with bias	$I_0 = \dfrac{G_0}{0.250}\left[V_A - b_A \right] + K_0$
Deviation—adjustable gain with bias	$I_0 = \dfrac{1}{0.250}\left[G_0\,(V_A - V_B) + V_C \right]$

Adjustable pots, 22 turn
Gain and bias pots, 1 turn graduated dials

Multiplier/divider	$I_0 = \dfrac{G_0}{0.250}\left[\dfrac{(G_A V_A \pm b_A)\,(G_C V_C \pm b_C)}{(G_B V_B \pm b_B)} \right] + 4\text{ mA}$

Other computation modules
 Lead/lag compensator
 Dual filter
 Ramp limiter
 Signal limiter
 Analog tracking
 Square-root extractor
 Signal selector (high/low)
 Signal selector (median)
 Peak picker
Signal converters
 Thermocouples: iron/constantan, Chromel/Alumel, copper/constantan, platinum/rhodium, Chromel/constantan
 Slidewire (nonisolated or isolated)
 Calculated relative humidity

 * Examples are for inputs of 1.5 volts dc and outputs of 4–20 mA.

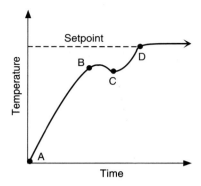

FIGURE 7 Self-tuning feature as available in one
SLC. Starting from a cold start *A,* the controller provides 100% output power until (in this case) the temperature reaches *B,* where the output power is cut off.
The controller then measures the system response *B*
to *C* and calculates the appropriate PID constants for
heat control. At point *C* the output is restored to bring
the system to the desired set point *D* with minimum
delay. Upon completion of the self-tuning routine, the
controller automatically stores and displays the PID
constants and starts using them in the control program. The operator can enter PID constants manually
if it is decided not to utilize the self-tuning feature.
(*LFE Instruments.*)

ations, as found in the metallurgical industries in particular, require bringing a product up to
temperature (as in annealing), holding the temperature for a given time interval, and then
lowering it in several steps. This is described further in the following article in this handbook.
Such schedules, of course, are not always limited to time and temperature, although this is the
most common application.

Other Features.   These include auto/manual, multi set point, self-diagnostics, and memory.

Networking.   Prior to the introduction of DCSs, single-loop controllers were commonly
linked to a central control room. There may be occasional instances when a user may opt for
networking the contemporary SLCs.

TIMERS AND COUNTERS

There are many industrial situations where the measurement of time and time-interval control are critical to the successful operation of a processing or manufacturing operation. Examples would include certain batching situations that must be controlled on an exact timing base
and time-variable sequencing situations, an example of which can be found in metallurgical

operations such as annealing, where several ramp and hold temperature-controlled segments of the process must be followed.

Counting of pieces, as notably encountered in the discrete-piece manufacturing industries, is important to packaging, determining production rates, and for inventory control.

TIMERS

Over the years, numerous methods have been used to achieve a reliable and accurate time reference base for scheduling one or more manufacturing events with reference to time. The requirement in most cases is one of elapsed time, that is, the time between two or more events, *not* simply in terms of the time of day. Early means used included spring-wound timers (as in an alarm clock), fluid dashpots, pneumatic devices, inertia devices, flea-power air motors, and bimetallic thermal devices. Commonly used for many years and widely used today are various types of synchronous motors (such as electric clocks) and various types of timing devices that utilize some electronic phenomenon. These include capacitor discharge and charge (*RC*) circuits, flux-decay (relays), and delay lines.

Control systems that incorporate some form of time-related logic [computers and programmable logic controls (PLCs)] have built-in timing functionality. This, in many cases, obviates the need for separately housed timing devices. However, there are literally thousands of industrial needs for timers where it is expedient to procure them as individual hardware elements to be mounted and used apart from other units of a system, or integrate them into the total system.

Functionality of Timers

Timers fall into certain classifications as determined by their intended use.

Time Switches. The time switch was one of the first time-base devices to be adopted for industrial and commercial applications, dating back many decades. A time switch is a 24-hour (usually) timing device used to open and close electric circuits in accordance with the time of day, on a continually repeating day-after-day schedule. Such timers are applied to daily automatic control of plant lighting, heating, and ventilating equipment; factory work signals, pumps, and compressors; and also for the commencement of heating processes prior to the start of the workday—as may apply to ovens and furnaces, lead and glue pots, dip tanks, soldering apparatus, molding press dies, and even testing equipment.

The greatest usefulness of time switches has been in the saving of electric power (lighting, for example) and labor costs (automatic versus manual connecting of numerous circuits). Time switches were adapted to electric time clocks for employees punching in and out before the turn of the century, and, in fact, the manufacture of these time clocks spawned one of today's leading computer and office automation firms.

Delay-on-Make Timers. The functionality of this type of timer is the same as a time-delay relay, that is, after an initiating signal is received, a timing action is commenced for a fixed (or adjustable) period during which no energy flows to the machine or process being controlled. On expiration of the timed period, current is permitted to flow through the timer contacts, continuing until such time that power to the timer is cut off and remains so for a short period prior to resetting for another cycle of operation (Fig. 1).

Interval Timers. Most timers are interval timers, although they may differ widely in such respects as means of initiating timing, range of intervals, contact operation, driving means,

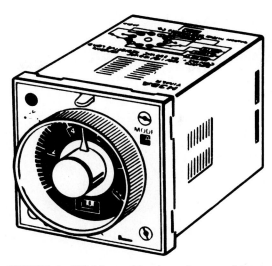

FIGURE 1 ON-delay multimode analog set solid-state timer with 16 time ranges (0.05 second to 100 hours) packaged in compact ⅟₁₆ DIN unit. Rotary switches select time unit, operation mode, and range. Plugs into standard 8-pin or 11-pin sockets. (*Omron Electronics, Inc.*)

arrangements for resetting, and whether or not timing is repeated continuously. An interval timer holds circuits open or closed for a predetermined interval and then returns them to their original positions. When the timer is energized, load contacts close or open and remain so until it has "timed out." Thus it controls the duration of an operation or process (Fig. 2).

A device of this type may be used to control the operating time of a variety of processes, such as those involving plastic molding presses, die casting machines, machine tools, conveyors, centrifuges, and pumps, among many others. Interval timers are also frequently used with bottle filling machines, photoprinters, mixers, bake ovens, resistance welding equipment, induction heaters, chemical feeders, and x-ray equipment; and in cooking, sterilizing, distillation, rubber curing, washing, electroplating, and heat-treating operations. In considering the many applications of control timers, one develops an enhanced appreciation of the great numbers of manufacturing operations that are time-dependent. The interdependence of time and temperature in a bake oven is one of the most obvious—time and temperature, taken together, actually define the variable of concern, namely, the quantity of heat available to be absorbed by the materials flowing through the oven. Within a fairly wide margin, crackers moving fast through a very hot oven can be satisfactorily browned or toasted; but the same result can be achieved by moving them more slowly through a less hot oven.

Percentage Timers.   Percentage timers are designed for applications that require a continuous cycling time that is remotely controlled and has an easily adjusted "on" or "off" setting as a percentage of full cycle time. These units cycle continuously whenever power is applied to the motor. When power to the motor is interrupted, the timer stops and will continue from that point when power is reestablished. Wiring through the common and normally open contacts of the switch gives the percentage of "on" with respect to the dial setting. Wiring through the common and the normally closed contacts of a switch gives the percentage of "off" with respect to the dial setting.

FIGURE 2 Push-button interval timer with automatic reset. Claimed repeat accuracy is ±½%. (*Cramer Company.*)

Repeat-Cycle Timers.   This is a timer that performs its function over and over until it is stopped. A timer of this type may be motor-driven or solid-state. In the electromechanical type, the control means may be a disk with trippers or a shaft with cams. Generally, a repeat-cycle timer is not designed for frequent changes because switch tripping means usually are not easily adjusted in the field. For frequent changes, split cams and other means designed for flexibility of adjustment are preferred. However, some units marketed as repeat-cycle timers incorporate a considerable degree of adjustment. Electromechanical repeat-cycle timers have been used for decades and are technically mature. They continue to be used widely. Electronic solid-state timers are also available to provide the repeat-cycle function, and many timer manufacturers offer both electromechanical and solid-state designs. A repeat-cycle timer can be wired, of course, to perform one cycle of the program and stop, but in such a case it would no longer meet the definition of repeat-cycle timer.

Reset Timers.   This type of timer is designed to reset itself automatically to the preselected setting once it has "timed out" (Fig. 3). The reset time is defined as the interval required for a timer to return its pointer to the position where it started, or how long it takes to prepare a solid-state timing circuit for a new cycle. It determines the closeness of timing cycles.

Programmable Timers.   A timer of this type can control several time intervals for various operations. The term time-cycle controller is also used. A timing diagram for a device of this kind is shown in Fig. 4. These actions can be affected mechanically by using a series of cams (Fig. 5), or electronically. Time-cycle control is encountered in such operations as tire manufacturing, which requires several mechanical operations to be performed, one after the other, in timed intervals. This is contrasted with time-schedule controllers described next.

FIGURE 3   Reset timer. (*Cramer Company.*)

Time-Schedule Controllers

These instruments link time to the control of a process variable. The value of the quantity or condition involved (commonly temperature) is regulated in accordance with a predetermined time schedule. Devices for accomplishing this action have three basic functions, (1) automatic control of the variable, (2) timing, and (3) changing the set point of the controller. In essence, a time-schedule controller is an automatic controller whose set point is automatically posi-

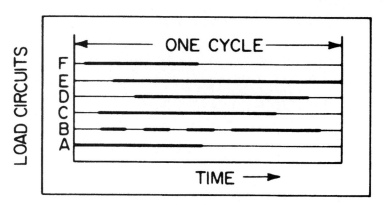

FIGURE 4   Timing diagram for a time-cycle controller.

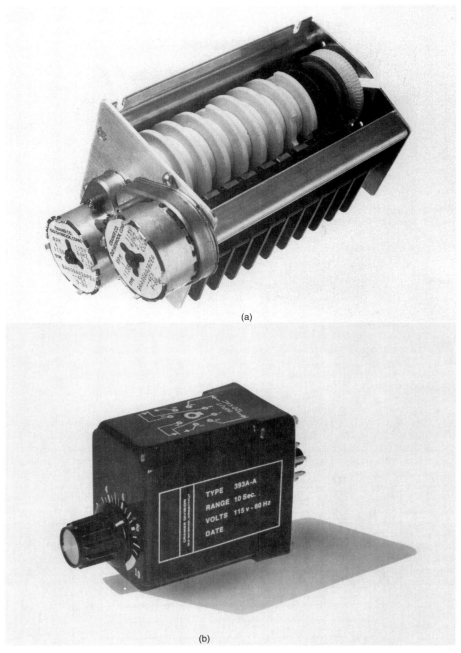

(a)

(b)

FIGURE 5 Programmable time-cycle controllers. (*a*) Mechanical type incorporating adjustable cams. (*b*) Solid-state electronic type. These are single multiple-switch timers that continuously repeat a preset sequential program. Time ranges are from 1 second to 24 hours or longer. Common applications include air-conditioning control, refrigeration system defrosting, lighting control, and machine sequencing. They are also used for appliance sequencing. (*Cramer Company.*)

tioned for a preselected time interval. It may control a variable at a number of rates of change, as well as at different fixed values, for a series of time intervals. In other words, a record of the variable may show a series of smooth rises and falls, or a relatively staccato, steplike variation. Other names for time-schedule controllers include contour, time-variable, time-pattern, and program controllers.

Single-loop controllers are available that provide two variable-versus-time profiles, with as many as 64 segments in a total control cycle. The latter number far exceeds the number of segments required normally. Rather than using cams (as in the earlier time-pattern instruments), which featured a link between the cam and the controller's set point, these actions are accomplished with function modules either built into the electronic controller per se or contained in a separate functional block. Thus time-variable controllers are approached today much as any other complex function, such as self-tuning (Fig. 6). Stand-alone dual-loop controllers also are obtainable with this feature, with a maximum of 16 profiles and 16 segments controlled for each loop. (See preceding article on stand-alone controllers.)

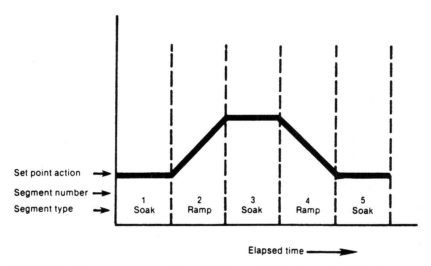

FIGURE 6   Time-temperature profile required in a metallurgical operation, indicating ramp and soak periods.

COUNTERS AND TOTALIZERS

Counters may be used for determining the number of parts or pieces, as in packaging, for counting the number of operations as in manufacturing, or for counting the number of certain kinds of events, as in a control system event counter.

Prior to modern electronic versions, counters depended on mechanical gears (as in an old-fashioned adding machine). Contemporary counters incorporate solid-state circuitry. They usually are small and have a light-emitting diode (LED) or a liquid-crystal display (LCD). These devices range from 0 to 99,999,999 counts. Counting speeds range up to 5 kilocycles per second. Memory is protected by battery back-up. They are available in UP or DOWN count configurations. A rather typical factory application is shown in Fig. 7. Counters are available which will cut off an operation upon reaching a certain high-limit count, in which case the counter may be termed a totalizer. This feature is particularly helpful in parts and piece packing (such as, so many screws per box) or in various boxing and bagging operations. A competing methodology is that of counting by weight.

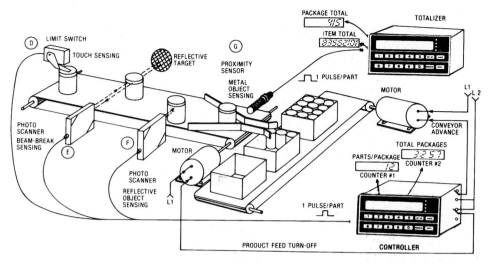

FIGURE 7 Representative operation in which a totalizer is used for a parts-boxing operation. (*LFE Instruments.*)

HYDRAULIC CONTROLLERS

by Raymond G. Reip*

A hydraulic controller is a device that uses a liquid control medium to provide an output signal which is a function of an input error signal. The error signal is the difference between a measured variable signal and a reference or set-point signal. Self-contained closed-loop hydraulic controllers continue to be used for certain types of process control problems, but as the use of the computer, with its electrical output, expands in process control applications, the electrohydraulic servo or proportional valve gains in usage. The combination adds the advantages of hydraulic control to the versatility of the computer. Also contributing to the expanding use of hydraulics is the steady improvement in fire-resistant fluids.

Since the servo or proportional valve does not accept low-level digital input directly, a digital-to-analog (D/A) converter and an amplifier are required. So where it can be used, a hydraulic controller which senses a controlled variable directly is preferred in the industrial environment for its easy maintainability.

* Consulting Mechanical Engineer, Shorewood Hills, Sawyer, Michigan 49125; formerly Chief Mechanical Engineer, L & J Engineering Inc., Crestwood, Illinois.

ELEMENTS OF HYDRAULIC CONTROLLERS

The major elements include (1) an amplifier, (2) an error detector, and (3) a signal-sensing section. As in all control-loop applications, a measured variable must be sensed and compared with a set point, and the difference or error signal manipulated by the controller to cause some final control element to correct the process. Depending on the particular construction of the hydraulic controller, these functions often can be incorporated physically into the hydraulic controller per se.

The hydraulic relay is common to all hydraulic controllers. The relay allows a small mechanical displacement to control the output from a hydraulic power supply in a fashion that will drive a work load. The jet pipe and spool valve are examples of these relays (Fig. 1). If the process variable is being sensed directly, such as pressure by means of a diaphragm, the signal sensing and error detection can become an integral part of the hydraulic controller. As shown in Fig. 1c, a force is generated from the signal system and compared with a set-point force operated by the spring, and the resulting mechanical displacement (error signal) moves the hydraulic relay (in this case a jet pipe) to control the work cylinder.

HYDRAULIC AMPLIFIERS

The three major traditional types of hydraulic amplifiers are (1) the jet pipe valve, (2) the flapper valve, and (3) the spool valve.

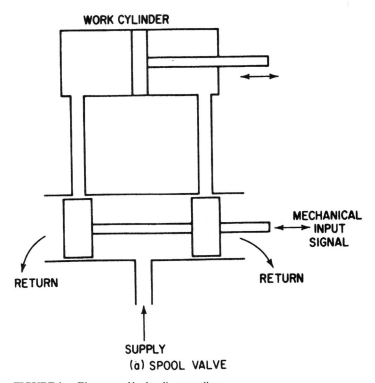

FIGURE 1a   Elements of hydraulic controllers.

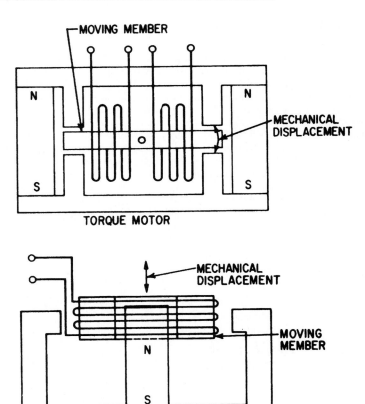

FIGURE 1b   Elements of hydraulic controllers.

Jet Pipe Valve

By pivoting a jet pipe, as shown in Fig. 1c, a fluid jet can be directed from one recovery port to another. The fluid energy is converted entirely to a velocity head as it leaves the jet pipe tip and then is reconverted to a pressure head as it is recovered by the recovery ports. The relationship between jet pipe motion and recovery pressure is approximately linear. The proportional operation of the jet pipe makes it very useful in proportional-speed floating systems (integral control), as shown in Fig. 2a, or as the first stage of a servo or proportional valve.

Position feedback can be provided by rebalancing the jet pipe from the work cylinder, as shown in Fig. 2b. A portional-plus-reset arrangement is shown in Fig. 2c. In the latter case, the proportional feedback is reduced to zero as the oil bleeds through the needle valve. The hydraulic flow obtainable from a jet pipe is a function of the pressure drop across the jet pipe.

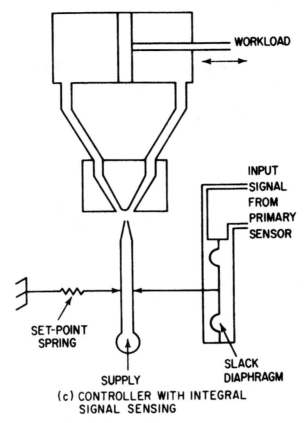

FIGURE 1c   Elements of hydraulic controllers.

Flapper Valve

This device makes use of two orifices in series, one of which is fixed and the other variable. The variable orifice consists of a flapper and a nozzle. The nozzle restriction is changed as the flapper is positioned closer to or farther from the nozzle. When the variable restriction is changed, the pressure drop across the fixed orifice changes, thus producing a variable output pressure with flapper position. Single- and double-flapper arrangements are available as diagramed in Fig. 3.

Spool Valve

This device, when used in automatic control or servo systems, usually is configured as a four-way valve, as previously shown in Fig. 1a. A four-way valve can put full supply pressure on one side of the work cylinder and drain the other side, or vice versa. If the widths of the spool lands are greater than those of their respective ports, the valve is overlapped and is referred to as a closed-center valve. An underlapped valve is referred to as an open-center valve. A

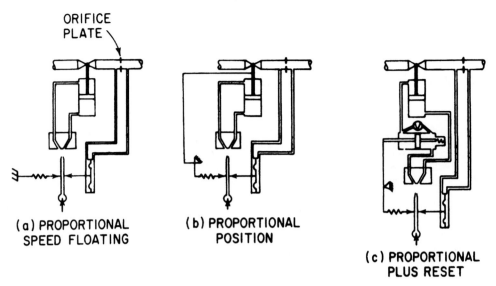

FIGURE 2 Hydraulic controllers arranged in different control modes.

line-to-line valve is one in which the spool lands are the same width as the porting. Flow through the spool valve is proportional to the square root of the differential pressure across the valve, as is the case with a jet pipe.

Two-Stage Valves

When higher flows and pressures are required, a two-stage valve may be used. In these cases the second stage usually is a spool valve, while the first stage may be a jet pipe, a spool valve,

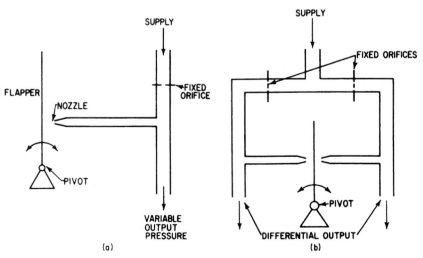

FIGURE 3 Flapper valves. (*a*) Single. (*b*) Double.

or a flapper valve. Different combinations of two-stage valves are shown in Fig. 4. In two-stage valves the second stage must be either positioned, such as with springs, or provided with a feedback to the first stage as required. Feedback may be mechanical, electrical, or hydraulic.

Electrohydraulic Amplifier

With reference to Fig. 4, frequently the process variable is converted to an electric signal by an appropriate transducer. In these cases generation of the set point and the error detection are performed electrically, and the error signal fed to the hydraulic amplifier is an electric signal. When this is done, an electromechanical transducer is required to generate the mechanical displacement. Two of the more popular schemes are shown in Fig. 4—the moving coil and the torque motor. The principal difference between these two transducers is in the direction of the mechanical displacement. The moving coil produces a linear mechanical displacement, while the torque motor output is rotational. Electromechanical transducers mechanically packaged together with a hydraulic amplifier commonly are referred to as servo or proportional valves.

Servo Valves

Figure 4 also illustrates one of the most widely used servo valve designs, a two-stage valve. One and three stages are also available. The single-stage design is used for the highest response rate; three-stage designs are used for the highest power output.

Typical dimensions and clearances for servo valves are given in Table 1. These figures apply to valves manufactured for aerospace or industrial use. The larger numbers shown apply to larger valves.

TABLE 1 Typical Dimensions and Clearances for Servo Valves

	Dimension	
Parameter	Inches	Millimeters
Air-gap spacing (each gap)	0.010–0.020	0.254–0.508
Maximum armature motion in air gap	0.003–0.006	0.076–0.152
Inlet orifice diameter	0.005–0.015	0.127–0.381
Nozzle diameter	0.010–0.025	0.254–0.635
Nozzle-flapper maximum opening	0.002–0.006	0.051–0.152
Drain orifice diameter	0.010–0.022	0.254–0.559
Spool stroke	±0.010–±0.060	±0.254–±1.524
Spool or bushing radial clearance	0.0001–0.00015	0.0025–0.0038

Most servo valve failures are due to sticking spools, usually directly traceable to contamination. Reliable servo valves depend on maintaining clean systems.

Proportional Valves

These valves grew out of a need for lower-cost servo valves capable of operating in a dirty environment. In many cases the basic designs are identical. However, there are some designs unique to proportional valves, such as proportional solenoids to drive the spool directly with a linear variable differential transformer (LVDT) feedback in a single stage. General frequency response of proportional valves is 2–10 Hz versus 10–300 Hz for servovalves.

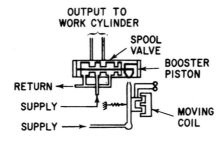

FIRST STAGE: JET PIPE
SECOND STAGE: SPOOL VALVE
FEEDBACK: HYDRAULIC

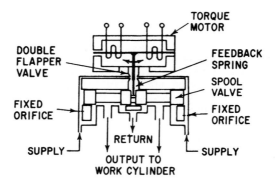

FIRST STAGE: DOUBLE FLAPPER
SECOND STAGE: SPOOL VALVE
FEEDBACK: MECHANICAL

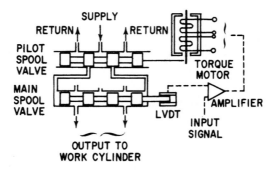

FIRST STAGE: SPOOL VALVE
SECOND STAGE: SPOOL VALVE
FEEDBACK: ELECTRIC

FIGURE 4   Various combinations of two-stage valves.

The filtration requirement of servo valves is generally 3 μm, while 10 μm is specified for proportional valves. However, both figures are on the very clean side. A value of 40 μm generally is specified for industrial hydraulics.

In servo valves the spool and the sleeve or body are matched for very close fit and zero overlap. Spools and bodies are interchangeable in proportional valves, thus there is a significant cost differential. However, overlap results in dead zone, which can be a problem.

HYDRAULIC FLUIDS

Petroleum oil is still the optimum hydraulic fluid unless fire resistance is required. With additives, petroleum oil does not have major shortcomings of the magnitude of those found in all fire-resistant fluids. In designing a hydraulic control system, the choice of fluid must be made early because a number of features of the system must be tailored to the fluid used.

Although there are numerous petroleum-based hydraulic oils designed specifically for various types of systems, automatic transmission fluid has been found to be a good engineering compromise for many systems because (1) it is available worldwide to the same specifications; (2) it contains viscosity index improvers which reduce the effects of viscosity change with temperature (like 10W-30 motor oil); (3) like other quality hydraulic oils, it has rust, oxidation, and foam inhibitors; and (4) it contains an antiwear zinc compound additive, although not as much as is contained in extreme pressure (EP) hydraulic oils.

Synthetic hydrocarbon oils have all the advantages of natural oils, plus complete freedom from wax content. This permits synthetic oils to remain fluid and thus be pumpable down to −65°F (−54°C). Although the cost of synthetic oils is relatively high, their use makes it possible to install hydraulic controls in adverse outside environments.

With increasing environmental awareness, the new vegetable-oil-based hydraulic fluids must be given consideration for many applications. These oils are advertised as readily biodegradable and nontoxic. Tests indicate better lubricity (less pump wear) and a higher viscosity index (less viscosity change with temperature) as compared with good petroleum hydraulic fluids. Also, the vegetable oils have shown performance characteristics comparable to those of petroleum oil at normal system operating temperatures. The same seal materials and metals can be used for both types of oils. On the negative side, the newer vegetable-based oils cost roughly 2½ times more than the conventional petroleum oils and are sensitive to low temperature.

Fire-Resistant Fluids

The principal available fire-resistant fluids are described briefly in Table 2. Their characteristics are compared in Table 3 with those of petroleum oils.

TABLE 2   Representative Types of Fire-Resistant Hydraulic Fluids*

Letter designation	Fluid description
HF-A	High-water-content emulsions and solutions are composed of formulations containing high percentages of water, typically greater than 80%. They include oil-in-water emulsions and solutions which are blends of selected additives in water.
HF-B	Water-in-oil emulsion fluids consist of petroleum oil, water emulsifiers, and selected additives.
HF-C	Water-glycol fluids are solutions of water, glycols, thickeners, and additives to provide viscosity and other properties desirable for satisfactory use in hydraulic systems.
HF-D	Synthetic fluids are nonwater fluids, such as phosphate esters or polyol esters.

* Extracted from ANSI B93.5M.

TABLE 3   Comparison of Hydraulic Fluids

Characteristic	Petroleum oil	HF-A, high water content, low viscosity	HF-B, water emulsion	HF-C, water glycol	HF-D synthetic	
					Phosphate ester	Polyol ester
Fire resistance	Poor	Excellent	Fair	Very good	Very good	Very good
Cost	1	0.1	1	2	5	5
Lubricity	Excellent	Fair	Good	Good	Excellent	Very good
High temperature	Excellent	Fair	Fair	Fair	Very good	Very good
Low temperature	Very good	Poor	Poor	Good	Fair	Good
Corrosion protection	Excellent	Good	Good	Good	Very good	Good
Standard hardware compatibility	Excellent	Fair	Fair	Fair	Good	Very good
Toxicity	Good	Very good	Good	Excellent	Fair	Very good

POWER CONSIDERATIONS

Hydraulic controllers are generally applied where either high-power or high-frequency response is required. The two requirements may or may not be needed for the same application. Frequency-response requirements are a function of the control loop dynamics and can be determined by appropriate automatic control theory techniques discussed elsewhere in this handbook. Power considerations will be discussed here.

Because hydraulic fluids can generally be considered incompressible, they can be handled under higher-pressure conditions than would be considered safe or practical with pneumatics. The amount of power delivered to the load is a function of the hydraulic power supply size, the pressure drop required by the hydraulic controller, and the pressure drop losses in the lines. With reference to Fig. 5 and neglecting any line losses, it can be said that the pressure drop available for the load is equal to the supply pressure less the hydraulic controller pressure drop. In equation form,

$$\Delta P_L = P_0 - \Delta P_c$$

If the oil flow under these conditions is Q, then

$$\text{Power available at load} = \Delta P_L Q$$

Q_{max} is determined by the hydraulic power supply capacity. As Q increases, the term $\Delta P_L Q$ tends to increase. However, increasing Q increases ΔP_c, thus reducing ΔP_L. It can be shown that for most hydraulic controllers the maximum $\Delta P_L Q$ occurs when $\Delta P_L = 2\Delta P_c$.

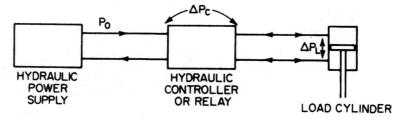

FIGURE 5   Hydraulic flow loop.

Figure 6 shows the force and power ranges normally encountered in hydraulics contrasted with the ranges commonly used with pneumatics. The load horsepower scales were calculated by selecting two representative load stroke lengths and three different full-stroke speeds, all corresponding to the force scale. The hydraulic power supply horsepower requirement is greater, however, since it must in addition supply the power for the hydraulic controller, line losses, and pump inefficiency. This can be expressed as

$$HP_{ps} = \frac{HP_L + HP_c + HP_{LL}}{\text{pump efficiency}}$$

The power lost across the hydraulic controller is a function of the controller capacity. This is generally expressed in terms of a flow–pressure-drop curve.

Conservation

Increasing energy costs have resulted in greater use of systems that load the pump motor only when there is a demand for hydraulic control power. Methods include (1) open-center control valves that bypass flow at low pressure, and (2) pressure-compensated variable-volume pumps that pump only a volume sufficient to maintain pressure in the system.

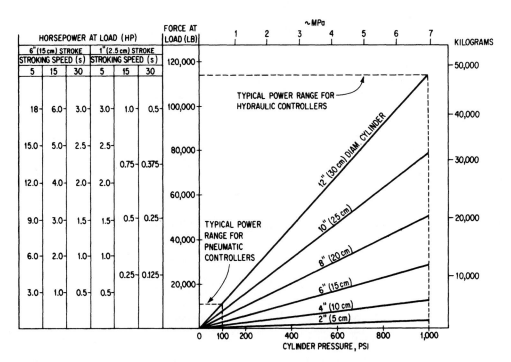

FIGURE 6   Force and power ranges normally encountered in hydraulic systems contrasted with ranges commonly used in pneumatic systems.

APPLICATIONS

Generally hydraulic control systems are selected based on more demanding control requirements rather than on lower initial costs. However, as power requirements increase, even the initial cost may favor hydraulics. Three typical hydraulic controller applications are shown in Figs. 7 to 9. The approximate power and performance requirements versus applications of servo valve control systems are illustrated in Fig. 10.

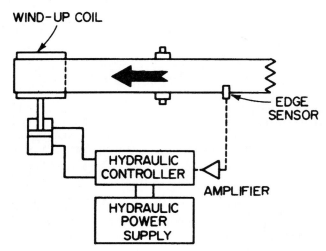

FIGURE 7   Hydraulic edge-guide control system. The windup coil is shifted in accordance with an edge sensor signal to provide an even coil.

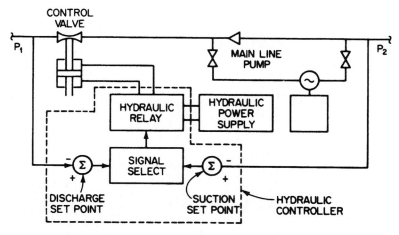

FIGURE 8   Hydraulic pipeline control system. If discharge pressure P_1 exceeds set point, or if suction pressure P_2 goes below set point, the control valve is throttled closed to correct the situation.

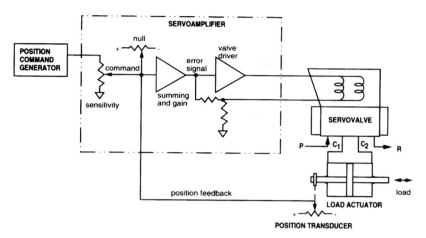

FIGURE 9   Hydraulic position control system, as used in machine tools.

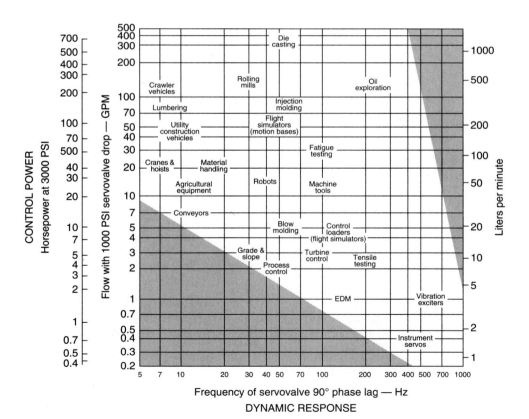

FIGURE 10   Spectrum of industrial applications for hydraulic servo valves. *Note:* 1000 psi = 6900 kPa.

Size-Selection Procedures

Size selection for servo or proportional valves proceeds as follows:

1. Size the actuator area to produce a stall force 30 percent greater than the total force required at the supply pressure available:

$$A = \frac{1.3F_r}{P_s}$$

where A = actuator area [in^2 (cm^2)]
F_r = total force required to move load [lb (kg)]
P_s = supply pressure [psi (kPa)]

2. From the maximum loaded velocity and the actuator force required at this velocity, determine the loaded flow:

$$Q_L = AV_L$$

where Q_L = loaded flow [in^2/s (L/s)]
V_L = maximum loaded velocity [in/s (cm/s)]

3. Calculate no-load flow:

$$Q_{NL} = Q_L\sqrt{\frac{P_s}{P_s - P_L}}$$

where Q_{NL} = no-load flow [in^3/s (L/s)]

4. Compute the rated flow required at the manufacturer's rated pressure drop. Increase 10 percent for margin.

$$Q_R = 1.1\,\frac{Q_{NL}}{C}\sqrt{\frac{P_R}{P_s}}$$

where 1.1 = margin
Q_R = rated flow on manufacturer's data sheets
P_R = pressure drop at which manufacturer rates flow [generally 1000 psi (6900 kPa) for servo valves; 150 psi (1000 kPa) for proportional valves]
C = conversion factor [in^3/s (3.85 gal/min); L/s (60 L/min)]

Relative Advantages and Limitations

Advantages of hydraulic controllers include the following:

1. High speed of response. The liquid control medium, being effectively incompressible, makes it possible for a load, such as a work cylinder, to respond very quickly to output changes from the hydraulic controller.

2. High power gain. Since liquids can be readily converted to high pressures or flows through the use of various types of pumps, hydraulic controllers can be built to pilot this high-energy fluid.

3. Simplicity of the final actuator. Most hydraulic controller outputs are two hydraulic lines that can be connected directly to a straight-type cylinder to provide a linear mechanical output.

4. Long life. The self-lubricating properties of most hydraulic controls are conducive to a long, useful life.

5. Relatively easy maintainability.

Some limitations include the following:

1. Maintenance of the hydraulic fluids. Depending on the type of hydraulic hardware used, filtration is required to keep the fluid clean; the use of fire-resistant fluids, because of poorer lubrication and possible corrosive effects, requires more careful maintenance.

2. Leakage. Care must be taken with seals and connections to prevent leakage of the hydraulic fluid.

3. Power supply. New hydraulic power supplies usually must accompany new hydraulic controller installations. This is contrasted with pneumatic and electrical or electronic systems where such power normally is available.

REFERENCES

1. Henke, R.: "Proportional Hydraulic Valves Offer Power, Flexibility," *Control Eng.,* vol. 28, no. 4, 1981, p. 68.
2. Maskrey, R. H., and W. J. Thayer: "A Brief History of Electrohydraulic Servomechanisms," *ASME J. Dyn. Syst. Meas. Control*, June 1978.
3. Niemas, F. J., Jr.: "Understanding Servo Valves and Using Them Properly," *Hydraul. Pneum.*, vol. 31, no. 13, October 1977, p. 152.
4. Staff: "High Water Content Systems for Profit-Making Designs," *Hydraul. Pneum.*, vol. 35, no. 4, April 1982, pp. HP1–HP18.
5. Totten, G. E., Ph.D.: "Novel Thickened Water-Glycol Hydraulic Fluids for Use at High Pressure," Paper No. 192-23.1, Proceedings of the International Fluid Power Applications Conference, March 1992.
6. Cheng, V. M., A. A. Wessol, and C. Wilks: "Environmentally Aware Hydraulic Oil," Paper No. 192-23.3, Proceedings of the International Fluid Power Applications Conference, March 1992.
7. McCloy, D., and H. R. Martin: *Control of Fluid Power, Analysis & Design*, Halsted Press, 1980.
8. Merritt, H. E.: *Hydraulic Control Systems*, J. Wiley, 1967.
9. Holzbock, W. E., P. E.: *Robotic Technology: Principles and Practice*, Van Nostrand Reinhold Co. Inc., 1986.
10. Parker Hannifin Corp.: "Industrial Hydraulic Technology," Bulletin No. 0232-B1.
11. Moog Controls Inc., "Servovalve Selection Guide."

PNEUMATIC CONTROLLERS

by J. D. Warnock*

Editor's Note: For almost a half-century (1920–1965) pneumatic operation was the preferred means of automatic control by large sectors of the industrial processing field. Commencing during the 1930s, pneumatic controls competed with electric analog controllers, which were generally preferred by a few sectors, notably the steel and metallurgical industries. In these applications, safe operation of equipment in explosive atmospheres generally was not a prime concern. A major advantage of pneumatic control was its innate safety. But this feature was substantially diminished by the appearance and acceptance of intrinsic safety. Several additional factors, however, made an even greater impact on the popularity of pneumatic control. These included the entry of developments in solid-state electronics, miniaturization, and computerization with their improved smaller packaging and the comparative ease with which digital technology could answer the needs of simple and complex systems. Changes in the control system architecture, which shifted the emphasis on highly centralized control to distributed instrumentation, were also a major factor in diminishing the status of pneumatic controllers.

Contemporary pneumatic controllers, which are products of a highly mature technology, are available and preferred by some users for some applications, mainly in connection with single or a limited number of control loops. Some earlier extensive installations also remain in operation. Ironically, control engineers continue to deal with pneumatically operated diaphragm control valves as final control elements, often with pneumatic valve positioners, in a high percentage of control applications. Because pneumatic technology remains active, selected portions of the article, "Pneumatic Transmission and Control" from the third edition (1986) of this handbook are reprinted here.

PNEUMATIC DEVICES

The design of pneumatic transmitters, controllers, and all other pneumatic instruments is based on a number of relatively simple pneumatic devices, which are combined in different mechanical arrangements to give a specific output-input relationship. These pneumatic devices can be considered the equivalent of certain passive or active electrical elements, and a useful analogy between a pneumatic system and electric circuits is often apparent.

Baffle-Nozzle Amplifier

A baffle-nozzle amplifier is the primary detector in almost all pneumatic transmitters and controllers. Its function is to convert mechanical motion to a pneumatic signal. Figure 1 shows a baffle nozzle actuated by pressure in an input bellows. Because the baffle is often mounted on a pivoting element, this instrument is also called a flapper-nozzle amplifier.

In principle, the operation of a baffle nozzle is quite simple. The output increases from a minimum value to supply pressure as decreasing nozzle clearance blocks the flow of air. The minimum output pressure is a function of the supply pressure and the relative sizes of the restriction and nozzle, and is usually less than 1 psig (6.9 kPa gage). Certain operating characteristics and limitations of the baffle nozzle are of particular interest in the design of pneumatic transmission instruments.

---

* Deceased; formerly Manager, Systems Engineering, Moore Products Company, Spring House, Pennsylvania.

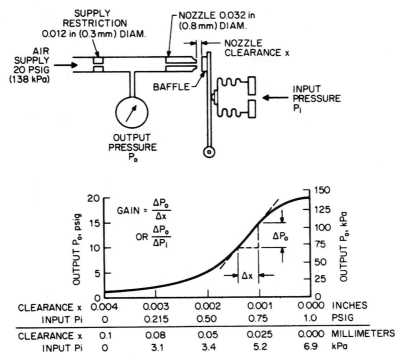

FIGURE 1   Baffle-nozzle amplifier, the primary detector used in pneumatic instruments.

1. The gain of a baffle nozzle, stated as the output pressure change per unit change in nozzle clearance, is high. In Fig. 1 a 3- to 15-psig (20- to 100-kPa gage) output requires only a 0.002-inch (0.05-mm) change in nozzle clearance. This gain is determined by the nozzle size and the ratio of the nozzle size to the restriction size. The gain can also be stated as the ratio of the output pressure change to the input pressure change. The pressure gain depends on the characteristics of the input element and can be made as high as necessary for a particular application. Increasing the baffle-nozzle gain improves the accuracy of the instrument in which it is used. Pressure gains as high as 1000 are used in pneumatic instruments.

2. The restriction and nozzle diameters shown in Fig. 1 are typical. These diameters are a compromise to meet different operating requirements. Small diameters increase the gain but they also increase the danger of clogging and the difficulty of aligning the baffle and the nozzle. Large diameters increase the continuous air consumption, which is equivalent to power consumption in an electrical device.

3. The output of a baffle-nozzle amplifier is nonlinear and subject to drift. Gradual accumulation of foreign matter in the supply restriction reduces the output for a given input and changes the gain. The output is also sensitive to changes in air supply pressure. For these reasons, a simple baffle-nozzle amplifier cannot be used as an accurate transmission instrument.

4. The output flow capacity of a baffle nozzle is low and is, therefore, not suitable for use as a transmission signal. All output flow is through the small supply restriction.

A baffle-nozzle amplifier, then, is a high-gain nonlinear drift-prone low-capacity element, which is not by itself useful as an accurate transmission device. It is equivalent to the active elements of electronic circuits—transistors and, in some respects, operational amplifiers. As with the electronic equivalents, negative-feedback techniques are used with a baffle-nozzle amplifier to design accurate linear instruments.

Baffle-Nozzle Controller

Despite its limitations, a baffle-nozzle amplifier is quite suitable for use as a simple proportional-only controller. Figure 2 shows a temperature controller with an adjustable gain. In pneumatic controllers the gain is usually described in terms of proportional band percentage, and in this kind of controller it may be adjustable to give a full-scale output change for 2 to 10 percent of the set-point adjustment span. Typical temperature set-point spans are from 100 to 1000°F (38 to 538°C). The action of this controller is reversible to suit the valve action (air-to-open or air-to-close).

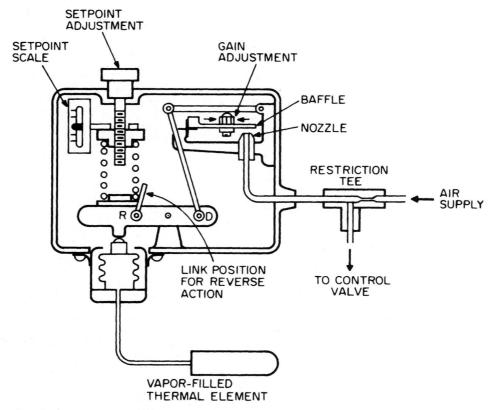

FIGURE 2   Baffle-nozzle temperature controller.

Normal baffle-nozzle drift errors are not serious in this kind of control because the controller acts over such a narrow temperature range; small output pressure changes represent very small temperature errors. Since this device is not intended for use as an accurate transmitter, linearity is not a primary concern. Controller output linearity, in any case, is not critical because of the larger nonlinearities in most control valve characteristics. Even the limited flow capacity of a baffle nozzle may be acceptable for short transmission lines and slow control loops. Where flow capacity is a problem, an inexpensive volume booster can be used.

Pilot Relays

These relays, also called pilot valves or booster relays, are used with baffle-nozzle amplifiers to provide a good flow capacity for transmission service. In addition to increased flow capacity, pilot relays usually provide some pressure gain. The pressure gain serves to increase the gain of the baffle-nozzle system for improvement in instrument accuracy. Additional gain permits the baffle nozzle to operate over a narrow more linear portion of the pressure characteristic and, by decreasing baffle-nozzle pressure, reduces air consumption.

In all the pilot relays shown in Fig. 3, an increase in nozzle pressure acts on the input diaphragm to push the valve plunger down, opening the supply port to increase the output pressure. A decrease in nozzle pressure causes the diaphragm to rise, opening the exhaust port to decrease the output pressure.

In the high-gain relay shown in Fig. 3a, for any output less than the supply pressure, both the supply and the exhaust ports must be open, and there is a continuous flow of air to the atmosphere. This kind of pilot is also called a bleed-type relay. Maximum steady-state air consumption occurs when the output is about half the supply pressure. It can be seen that increasing the size of the supply and exhaust ports to increase the flow capacity also increases air consumption. Thus a high-gain relay with a flow capacity of 2 std ft³/min (56.4 L/min) may have a steady-state air consumption of 1 std ft³/min (28.2 L/min) at midscale output. A typical pressure gain for this kind of relay is 10, and the output-input linearity is moderate. The gain is determined primarily by the stiffness of the input diaphragm and spring and by the travel of the plunger between the supply and exhaust seats. Stiffer elements and longer travel decrease the gain. Since a pilot relay is usually enclosed in a feedback circuit with a baffle nozzle, the exact gain and linearity are not critical. The spring below the input diaphragm determines the nozzle pressure required for a particular output and is used to fix the operating range of the

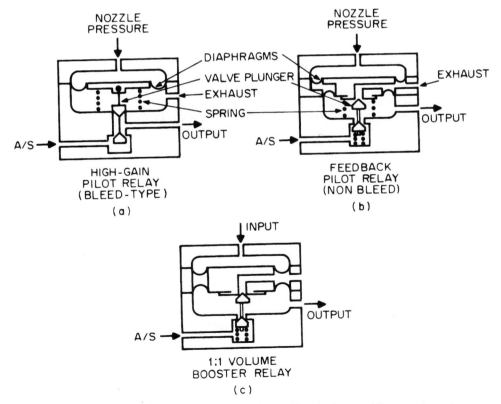

FIGURE 3 Pilot relays. (*a*), (*b*) Used with baffle-nozzle amplifiers for increased flow capacity and pressure gain. (*c*) Used as a volume booster on transmission lines. A/S—air supply.

baffle nozzle on a more linear part of its characteristic curve. With a high-gain relay, very little change in nozzle pressure is required to provide a 3- to 15-psi (20- to 100-kPa) output.

The feedback pilot relay shown in Fig. 3*b* differs from a high-gain relay because an output feedback diaphragm is used and the supply and exhaust ports can be closed at the same time. The gain of the relay is determined by the ratio of the input to the output diaphragm area. Gains of from 2 to 6 are commonly used. Higher gains can be obtained by reducing the area of the output diaphragm but are not often used because of problems in maintaining sensitivity and adequate valve plunger travel with smaller diaphragms. With a diaphragm ratio of 2, a 1-unit increase in nozzle pressure opens the supply port until the output increases to 2 units to close it. Similarly, a 1-unit decrease in nozzle pressure opens the exhaust port until the output decreases 2 units. In effect, the feedback diaphragm is used to close the supply and exhaust ports when the output reaches the correct value. Theoretically there is no flow in the steady state, and this is called a nonbleed relay. Most manufacturers design a small bleed into the supply or exhaust port so that the other port is always slightly open. This prevents a dead spot or hysteresis in responding to small input changes. Actual air consumption is on the order of 0.1 std ft^3/min (2.8 L/min) for a typical feedback relay.

Another result of using output feedback is that the output-input pressure relationship is linear. This is not important when used in a baffle-nozzle circuit but means that this kind of relay can be used in other applications as a linear amplifier. As with a high-gain relay, a bias spring is used to determine the optimum nozzle pressure range for the 3- to 15-psig (20- to 100-kPa) output. This spring can be made adjustable to give specific output-input relationships when the relay is used as a linear amplifier.

The 1:1 booster relay shown in Fig. 3*c* is not often used in baffle-nozzle circuits, but is shown here as another example of a feedback relay. A booster relay provides increased flow capacity, but no pressure gain. The relay is used with low-capacity transmitters and for small-volume termination on transmission lines.

Pressure Divider Circuit

In pneumatic instruments, a pressure divider circuit is used for simple gain adjustments. A graduated needle valve is used to adjust the output as a fraction of the input. In Fig. 4*a*, closing the needle valve gives a 0 gain, and opening the needle valve fully gives a gain of nearly 1. The electrical equivalent of a pressure divider is sometimes used for analysis, but the analogy is not exact. Because airflow is not a linear function of differential pressure, the output of a pressure divider is nonlinear. The nonlinearity is not severe [with a nominal gain of 0.5, the nonlinearity for a 3- to 15-psig (20- to 100-kPa gage) input range is ~4 percent], but a pressure divider is normally not used in transmitter circuits where better linearity and drift stability are required. A pressure divider is employed primarily in the gain adjustment circuit of a controller. Note that a pressure divider is a low-capacity device; any flow from the output causes a drop in output pressure. Another kind of pressure divider in which both the input and vent restrictions are adjusted is shown in Fig. 4*b*.

Feedback Amplifier

Force Balance. If the baffle nozzle is enclosed in a feedback circuit, the problems associated with nonlinearity and drift are eliminated. This is the principle on which most pneumatic transmitters and controllers operate (Fig. 5). This device is called a force-balance amplifier because the baffle nozzle changes the pressure in the feedback bellows to balance the force generated by the input, whether the input force is generated by another bellows or by any other kind of input element. An increase in input force reduces the nozzle clearance and increases the nozzle pressure and the output of the pilot relay until the feedback pressure, which is also the output, balances the input force. The output pressure, being then proportional to the input, can be used as an accurate transmission signal for measurement and control.

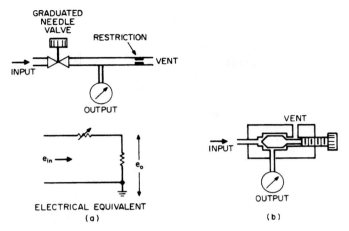

FIGURE 4 Pressure divider circuit used for gain adjustments in pneumatic instruments. (*a*) Graduated needle valve, used to adjust output as a fraction of input; also, electrical equivalent. (*b*) Pressure divider in which both input and vent restrictions are adjusted.

Motion Balance. Some input elements produce a displacement rather than a change in force, and a motion- or displacement-balance amplifier is used to convert the displacement to a proportional pneumatic pressure. An advantage of this arrangement is that the input displacement also can be used for direct mechanical indication of the measurement.

Increasing pressure in the bourdon tube (Fig. 6) raises the left end of the baffle lever and reduces the nozzle clearance. Increasing the nozzle pressure through the pilot valve increases the feedback pressure to reposition the nozzle lever. Since the nozzle clearance changes only a few thousandths of an inch (fractions of a millimeter) over the full output range, the baffle lever and the nozzle lever remain parallel. The feedback bellows displacement and output pressure are, therefore, proportional to the input displacement. In effect, the nozzle is moved to follow the input motion, and the pressure change required to reposition the nozzle is used to measure the input motion. Use of the feedback circuit results in a linear stable output signal, as in a force-balance amplifier.

The force required to reposition the motion-balance mechanism is derived primarily from the output pressure. Little force is required from the input element, so very sensitive input elements can be used without loss of accuracy. The input displacement for a motion-balance amplifier ranges from 0.02 to 3.00 inches (0.5 mm to 7.6 cm). With different input elements, motion-balance amplifiers are used in pressure, temperature, and level transmitters, for direct dimensional measurement, and in some kinds of pneumatic controllers.

PNEUMATIC CONTROL ACTIONS

Proportional Action

The proportional controller shown in Fig. 7 is a force-balance feedback amplifier with a differential-input bellows for the measurement and the set point. The resultant input to the controller is the difference between the measurement and the set point, called the error or deviation. The gain of the controller is changed by adjusting the pivot position. For a high gain, the pivot is at the right-hand end of the lever, so that a large change in output is required to balance a small change in input.

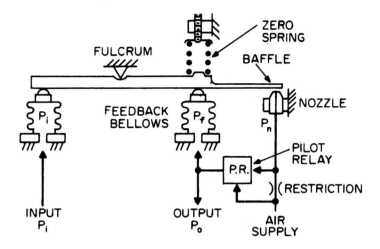

AMPLIFIER BLOCK DIAGRAM

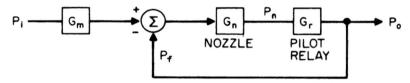

SIMPLIFIED BLOCK DIAGRAM

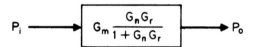

FIGURE 5   Force-balance feedback amplifier. Block diagram illustrates operation of a feedback amplifier schematically. Negative sign in the summing junction shows that an increase in feedback bellows pressure causes a decrease in nozzle pressure and that this is a negative-feedback device. G_m—mechanical gain of amplifier; G_n—nozzle gain.

In most pneumatic controllers the gain adjustment is graduated in proportional band units. The proportional band is the percent input change that will produce a 100 percent [usually 3 to 15 psi (20 to 100 kPa)] output change. The proportional band can be viewed as a band, on the indicator or recorder, through which a measurement must change to produce a full-scale valve position change. It should be noted that a receiver controller has no inherent pressure range. The pressure range of the system is determined by the calibration of the valve and the transmitter. A receiver controller can be used for any pressure range up to the pressure limits of the input and output elements and can, in fact, be used for different input and output ranges.

As shown in Fig. 7, the controller is direct acting, that is, the output increases as the measurement increases. The reversing switch permits interchanging of the position of the measurement and set-point bellows to achieve reverse action. The controller action must be set for each control loop so that an increase in the measurement changes the controller output in the direction required to decrease the measurement—as is required in any negative-

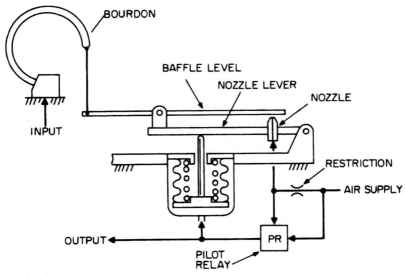

FIGURE 6   Motion-balance feedback amplifier.

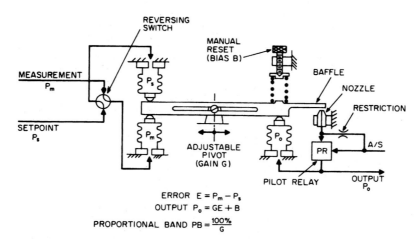

ERROR $E = P_m - P_s$

OUTPUT $P_o = GE + B$

PROPORTIONAL BAND $PB = \dfrac{100\%}{G}$

FIGURE 7   Force-balance proportional controller. A/S—air supply.

feedback loop. Because of operating considerations in selecting the valve action (most control valves are air-to-open), most process control loops are set for reverse action.

The manual reset or bias adjustment determines the controller output when the measurement is at the set point (error = 0). This adjustment is used to set the valve opening required to bring the measurement to the set point. With proportional control, a load disturbance (after the manual reset has been set) will cause an offset between the set point, that is, a different valve opening is then required to bring the measurement to the set point. If the offset is large enough to be objectionable, the manual reset must be readjusted for the new load condition. The offset can be reduced by reducing the proportional band (increasing the controller gain), but the stability of the control loop determines the minimum proportional band that can be used. A proportional band too narrow for the loop will cause control oscillation.

Proportional-plus-Reset Action

Automatic reset action causes the controller output to change as long as there is an error between the measurement and the set point. This eliminates the offset, which is characteristic of simple proportional control. Proportional-plus-reset action is also called proportional-plus-integral (PI) action.

In a proportional-plus-reset controller, as shown in Fig. 8, the output is connected to an opposing reset bellows through a needle valve and volume. From the output pressure equation for a proportional controller, it can be seen that the output is equal to the proportional output GE plus the pressure P_r in the reset volume. How this arrangement causes a continuous output change can be seen by considering the differential pressure across the reset needle valve. From the output pressure equation, the needle valve differential is equal to the product of the gain and the error GE. Therefore, as long as there is an error, there is airflow into or out of the reset volume, which changes the reset pressure P_r and the controller output. The controller output continues to change until the error returns to zero. At zero error, the controller output is equal to the reset pressure P_r.

The pressure in the reset volume is proportional to the integral of the airflow into the volume. If the airflow is, in turn, proportional to the differential pressure (which is approximately so for small differentials and for the small flows in the controller), then the reset pressure is proportional to the integral of GE. The response of the proportional-plus-reset controller can then be expressed in the equation showing integration of the error.

Mechanical Gain Adjustment

In Fig. 8 the controller gain or proportional band is changed by repositioning the pivot. Another design for changing the gain mechanically is shown in Fig. 9. In this design a floating disk acts as the balance beam and as a nozzle baffle. Rotating the adjusting lever changes the position of the fulcrum rollers for a 5 to 500 percent proportional band adjustment. This

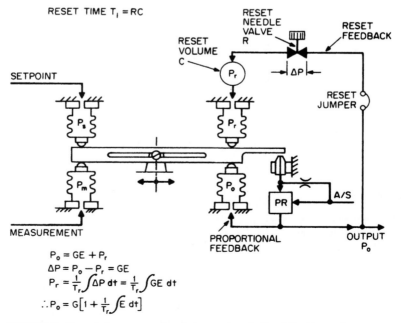

$$P_o = GE + P_r$$
$$\Delta P = P_o - P_r = GE$$
$$P_r = \frac{1}{T_r}\int \Delta P\, dt = \frac{1}{T_r}\int GE\, dt$$
$$\therefore P_o = G\left[1 + \frac{1}{T_r}\int E\, dt\right]$$

FIGURE 8   Force-balance proportional-plus-reset controller. A/S—air supply.

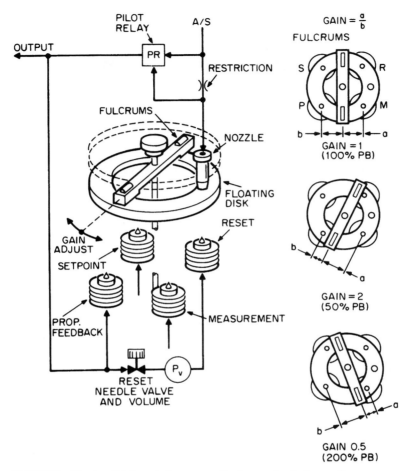

FIGURE 9 Floating disk force-balance controller (proportional-plus-reset).

arrangement permits compact design of the controller and gives smooth adjustment of the proportional band while the controller is in operation. Although mechanical details differ from those shown in Fig. 8, the principles of operation are the same.

Pneumatic Gain Adjustment—Stack Controller

A proportional-plus-reset controller in which the proportional band adjustment is made with a needle valve is shown in Fig. 10. Input elements are elastomer diaphragms separated by metal rings. This arrangement is referred to as a diaphragm stack controller.

In the error detector section, the area of the large diaphragms and the area of the smaller center diaphragm are in a ratio of 5:4. The measurement pressure, acting downward on the large diaphragm and upward on the small diaphragm, produces a resultant force downward, which is balanced by the output. Because the output acts on the full diaphragm area and the measurement acts only on the difference in areas, a change of 5 psi (34.5 kPa) in the measurement can be balanced by a change of 1 psi (6.9 kPa) in the output. Equation (1) in Fig. 10

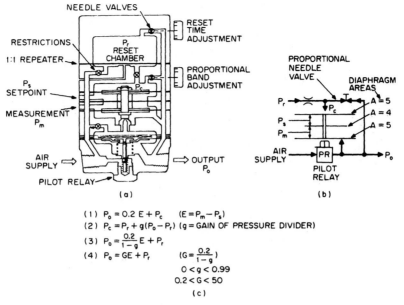

(1) $P_o = 0.2 E + P_c$ $(E = P_m - P_s)$
(2) $P_c = P_r + g(P_o - P_r)$ $(g = \text{GAIN OF PRESSURE DIVIDER})$
(3) $P_o = \dfrac{0.2}{1-g} E + P_r$
(4) $P_o = GE + P_r$ $(G = \dfrac{0.2}{1-g})$

$$0 < g < 0.99$$
$$0.2 < G < 50$$

(c)

FIGURE 10 Diaphragm stack controller with pneumatic gain adjustments. (*a*) Sectional view. (*b*) Schematic of proportional section. (*c*) Relationship.

shows the output required to balance the two forces produced by the other three pressures acting on the different diaphragm areas.

Gain adjustment is achieved by feedback of the output through a pressure divider. The gain *g* of the pressure divider is adjustable from 0 to 0.99 by means of the needle valve. This is a positive feedback and acts to increase the overall gain *G* of the controller. Substitution of the expression for P_c—the output of the pressure divider—into Eq. (1) and rearranging gives Eq. (3). This shows the controller output as a function of the pressure divider gain *g*. As *g* is adjusted from 0 to 0.99, the controller gain changes from 0.2 to 20 and the proportional band from 500 to 5 percent. With the proportional needle valve closed, there is no positive feedback. The gain is 0.2, as determined by the ratio of the diaphragm areas.

Proportional-plus-Derivative Action

Derivative action adds to the proportional response a component that is proportional to the rate of change in input. The derivative time T_d is usually defined as the time separating the proportional and the proportional-plus-derivative responses—with a straight-line ramp input, after transients have subsided. This is the time with which the derivative adjustment is graduated. A proportional-plus-derivative response is often described simply as a derivative action, although, in fact, a derivative response without the proportional component is seldom used in process control. Derivative action is also called rate action or preact.

One method of generating derivative action is by restricting the feedback pressure P_f in a force-balance amplifier, as shown in Fig. 11. The amplifier is designed so that $P_f = P_i$ when the amplifier is in balance. With a ramp input P_i the output must change to maintain the same pressure ramp in P_f. Since P_f is the output of a first-order lag, output P_o must lead the input by the lag time constant, as indicated in Fig. 11. The derivative component $P_o - P_i$ is then equal to the slope or derivative of P_i multiplied by the derivative time.

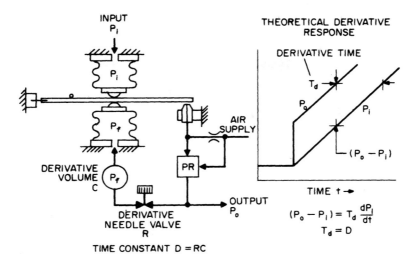

FIGURE 11 Proportional-plus-derivative amplifier showing a theoretical derivative response with infinite nozzle gain.

The derivative gain is limited by providing a direct proportional feedback from the output, as shown in Fig. 12. A change in input is balanced immediately by a change in the small proportional feedback bellows. Since P_c is the first-order lag response of the output, the pressure equation can be represented by the feedback block diagram, which can be simplified to show the proportional-plus-derivative transfer function. The derivative gain is determined by the ratio of the bellows areas. The step response graph shows clearly the effect of a limited gain.

A diaphragm stack proportional-plus-derivative unit is shown in Fig. 13. The pressure equation is the same as that shown in Fig. 12, and the output response is the same. This device is available as a separate unit for adding the derivative response to an existing controller or for dynamic compensation in feedforward systems. It is more commonly used as an integral component in a proportional-plus-reset-plus-derivative controller.

Direct-Connected Controllers

The direct-connected temperature controller shown as an example in Fig. 14 is a motion-balance instrument with proportional-plus-reset action. An increase in temperature in the gas-filled thermal element increases pressure in the bourdon tube, raising the left end of the flapper beam and increasing the nozzle pressure. Increasing the nozzle pressure through the pilot valve increases the output. The output is fed back to the proportional bellows, which lowers the left end of the flapper beam to restore nozzle clearance. Pneumatic gain adjustment is by a pressure divider, which feeds back some fraction of the output change to the proportional bellows. The proportional gain is increased by reducing the feedback to the proportional bellows so that a larger change in output is required to reposition the flapper beam. Reset action is generated, as shown in Fig. 8, by connecting the proportional feedback to an opposing bellows through a first-order lag.

The controller shown is direct acting. Action can be reversed by relocating the nozzle as shown and by interchanging the positions of the proportional-plus-reset bellows. The set point is adjusted by changing the position of the nozzle with the cam supporting the nozzle beam.

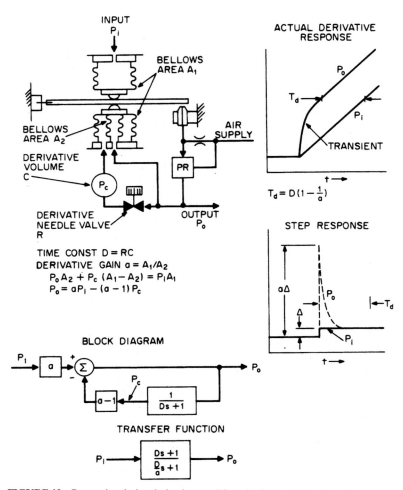

FIGURE 12   Proportional-plus-derivative amplifier with limited derivative gain.

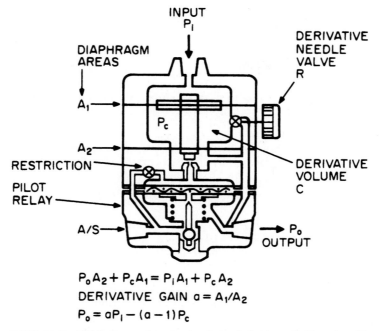

$$P_o A_2 + P_c A_1 = P_i A_1 + P_c A_2$$
$$\text{DERIVATIVE GAIN } a = A_1/A_2$$
$$P_o = aP_i - (a-1)P_c$$

FIGURE 13   Diaphragm stack proportional-plus-derivative unit. Note that this device has the same pressure equation as that given in Fig. 12.

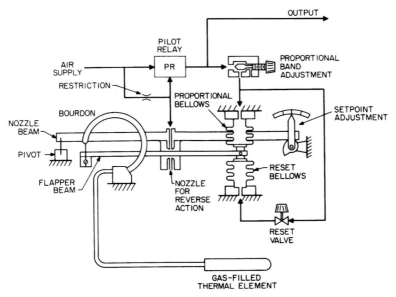

FIGURE 14   Direct-connected temperature controller.

BATCH PROCESS CONTROL

During the last few years much attention has been directed toward a better understanding of the dynamics of batch processes in an effort to achieve greater automation by applying advanced control knowledge from experience with continuous process controls and computers. This has proved to be more difficult and to require more time than had been contemplated originally. This study is far from complete, as witnessed by the numerous papers and articles in contemporary literature. However, the Instrument Society of America (ISA) plans to release a tentative "batch control model" sometime in 1993.

BATCH MODE

Prior to the late 1930s practically all chemical and petroleum production processes essentially were of the batch mode. When crude oil, for example, was first cracked into lighter hydrocarbon fractions, thermal cracking was of the batch mode. Where inherently suited (chemically or physically) and where a large and continuous market demand for a single product existed, manufacturers found that continuous production warranted the costs of scaling up equipment and integrating the flow to and from various unit operations in a continuous production system. Because these prerequisites do not always exist, or the return on investment for going "continuous" may be insufficient, the batch mode continues to be quite extensive. In addition, the batch mode has certain innate values, including multiple use of equipment and greater safety with hazardous products because of small runs.

Because of tremendous variations in raw materials, final products, and operational equipment used, it is difficult to characterize a "typical" batch process. The opinions of a few experts are along these lines:

1. Every process is a batch process; the difference is how long the process runs between startup and shutdown. (While this may be true theoretically, it is indeed an oversimplification.)

2. A batch process is discontinuous, in which ingredients are sequentially prepared, mixed, reacted, cooked, heated, or cooled, finished, and packed—all in accordance with a time-sequenced schedule worked out by a process engineer.

3. A batch process seldom is purely batch; a continuous process seldom is purely continuous. Most batch processes require continuous control over one or more variables throughout the entire schedule, and thus the overall batch control system must be a form of hybrid. Many products made continuously will, at the end of the line, require batch packaging, inspection, handling, storing, and shipping.

Although precise definitions are difficult, some generalities may be stated. A batch process usually is completed within a matter of minutes to a few hours, as contrasted with a few seconds for what may be termed a discrete operation or weeks or months for a continuous process from start-up to shutdown. However, in the latter case, for any given molecule of substance entering the process, the interval of time within the process (residence time or throughput rate) before it exits as part of a finished product usually can be measured in seconds to very few minutes.

Simplistic Model.   For making soup prior to canning, the steps may be as follows:

1. Fill a kettle with a given volume of water.
2. Add ingredients A, B, and C.
3. Turn on the mixer and raise the kettle temperature to a near-boil and maintain for 1 hour.

4. Allow kettle and ingredients to cool (*a*) for a programmed length of time or (*b*) until a given temperature is reached—whichever criterion is shorter.

5. Add ingredient D.

6. Stir for 10 minutes.

7. Open dump valve to transfer contents.

8. Close dump valve (*a*) after programmed 3 minutes or (*b*) when kettle level indicates zero—whichever criterion is shorter.

9. Start rinse cycle, turning on water spray and actuating stirrer for 4 minutes. (A second rinse might be scheduled.)

10. Close dump valve.

11. Indicate "Ready for Next Batch" on display.

If self-diagnostics for the system were incorporated, all conditions would be checked for readiness for the next batch. In the foregoing, note the combined need for continuous and discrete control.

This description minimizes the number of actions that usually must be taken and the time-sequencing required for more complex processes, which may involve many more ingredients, intricate time relationships, and measurement of several additional variables that require continuous or discontinuous control, safety interlocks, and, in modern systems, the tabulation of historical and quality control parameters.

Packaged or customized batch control systems are available today that will handle up to hundreds of steps for a single recipe and memory for storing up to several thousand recipes that may use the same group of batching operations. On an average, however, a control system will need to accommodate only a few hundred recipes. Table 1 lists some characteristics of batch operations versus discrete and continuous operations.

BATCHING NOMENCLATURE

Recipe.   A list of ingredients, amounts, or proportions of each ingredient. Depending on the product, the recipe may be created by a chemist or other specialist as the result of formulating the desired product in a laboratory (possibly pilot plant) environment. The chemist specifies the types of equipment that will be used. Each type of equipment is referred to as a unit. A multipurpose, multiproduct batching system will have the capability of producing numerous products and using several pieces of equipment. Also, from historical records of equipment (units) available and of other recipes in storage, in modern procedures the chemist will configure the batch sequence interactively using a sequential function chart (SFC) editor to select the necessary unit sequence from among a group, while simultaneously entering parameters for that sequence in accordance with prompt messages. As pointed out by Adams [1], new recipes can be prepared easily by repeating this series of operations.

Grade.   A variation in a recipe or formula, usually achieved through the use of different product data parameters such as temperature, times, and amounts. This may constitute a new recipe.

Unit.   A vessel and its associated process equipment that acts independently of other process units.

Operation.   A time- and event-based sequence that defines the actions of a process unit in performing a process function. A typical operation may be the sequence to charge, heat, or dump a given unit.

Procedure.   A sequence of unit operations with specific product data that constitutes the batch cycle; also commonly referred to as a recipe or formula.

TABLE 1 Process Types versus Process Characteristics*

Process type	Duration of process	Size of lot or run	Labor content	Process efficiency	Control type preference	Input-output content	
						Discrete	Analog
Discrete operations	Seconds	†	Medium to high	Low to medium	PLC	95%	5%
Batch operations	Minutes to hours	Small to medium	Medium	Medium	Various‡	60%	40%
Continuous operations	Weeks to months	Usually very large	Small	High	DCS	5%	95%

* Adapted from a chart originally presented by W. J. Loner, Allen-Bradley Co., Inc.

† Discrete operations to a large extent are encountered mainly in the discrete-piece manufacturing industries for the control of machines, conveyors, and so on. However, there are numerous discrete operations in process batching systems.

‡ In 1991 a survey indicated that programmable logic controllers (PLCs) were preferred by nearly 30% of users; approximately 45% preferred distributed control systems (DCSs); the remainder were various hybrid systems, some involving personal computers (PCs).

Discrete Control Device.   An on-off device, such as a solenoid valve, pump, or motor.

Regulatory Control.   A general term encompassing PID (proportional plus integral plus derivative) loops and continuous control functionality.

Sequenced Batch.   The most basic type of batching control. This can become quite complex because of the large number of operations in some cases. It may be considered as logic planning and incorporates all on-off functions and control of other discrete devices. In the pure form, sequence control involves no process feedback or regulatory control. It is quite amenable to ladder-logic-based programmable logic controllers (PLCs).

Programmed Batch.   Elements of sequenced batch plus the addition of some regulatory control. This system requires little operator intervention. Sophisticated regulatory controls (beyond PID) seldom are needed. The system is used frequently where a reactor or other batching equipment assembly is dedicated to a single product (or very few products) with limited process variations. The system also is amenable to PLCs in most cases. However, if there are several continuous control operations in proportion to the number of discrete operations, a hybrid, host, or distributed batch control system may be indicated.

High-Level Batch.   Elements of programmed batching with external and process feedback. Optimization routines may be used as, for example, in the production of polyvinyl chloride (PVC) and other polymeric emulsions.

DEFINING AND OUTLINING THE BATCH PROCESS

Since the late 1980s considerable software has become available for assisting in creating batch control programs. ISA standard SP88 is geared toward a high degree of advanced control and will be described briefly later in this article. For those readers who may not be fully versed in these more advanced concepts, it may be in order to include the following "plain language" description.[1]

Implementing a batch control system can be very complex and difficult without the proper definition. It is very important to determine the operation of individual process units and the associated equipment that makes the unit work as an entity. Careful consideration must also be given to a unit's interaction with other units and all common resources that may be used concurrently by multiple units. Contention for common resources by multiple units, without the proper mechanism for ensuring their use, can lead to costly coordination problems when operating the process automatically from a procedure level.

The definition phase of a batch control process can be difficult. Ensuring the proper equipment states for any possible combination of process conditions generally requires multiple iterations until a final control strategy is reached. This usually continues well into the process start-up and must be updated periodically during plant operation, as process optimization takes place and the overall strategies change. Therefore a structured approach to defining and implementing batch control strategies is offered in the batch formalism (Fig. 1).

The batch formalism described here lays out a set of guidelines by which a process designer may follow in the route to the final batch control strategy. Each step of the formalism will build upon the previous step, forcing the designer to complete each step before proceeding to the next.

Step 1.   The starting point is that of defining the control elements from the piping and instrumentation drawings (P&IDs). Using the P&ID, an equipment tag list can be constructed, including various regulatory and discrete devices.

Step 2.   Following the identification of the control elements, the individual strategies must be defined. Regulatory control strategies may be specified, using ISA symbology, or, for

---

[1] From information furnished by W. C. Thompson, Fisher Controls International, Inc., Austin, Texas.

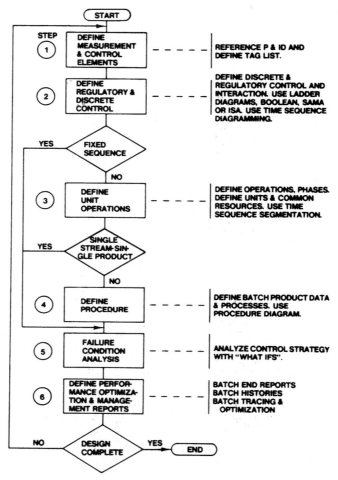

FIGURE 1   Batch formalism.

complex interactive loops, the Scientific Apparatus Makers Association (SAMA) standard RC22-11 may be useful. Due to the nature of batch control processes, attention should be given to loop mode changes, alarm status changes, and adaptive tuning requirements.

Traditionally the interlocking and sequencing of discrete control devices has been centered around ladder diagrams or in boolean logic form. These forms of diagraming, however accustomed designers are to them, are hard to follow and quite cumbersome when dealing with complex batch strategies. Time-sequence diagrams provide for a more organized time and event sequence and are considerably better for diagraming batch control processes.

While excellent for circuit wiring and fixed-sequence strategies, ladder logic does not indicate the time relationship required with most batch strategies. Time-sequence diagraming is essential in batch control due to the variable sequences, recipe changes, and common resource contentions, as well as the high degree of interaction between discrete and regulatory controls. This type of diagram is typified by Fig. 2. The symbology is shown later in Figs. 6 and 7.

Additional considerations in Step 2 of the batch formalism are more complex strategies, such as split ranging, weighing and metering, and changing control strategies depending on the transient nature of the process.

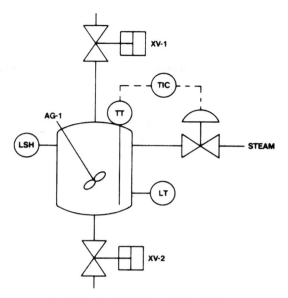

REACTOR AND ASSOCIATED INSTRUMENTATION

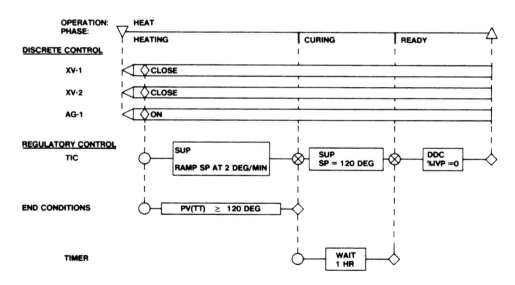

TIME SEQUENCE DIAGRAM

FIGURE 2   Representative reactor unit and associate time-sequence diagram.

If at this point in the procedure it is determined that the process sequence is fixed, the designer may proceed to Step 5, where the failure sequences based on analysis of the process conditions upon control element failure may be defined. However, if there is more than one time sequence for a process unit and the order of those different time sequences may change from batch to batch, the designer should consider unit operations.

Step 3.   Individual sets of time and event sequences are called operations. The actions within an operation include sequencing, interlocking, profiling, failure monitoring with emergency shutdown, calculations, integrators and timers, and parallel operations in addition to the discrete and regulatory functions. In addition the operation provides for manual operator entry and convenient entry and reentry points, all for coordinated control of the total unit.

In some operations, such as material transfers, unit equipment boundaries may overlap. In such cases, transfer headers may actually belong to different units at different times. This type of equipment is referred to as a common resource and must be coordinated between operations.

Figure 3 shows fill, heat, and dump operations for the unit shown in Fig. 2. It should be noted that the operations are subdivided into phases for satisfying operator interface and branching requirements.

Once the definition of the individual time sequences has been completed, the designer should define the operation or phase relationship and segment the process equipment by unit. Since this definition phase is highly application-dependent, defining the units and operation or phases is more easily done after the time-sequence diagrams are completed. Note that operations and phases are generally triggered at safe process conditions. When the designer is satisfied that the time sequences and units or operations are complete, the procedures should be defined.

Step 4: Procedure Definition.   As stated earlier, the procedure consists of a sequence of operations and sets of product parameters, because most batch processes are characterized by a series of small batches flowing sequentially through multiple units. Each batch follows some path, using some or all of the units or operations along that path. By treating each operation as a single control action in the procedure, a time-sequence-type diagram can be designed for the procedure (Fig. 4).

As with the operation or phase relationship, a procedure or process relationship is also established providing for safe hold points and better procedure definition for operator interface.

Just as common resources are contended for within operations, procedures may contend for the services of each unit in a multistream situation. The unit used in making the batch must be kept track of, so that cross contamination and improper mixes do not take place.

During procedure definition the designer should list all product parameters (values with engineering units), such as times, temperatures, ingredient amounts, and flow rates that are unique to the batch products and their various grades.

Step 5: Defining Failure Conditions.   After the first four steps of the formalism are complete, the designer should pass through the hierarchy once again to establish an analysis of the failure conditions of the process. Time sequences should then be defined for the application-dependent failure steps. These failure sequences should be incorporated by operation and should allow for reentry into the operation at a position that does not jeopardize personnel, equipment, or product.

Step 6: Optimization and Reports.   Finally, once all phases of the definition are complete, optimization of the performance of equipment and data measurement should be done and the determination of end report and batch history data completed. The format of batch end data should also be considered for printed reports.

The batch strategy is now complete and ready for implementation. Note that multiple iterations of this procedure may be necessary before implementation can be started.

SYSTEM SELECTION FACTORS

The actual implementation of the completed batch strategy is both application- and control-system-dependent. The application and user preference usually dictate the architecture and

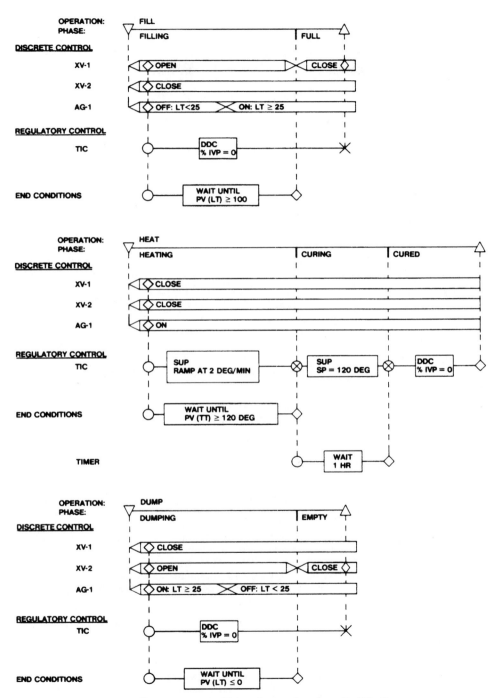

FIGURE 3 Time-sequence diagrams segmented into unit operations for unit of Fig. 2.

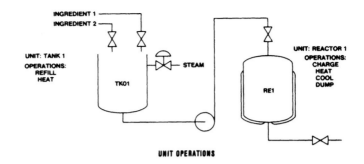

UNIT OPERATIONS

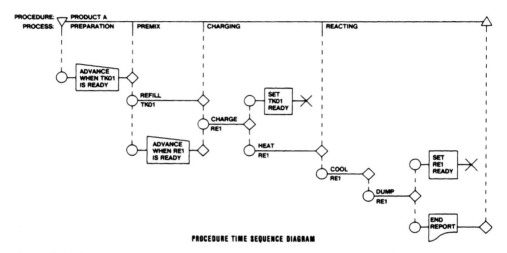

PROCEDURE TIME SEQUENCE DIAGRAM

FIGURE 4   Time-sequence diagram for batch procedure.

selection of the control equipment. There are several different control system architectures available. One of the most popular utilizes distributed devices on local area networks, or highways. As stated previously, batch control packages are still available commercially. However, redundancy becomes an expensive issue compared with the selective redundancy of a distributed system. In addition, the total installed cost of centralized computers makes them less cost-effective due to wiring costs and the programming efforts required with most centralized systems. Truly distributed control systems with advanced batch control software have been commercially available since the early 1980s. They incorporate sophisticated batch languages with hierarchical control that allow for easier implementation of unit operations, failure sequences, recipe level control, and effective operator interface. Depending on the process requirements, PCs, continuous controllers, and process management computers can easily be integrated with the batch (unit operations) controllers and consoles (for operator interface) to provide fully integrated plantwide control systems (Fig. 5). Subject to further confirmation, the symbology is shown in Figs. 6 and 7.

The system must meet certain minimum requirements.

1. The controllers must be able to handle the designer's control strategies. It is imperative in advanced batch control systems that the controller handle multiple unit operations for a number of units. Techniques such as tag aliasing (generating generic process tag names for use in unit operations) can greatly reduce configuration or programming time of the controllers.

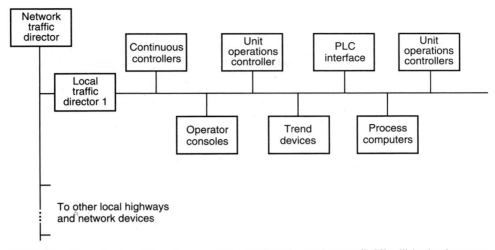

FIGURE 5   Generalized depiction of representative distributed control system (DCS) utilizing local area networks (LANs).

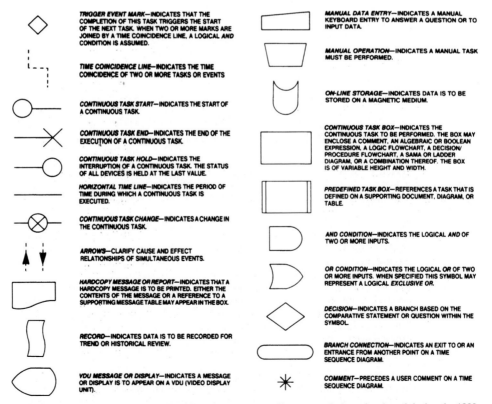

FIGURE 6   Subject to further confirmation, time-sequence diagram symbology developed during the 1980s.

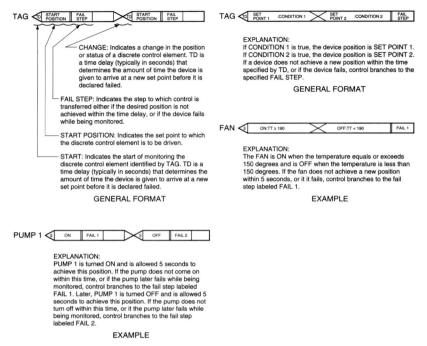

FIGURE 7   Time-sequence diagram symbology for discrete device control. Single condition shown at left, multiple conditions at right.

2. A clean operator interface with interactive color graphics is essential. Although continuous or steady-state control can be handled from panel-mounted operator stations or consoles with preformatted graphics, batch control generally requires custom graphic capability due to the complex interactions between units, loops, discrete devices, and the like.

3. Communications paths between controllers and consoles as well as peer-to-peer controller communication are usually essential. Because of the risk-distribution philosophy behind distributed control systems, most batch processes utilize more than one controller. It is generally required that most systems provide for standard interfaces to programmable controllers, continuous controllers, historical trend devices, and process or plant management computers, as well as providing a fully integrated package.

Control languages are available in many forms. Digital centralized computers generally require some knowledge of Fortran, Pascal, or other comparable languages by the designer. However, well-designed distributed control systems have taken the direction of configurability rather than programmability, allowing the designer of the process to perform the implementation. By using a high-level engineering language with configurable expressions and mnemonics, the engineer can virtually configure the control system directly from the time-sequence diagrams. Consequently time-sequence diagraming, although it may increase the time for design phases, can substantially reduce the implementation phase of a project and reduce errors and redesign after start-up.

ISA SP88.   This document [2] features a batch management model (Fig. 8), a control activity model (Figs. 9 and 10), and control characterization (Fig. 11). Tentative sequence instruction steps are tabulated in Fig. 12. The new standard is intended to provide flexibility for the user while including a comprehensive set of tools to accomplish a simple or a complex sequence control solution.

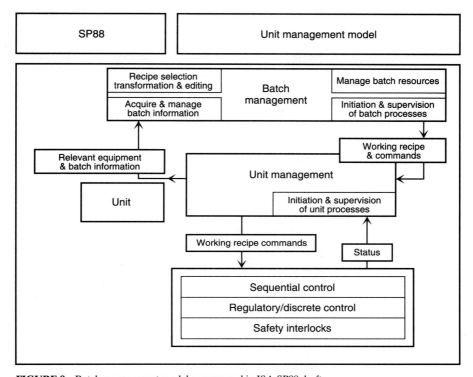

FIGURE 8 Batch management model as proposed in ISA SP88 draft.

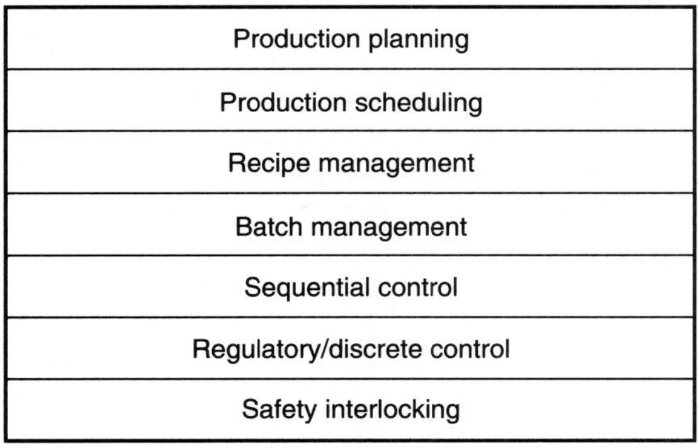

Production planning
Production scheduling
Recipe management
Batch management
Sequential control
Regulatory/discrete control
Safety interlocking

FIGURE 9 Control activity model.

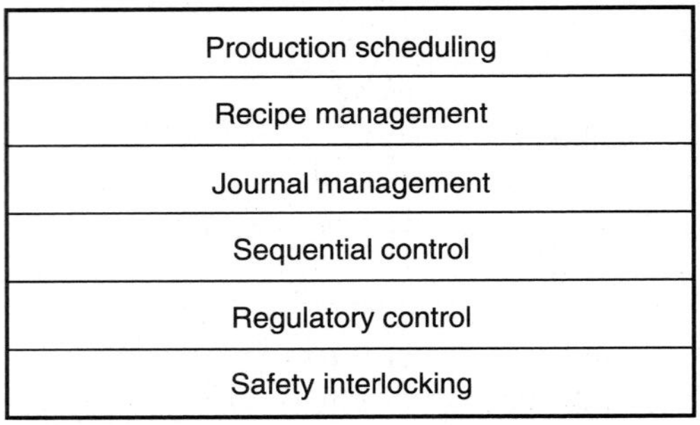

Production scheduling
Recipe management
Journal management
Sequential control
Regulatory control
Safety interlocking

FIGURE 10 Alternate control activity model.

Sequence controller model
Recipe selection transformation & editing
Recipe management
Production scheduling
Acquire & manage batch information
Initiation & supervision of batch processes
Relevant equipment & batch information
Working recipe & commands
Journal management
Sequential control
Working recipe commands
Status
Regulatory control
Safety interlocking

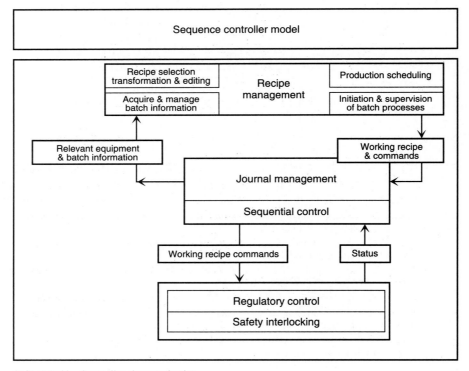

FIGURE 11 Controller characterization.

ACTIVATE	GO	RESTORE_RULES
BRANCH	HOLD	RESUME
CALL	IF	RULE
CAPTURE	INTERLOCK	SET_ABORT_KEY
CHAIN	LOG	SET_BIT
COMPARE	LOGIC	SET_HOLD_KEY
CONTINUE	MASK	STOP
CONTROL	MATH	SUSPEND_RULES
DEACTIVATE	MESSAGE	TREND
DELAY	ON_ABORT	UNLOCK
DEVICE	ON_ERROR	UNTIL
ELSE	OP_ENTRY	WAIT
END	PARALLEL	WHILE
END_WHILE	RAMP	COMMENT
FORMULA	REFER	DEFINE
FUNCTION	RELEASE	INCLUDE
GET_STATUS	REPEAT	NAME
	RESERVE	

FIGURE 12 Standardized names for sequence instruction steps.

REFERENCES

1. Adams, Arthur: "A Simplifying Batch Control and Recipe Management," *In-Tech,* January 1991, p. 41.
2. ISA SP88, "Batch Control Model," Instrument Society of America, final draft in process, 1993.
3. Wilkins, M. J.: "Simplify Batch Automation Projects," *Chem. Eng. Progress*, 61, April 1992.
4. Doerr, W. W. and R.T. Hessian, Jr.: "Control Toxic Emissions from Batch Operations," *Chem. Eng. Progress*, 57, September 1991.
5. Blankenstein, L. S.: "Batch Tracking: A Key for Process Validation," *Chem. Eng. Progress*, 63, August 1991.
6. Concordia, J. J.: "Batch Catalytic Gas/Liquid Reactions: Types and Performance Characteristics," *Chem. Eng. Progress*, 50, March 1990.
7. Musier, R. F. H.: "Batch Process Management," *Chem. Eng. Progress*, 66, June 1990.
8. Fisher, T. G.: "Batch Control Systems: Design, Application, and Instrumentation," Instrument Society of America, Research Triangle Park, North Carolina, 1990.
9. Staff: "Batch Control Service" (Videotapes), Instrument Society of America, Research Triangle Park, North Carolina, 1990.

AUTOMATIC BLENDING SYSTEMS

by Stan Stoddard*

The need to blend various ingredients in pots or vats dates back to antiquity. Over the years, various means of combining liquid or powder components in preprogrammed sequences and amounts were devised, including bucket trains, sprocket gears, water shells, and variable-speed gearing techniques. Later came pneumatic controllers, which could control the flow rate of a particular component. Proportioning techniques were developed that utilized mechanical and pneumatic devices. Electronics introduced analog amplifier comparators coupled with accurate metering devices and preset electromechanical counters for multiple-component flow control and metering. Further developments brought forth digital techniques with integrated circuits that performed precise measurement, comparisons, and control of multistream processes. Present microcomputer technologies and sophisticated programmable controllers find wide application in a variety of industries for bringing together multiple components in a precisely controlled manner. Applications are found in the petroleum, petrochemical, food and beverage, building materials, pharmaceutical, automotive, and chemical fields among others.

BATCH VERSUS CONTINUOUS BLENDING

In a batch-type process, a recipe is followed by adding specific amounts of ingredients in a predefined sequence with mixing, stirring, and brewing, or other processing times between the addition of each component. This practice, for example, was followed for many years in the manufacture of lubricating oils through programmed mixing, stirring, and heating of various hydrocarbon components and additives in large batch vats and allied equipment. This procedure has largely been replaced in many industries by in-line blending, which essentially refers to a process whereby component streams (liquids, gases, powders, or aggregates) are measured and controlled in a precise relationship or ratio to each other. All components flow together simultaneously to a central collection point, where they combine to form the finished product, such as a lubricating oil. The obvious advantage of in-line blending over the batch-type process is that large vessels for the mixing of components are eliminated. The blend heater, augmented by static or active in-line mixers, is all that is required to form the final product. The finished product can go directly to a canning line, to a finished product storage tank, or into a pipeline for distribution.

TYPICAL BLEND CONFIGURATION

A modern in-line blending scheme is shown in Fig. 1. The blend controller nominally utilizes microprocessor technology with a cathode-ray-tube (CRT) display. Each fluid component is pumped from a storage tank through a strainer and then through a flowmeter, with meters

---

* Marketing Manager, Waugh Controls Corporation, Chatsworth, California.

and valves selected for prevailing process conditions (viscosity, temperature, pressure, flow rates, and so on). The signal from the flowmeter is fed to the blend controller, which compares the actual flow rate to the desired flow rate. If the actual flow rate is incorrect, the blend controller will adjust the control valve via the 4- to 20-mA signal to the valve. In this way each component is controlled in a closed-loop fashion. Sometimes it is most practical to control the flow rate by means of proportioning pumps, which inject a precise amount of a specific fluid when a pulse signal from the blend controller is received. This type of open-loop control is cost-efficient, but some means for ensuring flow (not a dry line) should be considered inasmuch as no direct fluid measurement device is used.

Other variations of measurement and control involve the use of variable-speed pump motor controllers (silicon-controlled rectifiers) for flow control, adding a flowmeter in series with an injection pump, and the use of weigh belt feeders with variable feed-speed control and tachometer-load cell outputs (for powders and aggregates).

Liquid or Powder Blending

Solid materials or combinations of solid and fluid material, such as feeds to cement kilns, asphalt aggregates, and concrete aggregates, are readily controlled through the use of techniques similar to those used for fluids. Flow input, in the form of pulses representing the weight of the material, is obtained from weigh-belt feeders, while the 4- to 20-mA control output is used to operate a gate regulating the amount of material fed from a hopper onto the weigh belt. Many weigh-belt feeders require multiplication of belt speed by the mass of the belt in order to obtain the mass flow rate. This computation is performed by external computing modules in a rack. Many food and chemical products require a combination of both liquid and powder ingredients, blended together to form the finished product or an intermediate feed to additional processing. A combination-type blending system used to make bread or pastry dough is shown in Fig. 2.

Sizing a Blending System

The first step in designing a blending system is to construct a list of all the components required to form the various products. Next, after each component, list the ratio ranges for these fluids as they fulfill all daily production needs in an 8- to 9-hour shift, thus providing time for maintenance when required. Once the overall maximum delivery rate has been determined, each component stream can be sized to provide its percentage range of the total blend. The rangeability of the component streams then must be considered. This should not exceed 10:1 if turbine meters are used, or 20:1 in the case of positive-displacement (PD) meters. Possible future production rate increases should enter into sizing the component streams. However, caution should be exercised to avoid oversizing control valves excessively for the current flow rates.

BLEND CONTROLLER

Any number of blending systems of any type, or in any combination of types, are usually controlled by a microprocessor-based instrument system using one or more color monitors at the operator's control station. Such systems incorporate software to provide monitoring, control, data storage, and alarm functions, and input-output circuitry to accept signals from flowmeters and from weight and temperature transmitters, and to drive control devices, such as control valves, line-up valves, and pumps.

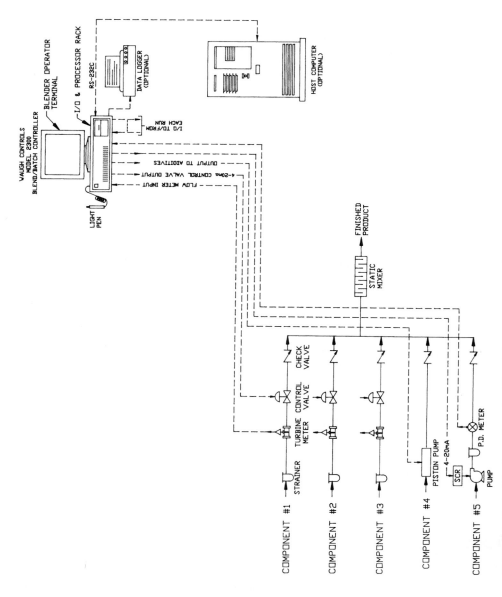

FIGURE 1 Typical blender configuration. (*Waugh Controls Corporation.*)

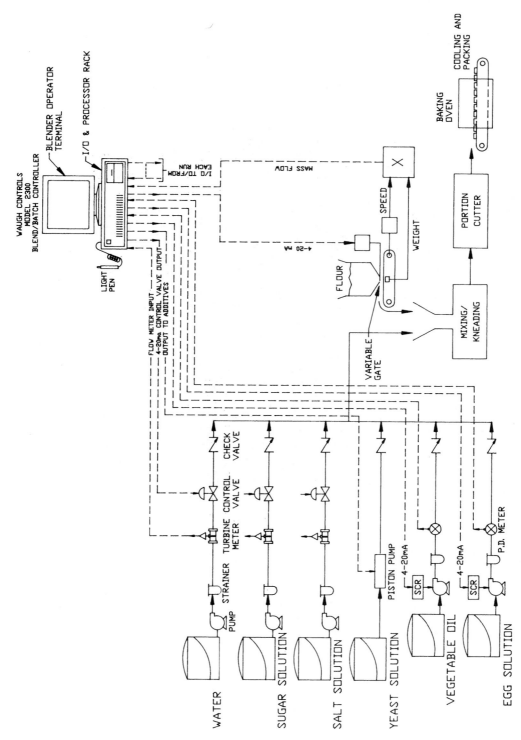

FIGURE 2 Blending system for preparing bread and pastry dough. (*Waugh Controls Corporation.*)

3.113

The system can use a distributed architecture, with monitoring and control instruments installed in the field near each blending unit and linked back to the central control system through a communications link, or a unit architecture, with all monitoring, control, and operator interfaces at the central control room. A typical modern blend control system is shown in Fig. 3.

FIGURE 3 Blending process management system. (*Waugh Controls Corporation.*)

Blend Stations

There will be a series of blend stations, usually skid-mounted, each controlling the flow of a single component of the blend. Each skid-mounted station may include a pump, a pressure-regulating valve, a strainer, an air eliminator, a flowmeter, a control valve, and a check valve. A typical skid assembly is shown in Fig. 4.

Blend Setup

To set up a blend, the operator will enter into the blend controller the actual percentages to be controlled at each blend station, the rate at which the blend is to be run, and the

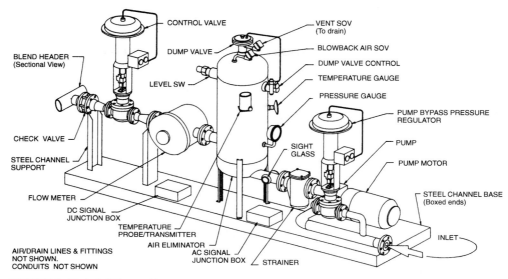

FIGURE 4 Typical skid assembly.

quantity to be blended. If these formulations are repetitive, they can be stored in memory or on a hard disk as numbered recipes. The recipe storage feature provides quick blender setup by simple entry of the recipe number. A blend controller[1] can store hundreds of recipes. Alternatively, the blender parameters may be set from a supervisory computer system.

Master Blend Control

With all in readiness, the "blend start" button is pressed, from which point the operation is entirely automatic. Pumps are started, after which the control valves are opened and the blend flow rates increased gradually until the preselected flow rate is reached. The master demand rate (which represents the total system throughput) is maintained until the demand total or measured total reaches a preshutdown value. At that point the demand rate is ramped down to a new holding value until the total batch size is reached. When the batch size is attained, the master demand rate immediately goes to zero and all loop control outputs go to zero. A master demand rate operation is shown in Fig. 5.

Blend Loop Control

For each component stream, closed-loop control is maintained by comparing the scaled pulse input from the flowmeter with the loop demand derived from the master demand total and

---

[1] For example, the Waugh 2300 blend controller.

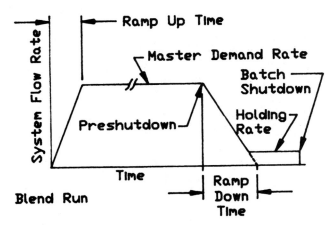

FIGURE 5   Master demand rate operation. (*Waugh Controls Corporation.*)

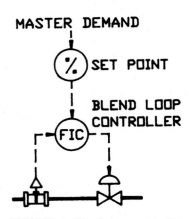

FIGURE 6   Blend loop controller. (*Waugh Controls Corporation.*)

the percent set point for that loop (Fig. 6). The difference between what is measured and what is demanded is the loop error or deviation error. The error from the present scan cycle is added to the previously accumulated errors. The blend loop controller then acts upon the integrated error and works to eliminate it. The output of the blend controller drives a final actuator to provide a fast response to error corrections. Blend loop control differs from conventional ratio control in that the accumulated loop error has a direct effect on the blend loop controller action. In contrast, ratio control uses only the error from the current cycle to bring the measurement back to the set point. It does not compensate for the error that accumulates while the controller is responding to the upset.

Automatic Rate Control

If, during the course of a blending operation, one or more streams are unable to maintain a sufficient flow rate to achieve the desired blend ratios, the blender will automatically ramp down to a rate at which the selected ratios can be maintained. Upon restoration of flow in the lagging streams, the blender will automatically return to its originally set blend flow rate. The system determines that a loop is unable to maintain its required flow when its associated control valve is 95 percent open. Thus the blender will always operate either at its set master demand flow rate or at the maximum rate allowed by the lagging stream, but, in either case, the blend ratios will be maintained automatically.

Time Control

The quality of the blended product may be adjusted on-line while the blend is in process via feedback from one or more analyzers which sample the blended stream. For example, in a lube oil blender an on-line viscometer may be used to control viscosity by adjusting the ratio of one of the lighter stocks. The analyzer signal acts as the process variable input to an internal PID controller function, whose set point is the desired product quality and whose output is then cascaded into the selected blend loop controller, serving to adjust the set point up or down. Changes in the flow of the selected stream affect the quality of the blend, thus closing the cascaded control loop.

Temperature Compensation

Each flow control loop possesses the capability to correct the volume as measured by the flowmeter to a standard volume at a selected reference temperature. Fluid temperature can be measured by a 100-ohm platinum probe with a 4- to 20-mA transmitter. Correction is made per American Petroleum Institute Standard 2540, Tables 6A, 6B, 6C, and 6D or a linear coefficient.

Programmable Logic Control

Blend controllers as described here contain integral PLC functions for interfacing with motor starters, solenoid valves, audible and visual alarms, valve limit switches, and other field equipment. For example, sequential starting of pumps to avoid sudden large increases in electrical load is readily accomplished by sequencing pump start relay closures at the beginning of each blend. A different time delay may be entered for each relay closure, thus permitting any desired time sequence of pump starting. Complex valve alignment procedures and interlocks can also be incorporated with these controllers.

DISTRIBUTED NUMERICAL CONTROL AND NETWORKING

by Terrence G. Cox*

Since the Industrial Revolution (around 1760) there has been a continuing search for more effective manufacturing methods. As part of this unrelenting process, numerically controlled (NC) machine tools were developed in the early 1960s and have been used widely through the present day.

NCs have been used to cut down job lot sizes, reduce the setup time, and trim machining time and the skill level of the direct labor force. Punched paper or mylar tape was the early medium used to feed stored feed instructions to the controllers. Although a major technological development of its day, this methodology was found to be time-consuming, error prone,

* Vice President, CAD/CAM Integration, Inc., Woburn, Massachusetts.

and limiting due to the cumbersome long tapes. Tapes often were difficult to locate by an operator from an ever-expanding inventory, and it was hard to make certain that the latest revised tape was in hand. Loading and unloading the tape required care. Also, the operator had to document any changes (edits) made during the machining process and to make certain that these changes were received by engineering.

Paper tape proved to be a poor medium for the shop floor. Subject to exposure to oil and solvent spills, the tape information would sometimes be altered to the point where scrap parts would be produced. Mylar tapes alleviated some of the problems, but tape reels also became longer and unhandy as the part size and complexity increased. This required splitting a single long tape into several spools, a time-consuming operation. Special instruction sets were required at the start and end of each tape, not to mention the additional time required for loading and unloading each tape.

Direct numerical control (DNC) was later developed to overcome these problems. A further stage of development yielded what is now called distributed numerical control, also commonly abbreviated as DNC.

DIRECT NUMERICAL CONTROL

To overcome the limitations of the earlier, simple NC, several large firms undertook the design of DNC systems in order to establish an interface between a computer and a machine tool wherein a computer could feed instructions directly to a single machine tool and which, in turn, would also control the servos and cut the part. Unfortunately when DNC was developed, the only computers available were costly mainframes. A further drawback was that only *one* machine tool could be controlled at a time. Consequently the general result was that the direct numerical approach was not cost-effective.

DISTRIBUTED NUMERICAL CONTROL

In DNC one computer feeds part programs to *multiple* machine tools. The advent of mini- and microcomputers, plus the ability to address multiple serial ports on a real-time basis, gained acceptance of this technology by manufacturers.

Basic DNC Configuration

In essence, a DNC system uploads and downloads part programs electronically to machine tools. In the initial use of this technology, the user does not need to have a CAD/CAM computer. Where a CAD/CAM system is not interfaced to the current DNC system, programs usually are entered through the machine tool controller in a manual data input (MDI) mode. Once in the controller, the program can be uploaded to the DNC system and, from that point, electronically downloaded to the required machine tool. The DNC system is used as (1) a file cabinet, (2) a traffic director, and (3) an interrupter.

File Cabinet.   It still is not unusual for a manufacturer today to have thousands of part programs stored on punched tape. With a DNC system, these programs are conserved electronically and are placed in organized directories with well-documented revision levels.

Traffic Director.   Base systems usually have a number of machine tools associated with a single DNC management station. Distributing the needed programs to the appropriate machine tool at the required time in an organized method is the primary job of the system.

Where a CAD/CAM computer is used, communications must be established and monitored between the host CAD/CAM and the DNC system regarding the system's ability to accept the transfer of jobs. The communications between the DNC system and each machine tool must be monitored in order for transmission to occur in a timely manner and to know when both systems are ready to accept information. The system also must take into account those instances where transfer cannot occur because the machine tool's memory buffers are full. All of this requires management and coordination from the DNC management station.

Interrupter.   There is much concern among control engineers today over the importance of standards for control system networks and interconnectability. Over the years the machine tool control field has been a major offender in this regard. Several different protocols persist. Consequently the particular control for a given system frequently must be interrupted and matched by the DNC system so that part programs can be distributed. Many different protocols are spread over large numbers of machine tool builders and, considering the introduction of new machine models periodically, the lack of standards has been a problem for several years.

In a modern DNC system the interrupter functionality must be of a user-friendly format so that end users can establish the needed communications as new machine tools are added.

DNC System Components

As shown in Fig. 1, a typical base DNC system will have four major facets: (1) a CAD/CAM system, (2) a DNC system, (3) the machine tools to be controlled, and (4) behind-the-tape readers (BTRs). The CAD/CAM system may be on a mainframe or on a personal com-puter (PC). The DNC system can run on any level of hardware, but usually will be PC-based. There usually are many machine tools that are not uniformly based as to standards and distributed in many locations on the shop floor. The BTRs are used mainly for two reasons:

1. A machine tool does not have an active RS-232 input port. In this case there is only one way to load information, namely, through the tape reader. The BTR is a hardware and software combination device which is located between the tape reader and the controller and is interfaced to the DNC system. (As far as the controller is concerned, it is accepting information from the tape reader.)

2. A machine tool may or may not have an active RS-232 input port, but it does *not* have sufficient on-board memory to store the entire part program. In this case a BTR will interface in the same manner, but the revolving buffer within the BTR will compensate for the needed memory. In this case, as the machine tool uses information, the BTR is "spoon feeding" additional information to the DNC computer, which in turn is "spoon feeding" the BTR. Under these circumstances a part program is limited only by the capacity of the hard disk in the DNC computer.

Advantages of Basic DNC System

The benefits of a DNC system are direct and measurable and include the following.

Designer Efficiency. Prior to direct numerical control it was the responsibility of the designer to punch out the tapes before they were run. If any edits were made after the tapes were punched, it was also the designer's responsibility to make the edits and to repunch the tapes. This process could interrupt the work of the designer by a factor ranging from 10 minutes to over an hour per day.

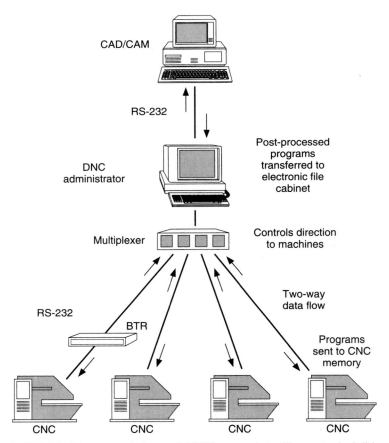

FIGURE 1 Direct numerical control (DNC) system that utilizes a standard disk operating system.

Basic system: (1) Entry level, (2) configured in 4, 8, or 16 ports (up to 64 devices), (3) uses off-the-shelf 286 and 386 PCs.

Features: (1) Centralized part programs, (2) eliminates paper tape, (3) full NC editor, (4) NC edit log, (5) electronic job transfer, (6) standard RS-232 or RS-422 communications.

Advantages: (1) Reduced machine downtime—no loading or unloading of paper tapes, (2) programs transferred to and from machine tools in seconds, (3) automatic update of job transfer and activity log, (4) immediate printout or screen view of any job, (5) reduced network due to NC revision control.

Hardware options: (1) Machine interface units (MIUs), (2) RS-422 communications, (3) behind-the-tape readers (BTRs), (4) communications box and queuing. (*CAD/CAM Integration, Inc.*)

Reduced Machine Downtime.   On average the associated tool downtime for paper tapes exceeds 15 minutes per job run. This time is consumed by punching a tape at the CAD/CAM system, transporting the tape to a particular machine tool, loading the tape on the machine tool, unloading the tape at the end of the run, and punching a new tape at the machine tool if any revisions were made through the editor on the machine tool controller.

Easier Job Transfers.   Many DNC systems incorporate a job transfer and activity log. This log updates reports whenever a part program transfer is made to or from the shop floor. When used properly, this can be very effective for flagging management on production trends, thereby catching production problems before they are out of control. In turn, machine scheduling problems can be handled, thereby reducing machine downtime.

Viewing an Entire Job.   With DNC an entire program can be viewed on the screen should questions arise. Prior to DNC, only one line at a time could be viewed on the controller screen. Viewing the entire job allows the operator a far better frame of reference.

Reduction of Rework.   The organized archiving procedure for part programs and their revision levels, made possible by DNC, eliminates scrap due to incorrect tapes being loaded onto a machine tool.

Additional Executable Functions

Within the DNC management system, a number of executable functions are required.

NC Editor.   An editor, written specifically for NC code, should be available. It is used to make minor changes to the part program, such as feeds and speeds, or changing G codes due to a different controller, all without going back to the CAD/CAM system. Common features include insert, overstrike, delete, resequencing, search and replace, cut and paste, and appending files. Advanced features found in some editors are mirroring, translate, and scaling.

Compare.   When a job is sent to a machine tool, the operator can use the controller's editor to alter the program. In order to manage these edits, the program designer must review the edits once uploaded from the machine tool and determine which should be kept, that is, either the original or the revised program. A DNC management station should have a compare feature to compare the two files efficiently and highlight the differences so that an intelligent decision can be made.

Activity Report.   This report should be available as a thumbnail sketch of shop floor activity. The report gives the manager the following information: what jobs have been downloaded or uploaded to which machine tools, at what time, and which jobs are waiting to be run at each machine tool.

DNC OPERATING SYSTEMS

There are several de facto standard operating systems available as of the early 1990s, including DOS (disk operating system), UNIX, VMS, OS/2, and various proprietary systems. In terms of trends, DOS and UNIX continue to have wide appeal. VMS appears to be giving way to ULTRAX (UNIX derivative). OS/2 is not receiving wide market acceptance. There is some consensus that proprietary systems will become virtually nonexistent by the early 2000s.

Disk Operating System.   A DOS environment provides a good platform. Because of the low cost of each license and minimum hardware requirements, a manufacturer can move into a tapeless environment for a modest cost. This is particularly important in connection with DNC because many thousands of small job shops constitute the major market. However, DOS is a single-tasking, single-user environment and, therefore, only one task can be accomplished at a time. This is limiting because the computer will feed information to only one machine at a time, even if a number of them are multiplexed from one computer. As an example, if a program is attempting to transfer from the CAD/CAM system, it must wait until all other tasks of that computer are completed.

UNIX Platform.   UNIX is much more powerful. It is a true multitasking, multiuser environment. A part program can be transferred from the CAD/CAM system. Part programs can be transferred to all interfaced machine tools, and the operator can edit jobs on the DNC management station—all simultaneously.

In the larger manufacturing firms the foregoing become critical capabilities because more machine tools are interfaced to a system and, therefore, real-time resource sharing is necessary. Another niche where this becomes critical occurs when long part programs are generated and must be fed to memory-insufficient multiple machine tools. Multiplexed systems cannot handle this situation effectively, but the multitasking features of UNIX easily can "drip feed" multiple machine tools.

Functionality Blur

As of the early 1990s a functionality "blur" or fuzziness is developing between the two operating systems just described. Third-party developers are writing multitasking features for DOS. Examples include windows and time-slice drives for serial ports. Although these are not true multitasking capabilities, they are sufficiently good for handling the requirements of multitasking DNC for a large share of existing needs.

Networks are also blurring the multiuser issue (Fig. 2). Networks such as Novell (described later) allow a number of DOS nodes to access a file server simultaneously. This allows a DOS environment to become functionally multiuser at a very reasonable cost.

Due to the complexities of the computing environment, a manufacturer must look carefully before implementing a tapeless factory environment. To determine which operating system may be best suited to a given situation, manufacturing management must evaluate the firm's current computing environment, its networking strategic plan, and the required functionality of the proposed DNC system. PCs and cost-effective networks are major factors in contemporary automated manufacturing technology.

Personal Computer DNC Systems

There are key economic, tactical, and strategic reasons why PCs have grown in popularity not only for DNC, but for many other manufacturing systems.

Economic Advantages of PCs.   Unlike most other areas of a business, manufacturing requires a number of specialized software programs to create the optimal automation

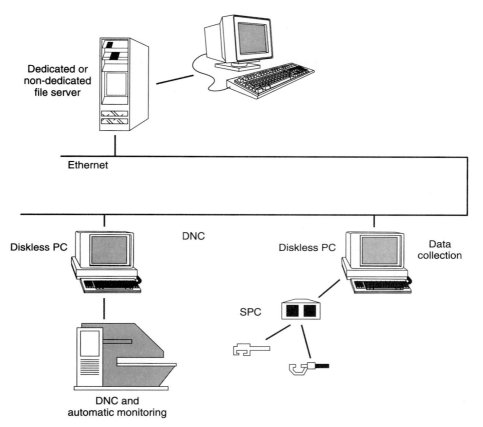

FIGURE 2 Entry-level network that is DOS-based.

Basic system: (1) Entry-level standard network, (2) dedicated or nondedicated file server, (3) based on Intel 286 microprocessor, (4) supports up to five active users, (5) controls application information over network, (6) applications can be integrated in seamless environment through factory network control system.

Features: (1) High performance—files are transferred at 10 Mb/s, (2) easy-to-use menu-driven utilities put the network supervisor in control, (3) security—allows supervisor to restrict network access, (4) single-source database for complete control of all files, (5) cost-effective network uses diskless PCs, (6) virtually unlimited expansion capabilities.

Applications: (1) Statistical process control (SPC), (2) direct numerical control (DNC), (3) data collection, (4) view graphics and documentation, (5) automatic monitoring. (*CAD/CAM Integration, Inc.*)

environment. In other departments a management resources planning (MRP) or CAD/CAM system can be procured which will handle the major share of the department's needs. Thus if these systems are on a mainframe and resources are being shared by a number of individuals, cost justification is easily defended. Furthermore, the software is written to function over a wide range of user applications, thereby reducing the need for customization.

Conversely, manufacturing applications tend to be more specialized, affecting a smaller realm of the corporate universe. Instead of having one MRP system to incorporate the needs of accounting planning, inventory control, traffic, and other departments, the manufacturing department requires many specialized programs, including DNC, statistical process control (SPC), scheduling, simulation, graphics and documentation, data collection, and cell control (Fig. 3). Each of these programs affects only a portion of manufacturing, each with its own specialized needs.

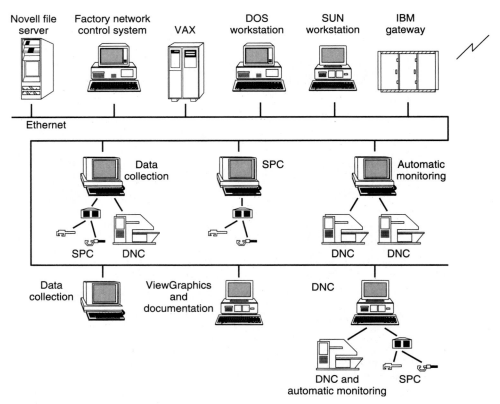

FIGURE 3 Advanced network that is DOS-based.

Basic system: (1) Controls application information of network, (2) applications can be integrated in seamless environment through factory network control system, (3) designed for medium to large business, (4) dedicated 386-based file server technology, (5) supports up to 100 active users on network, (6) high-performance network operating system.

Features: (1) Provides high functionality while maintaining high performance levels, (2) multiuser, multi-tasking architecture allows user to perform many operations simultaneously, (3) enhanced features for security, system reliability, and network management, (4) cell configurations decrease exposure to manufacturing down-time, (5) single-source database for complete control of files, (6) cost-effective network—uses diskless PCs, (7) simple-to-use menu-driven utilities make learning and adding applications easy.

Applications: (1) Statistical process control (SPC), (2) direct numerical control (DNC), (3) data collection, (4) view graphics and documentation, (5) automatic monitoring. (*CAD/CAM Integration, Inc.*)

Due to this diversity of applications and specialization within each application, it is obviously not economically justifiable to procure a mainframe with, for example, an SPC application. The cost of a mainframe and the associated software would be too great in comparison to the payback.

A review of the market in the 1980s substantiates this point, as there was not much specialized manufacturing mainframe software available. What was available did not integrate easily. Melding of software applications could not justify the total purchase.

Tactical Advantages of PCs.   Of all the departments within a company that are dependent on computers, manufacturing would be the most seriously affected by computer downtime. It is evident that production lines would be shut down, inspection would be crippled, work in progress would not be trackable, and so on. In a worst-case scenario, if a mainframe controlled the manufacturing process and the mainframe malfunctioned, the entire manufacturing process would have to shut down. In numerous instances the cost of such downtime is many thousands of dollars per minute. At best, repair time for the mainframe would be an hour or so, and at worst, a full day or more.

In a distributed PC network environment a hardware or software problem does not have the same rippling effect throughout manufacturing as would a mainframe. If a software application has a problem, a new node can be substituted by company personnel in a matter of minutes.

The majority of specialized manufacturing applications has been written for the PC-networked environment. Thus it is prudent for a manufacturer to establish a networked environment. PC interfaces tend to be more user-friendly than the mainframe counterparts. Less training and better acceptance of PC systems by employees have been demonstrated.

Strategic Advantages of PCs.   After a long period of trial and error the corporate universe essentially has embraced the concept of distributed networking. Firms are now requiring that proprietary systems must run over a single network and communicate with one another. The best hardware-software combinations are being purchased for specific applications, but with the proviso that the information generated must be capable of being transferred, accepted, and messaged by other applications of the network. (This strategy is better served through a distributed open architecture than a one-vendor mainframe mentality.)

UNIX-Based Network

The key feature of a UNIX-based network is that it allows the user to enter the realm of a true multitasking, multiuser environment. The power of UNIX is that it allows multiple users to perform multiple processes independently at each node, thereby reducing the network usage while simultaneously executing all needed tasks.

Because of the adherence to standards, a number of different UNIX machines can be interfaced to the same network, and any node can become a remote user of any other node. This facilitates the real-time exchange of information between departments which do not share the same application software. X-windows technology enhances this capability by standardizing the interface between all computers on the network, whether they be UNIX, DOS, or VMS, for example, so long as these computers adhere to the X-Window standard.

Novell Network

The strong points of Novell are threefold: (1) a manufacturer can install a network at a more reasonable cost than a full UNIX system, (2) as a firm's networking strategy develops, a Novell network will make changes easier because of the wide variety of topologies that it supports, and (3) DOS applications become available in a multiuser environment.

As of the early 1990s Novell networks can be procured in terms of a few hundred U.S. dollars. This permits users to justify the expense of installing a network and allows the network to expand as the firm becomes more comfortable with the networking concept. The justification argument, however, is not as great as in the one of Ethernet-based UNIX nodes.

When one couples lower cost with Novell's multiple topologies, there is a strong strategic case. As a firm develops the network strategy in both the manufacturing and the business areas, Novell can adapt and accommodate the different networks that emerge. Thus the investment is protected.

Considerable manufacturing software has been written for PC use. Novell allows single-user programs to share data and operate in a multiuser environment.

DNC System Trends

Standards-based PC and associated networks will become increasingly important. Communication standards, such as MMS and X-Windows, will play an integral part in the growing use of PCs and LANs in manufacturing. The degree of sophistication will continue to increase. Examples of contemporary DNC systems are shown in Figs. 4 and 5.

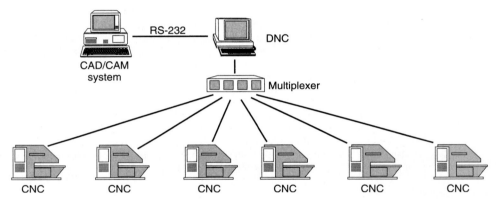

FIGURE 4   DNC system design for small workshop. Simple communications needed between CAM product and seven DNC machine tools with appropriate internal memory. There is a simple RS-232 interface between CAM product and DNC system. 8-port intelligent multiplexer interface to DNC system to communicate to seven machine tools via RS-232. This system allows job shop management to queue up jobs for each machine tool, compare original and revised part programs, and perform minor editing on part programs. System is fully upgradable to a LAN, and other third-party software packages can be added. (*CAD/CAM Integration, Inc.*)

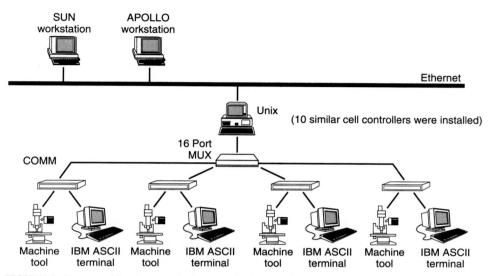

FIGURE 5   Large machine shop required communications to 10 machining cells with four machine tools in each cell. An existing Ethernet backbone was available. The management required that ASCII terminals be located at each machine tool, using a simple network that would minimize the number of wires running through the plant. It was desired to access the DNC system from the ASCII terminals to start job transfers to the machine tools. A UNIX node with star network configurations was installed at each machining cell. The star configuration provided machine tool interfacing to a central workstation hub, using a standard multiplexer. Each cell has additional multiplexer ports that could support additional machine tools if added at a later date. Each central hub attached to the Ethernet backbone provided closed-loop communications to existing workstations that already were making use of the network. A communications module was installed at each machine tool so that only one wire was drawn from the multiplexer to the machine tool. A communications module has two ports—one for interfacing to the machine tool controller and one for interfacing to the ASCII terminal. The installation is easily expandable both at the cell level and around the facility.

COMPUTERS AND CONTROLS

Ever since automatic controllers advanced from simple on-off devices, some form of computing has been required to implement algorithmic relationships in an effort to improve process and machine controllability. Over many decades, ingenious, essentially macroscale mechanical, pneumatic, hydraulic, and traditional electrical means were introduced and, considering their various vintage periods, operated quite satisfactorily. The computing capabilities required to achieve controllability that is considered acceptable in terms of the expectations of the early 1990s were enhanced by several factors, including the appearance of miniaturized solid-state electronic circuitry components with their high speed, their comparative simplicity, their progressively decreasing cost, and their inherent proclivity for handling instrumentation data in digital form, in contrast with the prior and essentially exclusive dependence on analog information.

The first electronic computer (although not solid-state) was developed by J. P. Eckert and J. W. Mauchly (Moore School of Electrical Engineering, University of Pennsylvania) with funds provided by the U.S. Army. The main purpose was that of calculating ballistic tables. Between 18,000 and 19,000 vacuum tubes replaced mechanical relays as switching elements. The machine weighed over 30 tons and occupied a very large room. Generally, the machine was not considered highly reliable. It existed until 1955 and logged some 80,000 hours of operation. The machine was known as the electronic numerical integrator and calculator (ENIAC).

Commencing essentially in the mid-1940s, digital computer development quickened in pace. IBM was active in the field and, in 1947, produced the selective sequence electronic calculator (SSEC). The Moore School developed a follow-up of the ENIAC, known as the electronic discrete variable computer (EDVAC), which was unveiled in 1952. Most historians now accept the fact that the EDVAC was the first stored-program computer. Unlike earlier machines, which were programmed at least partially by setting switches or using patch boards, the EDVAC and others to follow stored instructions and data in identical fashion. EDVAC used acoustical delay lines, which were simply columns of mercury through which data passed at the speed of sound, as the main memory. This type of storage soon gave way to magnetic-core memory, the antecedent of contemporary semiconductor memory.

In 1946 John Von Neumann, a mathematician at the Institute of Advanced Study in Princeton, New Jersey, prepared the paper, "Preliminary Discussion of the Logical Design of an Electronic Computing Instrument." The project, financed by the U.S. Army, suggested the principles of design for digital computers that were to be built over the next several decades. The principles outlined by Von Neumann included internal program storage, relocatable instructions, memory addressing, conditional transfer, parallel arithmetic, internal number-base conversion, synchronous internal timing, simultaneous computing while doing input-output, and magnetic tape for external storage. A computer built along these lines went into operation in the early 1950s. UNIVAC (universal automatic computer) was the first commercially available digital computer to use these principles. The UNIVAC was a descendent of ENIAC and EDVAC, having been built by Remington-Rand following the acquisition of the Eckert-Mauchly Computer Corporation. Eventually 48 UNIVACs were built, making Remington-Rand the number one computer manufacturer—until IBM commenced in earnest in 1954 with the introduction of the 700 line.

Early computer history would not be complete without mentioning the EDSAC (electronic data storage automatic calculator), developed by W. Renwick and M. V. Wilkes in 1949 at Cambridge, England; the Ferranti Mark I computer, developed at Manchester, United Kingdom, in 1950; the Pilot ACE (automatic computing engine), also developed at Manchester; the SEAC (standards eastern automatic computer), developed by the U.S. National Bureau of Standards (now NIST) in 1950; and WHIRLWIND I, developed by J. W. Forester at the Massachusetts Institute of Technology in 1950. It is interesting to note that WHIRLWIND I used an ultrasonic memory, later to be replaced by a Williams tube memory.

For nearly 3½ decades, commencing in the 1950s, computer architecture was largely based on the Von Neumann model. New concepts began to take hold in the 1980s, including parallelism, vector processing, multiprocessing, and memory access and interconnection schemes, such as the nearest-neighbor mesh, linear array, and hypercube schemes. Of course, a major event that has had a profound effect upon computers in instrumentation data acquisition and control systems was the introduction, a few years ago, of the so-called personal computer (PC), described later in this article.

INITIAL CONSIDERATIONS OF COMPUTERS IN CONTROL SYSTEMS

A relatively few instrument and control engineers were early in their recognition of how digital computers could be used to advantage in control and data acquisition.

Direct Digital Control.   In early systems a single mainframe computer accomplished data acquisition, control, reporting to the operator, and higher-level computation. In one of the first systems, the computer directly controlled the process, without intervening controllers. This architecture became known as direct digital control (DDC). Only one computer was used in the first system because of the high cost per processor and because of the general absence of computer-to-computer communications. Later systems used an auxiliary (redundant) computer. Early DDC system applications had a number of drawbacks. For example, if only one processor was used, a single failure could affect a large number of controlled variables and possibly disable an entire process. Although redundant processors were introduced to provide backup for hardware failure, additional processors for functional distribution did not appear until much later (Fig. 1).

Supervisory Control.   This concept, shown schematically in Fig. 2, developed over a period of years as an effort to use the many benefits available from the computer, but without the drawbacks of DDC. This concept had several advantages. In particular, it preserved the traditional panelboard control room while adding the capability of a digital computer and associated cathode-ray-tube (CRT) displays. This minimized the disruption of operations that could accompany a transition to full DDC. It also offered a buffer between the computer and the process, so that the operator had the option of selectively disabling the computer for troublesome loops. Although the impact of a computer failure was substantially reduced, it, too, used a few processors. Thus considerable functionality could still be disabled by a single failure.

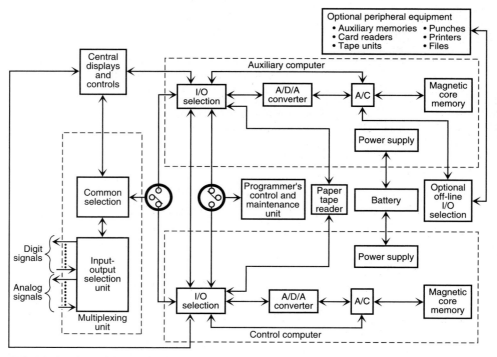

FIGURE 1   Direct digital control (DDC) system (with auxiliary computer) developed by D. Considine, D. McArthur, and D. Wilson. (*Hughes Aircraft Company.*)

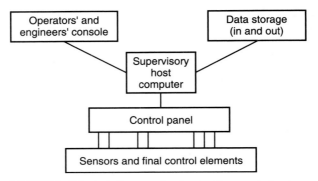

FIGURE 2   Schematic diagram of early supervisory control system.

Distributed Control Systems.   Introduced during the mid-1970s, distributed control systems (DCSs) combined three technologies, most of which either appeared for the first time or were much improved during the interim that dated back to DDC. These included (1) microprocessors, (2) data communications, and (3) CRT displays. Multiple microprocessors were used in some systems as dedicated control loop processors, CRT operating stations, and a variety of other control and communication functions. Although these systems could not accomplish all of the functions of the prior computer-based architectures in all instances, they did achieve considerable fault tolerance by minimizing the effect of a single failure. DCS is discussed in considerable detail in the first article of this handbook section (Fig. 3).

Hybrid Distributed Control Systems.   Because DCSs are limited in computer functionality and in logic or sequential control, they can be combined with supervisory host computers and with programmable controllers in hybrid architecture (Fig. 4). These systems use data communications channels between the host computer and both the programmable controllers and the DCS. Although these systems offer the strength of each subsystem, it is sometimes difficult to establish communication between the subsystems and confusing to provide software inasmuch as each subsystem may use different programming techniques and organizes data in separate databases.

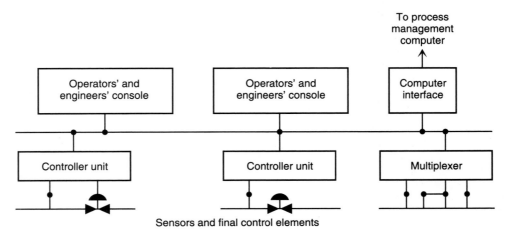

FIGURE 3   Schematic diagram of early distributed control system (DCS).

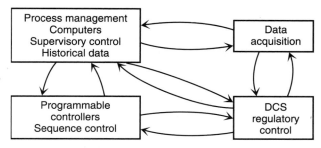

FIGURE 4 Schematic diagram of early hybrid distributed control system (DCS).

In another version of the hybrid approach, the DCS is replaced by single-loop digital controllers. These offer some advantage, especially in some cases where a user may prefer a conventional panel as the primary operator interface rather than one or more CRTs.

PERSONAL COMPUTERS IMPACT INDUSTRIAL INSTRUMENTATION[1]

Relatively early in the 1980s the PC enjoyed a multibillion dollar market and attracted the attention of pioneering professionals in the instrumentation and control field. It was recognized at an early date that one could develop a reasonably low-cost scan, control, alarm, and data acquisition (SCADA) platform and integrate it with general-purpose (shrink wrap[2]) software packages, such as spreadsheet and database management, to provide, in select cases, a cost-effective alternative to DCS. In a quantitative sense, prospective users of the PC tended to be cautious, and early applications occurred most often in cost-sensitive noncritical cases.

The initial acceptance of the PC was not universal, particularly with regard to using a PC as an industrial controller. Conservatives persisted in their opinions that PCs were limited to (1) off-line programming and downloading to real-time control computers, (2) for supervisor or operator advisory applications with artificial intelligence (AI) programs, (3) laboratory and small-scale pilot plant experiments, and (4) noncritical control applications, provided that programmable controller safety interlocks were also used. They were wary of the use of PCs in closed-loop situations but applauded their application for on-line engineering data acquisition and process performance analysis. Despite the conservative and limited acceptance by some, the PC has become very popular for a variety of control situations.

Basic PC

The advent of the modern PC makes it possible for virtually everyone to take advantage of the flexibility, power, and efficiency of computerized data acquisition and control. PCs offer high performance and low cost, coupled with an ease of use that is unprecedented. This con-

---

[1] Personal computers are discussed in several other articles in this handbook. Check Index.

[2] Feature-laden high-quality inexpensive software spawned by enterprising software houses and system integrators. This software was economically viable because the cost of the "intellectual property" could be spread over thousands of users.

trasts with the implementation of data acquisition and control tasks that once required the power of a 16-bit mainframe computer, the cost of which lead to timesharing among several users. The mainframe essentially was a centralized utility available to numerous parties. Consequently, small or remote jobs were often relegated to manual or, at best, simple electronic data-logging techniques. Thus smaller tasks could not benefit from the flexibility and power of a computerized solution.

Because of a significant (but not yet complete) degree of standardization among PC and data acquisition and control equipment manufacturers, a large selection of hardware and software tools and application packages has evolved. The end result is that an individual engineer or scientist can now implement a custom data acquisition and control system within a fraction of the time and expense formerly required. It has become practical to tailor highly efficient solutions to unique applications. The nature of the PC encourages innovation, which in just a few years has revolutionized office automation and currently is producing similar results in manufacturing facilities, production lines, testing, laboratories, and pilot plants.

Although subject to change with additional technological progress, as of 1992, in scientific, technical, and industrial environments, the popularity of IBM's PCs has made them a de facto standard. The PCX/AT bus architectures used in these computers also have become industry standards and have been incorporated in a vast array of IBM PC work-alikes. Some of these compatible PC models offer advantages in features, performance, or cost.

In addition, PC systems based on several other types of bus architectures are being used with increasing frequency in data acquisition applications. These alternate bus architectures include, for example, micro channel, used in the IBM PS/2 series; NuBus, used in Apple's Macintosh II; and extended industry standard architecture (EISA). The latter PCs are manufactured by Compaq and Hewlett-Packard, among others.

In the context of data acquisition and control applications, "true" IBM PC compatibility includes both hardware and software requirements. Only those computers that can run, without modification, the same software written for the IBM PC are compatible with it. Similarly, a compatible machine must accept the same range of add-on (or add-in) boards that plug directly inside the IBM PC.

Major Components of the PC

A complete PC consists of a system unit containing (1) a microprocessor, (2) a memory, (3) a power supply, (4) a keyboard, and (5) one or more output devices, including monitors, printers, and plotters. The system unit usually is housed in an enclosure, separate from other major components such as the monitor. An exception would be a portable PC, in which all devices are integrated into an easy-to-carry case.

Laboratory top and other miniature computers usually are *not* included in the true PC-compatible category because they lack expansion slots. However, growing numbers of small computers with expansion capabilities are appearing on the market.

An expansion slot is a physical and electrical space set aside for the connection of accessory hardware items to the PC. Electrical connection is made directly to the internal microcomputer bus. These accessory items usually take the form of a plug-in printed-circuit board (such as a graphics interface, a memory expansion module, or a data acquisition device).

Plug-in boards can be designed to be addressed by the microcomputer in two different ways, either as input-output (I/O) ports or as memory locations. Each method has its own advantages. However, memory addressing offers a higher level of performance that includes improved speed, extended address space, and the full use of the processor's instruction set.

Some of the computer I/O and memory addresses are reserved by the computer manufacturers for standard functions, such as graphics cards, RS-232 ports, memory, disk, and controllers. Most other types of plug-in boards are equipped with a bank of switches allowing the user to select an appropriate address location.

PC Memory Types

Several distinct types of memory are used in PCs. These include RAM (random-access memory), ROM (read-only memory), floppy disk, and hard disk. Other memory technologies include magnetic tape, optical disk, and bubble. RAM and ROM are semiconductor devices that offer a very high-speed operation. RAM has both read and write capabilities that are accessed by the microcomputer. ROM, in contrast, contains a fixed set of information that only can be read by the microcomputer. As with other computers, there is the central processing unit (CPU)—in this case, the microcomputer. ROM is preprogrammed at the factory and includes most of the basic CPU operating instructions, including the code that is required to start, or "boot," the computer. This is a special ROM or BIOS (basic I/O system). Other active program information is in RAM. The portion of RAM that can be used is a factor of the particular microprocessor chip used and, of course, the available software. The 20-bit address bus of the 8088 limits memory locations to about 1 Mbyte. The 80286 has a 24-bit bus which can address about 16 Mbytes. RAM is normally considered volatile, because in most systems its data will be destroyed with loss of power. Personal storage of data and programs usually is provided by the disk drives.

The complexities encountered with contemporary PCs usually discourage most users from attempting to "talk" directly with the CPU, BIOS, or disk drives. Interface software has been developed to bring the power of the PC within reach of nonspecialized users. Such software sometimes is referred to as the operating system. Among the widely used operating systems are PC-DOS and MS-DOS. Additional operating systems include UNIX and OS/2.

PC Speed

The time required to run a program (that is, to process instructions) depends on several factors, including factors under the user's control, such as choice of language and the efficiency of the resulting code. Software efficiency partly refers to how many machine cycles are needed to execute the desired instructions. A so-called tight program will require the minimum number of machine cycles and thus, from a time standpoint, is most efficient. Other factors affecting speed include basic PC design factors, including the microprocessor chip selected, additional logic, circuit configuration, and clock frequency. The 80326 processor increases speed and the amount of addressable RAM. Machines incorporating this chip and running at 33 MHz are available.

Regardless of the computer's CPU speed, most machines limit their expansion bus speed to 8 to 10 MHz to ensure compatibility with other add-on products. This results in a net effect such that while faster machines will process acquired data in proportionately less time, they usually will not acquire data at a faster rate.

Data Conversion and Handling

Related topics, such as signal types, signal conditioning, sampled data systems, analog and digital I/O methodologies, and common-mode rejection are discussed in Section 7, Article 1 of this handbook.

Interfacing the PC

A PC may be connected to a data acquisition (and control) system in one of two basic ways: (1) by direct connection to the PC bus (internal bus products) or (2) by way of a standard communications channel (external bus products), such as RS-232, RS-422, or IEEE-488. There are advantages and limitations of each method.

Internal Bus Products.   Direct connection (Fig. 5) generally yields three advantages.

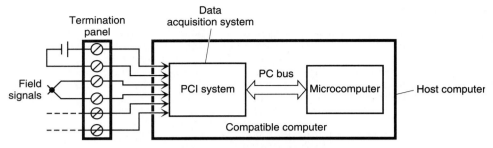

FIGURE 5   Block diagram of internal PC bus system. (*Intelligent Instrumentation.*)

1. *Higher speed* results from avoiding the relatively slow, external communications-channel protocol. For example, the data acquisition rate using RS-232 communication at 9600 baud is limited to about 20 analog readings per second. By contrast, some direct PC bus products can receive data at a rate faster than one million samples per second.

2. *Lower cost* results because no separate enclosure or power supply is required. Power is obtained from the PC.

3. *Smaller size* results from the more efficient utilization of space.

Limitations of internal bus products include:

1. Lack of channel expansion capability.

2. Inability to add functions not originally procured.

Internal bus products are board-level systems that make direct connection to the computer expansion bus. These boards may have:

1. *Fixed arrangement* of analog and digital I/O, which is not conducive to changing for future needs, but which is usually available at a comparatively lower cost.

2. *Modular design* that permits the user to select the number and configuration of the I/O functions needed. This is accomplished by using a grouping of function modules. (A modular board-level system offers some of the advantages of a box system.)

Fixed products include single-function and general-purpose configurations. The focused single-function boards often are effective in small applications or in well-defined instances where the board is embedded in a larger end product, as may be offered by an original equipment manufacturer (OEM).

Compromise is sometimes required in connection with general-purpose fixed I/O configurations. For example, the desired number of channels may not be obtainable (or the user may have to procure some unnecessary functions). Because of future developing needs, a mismatching may be inevitable. External add-on boards or boxes may be needed at a later date, frequently an undesirable situation because of space and cost considerations.

External Bus Products.   External PC connections (Fig. 6) frequently are advantageous for a number of reasons.

1. *Remote location* (from the host computer) permits the data acquisition portion of the system to be nearer to the sources of field signals and enables the construction of a distributed system. This allows many parameters to be monitored or controlled even if they are physically separated by an extensive distance from each other and from the host computer. For example, data originating from separate production lines in discrete-piece manufacturing or from

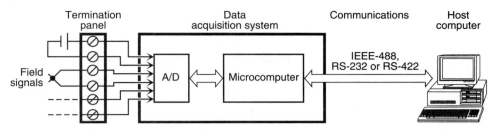

FIGURE 6   Block diagram of external bus data acquisition and control system. (*Intelligent Instrumentation.*)

separate units in a continuous chemical process—each with separate data acquisition and control systems—can be connected via RS-422. This permits monitoring by a single PC located several thousand feet (hundreds of meters) distant. For some systems this advantage can increase productivity and reduce overall costs.

2. *Offloading some data acquisition tasks* from the host computer through the use of data acquisition subsystems as described in 1.

3. *PC selection flexibility* because external bus products can be interfaced with virtually any kind of PC.

4. *System design flexibility* because the PC system need not be configured permanently from the start.

PC Software

In dealing with computers, including PCs, it is in order to frequently remind oneself that all communication with a computer must be in terms of digital 1's and 0's. At the outset of what has become computer science, only the machine language (a given voltage is present or is not present at an input) existed. Early computer programming was slow, error prone, difficult to maintain, and appreciated only by specialists.

Invention of the compiler was a great step forward in computer technology because a compiler understands (accepts) alphanumeric inputs and translates them into machine-readable code. Thus the compiler, which may be defined as an intermediate program, is a much better interface with human operators. Assembly language is the first step above machine code. An assembler is a low-level compiler that converts assembly language into machine code. The resulting code works with the computer's operating system to further simplify a specific programming task. Although more manageable than machine code, assembly language is not considered simple by the average computer user.

So-called high-level languages, such as BASIC, perform still more complex operations, while retaining a degree of recognizable user language, but still remain dominated by jargon and special syntax. C and Pascal are among several other high-level languages which vary in their ability in terms of execution speed and program maintainability. Compiled languages must be debugged as an entity. So-called interpreted languages are compiled incrementally, that is, line by line. This enables a single program line to be written and tested independently.

PC-based data acquisition systems have been designed so that users may write special programs for data acquisition, storage, display, logging, and control in high-level languages. It is the purpose of software to make these tasks as simple as possible. There are three classes of software available for PC-based data acquisition and control systems: (1) tutorial and program development tools, (2) function subroutine libraries, and (3) complete turnkey, menu-driven application packages.

In the case of program development tools, users may write their own unique application software, which usually includes "drivers" that provide the interface to the I/O hardware.

Such packages ease the task of writing programs in high-level languages. This type of programming is considered quite flexible and generally useful.

Some application packages are designed for immediate system start-up, with little or no programming needed. These packages frequently are structured and less flexible than other types of software.

A large selection of third-party software is available, as prepared by independent software firms, or houses. Sometimes the term "genetic software" is used for such programs. LABTECH CONTROL[3] is an example of a third-party program. This is a SCADA (supervisory control and data acquisition) type of program for use with some PCs. This software integrates process monitoring, data acquisition, output, test, measurement, analysis, and display. The program is compatible with most boards, carriers, modules, and IEEE-488 devices within the PC hardware system. The user interface is designed so that no programming is required. All features are selected with easy-to-use icons and menus. As a result, minimum computer skills are required to perform complex operations. The system includes fault-tolerant features, such as operating system error trapping and sensor voting. PID control and networking are optional capabilities. Custom color-graphic flow diagrams, including ISA symbols, provide display of real-time and trend information to the operator. Animation features can provide immediate visual feedback (such as the tank is filling or the door is open). Extensive alarm-processing capabilities are available. Alarms can be set to get the operator's attention, or they can automatically trigger a desired action. Data can be stored on floppy or hard disk, or written in a printer to provide historical records.

The program supports expanded memory beyond the traditional 640-kbyte limit for storing setup structures (such as channel configurations) and to house data buffers. This feature not only allows larger and a greater number of setups, but it also frees the DOS memory for other applications, such as terminate-and-stay-resident programs.

The LABTECH CONTROL[4] program is well suited for many real-world process control tasks in manufacturing and laboratory environments. It can automate experiments, control test sequences, perform calculations, display results in a graphic format, and generate reports. Applications include direct machine control, supervisory control, process monitoring, data logging, statistical process control (SPC), pilot plant production, and sequencing control. Specific areas of use include petrochemical and pharmaceutical processes, wave soldering, automotive machining, food processing, water treatment, and plastic extrusion. The program is menu-driven and icon-based. The conditions that define a current run are displayed on the screen and are readily modified. Conditions pertaining to a run can be stored and recalled as a group.

The program is flexible, permitting each channel to be set up with different characteristics. Sampling rates may vary from channel to channel and may also change at different times during a run. Stored data can be played back as though they were being acquired in real time, thus permitting comparisons of current data with prior data.

With the program, channels can be used for purposes other than simple inputs or outputs. The program has the ability to derive channels from other channels as, for example, channels can "operate" on others by calculating averages, derivatives, integrals, and so on, in real time. The group of mathematical, logical, statistical, and signal-processing functions also includes trigonometric functions, EXP, LN, LOG, OR, XOR, AND, filter, and FFT. These derived channels also can be used in determining triggers or as inputs to control loops.

Open- and closed-loop control algorithms may be implemented. In the open-loop mode, the user defines one period of any imaginable waveform, and the signal is then clocked out automatically during the run. For closed-loop control, both PID and on-off loops can be set up.

---

[3] Name proprietary to Intelligent Instrumentation, a Burr-Brown Company.

[4] If space permitted, numerous other PC software programs could be described. Programs like this typify the ingenious and rapid response of third-party software houses to the ever-increasing needs of the manufacturing and process industries.

The program also includes a curve-fitting function. It uses an iterative routine to fit an arbitrarily complex model (up to 10 parameters) to the collected data. This and other routines, such as PID and thermocouple linearization, take advantage of the PC's optional coprocessor if available. This offers 80-bit real number processing, reduces round-off error, and permits faster computation.

Software Control Techniques

Many, if not most, data acquisition and control applications depend on the timely execution of read and write operations. When speed or timing is critical, three techniques for software control are available for consideration: (1) polling, (2) interrupts, and (3) direct memory access (DMA).

Polling.   This is the simplest method for detecting a unique condition and then taking action. Polling involves a software loop that contains all of the required measurement, analysis, decision-making algorithms, and planned actions. The data acquisition program periodically tests the system's clock or external trigger input to sense a transition. Whenever a transition occurs, the program then samples each of the inputs and stores their values in a buffer. A buffer is simply a storage location that contains the values representing the specified inputs at a given time. The buffer can be stored in RAM, disk, or other types of memory. Each time the program senses a clock "tick," the inputs are scanned and converted, and a new value is added to the buffer. In this mode the PC/AT can support a data acquisition rate of about 180 kHz, depending on the CPU clock rate. Conversely, the design of the PC is such that potentially significant variations (or jitter) in timing can occur. In the IBM PC, jitter of approximately 12 μs is not uncommon. In addition, the PC is continuously busy when the polling loop is operational and hence no other tasks can be serviced. When an application cannot tolerate these characteristics, interrupt techniques may be indicated.

Interrupts.   These provide a means of tightly controlling the timing of events while allowing the processing of more than one task. Multitasking systems are also known as foreground/background systems. A method of putting data acquisition in the background is to relegate it to an interrupt routine. The clock or external timing signal, rather than being polled continuously, is used to generate an interrupt to the computer. Whenever the interrupt occurs, the computer suspends current activity and executes an interrupt service routine. The routine in this case might be a short program which acquires one frame of data and stores it in a buffer. The computer can perform other operations in the foreground while collecting data in the background. Whenever a clock tick or external interrupt occurs, the computer will automatically stop the foreground process, acquire the data, and then resume where it left off.

The reaction speed of the interrupt system is much slower than that of a well-prepared polling loop. This results because the interrupt mechanism in most computers involves a significant amount of software overhead. Speed for an IBM PC is about 4 kHz in the interrupt mode. Also, the software complexity of interrupts can be significant. In most cases the programmer must be prepared to write assembly language code. In contrast, most polled systems can be written in a high-level language. Interrupts are useful in situations where the acquisition rate is slow, timing accuracy is not a priority, and background operation is important. When the time required to service an interrupt is small, compared to the rate at which the interrupts can occur, then this technique yields good results.

The foregoing factors illustrate that careful analysis is required before making a polling or interrupt decision.

Direct Memory Access.   This is a hardware technique that often allows the highest-speed transfer of data to or from RAM. Given the potentially more expensive hardware, DMA can provide the means to read or write data at precise times without significantly restricting the microprocessor's tasks. For example, one system,[5] under DMA control, can read or write any combination of analog, digital, or counter/timer data to or from RAM at a maximum rate of 300 kbyte/s. This is accomplished by taking minimal time from the other tasks of the microprocessor. The amount of time required to respond to a DMA request is very small compared with the time required to service an interrupt. This makes the goal of high-speed foreground/background operation possible.

Progressive Enhancement of the PC

In examining the progress that has been made in applying PCs to industrial data acquisition and control systems, one must recall that PC systems originally were designed for office automation, that is, for single-user, single-tasking, simple, comparatively low-cost machines. Thus PCs, as originally conceived, excel at word processing and spreadsheet applications. Often they are not capable of handling the real-time requirements of high-performance data acquisition and control. Thus it is not surprising that most of the technical advances found in the data acquisition and control boards available today are directed toward overcoming this initial limitation.

One excellent solution involves putting a high-performance processor right on the data acquisition board itself. This processor can assume control of the data acquisition functions, allowing the PC to act in a supervisory role. The PC can download programs to the data acquisition system and leave all the time-critical operations where they are most effectively handled. The architecture of one available system is exemplary of this approach inasmuch as the data acquisition functions already are on modules. This permits the limited available carrier board space to be dedicated to the on-board processor's functions.

Almost immediately after the IBM PC was introduced, data acquisition boards to plug into them became available. The early boards were basic, that is, they handled analog-to-digital (A/D) conversion, digital-to-analog (D/A) conversion, and digital I/O. This was followed by a period of increasing sophistication. Faster A/D converters with programmable gain instrumentation amplifier front ends were incorporated. Powerful features, such as on-board memory and exotic DMA techniques, appeared. The number of competitive sources increased; existing products often became obsolete within short periods as newer products with more powerful features became available. Thus a wide array of contemporary products has been developed, ranging from low-end digital I/O boards to array processing systems, the cost of which may be several times that of the PC itself.

Most manufacturers offer fixed-configuration multifunction data acquisition boards. A typical board will contain 16 single-ended or eight differential analog inputs, two analog outputs, 16 digital I/O bits, some sort of pacer clock or timer for generating A/D conversions, and often one or more event counters. The main differentiating factor among these boards generally is either the speed or the resolution of the A/D converter, or the mix of I/Os offered.

A problem with the fixed multifunction board approach is the difficulty of expanding or upgrading a system, as is frequently the case in research and development applications where a system may be used for numerous applications over a period of time. Even in fixed-configuration production systems, the number of channels provided by the board may be insufficient to handle a given problem. To expand the number of analog or digital I/O channels, the user is faced with purchasing a complete second board. This can be a significant penalty if all that is required is a few more digital I/O points. And to upgrade speed, a user must procure an entirely different board with a higher-performance A/D converter. This also

---

[5] PCI system of Intelligent Instrumentation.

may involve different software inasmuch as an earlier program written for a slower board may require rewriting to accommodate the different architecture of the faster board. For other applications there may be functions provided by the multifunction boards which are not needed. (Typically, only about 30 percent of data acquisition users require analog outputs. If they are provided, the user pays for them whether or not they are used.) Some manufacturers have provided solutions for most of the aforementioned problems.

TERMINOLOGY

During the last quarter century a massive transfer of computer and digital technology has infused into traditional instrumentation and control engineering. This was addressed in the mid-1980s by the editors of the third edition of this handbook by extensive discussions of computers and digital techniques. Because of the growing acceptance of the PC for data acquisition and control applications, and a deeper understanding gained by the profession during the last few years, there is less need for tutorial information. Thus the earlier glossary of terms has been reduced. However, some holders of the third edition may opt to retain their copy to augment this current, fourth edition.

Adder.   A digital circuit which provides the sum of two or more input numbers as an output. A 1-bit binary adder is shown in Fig. 7. In this diagram, A and B are the input bits and C and $\bar{C}$ are the carry and no-carry bits from the previous position. There are both serial and parallel adders. In a *serial adder,* only one adder position is required and the bits to be added are sequentially gated to the input. The carry or no-carry from the prior position is remembered and provided as an input along with the bits from the next position. In a *parallel adder,* all the bits are added simultaneously, with the carry or no-carry from the lower-order position propagated to the higher position. In a parallel adder, there may be a delay due to the carry propagation time.

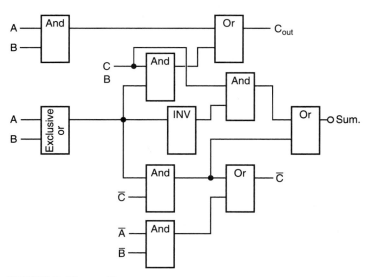

FIGURE 7   Binary adder.

An adder may perform the subtraction as well as the addition of two numbers. Generally, this is effected by complementing one of the numbers and then adding the two factors. The following is an example of a two's complement binary subtraction operation:

$$
\begin{array}{lll}
(a) & 0110 & +6 \;\;\text{(true)} \\
    & (+)\,\underline{1010} & -6 \;\;\text{(complement)} \\
    & 10000 & 0 \;\;\text{(true)}
\end{array}
$$

$$
\begin{array}{lll}
(b) & 0101 & +5 \;\;\text{(true)} \\
    & (+)\,\underline{1010} & -6 \;\;\text{(complement)} \\
    & 1111 & -1 \;\;\text{(complement)}
\end{array}
$$

$$
(c) \qquad 1111 \;\text{(complement)} = -0001 \;\text{(true)}
$$

The two's complement of a binary number is obtained by replacing all 1's with 0's, replacing all 0's with 1's, and adding 1 to the units position. In (a), 6 is subtracted from 6, and the result is all 0's; the carry implies that the answer is in true form. In (b), 6 is subtracted from 5 and the result is all 1's with no carry. The no-carry indicates the result is in complement form and that the result must be recomplemented as shown in (c).

Address.   An identification, represented by a name, label, or number, for a digital computer register, device, or location in storage. Addresses are also parts of an instruction word along with commands, tags, and other symbols. The part of an instruction which specifies an operand for the instruction may be an address.

Absolute address or *specific address* indicates the exact physical storage location where the referenced operand is to be found or stored in the actual machine code address numbering system.

Direct address or *first-level address* indicates the location where the referenced operand is to be found or stored with no reference to an index register.

Indirect address or *second-level address* in a computer instruction indicates a location where the address of the referenced operand is to be found. In some computers, the machine address indicated can in itself be indirect. Such multiple levels of addressing are terminated either by prior control or by a termination symbol.

Machine address is an absolute, direct, unindexed address expressed as such or resulting after indexing and other processing has been completed.

Symbolic address is a label, alphabetic or alphanumeric, used to specify a storage location in the context of a particular program. Sometimes programs may be written using symbolic addresses in some convenient code, which then are translated into absolute addresses by an assembly program.

Base address permits derivation of an absolute address from a relative address.

Effective address is derived from applying specific indexing or indirect addressing rules to a specified address.

Four-plus-one address incorporates four operand addresses and a control address.

Immediate address incorporates the value of the operand in the address portion instead of the address of the operand.

N-level address is a multilevel address in which N levels of addressing are specified.

One-level address directly indicates the location of an instruction.

One-plus-one address contains two address portions. One address may indicate the operand required in the operation, and the other may indicate the following instruction to be executed.

Relative address is the numerical difference between a desired address and a known reference address.

Three-plus-one address incorporates an operation code, three operand address parts, and a control address.

Zero-level address permits immediate use of the operand.

*AND **Circuit.*** A computer logical decision element which provides an output if and only if all the input functions are satisfied. A three-variable AND element is shown in Fig. 8. The function F is a binary 1 if and only if A and B and C are all 1's. When any of the input functions is 0, the output function is 0. This may be represented in boolean algebra by $F = A \cdot B \cdot C$ or $F = ABC$. Diode and transistor circuit schematics for two-variable AND functions also are shown in Fig. 8. In integrated circuits, the function of the two transistors or diodes may be fabricated as a single active device. In the diode AND circuit, output F is positive only when both inputs A and B are positive. If one or both inputs are negative, one or both diodes will be forward-biased and the output will be negative. The transistor AND circuit operates in a similar manner, that is, if an input is negative, the associated transistor will be conducting and the output will be negative.

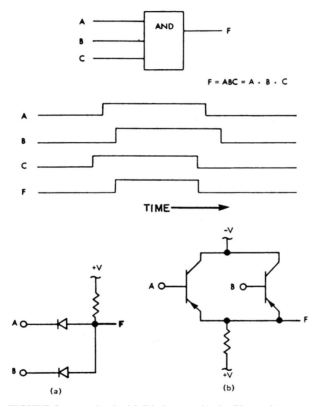

FIGURE 8 AND circuit: (*a*) Diode-type circuit; (*b*) transistor-type circuit.

Generally referred to as "fan in," the maximum number of input functions of which a given circuit configuration is capable is determined by the leakage current of the active element. Termed "fan out," the number of circuits which can be driven by the output is a function of current that can be supplied by the AND circuit.

Assembler. A computer program which operates on symbolic input data to produce machine instructions by carrying out such functions as (1) translating symbolic operation codes into computer instructions, (2) assigning locations in storage for successive instructions, and (3) assigning absolute addresses for symbolic addresses. An assembler generally translates input symbolic codes into machine instructions item for item and produces as output the same number of instructions or constants that were defined in the input symbolic codes.

Assembly language may be defined as computer language characterized by a one-to-one relationship between the statements written by the programmer and the actual machine instructions performed. The programmer thus has direct control over the efficiency and speed of the program. Usually the language allows the use of mnemonic names instead of numerical values for the operation codes of the instructions and similarly allows the user to assign symbolic names to the locations of the instructions and data. For the first feature, the assembler contains a table of the permissible mnemonic names and their numerical equivalents. For the second feature, the assembler builds such a table on a first pass through the program statements. Then the table is used to replace the symbolic names by their numerical values on a second pass through the program. Usually dummy operation codes (or pseudocodes) are needed by the assembler to pass control information to it. As an example, an origin statement is usually required as the first statement in the program. This gives the numerical value of the desired location of the first instruction or piece of data so that the assembler can, by counting the instructions and data, assign numerical values for their symbolic names.

The format of the program statements is usually rigidly specified, and only one statement per input record to the assembler is permitted. A representative statement is: symbolic name, operation code (or pseudocode), modifiers and/or register addresses, symbolic name of data. The mnemonic names used for the operation codes usually are defined uniquely for a particular computer type, with little standardization among computer manufacturers even for the most common operations. The programmer must learn a new language for each new machine.

Asynchronous.   A term used to designate the property of a device or action whose timing is not a direct function of the clock cycles in the system. In an asynchronous situation, the time of occurrence or duration of an event or operation is unpredictable because of factors such as variable signal propagation delay or a stimulus which is not under the control of the computer.

In terms of a computer channel, an asynchronous channel does not depend on computer clock pulses to control the transmission of information to and from the input or output device. Transmission of the information is under the control of interlocked control signals. Thus when a device has data to send to the channel, the device activates a service request signal. Responding to this signal, the channel activates a SERVICE OUT signal. The latter, in turn, activates a SERVICE IN signal in the device and also deactivates the request signal. Information is then transferred to the channel in coincidence with SERVICE IN, and the channel acknowledges receipt of the data by deactivating SERVICE OUT.

Asynchronous operation also occurs in the operation of analog-to-digital (A/D) subsystems. The systems may issue a command to the subsystem to read an analog point and then proceed to the next sequential operation. The analog subsystem carries out the A/D conversion. When the conversion is complete, the subsystem interrupts the system to signal the completion.

Asynchronous also has a broader meaning—specifically unexpected or unpredictable occurrences with respect to a program's instructions.

BASIC.   An acronym for *b*eginner's *a*ll-purpose *s*ymbolic *i*nstruction *c*ode. This language was developed at Dartmouth College by Kemeny and Kurz for timeshared use by students and other nonprofessional programmers. The language has been adopted by most computer firms and has been offered via timesharing services. BASIC is extensively used for small desktop computer systems and by computer hobbyists. The language is characterized by a very simple statement form. It normally requires that the first word in the statement be one of the small number of keywords and that a restricted naming convention for variables be used. An example of a program prepared in BASIC is

```
 05 LET E = -1
 10 READ A, B, C
 15 DATA 1, 2, 3
 20 LET D = A/B*C
 30 IF D = 4 GO TO 60
 40 IF D = 3/2 GO TO 80
 45 PRINT A, B, C, D, E
 50 STOP
 60 LET E = +1
 70 GO TO 45
 80 LET E = 0
 90 GO TO 45
100 END
```

The nucleus of the language is described by current ANSI specifications.

Baud. A traditional unit of telegraph signaling speed derived from the duration of the shortest signaling pulse. A telegraphic speed of 1 baud is one pulse per second. The term *unit pulse* sometimes has the same meaning. A related term, *dot cycle,* refers to an on-off or mark-space cycle in which both mark and space intervals have the same length as the unit pulse.

Bit. A contraction of *bi*nary dig*it*. A single character in a binary numeral, that is, a 1 or 0. A single pulse in a group of pulses also may be referred to as a bit. The bit is a unit of information capacity of a storage device. The capacity in bits is the logarithm to the base 2 of the number of possible states of the device.

   Parity Bit. A check bit that indicates whether the total number of binary 1 digits in a character or word (excluding the parity bit) is odd or even. If a 1 parity bit indicates an odd number of 1 digits, then a 0 bit indicates an even number of 1 digits. If the total number of 1 bits, including the parity bit, is always even, the system is called an even-parity system. In an odd-parity system, the total number of 1 bits, including the parity bit, is always odd.

   Zone Bit. (1) One of the two leftmost bits in a system in which 6 bits are used for each character; related to overpunch. (2) Any bit in a group of bit positions used to indicate a specific class of items, such as numbers, letters, special signs, and commands.

Boolean Algebra. Originated by George Boole (1815–1864), boolean algebra is a mathematical method of manipulating logical relations in symbolic form. Boolean variables are restricted to two possible values or states. Possible pairs of values for the boolean algebra variable are YES and NO, ON and OFF, TRUE and FALSE, and so forth. It is common practice to use the symbols 1 and 0 as the boolean variables. Since a digital computer typically uses signals having only two possible values or states, boolean algebra enables the computer designer to combine mathematically these variables and manipulate them in order to obtain the minimum design which realizes a desired logical function. A table of definitions and symbols for some of the logical operations defined in boolean algebra is shown in Fig. 9.

Logical Operation	Symbol	Definition
AND	$\cdot$	$A \cdot A = A; A \cdot 0 = 0; A \cdot 1 = 1; A \cdot \bar{A} = 0$
OR	$+$	$A + A = A; A + 0 = A; A + 1 = 1; A + \bar{A} = 1$
NOT	$-$	$A\bar{A} = 0; A + \bar{A} = 1$
EXCLUSIVE OR	$\oplus$	$A \oplus B = \bar{A}B + A\bar{B} = \overline{A \odot B}$
COINCIDENCE	$\odot$	$A \odot B = A\bar{B} + AB = \overline{A \oplus B}$
NAND (or Sheffer stroke)	$/$	$A/B = \bar{A} + \bar{B} = \overline{AB}$
NOR (or Peirce)		$A \ B = \bar{A}\bar{B} = \overline{A+B}$

FIGURE 9   Boolean algebra symbols.

Branch.   A set of instructions that may be executed between a couple of successive decision instructions. Branching allows parts of a program to be worked on to the exclusion of other parts and provides a computer with considerable flexibility. The branch point is a junction in a computer routine where one or both of two choices are selected under control of the routine. Also refers to one instruction that controls branching.

Buffer.   An internal portion of a digital computer or data-processing system which serves as intermediary storage between two storage or data handling systems with different access times or formats. The word *buffer* is also used to describe a routine for compensating for a difference in data rate, or time of occurrence of events, when transferring data from one device or task to another. For example, a buffer usually is used to connect an input or output device with the main or internal high-speed storage. The term *buffer amplifier* applies to an amplifier which provides impedance transformation between two analog circuits or devices. A buffer amplifier may be used at the input of an analog-to-digital (A/D) converter to provide a low source impedance for the A/D converter and a high load impedance for the signal source.

Byte.   A group of binary digits, usually shorter than a word and usually operated on as a unit. Through common usage, the word most often describes an 8-bit unit. This is a convenient information size to represent an alphanumeric character. Particularly in connection with communications, the trend has been toward the use of *octet* for an 8-bit byte, *sextet* for a 6-bit byte, and so on. Computer designs commonly provide for instructions that operate on bytes, as well as word-oriented instructions. The capacity of storage units is often specified in terms of bytes.

Central Processing Unit.   Also called the *mainframe,* the central processing unit (CPU) is the part of a computing system exclusive of input or output devices and, sometimes, main storage. As indicated in Fig. 10, the CPU includes the arithmetic and logic unit (ALU), channels, storage and associated registers, and controls. Information is transmitted to and from the input or output devices via the channel. Within the CPU, data are transmitted between storage and the channel and between each of these and the ALU. In some computing systems, if the channel and storage are separate assemblages of equipment, the CPU includes only the ALU and the instruction interpretation and execution controls.

A simplified example of CPU data flow was also given in Fig. 10. Generally, the operations performed in the CPU can be divided into (1) fetching instructions, (2) indexing (if appropriate to the instruction format), and (3) execution (Fig. 11). In the example shown, to initiate an operation, the contents of the instruction address register are transmitted to the storage address register (SAR), and the addressed instruction is fetched from storage and transferred to the storage

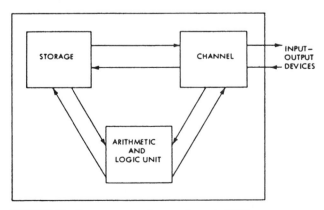

FIGURE 10   Computer central processing unit (CPU).

data register. The operation code portion of the instruction is set into the operation code register (OP), and the data address part is set into the data address register. If indexing is defined in
the instruction format, the indexing address is set into the index address register.

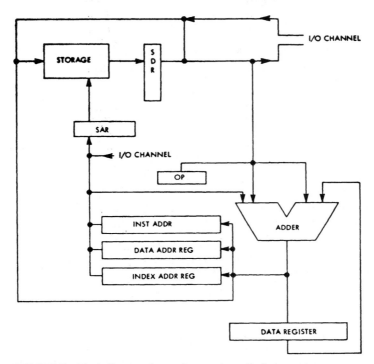

FIGURE 11   Block diagram of central processing unit of a computer.

During the indexing phase of the operation, the index address register is transmitted to the
SAR and the contents of the storage location are set into the storage data register. The contents of the storage data register and of the data address register are summed in the adder,
and the result is placed in the data address register.

In connection with an arithmetic operation, the data address register is transmitted to the
SAR, and the data in the referenced location are gated through the adder along with the contents of the data register. The result replaces the contents of the data register. If the arithmetic
operation causes a carry-out of the high-order position of the word being operated on, the
overflow indication is saved in a condition code trigger. This condition may be tested on a subsequent operation, such as a transfer on condition. On completion of the specified operation,
the next instruction is fetched from storage and the previously described phases are repeated.

Channel.   The portion of the central processing unit (CPU) of a computer which connects
input and output devices to the CPU. It may also execute instructions relating to the input or
output devices. A channel also provides the interfaces and associated controls for the transfer of data between storage and the input-output devices attached to the CPU. The channel
generally maintains the storage address for the device in operation and includes the buffer
registers needed for synchronizing with storage. A serial or selector channel may be used for
slower devices.

A *multiplex channel* has the capability of concurrently servicing several devices. The storage address for each operating device is controlled by the channel and is maintained in main storage or in the logic of the channel. If the address is maintained in the logic, this increases the maximum data rate of the channel and reduces the number of storage references needed to service the device. The storage-access channel is an adaptation of the multiplex channel. With this type of channel, the device transmits both the data and the storage address to the central processor when it requires servicing.

Where a *series channel* is used, one or more devices may be physically attached to the channel interface, but only one of them is logically connected at any given time. Thus for the selected device, the full channel data rate capacity is available. For a given device data rate requirement, a series channel is less costly than either dedicated channels per device or a multiplex channel. In the latter instance, the saving is realized from the fact that the maximum required data rate is determined by the data rate of a single attached device. The maximum rate of a multiplex channel is the sum of the data rate requirements of several devices, plus the data rate required for device addressing.

Channel also may be used to describe other portions of some computers. For example, *analog input channel* refers to the path between the input terminals of an analog input subsystem and an analog-to-digital (A/D) converter. Similarly, it may describe the logical or physical path between the source and the destination of a message in a communications system. In some applications relating to data acquisition in physics experiments, it is synonymous with the quantization interval in an A/D converter. The tracks along the length of magnetic tape used for storing digital data are also referred to as channels.

Character. One symbol of a set of elementary symbols, such as those corresponding to the keys of a typewriter, that is used for organization, representation, or control of data. Symbols may include the decimal digits 0 through 9, the letters A through Z, and any other symbol a computer can read, store, or write. Thus such symbols as @, #, $, and / are commonly used to expand character availability.

A *blank character* signifies an empty space on an output medium; or a lack of data on an input medium, such as an unpunched column on a punched card.

A *check character* signifies a checking operation. Such a character contains only the data needed to verify that a group of preceding characters is correct.

A *control character* controls an action rather than conveying information. A control character may initiate, modify, or stop a control operation. Actions may include the line spacing of a printer, the output hopper selection in a card punch, and so on.

An *escape character* indicates that the succeeding character is in a code that differs from the prior code in use.

A *special character* is not alphabetic, numeric, or blank. Thus @, #, . . . , are special characters.

Check. A process of partial or complete testing of the correctness of computer or other data-processing machine operations. Checks also may be run to verify the existence of certain prescribed conditions within a computer or the correctness of the results produced by a program. A check of any of these conditions usually may be made automatically by the equipment; or checks may be programmed.

Automatic Check. A procedure for detecting errors that is a built-in or integral part of the normal operation of a device. Until an error is detected, automatic checking normally does not require operating system or programmer attention. For example, if the product of a multiplication is too large for the space allocated, an error condition (overflow) will be signaled.

Built-in Check. An error detecting mechanism that requires no program or operator attention until an error is detected. The mechanism is built into the computer hardware.

Checkpoint. A point in time in a machine run at which processing is momentarily halted to perform a check or to make a magnetic tape or disk record (or equivalent) of the condition of all the variables of the machine run, such as the status of input and output devices and a copy of working storage. Checkpoints are used in conjunction with a restart routine to minimize reprocessing time occasioned by functional failures.

Duplication Check. Two independent performances of the same task are completed and the results compared. This is illustrated by the following operations:

12	31	84	127
9	14	43	66
21	45	127	193

Parity Check. A summation check in which the binary digits in a character or word are added, modulo 2, and the sum checked against a single, previously computed parity digit, that is, a check which tests whether the number of 1's in a word is odd or even.

Programmed Check. A system of determining the correct program and machine functioning either by running a sample problem with similar programming and a known answer or by using mathematical or logic checks, such as comparing $A \times B$ with $B \times A$. Also, a check system built into the program for computers that do not have automatic checking.

Reasonableness Check. Same as *validity check.* See below.

Residue Check. An error detection system in which a number is divided by a quantity n and the remainder compared with the original computer remainder.

Sequence Check. A data-processing operation designed to check the sequence of the items in a file assumed to be already in sequence.

Summation Check. A check in which groups of digits are summed, usually without regard for overflow, and that sum is checked against a previously computed sum to verify that no digits have been changed since the last summation.

Validity Check. A check based on known limits or on given information or computer results; for example, a calendar month will not be numbered greater than 12, or a week does not have more than 168 hours.

Code. A system of symbols for representing data or instructions in a computer or data-processing machine. A machine language program sometimes is referred to as a *code.*

Alphanumeric Code. A set of symbols consisting of the alphabet characters A through Z and the digits 0 through 9. Sometimes the definition is extended to include special characters. A programming system commonly restricts user-defined symbols to only those using alphanumeric characters and for the system to take special action on the occurrence of a nonalphanumeric character, such as $, %, or &. For example, a job currently in progress may be stopped should a given special character be encountered.

Binary Code. (1) A coding system in which the encoding of any data is done through the use of bits, that is, 0 or 1. (2) A code for the 10 decimal digits 0, 1, . . . , 9 in which each is represented by its binary, radix 2, equivalent, that is, straight binary.

Biquinary Code. A two-part code in which each decimal digit is represented by the sum of the two parts, one of which has the value of decimal 0 or 5, and the other the values 0 through 4. The abacus and soroban both use biquinary codes. An example follows:

Decimal	Biquinary	Interpretation
0	0 000	0 + 0
1	0 001	0 + 1
2	0 010	0 + 2
3	0 011	0 + 3
4	0 100	0 + 4
5	1 000	5 + 0
6	1 001	5 + 1
7	1 010	5 + 2
8	1 011	5 + 3
9	1 100	5 + 4

Column-Binary Code.   A code used with punch cards in which successive bits are represented by the presence or absence of punches in contiguous positions in successive columns as opposed to rows. Column binary code is widely used in connection with 36-bit-word computers, where each group of three columns is used to represent a single word.

Computer Code or Machine Language Code.   A system of combinations of binary digits used by a given computer.

Excess-Three Code.   A binary-coded decimal code in which each digit is represented by the binary equivalent of that number plus 3; for example:

Decimal digit	Excess-3 code	Binary value
0	0011	3
1	0100	4
2	0101	5
3	0110	6
4	0111	7
5	1000	8
6	1001	9
7	1010	10
8	1011	11
9	1100	12

Gray Code.   A binary code in which sequential numbers are represented by expressions which are the same except in one place and in that place differ by one unit:

Decimal	Binary	Gray
0	000	000
1	001	001
2	010	011
3	011	010
4	100	110
5	101	111

Thus in going from one decimal digit to the next sequential digit, only one binary digit changes its value; synonymous with cyclic code.

Instruction Code.   The list of symbols, names, and definitions of the instructions which are intelligible to a given computer or computing system.

Mnemonic Operation Code.   An operation code in which the names of operations are abbreviated and expressed mnemonically to facilitate remembering the operations they represent. A mnemonic code normally needs to be converted to an actual operation code by an assembler before execution by the computer. Examples of mnemonic codes are ADD for addition, CLR for clear storage, and SQR for square root.

Numeric Code.   A system of numerical abbreviations used in the preparation of information for input into a machine; that is, all information is reduced to numerical quantities. Contrasted with *alphabetic code.*

Symbolic Code or Pseudocode.   A code which expresses programs in source language; that is, by referring to storage locations and machine operations by symbolic names and addresses which are independent of their hardware-determined names and addresses.

Two-out-of-Five Code.   A system of encoding the decimal digits 0, 1, . . . , 9 where each digit is represented by binary digits of which two are 0's and three are 1's, or vice versa.

Column.   A character or digit position in a positional information format, particularly one in which characters appear in rows and the rows are placed one above another; for example, the rightmost column in a five-decimal-place table or in a list of data.

Command.   The specification of an operation to be performed. In terms of a control signal, a command usually takes the form of YES (go ahead) or NO (do not proceed). A command should not be confused with an instruction. In most computers, an instruction is given to the central processing unit (CPU), as contrasted with a command, which is an instruction to be followed by a data channel. An input command, for example, may be READ and an output command, WRITE.

Compiler.   A program designed to translate a higher-level language into machine language. In addition to its translating function, which is similar to the process used in an assembler, a compiler program is able to replace certain items of input with a series of instructions, usually called subroutines. Thus where an assembler translates item for item and produces as output the same number of instructions or constants that were put into it, a compiler typically produces multiple output instructions for each input instruction or statement. The program which results from compiling is a translated and expanded version of the original.

Compiler language is characterized by a one-to-many relationship between the statements written by a programmer and the actual machine instructions executed. The programmer typically has little control over the number of machine instructions executed to perform a particular function and is dependent on the particular compiler implementation. Typically, the language is very nearly machine-independent and may be biased in its statements and features to a particular group of users with similar problems. Thus these languages sometimes are referred to as problem-oriented languages (POLs). Slightly different implementations of a given language are sometimes called *dialects.*

Computer Operating System.   Generally defined as a group of interrelated programs to be used on a computer system in order to increase the utility of the hardware and software. There is a wide range in the size, complexity, and application of operating systems. The need for operating systems arose from the desire to obtain the maximum amount of service from a computer. A first step was simple monitor systems providing a smooth job-to-job transition. Modern operating systems contain coordinate programs to control input-output scheduling, task scheduling, error detection and recovery, data management, debugging, multiprogramming, and on-line diagnostics.

Operating system programs fall into two main categories: (1) control programs and (2) processing programs:

1. Control programs
   a. Data management, including all input-output
   b. Job management
   c. Task management
2. Processing programs
   a. Language translators
   b. Service programs (linking programs, sort or merge, utilities)
      System library routines
   c. User-written programs

Control programs provide the structure and environment in which work may be accomplished more efficiently. They are the components of the supervisory portion of the system. Processing programs are programs that have some specific objective that is unrelated to controlling the system work. These programs use the services of supervisory programs rather than operating as part of them.

Data management is involved with the movement of data to and from all input-output devices and all storage facilities. This area of the system embraces the functions referred to as the input-output control system (IOCS), which frequently is segmented into two parts, the physical IOCS and the logical IOCS.

Physical IOCS is concerned with device and channel operations, error procedures, queue processing, and, generally, all operations concerned with transmitting physical data segments from storage to external devices. *Logical IOCS* is concerned with data organization, buffer handling, data referencing mechanisms, logical device reference, and device independence.

Job management involves the movement of control cards or commands through the system input device, their initial interpretation, and the scheduling of jobs so indicated. Other concerns are job queues, priority scheduling of jobs, and job accounting functions.

Task management is concerned with the order in which work is performed in the system. This includes management of the control facilities, namely, central processing unit, storage, input-output channels, and devices, in accordance with some task priority scheme.

Language translators allow users to concentrate on solving logical problems with a minimum of thought given to detailed hardware requirements. In this category are higher-level languages, report generators, and special translation programs.

Service programs are needed to facilitate system operation or to provide auxiliary functions. Sort programs, program linking functions, and file duplication, as well as system libraries are within this group.

User-written programs are programs prepared specifically to assist the user in the accomplishment of his or her objectives.

Concurrent Operation.   The performance of several actions during the same interval of time, although not necessarily simultaneously. Compare with *parallel operation,* described later in this section. Multiprogramming is a technique that provides concurrent execution of several tasks in a computer. See also *multiprogramming* in this section.

Conversion Time.   The interval of time between initiation and completion of a single analog-to-digital (A/D) or digital-to-analog (D/A) conversion operation. Also, the reciprocal of the conversion rate. In practice, the term usually refers to A/D converters or digital voltmeters. The conversion time required by an A/D converter is comprised of (1) the time needed to reset and condition the logic, (2) a delay to allow for the settling time of the input buffer amplifier, (3) a polarity determination time, (4) the actual A/D conversion operation, and (5) any time required to transfer the digital result to an output register. Not all factors are always present. A unipolar A/D converter, for example, involves no polarity determination time.

Conversion time also is used in connection with data acquisition or analog input subsystems for process control computers. The more precise term in this case is *measurement time.* In this case the measurement rate may not be the reciprocal of the measurement time, inasmuch as some of the operations performed may overlap with other operations. With some types of A/D converters it is possible to select the next multiplexer point during the time the previous value is being converted from analog to digital form.

Conversion or measurement time also may include the time required for such operations as multiplexer address decoding, multiplexer switch selection, and settling time, and the time required to deactivate the multiplexer switches and permit the subsystem to return to an initial state.

Counter.   A physical or logical device capable of maintaining numeric values which can be incremented or decremented by the value of another number. A counter, located in storage, may be incremented or decremented under control of the program. An example of such use is recording the number of times a program loop has been executed. The counter location is set to the value of the number of times the sequence of instructions is to be performed. On completion of the sequence, the counter is decremented by 1 and then tested for zero. If the answer is nonzero, the sequence of instructions may be repeated.

A counter also may be in the form of a circuit that records the number of times an event occurs. A counter which counts according to the binary number system is illustrated by the truth table shown in Fig. 12. Counters also may be used to accumulate the number of times an external event takes place. The counter may be a storage location or a counter circuit which is incremented as the result of an external stimulus. Specific counter definitions include the following.

	A_3	A_2	A_1
P_1	0	0	1
P_2	0	1	0
P_3	0	1	1
P_4	1	0	0
P_5	1	0	1
P_6	1	1	0
P_7	1	1	1
P_8	0	0	0

FIGURE 12 Truth table for binary counter. The situation illustrates a binary counter of three stages capable of counting up to eight pulses. Each trigger changes state when a pulse is gated to its input. In the instance of trigger A_2, it receives an input pulse only when trigger A_1 and the input are both 1. Subsequently, trigger A_3 changes its state only when both triggers A_1 and A_2 and the input pulse are all 1's. This table shows the values of each trigger after each input pulse. At the eighth pulse, the counter resets to zero.

A *binary counter* is (1) a counter which counts according to the binary number system, or (2) a counter capable of assuming one of two stable states.

A *control counter* records the storage location of the instruction word which is to be operated on following the instruction word in current use. The control counter may select storage locations in sequence, thus obtaining the next instruction word from the subsequent storage location, unless a transfer or special instruction is encountered.

A *location counter* or *instruction counter* is (1) the control section register which contains the address of the instruction currently being executed, or (2) a register in which the address of the current instruction is recorded. It is synonymous with *program address counter.*

Diagnostics. Programs provided for the maintenance engineer or operator to assist in discovering the source of a particular computer system malfunction. Diagnostics generally consist of programs which force extreme conditions (the worst patterns) on the suspected unit with the expectation of exaggerating the symptoms sufficiently for the engineer to readily discriminate among possible faults and to identify the particular fault. In addition, diagnostics may provide assistance in localizing the cause of a malfunction to a particular card or component in the system.

There are two basic types of diagnostic programs: (1) off-line and (2) on-line. Off-line diagnostic programs are those which require that there be no other program active in the computer system, sometimes requiring that there be no executive program in the system. Off-line diagnostics are typically used for central processing unit (CPU) malfunctions, very obscure and persistent peripheral device errors, or critically time-dependent testing. For example, it may be suspected or known that a harmonic frequency is contributing to the malfunction. Thus it may be desirable to drive the unit continuously at various precise frequencies close to the suspected frequency to confirm the diagnosis and then to confirm the cure. Interference from other activities may well make such a test meaningless. Thus all other activity on the system must cease.

On-line diagnostics are used mainly in a multiprogramming environment and are vital to the success of real-time systems. The basic concept is that of logically isolating the malfunctioning unit from all problem programs and allowing the diagnostic program to perform any and all functions on the unit. Many of the more common malfunctions can be isolated by such diagnostics, but there are limitations imposed by interference from other programs also using the CPU.

Digit. A symbol used to convey a specific quantity either by itself or with other numbers of its set; for example, 2, 3, 4, and 5 are digits. The base or radix must be specified and each digit's value assigned.

Binary Digit. A number on the binary scale of notation. This digit may be 0 or 1. It may be equivalent to an ON or an OFF condition or a YES or NO condition. Often abbreviated *bit.*

Check Digit. One or more redundant digits carried along with a machine word and used in relation to the other digits in the word as a self-checking or error detecting code to detect malfunctions of equipment in data transfer operations.

Equivalent Binary Digits. The number of binary positions needed to enumerate the elements of a specific set. In the case of a set with five elements, it will be found that three equivalent binary digits are needed to enumerate the five members of the set 1, 10, 11, 100, and 101. Where a word consists of three decimal digits and a plus or minus sign, 1999 different combinations are possible. This set would require 11 equivalent binary digits in order to enumerate all its elements.

Octal Digit. The symbols 0, 1, 2, 3, 4, 5, 6, or 7 used as a digit in the system of notation which uses 8 as the base or radix.

Sign Digit. A digit incorporating 1 to 4 binary bits, which is associated with a data item for the purpose of denoting an algebraic sign. In most binary, word-organized computers, a 1-bit sign is used: 0 = + (plus); and 1 = − (minus). Although not strictly a digit by the foregoing definition, it occupies the first digit position and it is common to consider it a digit.

EXCLUSIVE OR *Circuit.* A logical element which has the properties that if either of the inputs is a binary 1, then the output is a binary 1. If both the inputs are a binary 1 or 0, the output is a binary 0. In terms of boolean algebra, this function is represented as $F = AB' + BA'$, where the prime denotes the NOT function. With reference to the transistor EXCLUSIVE OR circuit shown in Fig. 13, the output is positive when either transistor is in saturation. When input A is positive and B is negative, transistor T_2 is in saturation. When B is positive and A is negative, transistor T_1 is in saturation. When A and B are either both positive or both negative, then both transistors are cut off and the output F is negative. Although shown as discrete devices in the figure, fabrication using large-scale integrated circuit technology may utilize other circuit and design configurations.

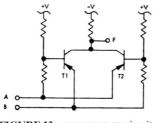

FIGURE 13 EXCLUSIVE OR circuit.

Fixed-Point Arithmetic. A method of storing numeric data in a computer such that the data are all stored in integer form (or all in fractional form) and the user postulates a radix point between a certain pair of digits. Consider a computer whose basic arithmetic is in decimal and in which each computer word consists of seven decimal digits in integer form. If it is desired to add 2.796512 to 4.873214, the data are stored in the computer as 2796512 and 4873214, the sum of which is 7669726. It is recalled that a decimal point between digits 1 and 2 has been postulated. The result, therefore, represents 7.669726. Input and output conversion routines often are provided for convenience. These routines can add or delete the radix point in the external representation and align the data as required internally.

Fixed-point operations are fast and thus preferred over floating-point operations. It is important, of course, that the magnitude of the numbers be much better known than for floating-point numbers, since the absolute magnitude is limited by word size and the availability of double-length operations. For many applications, fixed-point calculations are practical and increase speed.

Fortran. An acronym standing for *for*mula *tran*slation. It is a programming language designed for problems which can be expressed in algebraic notation, allowing for exponentiation, subscripting, and other mathematical functions.

Fortran was introduced by IBM in 1957 after development by a working group headed by John W. Backus. It was the first computer language to be used widely for solving numerical problems and was the first to become an American National Standard. Numerous enhancements have been added to the language. The current version is described in ANSI/ISA publications.

Index Register. The contents of the index register of a computer are generally used to modify the data address of the instruction as the instruction is being read from storage. The modified address is called the *effective data address*. A particular index register is addressed by a specified field in the format of the instruction. The data address of the instructions would contain the address of the required data with reference to the start of the table. All instructions which reference table data are indexed by the specified index register which contains the address of the start of the table. Thus when the program is to perform these operations on another table of data, the value in the index register is changed to the start address of the new data table. This effectively modifies all the indexed instructions in the sequence.

Instruction.   (1) A set of characters which defines an operation together with one or more addresses, or no address, and which, as a unit, causes the computer to perform the operation on the indicated quantities. The term *instruction* is preferable to the terms *command* and *order. Command* is reserved for a specified portion of the instruction word, that is, the part which specifies the operation to be performed. *Order* is reserved for the ordering of the characters, implying sequence, or the order of the interpolation, or the order of the differential equation. (2) The operation or command to be executed by a computer, together with associated addresses, tags, and indices.

Interrupt.   A signal which causes the central processing unit (CPU) to change state as the result of a specified condition. An interrupt represents a temporary suspension of normal program execution and arises from an external condition, from an input or output device, or from the program currently being processed in the CPU. See earlier text in this article.

Language.   A communications means for transmitting information between human operators and computers. The human programmer describes how the problem is to be solved using the computer language. A computer language consists of a well-defined set of characters and words, coupled with a series of rules (*syntax*) for combining them into computer instructions or statements. There is a wide variety of computer languages, particularly in terms of flexibility and ease of use. There are three levels in the hierarchy of computer languages:
   Machine Language.   (1) A language designed for interpretation and use by a machine without translation. (2) A system for expressing information which is intelligible to a specific machine, such as a computer or class of computers. Such a language may include instructions which define and direct machine operations, and information to be recorded by or acted on by these machine operations. (3) The set of instructions expressed in the number system basic to a computer, together with symbolic operation codes with absolute addresses, relative addresses, or symbolic addresses. In this case, it is known as an *assembly language.*
   Problem-Oriented Language.   A language designed for the convenience of program specification in a general problem area. The components of such a language may bear little resemblance to machine instructions and often incorporate terminology and functions unique to an application. This type of language is also known as an *applications language.*
   Procedure-Oriented Language.   A machine-independent language which describes how the process of solving the problem is to be carried out. For example, Fortran, Algol, PL/1, and Cobol.
   Other computer languages include:
   Algorithmic Language.   An arithmetic language by which numerical procedures may be precisely presented to a computer in a standard form. The language is intended not only as a means of directly presenting any numerical procedure to any appropriate computer for which a compiler exists but also as a means of communicating numerical procedures among individuals.
   Artificial Language.   A language specifically designed for ease of communication in a particular area of endeavor, but one that is not yet "natural" to that area. This is contrasted with a natural language which has evolved through long usage.
   Common Business-Oriented Language.   A specific language by which business data-processing procedures may be precisely described in a standard form. The language is intended not only as a means for directly presenting a business program to any appropriate computer for which a compiler exists but also as a means of communicating such procedures among individuals.

Common Machine Language. A machine-sensible information representation which is common to a related group of data-processing machines.

Object Language. A language which is the output of an automatic coding routine. Usually object language and machine language are the same. However, a series of steps in an automatic coding system may involve the object language of one step serving as a source language for the next step, and so forth.

Logic. In hardware, a term referring to the circuits which perform the arithmetic and control operations in a computer. In designing digital computers, the principles of boolean algebra are employed. The logical elements of AND, OR, INVERTER, EXCLUSIVE OR, NOR, NAND, NOT, and so forth, are combined to perform a specified function. Each of the logical elements is implemented as an electronic circuit which in turn is connected to other circuits to achieve the desired result. The word *logic* is also used in computer programming to refer to the procedure or algorithm necessary to achieve a result.

Logical Operation. (1) A logical or boolean operation on N-state variables which yields a single N-state variable, such as a comparison of the three-state variables A and B, each represented by $-$, 0, or $+$, which yields $-$ when A is less than B, 0 when A equals B, and $+$ when A is greater than B. Specifically, operations such as AND, OR, and NOT on two-state variables which occur in the algebra of logic, that is, boolean algebra. (2) Logical shifting, masking, and other nonarithmetic operations of a computer.

Loop. A sequence of instructions that may be executed repeatedly while a certain condition prevails. The productive instructions in a loop generally manipulate the operands, while bookkeeping instructions may modify the productive instructions and keep count of the number of repetitions. A loop may contain any number of conditions for termination, such as the number of repetitions or the requirement that an operand be nonnegative. The equivalent of a loop can be achieved by the technique of straight-line coding, whereby the repetition of productive and bookkeeping operations is accomplished by explicitly writing the instructions for each repetition.

Macroassembler. An assembler which permits the user to define pseudo computer instructions which may generate multiple computer instructions when assembled. Source statements which may generate multiple computer instructions are termed *macrostatements* or *macroinstructions*. With a process control digital computer, a macroassembler can be a most significant tool. By defining a set of macrostatements, for example, a process control engineer can define a process control programming language specifically oriented to the process.

Mask. A pattern of digits used to control the retention or elimination of portions of another pattern of digits. Also, the use of such a pattern. For example, an 8-bit mask having a single i bit in the ith position, when added with another 8-bit pattern, can be used to determine whether the ith bit is a 1 or a 0; that is, the ith bit in the pattern will be retained and all other bits will be 0's.

As another example, there are situations where it is desirable to delay the recognition of a process interrupt by a digital computer. A mask instruction permits the recognition of specific interrupts to be inhibited until it is convenient to service them.

Memory. Although the word *memory* is widely used, international standards list the word as a deprecated term and prefer the word *storage*. Memory or storage is a means of storing information. A complete memory system incorporates means of placing information into storage and of retrieving it. See earlier text in this article.

Microcomputer. A computer which utilizes a microprocessor as its central processing unit (CPU). This CPU must perform two functions: (1) sequence through the instructions and (2) execute each instruction. A microcomputer requires two other fundamental elements: (1) memory for the program—a sequence of instructions to be performed, and (2) input-output circuits to tie it to external devices. A microcomputer is a digital logic device, that is, all input-output signals are at digital logic levels, either 0 or 5 V. A significant amount of interface circuitry is required between the microcomputer and external devices.

Microprocessor.   A program-controlled component, normally contained on a single crystal of silicon on which over one thousand transistors may be implemented. The microprocessor functions to control the sequence of instructions making up a program. It sequences through these instructions, examining each instruction in turn and executing it. This means that the microprocessor performs some manipulation of data. See text in earlier part of this article.

Microprogram.   Microprogramming is a technique of using a special set of instructions for an automatic computer. Elementary logical and control commands are used to simulate a higher-level instruction set for a digital computer. The basic machine-oriented instruction set in many computers is comprised of commands, such as ADD, DIVIDE, SUBTRACT, and MULTIPLY, which are executed directly by the hardware. The hardware actually implements each function as a combination of elementary logical functions, such as AND, OR, and EXCLUSIVE OR. The manner of exact implementation usually is not of concern to the programmer. Compared with the elementary logical functions, the ADD, SUBTRACT, MULTIPLY, and DIVIDE commands are a high-level language set in the same sense that a macrostatement is a higher-level instruction set when compared with the machine-oriented language instruction set.

In a microprogrammed computer, the executable (micro) instructions which may be used by the (micro) programmer are comprised of a logical function, such as AND and OR, and some elementary control functions, such as shift and branch. The (micro) programmer then defines microprograms which implement an instruction set analogous to the machine-oriented language instruction set in terms of these microinstructions. Using this derived instruction set, the systems or applications programmer writes programs for the solution of a problem. When the program is executed, each derived instruction is executed by transferring control to a microprogram. The microprogram is then executed in terms of the microinstructions of the computer. After execution of the microprogram, control is returned to the program written in the derived instruction set. The microprogram typically is stored in read-only storage (ROS) and thus is permanent and cannot be changed without physical replacement.

An advantage of microprogramming is increased flexibility. This can be realized by adapting the derived machine-oriented instruction set to a particular application. The technique enables the programmer to implement a function, such as a square root, directly without subroutines or macrostatements. Thus significant programming and execution efficiency is realized if the square root function is commonly required. Also, the instruction set of a character-oriented computer can be implemented on a word-oriented machine where adequate microprogramming is provided.

Minicomputer.   There is no widely accepted definition of this term, although it is frequently used in the information processing industry, as are other computer classifications, such as microcomputer, supermini, supercomputer, and mainframe computer. Nevertheless, computers classified as minicomputers often have one or more of the following characteristics: (1) They are called minicomputers by their manufacturers; (2) they utilize a 16-bit instruction and data word, often divided into two separately addressable 8-bit bytes; the supermini utilizes a 32-bit word, usually consisting of four separately addressable bytes; (3) they are often packaged using a rack-and-panel construction, although some may consist of only a single printed circuit board; (4) they are highly modular, having a wide variety of optional features, peripheral equipment, and adapters for attaching specialized peripheral devices, such as data acquisition equipment; (5) they are often utilized by original equipment manufacturers (OEMs) to provide computing capability in specialized equipment, such as electronic assembly testers; and (6) they often provide as standard or optional features devices such as time-of-day clocks, interval timers, and hardware-implemented priority interrupts which facilitate their use in real-time applications. To a considerable extent, personal computers have markedly impacted on the use of the minicomputer. See text in earlier part of this article.

Module.   In a computer system, particularly in connection with data acquisition and personal computers, a module or a series of modules may be used to expand or, in some cases, alter the ability and performance of a computer. See text in earlier portion of this article.

Multiprogramming.   The essentials of a multiprogramming system in connection with digital computer operations are: (1) several programs are resident in main storage simultaneously; and (2) the central processing unit (CPU) is timeshared among the programs. This makes for better utilization of a computer system. Where only one program may reside in the main storage at any given time, inefficient use of the CPU time results when a program requests data from an input-output device. The operation is delayed until the requested information is received. In some applications, such delays can constitute a large portion of the program execution time. In the multiprogramming approach, other programs resident in storage may use the CPU while a preceding program is awaiting new information. Multiprogramming practically eliminates CPU lost time due to input-output delays. Multiprogramming is particularly useful in process control or interactive applications which involve large amounts of data input and output.

Input-output delay-time control is a basic method for controlling the interplay between multiple programs. Where multiprogramming is controlled in this manner, the various programs resident in storage are normally structured in a hierarchy. When a given program in the hierarchy initiates an input-output operation, the program is suspended until such time as the input-output operation is completed. A lower-priority program is permitted to execute during the delay time.

In a *time-slice* multiprogramming system, each program resident in storage is given a certain fixed interval of CPU time. Multiprogramming systems for applications where much more computation is done than input-output operation usually use the time-slice approach.

Multiprogramming systems allow multiple functions to be controlled simultaneously by a single process control digital computer. A multiprogramming system allows a portion of the storage to be dedicated to each type of function required in the control of the process and thus eliminates interference among the various types of functions. In addition, it provides the means whereby asynchronous external interrupts can be serviced effectively on a timely basis.

NAND Circuit.   A computer logical decision element which has the characteristic that the output F is 0 if and only if all the inputs are 1's. Conversely, if any one of the input signals A or B or C or the three-input NAND element shown in Fig. 14 is not a 1, the output F is a binary 1. Although the NAND function can be achieved by inverting the output of an AND circuit, the specific NAND circuit requires fewer circuit elements. A two-input transistor NAND circuit is shown in Fig. 15. The output F is negative only when both transistors are cut off. This occurs when both inputs are positive. The number of inputs, or fan-in, is a function of the components and circuit design. NAND is a contraction of *not and.*

NOR Circuit.   A computer logical decision element which provides a binary 1 output if all the input signals are a binary 0. This is the overall NOT of the logical OR operation. A boolean algebra expression of the NOR circuit is $F = (AB)'$, where the prime denotes the NOT function. A two-input transistor NOR circuit is shown in Fig. 16. Output F is positive only when both transistors are cut off. This occurs when both inputs A and B are negative.

NOT Circuit.   Also known as an inverter circuit, this is a circuit which provides a logical NOT of the input signal. If the input signal is a binary 1, the output is a binary 0. If the input signal is in the 0 state, the output is in the 1 state. In reference to Fig. 17, if A is positive, the output F is at 0 V inasmuch as the transistor is biased into conduction. If A is at 0 V, the output is at + V because the transistor is cut off. Expressed in boolean algebra, $F = A'$, where the prime denotes the NOT function.

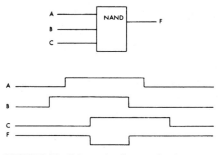

FIGURE 14   Schematic of NAND circuit.

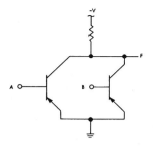

FIGURE 15   Transistor-type
NAND circuit.

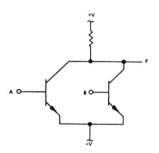

FIGURE 16   Transistor-type
NOR circuit.

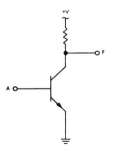

FIGURE 17   Inverter or NOT
circuit.

OR Circuit.   A computer logical decision element which has the characteristic of providing a binary 1 output if any of the input signals are in a binary 1 state. This is expressed in terms of boolean algebra by $F = A + B$. A diode and a transistor representation of this circuit is shown in Fig. 18. In a diode-type OR circuit, if either input signal A or B or both are positive, the respective diode is forward-biased and the output F is positive. The number of allowable input signals to the diode OR gate is a function of the back resistance of the diodes. The input transistors of a transistor-type OR circuit are forced into higher conductivity when the respective input signal becomes positive. Thus the output signal becomes positive when either or both inputs are positive.

Parallel Operation.   The simultaneous performance of several actions, usually of a similar nature, through provision of individual similar or identical devices for each such action, particularly flow or processing of information. Parallel operation is performed to save time over serial operation. The decrease in the cost of multiple high-function integrated circuits has made parallel operation much more common in recent years. Because computer logic speeds already have approached a very high limit, the use of parallel execution holds the best promise for increasing overall computing speeds. Multiprocessors represent a case where parallel operation is achieved by providing two or more complete processors capable of simultaneous operation. Compare with *concurrent operation.*

Pascal.   Invented by N. Wirth in 1971, Pascal is a comparatively recent high-level programming language named in honor of Blaise Pascal, the seventeenth-century French philosopher who invented the first workable mechanical adding machine at age 19. The language was designed for the systematic teaching of programming as a discipline based on fundamental concepts of structure and integrity. Although taught widely, Pascal is increasingly being used outside the classroom in non-business-oriented applications. The language is characterized by strong typing and a syntax which encourages readable programs and good programming practices. It is a descendant of Algol, although not a strict superset of it. Pascal compilers are available on many computers, ranging from microcomputers to mainframes. Efforts to develop national and international standards for this language commenced in 1981.

Program.   (1) The complete plan for the computer solution of a problem, more specifically the complete sequence of instructions and routines necessary to solve a problem. (2) To plan the procedures for solving a problem. This may involve, among other things, analysis of the problem, preparation of a flow diagram, preparing details, texting and developing subroutines, allocation of storage locations, specification of input and output formats, and the incorporation of a computer run into a complete data processing system.

   Internally Stored Program.   A sequence of instructions stored inside the computer in the same storage facilities as the computer data, as opposed to external storage on punched paper tape and pinboards.

   Object Program.   The program which is the output of an automatic coding system, such as an assembler or compiler. Often the object program is a machine-language program ready for execution, but it may well be in an intermediate language.

   Source Program.   A computer program written in a language designed for ease of expression of a class of problems or procedures by humans, such as symbolic or algebraic. A generator, assembler, translator, or compiler routine is used to perform the mechanics of translating the source program into an object program in machine language.

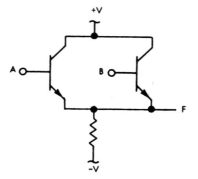

FIGURE 18   OR circuit.

Program Generator.   A program that permits a computer to write other programs automatically. Generators are of two types: (1) the *character-controlled generator,* which is like a compiler in that it takes entries from a library of functions but is unlike a simple compiler in that it examines control characters associated with each entry and alters instructions found in the library according to the directions contained in the control characteristics; (2) the *pure generator,* which is a program that writes another program. When associated with an assembler, a pure generator is usually a section of program which is called into storage by the assembler from a library and then writes one or more entries in another program. Most assemblers are also compilers and generators. In this case the entire system is usually referred to as an *assembly program.*

Programming Flowchart.   A graphic representation of a program, in which symbols are used to represent data, flow, operations, equipment, and so forth. A digital computer program may be charted for two primary reasons: (1) ease of initial program design and (2) program documentation. By coding from a flowchart, instead of coding without any preliminary design, a programmer usually conserves time and effort in developing a program. In addition, a flowchart is an effective means of transmitting an understanding of the program to another person.

A programming flowchart is comprised of function blocks with connectors between the blocks. A specific function box may represent an input-output operation, a numerical computation, or a logic decision. The program chart shown in Fig. 19 is of a program that reads values from the process, converts the values to engineering units, limit-checks the converted values, and, if there is a violation, prints an alarm message on the process operator's display. A set of symbols used in flowcharting is described by the American National Standards Institute (ANSI).

Various levels of detail are presented in programming flowcharts. A functional block in a low-level flowchart may represent only a few computer instructions, whereas a functional block in a high-level flowchart may represent many computer instructions. A high-level flowchart is used mainly for initial program design and as a way of informing a nonprogrammer of what the program does. A low-level chart usually appears in the last stage of flowcharting before a program is actually coded. It is used for documentation that may be needed for later modifications and corrections.

Program Relocation.   A process of modifying the address in a unit of code, that is, a program subroutine, or procedure to operate from a different location in storage than that for which it was originally prepared. Program relocation is an integral part of nearly all programming systems. The technique allows a library of subroutines to be maintained in object form and made a part of a program by relocation and appropriate linking. When a program is prepared for execution by relocating the main routine and any included routines to occupy a certain part of the storage on a one-time basis, the process usually is termed *static relocation.* The resulting relocated program may be resident in a library and may be loaded into the same location in storage each time it is executed. Dynamic storage allocation schemes also may be set up so that each time a program is loaded, it is relocated to an available space in storage. This process is termed *dynamic relocation.* Timesharing systems may temporarily stop the execution of programs, store them in auxiliary storage, and later reload them into a different location for continued execution. This process is also termed dynamic relocation.

Software relocation commonly refers to a method whereby a program loader processes all the code of a program as it is loaded and modifies any required portions. Auxiliary information is carried in the code to indicate which parts must be altered. Inasmuch as all code must be examined, this method can be time-consuming.

In *hardware relocation* special machine components are used, as a base or relocation register, to alter addresses automatically at execution time in order to achieve the desired

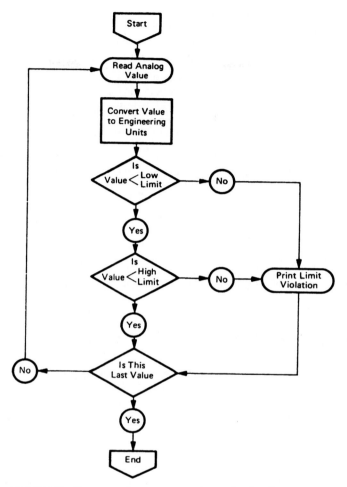

FIGURE 19   Representative computer programming flowchart.

results. In dynamic relocation situations this is a fast method. Coding techniques also are used for relocation. The resulting code is caused to be self-relocating and executes in any storage location into which it is loaded. An index register may be used to make all references to storage. Double indexing is required to perform both the relocation and the normal indexing operations. Or the code may actually modify all addresses as it executes to provide the correct reference. This method is slow compared with other methods and requires additional storage.

Queue.   When events occur at a faster rate than they can be handled in a computer system, a waiting line, or *queue,* must be formed. The elements of a queue typically are pointers (addresses) which refer to the items waiting for service. The items may be tasks to be executed or messages to be sent over communications facilities. The term is also used as a verb, meaning to place an item in a queue. Several methods of organizing queues are used. The *sequential queue* is common. As new elements arrive, they are placed at the end; as elements are processed, they are taken from the front. This is the first in–first out (FIFO) organization.

In the *pushdown queue* the last one in is the first one out (LIFO). The *multipriority queue* processes from the front, but in terms of priority of the elements waiting. Essentially, this is a modified sequential queue containing subsequences and is sometimes referred to as priority in–first out (PIFO).

Resolution.   In systems where either the input or the output of the subsystem is expressed in digital form, the resolution is determined by the number of digits used to express the numerical value. In a digital-to-analog converter, the output analog signal takes on a finite number of discrete values which correspond to the discrete numerical input. The output of an analog-to-digital converter is discrete, although the analog input signal is continuous.

In digital equipment, resolution is typically expressed in terms of the number of digits in the input or output digital representation. In the binary system, a typical specification is that "resolution is x bits." As an example, if V_{fs} is the full-scale input or output voltage range, this specification states that the resolution is $V_{fs}/2^x$. If $x = 10$ and $V_{fs} = 5$ V, the resolution is $5/2^{10}$, or 0.00488 V. It is also common to express resolution in terms of parts. A four-digit decimal converter may be said to have a resolution of 1 part in 10,000, and a 10-bit binary converter may be said to have a resolution of 1 part in 1024. The term *least-significant bit* (LSB) also is used. It may be stated, for example, that the binary resolution is $\pm\frac{1}{2}$ LSB. Also used is the term *least-significant digit* (LSD). This term is used in relation to decimal or other nonbinary digital equipment.

Routine.   A set of coded instructions arranged in proper sequence to direct a computer to perform a desired operation or sequence of operations. A subdivision of a program consisting of two or more instructions that are functionally related, hence a program.

Diagnostic Routine.   A routine used to locate a malfunction in a computer or to aid in locating mistakes in a computer program. Thus, in general, any routine specifically designed to aid in debugging or troubleshooting.

Executive Routine.   A routine which controls loading and relocation of routines and in some cases makes use of instructions not available to the general programmer. Effectively, an executive routine is part of the machine itself. Synonymous with *monitor routine, supervisory routine,* and *supervisory program.*

Heuristic Routine.   A routine by which the computer attacks a problem not by a direct algorithmic procedure but by a trial-and-error approach frequently associated with the act of learning.

Interpretive Routine.   A routine which decodes and immediately executes instructions written as pseudocodes. This is contrasted with a compiler which decodes pseudocodes into a machine language routine to be executed at a later time. The essential characteristic of an interpretive routine is that a particular pseudocode operation must be decoded each time it is executed. Synonymous with interpretive code.

Service Routine.   A broad class of routines provided at a particular installation for the purpose of assisting in the maintenance and operation of a computer as well as the preparation of programs as opposed to routines for the actual solution of production problems. This class includes monitoring or supervisory routines, assemblers, compilers, diagnostics for computer malfunctions, simulations of peripheral equipment, general diagnostics, and input data. The distinguishing quality of service routines is that they are generally tailored to meet the servicing needs at a particular installation, independent of any specific production-type routine requiring such services.

Tracing Routine.   A diagnostic routine used to provide a time history of one or more machine registers and controls during execution of the object routine. A complete tracing routine reveals the status of all registers and locations affected by each instruction each time the instruction is executed. Since such a trace is prohibitive in machine time, traces which provide information only following the execution of certain types of instructions are more frequently used. Furthermore, a tracing routine may be under the control of the processor or may be called in by means of a trapping feature.

Scale Factor.   In digital computing, an arbitrary factor which may be associated with numbers in a computer to adjust the position of the radix so that the significant digits occupy specified columns. In analog computing, a proportionality factor which relates the magnitude of a variable to its representation within a computer.

Serial Operation.   The flow of information through a computer in time sequence using only one digit, word, line, or channel at a time. Serial addition in character-oriented computers permits the formation of sums with low-cost hardware. Addition occurs from right to left. Parallel addition is used in faster word- or byte-organized computers.

Magnetic disk and drum storage units may access and record data on a serial-by-bit basis. Conversion to (or from) the parallel form utilized in the central processing unit (CPU) is performed in the associated control unit. Except in the case of short distances, most communications between computers or between computers and many types of terminals take place on a serial-by-bit basis.

Software.   The totality of programs, procedures, rules, and (possibly) documentation used in conjunction with computers, such as compilers, assemblers, narrators, routines, and subroutines. References are made to the software and hardware parts of a system, where the hardware comprises the physical (mechanical and electronic) components of the system. In some machines the instructions are microprogrammed in a special control storage section of the machine, using a more basic code actually wired into the machine. This is contrasted with the situation where the instructions are wired into the control unit. The microprogram technique permits the economic construction of various size machines which appear to have identical instruction sets. However, microprograms generally are not considered software and are sometimes called *firmware*. For additional information on software, see Index.

Storage.   Any medium capable of storing information. As generally defined, however, a storage unit is a device on or in which data can be stored, read, and erased. The major classifications of storage devices associated with computer systems are (1) immediate access, (2) random access, and (3) sequential access. As a general rule, the cost per bit of information is greater for immediate-access storage devices, but the access time is considerably faster than for the other two types.

Immediate-Access Storage.   Devices in which information can be read in a microsecond or less. Usually an array of storage elements can be directly addressed, and thus all information in the array requires the same amount of time to be read.

Random-Access Storage.   Devices in which the time required to obtain information is independent of the location of the information most recently obtained. This strict definition must be qualified by the observation that what is meant is *relatively* random. Thus magnetic drums are relatively nonrandom access when compared with monolithic storage, but are relatively random access when compared with magnetic tapes for file storage. Disk storage and drum storage units usually are referred to as random-access storage devices. The time required to read or write information on these units generally is in the 10- to 200-ms range, but is dependent on where the information is recorded with respect to the read-write head at the time the data are addressed.

Sequential-Access Storage.   Devices in which the items of information stored become available only in a one-after-the-other sequence, whether or not all the information or only some of it is desired. Storage on magnetic tape is an example. Also see text in earlier portion of this article.

Storage Protect.   Several methods are used to effect storage protection in digital computers. The objective is to protect certain areas of code from alteration by other areas of code. Storage may be protected by areas, each area being given a different program-settable code or key. A master key is used to permit the control program to refer to all areas. In another system, the storage is protected on an individual word basis, providing finer resolution but increasing the time required to protect a given area size. In both cases, instructions are provided to set the protect status by programming. Some systems provide for a disable of the feature from the console. This permits a preset protection pattern to be established, but with the program capability to disable selected protection status. Protection is required so that stored information will not be altered accidentally by store-type instructions. In single-word schemes, any storage modification is prevented. In the multikey area systems storage approach, modification is limited to operation in areas with the same key.

Synchronous.   A synchronous operation takes place in a fixed time relation to another operation or event, such as a clock pulse. When a set of contacts is sampled at a fixed time interval, the operation is termed synchronous. This situation is to be contrasted with that where the contacts may be sampled randomly under the control of an external signal. Generally, the read operation of a main storage unit is synchronous. The turning on of the X and Y selection drivers and the sampling of the storage output on the sense line are controlled by a fixed frequency. Contrast with *asynchronous.*

Timesharing.   The use of a device, particularly a computer or data-processing machine, for two or more purposes during the same overall time interval, accomplished by interspersing component and subsystem actions in time. In the case of a digital computer, timesharing generally connotes the process of using main storage for the concurrent execution of more than one job by temporarily storing all jobs in auxiliary storage except the one in control. This technique allows a computer to be used by several independent program users. The method most often is associated with a computer-controlled terminal system used in an interactive mode.

Time Slicing.   A technique that allows several users to utilize a computer facility as though each had complete control of the machine. Several users can be serviced one at a time, unaware of each other, because of the relative speed between computer operation and human response. Essentially, time slicing is used by a software control system to control the allocation of facilities of a computer to tasks requesting service. The allocation basis is a fixed time interval. Each job is executed for the time period used in the time slice. The job is then temporarily stored on an auxiliary storage device (timesharing) or suspended (multiprogramming) while another job is being run. Each job, therefore, is run in an incremental fashion until complete.

Word.   A character or bit string that traditionally has been an entity in computer technology. A word typically consists of one or more bytes. In small computers, there are usually 2 bytes per word. In large machines, there may be up to 8 bytes or more. Instructions are provided for manipulating words of data and, typically, most instructions occupy one word of storage. In addition, the internal data paths in a computer (parallel) are designed to transfer one word of data at a time.

SECTION 4
PROCESS VARIABLES—FIELD INSTRUMENTATION*

L. Arnold
Johnson Yokogawa Corporation, Newnan, Georgia (Flow Systems)

S. Barrows
The Foxboro Company (A Siebe Company), Norcross, Georgia (Temperature Systems)

W. H. Burtt
The Foxboro Company (A Siebe Company), Foxboro, Massachusetts (Resonant-Wire Pressure Transducers)

H. R. Cantrell
Delta m Corporation, Oak Ridge, Tennessee (Thermal Level System)

W. A. Clayton
Hy-Cal Engineering (A Unit of General Signal), El Monte, California (Temperature Systems—prior edition)

R. Collier
ABS-Kent-Taylor (ASEA Brown Boveri), Rochester, New York (Temperature Systems—prior edition)

L. E. Cuckler
(Deceased) formerly, Robertshaw Controls Company, Anaheim, California (Moisture Measurement)

Z. C. Dobrowolski
(Deceased) formerly Chief Development Engineer, Kinney Vacuum Company (A Unit of General Signal), Cannon, Massachusetts (High Vacuum Measurement—prior edition)

G. L. Dyke
Bailey Controls Company, Wickliffe, Ohio (Pressure Transducers and Transmitters)

C. J. Easton
President, Sensotec, Inc., Columbus, Ohio (Strain-Pressure Transducers—prior edition)

---

* Persons who authored complete articles or subsections of articles, or who otherwise cooperated in an outstanding manner in furnishing information and helpful counsel to the editorial staff.

4.1

C. E. Fees
Fischer & Porter Company, Warminster, Pennsylvania (Rotameters—prior edition)

H. Grekksa
Square D Company, Infrared Measurement Division, Niles, Illinois (Temperature Systems)

E. H. Higham
Foxboro Great Britain Limited, Redhill, Surrey, England (Magnetic Flowmeter—prior edition)

D. A. Jackson
Omega Engineering, Inc., Stamford, Connecticut (Temperature Systems—prior edition)

J. Kortright
Leeds & Northrup (A Unit of General Signal), North Wales, Pennsylvania (Flow Systems; Temperature Systems)

G. Kuebler
Great Lakes Instruments, Inc., Milwaukee, Wisconsin (Flow Systems; Fluid Level Systems)

A. J. Kurylchek
Consultant, Wayne, New Jersey (Industrial Weighing and Density Systems)

R. W. Lally
Engineering Department, PCB Piezotronics, Inc., Depew, New York (Pressure Sensors)

G. Leavitt
ABB-Kent-Taylor (ASEA Brown Boveri), Rochester, New York (Temperature Systems—prior edition)

M. Levine
Process Instrument Division, Panametrics, Inc., Waltham, Massachusetts (Moisture Sensors)

C. L. Mamzic
Manager, Systems and Application Engineering, Moore Products Company, Spring House, Pennsylvania (Flow Systems—prior edition)

John Masters
Great Lakes Instruments, Inc., Milwaukee, Wisconsin (RF Level Systems)

G. R. McFarland
Senior Project Manager, ABB-Kent-Taylor, Inc., Rochester, New York (Wedge-Type Flow Element)

S. Milant
Managing Director, Ryan Instruments, Redmond, Washington (Thermistors)

R. W. Miller
Consultant, Foxboro, Massachusetts (Flow Differential Producers—prior edition)

A. E. Mushin
Omega Engineering, Inc., Stamford, Connecticut (Thermocouple Systems—prior edition)

R. Peacock
Vice-President, Engineering, Land Instruments, Inc., Tullytown, Pennsylvania (Radiation Thermometers—prior edition)

B. Pelletier
Rosemount Inc., Measurement Division, Eden Prairie, Minnesota (Temperature Systems; Hydrostatic Level Gages)

G. Rebucci
Schenck Weighing Systems, Totowa, New Jersey (Weighing Systems—prior edition)

W. L. Ricketson
Toledo Scale Division, Reliance Electric Company, Atlanta, Georgia (Weighing Systems)

Bill Roeber
Great Lakes Instruments, Inc., Milwaukee, Wisconsin (Target Flowmeters)

Staff
AMETEK, Process and Analytical Division, Newark, Delaware (Quartz Crystal Moisture Detector)

Staff
Brooks Instrument Division, Emerson Electric Company, Hatfield, Pennsylvania (Positive-Displacement Meters)

Staff
Lucas Schaevitz, Pennsauken, New Jersey (Gage Pressure Transducers)

Staff
Raytek Incorporated, Santa Cruz, California (Temperature Systems)

Technical Staff
Moisture Systems Corporation, Hopkinstown, Massachusetts (IR Moisture Analyzer)

R. D. Thompson
ABB-Kent-Taylor (ASEA Brown Boveri), Rochester, New York (Temperature Systems—prior edition)

Peter E. Wiederhold
President, General Eastern Instruments Corporation, Watertown, Massachusetts (Humidity—prior edition)

J. A. Wise
National Institute of Standards and Technology, Gaithersburg, Maryland (Temperature Systems—prior edition)

Gene Yazbak
Senior Applications Engineer, MetriCor, Inc., Monument Beach, Massachusetts (Temperature Systems)

TEMPERATURE SYSTEMS

Over many decades the demand for temperature sensors and controllers has shown that temperature is the principal process variable of serious concern to the process industries, that is, those industries that handle and convert gases, liquids, and bulk solids into products and by-products. Chemical, petroleum, petrochemical, polymer, plastic, and large segments of metallurgical and food processors are examples. Temperature control is critical to such processes and operations as chemical reactions and in materials separations, such as distillation, drying, evaporation, absorbing, crystallizing, baking, and extruding. Temperature control also plays a critical role in the safe operation of such facilities.

Although critical temperature control applications occur less frequently in the discrete-piece manufacturing industries, there are numerous examples. The extensive demand for temperature controllers in the air-conditioning field is self-evident.

TEMPERATURE DEFINED

Although temperature fundamentally relates to the kinetic energy of the molecules of a substance (as reflected in the definition of the absolute, thermodynamic, or Kelvin temperature scale), temperature may be defined in a less academic fashion as "the condition of a body which determines the transfer of heat *to* or *from* other bodies," or even more practically, as "the degree of 'hotness' or 'coldness' as referenced to a specific scale of temperature measurement."

TEMPERATURE SCALES

Thermodynamic Kelvin Scale. The currently accepted theoretical scale is named for Lord Kelvin, who first enunciated the principle on which it is based. Thermodynamic temperature is denoted by T and the unit is the kelvin (K) (no degree sign is used). The kelvin is the fraction $1/273.16$ of the thermodynamic temperature of the triple point of water. The triple point is realized when ice, water, and water vapor are in equilibrium. It is the sole defining fixed point of the thermodynamic Kelvin scale and has the assigned value of 273.16 K.

Celsius (Centigrade) Scale. In 1742 Anders Celsius of Uppsala University, Sweden, reported on the use of thermometers in which the fundamental interval, ice point to steam point, was 100°. Celsius designated the ice point at 100° and the steam point at 0°. Subsequently Christin (1743) in Lyon, France, and Linnaeus (1745) at Uppsala independently interchanged the designations. For many years prior to 1948 it was known as the centigrade scale. In 1948, by international agreement, it was renamed in honor of its inventor. Used worldwide, temperatures are denoted as degrees Celsius (°C). By personal choice, the degree symbol is sometimes eliminated.

Fahrenheit Scale. Daniel Gabriel Fahrenheit (1724) first defined the Fahrenheit scale, using the ice point (32°) and the human body temperature (96°) as the fixed points of the scale. The fundamental interval (ice point to steam point) turned out to be 180 degrees (212 − 32 = 180). Although very serious attempts have been and are being made to convert to the Celsius scale, the Fahrenheit scale remains popular in English-speaking countries. Scientific and engineering publications largely have converted to the Celsius scale, but because the conversion still is far from complete, many technical publications usually follow a value in °C by the equivalent value in °F. Again, by personal choice the degree symbol is sometimes eliminated, such as in 100 F.

Réaumur Scale.   Invented prior to 1730 by René-Antoine Ferchalt de Réaumur, the scale today is essentially limited to the brewing and liquor industries. The fundamental temperature interval is defined by the ice point (0°) and a steam-point designation of 80°. The symbol is °R.

Rankine Scale.   This scale is the equivalent of the thermodynamic Kelvin scale, but is expressed in terms of Fahrenheit degrees. Thus the temperature of the triple point of water on the Rankine scale, corresponding to 273.16 K, is very nearly 491.69° Rankine.

International Practical Temperature Scale.   For precision calibration needs, the concept of an international temperature scale with fixed reference points in addition to the ice point and the steam point was proposed as early as 1887. The last revisions to this scale of any note occurred with the publication of the fixed points for the International Practical Temperature Scale (IPTS) of 1968. In the usual applications of thermometry, this scale is not frequently used. Some of the intermediate reference points on the scale include the triple point of equilibrium of hydrogen, the boiling point of neon, the triple point of oxygen, and the freezing points of zinc, silver, and gold. The IPTS is reviewed periodically, as recently as 1990.

Temperature Scale Conversion

A convenient chart for converting degrees Celsius to degrees Fahrenheit and vice versa is given in Table 1.

TEMPERATURE SENSORS[1]

All materials are affected by temperature, and thus it is not surprising that there are so many means available for inferring temperature from some physical effect. Early thermometers depended on volumetric changes of gases and liquids with temperature change, and, of course, this principle still is exploited, as encountered in industrial gas- and liquid-filled thermal systems and in the familiar liquid-column fever thermometer. Although these instruments were accepted widely for many years, the filled-system thermometer has been significantly displaced by other simpler and more convenient approaches, including the thermocouple and the resistance temperature detector (RTD). The contraction and expansion of solids, notably metals, is a phenomenon that has been applied widely in thermometry as, for example, in bimetallic metallic temperature controllers commonly found in the air-conditioning field. Thermoelectric methods, such as the thermocouple, and thermoresistive effects, such as the change of electrical resistance with temperature change, as found in RTDs and thermistors, also have been known and applied for many decades. Thermal radiation of hot

---

[1] Temperature sensors are described in much more detail in the forthcoming *Industrial Sensors and Measurements Handbook,* D. M. Considine, Editor-in-Chief, McGraw-Hill, New York (circa 1995).

bodies has served as the basis for radiation thermometers [once commonly referred to as radiation pyrometers and now called infrared (IR) thermometers] and has also been known and practiced for many decades. Through technological advancements IR thermometers have grown in acceptance in recent years and displaced other measurement means for a number of temperature measurement applications.

Thus as is the case with other process variables, there is a wide selection of thermal sensors. An effort is made in this article to summarize the relative advantages and limitations of easily available temperature sensors, but for some uses, the justification of one method over another is sometimes difficult.

THERMOCOUPLES

For many years the thermocouple was the clear-cut choice of instrumentation and control engineers in the process industries, but in recent years the position of the thermocouple has been increasingly challenged by the RTD. Nevertheless, the thermocouple still is used widely.

Thermocouple Principles

Seebeck Effect.   As early as 1821, Seebeck observed the existence of thermoelectric circuits while studying the electromagnetic effects of metals. He found that bonding wires of two dissimilar metals together to form a closed circuit caused an electric current to flow in the circuit whenever a difference in temperature was imposed between the end junctions (Fig. 1).

Peltier Effect.   Jean Peltier (1834) discovered that when an electric current flows across a junction of two dissimilar metals, heat is liberated or absorbed. When the electric current flows in the same direction as the Seebeck current, heat is absorbed at the hotter junction and liberated at the colder junction. The Peltier effect may be defined as the change in heat content when a quantity of charge (1 coulomb) crosses the junction (Fig. 2). The Peltier effect is the fundamental basis for thermoelectric cooling and heating.

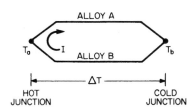

FIGURE 1 Seebeck's circuit. The direction and magnitude of the Seebeck voltage E_s depend on the temperature of the junctions and on the materials making up the thermocouple. For a particular combination of materials A and B over a small temperature difference, $dE_s = \alpha A,B\ dT$, where $\alpha A,B$ is a coefficient of proportionality called the Seebeck coefficient.

Thomson Effect.   Sir William Thomson (Lord Kelvin) discovered in 1851 that a temperature gradient in a metallic conductor is accompanied by a small voltage gradient whose magnitude and direction depend on the particular metal. When an electric current flows, there is an evolution or absorption of heat due to the presence of the thermoelectric gradient, with the net result that the heat evolved in an interval bounded by different temperatures is slightly greater or less than that accounted for by the resistance of the conductor. The Thomson effects are equal and opposite and thus cancel each other, thus allowing the use of extension wires with thermocouples because no electromotive force (EMF) is added to the circuit.

Practical Rules Applying to Thermocouples.   Based on decades of practical experience, the following rules apply:

TABLE 1 Temperature Conversion Table*†

°C		°F	°C		°F	°C		°F	°C		°F	°C		°F	°C		°F	°C		°F	°C		°F	°C		°F
-273.1	**-459.4**		-17.8	**0**	32	10.0	**50**	122.0	38	**100**	212	260	**500**	932	538	**1000**	1832	816	**1500**	2732	1093	**2000**	3632	1371	**2500**	4532
-268	**-450**		-17.2	**1**	33.8	10.6	**51**	123.8	43	**110**	230	266	**510**	950	543	**1010**	1850	821	**1510**	2750	1099	**2010**	3650	1377	**2510**	4550
-262	**-440**		-16.7	**2**	35.6	11.1	**52**	125.6	49	**120**	248	271	**520**	968	549	**1020**	1868	827	**1520**	2768	1104	**2020**	3668	1382	**2520**	4568
-257	**-430**		-16.1	**3**	37.4	11.7	**53**	127.4	54	**130**	266	277	**530**	986	554	**1030**	1886	832	**1530**	2786	1110	**2030**	3686	1388	**2530**	4586
-251	**-420**		-15.6	**4**	39.2	12.2	**54**	129.2	60	**140**	284	282	**540**	1004	560	**1040**	1904	838	**1540**	2804	1116	**2040**	3704	1393	**2540**	4604
-246	**-410**		-15.0	**5**	41.0	12.8	**55**	131.0	66	**150**	302	288	**550**	1022	566	**1050**	1922	843	**1550**	2822	1121	**2050**	3722	1399	**2550**	4622
-240	**-400**		-14.4	**6**	42.8	13.3	**56**	132.8	71	**160**	320	293	**560**	1040	571	**1060**	1940	849	**1560**	2840	1127	**2060**	3740	1404	**2560**	4640
-234	**-390**		-13.9	**7**	44.6	13.9	**57**	134.6	77	**170**	338	299	**570**	1058	577	**1070**	1958	854	**1570**	2858	1132	**2070**	3758	1410	**2570**	4658
-229	**-380**		-13.3	**8**	46.4	14.4	**58**	136.4	82	**180**	356	304	**580**	1076	582	**1080**	1976	860	**1580**	2876	1138	**2080**	3776	1416	**2580**	4676
-223	**-370**		-12.8	**9**	48.2	15.0	**59**	138.2	88	**190**	374	310	**590**	1094	588	**1090**	1994	866	**1590**	2894	1143	**2090**	3794	1421	**2590**	4694
-218	**-360**		-12.2	**10**	50.0	15.6	**60**	140.0	93	**200**	392	316	**600**	1112	593	**1100**	2012	871	**1600**	2912	1149	**2100**	3812	1427	**2600**	4712
-212	**-350**		-11.7	**11**	51.8	16.1	**61**	141.8	99	**210**	410	321	**610**	1130	599	**1110**	2030	877	**1610**	2930	1154	**2110**	3830	1432	**2610**	4730
-207	**-340**		-11.1	**12**	53.6	16.7	**62**	143.6	100	**212**	413	327	**620**	1148	604	**1120**	2048	882	**1620**	2948	1160	**2120**	3848	1438	**2620**	4748
-201	**-330**		-10.6	**13**	55.4	17.2	**63**	145.4	104	**220**	428	332	**630**	1166	610	**1130**	2066	888	**1630**	2966	1166	**2130**	3866	1443	**2630**	4766
-196	**-320**		-10.0	**14**	57.2	17.8	**64**	147.2	110	**230**	446	338	**640**	1184	616	**1140**	2084	893	**1640**	2984	1171	**2140**	3884	1449	**2640**	4784
-190	**-310**		-9.44	**15**	59.0	18.3	**65**	149.0	116	**240**	464	343	**650**	1202	621	**1150**	2102	899	**1650**	3002	1177	**2150**	3902	1454	**2650**	4802
-184	**-300**		-8.89	**16**	60.8	18.9	**66**	150.8	121	**250**	482	349	**660**	1220	627	**1160**	2120	904	**1660**	3020	1182	**2160**	3920	1460	**2660**	4820
-179	**-290**		-8.33	**17**	62.6	19.4	**67**	152.6	127	**260**	500	354	**670**	1238	632	**1170**	2138	910	**1670**	3038	1188	**2170**	3938	1466	**2670**	4838
-173	**-280**		-7.78	**18**	64.4	20.0	**68**	154.4	132	**270**	518	360	**680**	1256	638	**1180**	2156	916	**1680**	3056	1193	**2180**	3956	1471	**2680**	4856
-169	**-273**	-459.4	-7.22	**19**	66.2	20.6	**69**	156.2	138	**280**	536	366	**690**	1274	643	**1190**	2174	921	**1690**	3074	1199	**2190**	3974	1477	**2690**	4874
-168	**-270**	-454	-6.67	**20**	68.0	21.1	**70**	158.0	143	**290**	554	371	**700**	1292	649	**1200**	2192	927	**1700**	3092	1204	**2200**	3992	1482	**2700**	4892
-162	**-260**	-436	-6.11	**21**	69.8	21.7	**71**	159.8	149	**300**	572	377	**710**	1310	654	**1210**	2210	932	**1710**	3110	1210	**2210**	4010	1488	**2710**	4910
-157	**-250**	-418	-5.56	**22**	71.6	22.2	**72**	161.6	154	**310**	590	382	**720**	1328	660	**1220**	2228	938	**1720**	3128	1216	**2220**	4028	1493	**2720**	4928
-151	**-240**	-400	-5.00	**23**	73.4	22.8	**73**	163.4	160	**320**	608	388	**730**	1346	666	**1230**	2246	943	**1730**	3146	1221	**2230**	4046	1499	**2730**	4946
-146	**-230**	-382	-4.44	**24**	75.2	23.3	**74**	165.2	166	**330**	626	393	**740**	1364	671	**1240**	2264	949	**1740**	3164	1227	**2240**	4064	1504	**2740**	4964
-140	**-220**	-364	-3.89	**25**	77.0	23.9	**75**	167.0	171	**340**	644	399	**750**	1382	677	**1250**	2282	954	**1750**	3182	1232	**2250**	4082	1510	**2750**	4982
-134	**-210**	-346	-3.33	**26**	78.8	24.4	**76**	168.8	177	**350**	662	404	**760**	1400	682	**1260**	2300	960	**1760**	3200	1238	**2260**	4100	1516	**2760**	5000
-129	**-200**	-328	-2.78	**27**	80.6	25.0	**77**	170.6	182	**360**	680	410	**770**	1418	688	**1270**	2318	966	**1770**	3218	1243	**2270**	4118	1521	**2770**	5018
-123	**-190**	-310	-2.22	**28**	82.4	25.6	**78**	172.4	188	**370**	698	416	**780**	1436	693	**1280**	2336	971	**1780**	3236	1249	**2280**	4136	1527	**2780**	5036
-118	**-180**	-292	-1.67	**29**	84.2	26.1	**79**	174.2	193	**380**	716	421	**790**	1454	699	**1290**	2354	977	**1790**	3254	1254	**2290**	4154	1532	**2790**	5054
-112	**-170**	-274	-1.11	**30**	86.0	26.7	**80**	176.0	199	**390**	734	427	**800**	1472	704	**1300**	2372	982	**1800**	3272	1260	**2300**	4172	1538	**2800**	5072
-107	**-160**	-256	-0.56	**31**	87.8	27.2	**81**	177.8	204	**400**	752	432	**810**	1490	710	**1310**	2390	988	**1810**	3290	1266	**2310**	4190	1543	**2810**	5090

Temperature Conversion Table*†

°C	°(convert)	°F
-101	**-150**	-238
-95.6	**-140**	-220
-90.0	**-130**	-202
-84.4	**-120**	-184
-78.9	**-110**	-166
-73.3	**-100**	-148
-67.8	**-90**	-130
-62.2	**-80**	-112
-56.7	**-70**	-94
-51.1	**-60**	-76
-45.6	**-50**	-58
-40.0	**-40**	-40
-34.4	**-30**	-22
-28.9	**-20**	-4
-23.3	**-10**	14
-17.8	**0**	32

°C	°(convert)	°F
0	**32**	89.6
0.56	**33**	91.4
1.11	**34**	93.2
1.67	**35**	95.0
2.22	**36**	96.8
2.78	**37**	98.6
3.33	**38**	100.4
3.89	**39**	102.2
4.44	**40**	104.0
5.00	**41**	105.8
5.56	**42**	107.6
6.11	**43**	109.4
6.67	**44**	111.2
7.22	**45**	113.0
7.78	**46**	114.8
8.33	**47**	116.6
8.89	**48**	118.4
9.44	**49**	120.2

°C	°(convert)	°F
27.8	**82**	179.6
28.3	**83**	181.4
28.9	**84**	183.2
29.4	**85**	185.0
30.0	**86**	186.8
30.6	**87**	188.6
31.1	**88**	190.4
31.7	**89**	192.2
32.2	**90**	194.0
32.8	**91**	195.8
33.3	**92**	197.6
33.9	**93**	199.4
34.4	**94**	201.2
35.0	**95**	203.0
35.6	**96**	204.8
36.1	**97**	206.6
36.7	**98**	208.4
37.2	**99**	210.2

°C	°(convert)	°F
210	**410**	770
216	**420**	788
221	**430**	806
227	**440**	824
232	**450**	842
238	**460**	860
243	**470**	878
249	**480**	896
254	**490**	914

°C	°(convert)	°F
438	**820**	1508
443	**830**	1526
449	**840**	1544
454	**850**	1562
460	**860**	1580
466	**870**	1598
471	**880**	1616
477	**890**	1634
482	**900**	1652
488	**910**	1670
493	**920**	1688
499	**930**	1706
504	**940**	1724
510	**950**	1742
516	**960**	1760
521	**970**	1778
527	**980**	1796
532	**990**	1814

°C	°(convert)	°F
716	**1320**	2408
721	**1330**	2426
727	**1340**	2444
732	**1350**	2462
738	**1360**	2480
743	**1370**	2498
749	**1380**	2516
754	**1390**	2534
760	**1400**	2552
766	**1410**	2570
771	**1420**	2588
777	**1430**	2606
782	**1440**	2624
788	**1450**	2642
793	**1460**	2660
799	**1470**	2678
804	**1480**	2696
810	**1490**	2714

°C	°(convert)	°F
993	**1820**	3308
999	**1830**	3326
1004	**1840**	3344
1010	**1850**	3362
1016	**1860**	3380
1021	**1870**	3398
1027	**1880**	3416
1032	**1890**	3434
1038	**1900**	3452
1043	**1910**	3470
1049	**1920**	3488
1054	**1930**	3506
1060	**1940**	3524
1066	**1950**	3542
1071	**1960**	3560
1077	**1970**	3578
1082	**1980**	3596
1088	**1990**	3614

°C	°(convert)	°F
1271	**2320**	4208
1277	**2330**	4226
1282	**2340**	4244
1288	**2350**	4262
1293	**2360**	4280
1299	**2370**	4298
1304	**2380**	4316
1310	**2390**	4334
1316	**2400**	4352
1321	**2410**	4370
1327	**2420**	4388
1332	**2430**	4406
1338	**2440**	4424
1343	**2450**	4442
1349	**2460**	4460
1354	**2470**	4478
1360	**2480**	4496
1366	**2490**	4514

°C	°(convert)	°F
1549	**2820**	5108
1554	**2830**	5126
1560	**2840**	5144
1566	**2850**	5162
1571	**2860**	5180
1577	**2870**	5198
1582	**2880**	5216
1588	**2890**	5234
1593	**2900**	5252
1599	**2910**	5270
1604	**2920**	5288
1610	**2930**	5306
1616	**2940**	5324
1621	**2950**	5342
1627	**2960**	5360
1632	**2970**	5378
1638	**2980**	5396
1643	**2990**	5414

* General formula: °F = (°C × 9/5) + 32; °C = (°F − 32) × 5/9.

† The numbers in **boldface** type refer to the temperature (in either Celsius or Fahrenheit degrees) which it is desired to convert into the other scale. If converting from degrees Fahrenheit to degrees Celsius, the equivalent temperature is in the left column, while if converting from degrees Celsius to degrees Fahrenheit, the equivalent temperature is in the column on the right.

Interpolation factors

°C		°F	°C		°F
0.56	**1**	1.8	3.33	**6**	10.8
1.11	**2**	3.6	3.89	**7**	12.6
1.67	**3**	5.4	4.44	**8**	14.4
2.22	**4**	7.2	5.00	**9**	16.2
2.78	**5**	9.0	5.56	**10**	18.0

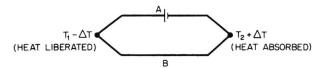

FIGURE 2   Peltier effect.

- A thermocouple current cannot be sustained in a circuit of a single homogeneous material, however varying in cross section, by the application of heat alone.
- The algebraic sum of the thermoelectromotive forces in a circuit composed of any number of dissimilar materials is zero if all of the circuit is at a uniform temperature. This means that a third homogeneous material always can be added to a circuit with no effect on the net EMF of the circuit so long as its extremities are at the same temperature. Therefore a device for measuring the thermal EMF may be introduced into a circuit at any point without affecting the resultant EMF, provided all the junctions added to the circuit are at the same temperature. It also follows that any junction whose temperature is uniform and which makes a good electrical contact does not affect the EMF of the thermocouple circuit regardless of the method used in forming the junction (Fig. 3).

- If two dissimilar homogeneous metals produce a thermal EMF of E_1 when the junctions are at temperatures T_1 and T_2, and a thermal EMF of E_2 when the junctions are at T_2 and T_3, the EMF generated when the junctions are at T_1 and T_3 will be $E_1 + E_2$.

The application of this law permits a thermocouple calibrated for a given reference temperature to be used with any other reference temperature through the use of a suitable correction. Figure 4 shows a schematic example. Another example of this law is that extension wires having the same thermoelectric characteristics as those of the thermocouple wires can be introduced into the thermocouple circuit (from region T_2 to region T_3 in Fig. 4) without affecting the net EMF of the thermocouple.

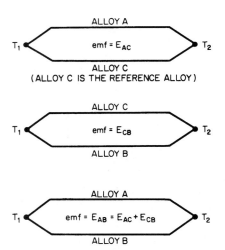

FIGURE 3   Thermocouple EMF algebraic sums.

Thermocouple Signal Conditioning

The thermocouple output voltage is quite nonlinear with respect to temperature. Further, changes in the reference junction temperature influence the output signal of thermocouples. Thermocouple signal conditioning is discussed in considerable detail in Section 7, Article 1, of this handbook.

Types of Thermocouples

In present times thermocouple users have the advantage of applying a very mature technology. Over many decades of research into finding metal and alloy combinations that provide an ample millivoltage per degree of temperature, and that resist corrosion in oxidizing and

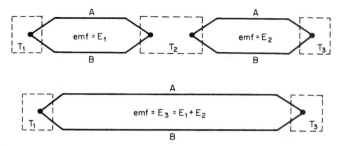

FIGURE 4   EMFs are additive for temperature intervals.

reducing atmospheres, a massive base of scientific and engineering information has developed. Much research also has gone into finding sheaths and protecting wells that will withstand both very high and very low temperatures. It should be pointed out, however, that the universal thermocouple has not been found. Thermocouple selection requires much experience to make the optimum choice for a given application.

This situation is evident from Table 2, which provides key specifying information on the five principally used base metal thermocouples and the three precious- or noble-metal thermocouples that essentially are standard for high-temperature and high-accuracy needs of some applications, and illustrates the wide diversity of thermocouples easily obtainable from reliable manufacturers.

Thermocouple Fabrication and Protection

A very limited number of users prefer to construct their own thermocouple assemblies. Accepted methods for joining the wires to make a junction are shown in Fig. 5. The main components of a complete thermocouple assembly are presented in Fig. 6. Measurement junctions may be exposed, ungrounded, or grounded (Fig. 7).

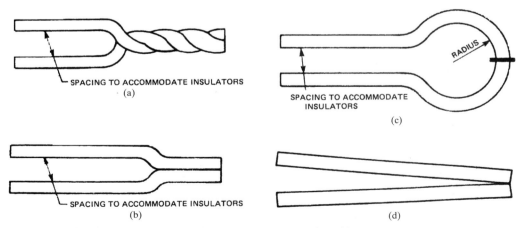

FIGURE 5   Methods for joining dissimilar wires in making thermocouples. (*a*) Twisting wires for gas and electric arc welding. (*b*) Forming wires for resistance welding. (*c*) Forming butt-welded thermocouple. (*d*) Forming wires for electric arc welding.

TABLE 2 Characteristics of Principal Industrial Thermocouples*

ANSI-ISA Types	Metal combinations†	Thermo-element codes	Chemical composition	Common temperature ranges	Limits of error — Standard grade	Limits of error — Premium grade	Atmosphere suitability
				Base-metal thermocouples			
J	Iron/	JP	Fe	−73°C (−100°F) to 427°C (800°F)	±2.2°C (±4°F) (±¾%)	±1/1°C (±2°F) (±⅜%)	Oxidizing
	constantan	JN	44Ni:55Cu	427°C (800°F) to 760°C (1400°F)			Reducing‡
K	*Chromel*	KP	90Ni:9Cr	0°C (32°F) to 277°C (530°F)	±2.2°C (±4°F) (±¾%)	±1.1°C (±2°F) (±⅜%)	Oxidizing,
	Alumel	KN	94Ni:Al:Mn:Fe	277°C (530°F) to 1149°C (2100°F)		(±1%)	inert‖
T	Copper/	TP	Cu	−101°C (−150°F) to −60°C (−75°F)	±1.7°C (±3°F) ±0.8°C (±1.5°F) (±¾%)	(±¾%)	Oxidizing,
	constantan	TN	44Ni:55Cu	−75°C (−103°F) to 93°C (200°F)		(±⅜%)	reducing
				99°C (200°F) to 371°C (700°F)			
E	*Chromel*	EP	90Ni:9Cr	0°C (32°F) to 316°C (600°F)	±1.7°C (±3°F) (±½%)	±1.1°C (±2°F) (±⅜%)	Oxidizing, inert
	constantan	EN	44Ni:55Cu	316°C (600°F) to 871°C (1600°F)			
N	*Nicrosil/*	NP	Ni:14.2Cr:1.4Si	0°C (32°F) to 277°C (530°F)	±2.2°C (±4°F) (±¾%)	—	Oxidizing, inert
	Nisil	NN	Ni:4Si:0.15Mg	277°C (530°F) to 1149°C (2100°F)			
				Precious-metal thermocouples			
R	Platinum-rhodium/	RP	87Pt:13Rh	Available up to 1480°C (2700°F), depending on sheath materials used	Check with supplier		Oxidizing, inert
	platinum	RN	Pt				
S	Platinum-rhodium/	SP	90Pt:10Rh	−18°C (0°F) to 538°C (1000°F)	±1.4°C (±2.5°F) (±¼%)	—	Oxidizing, inert
	platinum	SN	Pt	538°C (1000°F) to 1149°C (2100°F)			
B	Platinum-rhodium/	BP	70Pt:30Rh	Available up to 1700°C (3100°F), depending on sheath materials used	Check with supplier		Oxidizing, inert, vacuum
	platinum-rhodium	BN	94Pt:6Rh				

* Specifications vary somewhat from one manufacturer to the next. Values in this table generally include temperature limitations of the type of sheath used.

† Terms in italics are registered trade names and proprietary compositions.

‡ Type J can be used in oxidizing and reducing atmospheres up to 760°C (1400°F), but above that point the iron oxidizes, causing accelerated deterioration of the accuracy of the sensor.

‖ Type K must not be used in reducing atmospheres, such as hydrogen, dissociated ammonia, carbon monoxide, and many reducing atmospheres as encountered in heat-treating applications. Under such conditions, the KP element, which contains chromium, forms green chromic oxide instead of a spinel-type nickel-chromium oxide. This condition sometimes is called "green rot."

General evaluation:

• Type J is the most widely used of all industrial thermocouples. It has high and fairly uniform sensitivity in millivolts per degree temperature change. Comparatively low-cost.

• Type K is a moderate-cost sensor, particularly for high-temperature measurements in oxidizing atmospheres.

• Type T is highly stable at subzero temperatures with a high conformity to published calibration data. These sensors are frequently the thermocouple of choice for cryogenic and ambient temperature conditions.

• Type E thermocouple provides the highest millivoltage per degree temperature change, providing maximum sensitivity, and is especially useful for short ranges or differential-temperature measurements.

• Type N provides superior thermal stability, longer life, and better accuracy for longer periods than type K. Depending upon manufacturer, this thermocouple can be somewhat limited to smaller outside diameters.

• Type S thermocouple is calibrated to IPTS 68 standard. Essentially, it is used for high-accuracy, high-temperature applications.

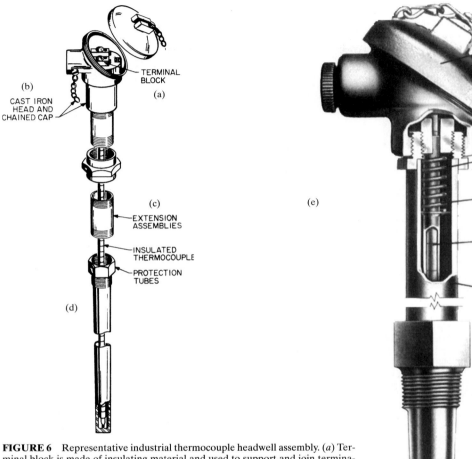

FIGURE 6 Representative industrial thermocouple headwell assembly. (*a*) Terminal block is made of insulating material and used to support and join termination of conductors. (*b*) Connection head is a housing that encloses the terminal block and usually is provided with threaded openings for attachment to a protection tube and for attachment of a conduit. (*c*) Connection head extension usually is a threaded fitting or an assembly of fittings extending between the thermowell or angle fitting and the connection head. Exact configuration depends on installation requirements. (*d*) Protection tube is used to protect sensor from damaging environmental effects. Ceramic materials, such as mullite, high-purity alumina, and some special ceramics, are used mainly in high-temperature applications. They also are used in lower-temperature applications for severe environmental protection. High-purity alumina tubes are required with platinum thermocouples above 1200°C (2200°F) because mullite contains impurities which can contaminate platinum above that temperature. (*e*) Spring-loaded thermocouple assemblies are particularly effective where a temperature measurement is made for control purposes. Spring loading not only improves response, but also protects the junction from the effects of severe vibration. In one design (*Leeds & Northrup*) a retaining ring is brazed to the tube or sheath close to the head end. A spring is compressed between the ring and a bushing assembly, forcing the thermocouple junction into contact with the tip of the well. A grounded junction is required. A silver plug contains the measuring junction. This results in superior response to a temperature change, with a time constant of about 12 seconds, including 3.5 seconds for initial response.

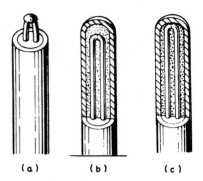

(a)   (b)   (c)

FIGURE 7 Thermocouple measuring junctions. (*a*) Exposed junction. (*b*) Ungrounded junction. (*c*) Grounded junction.

The exposed junction is often used for the measurement of static or flowing non-corrosive gas temperatures where the response time must be minimal. The junction extends beyond the protective metallic sheath to provide better response. The sheath insulation is sealed at the point of entry to prevent penetration of moisture or gas.

The ungrounded junction often is used for the measurement of static or flowing corrosive gas and liquid temperatures in critical electrical applications. The welded wire thermocouple is physically insulated from the thermocouple sheath by soft magnesium oxide (MgO) powder or equivalent.

The grounded junction often is used for the measurement of static or flowing corrosive gas and liquid temperatures and for high-pressure applications. The junction is welded to the protective sheath, providing faster response than an ungrounded junction.

Thermocouple Installation

Thermocouples with sheaths or placed in thermowells are subject to a temperature gradient along the length of the sheath. Such errors can be minimized by specifying long, small-diameter sensors, by using sheath materials with low thermal conductivity, and by providing high convective heat transfer coefficients between the fluid and the thermocouple. Mounting of the thermocouple also plays a large role in minimizing errors. Some users follow a long-regarded rule that the sensor immersion depth should be equal to 10 times the sheath diameter. But according to W. C. Behrmann of Exxon Research and Development Laboratories [1] (in an *InTech* article of August 1990, p. 36) the problem is more complex. In a mathematical study of thermocouple location geometry it has been found that 90° bend couples and curved couples are less prone to error than the common side-entering couples. Of significance, however, the installation difficulties are least with the side-entering geometry.

Thermocouple sheath materials fall into two major categories (Table 3):

1. Metals, such as Inconel 600 [maximum air temperature of 1150°C (2100°F)] and a number of stainless steels: 310SS [1150°C (2100°F)], 304SS, 316SS, and 347SS [900°C (1650°F)]
2. Ceramic materials, such as silicon carbide, Frystan, alumina, and porcelains of various types and brands

Thermocouple Wire Insulators

The dissimilar wires of a thermocouple must be insulated. Traditionally, ceramics have been used (Fig. 8).

Special Thermocouples

In addition to the traditional industrial thermocouples just described, there also are surface probes and cement-on styles. Thermocouples can be drawn in metal-sheathed form to as small as 0.25 mm (0.01 inch) OD. In wire form, 0.013 mm (0.0005-inch) thermocouples can be made.

Types K and E surface probes are commercially available. Type E usually is preferred because of its high accuracy in most low-temperature applications. Type K is used where high temperatures must be measured. Types J and T are not commonly used as probes.

Cement-on style thermocouples have grown in popularity in recent years. Special fast-responding techniques include thin-foil couples with plastic laminates for cementing directly on equipment. The full sensor is embedded between two thin glass-reinforced, high-temperature polymer laminates which both support and electrically insulate the foil section as well as providing a flat surface for cementing. The polymer-glass laminate, in general, determines the maximum temperature of the construction, which is 260°C (500°F) in continuous service (Fig. 9).

TABLE 3   High-Temperature Sheath Materials

Sheath material	Maximum operating temperature	Workability	Working environment	Approximate melting point	Remarks
Molybdenum*	2205°C (4000°F)	Brittle	Inert, vacuum, reducing	2610°C (4730°F)	Relatively good hot strength; sensitive to oxidation above 500°C (930°F); resists many liquid metals and most molten glasses
Tantalum†	2482°C (4500°F)	Malleable	Inert, vacuum	3000°C (5425°F)	Resists most acids and weak alkalies; very sensitive to oxidation above 300°C (570°F)
Platinum-rhodium alloy	1677°C (3050°F)	Malleable	Oxidizing, inert, vacuum	1875°C (3400°F)	No attack by SO_2 at 1093°C (2000°F); silica is detrimental; halogens attack at high temperatures
Inconel 600	1149°C (2100°F)	Malleable	Oxidizing, inert, vacuum	1410°C (2570°F)	Excellent resistance to oxidation at high temperature; do not use in presence of sulfur above 538°C (1000°F); hydrogen tends to embrittle

* Refractory metals are extremely sensitive to any trace of oxygen above approximately 260°C (500°F). They must be used in vacuum or in very pure inert gases such as helium and argon.
† Suitable for exposure to certain reducing atmospheres as well as inert gases and vacuum.

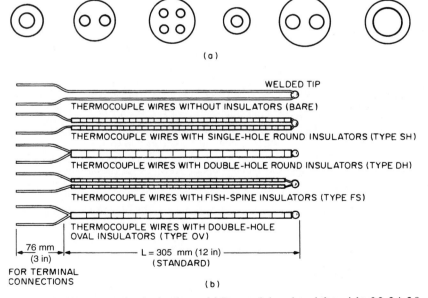

FIGURE 8   Thermocouple wire insulators. (*a*) Range of sizes, from left to right: 3.2, 2.4, 2.0, 1.6, 1.2, 0.8, and 0.4 mm (⅛, 3⁄32, 5⁄64, 1⁄16, 3⁄64, 1⁄32, and 1⁄64 inch) bore diameter. (*b*) Application of insulators to various styles of thermocouples.

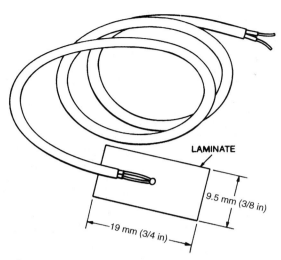

FIGURE 9   Cement-on-style thermocouples.

Thermocouple Circuit Flexibility

Normally one envisions the use of thermocouples one at a time for single temperature measurements. As shown in Fig. 10, thermocouples may be used in parallel, in series, and in switching and differential circuits.

Checking Thermocouples

Authorities suggest that the calibration of thermocouples be checked annually, or at least semiannually. In accordance with a maintenance schedule, past experience should be used for determining how often a thermocouple should be replaced prior to serious changes in calibration. This is particularly true of the base-metal sensors. Subsequent checking, that is, checking after initial installation, should be done in situ. Portable calibrators that can measure or simulate a thermocouple signal are readily available and simplify field calibration.

Resistance checks of thermocouples are effective for determining the condition of a sensor. For example, a low resistance generally indicates a satisfactory situation, whereas a high resistance may indicate that a thermocouple is approaching the end of its useful life. To determine correct installation, particularly in connection with thermocouples made "in house," a small magnet can be useful. For example, the positive thermoelement of a type J couple is iron, which is magnetic, whereas the negative element, constantan, is nonmagnetic. In the type K couple, the alumel element is magnetic and the other thermoelement is not.

RESISTANCE TEMPERATURE DETECTORS

The science of measuring temperature by utilizing the characteristic relationship of electrical resistance to temperature has been advanced periodically since the early work of Faraday (circa 1835). Certain suitably chosen and prepared materials which vary in resistance in a well-defined and calibrated manner with temperature became readily available around 1925, prompting the use of resistance thermometers as primary sensors for industrial applications

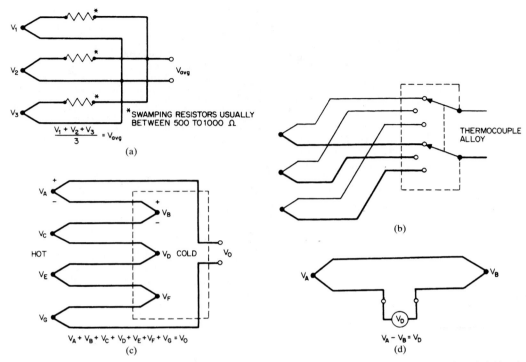

FIGURE 10 Use of thermocouples in multiples. (*a*) Thermocouples in parallel. (*b*) Thermocouples in switch circuit. Switch must be isothermal or made of the same thermocouple alloy material. (*c*) Thermocouples in series (thermopile). *Note:* V_B, V_D, and V_F are negative thermoelectric voltages compared with V_A, V_C, V_E, and V_G. However, the alloys are also reversed, thus creating a net voltage. (*d*) Differential circuit. *Note:* Output voltage cannot be accurately cold-junction compensated because of the nonlinearity of thermocouple EMF versus temperature. Approximations can be made if the absolute temperature is known.

where reproducibility and stability are of critical importance. Platinum resistance thermometers became the international standard for temperature measurements between the triple point of hydrogen at 13.81 K and the freezing point of antimony at 730.75°C. Since the 1970s RTDs have made very serious inroads on the thermocouple for very broad usage in industry—for practical industrial use, not just for applications requiring exceptional accuracy. The advantages and limitations of RTDs as compared with thermocouples in this present time span are presented later in this article.

Principles of Resistance Thermometry

For pure metals, the characteristic relationship that governs resistance thermometry is given by

$$R_t = R_0\,(1 + at + bt^2 + ct^3 + \cdots)$$

where R_0 = resistance at reference temperature (usually at ice point, 0°C), Ω
            R_t = resistance at temperature t, Ω
            a = temperature coefficient of resistance, Ω/Ω (°C)
     b, c = coefficients calculated on the basis of two or more known resistance-temperature (calibration) points

For alloys and semiconductors, the relationship follows a unique equation dependent on the specific material involved. Whereas most elements constructed from metal conductors generally display positive temperature coefficients, with an increase in temperature resulting in increased resistance, most semiconductors display a characteristic negative temperature coefficient of resistance.

Only a few pure metals have a characteristic relationship suitable for the fabrication of sensing elements used in resistance thermometers. The metal must have an extremely stable resistance-temperature relationship so that neither the absolute value of the resistance R_0 nor the coefficients a and b drift with repeated heating and cooling within the thermometer's specified temperature range of operation. The material's specific resistance in ohms per cubic centimeter must be within limits that will permit fabrication of practical-size elements. The material must exhibit relatively small resistance changes for nontemperature effects, such as strain and possible contamination which may not be totally eliminated from a controlled manufacturing environment. The material's change in resistance with temperature must be relatively large in order to produce a resultant thermometer with inherent sensitivity. The metal must not undergo any change of phase or state within a reasonable temperature range. Finally, the metal must be commercially available with essentially a consistent resistance-temperature relationship to provide reliable uniformity.

Industrial resistance thermometers, often referred to as RTDs, are commonly available with elements of platinum, nickel, 70% nickel–30% iron (Balco), or copper. The entire resistance thermometer is an assembly of parts, which include the sensing element, internal leadwires, internal supporting and insulating materials, and protection tube or case (Fig. 11 and Table 4).

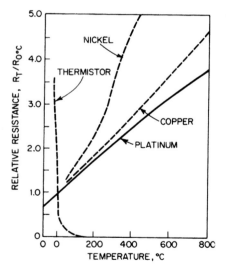

FIGURE 11   Resistance-temperature characteristics of thermoresistive materials at elevated temperatures. Platinum and nickel are the most commonly used metals for industrial applications.

Platinum RTDs

Of all materials currently utilized in the fabrication of thermoresistive elements, platinum has the optimum characteristics for service over a wide temperature range. Although platinum is a noble metal and does not oxidize, it is subject to contamination at elevated temperatures by some gases, such as carbon monoxide and other reducing atmospheres, and by metallic oxides.

The metal is available commercially in pure form, providing a reproducible resistance-temperature characteristic. Platinum with a temperature coefficient of resistance equal to 0.00385 Ω/Ω (°C) (from 0 to 100°C) has been used as a standard for industrial thermometers throughout the United Kingdom and Western Europe since World War II and has gained prominence in recent years in the United States in the absence of a defined and commonly accepted standard coefficient. Platinum has a high melting point and does not volatilize appreciably at temperatures below 1200°C. It has a tensile strength of 18,000 psi (124 MPa) and a resistivity of 60.0 Ω/cmil · ft at 0°C (9.83 $\mu\Omega$ · cm).

Platinum is the material most generally used in the construction of precision laboratory standard thermometers for calibration work. In fact, the laboratory-grade platinum resistance thermometer (usually with a basic resistance equal to 25.5 Ω at 0°C) is the defining standard for the temperature range from the liquid oxygen point (−182.96°C) to the antimony point (630.74°C) as defined by the International Practical Temperature Scale.

TABLE 4   Resistance versus Temperature for Various Metals

Metal	Resistivity, gΩ · cm	Relative resistance R_t/R_0 at 0°C											
		−200	−100	0	100	200	300	400	500	600	700	800	900
Alumel*	28.1			1.000	1.239	1.428	1.537	1.637	1.726	1.814	1.899	1.982	2.066
Copper	1.56	0.117	0.557	1.000	1.431	0.862	2.299	2.747	3.210	3.695	4.208	4.752	5.334
Iron	8.57			1.000	1.650	2.464	3.485	4.716	6.162	7.839	9.790	12.009	12.790
Nickel	6.38			1.000	1.663	2.501	3.611	4.847	5.398	5.882	6.327	6.751	7.156
Platinum	9.83	0.177	0.599	1.000	1.392	1.773	2.142	2.499	3.178	3.178	3.500	3.810	4.109
Silver	1.50	0.176	0.596	1.000	1.408	1.827	2.256	2.698	3.616	3.616	4.094	5.586	5.091

* Registered trademark of Hoskins Manufacturing Co., Detroit, Michigan.

The resistance-temperature relationship for platinum resistance elements is determined from the Callendar equation above 0°C,

$$t = \frac{100(R_t - R_0)}{R_{100} - R_0} + \delta \left(\frac{t}{100} - 1 \right) \frac{t}{100}$$

where

t = temperature, °C
R_t = resistance at temperature t, Ω
R_0 = resistance at 0°C, Ω
R_{100} = resistance at 100°C, Ω
δ = Callendar constant (approximately 1.50)

The fundamental coefficient (temperature coefficient of resistance) α is defined over the fundamental interval of 0 to 100°C,

$$\alpha = \frac{R_{100} - R_0}{100R_0}$$

Thin-Film Platinum RTDs.   Processing techniques developed during the past decade or two have provided the capability of producing thin-film platinum RTD elements that essentially are indistinguishable from wire-wound elements in reproduction and stability. An industrial RTD thermowell assembly is shown in Fig. 12. Thin-film platinum RTD elements are trimmed to the final desired resistance value, usually by automated equipment. These high-resistance elements have, in numerous instances, made obsolete the need to consider base-metal wire-wound sensors of nickel or nickel-iron. Thermistors, too, are sometimes replaced with a sensor of significantly greater stability and temperature range.

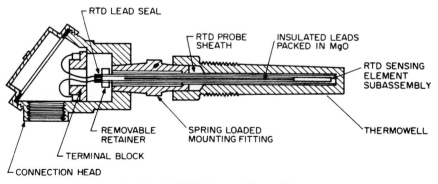

FIGURE 12   Representative industrial RTD thermowell assembly.

The use of sheaths and thermowells, as described previously for thermocouples, applies similarly to RTDs.

Wire-Wound Platinum RTDs.   Some users for certain applications prefer the wire-sound RTD. In the fully encapsulated RTD the platinum wire, usually 0.025 mm (0.001 inch) OD or less, is wound into a coil and inserted into a multibore high-purity ceramic tube, or may be wound directly on the outside of a ceramic tube. The most commonly used ceramic material is aluminum oxide (99.7% Al_2O_3). The winding is completely embedded and fused within or on the ceramic tube utilizing extremely fine granular powder. The resultant fully encapsulated element, with only two noble-metal leadwires exposed, provides maximum protection for the platinum resistance coil. Although special fusing techniques are used, such elements are not completely strain-free, but the effects of existing strains are fairly constant with resulting errors well within the permissible limits for industrial applications. The intimate contact between the platinum winding and the ceramic encapsulation permits a rapid speed of response with the thermal conductivity of ceramic adequate for heat transmission through the protecting layer. A platinum industrial RTD assembly with a maximum temperature range is shown in Fig. 13. For applications where high-temperature requirements are combined with high pressure, high flow, and high vibration, the assembly includes a thermowell protecting tube.

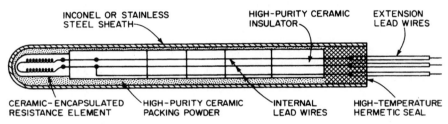

FIGURE 13   Representative platinum RTD assembly.

Nickel RTDs.   During recent years, widespread use of improved platinum RTDs with their superior performance characteristics, often at lower cost, has taken precedence over the use of nickel RTDs in a wide range of industrial applications. Current availability of nickel sensors has continued primarily as a component replacement for already existing industrial systems.

Copper RTDs.   The observations pertaining to nickel RTDs are also essentially applicable to copper RTDs. The straight-line characteristics of copper have in the past been useful in allowing two sensors to be applied directly for temperature-difference measurements.

Ranges and Performance Characteristics of RTDs

The most common RTD temperature ranges commercially available and some additional performance characteristics are summarized in Table 5.

RTD Circuitry

Traditionally three methods have been used for making electric connections from the resistance thermometer assembly to the measuring instrument. Diagrams and their equations are presented in Fig. 14. When highly stable constant-current-source (CCS) power supplies

TABLE 5 Representative Standard RTDs Available Commercially

	Platinum	Platinum	Platinum	Platinum	Platinum	Nickel
Temperature	−200 to 200°C (−328 to 392°F)	−50 to 200°C (−58 to 392°F)	−100 to 260°C (−148 to 500°F)	−65 to 200°C (−85 to 392°F)	−200 to 650°C (−328 to 1200°F)	−130 to 315°C (−200 to 600°F)
Configuration	316 SS sheath 200°C (392°F) 17.2 MPa (2500 psig)	316 SS sheath 200°C (392°F) 15.2 MPa (2200 psig)	Surface mount	Surface mount	316 SS sheath 480°C (900°F); *Inconel* to 650°C (1200°F)	Stainless steel; other metals
Repeatability	±0.05% max ice-point resistance 0.13°C (0.23°F)	±0.025% max ice-point resistance	±0.04% max ice-point resistance 0.1°C (0.18°F)	±0.08% max ice-point resistance 0.2°C (0.36°F)	±0.26°C (0.47°F) up to 480°C (896°F); ±0.5% (reading) up to 650°C (1200°F)	Limit of error from ±0.3°C (±0.5°F) to ±1.7°C (3°F) at cryogenic temperatures
Stability	±0.08% max ice-point resistance	±0.035% max ice-point resistance	±0.05% max ice-point resistance	±0.15% max ice-point resistance	—	—
Time constant (to reach 63.2% of sensor response)	8 seconds	7 seconds	1.25 seconds	2.5 seconds	—	—
Immersion length	0.6 to 6 m (2 to 200 ft)	31, 61, 91, 122 cm (12, 24, 36, 48 in)	—	—	89 to 914 mm (3.5 to 36 in)	89 to 914 mm (3.5 to 36 in)
Leadwire	Teflon insulated, nickel-coated 22-gauge standard copper wire	Teflon insulated, nickel-coated 22-gauge standard copper wire	0.25 mm (0.01 in)-diameter platinum wire	Teflon insulated, 24 AWG standard copper wire	—	—

4.23

became available in miniature packages at relatively low cost, these offered an effective alternative to null-balance bridge-type instruments, particularly for industrial process systems that require scanning (both manual and automatic) of up to 100 or more individual RTD points, often located at varying distances from a central point. The circuit shown in Fig. 15 is the basic four-lead circuit with two leads joined in close proximity to each side of the resistance element. The CCS can be connected across leads t and c supplying a constant current I_c across the resistance element X. The value for i_c may be kept to a minimum of 1 mA or less to avoid excessive self-heating errors. The voltage drop across the resistance element then is measured between T and C. The resultant voltage drops across the thermometer element, in the constant-current mode, then varies with resistance directly as a function of temperature. The advantage of CCS circuits becomes apparent to the user already versed in bridge measuring techniques. The CCS power supply continues to maintain the fixed constant current (within its compliance voltage limitations) across the thermometer element, thus making costly matching techniques associated with leadwires unnecessary. In addition, leadwire contact resistances associated with high-speed automatic switching are reduced to a minimum. An added feature of the CCS measuring circuit is its ability to interface directly with a wide variety of voltage measuring instruments. Digital linearizers can be applied to operate on the nonlinear platinum resistance thermometer function to display directly in engineering units with linearization conformities of a fraction of a degree (temperature). CCS measuring circuits provide added binary-coded decimal (BCD) output for interface with digital printers and process control computers.

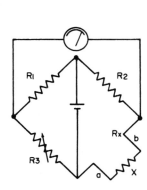

Circuit equations:

$$R_1 + R_3 = R_2 + a + b + X \quad (8)$$

$$R_1 = R_2$$

$$\therefore \quad R_3 = a + b + X \quad (9)$$

(a)

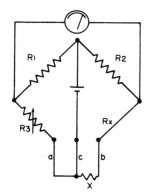

Circuit equations:

$$R_1 + R_3 + a + c = R_2 + b + X + c \quad (10)$$

$$R_1 = R_2$$

If $a = b$ (lead resistance equal)

Then $R_3 = X \quad (11)$

(b)

FIGURE 14 Traditional RTD circuits presented as backdrop to contemporary circuit designs. (*a*) Two-lead circuit permissible only when leadwire resistance can be kept to a minimum and only where a moderate degree of accuracy is required. (*b*) Three-lead circuit. Two leads are connected in close proximity to the resistance element at a common node. Third lead is connected to the opposite resistance leg of the element. Resistance of lead a is added to bridge arm R_3, while resistance of lead b remains on bridge arm R_X, thereby dividing the lead resistance and retaining a balance in the bridge circuit. Lead resistance c is common to both left and right loops of the bridge circuit. Although this method compensates for the effect of lead resistance, the ultimate accuracy of the circuit depends on leads a and b being of equal resistance. Special matching techniques must be used on leads a and b, particularly when distance between sensor and measuring equipment is relatively large. A four-lead circuit, not shown, is used only when the highest degree of accuracy, as in laboratory temperature standards, is required.

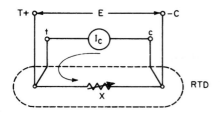

Circuit equations:

$E = IX$

$\quad I \cong constant$

$E = f(X) = f'(temperature)$ (18)

FIGURE 15 Four-lead constant-current measuring circuit.

COMPARISON OF THERMOCOUPLE AND RTD SELECTION

The choice of a thermocouple or a resistance temperature sensor for a given application is not always clear-cut. In making an optimum choice, the user must assign relative priorities to the most and the least important characteristics needed. As an example, for a specific application, vibration may be an outstanding problem and a thermocouple may be selected over an RTD because the thermocouple is less likely to be affected by vibration, even though most other criteria may favor the RTD. Or short-term reproducibility may be a key criterion that favors the RTD over the thermocouple.

Where special requirements are not a factor, then the principal criteria include accuracy, response time, size, original and installed cost (an important factor where many points of measurement are involved), sealing the detector from the process environment without significantly reducing response time, and the user's experience factor. When all other evaluating criteria are about equal, the user simply may prefer one type of detector over the other because in-house engineers, operators, and maintenance personnel have much more experience and know-how with a given type of temperature detector.

Numerous professionals have expressed their opinions. These have been distilled in Table 6. Thus a few inconsistencies may be expected.

THERMOCOUPLE AND RTD TRANSMITTERS

Data signal handling in computerized instrumentation systems is the topic of in Section 7, Article 1, of this handbook. Thermocouple, RTD, and solid-state temperature detectors are specifically mentioned in that article.

Although variously termed "smart" or "intelligent" temperature (and other process variables) transmitters have numerous electronically built-in advantages, the basic purpose of the transmitter must not be minimized: the ultimate and self-evident function of the transmitter is that of ensuring data integrity. This is particularly important in noisy electrical environments as encountered in numerous processing and manufacturing situations. Other transmitter capabilities, in essence, are extra dividends that can be derived conveniently and at reasonably low cost from associated electronics. These include, among other advantages, self-diagnostics, elimination of hardware changes or recalibration, and bidirectional communication with a distributed control system. Because so many industrial variable sensors, such as pressure, flow, and liquid level, also generate voltage and current signals, some available single units can be programmed electronically to replace any of the other units in a system, thus reducing replacement inventory costs. This is particularly true in the temperature field.

Evolution of Temperature Transmitters

The traditional practice over many decades (still present in numerous systems) was that of using long wire runs to deliver low-level thermocouple and RTD signals to some operator-based central location. For reasons of cost and vulnerability to electrical noise, such systems were rather severely distance-limited, not to mention their high installation and maintenance costs.

TABLE 6 Relative Advantages and Limitations of Thermocouples and Resistance Temperature Detectors

Advantages	Limitations
Thermocouples	
Cost: Although cost of RTDs is trending downward, thermocouples generally continue to be less expensive. Ruggedness: In terms of process environmental conditions, including high temperature and vibration, thermocouples are regarded highly. Higher temperature range: Extends to about 1150°C (2100°F) or higher. However, the thermocouple's cryogenic range is somewhat less than that of the RTD. Mounting cost: Generally considered to be somewhat lower than for RTDs.	Accuracy: Generally expected, after installation, ±4°C (±7.2°F), but there are exceptions. Thermocouples require a reference junction or special extension wire. The overall thermocouple system includes the inaccuracies associated with two separate temperature measurements—the measuring junction and the cold or reference junction. Stability: Less than for the RTD. Estimated at 0.6°C (1°F) per year. Inherently small millivolt output: Can be affected by electrical noise. Calibration: Nonlinear over normal spans. Signal requires linearizing. Calibration can be changed by contamination.
Resistance temperature detectors	
Accuracy: Generally expected, after installation, ±0.5°C (±0.9°F). The platinum RTD, for example, is used to define the IPTS at the oxygen point (−182.97°C) and the antimony point (+630.74°C). No reference junction is required. Repeatability: Within a few hundredths of a degree; can often be achieved with platinum RTDs. Less than 0.1% drift in 5 years. Substantial output voltage (1 to 6 volts): This is an advantage in that the output can be controlled by adjusting the current and the bridge design in the RTD signal conversion module. Because a higher output voltage to the RTD usually can be achieved, the recording, monitoring, and controlling of temperature signals is simpler. This permits more accurate measurements without requiring complex calculations for large spans. Short-term reproducibility: Superior to that of thermocouples. Reproducibility not affected by temperature changes. Relatively narrow spans: Some versions may have spans as narrow as 5.6°C (10°F). Compensation: Not required. Suppressed ranges: Available. Size: Generally smaller than thermocouples.	Cost: Generally higher than that of thermocouples. However, RTDs do not require compensation, special leadwires, and special signal conditioning for long runs. With platinum-film technology, cost is trending downward. Less rugged: For adverse process environments, including high temperatures and vibration, RTDs are not regarded as highly as thermocouples. Lower temperature range: Limited to about 870°C (1600°F). Note, however, that RTDs extend to lower cryogenic ranges than do thermocouples. Self-heating errors: This may be a problem unless corrected in the transmitter's electronics.

Two-Wire Analog Temperature Transmitter. This was widely accepted as a means for improving measurement stability and accuracy—accomplished by replacing low-level millivolts or ohms with high-level current signals. This enabled the replacement of high-cost shielded cables and conduits with twisted-pair copper wire in the case of RTDs, and of costly extension wires in the case of thermocouples.

Microprocessor-Based Transmitters

These smart devices were introduced in the late 1980s to take advantage of the large advances in digital electronics that had occurred just a few years previously. These transmitters have achieved wide acclaim because of their versatility. Generally the units are preprogrammed by the manufacturer with the information necessary to linearize a given sensor signal, including a given thermocouple, RTD, or other millivolt or ohm device. An important advantage of these transmitters is their potential for reconfiguration for accepting other sensor inputs without having to make hardware changes and recalibrate. Also, they offer a variety of self-diagnostics for power supply and loop problems (Fig. 16). The specifications of available transmitters are compared in Table 7.

THERMISTORS

The name thermistor is derived from thermally sensitive resistor, since the resistance of a thermistor varies as a function of temperature. Although the fundamental principles of the thermistor were established several decades ago, industrial and commercial utilization of thermistors in temperature measurement systems was very slow in developing. Since the early 1980s a proliferation of relatively low-cost, usually portable thermistor thermometers has become available. However, the thermistor to date has not been widely applied in the traditional industrial and process control temperature measurement field because of several inherent problems.

A thermistor is an electrical device made of a solid semiconductor with a high temperature coefficient of resistivity which would exhibit a linear voltage-current characteristic if its temperature were held constant. When a thermistor is used as a temperature sensing element, the relationship between resistance and temperature is of primary concern. The approximate relationship applying to most thermistors is

$$R_t = R_0 \exp B \left(\frac{1}{T} - \frac{1}{T_0} \right)$$

TABLE 7   Comparison of Specifications of Main Types of Temperature Transmitters

Traditional type	Smart or intelligent type	Throwaway type (potted)
Analog	Digital	Analog
Thermocouple or RTD must be specified, changes require hardware reconfiguring	One model handles all thermocouple, RTD, mV, and ohm sensors	Limited to one input
Moderate cost	Premium cost, downward trend	Low cost
Variable span; moderate range	Variable span; wide range	Single span; limited range
Sometimes isolated	Isolated	Not isolated
Variety of applications	Almost every application	Only one application
Sometimes intrinsically safe	Nearly always intrinsically safe	Usually not intrinsically safe
Can be reconfigured on site or in shop	Can be reconfigured remotely	Very limited reconfiguration
	Remote diagnostics	
Moderate to good performance	Superior accuracy	Performance accuracy limited
RTD models easily linearized; thermocouples sometimes linearized	Linearization is selectable	Must linearize at receiver
Stability depends on manufacture and application	Quite stable with ambient temperature and time; infrequent needs to recalibrate	Stability depends on manufacture and application

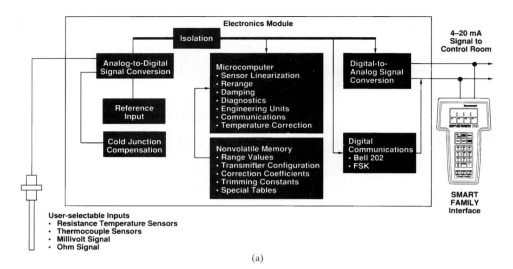

(a)

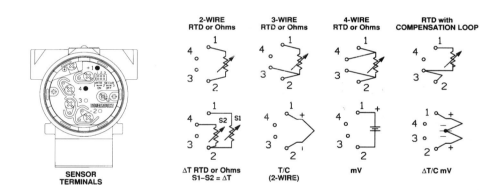

(b)

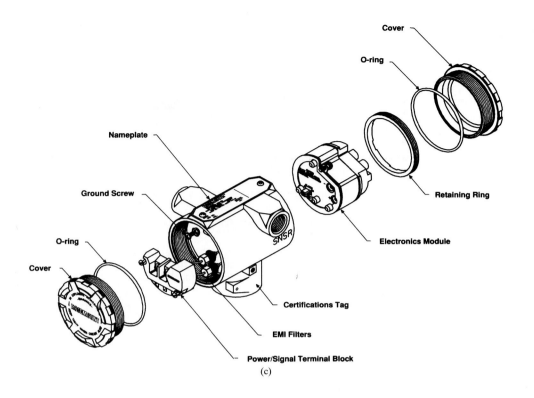

Cover

O-ring

Retaining Ring

Electronics Module

Nameplate

Ground Screw

O-ring

Cover

Certifications Tag

EMI Filters

Power/Signal Terminal Block

(c)

FIGURE 16 Example of contemporary smart temperature transmitter designed for use with a variety of sensors, such as RTDs, thermocouples, and other millivolt and ohm signals.

(*a*) Block diagram of electronics module. The interface can be connected at any termination point in the 4- to 20-mA signal loop. The electronics module consists of two circuit boards sealed in an enclosure with an integral sensor terminal block. The module utilizes digital ASIC (application-specific integrated circuit), microcomputer, and surface-mount technologies. The electronics digitize the input signal from the sensor and apply correction coefficients selected from nonvolatile memory. The output section of the electronics module converts the digital signal to a 4- to 20-mA output and handles the communication with other interfaces of the control system. An LCD (liquid-crystal display) meter may plug into the electronics module to display the digital output in user-configured units.

Configuration data are stored in nonvolatile EEPROM (electrically erasable programmable read-only memory) in the electronics module. These data are retained in the transmitter when power is interrupted—so the transmitter is functional immediately upon power-up.

The process variable (temperature) is stored as digital data and engineering unit conversion. The corrected data then are converted to a standard 4- to 20-mA current applied to the output loop. The control system can access the sensor reading directly as digital signal, thus bypassing the digital-to-analog conversion process for higher accuracy.

(*b*) Transmitter field wiring. The meter interface previously described can be connected at any termination point in the signal loop. In this system the signal loop must have 250-ohm minimum load for communications. In addition to the initial setting of the transmitter's operational parameters (sensor type, number of wires, 4- and 20-mA points, damping, and unit selection), informational data can be entered into the transmitter to allow identification and physical description of the transmitter. These data include Tag (8 alphanumeric characters), descriptor (16 alphanumeric characters), message (32 alphanumeric characters), date, and integral meter. In addition to the configurable parameters, the system's software contains several kinds of information that are *not* user-changeable, namely, transmitter type, sensor limits, and transmitter software revision levels.

The system performs continuous self-tests. In the event of a problem, the transmitter activates the user-selected analog output warning. Then the meter interface (or other designated control component) can interrogate the transmitter to determine the problem. The transmitter outputs specific information to the interface or control system component to identify the problem for fast and convenient corrective action. If an operator believes there may be a loop problem, the transmitter can be directed to provide specific outputs for loop testing.

During initial setup of the transmitter and for maintenance of the digital electronics, the format function is used. The top-level format menu offers two functions: characterize and digital trim. These functions allow the user to select the sensor type and adjust the transmitter's digital electronics to user plant standards.

(*c*) Exploded view of smart transmitter. (*Rosemount Inc., Measurement Division.*)

where R_0 = resistance value at reference temperature T_0 K, Ω
R_T = resistance at temperature T K, Ω
B = constant over temperature range, dependent on manufacturing process and construction characteristics (specified by supplier),

$$B \cong \frac{E}{K}$$

where E = electronvolt energy level
K = Boltzmann's constant, = 8.625×10^{-5} eV/K

A second form of the approximate resistance-temperature relationship is written in the form

$$R_T = R_\infty e^{B/T}$$

where R_∞ is the thermistor resistance as the temperature approaches infinity, in ohms.

These equations are only best approximations and, therefore, are of limited use in making highly accurate temperature measurements. However, they do serve to compare thermistor characteristics and thermistor types.

The temperature coefficient usually is expressed as a percent change in resistance per degree of temperature change and is approximately related to B by

$$a = \frac{dR}{dT} \left(\frac{1}{R} \right) = \frac{-B}{T_0^2}$$

where T_0 is in kelvin. It should be noted that the resistance of the thermometer is solely a function of its absolute temperature. Furthermore, it is apparent that the thermistor's resistance-temperature function has a characteristic high negative coefficient as well as a high degree of nonlinearity. The value of the coefficient a for common commercial thermistors is on the order of 2 to 6 percent per kelvin at room temperature. This value is approximately 10 times that of metals used in the manufacture of resistance thermometers.

Resultant considerations due to the high coefficient characteristic of thermistors include inherent high sensitivity and high level of output, eliminating the need for extremely sensitive readout devices and leadwire matching techniques, respectively. However, limitations on interchangeability (particularly over wide temperature ranges), calibration, and stability—also inherent to thermistors—are quite restrictive. The high degree of nonlinearity in the resistance-temperature function usually limits the range of the readout instrumentation. In many applications, special prelinearization circuits must be used before interfacing with related system instrumentation. The negative temperature coefficient also may require an inversion (to positive form) when interfacing with some analog or digital instrumentation.

A number of metal oxides and their mixtures, including the oxides of cobalt, copper, iron, magnesium, manganese, nickel, tin, titanium, uranium, and zinc, are among the most common semiconducting materials used in the construction of thermistors. Usually compressed into the desired shape from the specially formulated powder, the oxides are then recrystallized by heat treatment, resulting in a dense ceramic body. The leadwires are then attached while electric contact is maintained, and the finished thermistor is then encapsulated. A thermistor can be made in numerous configurations, as illustrated in Fig. 17.

Thermistor Performance

The evaluation of thermistor performance characteristics is in many cases similar to that of resistance thermometers. Figure 18 shows the logarithm of the specific-resistance-verus-temperature relationship for three thermistor materials as compared with platinum metal. The

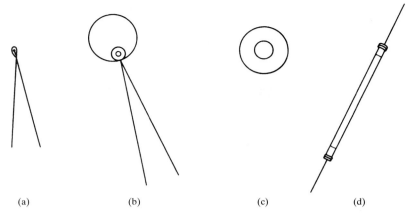

(a)	(b)	(c)	(d)

FIGURE 17 Various configurations of thermistors. (*a*) Beads may be made by forming small ellipsoids of material suspended on two fine leadwires. The wires may be approximately 2.5 mm (0.1 inch) apart or even closer. If the material is sintered at an elevated temperature, the leadwires become tightly embedded within the bead, making electric contact with the thermistor material. For more rugged applications, the bead thermistor may be encapsulated or placed within a suitable metal sheath. (*b*) Disk configurations are manufactured by pressing the semiconductor material into a round die to produce a flat, circular probe. After sintering, the pieces may be silvered on the two flat surfaces. Thermistor disks may range from 2.5 to 25 mm (0.1 to 1 inch) or less in diameter, and from 0.5 to 13 mm (0.02 to 0.5 inch) or less in thickness. The disk configuration is well suited where a moderate degree of power dissipation is needed. (*c*) Washer-type thermistors are like the disk types, but provide a hole for bolt mounting. Often they are applied where high power dissipation is a primary requirement. (*d*) Rod-type thermistors are extruded through dies, resulting in long, cylindrical probes. Rod configurations generally are of high thermal resistance and are applied wherever power dissipation is not a major concern.

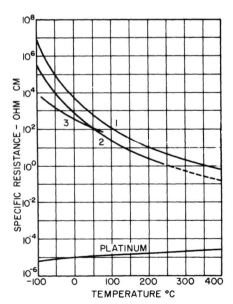

FIGURE 18 Resistance-temperature characteristics of three thermistors compared with platinum.

specific resistance of the thermistor represented by curve 1 *decreases* by a factor of 50 as the temperature is increased from 0 to 100°C. Over the same temperature range, the resistivity of platinum will *increase* by a factor of approximately 1.39.

Thermistors range in terminal resistance at room temperature from about 1 ohm to the order of 10^8 ohms, depending on composition, shape, and size. Within a given type, they commonly vary 10 to 30 percent in resistance from the nominal at the reference temperature. Advanced manufacturing techniques and careful selection of suitable units can bring resistance values within closer limits.

Thermistor Applications

The application of thermistors as primary temperature elements follows the usual principle of resistance thermometry. Conventional bridge or other resistance measuring circuits as well as constant-current circuits are used. Special application considerations must be given to the negative and highly nonlinear resistance-temperature relationship. Common to resistance thermometers, consideration must be given to keeping the measuring circuit small enough to avoid significant heating in order that the element resistance will be solely dependent on the measured medium.

A current very extensive use of thermistors is in trip thermometers, which have been well accepted in the food transportation industry. Small, portable recording-monitoring temperature instruments of the type shown in Fig. 19 initially were developed to provide a temperature history of foodstuffs (vegetables, fruits, seafoods) during transit from producer to final distributor by truck, rail, ship, and air. Such units not only provide information required by regulatory agencies, but also for the improvement of preparation and handling processes. A few examples of the more critical needs for such temperature information would include fresh lettuce, bananas, lobster, and pizza dough. Shipping time spans range from 4 to 16 shipping days. Temperature span covers −29 to 38°C (−20 to 100°F), with a claimed accuracy of ±2 percent (scale). Extended temperature ranges and longer shipment intervals (up to 75 days) are available. Both analog and digital versions are available. Some of the units also incorporate a humidity (capacitive) detector with scales ranging from 10 to 90 percent relative humidity.

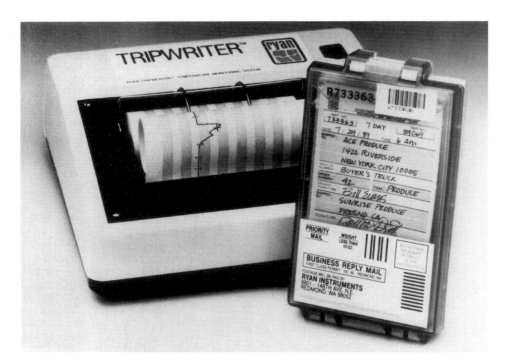

FIGURE 19   Thermistor-based temperature recorder (right) used for monitoring environmental conditions during the transport of various perishable foodstuffs, such as fruits and vegetables, for up to 16 days or more of travel time. These battery-operated units are packaged for mailing. Upon receipt of mailable units, such as *Tripmentor*™ and *Tripwriter*™, user may print out the information on the recorder (left). (*Ryan Instruments.*)

Another rapidly developing market for thermistor thermometers where portability and convenience are important factors is found in laboratories, pilot plants, and for checking the temperature of flow and return pipes in heating and air-conditioning systems. One design makes a 3-second calibration from a special microprocessor each time it is turned on. A typical measuring range is from −51 to 150°C (−60 to 300°F) with a resolution of 0.6°C (1°F) and an accuracy of 0.2°C (0.4°F). Power supply is a 9-volt battery. Depending on the model, the instrument weighs from 150 to 250 grams (5.3 to 7.8 oz).

SOLID-STATE TEMPERATURE SENSORS

Of limited, but expanding industrial use is the silicon or quantum detector. In one form, incident infrared radiation interacts with a bound electron within the semiconductor crystal lattice. The energy of a photon, if sufficiently large, is transferred to an electron to free it from its immobile state and permit it to move through the crystal. During the time it is free, the electron can produce a signal voltage in the detector. After a short interval, the electron will return to its bound state. This interval generally is far shorter than the thermal time constant of a thermal detector.

In essence, the quantum detector is a photon counter which is equally sensitive to all photons having the minimum energy to free a bound electron. A detector of this type will exhibit a fairly uniform response to all photons up to a particular wavelength. The practical advantage of these detectors lies in their ability to produce electrical signals which faithfully measure the incident photon flux. This permits a method of continuous temperature measurement without contact. Thus solid-state detectors are used in some infrared thermometer designs.

RADIATION THERMOMETERS

Radiation thermometry represents a practical application of the Planck law and Planck radiation formula (1900) and makes it possible to measure the temperature of an object without making physical contact with the object. Although these instruments are widely used because of the contactless feature, this is by no means the exclusive advantage of the method. Another important advantage is the wide useful temperature range—from subzero temperatures to extremely high, virtually unlimited values. Carefully constructed instruments have been used for many years to maintain laboratory primary and secondary standards above the gold point (1064.43°C) within an accuracy of ±0.01°C. Representative industrial designs generally have a precision of ±0.5 to ±1 percent, even under rather adverse conditions.

Radiation thermometers, although much improved during the past decade or two through the incorporation of technological advancements in electronic circuitry and optics, have been used for many decades. Originally called radiation pyrometers, then radiation thermometers, and, more recently, infrared (IR) thermometers, the initial applications of the method usually were in those processing and manufacturing applications where great amounts of heat were required, often in small spaces, thus creating very high temperatures. Often such materials were moving and could not be contacted by a temperature detector.

Thus early applications generally involved monitoring such operations as glass, metal, chemical, cement, lime, and refractory materials. During the last few decades, radiation thermometry has extended into lower temperature regions, including subzero measurements, as encountered in the foods, electronics, paper, pharmaceutical, plastics, rubber, and textile industries, among others. Portable IR thermometers also find wide application in industrial energy conservation strategies for checking large lines, vessels, steam traps, and so on, for faulty operation. Relatively recent increased usage of IR thermometers now gives it well over 10 percent of the industrial temperature measurement field.

IR Thermometry Principles

Planck's law predicts very accurately the radiant power emitted by a blackbody per unit area per unit wavelength, or complete radiation. It is written

$$M^b(\lambda, T) = \frac{C_1}{\lambda^5} \frac{1}{e^{C_2/\lambda T} - 1} \quad W \cdot m^{-3}$$

where $C_1 = 2\pi h C^2 = 3.7415 \times 10^{16}$ W $\cdot$ m^2, called the first radiation constant
$C_2 = Ch/k = 1.43879 \times 10^{-2}$ m $\cdot$ k, called the second radiation constant

This radiation formula can be written in other forms, such as using wavenumbers instead of wavelengths. In this article, the above form expresses the radiant exitance[2] in terms of wavelength and absolute temperature. The units used are SI, and the nomenclature is that recommended by the Optical Society of America.

Blackbody Concept.   This is central to radiation thermometer technology. The energy radiated by an object as a result of its temperature is quantitatively expressed in terms of a perfect radiating body, which is traditionally designated a blackbody. This concept has been described in several ways, such as a body which absorbs all the radiation it intercepts, and a body that radiates more thermal energy for all wavelength intervals than any other body of the same area, at the same temperature. Physical realization of the blackbody, necessary for instrument calibration purposes, includes the spherical cavity and the wedge-shaped cavity as described by Mendenhall (Fig. 20). Numerous other blackbody source designs have been made and are available commercially (Fig. 21).

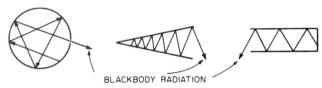

BLACKBODY RADIATION

FIGURE 20   Classical blackbody models. The radiation leaving a small hole in the sphere or the inside of the wedge, for example, will fulfill the definition of blackbody radiation, provided the walls of the cavity are opaque and the cavity is uniform in temperature. Because the geometry of these models promotes multiple internal reflection of the radiated energy, the wall material may be an imperfect radiator and thus a practical substance. Laboratory blackbody sources must be stable and uniform when used as reproducible sources in the calibration of secondary standards for radiation thermometers.

Emissivity.   This is a measure of the ratio of thermal radiation emitted by a nonblackbody at the same temperature.

Total Emissivity.   This is the ratio of the total amount of radiation emitted. By definition, a blackbody has maximum emissivity, the value of which is unity. Other bodies have an emissivity less than unity. The ratio is designated as total emissivity.

Gray Body.   This refers to an object which has the same spectral emissivity at every wavelength, or one which has its spectral emissivity equal to its total emissivity.

Stefan-Boltzmann Law.   As mentioned previously, Planck's radiation law predicts precise levels of radiation emitted per unit surface area of a blackbody at each wavelength. This may be written as

$$M^b(\lambda, T) = c_1 \lambda^{-5}(e^{c_2/\lambda T} - 1)^{-1}$$

The total radiation emitted per unit surface area is the integral of this equation over all wavelengths,

$$M^b(T) = c_1 \int_0^{+\infty} \lambda^{-5}(e^{c_2/\lambda T} - 1)^{-1} \, d\lambda$$

or, simply,

$$M^b(T) = \tau T^4$$

This is known as the Stefan-Boltzmann law. It is used extensively in the calculation of radiant heat transfer.

---

[2] A coined word not to be confused with excitance.

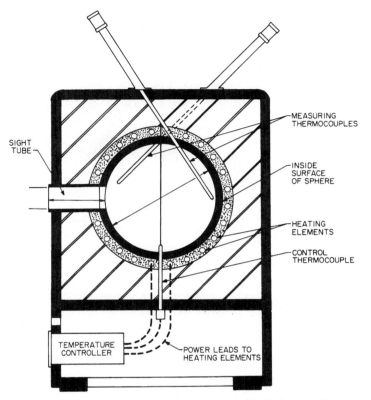

FIGURE 21 Schematic view of a commercial blackbody furnace. Entrance sight tube is shown to the left of the hollow sphere. One thermocouple is used to determine the sphere's temperature, another to determine temperature gradients. Blackbody source designs also may take the form of conical or cylindrical chambers where the length-to-diameter ratio is large. The emissivity of laboratory sources ranges from 0.98 to 0.9998. In some sources a temperature uniformity of better than 1 K is obtainable. Thus, reproducible calibration conditions of ±1 percent are readily obtainable, and ±0.1 percent can be achieved for industrial instruments.

In situations where the exponential term in Planck's law is much greater than 1, that is, $e^{c_2/\lambda T} \gg 1$, Planck's equation can be approximated by

$$M^b(\lambda,\ T) = c_1 \lambda^{-5} e^{-c_2/\lambda T}$$

which is known as Wein's law.

A graphic representation of the radiant exitance as predicted by Planck's law, for several temperatures, is given in Fig. 22. Note that for each temperature there is a peak value of emission and that the peak value shifts to shorter wavelengths as the temperature is increased. This shift can be expressed as

$$\lambda_m T = b$$

which is known as Wein's displacement law.

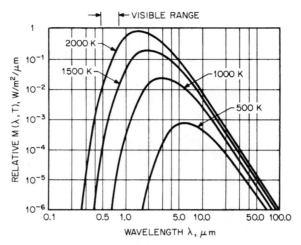

FIGURE 22   Radiant emission of thermal radiation from a blackbody at various temperatures (in kelvin).

Measuring Temperatures of Nonblackbodies

Temperature measurements of real nonblackbody objects can be complicated by three major factors:

1. Nonblackbodies emit less radiation than blackbodies, and often this difference is wavelength-dependent. Often, but not always, an emissivity correction must be made. Without the needed emissivity correction of the measured signal, the apparent temperature will be lower than the actual temperature.

2. Extra radiation from other radiant sources may be reflected from the object's surface, thus adding to the measured radiation and thereby increasing the apparent temperature. In certain cases the reflected amount may compensate for that needed to correct the object's emissivity.

3. The intensity of the emitted radiation may be modified in passing through media between the object and the instrument, thus resulting in a change in the apparent temperature. If radiation is lost, the apparent temperature will be low; if radiation is added, the temperature will read high.

In addition to these practical problems, there is often a need to measure an object that is transparent. Additional problems include added radiation produced by hot objects behind the object of measurement.

Effective Wavelength.   The calibration function, or output, of a radiation thermometer is a nonlinear voltage or current. Mathematically it is an equation involving the spectral characteristics of the optical system and the detector response, integrated over all wavelengths. Once an instrument design is fixed, the relation between the thermometer output and a blackbody source temperature can be written,

$$V(T) = K \int_{0}^{+\infty} M^{b}(\lambda,T)S(\lambda)\, d\lambda$$

where $S(\lambda)$ = net thermometer wavelength sensitivity
K = calibration constant

The calibration function is generated simply by aligning a unit with a blackbody source and measuring the output at different temperatures. Under conditions where Wein's law may be substituted for Planck's law, this equation becomes

$$V(T) = Kc_1 \int_{\lambda_1}^{\lambda_2} \lambda^{-5} e^{-c_2/\lambda T} \, d\lambda$$

and a wavelength λ_e can be found such that

$$\lambda_e^{-5} e^{-c_2/\lambda_e T} = \int_{\lambda_1}^{\lambda_2} \lambda^{-5} e^{-c_2/\lambda T} \, d\lambda$$

The single wavelength λ_e, the effective wavelength, is representative of an instrument at temperature T. Thus the calibration function $V(T)$ can be written as

$$V(T) = K' e^{-c_2/\lambda_e T}$$

from which it can be shown that the effective wavelength can be expressed in terms of the rate of change of the thermometer calibration function at T as

$$\lambda_e = \frac{c_2}{T_2} \left[\frac{\Delta V(T)}{V(T) \Delta T} \right]^{-1}$$

N Values.   Over a range of temperatures, say from T_1 to T_2, the effective wavelength will change. Also, at a given temperature the calibration function can be approximated as a single-term power function of temperature:

$$V(T) = KT^N$$

Making use of the effective wavelength concept, it can also be shown that

$$N = \frac{c_2}{\lambda_e T}$$

where the power N is called the N value of the thermometer. A ratio thermometer can be described in terms of effective (or equivalent) wavelength, even though it has two distinct and possibly widely separated wavelength response bands.

At high temperatures, $c_2/\lambda T \gg 1$, the calibration function of a ratio thermometer is essentially the ratio of the two calibration functions of the individual "channels" $V_1(T)$ and $V_2(T)$,

$$V_R(T) = \frac{V_1(T)}{V_2(T)} = A \, \frac{e^{-c_2/\lambda_1 T}}{e^{-c_2/\lambda_2 T}}$$

where λ_1 and λ_2 are the effective wavelengths of the two channels. When a net effective or equivalent wavelength λ_E is defined as

$$\frac{1}{\lambda_E} = \frac{1}{\lambda_1} - \frac{1}{\lambda_2}$$

then the *form* of the calibration function $V_R(T)$ is nearly identical to that of a single-wavelength thermometer with an effective wavelength of λ_e. The nearer the two effective wavelengths of the ratio thermometer channels are, the longer the equivalent wavelength λ_E.

A mean equivalent wavelength and N value can also be used to characterize a given ratio thermometer. It should be noted that, in using the foregoing equations, the temperatures must be expressed as absolute temperatures.

Optical Field of View.   Inasmuch as Planck's law deals with the radiation emitted per unit surface area, radiation from a known area must be measured in order to establish a temperature measurement. This property, called the *field of view* of the instrument, is simply the target size that the thermometer "sees" at a given distance. Thus a target-size-versus-distance table, chart, or formula is essential for the correct use of a radiation thermometer. At a minimum, the object of measurement must fill the required target size at a given distance. The optical field of view of a radiation thermometer is shown in Fig. 23. Similarly, in calibrating a thermometer, the radiation source must fill the field of view in order to generate or check the calibration output. If the field of view is not filled, the thermometer will read low. If a thermometer does not have a well-defined field of view, the output of the instrument will increase if the object of measurement is larger than the minimum size.

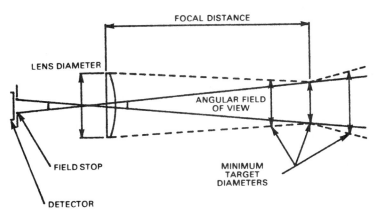

FIGURE 23   Schematic representation of the field of view of a radiation thermometer.

Since a ratio thermometer essentially measures the ratio of two thermometer signals, anything that tends to reduce the actual target size will not upset the ratio. In principle, ratio thermometers should be immune to changes in target size or not seriously affected if the object does not fill the field of view. There are, however, two limitations to this: (1) the optical fields of view in the two wavebands must be the same, and (2) the net signals to be ratioed must be larger than any internal amplifier drifts, offsets, and noise.

Transparent Objects.   The law of conservation of energy requires that, at every wavelength, the coefficients of transmission, reflection, and emission (absorption) of radiation add up to 1, as shown by

$$\varepsilon(\lambda) + r(\lambda) + t(\lambda) = 1$$

There are several ways to deal with transparent objects such as glass, plastic, semiconductor materials, and gases. The first and simplest is to select a waveband in which the object is opaque. For example, nearly all ordinary glasses are opaque at wavelengths longer than about 5.0 μm, provided the glass thickness is 3.0 mm or more. Similarly, most thin polyethylene plastics are opaque in a narrow wavelength band centered at 3.43 μm. As a simple guide in determining if an object is opaque, one must examine the spectral absorption coefficient which can be roughly deduced from spectral transmission curves using the relationship

$$\alpha = -\frac{1}{x} \ln \frac{T}{[1 - r(\lambda)]^2}$$

where $r(\lambda)$ = surface reflection coefficient which, at normal incidence, is obtained from the
index of refraction n
x = thickness
T = transmission

$$r(\lambda) = \left(\frac{n-1}{n+1} \right)^{2}$$

Other Factors Affecting Performance.   These include (1) the object has unknown or varying emissivity and (2) attenuation or enhancement of radiation occurs in the instrument's sighting path. Still another factor is ambient temperature stability.

Emissivity data are obtainable from numerous handbooks on materials, but where these data cannot be located, as in the case of a brand new material for which data have not yet been published (or may be proprietary) or where the instrument supplier may not have any specific data, some laboratory work may be needed. In such cases the user either should measure the object's temperature, or at least that of a representative sample, by another accurate and reliable means and adjust the thermometer emissivity control to make the thermometer match; or should have such tests conducted by an approved outside laboratory.

It should be stressed that the angle of viewing the target affects the target size and shape, as shown in Fig. 24.

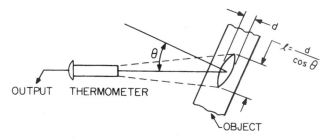

FIGURE 24   Effect of angle of viewing on target size and shape.

The user should determine which of the following situations may apply to a given application: (1) object and surroundings are at about the same temperature; (2) ambient conditions are cooler; or (3) ambient conditions are hotter. Measurements can be made which describe a given situation by using the following equation, which accounts for both the emissivity of the surface and the effect of thermal radiation from other sources reflected from the surface into the thermometer:

$$V(T_A) = \varepsilon(\lambda)V(T_O) + [1 - \varepsilon(\lambda)]V(T_B)$$

where $V(T_A)$ = indicated apparent blackbody temperature
$V(T_O)$ = blackbody temperature of object
$V(T_B)$ = blackbody temperature of net surroundings (assuming for simplicity that a reasonably uniform effective background exists)
$\varepsilon(\lambda)$ = spectral emissivity
$r(\lambda)$ = average reflectivity of object over same wavelength span, $= 1 - \varepsilon(\lambda)$

These problems are particularly important where the object is in hotter surroundings. What might be termed the specular solution to the problem can be used with many materials that have a relatively smooth surface, as, for example, flat glas in tempering furnaces. Rolled steel sheets for automotive and appliance uses are often processed in high-temperature ovens for annealing or cleaning. Such products often are highly specular, sometimes amounting to 90 percent or more, and can be measured successfully with very high N-value thermometers. Further, dual thermometers, as shown in Fig. 25, may be used.

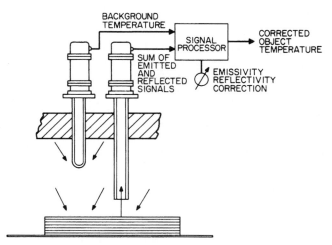

FIGURE 25   Dual-thermometer method for measuring the temperature of an object in surroundings that are hotter.

In terms of attenuating the radiation received by the thermometer, water vapor is a major cause of error, particularly with wideband thermometers because of the overlapping wavelengths involved. Radiation also may be attenuated by dust, smoke, condensing steam, radiation-absorptive gases, and simply by any object that may block the field of view of the instrument. In difficult situations of this kind, a closed or open sighting tube may be used as part of the instrument (Fig. 26), or, for example, in the case of gas turbine measurements, a two-wavelength thermometer may be the solution.

IR Detectors

Thermal detectors are most frequently used in contemporary IR thermometers. These include thermocouples (types J, K, N, R, and S) and RTDs. Quantum detectors, which consist

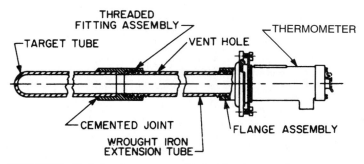

FIGURE 26   Radiation thermometer with closed-end target tube.

of a semiconductor crystal, are favored for some applications that require exceptional response speed. Both types of detectors have been described earlier in this article.

Classes of IR Thermometers

Generally IR thermometers are classified into five categories.

Wideband Instruments. These devices are the simplest and of lowest cost. They respond to radiation with wavelengths from 0.3 μm to between 2.5 and 20 μm, depending on the lens or window material used. These instruments also have been called broadband or total radiation pyrometers because of their relatively wide wavelength response and the fact that they measure a significant fraction of the total radiation emitted by the object of measurement. Historically these devices were the earliest fixed or automatic units. Standard ranges include 0 to 1000°C (32 to 1832°F) and 600 to 1900°C (932 to 1652°F).

Narrowband Instruments. These instruments usually have a carefully selected, relatively narrow wavelength response and are often selected to provide the particular wavelength response required by an application. Development of many different narrowband thermometers in recent years has been prompted by an increased need to match the IR thermometer with specific applications, as well as overcoming prior deficiencies of wideband instruments. Standard ranges are many, varying from one supplier to the next. Examples would include −50 to 600°C (−36 to 1112°F), 0 to 1000°C (32 to 1832°F), 600 to 3000°C (1112 to 5432°F), and 500 to 2000°C (932 to 3632°F).

Ratio (Two-Color) Instruments. These instruments essentially consist of two radiation thermometers contained within a single housing. Some internal components, such as lens and detector, may be shared. The unique characteristic is that the output from the two thermometers, each having a separate wavelength response, is ratioed. The concept behind the ratio thermometer is that the ratio signal is also a function of temperature, and so long as the ratio value is unchanged, the temperature measurement is accurate. Target, so long as it is sufficient, is not critical because the ratio of the signals from a small target is the same as that from a large target. These instruments cover wide temperature ranges. Examples would include 500 to 1800°C (932 to 3272°F), 0 to 1000°C (32 to 1832°F), and 825 to 1800°C (1441 to 3272°F).

Fiber-Optic Thermometers. These instruments enable near-infrared and visible radiation to be transmitted around corners and away from hot, hazardous environments to locations more suitable for the electronics associated with contemporary radiation thermometers. Fiber-optics also make possible measurements in regions where access is restricted to larger instruments and where large electric or radio frequency fields would seriously affect an ordinary sensor. Conceptually, a fiber-optic system differs from an ordinary IR thermometer by the addition of a fiber-optic light guide, with or without a lens. The optics of the light guide define the field of view of the instrument, while the optical transmission properties of the fiber-optic elements form an integral part of the thermometer spectral response (Fig. 27).

Disappearing-Filament IR Thermometer (Optical Pyrometer). Also sometimes referred to as a brightness thermometer, this instrument makes a photometric match between the brightness of an object and an internal lamp. These pyrometers are sensitive only in a very narrow wavelength range. The instrument is particularly adapted to manual operation and portability. One of the earliest noncontact temperature-measuring devices, its reputation for high accuracy was established a number of decades ago. Optical pyrometers differ from other IR thermometers in both the type of reference source used and the method of achieving the brightness match between object and reference. Figure 28 is a schematic view of a typical visual instrument and the indicators for over, under, and matching conditions as seen by a

FIGURE 27 Fiber-optic IR thermometer. Instruments like this cover three spectral regions and can measure temperatures from 177 to 3500°C (350 to 6500°F). A two-color version is also available (*Fiber Optic Mirage*™). Claimed accuracy is within ±1 percent and repeatability, ±0.3 percent (full-scale). (*Square D. Company (Ircon), Infrared Measurement Division.*)

manual operator. The temperature range is limited at the lower end by the need for an incandescent image of the filament to about 800°C (1472°F). The upper temperatures are limited only by the needs of the application.

FILLED-SYSTEM THERMOMETERS

An industrial filled-system thermometer is shown schematically in Fig. 29. These instruments may be separated into two fundamental types. The measuring element (a bourdon, bellows, or helix) responds either to volume changes or to pressure changes with changes in temperature. The liquid expansivity with temperature is greater than that of the bulb metal, the net volume change being communicated to the bourdon. An internal-system pressure change is always associated with the bourdon volume change, but this effect is not of primary importance.

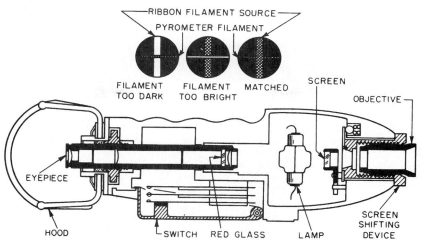

FIGURE 28 Optical pyrometer telescope and principle of operation. The reference source is an aged and calibrated tungsten strip lamp. In the manual version, the operator views the object to be measured and the lamp filament simultaneously through an optical filter. The combination of filter characteristics and the response of the average human eye produce a net instrument wavelength response band that is very narrow and centered near 0.65 µm. By adjusting the current flow through the lamp, or varying the intensity of the object radiation, the operator can produce a brightness match over at least a portion of the lamp filament, according to relative target size. Under matched conditions, the two "scenes" merge into one another, that is, the filament appears to vanish. Automatic optical pyrometers function similarly, in the same waveband, but utilize a photomultiplier tube or other detector element in place of the human eye. Advantage is taken of electronic feedback circuitry. (*Leeds & Northrup.*)

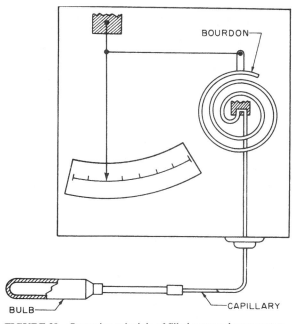

FIGURE 29 Operating principle of filled-system thermometer.

The following systems are generally available:

- *Class IA.*   The thermal system is completely filled with an incompressible liquid under pressure. The system is fully compensated for ambient temperature variations at the case and along the tubing (Fig. 30). Standard ranges, spans, and range limits are indicated in Table 8.

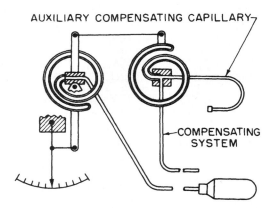

FIGURE 30   Fully compensated system as used in Class IA thermal systems.

- *Class IIA.*   The thermal system is evacuated and partially filled with a volatile liquid, such as methyl chloride (CL), ether (ET), butane (BU), or toluene (T) (Table 8).
- *Class IIIB.*   The thermal system is filled under pressure with purified nitrogen (Table 8).

In all systems bulbs may be flange-mounted for direct insertion into the process, or contained in a thermowell. On special order, longer-capillary lengths, sanitary fittings, and other ranges and spans (within limitations) are commercially available.

Advantages and Limitations of Filled Systems.   The fundamental simplicity of these systems favors lower cost, easy installation, and low maintenance. (However, in case of failure, the entire bulb and capillary assembly may have to be replaced.) The measuring system is self-contained, requiring no auxiliary power unless it is combined with an electronic or pneumatic transmission system. The capillary allows considerable separation between point of measurement and point of indication. [However, beyond about 120 meters (400 ft) it becomes more economic to use a transmission system.] In some cases the system can be designed to deliver significant power, if needed, to drive indicating or controlling mechanisms, including valves. Filled systems are limited to temperatures between −73 and 260°C (−100 and 500°F).

FIBER-OPTIC TEMPERATURE SENSOR[3]

A miniature fiber-optic sensor infers temperature from the refractive index (RI) of a thin layer of silicon that is "sandwiched" between two pieces of Pyrex glass. Rather than measuring the angle of refraction as is commonly done in many refractometers, an extrinsic Fabry-Perot interferometer spectrally modulates a light beam (wavefront) in proportion to temperature. Inasmuch as the units are based on spectral rather than amplitude modulation,

---

[3] Gene Yazbak, Senior Applications Engineer, MetriCor Inc., Monument Beach, Massachusetts.

TABLE 8 General Specifications of Filled-System Thermometers

	Class IA				Class IIA	Class IIIB		
	Bulb length		Range limits				Range limits	
Standard spans	mm	in	Minimum	Maximum	Standard ranges (fills)	Standard spans	Minimum	Maximum
°C			°C	°C	°C	°C	°C	°C
25	122	4.8	13	120	38 to 105 (BU)	100	−40	170
50	69	2.7	−15	200	105 to 150 (ET)	150	−15	400
75	53	2.1	−38	200	50 to 205 (T)	200	−15	500
100	43	1.7	−63	200	38 to 150 (ET)	250	−15	500
150	36	1.4	−73	250	50 to 250 (T)	300	−15	500
200	30	1.2	−73	260		400	−15	500
250	28	1.1	−73	260		500	−15	500
°F			°F	°F	°F	°F	°F	°F
50	112	4.4	50	250	100 to 180 (CL)	150	−50	240
100	69	2.7	0	400	100 to 220 (BU)	200	−50	425
150	51	2.0	−50	400	100 to 250 (BU)	300	0	825
200	43	1.7	−100	400	100 to 270 (BU)	400	0	1000
250	36	1.4	−100	400	100 to 300 (ET)	500	0	1000
300	33	1.3	−100	500	100 to 350 (ET)	600	0	1000
400	28	1.1	−100	500	100 to 400 (T)	700	0	1000
500	25	1.0	−100	500	200 to 500 (T)	800	0	1000
						1000	0	1000

Accuracy		
±0.5% of calibrated span for spans up to 215°C (400°F); ±0.75% of calibrated span for spans between 215 and 330°C (400 and 600°F).	±0.5% of calibrated span over upper two-thirds of scale.	±0.5% of calibrated span for spans up to 330°C (600°F); ±0.75% of upper range value for upper range value above 330°C (600°F); ±0.75% of lower range value for lower range value below −45°C (−50°F).

they are not affected by such common system phenomena as fiber bending, connector losses, and source and detector aging (Fig. 31).

Although the temperature probes can be made as small as 0.032 inch (0.8 mm), they usually are packaged as ⅛-inch (3.2-mm) stainless-steel probe assemblies which are interfaced directly to process vessels or pipes, using off-the-shelf compression fittings. Claimed advantages include immunity to electromagnetic and radio frequency; installation and maintenance are simple.

The sensors have been found effective in chemical and food processing and notably in microwavable food packaging research for gathering data on bumping and splattering of foods during heating. Nonmetallic temperature probes available for such applications have been found to minimize error because of their low thermal conductivity and small size. Sensors can be located up to 914 meters (3000 ft) distant from the optoelectronics module.

Similarly conceived sensors are available for pressure measurement and direct refractive index measurement.

ACOUSTIC PYROMETER

The modulating effect of temperature on the speed of sound dates back to the early studies of Laplace over two centuries ago. Thus temperature can be measured inferentially by measuring

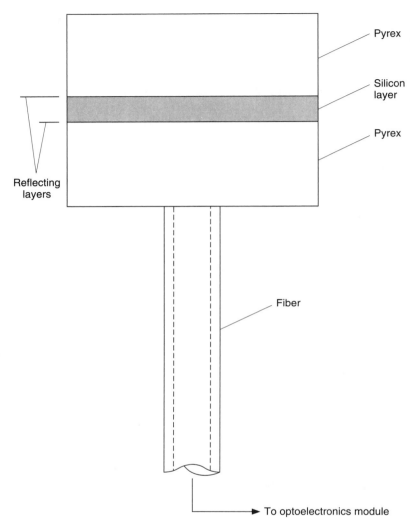

FIGURE 31   Fiber-optic temperature sensor. (*MetriCor, Inc.*)

the speed of an acoustic wave in a given environment. This principle has been applied to the measurement of furnace exit-gas temperatures. Traditionally such measurements are made by optical and radiation pyrometry or wall thermocouples, or estimated on the basis of heat balance calculations. Exit-gas temperatures are particularly important during boiler start-up so that superheater tubes will not be overheated prior to the establishment of steam flow.

If the speed of sound can be determined by measuring the flight time of an acoustic wave between two locations, then the temperature of the measured medium can be determined. Experiments with an acoustic pyrometer for furnace exit gas indicate that temperatures up to 1650°C (3000°F) may be encountered and that acoustic velocities may be in excess of 879 m/s (2880 ft/s) and wavelengths on the order of 1 meter (3.28 ft). As pointed out by some researchers, there still is much work to be done before a sound source is developed that will satisfy all of the conflicting requirements generated by the possible applications for acoustic pyrometers.

FLUID PRESSURE SYSTEMS

Pressure measurement not only is critical to the safe and optimum operation of such industrial processes as air and other gas compression; hydraulic equipment operation; separating operations, such as absorption, desorption, distillation, and filtration; steam generation; and vacuum processing—but other process variables, such as the contents level of tanks (hydrostatic pressure) and flow (differential pressure) can be inferred from pressure measurements. Pressure, of course, is the key to applying pneumatic transmitters and controllers. Further, pressure sometimes is selected over temperature as the variable to control. Air pressure also activates diaphragm motor valves.

Pressure instrumentation ranges widely from the comparative simplicity of low-cost bourdon- and bellows-actuated gages to some of the contemporary complex and sophisticated pressure sensors-transducers (transmitters) that have appeared in very recent years. Modern pressure transmitters differ from their historical counterparts mainly in two design respects:

1. Mechanical transducers utilizing links, levers, and pivots having been replaced by electric and electrooptic transducers, which permits varying degrees of miniaturization of the force-receiving sensors

2. Introduction of what is commonly termed "smart" or "intelligent" electronics into transmitter design, notably the incorporation of a microprocessor along with other ingenious electronic circuitry

A 1991 survey of users indicates that about one-third of the modernized electronic pressure transmitters sold today are of the smart variety, and with an increasing rate of acceptance, even though costs run somewhat higher than for the simpler versions.

It is interesting to note that progress in the pressure transmitter field has stemmed mainly from discoveries and developments in the electronics and computer industries. Because of the dynamics of these industries, it would be unrealistic to assume that pressure transmitters or, in fact, any other areas of industrial instrumentation have reached a status that could be identified as mature. Thus the field will remain in a transitory phase for some time to come, which makes selection a continuing difficult task.

Because a substantial market remains for the historically less sophisticated pressure gages that incorporate simple mechanics and design, these devices are described in the early portions of this article.

MANOMETERS

Because of their inherent accuracy, manometers are used for the direct measurement of pressure and vacuum. Although some rugged designs can be used in the field and on-line, manometers largely serve as standards for calibrating other pressure-measuring instruments.

U-Tube Manometer.   A glass U-tube is partially filled with liquid, and both ends are initially open to the atmosphere (Fig. 1).

When a gage pressure P_2 is to be measured, it is applied to the top of one of the columns and the top of the other column remains open. When the liquid in the tube is mercury, for example, the indicated pressure h is usually expressed in inches (or millimeters) of mercury. To convert to pounds per square inch (or kilograms per square centimeter),

$$P_2 = dh$$

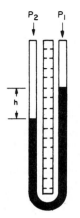

FIGURE 1   U-tube manometer.

where P_2 = pressure, psig (kg/cm^2)
$\quad\quad d$ = density, lb/in^3 (kg/cm^3)
$\quad\quad h$ = height, inches (cm)

For mercury, the density is 0.490 lb/in^3 at 60°F (15.6°C), and the conversion of inches of mercury to pounds per square inch becomes

$$P_2 = 0.490h$$

The density of water at 60°F (15.6°C) is 0.0361 lb/in^3, and if water is used in a manometer, the conversion of inches of water to pounds per square inch becomes

$$P_2 = 0.0361h$$

The same principles apply when metric units are used. For example, the density of mercury at 15.6°C (60°F) may also be expressed as 0.0136 kg/cm^3, and the conversion of centimeters of mercury to kilograms per square centimeters

$$P_2 = 0.0136h$$

For measuring differential pressure and for static balance,

$$P_2 - P_1 = dh$$

The U-tube manometer principle has also been utilized in industry in an instrument usually called a differential-pressure manometer. In this device the tubes are expanded into chambers and a float rides on top of the liquid in one of the chambers. The float positions an outside pointer through a pressuretight bearing or torque tube.

Well Manometer.   In this design one leg is replaced by a large-diameter well so that the pressure differential is indicated only by the height of the column in the single leg. The ratio of diameters is important and should be as great as possible to reduce the errors resulting from the change in level in the large-diameter well (Fig. 2).

The pressure difference can be read directly on a single scale. For static balance,

$$P_2 - P_1 = d\left(1 + \frac{A_1}{A_2}\right)h$$

where A_1 = area of smaller-diameter leg
$\quad\quad A_2$ = area of well

If the ratio of A_1/A_2 is small compared with unity, then the error in neglecting this term becomes negligible, and the static balance relation becomes

$$P_2 - P_1 = dh$$

On some manometers this error is eliminated by reducing the spacing between scale graduations by the required amount.

Inclined-Tube Manometer.   In this device, so as to read small pressure differentials more accurately, the smaller-diameter leg is inclined (Fig. 3). This produces a longer scale so that

$$h = L \sin \alpha$$

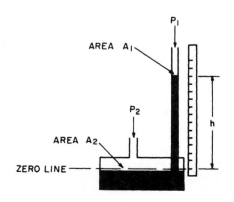

FIGURE 2   Well manometer.

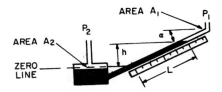

FIGURE 3   Inclined-tube manometer.

Bell-Type Manometer.   This device utilizes a container immersed in a sealing liquid. The pressure to be measured is applied to the inside of the bell, the motion of which is opposed by a restricting spring (Fig. 4a). In the bell-type differential-pressure gage, pressures are applied to both the outside and the inside of the bell. Motion is restricted by an opposing spring (Fig. 4b).

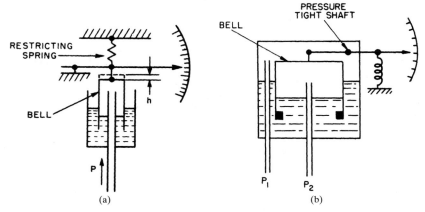

(a)   (b)

FIGURE 4   Bell-type manometers. (a) Liquid-sealed bell. (b) Differential-pressure gage.

Liquid Barometer.   A simple barometer may be constructed from a glass tube which is closed at one end and open at the other. The length of the tube must be greater than 30 inches (76.2 cm). The tube is first completely filled with mercury, the open end temporarily plugged, and then the plugged end placed in a container partially filled with mercury.

When the plug is removed, the mercury in the tube will drop by a certain amount, creating a vacuum at the top of the tube. The height of the column, as measured in Fig. 5 and expressed in inches or millimeters of mercury, will then be proportional to the atmospheric pressure.

Absolute-Pressure Manometer.   This type of gage is comprised of a glass U-tube partially filled with mercury, with the top of one leg evacuated and sealed (Fig. 6). The pressure to be measured is applied to the other leg, and h may be read in units of mercury absolute. To convert to pounds per square inch absolute (psia),

$$P = 0.490h$$

where P is absolute pressure in psia. If h is indicated in centimeters, this value may be converted to kilograms per square centimeter absolute by multiplying by 0.0136.

McLeod Gage (Liquid Manometer).   This device is designed for vacuum measurement. It functions essentially as a pressure amplifier. If, for example, 100 cm³ (volume V in Fig. 7) of permanent gas is compressed into a section of capillary having a volume of 0.1 cm³, the resulting pressure reading is amplified 1000 times. This princi-

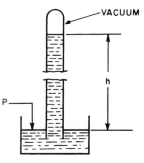

FIGURE 5   Liquid barometer.

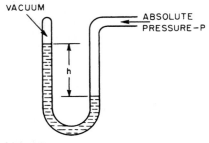

FIGURE 6 Absolute-pressure gage (manometer).

ple allows pressure measurements into the 10^{-6}-torr region, considerably below the 10^{-2}-torr range of precision manometers.

If we assume that the volume V of gas trapped at the unknown pressure p (in centimeters of mercury for convenience) obeys Boyle's law, then $pV = (p + H)A$, where A is the cross section of the closed capillary in square centimeters, and

$$P = \frac{AH^2}{V - HA}$$

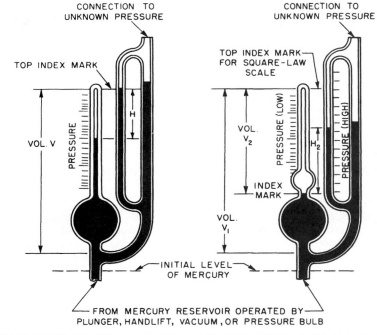

(a) SINGLE-RANGE, SQUARE-LAW SCALE

$$P = \frac{AH^2}{V} \text{ (VERY NEARLY)}$$

(b) DOUBLE-RANGE, LINEAR AND SQUARE-LAW SCALES

$$P = \frac{V_2}{V_1} H_2 \text{ (VERY NEARLY)}$$

FIGURE 7 Two versions of McLeod gage.

In practice, HA is quite negligible when compared with volume V, with $p = 10AH^2/V$ torr, and with other values expressed in centimeters.

A conveniently small McLeod gage may have a volume V of 200 cm^3, with a capillary cross section A of 0.02 cm^2 and a length of 10 cm. Thus for $H = 0.1$ cm, $p = 5 \times 10^{-6}$ torr, which would be the limit of unaided visual resolution and the reading could be wrong by 100 percent. At $H = 1$ cm, $p = 5 \times 10^{-4}$ torr, the possible error becomes 10 percent. For various reasons, the only significant improvement in accuracy can be achieved by an increase in volume V.

A carefully constructed nonportable gage with a 1300-cm³ volume gives reproducible readings of ±0.5, ±0.6, ±2, and ±6 percent at pressures of 10^{-2}, 10^{-3}, 10^{-4}, and 10^{-5} torr, respectively. The errors for other volumes can be estimated to be no lower than those based on volume proportionality. Thus in the previous example with $V = 200$ cm³ and $p = 10^{-4}$ torr, percent error = $(1300/200) \times 2 = 13$ percent, which is in good agreement with the previous rough estimate of 10 percent.

Since the measured pressure in a McLeod gage is derived from basic (linear) dimensions, it is the industrial pressure standard with reference to which all other vacuum gages are calibrated. However, it should be emphasized that only the pressure of permanent gases is measured correctly. On account of the high compression ratio employed, vapor pressure of a substance of several tenths of a torr would not be detected with condensed liquid occupying negligible volume and not being visible to the eye. A highly portable version of the McLeod gage is shown in Fig. 8.

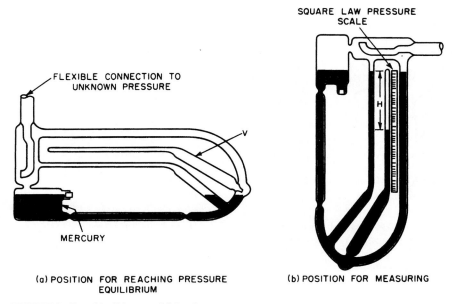

FIGURE 8   Portable tilting-type McLeod gage.

ELASTIC-ELEMENT MECHANICAL PRESSURE GAGES

Dating back to the early years of steam power and compressed air and hydraulic technologies, this class of pressure sensors uses some form of elastic element whose geometry is altered by changes in pressure. These elements are of four principal types: bellows, bourdon tube, diaphragm, and capsule.

Bellows.   This is a thin-wall metal tube with deeply convoluted sidewalls which permit axial expansion and contraction (Fig. 9). Most bellows are made from seamless tubes, and the convolutions either are hydraulically formed or mechanically rolled. Materials used include brass, phosphor bronze, beryllium copper, Monel, stainless steel, and Inconel. Bellows elements are well adapted to use in applications that require long strokes and high developed forces. They are well suited for input elements for large-case recorders and indicators and for feedback elements in pneumatic controllers.

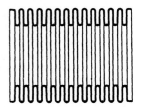

FIGURE 9   Common form of bellows used in pressure gage.

Bourdon Tube.   In the 1852 patent its inventor E. Bourdon described the bourdon tube as a curved or twisted tube whose transfer section differs from a circular form. In principle, it is a tube closed at one end, with an internal cross section that is not a perfect circle and, if bent or distorted, has the property of changing its shape with internal pressure variations. An internal pressure increase causes the cross section to become more circular and the shape to straighten, resulting in motion of the closed end of the tube, a motion commonly called tip travel. Common forms of bourdon tubes are shown in Fig. 10.

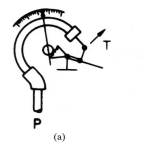

(a)

(b)

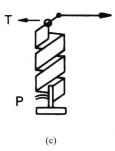

(c)

FIGURE 10   Types of bourdon springs. (*a*) C-type tube. (*b*) Spiral tube. (*c*) Helical tube.

A wide range of alloys can be used for making bourdon elements, including brass, phosphor bronze, beryllium copper, Monel, Ni-Span C, and various stainless-steel alloys.

Diaphragm.   This is a flexible disk, usually with concentric corrugations, that is used to convert pressure to deflection. (In addition to use in pressure sensors, diaphragms can serve as fluid barriers in transmitters, as seal assemblies, and also as building blocks for capsules.) A diaphragm usually is designed so that the deflection-versus-pressure characteristics are as linear as possible over a specified pressure range, and with a minimum of hysteresis and minimum shift in the zero point. However, when required, as in the case of an altitude sensor, a diaphragm can be purposely designed to have a nonlinear characteristic.

Metals commonly used for making diaphragms are trumpet brass, phosphor bronze, beryllium copper, stainless steel, NiSpan C, Monel, Hastelloy, titanium, and tantalum. Both linearity and sensitivity are determined mainly by the depth and number of corrugations and by the angle of formation of the diaphragm face.

In many pressure-measuring applications, the process fluid must not contact or seep into the pressure element in order to prevent errors due to effects of static head, to isolate the pressure element from corrosive and otherwise fouling fluids, to assist in cleaning (as in the food industry), and to prevent loss of costly or hazardous process fluids. Thus diaphragm seals are commonly used.

Capsule.   A capsule is formed by joining the peripheries of two diaphragms through soldering or welding. Two or more capsules can be joined together (Fig. 11), and thus the total deflection of the assembly is equal to the sum of the deflections of the individual capsules.

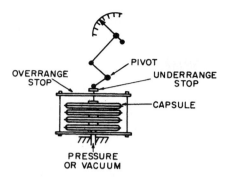

FIGURE 11 Use of capsule element in pressure gage.

Such elements are used in some absolute pressure gages. These configurations also are used in aircraft applications.

Range of Measurement

The minimum and maximum pressure ranges of elastic-element mechanical pressure gages are given in Table 1.

FORCE OR PRESSURE SENSORS, TRANSDUCERS, AND TRANSMITTERS

There is a marked distinction between a pressure sensor and a pressure transducer. The sensor provides the basis of measurement; the transducer converts energy from one form to another.

In the fully mechanical pressure instruments described previously, a spring may furnish the restoring force and, by means of links and lever, amplify and transmit the sensor value to a mechanically operated indicator, recorder, or controller.

In pneumatic pressure transducers, a counterpressure of air acts on the diaphragm, bellows, bourdon, or other elastic element to equalize the sensed (process) pressure. A force- or position-balance system may be used in pneumatic instruments, which are described in Section 3, Article 6, of this handbook. Current-to-pressure transducers used for the operation of pneumatic diaphragm control valves are described in Section 9 of this handbook.

In electronic or electrooptical transducers, sensor values are converted to electrical quantities—current, resistance, capacitance, reluctance, and alterations in piezoelectric and optical outputs.

Invention of the strain gage served as the initial impetus to use electrical transducers. There are numerous advantages for a large number of applications to be derived from some form of

TABLE 1 Ranges of Elastic-Element Pressure Gages

Element	Application	Minimum range	Maximum range
Capsule	Pressure	0–0.2 in (0.5 cm) H_2O	0–1000 psig (70.3 kg/cm^2)
	Vacuum	0–0.2 in (0.5 cm) H_2O	0–30 in (76.2 cm) Hg vacuum
	Compound vacuum and pressure	Any span within pressure and vacuum ranges, with a total span of 0.2 in (0.5 cm) H_2O	—
Bellows	Pressure	0–5 in (12.7 cm) H_2O	0–2000 psig (141 kg/cm^2)
	Vacuum	0–5 in (12.7 cm) H_2O	0–30 in (76.2 cm) Hg vacuum
	Compound vacuum and pressure	Any span within pressure and vacuum ranges, with a total span of 5 in (12.7 cm) H_2O	—
Bourdon	Pressure	0–5 psig (0.35 kg/cm^2)	0–100,000 psig (7030 kg/cm^2)
	Vacuum	0–30 in (76.2 cm) Hg vacuum	—
	Compound vacuum and pressure	Any span within pressure and vacuum ranges, with a total span of 12 psi (0.84 kg/cm^2)	—

electronic transduction. Such units are quite small, they are easy to integrate into electrical networks, and numerous other electronic features can be added to transducers and transmitters, including built-in calibration checks, temperature compensation, self-diagnostics, signal conditioning, and other features, which may be derived from integrating a microprocessor in the sensor-transducer-transmitter unit.

Strain-Gage Transducers

These devices have been used extensively in pressure and weighing load cells for several years. Strain gages usually are mounted directly on the pressure sensor or force summing element. They may be supported directly by sensing diaphragms or bonded to cantilever springs, which act as a restoring force. The operating principle of a resistance-type strain gage is illustrated in Fig. 12; a contemporary unit is shown in Fig. 13. The performance and range characteristics of some commercially available units are summarized in Table 2.

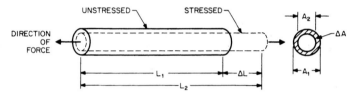

FIGURE 12   Basic relation between resistance change and strain in resistance-type strain gage. When under stress, the wire changes in length from L_1 to L_2 and in area from A_1 to A_2. The resistance is

$$R = p\,\frac{L}{A}$$

where L = conductor length
A = cross section area
p = resistivity constant

and the strain sensitivity (gage factor) is

$$S = \frac{\Delta R/R}{\Delta L/L}$$

where $\dfrac{\Delta R}{R}$ = resistance change

$\dfrac{\Delta L}{L}$ = strain

The strain sensitivity is commonly called the gage factor when referring to a specific strain-gage material. Poisson's ratio for most wire is approximately 0.3. The strain sensitivity or gage factor is approximately 1.6 when considering only the dimensional change aspect. This means that a 0.1 percent increase in length within the elastic range should produce a resistance increase of 0.16 percent. When actual tests are performed, a metal or alloy exhibits different values of strain sensitivity for different temperatures.

The ideal strain gage would change resistance in accordance with deformations of the surface to which it is bonded and for no other reason. However, gage resistance is affected by other factors, including temperature. Any resistive change in the gage not caused by strain is referred to as apparent strain. Apparent strain may be caused by a change in the gage factor due to temperature (thermal coefficient of the gage factor), by a change in resistance due to temperature (ther-

TABLE 2 Representative Characteristics of Strain-Gage Pressure Transducers

Pressure range	Pressure reference	Performance, %FRO	Accuracy (static error band)	Input	Output	Electrical connection
Low pressure:						
0–2 in. H$_2$O to 0–100 psi (0–5 mbar to 0–7 bar)	Vented gage; absolute; differential; vacuum	Very low range; high performance	±0.50	10–32 V dc; ±15 V dc	5V; 4–20 mA	Cable
0–1.5 to 0–5 psi (0–100 to 0–350 mbar)	Vented gage; absolute	Low range; high performance	±0.35	10; 11–18; 18–32; 10–36 V dc; ±15 V dc	25 mV; 5 V; 2.5 V; 4–20 mA	Cable or connector
0–1.5 to 0–5 psi (0–100 to 0–350 mbar)	Wet/dry unidirectional differential	Low range; high performance	±0.30	10; 11–18; 18–32; 10–36 V dc; ±15 V dc	25 mV; 5 V; 2.5 V; 4–20 mA	Cable or connector
0–1.5 to 0–30 psi (0–100 mbar to 0–2 bar)	Wet/wet unidirectional differential	Low range; differential measurement	±0.50	10; 11–18; 18–32; 9–36 V dc; ±15 V dc	25 mV; 5 V; 2.5 V; 4–20 mA	Cable or connector
Medium to high pressure:						
0–10 to 0–1000 psi (0–700 mbar to 0–70 bar)	Unidirectional wet/dry differential	Medium range; high performance	±0.30	10; 11–18; 18–32; 10–36 V dc; ±15 V dc	25 mV; 5 V; 2.5 V; 4–20 mA	Cable or connector
0–10 to 0–3500 psi (0–700 mbar to 0–250 bar)	Wet/wet differential	High range; differential measurement	±0.35	10; 11–18; 18–32; 10–36 V dc; ±15 V dc	25 mV; 5 V; 2.5 V; 4–20 mA	Cable or connector
0–10 to 0–3500 psi (0–700 mbar to 0–250 bar)	Vented gage; sealed gage; absolute	Flush diaphragm	±0.30	10; 11–18; 18–32; 10–36 V dc; ±15 V dc	25 mV; 5 V; 2.5 V; 4–20 mA	Cable or connector
0–10 to 0–10,000 psi (0–700 mbar to 0–700 bar)	Vented gage; sealed gage; absolute	Premium performance	±0.15	10; 11–18; 18–32; 10–36 V dc; ±15 V dc	25 mV; 5 V; 2.5 V; 4–20 mA	Cable or connector
0–75 to 0–10,000 psi (0–5 mbar to 0–700 bar)	Vented gage; sealed gage; absolute	High performance	±0.30	10; 10–36 V dc	24–32 mV; 5 V; 4–20 mA	Cable
0–15 to 0–5000 psi (0–1 bar to 0–350 bar)	Vented gage; sealed gage; absolute	Economical, general performance	±1.0	5; 9–20; 12–45 V dc	50 mV; 1–6 V; 4–20 mA	Cable or pins
Special configurations: 13–25 to 13–80 psi (900–1700 to 900–5500 mbar)	Absolute	Configured for tank content	±0.20	10–36 V dc	4–20 mA	Connector
0–75 to 0–500 psi (0–5 to 0–35 bar)	Vented gage; sealed gage; absolute	Meets hygienic requirements for aseptic facilities	±0.50	10; 11–18; 18–32; 10–36 V dc; ±15 V dc	25 mV; 5 V; 4–20 mA	Cable or connector
0–150 to 0–6000 psi (0–10 to 0–400 bar)	Sealed gage; absolute	Meets long-term operations in subsea conditions	±0.30	10–36 V dc	4–20 mA	Cable or connector

Note: In addition to traditional process applications, strain-gage pressure transducers find wide application in other fields.

- *Aerospace:* Hydraulic pressure measurement on aircraft; in-flight pressure control; measurement of altitude (atmospheric pressure and air speed).
- *Energy applications:* Monitoring of nuclear reactor core pressure; measurement of water level behind hydroelectric dams; fossil-fuel power plants.
- *Marine applications:* Control of submersibles, such as measuring ballast pressure on each leg of an oil-rig platform as it is submersed to the ocean floor; content measurement on chemical tankers; monitoring hydraulic oil and fuel pressures of shipborne equipment.
- *Special process applications:* Measurement of wet/dry and wet/wet unidirectional differential pressures; tank contents pressure transmission.

  Source: Lucas Schaevitz.

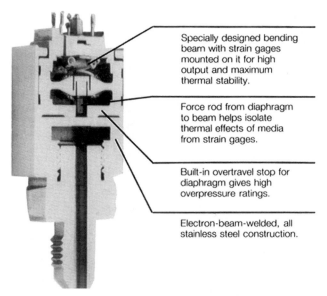

Specially designed bending beam with strain gages mounted on it for high output and maximum thermal stability.

Force rod from diaphragm to beam helps isolate thermal effects of media from strain gages.

Built-in overtravel stop for diaphragm gives high overpressure ratings.

Electron-beam-welded, all stainless steel construction.

FIGURE 13 Strain-gage pressure transducer incorporating a force transfer rod between diaphragm and double cantilever beam with four foil strain gages on beam. Units are calibrated and temperature-compensated to ensure stability over specified ranges, as well as unit interchangeability. Units are available with required signal conditioning. (*Lucas Schaevitz.*)

mal coefficient of the resistance), by the stability of the metal, and even by the properties of the adhesive that bonds the gage to the surface being measured. Many improvements in strain-gage materials have been made in recent years, thus reducing the effects of apparent strain. Some common types of resistance strain gages and their characteristics are listed in Table 3.

Bonded Strain-Gage Systems.   Although increasing the gage factor makes the gage more sensitive to strain, this also increases the undesirable effects of temperature. Thus small size is preferred so that the gage can be placed close to the high-strain area. A high resistance

TABLE 3 Major Types of Resistance Strain Gages

Common name	Basic material of strain gage	Method of attachment of strain gage to surface	General application
Unbonded	Wire	Connected at ends	Transducer
Bonded metallic	Wire or foil	Epoxy	Stress analysis and transducer
Flame spray	Wire	Spray-coated	Stress analysis
Welded	Foil	Spot-welded	Stress analysis
Bonded semiconductor	Silicon or germanium	Epoxy	Stress analysis and transducer
Diffused semiconductor	Silicon	Semiconductor diffusion	Transducer
Thin film	Metal alloy	Sputtering or deposition	Transducer

permits larger input voltage excitation and thus a larger millivolt output with a lower power consumption.

Bonded foil strain gages are made using special metal alloy conductors with high resistivities, high gage factors, and low temperature coefficients. Wire strain gages are not used widely because in order to obtain 350 ohms by using no. 28 copper wire [0.000126 in² (0.08 mm²) in cross section], 5360 ft (1633.7 meters) of wire would be needed [350 Ω/(65 Ω/1000 ft)]. The metal alloy of a bonded foil strain gage is formed into a back-and-forth grid to decrease the overall length of the strain gage system (Fig. 14). The length of the grid versus the width is designed to concentrate the strain-sensing grid over the high-strain area. Foil strain gages with gage resistance values of 120, 350, and 1000 ohms are common, with special gages for use with 4- to 20-mA electronic transmitters having resistances as high as 5000 ohms.

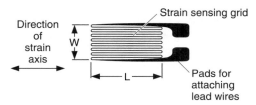

FIGURE 14   Representative single bonded-foil strain gage. Dimensions can be of the order of 0.021 X 0.062 in (0.79 X 1.57 mm). Thickness of a single gage may be on the order of 0.0022 in (0.056 mm). With improved manufacturing techniques, the trend is toward smaller dimensions.

The sensing grid is tiny and fragile in comparison to the structure to which it is usually attached. Therefore pads for connecting leadwires must be manufactured as part of the strain gage. The strain gage is bonded to the specimen surface by a thin layer of epoxy adhesive (Fig. 15), and care must be taken to ensure a thin, uniform, strong bond. A uniform bonding force applied by a contoured gaging block is used to exert a constant, even pressure against the strain gage. In summary, when installed and ready for use, the strain-gage system consists of the specimen surface, an effective bond between gage and specimen, the strain gage, appropriate leads and connectors, and, if needed, a protective waterproof coating.

Metallic Strain-Gage Materials.   All electrical conductors exhibit a strain-gage effect, but only a few meet the necessary requirements to be useful as strain gages. The major properties of concern are (1) gage factor, (2) resistance, (3) temperature coefficient of gage factor, (4) thermal coefficient of resistivity, and (5) stability. High-gage-factor materials tend to be more sensitive to temperature and less stable than the lower-gage-factor materials.

Strain-gage materials that have been commonly used in the past include the following:

Constantan.   Constantan or Advance (copper-nickel alloy) is primarily used for static strain measurement because of its low and controllable temperature coefficient. For static measurements, under ideal compensation conditions, or for dynamic measurements the alloy may be used from −100 to +460°F (−73.3 to +283°C). Conservative limits are 50 to 400°F (10 to 204°C).

Karma.   Karma (nickel-chrome alloy with precipitation-forming additives) provides a wider temperature compensation range than Constantan. Special treatment of this alloy gives minimum drift to 600°F (316°C) and excellent self-temperature compensation characteristics to ~800°F (427°C).

Nichrome V.   Nichrome V (nickel-chrome alloy) is commonly used for high-temperature static and dynamic strain measurements. Under ideal conditions, this alloy may be used for static measurements to 1200°F (649°C) and for dynamic measurements to 1800°F (982°C).

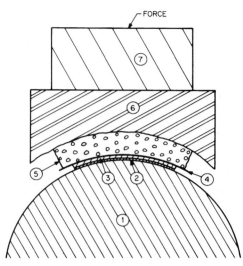

FIGURE 15 Installation of foil strain gage on non-planar surface. (1) Cylindrical specimen surface; (2) thin adhesive layer [0.001 inch (0.025 mm)]; (3) strain gage; (4) polymer pad sheet to prevent pressure pad from sticking; (5) pressure pad; (6) metal gaging block that conforms to specimen surface; (7) weight or clamp to apply pressure while adhesive is curing.

Isoelastic. Isoelastic (nickel-iron alloy plus other ingredients) is used for dynamic tests where its larger temperature coefficient is of no consequence. The higher gage factor is a distinct advantage where dynamic strains of small magnitude are measured.

479PT. 479PT (platinum-tungsten alloy) shows an unusually high stability at elevated temperatures. It also has a relatively high gage factor for an alloy. A gage of this material is recommended for dynamic tests to 1500°F (816°C) and static tests to 1200°F (649°C).

Semiconductor (Silicon) Strain-Gage Materials. Semiconductor material has an advantage over metals because its gage factor is approximately 50 to 70 times higher. However, the desirable increase in gage factor is partially offset by its greater thermal coefficient of resistivity (the common term is temperature effect). Comparatively recently, semiconductor strain gages are gaining in importance, particularly in the manufacture of miniature pressure and force transducers. Micromachined silicon assemblies also permit the integration of numerous other functions in a pressure transmitter. Representative properties of strain-gage materials, including semiconductors, are given in Table 4.

Strain-Gage Bonding Agents. The importance of the adhesive which bonds the strain gage to the metal structure under test or as part of a transducer cannot be overemphasized. An ideal adhesive should be suited to its intended environment, transmit all strain from the surface to the gage, have high mechanical strength, high electrical isolation, low thermal insulation, and be very thin. Also, it should not be affected by temperature changes. The adhesive must provide a strong bond while electrically isolating the gage from the surface to which it is attached. Electrical isolation is needed because most of the structures to which gages are bonded would electrically short out the elements if no electrical isolation existed. In a typical strain-gage installation, the electrical isolation between the gage and the specimen surface should be at least 1000 MΩ at room temperature and 50 volts dc. Electrical isolation (leakage) becomes a problem with bonding agents at high temperatures and in high-moisture environ-

TABLE 4 Properties of Strain-Gage Materials

Material	Composition, %	Gage factor	Thermal coefficient of resistivity, $°C^{-1} \times 10^{-5}$
Constantan (Advance)	Ni 45, Cu 55	2.1	±2
Isoelastic	Ni 36, Cr 8, Mn-Si-Mo 4, Fe 52	3.52 to 3.6	+17
Karma	Ni 74, Cr 20, Fe 3, Cu 3	2.1	+2
Manganin	Cu 84, Mn 12, Ni 4	0.3 to 0.47	±2
Alloy 479	Pt 92, W 8	3.6 to 4.4	+24
Nickel	Pure	−12 to −20	670
Nichrome V	Ni 80, Cr 20	2.1 to 2.63	10
Silicon	p-Type	100 to 170	70 to 700
Silicon	n-Type	−100 to −140	70 to 700

ments. At high temperatures, even ceramic materials begin to exhibit a loss of electrical isolation. This is one of the most severe limitations on strain-gage performance at temperatures above 1200°F (649°C).

Because of the wide variation in properties obtainable with different resin and hardener combinations, epoxy resins are an important class of strain-gage adhesives. Alternate attachment methods, such as the flame spray technique, have been used.

Basic Strain-Gage Bridge Circuit.   In order to make use of the basic operating principle of the bonded resistance strain gage (that is, change in resistance proportional to strain), the strain-gage input must be connected to an electric circuit capable of measuring small changes in resistance. Since the strain-induced resistance changes are small (typically 0.2 percent for full-scale output in one active gage), the gages are wired into a Wheatstone bridge. A Wheatstone bridge is a circuit designed to accurately measure small changes. It can be used to determine both dynamic and static strain-gage readings. The Wheatstone bridge also has certain compensation properties.

The Wheatstone bridge detects small changes in a variable by comparing its value to that of a similar variable and then measuring the difference in magnitude, instead of measuring the magnitude directly. For instance, if four equal-resistance gages are wired into the bridge (Fig. 16) and a voltage is applied between points A and C (input), then there will be no potential difference between points B and D (output). However, any small change in any one of these resistances will cause the bridge to become unbalanced, and a voltage will exist at the output in proportion to the imbalance.

In the simplest Wheatstone bridge configuration, a strain-sensing grid is wired in as resistance R_1. For this circuit, the output voltage E_o can be derived easily. In reference to the circuit shown in Fig. 16, the voltage drop across R_1 is denoted by V_{ab} and given as

$$V_{ab} = \frac{R_1}{R_1 + R_2} \text{ volts} \tag{1}$$

Similarly, the voltage drop across R_4 is denoted by V_{ad} and given by

$$V_{ad} = \frac{R_4}{R_3 + R_4} \text{ volts} \tag{2}$$

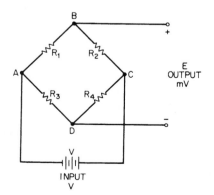

FIGURE 16 Four-arm Wheatstone bridge circuit. Strain gages are inserted at R_1, R_2, R_3, and R_4.

The output voltage from the bridge E is equivalent to V_{bd}, which is given by

$$E = V_{bd} = V_{ab} - V_{ad} \tag{3}$$

Substituting Eqs. (1) and (2) into Eq. (3) and simplifying gives

$$
\begin{aligned}
E &= \frac{R_1}{R_1 + R_2} - \frac{R_4}{R_3 + R_4} \\[2mm]
&= \frac{R_1 R_3 - R_2 R_4}{(R_1 + R_2)(R_3 + R_4)} \quad \text{volts}
\end{aligned}
\tag{4}
$$

The voltage E will go to zero, and thus the bridge will be considered to be balanced when

$$R_1 R_3 - R_2 R_4 = 0$$

or

$$R_1 R_3 = R_2 R_4 \tag{5}$$

Therefore the general equation for bridge balance and zero potential difference between points B and D is

$$\frac{R_1}{R_4} = \frac{R_2}{R_3} \tag{6}$$

Any small change in the resistance of the sensing grid will throw the bridge out of balance and can be detected by a voltmeter.

When the bridge is set up so that the only source of imbalance is a resistance change in the gage (resulting from strain), then the output voltage becomes a measure of the strain. From Eq. (4), with a small change in R_1,

$$E = \frac{(R_1 + \Delta R_1)R_3 - R_2 R_4}{[(R_1 + \Delta R_1) + R_2](R_3 + R_4)} \quad \text{volts}$$

Most Wheatstone bridge circuits are produced with all four arms serving as active strain gages.

Types of Resistance Strain Gages.   Many different types of resistance strain gages have been developed since the first bonded strain gage was introduced in 1936. Bonded gages have been used widely, but there have been numerous changes in technology, as described later, in the assembly of strain-gage sensors, notably as they are used in process-type pressure sensors and transducers. These alterations include bonded semiconductor strain gages and diffused semiconductor strain gages.

Bonded Foil Strain Gages.   Serious commercial attention to the strain gage commenced in the mid-1950s, at which time foil strain gages were produced by a printed-circuit process or by being stamped from selected alloys that had been rolled into a thin foil. Foil thicknesses ranged from 0.0001 to 0.002 inch (0.00254 to 0.00508 mm). The foil usually was heat-treated before use in order to optimize mechanical properties and the thermal coefficient of resistivity. For a given cross-sectional area, a foil conductor displays a large surface area. The large ratio of surface area to cross section provides superior mechanical stability under prolonged strain and high-temperature conditions. The large surface area also provides a good heat-transfer surface between grid and specimen and, therefore, high input-voltage levels are possible without developing severe temperature gradients across the insulating matrix.

Photoetching permits the manufacture of sensing grids in virtually any two-dimensional pattern. Good practice is to develop a geometric pattern that provides maximum electrical and mechanical efficiency from the sensing element. Some common configurations that have been used are shown in Fig. 17.

 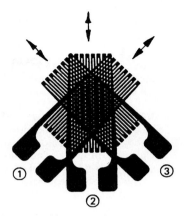

FIGURE 17   Strain-gage patterns that have been used successfully. (*a*) For measurement of strain in a diaphragm, elements 1 and 4 are subjected to compressive radial strains, while elements 2 and 4 are subjected to tensile tangential strains. (*b*) Rosette gage that measures strain in three directions simultaneously.

Bonded Semiconductor Strain Gages.   The principal difference between foil and semiconductor gages is the greater response of semiconductor gages to both strain and temperature. The large resistance-versus-strain characteristic of a properly oriented semiconductor crystal is due primarily to the piezoresistive effect. Gage behavior may be accurately described by

$$\frac{\Delta R}{R_0} = \left(\frac{T_0}{T}\right) E(GF) + \left(\frac{T_0}{T}\right)^2 E^2 C_2$$

where
R_0 = unstressed gage resistance at T (changes as T changes)
ΔR = change in gage resistance from R_0
T = temperature, K
T_0 = 298 K (24.9°C)
E = strain
GF, C_2 = constants of particular gage in question

The resistance change due to strain is a parabola for high-resistivity *p*-type silicon. Pure material of this resistivity is not used to produce gages because of this severe nonlinearity. As can be seen in the equation, the linearity can be improved by reducing the nonlinearity constant C_2. Figure 18 shows the behavior of a typical *p*-type semiconductor strain gage for a material that has been doped so that C_2 is low and the slope is more linear. The equation also shows that large tensile strains on the gage filament and higher temperatures increase gage linearity. As temperature T rises, the value of both terms on the right-hand side of the equation decrease, as does gage sensitivity. The nonlinearity coefficient, however, decreases faster than the gage factor coefficient, thus improving the linearity.

n-Type semiconductor strain gages are similar in behavior to *p*-type gages, except that the gage factor is negative.

The high output obtainable from semiconductor piezoresistive elements makes them particularly attractive for transducers that are ⅛ inch (3.2 mm) and smaller. Miniature transducers are formed by attaching

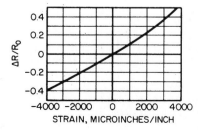

FIGURE 18   Gage sensitivity versus strain level for *p*-type semiconductor gage.

individual silicon strain gages to a force-collecting diaphragm or beam. When the diaphragm or beam deflects, the surface strains are sensed by the semiconducting elements. Output levels typically ~100 mV full-scale are available from full bridge transducers with ~10-volt input.

A typical semiconductor gage, unlike a bonded foil gage, is not provided with a backing or carrier. Therefore, bonding the gage to a surface requires extreme care in order to obtain a thin epoxy bond. The same epoxies used for foil gages are used for semiconductor gages.

Diffused Semiconductor Strain Gages.   A major advance in transducer technology was achieved with the introduction of diffused semiconductor strain gages. The gages are diffused directly into the surface of a diaphragm, utilizing photolithographic masking techniques and solid-state diffusion of an impurity element, such as boron. Since the bonding does not use an adhesive, no creep or hysteresis occurs.

The diffusion process does not lend itself to the production of individual strain gages and also requires that the strained member (diaphragm or beam) be made from silicon. Therefore, diffused semiconductors are used for manufacturing transducers (primarily pressure) instead of for stress analysis. Typically, a slice of silicon 2 to 3 inches (5 to 7.5 cm) in diameter is selected as the main substrate. From this substrate, hundreds of transducer diaphragms 0.1 to 0.5 inch (2.5 to 21.7 mm) in diameter with full four-arm Wheatstone bridges can be produced. A silicon pressure transducer diaphragm with a diffused semiconductor strain gage is shown in Fig. 19. The entire Wheatstone bridge circuitry is diffused into the diaphragm (strain gages and connection solder areas for leadwires).

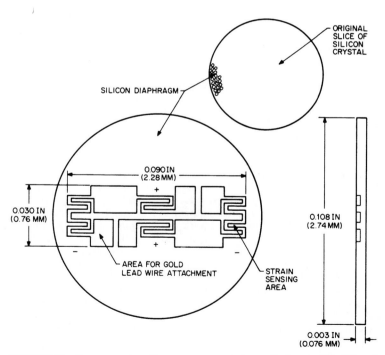

FIGURE 19   Representative silicon pressure transducer diaphragm with diffused Wheatstone bridge circuit. Many diaphragms are produced from a single slice of silicon crystal. The strain-gage elements are situated to measure compressive radial strains and tensile tangential strains.

A diffused silicon sensor is unsuitable for high-temperature measurements because the gage-to-gage resistance decreases sharply as a function of temperature. Two mechanisms combine to produce this effect. First, isolation from the substrate is accomplished by a *pn* junction, and its effectiveness is sensitive to increased heat. Second, the diaphragm (essentially an insulator) becomes increasingly conductive as the temperature is raised.

Refinements in the semiconductor diffusion process have allowed manufacturers to produce the entire strain-gage transducer diaphragm, strain gage, temperature compensation elements (that is, thermistors), and amplifier circuits with semiconductor technology. The introduction of very-high-volume, extremely low-cost transducers is now practical.

Thin-Film Strain Gages.   Another relatively recent development of strain-gage technology is the thin-film process. This technique controls the major strain-gage properties independently (that is, strain-gage and electrical isolation). It uses the inherent advantages of metal strain gages (low-temperature effects and high gage factors) and the process advantages available with the diffused semiconductor technique (no adhesive bonding). The thin-film strain gage is potentially capable of producing an ideal strain-gage system.

A thin-film strain gage is produced by depositing a thin layer of metal alloy on a metal specimen by means of vacuum deposition or sputtering. This technique produces a strain gage that is molecularly bonded to the specimen, so the disadvantages of the epoxy adhesive bond are eliminated. Like the diffused semiconductor process, the thin-film technique is used almost exclusively for transducer applications.

To produce thin-film strain-gage transducers, first an electrical insulation (such as a ceramic) is deposited on the stressed metal member (diaphragm or beam). Next the strain-gage alloy is deposited on top of the isolation layer. Both layers may be deposited either by vacuum deposition or by sputtering.

In vacuum deposition, the material to be deposited is heated in a vacuum and vapor is emitted. The vapor deposits on the transducer diaphragm in a pattern determined by substrate masks.

The sputtering technique also employs a vacuum chamber. With this method, the gage or insulating material is held at a negative potential and the target (transducer diaphragm or beam) is held at a positive potential. Molecules of the gage or insulating material are ejected from the negative electrode by the impact of positive gas ions (argon) bombarding the surface. The ejected molecules are accelerated toward the transducer diaphragm or beam and strike the target area with kinetic energy several orders of magnitude greater than that possible with any other deposition method. This produces superior adherence to the specimen.

In order to obtain maximum bridge sensitivity (millivolt output) to minimize heating effects, and to obtain stability, the four strain gages, the wiring between the gages, and the balance and temperature compensation components are all integrally formed during the deposition process. This ensures the same composition and thickness throughout.

The thin-film strain-gage transducer has many advantages over other types of strain-gage transducers. The principal advantage is long-term stability. The thin-film strain-gage circuit is molecularly bonded to the specimen, and no organic adhesives are used which could cause drift with temperature or stress creep. The thin-film technique also allows control of the strain-gage resistance value. A resistance as high as 5000 ohms can be produced in order to allow increased input and output voltages with low power consumption.

Strain-Gage Bridge Correction Circuits.   When static strains or the static component of a varying strain are to be measured, the most convenient circuit is the Wheatstone bridge, previously shown in Fig. 16. The bridge is balanced ($E = 0$) when

$$\frac{R_1}{R_4} = \frac{R_2}{R_3}$$

Now consider a bridge in which all four arms are separate strain gages. Assume that the bridge is initially balanced, so that $R_1 R_3 = R_2 R_4$ and $E = 0$. A strain in the gages will cause a change in each value of resistance R_1, R_2, R_3, and R_4 by incremental amounts ΔR_1, ΔR_2, ΔR_3, and ΔR_4, respectively. The voltage output ΔE of the bridge can be obtained from

$$E = \frac{R_1 R_3 - R_2 R_4}{(R_1 + R_2)(R_3 + R_4)} \quad \text{volts}$$

which becomes

$$\Delta E = \frac{(R_1 + \Delta R_1)(R_3 + \Delta R_3) - (R_2 + \Delta R_2)(R_4 + \Delta R_4)}{[(R_1 + \Delta R_1) + (R_2 + \Delta R_2)][(R_3 + \Delta R_3) + (R_4 + \Delta R_4)]} \quad \text{volts}$$

After considerable simplification, this becomes

$$E = \frac{R_2/R_1}{(1 + R_2/R_1)^2} \left(\frac{\Delta R_1}{R_1} - \frac{\Delta R_2}{R_2} + \frac{\Delta R_3}{R_3} - \frac{\Delta R_4}{R_4} \right) \quad \text{volts}$$

This latter equation shows that if all four gages experience the same strain, the resistance changes will cancel out and the voltage change ΔE will equal zero. On the other hand, if gages R_1 and R_3 are in tension (ΔR_1 and ΔR_3 positive) and gages R_2 and R_4 are in compression (ΔR_2 and ΔR_4 negative), then the output will be proportional to the sum of all the strains measured separately. All four-arm Wheatstone bridge transducers are wired to give two gages in tension and two gages in compression. An example of a four-gage setup for the diaphragm of a pressure transducer is shown in Fig. 20. This design takes full advantage of the tensile tangential strains developed at the center of the diaphragm and the compressive radial strains present at the edge.

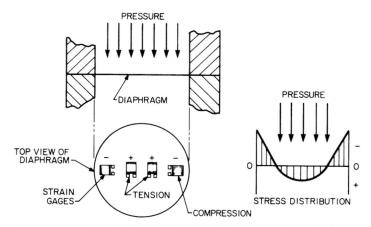

FIGURE 20 Representative strain-gage positions on a pressure diaphragm. Orientations take advantage of the stress distribution. The gages are wired into a Wheatstone bridge with two gages in tension and two in compression.

Another advantage of using a four-gage bridge, besides the increased output, is the effect on the temperature sensitivity. If the gages are located close together, as on a pressure transducer diaphragm, they will be subjected to the same temperature. Therefore the resistance change due to temperature will be the same for each arm of the Wheatstone bridge. If the gage resistance changes due to temperature are identical, the temperature effects will all cancel out and the output voltage of the circuit will not increase or decrease due to temperature.

The output voltage of the Wheatstone bridge is expressed in millivolts output per volt input. For example, a transducer rated at 3.0 mV/V at 500 psi (~73 kPa) will have an output signal of 30.00 mV for a 10-volt input at 500 psi (~73 kPa) or 36.00 mV for a 12-volt input. Any variation in the power supply will directly change the output of the bridge. Generally, power supply regulation should be 0.05 percent or better.

In production the four strain gages in a Wheatstone bridge never come out to be exactly equal for all conditions of strain and temperature (even in the diffused semiconductor process). Therefore various techniques have been developed to correct the differences in the individual strain gages and to make the strain-gage bridge easier to use with electronic instrumentation. Four main values normally need adjusting (Fig. 21): (1) electrical bridge imbalance, (2) balance shift with temperature, (3) span or sensitivity shift of bridge output with temperature, and (4) standardization of the bridge output to a given millivolts-per-volt value. Other transducer characteristics such as accuracy, linearity, hysteresis, acceleration effect, and drift are part of the transducer element design (beam or diaphragm) and cannot be corrected after the strain-gage bridge has been produced.

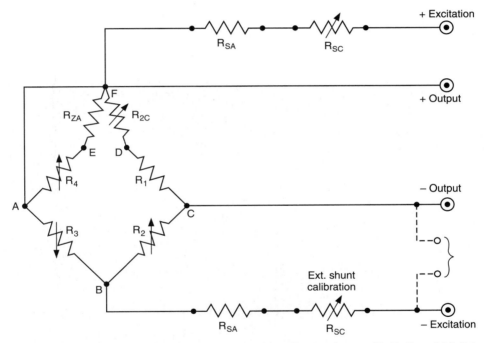

FIGURE 21   Strain-gage transducer circuit with four active strain-gage elements (R_1, R_2, R_3, and R_4). Balance, sensitivity, and thermal compensation resistors are also shown.

Figure 21 shows the circuit diagram of a Wheatstone bridge circuit with adjusting resistors. One corner of the bridge (points D and E) remains "open," so that the bridge can be adjusted electrically. This means that five leads come from the four gages. The zero-balance adjustment compensates for the electrical imbalance in the bridge caused by unequal resistances of the strain gages. Depending on which leg is unbalanced, R_{za} is placed between points E and F or between points D and F. The zero balance changes with temperature, and R_{zc} is inserted inside the bridge to correct for this change. A small piece of nickel wire is selected to provide a resistance change opposite the resistance change of the bridge. R_{sc} is also a temperature thermistor or sensor which changes resistance with temperature to adjust the excitation to the bridge. The values for R_{zc} and R_{sc} have to be selected by running each bridge over its desired temperature range [usually −65 to +300°F (−54 to +149°C)]. R_{sa} is a non-temperature-sensitive resistor, and it is used to adjust the output to a precise millivolts-per-volt value once all the balance and temperature-sensitive resistors have been inserted within the bridge. The user of transducers is not affected because all this circuitry is contained within the transducer and does not interfere with connections to amplifiers, power supplies, or computers.

A Wheatstone bridge can also be used in applications that require only one or two active strain gages. To compensate for temperature in two-gage applications, the gages must be located in adjacent arms of the bridge, as shown in Fig. 22. In placing gages, one must only recognize that the bridge is unbalanced in proportion to the difference in the strains of the gages located in adjacent arms and in proportion to the sum of the strains of gages located in opposite arms.

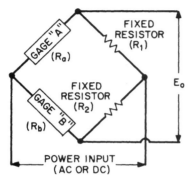

FIGURE 22 Wheatstone bridge circuit utilizing two strain gages.

Electronics for Strain-Gage Transducers and Transmitters. The full-scale output of a typical bonded-foil, four-active-element strain-gage bridge with all compensating and adjusting resistors connected is ~20 to 30 mV at 10-volt excitation. An amplifier must be used to obtain the 0- to 5-volt or 4- to 20-mA outputs used in control instrumentation. As a result of the advances in integrated circuitry, many transducers now have amplifiers that are internally installed within the transducer body (Fig. 23).

High-gain, low-noise, instrumentation-quality differential operational amplifiers such as the OP-07 make amplification of the strain-gage bridge output for standard 0- to 5-volt transducers and 4- to 20-mA transmitters reliable and easy. These integrated-

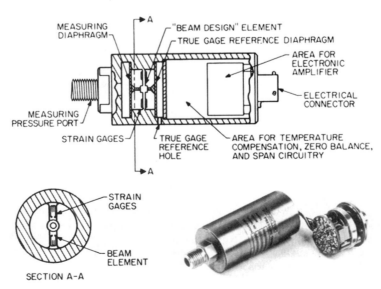

FIGURE 23 Pressure transducer (diaphragm-beam design) that compares measured pressure to atmospheric (reference) pressure. Welded stainless-steel diaphragms permit use in corrosive environments because strain gages are in environmentally protected chamber. (*Sensotec.*)

circuit amplifiers have a high common-mode rejection ratio and are thus well suited for use with Wheatstone bridge circuits. They are also inherently well compensated in order to deliver a constant output irrespective of temperature changes. The operational amplifiers used in instruments have controllable gains and zero-balance adjustments. Since the offset of the instrumentation channel's output is equal to the sum of the offsets in the bridge and in the amplifier, the combined offset can be adjusted at the amplifier so that the channel delivers 0 volts at zero stimulus (pressure, load, torque, and so on) for 0- to 5-volt output transducers or 4-mA at zero stimulus for 4- to 20-mA output transmitters.

The shear-web-element load cell is of a somewhat different configuration. This design is used in making high-capacity [50,000-lb (22,650-kg)] load cells for industrial weighing, crane scales, and so on. A shear web connects an outer, stationary hub to an inner, loaded hub. Strain gages detect the shear strain produced in the web. The large diameter of a very high-capacity shear element requires that the strain be measured at more than one location. Note the three webs in Fig. 24. This is common practice in all types of large transducer elements in order to obtain an average of the total strains on the element and to eliminate errors caused by minor off-center balancing. The strain gages are wired so that the user sees only one 350-ohm bridge.

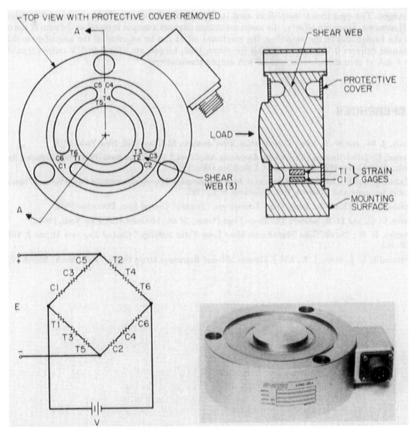

FIGURE 24 Pancake (shear-web) load cell. Each arm of the Wheatstone bridge circuit contains one strain gage from each of the three shear webs. The microstrains from the three webs are added together in one bridge circuit to determine the load. (*Sensotec.*)

Capacitive Pressure Transducers

In a traditional capacitance-type (capacitive) transducer, a measuring diaphragm (elastic element) moves relative to one or two fixed plates. Changes in capacitance are detected by an oscillator or bridge circuit. Generally, capacitive transducers are of low mass and high resolution, and they have good frequency response. Limitations have included a requirement for sophsticated signal conditioning, some sensitivity to temperature, and the effects of stray noise on sensor leads. As will be pointed out, much research and development during the past few years has gone forward to improve capacitive transducer performance—and generally with excellent results. These improvements largely have been made by way of testing and substituting new materials and through taking advantage of increasing ingenuity in the design and miniaturization of electronic circuitry, notably in the increasing use of microprocessors.

Transducer design concentration has centered on two classes of error sources. (1) Deficiencies such as in long-term stability, that is, essentially those error sources which cannot be corrected by the built-in electronics. Improved answers to these error sources have been derived essentially by testing and utilizing new materials as, for example, micromachined silicon, ceramics, quartz, and sapphire which, by their nature, exhibit minimal hysteresis. (2) Deficiencies which are amenable to improvement by electronic measures, including signal conditioning, calibration, and error self-diagnosis.

In a typical capacitive pressure transducer, as pressure is applied and changes, the distance between two parallel plates varies—hence altering the electric capacitance. This capacitive change can be amplified and used to operate into phase-, amplitude-, or frequency-modulated carrier systems. A frequency-modulated system using a tuned resonant circuit is shown in simple form in Fig. 25. In this electric circuit, the capacitance C_3 is part of a tuned resonant circuit $L_2C_2C_3$. L_1C_1 forms part of a stable high-frequency oscillator circuit. The tuned circuit $L_2C_2C_3$ is loosely coupled to the circuit L_1C_1. The high-frequency potential induced in circuit $L_2C_2C_3$ is rectified, and the dc output current of the rectifier is indicated on a microammeter. The response of the tuned circuit $L_2C_2C_3$ to a constant frequency is shown in Fig. 26 as a function of the capacitance $C_2 + C_3$ of this circuit. Peak output occurs at point A when the circuit is tuned to resonate at the oscillator frequency. This circuit is tuned to its operating point B by increasing capacitor C_2 until the rectifier meter reads approximately 70 percent of maximum. Any small change in pressure transducer capacitance C_3, due to pressure on the diaphragm, affects the response of the circuit according to Fig. 26.

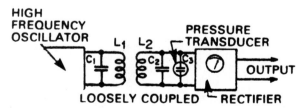

FIGURE 25 Schematic of tuned resonant circuit used in some capacitive pressure transducers.

In order to eliminate the effect of cable capacity between the transducer C_3 and the tuned circuit L_2C_2, a circuit of the type shown in Fig. 27 can be used. In this circuit, a coil L_3 is built as an integral part of the capacitive-transducer assembly. The coil L_3 is connected in parallel with the transducer capacity C_3 to form a tuned circuit with a resonant frequency (for example, 600 kHz). The tuned circuit L_3C_3 is close-coupled to the tuned circuit L_2C_2 by means of the link coils L_4 and L_5, which are connected by means of a low-impedance (70-ohm) untuned cable. The change in cable capacity, such as that produced by vibration, is negligible when

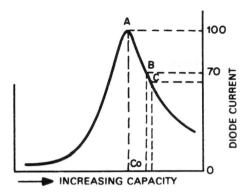

FIGURE 26 Response of resonant circuit to constant frequency shown as a function of circuit capacity.

reflected into the high-impedance tuned circuit. In this way, long cables can be used between the transducer and the electronic unit. The tuning characteristics of a link-coupled circuit are shown in Fig. 28. The operating range is the linear section noted midway between the maximum and minimum readings obtained with changing capacity.

A phase-modulated carrier system can be used in combination with transducers which incorporate a radio-frequency matching transformer which is tuned with the fixed condenser plate and stray capacitances in the pickup to approximately the oscillator frequency and is properly matched to the transducer connecting coaxial cable. When pressure is applied to the diaphragm, the increase in capacity lowers the resonant frequency of the circuit. The resulting change in reactance is coupled back to the indicator by a suitable transmission line, producing a phase change in the discriminator. This, in turn, produces an output voltage which is a function of the pressure on the diaphragm. The voltage can be indicated or recorded by the usual methods.

The measuring circuit of a capacitive differential-pressure transducer transmitter is shown schematically in Fig. 29. In this system two process diaphragms (high side and low side) are mechanically attached to a connecting rod. In the middle of the connecting rod, the movable electrode is attached and held in position by a spring diaphragm. The differential pressure is balanced by the restoring force of the spring diaphragm. Hence the spring diaphragm represents the measuring element. When applying a differential pressure on the system, the movable electrode is shifted and the distances d_1 and d_2 to the fixed electrodes are changed simultaneously. As a result of the change in distance between the fixed and movable electrodes, the capacitances of the differential capacitor are also changed. This

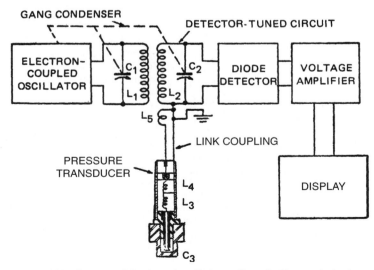

FIGURE 27 One type of circuit used to eliminate effect of cable capacity between transducer C_1 and tuned circuit L_2C_2.

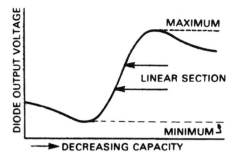

FIGURE 28 Tuning characteristic of a link-coupled circuit.

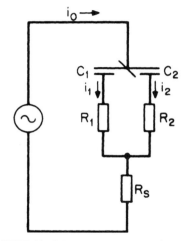

FIGURE 29 Schematic of a type of measuring circuit used in capacitive differential-pressure transducer transmitter.

change is amplified electronically and transduced to a 4- to 20-mA dc output signal directly proportional to the differential pressure.

Assume the gaps between the movable electrode and two fixed electrodes are both equal to d_0. When differential pressure $P_1 - P_2$ is applied, the connecting rod moves a distance of Δd. Then

$$d_1 = d_0 + \Delta d$$

$$d_2 = d_0 - \Delta d$$

$$\Delta d = K_1(P_1 - P_2)$$

where d_1 and d_2 represent the interelectrode gaps on the high and low sides, respectively; K_1 is a proportional constant.

The capacitances C_1 and C_2 of the gaps are, respectively,

$$C_1 = \frac{K_2}{d_1} = \frac{K_2}{d_0 + \Delta d}$$

$$C_2 = \frac{K_2}{d_2} = \frac{K_2}{d_0 - \Delta d}$$

where K_2 is a proportional constant and depends on the electrode area and the dielectric constant of the material filling the gaps.

In the conventional capacitive pressure sensor, the compliance of the diaphragm is selected so that the device will produce about a 25 percent change in capacitance for a full-scale pressure change. This large change gives the device an advantage for the measurement of comparatively low pressures. The device also permits the designer to include backstops on either side of the diaphragm for overpressure protection. The relatively high sensitivity of the device also allows it to generate digital or analog outputs.

All-Silicon Capacitive Sensor. In the recent past, performance of capacitive pressure sensors which utilize an all-silicon sandwich design provide better thermal stability because material mismatching (a major source of thermal effects) is eliminated. This provides an advantage over piezoresistive sensors because the latter normally use a sandwich constructed of different materials and hence different thermal coefficients of expansion.

As of early 1992 all-silicon sensors have been designed that displace only by 0.5 μm at a full-scale pressure of a 10-inch (25.4-cm) water column. The silicon diaphragm is sandwiched between two other pieces of silicon, the top piece serving as a mechanical overpressure stop, the bottom piece containing CMOS circuitry and the backplate of the capacitor, in addition to serving as an overpressure stop for pressure change in the opposite direction. The CMOS die contains both analog and digital circuitry, 13-bit digital output, an 8-bit parallel sensor interface, and an ASIC capability for local control or decision-point circuitry as well as for special communication protocols.

Piezoresistive Pressure Transducers

Stemming from research[1] in the 1950s on the piezoresistive properties of silicon-diffused layers and the development of a piezoresistive device for a solid-state accelerometer, the first piezoresistive pressure transducers were developed as pressure inputs for a commercial airliner in the 1960s. Although piezoresistive transducers have been available for other applications over an input pressure range of 1 to 680 atm, the principal application developed in the early 1970s was in the automotive field. Since that time, uses for piezoresistive pressure transducers in process control and industrial applications have increased.

The sensing element consists of four nearly identical piezoresistors buried in the surface of a thin, circular silicon diaphragm. Gold pads attached to the silicon diaphragm surface provide connections to the piezoresistors and serve as pads for probe-type resistance measurements or for bonding of wire leads. The thin diaphragm is formed by chemically etching a circular cavity into the surface opposite the piezoresistors. The unetched portion of the silicon slice provides a rigid boundary constraint for the diaphragm and a surface for mounting to some other member. A cross-sectional view of the sensing element with wire leads bonded to the metal contacts is shown in Fig. 30.

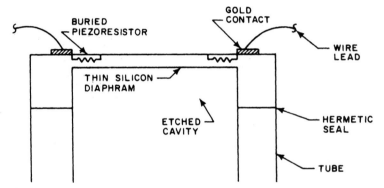

FIGURE 30 Cross section of piezoresistive sensing element with wire leads bonded to metal contacts.

Pressure causes the thin diaphragm to bend, inducing a stress or strain in the diaphragm and also in the buried resistor. The resistor values will change, depending on the amount of strain they undergo, which depends on the amount of pressure applied to the diaphragm. Hence a change in pressure (mechanical input) is converted to a change in resistance (electrical output). The sensing element converts energy from one form to another. The resistor can be connected to either a half-bridge or a full Wheatstone bridge arrangement. For pressure applied to the diaphragm using a full bridge, the resistors can theoretically be approximated as shown in Fig. 31 (nonamplified units). The signal voltage generated by the full-bridge arrangement is proportional to the amount of supply voltage V_{cc} and the amount of pressure applied, which generates the resistance change ΔR.

A half-bridge configuration used in a signal-conditioned version of the piezoresistive pressure transducer is shown in Fig. 32.

Among the pressure ranges of the transducers most frequently used are 0 to 1, 0 to 15, 3 to 15, 0 to 30, 0 to 100, and 0 to 250 psi (1 psi ≈ 6.9 kPa). Repeatability and hysteresis effects are

---

[1] Honeywell Inc.

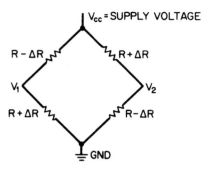

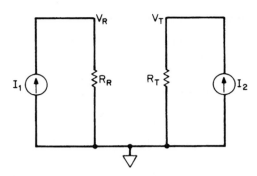

FIGURE 31 Full-bridge arrangement of piezoresistive transducer. $R + \Delta R$ and $R - \Delta R$ represent actual resistor values at applied pressure. R represents resistor value for undeflected diaphragm ($P = 0$) where all four resistors are nearly equal in value. ΔR represents change in resistance due to applied pressure. All four resistors will change by approximately the same value. Note that two resistors increase and two decrease, depending on their orientation with respect to the crystalline direction of the silicon material.

FIGURE 32 Half-bridge configuration used in signal-conditioned version of piezoresistive pressure transducer. Voltage across piezoresistors: $V_R = L_1R_R = I_1(R_{RO} + kP)$ for radial resistor; and $V_T = I_2R_T = I_2(R_{TO} - kP)$ across tangential resistor. I_1 and I_2 are adjusted at zero pressure to obtain $V_R = V_T$ or $I_1R_{RO} = I_2R_{TO}$. At other temperatures, when R_{RO} and R_{TO} vary, the equality will hold provided that the temperature coefficients of R_{RO} and R_{TO} are equal. I_1 and I_2 increase with temperature to compensate for the chip's negative temperature coefficient of span. The temperature coefficient of null, which may be of either polarity, is compensated for by summing a temperature-dependent voltage $V_N(T)$ with the piezoresistor voltage so that the output $V_O = V_R - V_T \pm V_N(T)$, with the polarity of $V_N(T)$ selected to provide compensation.

typically less than 0.1 percent of full-scale, and combined linearity and hysteresis do not exceed ±1 percent of full-scale output. The operating temperature range for standard units is from −40 to 125°C (−40 to 252°F).

Piezoelectric Pressure Transducers[2]

When certain asymmetrical crystals are elastically deformed along specific axes, an electric potential produced within the crystal causes a flow of electric charge in external circuits. Called the piezoelectric effect, this principle is widely used in transducers for measuring dynamic pressure, force, and shock or vibratory motion. In a piezoelectric pressure transducer, as shown in Fig. 33, the crystal elements form an elastic structure which functions to transfer displacement caused by force into an electric signal proportional to the pressure applied. Pressure acting on a flush diaphragm generates the force.

Piezoelectric pressure transducers historically have used two different types of crystals: (1) natural single crystals, such as quartz and tourmaline, and (2) synthetic polycrystalline ceramic materials, such as barium titanate and lead zirconate. With the relatively recent development of artificially cultured quartz crystals, the foregoing distinction is no longer clear-cut.

Cultured quartz has the advantage of being readily available and reasonably priced. Near-perfect elasticity and stability, combined with an insensitivity to temperature, make quartz an ideal transduction element. Ultrahigh insulation resistance and low leakage allow static calibration, accounting for the popularity of quartz in pressure transducers.

Natural tourmaline, because of its rigid, anisotropic nature, offers submicrosecond response in pressure-bar-type blast transducers. Artificial ceramic piezoelectric crystals and

[2] Based on R. W. Lally, PCB Piezotronics, Inc., Depew, New York.

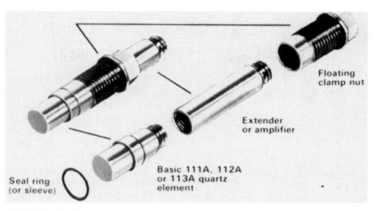

FIGURE 33 Modular assembly of piezoelectric dynamic pressure transducer. (*PCB Piezotronics, Inc.*)

electret (permanently polarized dielectric material, the analog of a magnet) materials are readily formed into compliant transducer structures for generating and measuring sound pressure.

The charge signal from a piezoelectric pressure transducer is usually converted into a voltage-type signal by means of a capacitor, according to the law of electrostatics: $E = Q/C$, where E is the voltage signal, Q is the charge, and C is the capacitance. This circuit is shown in Fig. 34.

In response to a step-function input, the charge signal stored in the capacitor will exponentially leak off through the always finite insulation resistance of the circuit components, precluding static measurements. The initial leakage rate is set by the circuit discharge time constant $R \times C$, where R is the leakage resistance value, which can be as high as 10^8 MΩ in quartz crystals.

Because of the automatic rezeroing action of the discharge circuit, piezoelectric sensors measure relative pressure, sometimes denoted as psir. They measure pressure relative to the initial level for transient events and relative to the average level for repetitive phenomena. Sometimes the slow action of these circuits is mistaken for zero drift by impatient operators.

To prevent the charge signal from quickly leaking off through the recorder or oscilloscope input resistance, a special isolation amplifier is required between the crystal and the recorder. If the charge-converting capacitance is located at the input of this isolation amplifier, the amplifier is called a voltage amplifier. If the capacitor is in the feedback path, it is called a charge amplifier. Amplifiers are further classified as electrostatic (dc-coupled) or vibration (ac-coupled). The ac coupling circuitry behaves similarly to the sensor discharge circuit.

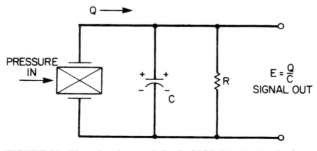

FIGURE 34 Piezoelectric crystal circuit. (*PCB Piezotronics, Inc.*)

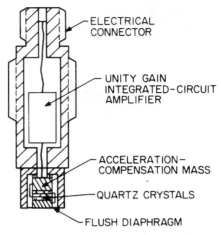

FIGURE 35 Acceleration-compensated quartz pressure sensor with built-in microelectronic unity-gain isolation amplifier. (*PCB Piezotronics, Inc.*)

The high-frequency response of piezoelectric sensor systems depends on the resonant behavior of the sensor's mechanical structure, or on electronic low-pass filters in the sensor, amplifier, or recorder.

The advent of microelectronics and charge-operated field-effect transistors (JFET and MOSFET) is continuing to change the design of piezoelectric sensors profoundly. The current practice is to package the isolation amplifier and signal-conditioning circuitry inside the sensor. These integrated circuit piezoelectric (ICP) sensors with built-in microelectronics, which operate over a simple two-wire cable, are called "smart" sensors.

To eliminate spurious signals caused by environmental effects, such as temperature and motion, the mechanical structures of some piezoelectric pressure sensors are quite sophisticated. A typical acceleration-compensated pressure sensor containing an integrated accelerometer to cancel out motion signals is shown in Fig. 35. Durable conformal coatings of the sensor case and diaphragm provide electrical and thermal insulation. Hermetic seals are electron-beam-welded.

Piezoelectric pressure sensors offer several advantages for measuring dynamic pressures. They are generally small in size, lightweight, and very rugged. One transducer may cover a measuring range of greater than 10,000:1 and a frequency range from less than 1 hertz to hundreds of kilohertz with little or no phase shift (time delay). As mentioned previously, piezoelectric sensors cannot measure static or absolute pressures for more than a few seconds, but this automatic elimination of static signal components allows unattended, drift-free operation.

Because of their unusual ruggedness, piezoelectric pressure sensors are widely used in difficult applications, such as ballistics, blasts, explosions, internal combustion, fuel injection, flow instabilities, high-intensity sound, and hydraulic or pneumatic pulsations—in connection with problems which may be encountered in connection with guns, shock tubes, closed bombs, rocket motors, internal-combustion engines, pumps, compressors, pipelines, mufflers, and oil exploration imploders.

Resonant-Wire Pressure Transducers[3]

A wire under tension is caused to oscillate at its resonant (or natural) frequency, and changes in pressure are converted to changes in this frequency. The application of an oscillating wire as a primary means of detecting force is based on fundamental principles initially outlined by Rayleigh's equations for a bar vibrating in a vacuum. Holst et al. (1979) modified Rayleigh's equations to fit an oscillating wire.

Their approximation of the resonant frequency f_n of a wire in a vacuum is

$$f_n = \frac{1}{2l} \sqrt{\frac{T}{\rho A} + (12 + \pi^2) \frac{EK^2}{\rho l^2}} + \frac{1}{l} \sqrt{\frac{EK^2}{\rho}}$$

---

[3] Based on W. H. Burtt, The Foxboro Company (A Siebe Company), Foxboro, Massachusetts.

where ρ = density of wire material
   A = cross-sectional area of wire
   T = tension in wire
   E = modulus of elasticity of wire material
   K = radius of gyration
   l = length of wire

Practical use of a resonant-wire pressure sensor requires that the wire be in a nonvacuum environment. Fairly rigorous modifications and refinements of the prior equation were made by Holst et al. (1979) to account for this. Simplifying and assuming that length, density, and area remain constant in the range of tension applied to the wire, the result of these modifications can be approximated as

$$f_n \propto T^2$$

A representative resonant-wire sensor for differential pressure or liquid-level measurement is shown schematically in Fig. 36; a resonant-wire sensor for measuring gage pressure is shown schematically in Fig. 37. This principle also is used in connection with liquid-level measurement. A block diagram of the electronic circuitry of a resonant-wire pressure transducer is presented in Fig. 38.

LVDT Pressure Transducers

Particularly for low-pressure measurement, some modern transducers utilize linear variable differential transformers (LVDTs) (Fig. 39). The LVDT measures the displacement of a pressure carrying capsule. It is claimed that it is well suited for measuring low pressures accurately in absolute, vented gage, and differential-pressure applications. LVDTs are described in greater detail in Section 5, Article 1, of this handbook.

Carbon-Resistive Elements

These transducers no longer are popular, but should be mentioned to demonstrate the many transducer approaches used in pressure measurements. Some of these methods are described in Fig. 40.

Inductive Elements

These methods also have been reduced in terms of usage. They are described briefly in Fig. 41.

Reluctive Elements

Variable-reluctance pressure transducers are distinguished mainly by the manner in which the exciting electric energy enters the system. In these devices, energy is introduced as a magnetomotive force which may be derived from either a permanent magnet or an electromagnet assembly. Electromagnetic induction, roughly speaking, is the production of an electric current (or voltage) by the movement of a conductor through a magnetic field, or by changing the strength of the field while the conductor is within it. In either case it is the interaction of a magnetic field and an electric conductor in such a manner as to produce a current, a change in position, or a change in flux that is always involved (Fig. 42).

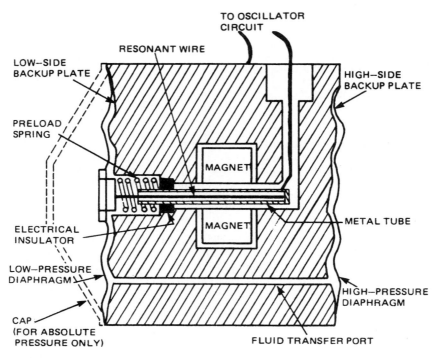

FIGURE 36 Schematic diagram of differential-pressure sensor that utilizes the resonant-wire principle. A wire under tension is located in the field of a permanent magnet. The wire is an integral part of an oscillator circuit that causes the wire to oscillate at its resonant frequency. One end of the wire is connected to the closed end of a metal tube, which is fixed to the sensor body by an electrical insulator. The other end of the wire is connected to the low-pressure diaphragm and loaded in tension by a preload spring. The spaces between the diaphragms and the backup plates, the fluid transfer port, and the metal tube are filled with fluid. An increasing pressure on the high-pressure diaphragm tends to move the diaphragm toward its backup plate. The fluid thus displaced moves through the fluid transfer port and tends to push the low-pressure diaphragm away from its backup plate. This increases the tension on the wire, raising its resonant frequency, and increasing the output signal of the transducer. Overrange protection for the wire is provided by an overrange spring (not shown) that is selected to limit the maximum tension on the wire to about two-thirds of its yield strength. The diaphragms are protected from overrange by the backup plates. (*Foxboro.*)

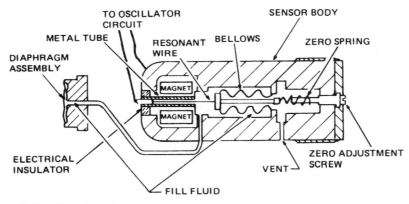

FIGURE 37 Schematic diagram of resonant-wire sensor for measuring gage pressure. (*Foxboro.*)

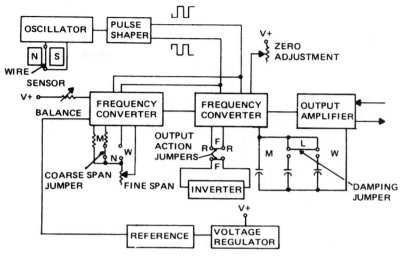

FIGURE 38   Block diagram of electronic circuitry of resonant-wire pressure transducer. (*Foxboro.*)

Optical Pressure Transducers

Over many years, optical methods have been used to measure the movement of diaphragms, bellows, or other summing elements in pressure sensors (Figs. 43 and 44). A very recent (1992) fiber-optic pressure transducer is described in Section 7, Article 4, of this handbook.

VACUUM MEASUREMENT[4]

Subatmospheric pressure usually is expressed in reference to perfect vacuum or absolute zero pressure. Like absolute zero temperature (the concept is analogous), absolute zero pressure cannot be achieved, but it does provide a convenient reference datum. Standard atmospheric pressure is 14.695 psi absolute, 30 inches of mercury absolute, or 760 mmHg of density 13.595 g/cm^3 where acceleration due to gravity is $g = 980.665$ cm/s^2. 1 mmHg, which equals 1 torr, is the most commonly used unit of absolute pressure. Derived units, the millitorr or micrometer, representing 1/1000 of 1 mmHg or 1 torr, are also used for subtorr pressures.

In the MKS system of units, standard atmospheric pressure is 750 torr and is expressed as 100,000 Pa (N/m^2) or 100 kPa. This means that 1 Pa is equivalent to 7.5 millitorr (1 torr = 133.3 pascal). Vacuum, usually expressed in inches of mercury, is the depression of pressure below the atmospheric level, with absolute zero pressure corresponding to a vacuum of 30 inches of mercury.

When specifying and using vacuum gages, one must constantly keep in mind that atmospheric pressure is *not* constant and that it also varies with elevation above sea level.

---

[4] Z. C. Dobrowolski, deceased; formerly Chief Development Engineer, Kinney Vacuum Company (A Unit of General Signal), Cannon, Massachusetts.

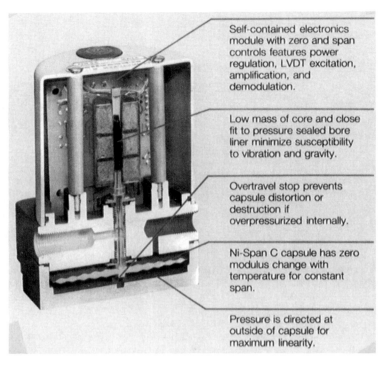

Self-contained electronics module with zero and span controls features power regulation, LVDT excitation, amplification, and demodulation.

Low mass of core and close fit to pressure sealed bore liner minimize susceptibility to vibration and gravity.

Overtravel stop prevents capsule distortion or destruction if overpressurized internally.

Ni-Span C capsule has zero modulus change with temperature for constant span.

Pressure is directed at outside of capsule for maximum linearity.

FIGURE 39   Sectional view of low-pressure LVDT-type transducer. Applicable pressure ranges are from as low as 2 inches (50 mm) of water to 100 psi (7 bar) full-scale. Typical applications include climate control and energy management systems, measurement of liquid levels in bulk storage tanks and closed pressure vessels. The unit incorporates a NiSpan C capsule, which offers low hysteresis and constant scale factor with temperature variation. Deflection of the capsule when pressurized is measured by an LVDT displacement sensor whose core is directly coupled to the capsule. The electrical output of the LVDT is directly proportional to core motion, which, in turn, is proportional to presssure applied to the capsule. (*Lucas Shaevitz.*)

Types of Vacuum Gages

Vacuum gages can be either direct or indirect reading. Those which measure pressure by calculating the force exerted by incident particles of gas are direct reading, while instruments which record pressure by measuring a gas property which changes in a predictable manner with gas density are indirect reading.

The range of operation for these two classes of vacuum instruments is given in Table 5. Since the pressure range of interest in present vacuum technology extends from 760 to 10^{-13} torr (over 16 orders of magnitude), there is no single gage capable of covering such a wide range. The ranges of vacuum where specific types of gages are most applicable are shown in Fig. 45; pertinent characteristics of these gages are given in Fig. 46.

The operating principles of some vacuum gages, such as liquid manometers, bourdon, bellows, and diaphragm gages involving elastic members, were described earlier in this article. The remaining vacuum measurement devices include the thermal conductivity (or Pirani and thermocouple-type gages), the hot-filament ionization gage, the cold-cathode ionization gage (Philips), the spinning-rotor friction gage, and the partial-pressure analyzer.

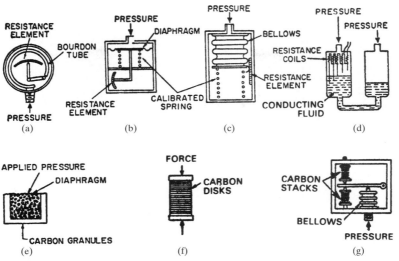

FIGURE 40 Group of resistive pressure transducers used for many years. Illustrations are highly schematic and essentially of historic interest. Other approaches have been miniaturized through the application of solid-state electronics. (*a*) Bourdon tube. (*b*) Diaphragm. (*c*) Bellows. (*d*) Differential coil. (*e*) Carbon pile. (*f*) Stacked carbon disk. (*g*) Carbon stacks with bellows coupling.

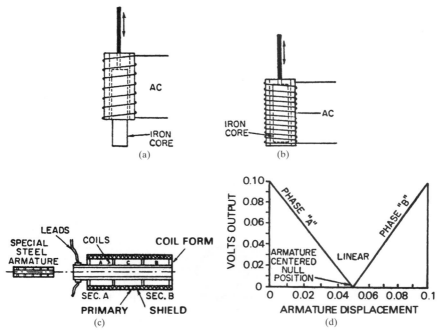

FIGURE 41 Various forms of inductive elements that are or have been used in pressure transducer designs. (*a*) Variable-inductance unit. (*b*) Inductance-ratio element. (*c*) Mutual-inductance element. (*d*) Phase relationship in mutual-inductance element.

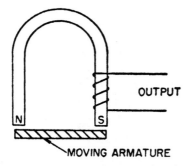

FIGURE 42 Schematic representation of variable-reluctance pressure transducer.

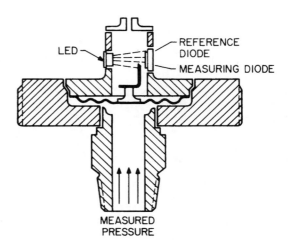

FIGURE 43 Cutaway view of Heise noncontacing optical sensor.

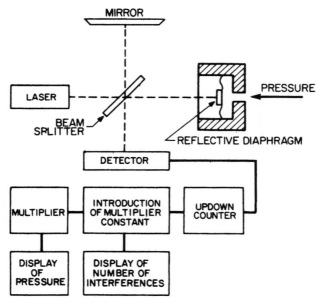

FIGURE 44 Supersensitive and accurate pressure sensor that utilizes optical interferometry.

Pirani or Thermocouple Vacuum Gage

Commercial thermal conductivity gages should not ordinarily be thought of as precision devices. Within their rather limited but industrially important pressure range they are outstandingly useful. The virtues of these gages include low cost, electrical indication readily adapted to remote readings, sturdiness, simplicity, and interchangeability of sensing elements. They are well adapted for uses where a single power supply and measuring circuit is used with several sensing elements located in different parts of the same vacuum system or in several different systems.

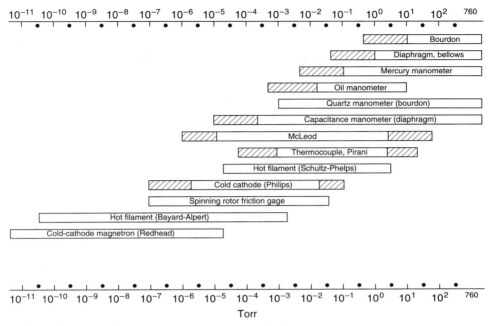

FIGURE 45   Ranges where certain vacuum gages are most suitable.

The working element of the gages consists of a metal wire or ribbon exposed to the unknown pressure and heated by an electric current (Fig. 47). The temperature attained by the heater is such that the total rate of heat loss by radiation, gas convection, gas thermal conduction, and thermal conduction through the supporting leads equals the electric power input to the element. Convection is unimportant and can be disregarded, but the heat loss by thermal conduction through the gas is a function of pressure. At pressures of approximately 10 torr and higher, the thermal conductivity of a gas is high and roughly independent of further pressure increases. Below about 1 torr, on the other hand, the thermal conductivity decreases with decreasing pressure, eventually in linear fashion, reaching zero at zero pressure. At pressures above a few torr, the cooling by thermal conduction limits the temperature attained by the heater to a relatively low value. As the pressure is reduced below a few

TABLE 5   Range of Operation of Major Vacuum Gages

Principle	Gage type	Range, torr
Direct reading	Force measuring:	
	Bourdon, bellows, manometer (oil and mercury),	$760–10^{-6}$
	McLeod capacitance (diaphragm)	760×10^{-6}
Indirect reading	Thermal conductivity:	
	Thermocouple (thermopile)	$10–10^{-3}$
	Pirani (thermistor)	$10–10^{-4}$
	Molecular friction	$10^{-2}–10^{-7}$
	Ionization:	
	Hot filament	$10–10^{-10}$
	Cold cathode	$10^{-2}–10^{-15}$

Key
- ☐ Yes
- Qualified Yes
- Qualified No
- No

	Bourdon	Diaphragm, bellows	Manometer	McLeod	Quartz manometer	Capacitance manometer	Thermocouple, Pirani	Ionization: cold cathode	Ionization: hot filament	Ionization: Schultz-Phelps
1 Composition independent	Yes	Yes	Yes	Qualified Yes	Yes	Yes	No	No	Qualified No	Qualified No
2 Continuous indicating	Yes	Yes	Yes	No	Yes	Yes	Yes	Yes	Yes	Yes
3 Remote indication and interfacing	Qualified Yes	Qualified Yes	Qualified Yes	No	Yes	Yes	Qualified Yes	Qualified Yes	Qualified Yes	Qualified Yes
4 Corrosion resistance	Qualified Yes	No	Qualified Yes	Qualified Yes	Yes	Yes	No	Qualified Yes	No	Qualified Yes
5 Accuracy better than 10%	Yes	Yes	Yes	Yes	Yes	Yes	Qualified No	Qualified No	Yes	Yes
6 Approximate cost	1-3	3-5	1-5	3-7	9	8-9	2-5	3-5	5-7	8
Comments	Common version inexpensive	Barometric compensation normal	Mercury vapor	Mercury vapor	Superior corrosion resistance	Widest useful pressure range	Convenient, inexpensive	Subject to oil contamination, rugged	Reaction with filament, burnout	Filament failure

Key	
1	S50 — 99
2	100 — 199
3	200 — 299
4	300 — 399
5	400 — 599
6	500 — 799
7	800 — 999
8	1000 — 4999
9	5000 and over

FIGURE 46 Vacuum-gage properties. Four different symbols are used in this chart, ranging from "yes" through "qualified yes" and "qualified no" to "no." For easy and uniform reference, these symbols are made to appear in shaded squares ranging from white (blank) to nearly black (heavy hash marks). This chart allows one to determine at a glance if the number of disadvantages or gage limitations is high or low. The assigned answers are unavoidably somewhat arbitrary. Reference to specific descriptions is suggested.

hundred millitorr, the heater temperature rises, and at the lowest pressures, the heater temperature reaches an upper value established by heat radiation and by thermal conduction through the supporting leads.

Hot-Filament Ionization Vacuum Gage

The hot-filament ionization gage is the most widely used pressure-measuring device for the region from 10^{-2} to 10^{-11} torr. The operating principle of this gage is illustrated in Fig. 48.

A regulated electron current (typically about 10 mA) is emitted from a heated filament. The electrons are attracted to the helical grid by a dc potential of about +150 volts. In their passage from filament to grid, the electrons collide with gas molecules in the gage envelope,

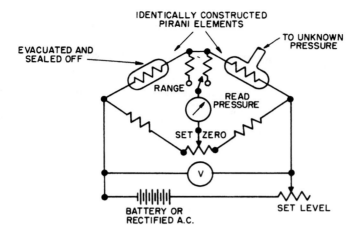

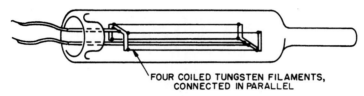

FIGURE 47   Pirani gage. (*Top*) Gage in fixed-voltage Wheatstone bridge. (*Bottom*) Sensing element.

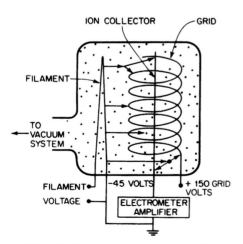

FIGURE 48   Hot-filament ionization gage (*Bayard-Alpert type*).

causing a fraction of them to be ionized. The gas ions formed by electron collisions are attracted to the central ion collector wire by the negative voltage on the collector (typically −30 volts). Ion currents collected are on the order of 100 mA/torr. This current is amplified and displayed using an electronic amplifier.

This ion current will differ for different gases at the same pressure, that is, a hot-filament ionization gage is composition-dependent. Over a wide range of molecular density, however, the ion current from a gas of constant composition will be directly proportional to the molecular density of the gas in the gage.

Cold-Cathode Ionization Vacuum Gage

This ingenious gage, invented by Penning, possesses many of the advantages of the hot-filament ionization gage without being susceptible to burnout. Ordinarily an electrical discharge between two electrodes in a gas cannot be sustained below a few millitorr pressure. To simplify a complicated set of relationships, this is because the "birthrate" of new electrons capable of sustaining ionization is smaller than the "death rate" of electrons and ions. In the Philips gage this difficulty is overcome by the use of a collimating magnetic field which forces the electrons to traverse a tremendously increased path length before they can reach the collecting electrode. In traversing this very long path, they have a correspondingly increased opportunity to encounter and ionize molecules of gas in the interelectrode region, even though this gas may be extremely rarefied. It has been found possible by this use of a magnetic field and appropriately designed electrodes, as indicated in Fig. 49, to maintain an electric discharge at pressures below 10^{-9} torr.

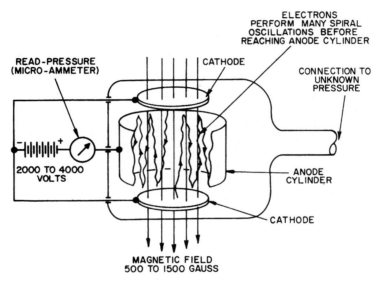

FIGURE 49   Philips cold-cathode ionization vacuum gage.

Comparison with the hot-filament ionization gage reveals that, in the hot-filament gage, the source of the inherently linear relationship between gas pressure (more exactly molecular density) and gage reading is the fact that the ionizing current is established and regulated independently of the resulting ion current. In the Philips gage this situation does not hold. Maintenance of the gas discharge current involves a complicated set of interactions in which electrons, positive ions, and photoelectrically effective x-rays all play a significant part. It is thus not surprising that the output current of the Philips gage is not perfectly linear with respect to pressure. Slight discontinuities in the calibration are also sometimes found, since the magnetic fields customarily used are too low to stabilize the gas discharge completely. Despite these objections, a Philips gage is a highly useful device, particularly where accuracy better than 10 or 20 percent is not required.

The Philips gage is composition-sensitive but, unlike the situation with the hot-filament ionization gage, the sensitivity relative to some reference gas such as air or argon is not independent of pressure. Leak hunting with a Philips gage and a probe gas or liquid is a useful technique. Unlike the hot-filament ionization gage, the Philips gage does not involve the use of a high-temperature filament and consequently does not subject the gas to thermal stress. The voltages applied in the Philips gage are on the order of a few thousand volts, which is sufficient to cause some sputtering at the high-pressure end of the range, resulting in a certain amount of gettering or enforced take-up of the gas by the electrodes and other parts of the gage. Various design refinements have been used to facilitate periodic cleaning of the vacuum chamber and electrodes, since polymerized organic molecules are an ever-present contaminant.

The conventional cold-cathode (Philips) gage is used in the range from 10^{-2} to 10^{-7} torr. Redhead has developed a modified cold-cathode gage useful in the 10^{-6} to 10^{-12} torr range (Fig. 50). The operating voltage is about 5000 volts in a 1-kG magnetic field.

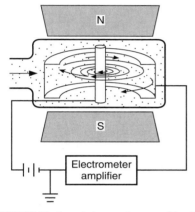

FIGURE 50   Inverted magnetron, a cold-cathode gage, produces electrons by applying a high voltage to unheated electrodes. Electrons spiraling in toward the central electrode ionize gas molecules, which are collected on the curved cathode.

Spinning-Rotor Friction Vacuum Gage

While liquid manometers (U-tube, McLeod) serve as pressure standards for subatmospheric measurements (760 to 10^{-5} torr) and capacitance (also quartz) manometers duplicate this range as useful transfer standards, calibration at lower pressures depends on volume expansion techniques and presumed linearity of the measuring system. The friction gage allows extension of the calibration range directly down to 10^{-7} torr. It measures pressure in a vacuum system by sensing the deceleration of a rotating steel ball levitating in a magnetic field (Fig. 51).

Partial-Pressure Analyzers (Vacuum)[5]

Many applications of high-vacuum technology are more concerned with the partial pressure of particular gas species than with total pressure. Also, "total" pressure gages generally give accurate readings only in pure gases. For these reasons partial-pressure analyzers are finding increasing application. These are basically low-resolution, high-sensitivity mass spectrometers which ionize a gas sample in a manner similar to that of a hot-filament ionization gage. The resulting ions are then separated in an analyzer section, depending on the mass-to-charge ratio of each ion. The ion current corresponding to one ion type is then collected, amplified, and displayed. Partial-pressure gages are very valuable diagnostic tools in both research and production work. The major types are listed in Table 6.

SMART (INTELLIGENT) PRESSURE TRANSMITTERS

The concept of the smart or intelligent transmitter applies to nearly all process variables today, not just to pressure transmitters, where it was applied initially. Although details differ for each variable measured, smart transmitter designs have more parallels than differences.

---

[5] Also known as residual gas analyzers (RGAs).

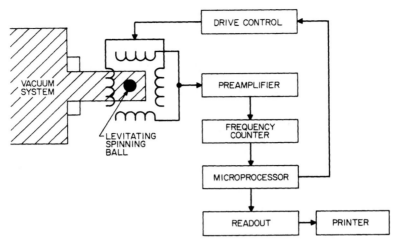

FIGURE 51   Schematic diagram of spinning-rotor friction gage.

Rather than duplicate this commonality, the reader is referred to Section 4, Article 1, of this handbook, where a smart RTD transmitter is described and illustrated in considerable detail.

Most smart transmitters (Fig. 52) are characterized by their inclusion of microprocessors and miniature electronics for storing important range, calibration, self-diagnostic troubleshooting, and other data with a remote capability for sending data to and receiving data from a measuring unit located at some site in the field.

Surveys as of 1991 indicate that approximately one-third of electronic-type pressure transmitters purchased are smart, thus indicating a rather rapidly rising rate of acceptance.

As of the early 1990s, some smart transducer connectivity problems remain. What is needed, of course, is a return to the convenient simplicity that reigned during that earlier period when 3- to 15-psi (20- to 100-kPa) pneumatic control systems operated successfully with components furnished by different instrument makers without protocol issues; or when devices operated with the universally accepted electronic standard of 4- to 20 mA control loops. The situation changed with the appearance of digital communications and smart transmitters. Fortunately, a solution may be in view (ISA SP50 Committee).

With the emphasis on statistical process control (SPC) and the general recognition that quality performance through instrumentation must receive closer attention than in the past, increasing efforts must be taken to maintain installed device accuracy. The ability of smart transmitters to be "drift-free" over long time periods cuts down on the time required to carry out routine calibration checks. But where a plant has performed well for a long

TABLE 6   Characteristics of Partial-Pressure Analyzers

Type	Minimum partial pressure, torr	Resolution,* au	Magnetic field
Magnetic sector	10^{-11}	20–150	Yes
Cycloidal	10^{-11}	100–150	Yes
Quadrupole	10^{-12}	100–300	No
Time of flight	10^{-12}	200	No

* Maximum mass number at which a mass number difference of 1 can be observed.

FIGURE 52 Smart transmitter terminal and transmitter incorporate modular electronics to enhance maintenance and upgrade capabilities to support SP50 protocols. Provides simultaneous communications up to 1 mile
(1.6 km) from field bus or current loop. Also features on-board liquid-crystal display scalable in engineering
units. (*Bailey Controls.*)

time span and where people are available to perform calibration checks and other routine
maintenance, there is a lesser tendency to replace existing devices with their smart counterparts.

As summarized by some experts (mid-1992), the smart pressure transmitter is a high-technology element of process control instrumentation and must be so regarded. In some user cir-
cles it is considered to be in a very early stage of maturity. Once a standard process industry
high-speed communication protocol is established, an increasing number of digital instruments will appear in the marketplace. Most likely an integration of not only smart transmitters, but also other devices in a common data highway will occur. Communication will not be
limited to dedicated distributed control systems and hand-held field calibrators.

PRESSURE SENSOR SELECTION GUIDELINES

Checklists of factors to consider when specifying a pressure sensor are given in Tables 7 and 8.

TABLE 7 Pressure Sensor Selection Guidelines

Accuracy (static error band): Combined effect of nonlinearity, hysteresis, and nonrepeatability, expressed as a percentage of full-range output (%FRO).

Best straight line: Line midway between two parallel straight lines enclosing all output-versus-pressure values on a calibration curve.

Burst pressure rating: Maximum pressure that may be applied to the transducer inlet port without rupture of the transducer case or diaphragm.

Critical damping: Lowest value of damping that renders a simple vibratory system nonoscillatory.

Damping ratio: For a simple vibratory system, ratio of actual damping to the damping required for critical damping.

Full-range output: Span of electrical output between maximum positive and maximum negative end points of the calibration curve.

Hysteresis: Maximum difference between electrical outputs obtained for both increasing and decreasing input, expressed as a percentage of full-range output (%FRO).

Line (base) pressure: Pressure to which the reference chamber is subjected. (This refers to differential-pressure transducers.)

Maximum electrical excitation: Maximum input voltage that may be applied without damage to the transducer.

Nonlinearity: Maximum deviation of any calibration point for both increasing and decreasing input from a best straight line and making the deviation a minimum, expressed as a percentage of full-range output (%FRO).

Nonlinearity and hysteresis: Maximum deviation of any calibration point obtained by either increasing or decreasing input, from a best straight line drawn through all calibration points, and making the deviation a minimum. Nonlinearity is expressed as a percentage of full-range output (%FRO).

Nonrepeatability: Maximum deviation of electrical output values obtained by applying the same input consecutively under the same conditions and in the same direction, expressed as a percentage of full-range output (%FRO).

Operational temperature range: Range of temperature over which the transducer may be used without damage.

Pressure limit (overpressure): Maximum allowable pressure difference between input and reference pressure, without affecting, in subsequent operation, the performance requirements over the rated pressure range.

Pressure range (rated pressure): Algebraic difference between maximum and minimum pressure over which the transducer is calibrated.

Pressure reference (operational pressure model): Reference pressure against which the input pressure is measured in the following devices:
- *Vented gage:* ambient atmospheric pressure
- *Sealed gage:* internal sealed atmospheric pressure
- *Absolute gage:* internal sealed vacuum
- *Differential gage:* difference of two unknown pressures
- *Vacuum:* pressure lower than atmospheric pressure

Rated electrical excitation: Applied input voltage at which the transducer calibration applies.

Rated temperature range: Range of temperature over which the thermal errors of the transducer have been compensated and to which specified values of thermal zero shift, thermal sensitivity shift, thermal zero repeatability, and thermal sensitivity repeatability apply.

Residual unbalance: Electrical output at rated electrical excitation and zero applied pressure at room temperature.

Secondary containment: Maximum pressure that may be contained in the event of a diaphragm failure without loss of pressure seal of the transducer case.

Sensitivity: Output per unit at rated excitation and room temperature.

Thermal sensitivity repeatability: Maximum difference between electrical outputs at full-range input obtained after cycling the transducer a specified number of times over the rated temperature range, expressed as a percentage of full-range output per degree temperature (%FRO/°C or %FRO/°F).

Thermal sensitivity shift: Difference between the electrical outputs at temperatures T_1 and T_2 with full-range applied pressure, expressed as a percentage of full-range output per degree temperature (%FRO/°C or %FRO/°F).

Thermal zero repeatability: Maximum difference between electrical outputs at zero applied input obtained after cycling the transducer a specified number of times over the rated temperature range, expressed as a percentage of full-range output per degree temperature (%FRO/°C or %FRO/°F).

Thermal zero shift: Difference between electrical outputs at temperatures T_1 and T_2 with zero applied pressure, expressed as a percentage of full-range output per degree temperature (%FRO/°C or %FRO/°F).

TABLE 8

Pressure medium and operating environment: Suitability of transducer for liquids or gases and likely ability to withstand corrosion. Shock and vibration also must be considered. Special stainless steels can withstand most dry gases and pressure media, such as lubrication oil, hydraulic oil, harsh chemicals, and temperature extremes. Subsea environments, some chemical processes, and food operations may require special shielding materials, such as Hasteloy C276 or Inconel 625.

Operating temperature: Most commercial pressure sensors incorporate temperature compensation up to 100°C (212°F). The temperature of the measured medium frequently is higher. For excessively high medium temperatures, the effect may be reduced by fitting a standoff pipe between medium and sensor.

Physical dimensions: Suppliers generally have attempted to reduce the size and weight of pressure sensors to a minimum. Some sensors are specifically designed for cleanability, as encountered in food processing.

Connections: The electrical termination of a pressure sensor will be designed for (*a*) a 6-pin electrical connector or (*b*) a cable-style connection (Fig. 53).

Overload protection: Where cycling, in particular involving occasional high-pressure spikes, of pressure of the measured medium may be contemplated, a premium-quality sensor/transducer may be required. Some suppliers offer units with 5× overpressure protection.

Excitation voltage required: A basic sensor may have a millivolt output from a 10-volt ac/dc-stabilized supply. However, using the same 10-volt supply, an amplifier can be added to give 5-volt dc output. From there a range of regulators can be added, operating from 11 to 18, 18 to 32, or 10 to 36 volts dc. All may provide a 4- to 20-mA output and have intrinsic safety approval. Units also are available for operating from ±15 volts dc.

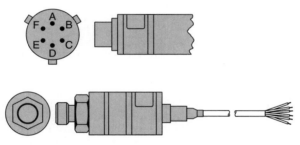

FIGURE 53   Electrical interface configurations. (*Top*) Connector style. (*Bottom*) Cable style. (*Lucas Schaevitz.*)

FLOW SYSTEMS

This article examines the numerous means for measuring the flow rate of fluids, including liquids, slurries, gases, and vapors, as these materials transit through pipelines, conduits, and open channels. The flow rate of bulk solids as they move on conveyor belts is also described.

Flow control establishes the material's feed rate to reactors and other chemical and mechanical operations so that optimum throughputs may be maintained. For example, the mathematics of chemical reactions can be precisely calculated in advance and the flow rates of reactants carefully flow-controlled to avoid excess wastes. In the control of other process variables, as encountered, for example, in separation operations, such as temperature, pressure, and level, the flow rate frequently is selected as the variable that will provide a rapid response. In a manufacturing plant, materials on the move must be accounted for in terms of production cost control, embracing incoming materials, inventory in process, and end products ready for shipment. Thus flowmeters are a key tool for accounting and financial control. Particularly in recent years, flowmeters have grown in importance for the measurement of pollutants and neutralizing materials in environmental protection programs.

The counterpart of fluid flow measurement in the discrete-piece manufacturing industries, such as parts counting and machine and assembly rates, is covered elsewhere in this handbook.

BROAD SELECTION OF FLOW SENSORS[1]

Because of the very marked differences in the chemical and physical properties of industrial fluids, numerous flow measurement methodologies have been developed over the years. Several physical parameters are used in flow sensors. Some flowmeters determine mass directly, but the majority of contemporary systems measure some quantitative dimension from which the flow rate can be inferred.

Exceptional progress has occurred since the early 1980s in the exploitation of several inferential principles that have been known for decades, but which previously had not been reduced to practice. In this category may be included ultrasonic approaches and use of the Doppler effect, of Coriolis forces, and of the Coanda effect. The use of magnetic induction also is relatively recent. However, it should be stressed that traditional flow sensors, such as differential producers, positive displacement, and turbine devices, among several others, have continued to improve and to compete well for user preference. While the appearance of additional flowmeters has made it possible to solve heretofore very difficult flow measurement problems, the process of selecting the right flowmeter also has become more complex.

DIFFERENTIAL-PRESSURE FLOWMETERS

Differential-pressure (DP) flowmeters have been used for many decades in numerous applications. In 1732 Henry Pitot invented the Pitot tube, and in 1791 Venturi did his basic work on the venturi tube. Venturi developed the mathematical basis and much of the theory now used in contemporary flowmeter computations. It was not until 1887 that Clemens Herschel developed a commercially satisfactory venturi tube.

Demands of the natural gas industry in 1903 led Thomas Weymouth to use a thin-plate, sharp-edged orifice to measure large volumes of gas. Throughout many years, DP flowmeters have enjoyed wide acceptance in the process and utility industries. Standards and vast amounts of engineering data have been developed by numerous technical organizations, including the American Gas Association, the American Society of Mechanical Engineers, the Instrument Society of America, the American National Standards Institute, and the National Engineering Laboratory (Germany). Some guidance for specifying DP flowmeters is given in Table 1, which lists the principal characteristics, including advantages and limitations.

Differential Producers

Flow is related to pressure by causing the flowing fluid to pass through some form of restriction in the transport pipe, which creates a momentary loss of pressure. As will be explained shortly, this pressure differential is related mathematically to the flow rate.

---

[1] Flow sensors are described in much more detail in the forthcoming *Industrial Sensors and Measurements Handbook,* D. M. Considine, Editor-in-Chief, McGraw-Hill, New York (circa 1995).

TABLE 1   Comparison of Differential Pressure and Target Flowmeters

Applications:
  Liquids and gases:
    Orifice, venturi, flow nozzle, flow tube, elbow, pitot, target
  Steam:
    Orifice, venturi, flow nozzle, flow tube, target

Flow range:
  Liquids (minimum):

Orifice	0.1 cm^3/min
Venturi	20 cm^3/min (5 gal/min)
Flow nozzle	20 cm^3/min (5 gal/min)
Flow tube	20 cm^2/min (5 gal/min)
Elbow	Depends on pipe size
Pitot	Depends on pipe size
Target	0.25 L/min (0.07 gal/min)

  Liquids (maximum):

All types	Depends on pipe size

  Gases (minimum):

Orifice	Gas equivalent of liquids
Venturi	35 m^3/h (20 scfm)
Flow nozzle	35 m^3/h (20 scfm)
Flow tube	35 m^3/h (20 scfm)
Elbow	Depends on pipe size
Pitot	Depends on pipe size
Target	0.5 m^3/h (0.3 scfm)

  Gases (maximum):

All types	Depends on pipe size

Operating pressure (maximum):
  Generally depends on transmitter selected

Target meter	68 MPa (10,000 psig)

Operating temperature (maximum):
  Depends on construction material selected

Target meter	398°C (750°F)

Scale:

All types	Square root

Nominal accuracy:

Orifice	±0.6% (maximum flow)
Venturi	±1% (maximum flow)
Flow nozzle	±1% (full-scale)
Flow tube	±1% (full scale)
Elbow	±5 to ±10% (full-scale)
Pitot	±5% or better (full-scale)
Target	±0.5 to ±5% (full-scale)

Rangeability:

Target meter	3:1
Other types	4:1

Signal:

All types	Analog electric or pneumatic

Relative advantages and limitations:
  Comparatively high cost:
    Venturi

TABLE 1   Comparison of Differential Pressure and Target Flowmeters (*Continued*)

Comparatively low cost:
  Orifice, flow nozzle, flow tube, Pitot, elbow, target
Comparatively easy to install:
  Orifice, elbow
Same transmitter can be used for variety of pipe sizes:
  Orifice, flow nozzle, flow tube, Pitot, elbow
Comparatively low pressure loss:
  Venturi, flow nozzle, flow tube
Good for dirty liquids and slurries:
  Orifice (if equipped with eccentric or segmental plates), venturi, target meter (especially for hot, tarry, and
    sediment-bearing materials)
Large and heavy:
  Venturi
Requires straight runs of piping (upstream, downstream):
  Orifice, target meter
Limited background engineering data:
  Flow nozzle, flow tube
Bidirectional flow measurement:
  Elbow
Minimum upstream straight piping:
  Elbow
Flow averaging:
  Pitot
Accuracy limited:
  Elbow

Orifice Restrictions

In Weymouth's early work he used flange pressure taps, one upstream and one downstream from the face of the orifice. These later became the principal pipe tap locations (Fig. 1).

The general operating principle of the orifice-type flowmeter is shown and defined in Fig. 2. Weymouth developed empirical coefficient data correlated with the ratio of orifice bore to pipe diameter, called the beta ratio β. It was the ability to predict a coefficient from measured dimensions that led to the first commercialization of the orifice flowmeter. For most applications this ratio should be between 0.2 and 0.75, depending on the desired upper range differential. A high beta ratio produces less differential for the same flow rate than does a low beta ratio.

Various orifice plate designs are shown and described in Fig. 3*a* through *e*. For small pipe sizes handling clean fluids, an orifice can be installed integrally with the DP transmitter and thus provide a compact installation with acceptable accuracy for many applications.

Wedge Primary Elements

Of proprietary design (ABB Kent-Taylor), these units (Fig. 3*f* and *g*) are available for replacing standard orifice configurations for some applications at a claimed lower cost. Wedges provide linear measurement down to low Reynolds numbers for extremely viscous fluids and slow flow rates. The units can be mounted vertically to handle two-phase flows and are available in increments from 1- to 4-inch (25- to 100-mm) pipe sizes.

Venturi Tube

Rather than depend on an invasive restriction in a transport pipe, the venturi tube, shown in Fig. 4, is a specially shaped pipe assembly that, in essence, becomes an integral part of the pip-

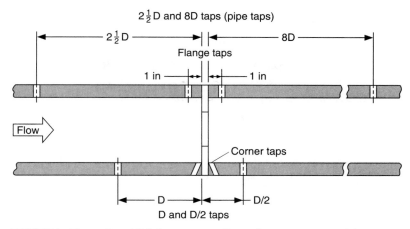

FIGURE 1 Flange, *D*, and *D*/2 flowmeter taps. Depending on upstream and downstream tap locations, a flowmeter may be referred to as a corner tap, a *D* and *D*/2 tap, a pipe tap, or a vena contracta tap orifice flowmeter. Corner and *D* and *D*/2 taps are widely used in Europe, while flange taps are usually preferred in the United States.

ing system. The flow-restricting effect thus is much less abrupt. Initially designed for large-line-size [6 inches (150 mm) or larger] water and waste applications, the venturi became popular for smaller line sizes with the introduction of the universal venturi tube, which has reduced weight and a shorter overall length (Fig. 5).

For pipe Reynolds numbers greater than 100,000, discharge coefficients for venturis are constant and predictable to within ±0.5 to 2 percent, depending on design. Venturis normally are not used for lower Reynolds numbers without accepting a decrease in accuracy. When pumping cost or shorter upstream installation lengths are important, the additional expense for a venturi design may be warranted.

Flow Nozzle

The flow nozzle shown in Fig. 6 has an elliptical or radius entrance and generally is used in steam (vapor) service operating at high pipeline velocities [100 ft/s (30.5 m/s)]. Because of improved rigidity, the flow nozzle is dimensionally more stable at higher temperatures and velocities than an orifice.

The initial cost is higher than that of an orifice, but markedly less than that of a venturi. When sized to create the same differential at the same flow rate, the pressure loss approximates that of an orifice. Flow nozzles or critical nozzle venturis (Fig. 6) are sometimes operated at critical (choked) flow for flow limiting, or as secondary flow standards. Widely used is the ASME elliptical-inlet long-radius wall tap nozzle, as well as the ASME throat tap nozzle for steam turbine testing. In Europe the ISA nozzle is used widely.

Numerous proprietary low-loss tubes have been introduced over the years. Operating data for these designs are not as extensive as in the case of most flow measurement devices.

Elbow Flowmeters

As a fluid passes through a pipe elbow, the pressure increases at the outside radius of the elbow because of centrifugal force. If pressure taps are located at the outside and inside of the elbow (at either 25 or 45°), a reproducible measurement can be made. Taps located at

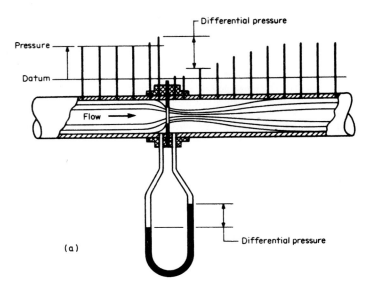

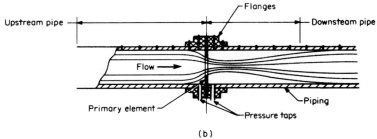

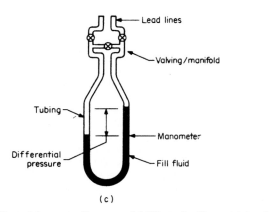

FIGURE 2 Differential pressure flowmeter. (*a*) Effect of orifice restriction in creating a pressure differential across plate and recovery of pressure loss downstream from plate is shown graphically in top view. A liquid manometer is used for illustrative purposes. Other DP transducers are used in practice. The pressure differential created depends on the type of restriction used. With an orifice plate, the flow contraction is abrupt; with a flow nozzle, it is more gradual. (*b*) Relationship between measured differential and flow rate is a function of tap locations, the particular type of restriction used, and the associated upstream and downstream piping. These factors are included in the discharge coefficient, which relates actual flow rate to theoretically calculated rate. (*c*) Commonly used piping arrangements used for installing a DP flowmeter.

angles greater than 45° are not recommended because flow separation may cause erratic readings.

Many flow piping systems have elbows. The system is relatively low cost. Elbows are not always consistently constructed and thus limit the accuracy, but precision (repeatability) is considered good. Some makes offer proprietary machined elbows for improved accuracy. A very low differential is produced, particularly for gas flows (Fig. 7).

Pitot Tube

A Pitot tube such as that shown in Fig. 8 is used for large pipe sizes when the fluid is a clean liquid or gas (vapor) and a relatively low-cost measurement is needed. The difference between total (stagnation) pressure and static pressure follows a square-root relationship, with velocity being sensed at the insertion depth only. By traversing, an average velocity point can be located and used as a measure of flow rate. A proprietary design incorporates a multiple-ported Pitot tube that spans the pipe. Pressure ports are located at mathematically defined positions based on published axisymmetric pipeline velocity profiles. These are claimed to average the differential, thereby eliminating the need to locate the average velocity point, as is the case with conventional Pitot tubes. A variety of multiported designs is now available.

Annular Orifice

The annular orifice (Fig. 9) was developed to overcome the problem of dirt buildup in front of an orifice in a liquid stream and of liquid buildup in a moist gas stream. Total (stagnation) pressure taps and rearward-facing taps produce a high differential for a given beta ratio β, redefined here as the ratio of disk diameter to pipe diameter. Data on line-size correlation are limited. Air data are available for the normally used beta ratios.

Target Flowmeter

The target flowmeter shown in Fig. 10 has the features of the annular orifice, but without the disadvantages of freezing or plugging lead lines. In the target meter the primary element consists of a sharp leading-edge disk (target) fastened to a bar. Differential pressure produced by the reduced annular area creates a disk drag force. This force is transmitted through the bar to an appropriate force-measuring secondary device. The flow rate is calculated as the square root of this output. Target flowmeters are well suited for liquids, gases, vapors, dirty fluids, light slurries, and high-viscosity liquids.

VARIABLE-AREA FLOWMETERS

The principle of using differential pressure as a measure of flow rate, as described in the case of the orifice, venturi, and other devices, is also applied in variable-area flowmeters, but with one important variation. In the orifice meter there is a fixed aperture (restriction) and the flow rate is indicated as a function of the differential pressure created. In the case of the variable-area meter there is a variable orifice and a relatively constant pressure drop. Thus the flow rate is indicated by the area of the annular opening through which the flow must pass. This area is generally read out as the position of a float or obstruction in the orifice.

Frequently referred to as glass-tube flowmeters or rotameters, variable-area flowmeters are designed in three general configurations:

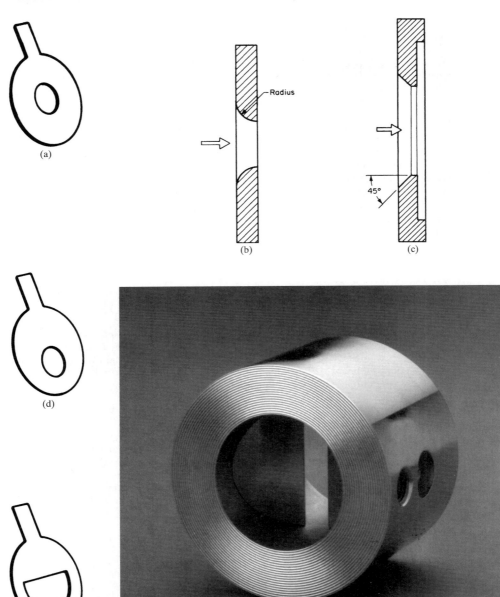

(a)

Radius

(b)

45°

(c)

(d)

(e)

(f)

(g)

FIGURE 3   Orifice designs. (*a*) Square-edged concentric orifice. Commonly used for clean liquids, gases, and low-velocity vapor (steam). It is a sharp, square-edged hole bored in a flat, thin plate. (*b*) Quadrant-edged orifice. This is suitable when the pipe Reynolds number is below 10,000. The upstream orifice edge is rounded. It is commonly used in the United States. (*c*) Conical orifice. This is popular in Europe. Both the quadrant-edged and conical orifices have contours that produce a more constant and predictable discharge coefficient at lower Reynolds numbers. By contrast, at low Reynolds numbers, the coefficient of a square-edged orifice may change by as much as 30%, whereas with the quadrant and conical geometries, the effect only may be 1 to 2%, making them more usable for viscous fluids. (*d*) Eccentric orifice, where the orifice hole is located at the bottom of the pipe for gases and at the top for liquids, where entrained water or gases flow through the plate rather than building up in front of it. (*e*) Segmental orifice, made by machining a segmental opening in the plate. This also permits the passage of liquids, air, or particulate matter. The data available for eccentric and segmental orifice plates are limited, although the orifice geometries may be attractive for handling troublesome fluids. (*f*) Bidirectional Wedge flow primary element with a cost less than that of an orifice plate, handles most types of flow: viscous, two-phase, entrained solids, erosive fluids, steam, or gas. Unit provides linear measurements down to low Reynolds numbers for extremely viscous flows or very low flow rates, mounted vertically for two-phase flows. Minimal pressure loss. Standard sizes 1.0, 1.5, 2.0, 3.0, and 4.0 inches (25, 37.5, 50, 75, 100 mm) standard pipe sizes. Wedge-to-diameter ratios of 0.2, 0.3, and 0.4 accommodate different flow rates. (*g*) Wedge flow element with remote seal connections measures slurries and high-viscosity fluids down to a Reynolds number of 500. Accuracy of ±0.5% (actual flow) can be achieved with optional calibration. (*Wedge flow elements, ABB Kent-Taylor, Inc.*)

FIGURE 4   Classical venturi.

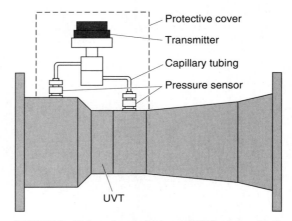

FIGURE 5   Universal venturi primary (UVT) equipped with an electronic flow transmitter. Differential pressure is statically sensed by units that are mounted flush with the inside of the venturi tube. Units like this are used in connection with solids-bearing fluids, such as waste water or light slurries. The system is designed to eliminate clogging or contaminant buildup. A 4- to 20-mA dc signal is produced. System accuracy is ±1.0% over a 10:1 range. (*Leeds & Northrup, BIF Products.*)

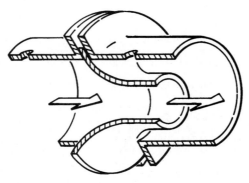

FIGURE 6   Flow nozzle.

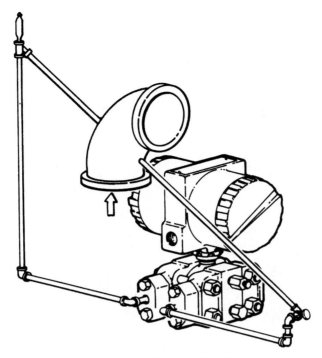

FIGURE 7 Highly schematic diagram of elbow flowmeter.

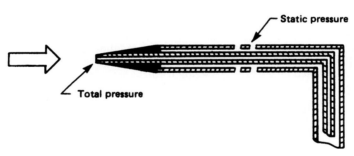

FIGURE 8 Pitot tube.

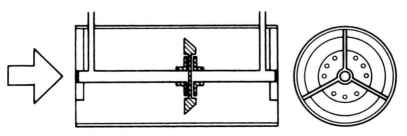

FIGURE 9 Annular orifice.

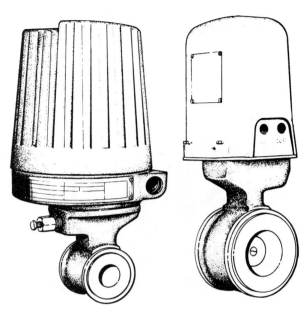

FIGURE 10   Target flowmeter. (*Foxboro.*)

1. The *tapered-tube meter,* in which a float (usually of a density greater than the fluid being measured) is contained in an upright glass tube, which is tapered, that is, the bottom of the tube is of a somewhat smaller diameter than the top. Thus the area or concentric orifice through which the fluid flows is greater as the float rises in the tube. The term rotameter arose from early technology when the rotor was slotted and thus rotated by the flowing fluid. The float is lifted to a position of equilibrium between the downward (gravity) force and the upward force caused by the flowing medium. Later technology introduced rod- or rib-guided nonrotating floats.

2. The *orifice and tapered-plug meter,* which is equipped with a fixed orifice mounted in an upright chamber. The float has a tapered body with the small end at the bottom and is allowed to move vertically through the orifice. The fluid flow causes the float to seek an equilibrium position.

3. The *piston-type meter,* in which a piston is fitted accurately inside a sleeve and is uncovered to permit the passage of fluid. The flow rate is indicated by the position of the piston.

Schematic diagrams of the different types of variable-area flowmeters are shown in Fig. 11.

Floats are made from numerous corrosion-resistant materials. For ordinary service, metering tubes are borosilicate glass. Metal tubes (partial or full) are used for high-pressure and hazardous applications, thus requiring a magnetic coupling to the float position to provide flow-rate indication. Float designs have varied over the years. Earlier meters were affected to some degree by viscosity changes of the metered fluid. Floats having sharp edges were found to be most immune to viscosity changes and are commonly used today for many applications. A majority of variable-area meters is used for direct reading, but they can be equipped with a variety of contemporary hardware for data transmission, alarming, and integration into sophisticated systems. The principal characteristics of variable-area flowmeters are summarized in Table 2.

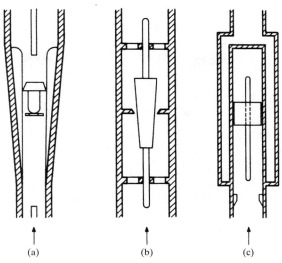

FIGURE 11 Schematic views of three basic types of variable-area flowmeters. (*a*) Tapered tube (rotameter). (*b*) Orifice and tapered plug. (*c*) Cylinder and piston.

TABLE 2 Characteristics of Variable-Area Flowmeters

Service	Liquids, gases, steam
Flow rate:	
Liquids	0.2 to 23 gal/min (0.6 to 87 L/min)
Gases	0.9 to 106 scfm (26 to 2993 L/min)
Rangeability	10:1
Accuracy	Depends on model
	High accuracy designs: ±1% of reading from 100 to 10% (full-scale)
	Purge meters: ±10% (full-scale)
Repeatability	Depends on model; average 0.5% (full-scale)
Configurations	Full view; armored for high pressure; glass and metal tubes (magnetically coupled indicator); alarms
Operating temperature (maximum)	Glass tube: 400°F (204°C)
	Metal tube: 1000°F (583°C)
Operating pressure (maximum)	Glass tube: 350 psig (2.4 MPa)
	Metal tube: 720 psig (4.9 MPa)
Floats	Rod-guided, rib guided, tube-guided glass or sapphire spheres
Construction materials	Metering tubes: borosilicate glass
	High pressure: stainless steel, Hastelloy, Mone
	Depending on service, floats are glass, sapphire, tantalum, Carboloy, stainless steel, aluminum, Hastelloy, CPVC, Teflon, PVC
Advantages	Relatively low cost; direct indicating; minimum piping required
Limitations	Must be vertically mounted; need minimum back pressure (gases); usually limited to direct viewing

MAGNETIC FLOWMETERS

The magnetic flowmeter, also referred to as magmeter, has numerous advantages for measuring liquids that are electrically conductive. This is also a factor, however, which limits its use in terms of nonconducting fluids. Where applicable, the magmeter is quite advantageous in terms of difficult fluid-measuring problems, including corrosives, viscose, sewage, rock and acid slurries, paper pulp stock, rosin size, detergents, tomato pulp, and beer, among other fluids encountered in the chemical and food-processing fields. The magmeter can be adapted as a controller for batching, as may be found in brewery and bottling-plant batching operations, where it operates with an accuracy of ±0.2 percent.

Early magmeter designs were full-line devices; they were quite costly, heavy, and bulky. In more recent years some insertion configurations have been developed. In such designs the sensor can be introduced through a hole in a large pipeline or by a T-fitting in a small line. Although the same accuracy generally is not achieved with in-line models, these designs do offer lower initial cost, ease of installation, and lower maintenance. As of the early 1990s some authorities consider that the Coriolis mass flowmeter will become a serious rival for traditional magmeter applications.

Magmeter Operating Principle

The magmeter comprises a tube through which the measured fluid passes, coils, a laminated iron core, a cover, and end connections (Fig. 12). Traditionally the tube has been constructed of nonmagnetic stainless steel or fiberglass-laminated plastic. Research on nonconducting materials to prevent the generation of EMF from short-circuiting continues. Contemporary ceramic liners approach the desired characteristics, that is, they satisfy the electrical requirements as well as coping with the corrosive and other characteristics of the measured fluid.

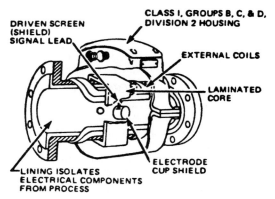

FIGURE 12   Layout of essential components of magnetic flowmeter.

Operation of the magmeter is based on Faraday's law of electromagnetic induction, which states that the voltage E induced in a conductor of length d moving through a magnetic field h is proportional to the velocity v of the conductor. Thus

$$E = Chdv$$

where C is a dimensional constant (Fig. 13).

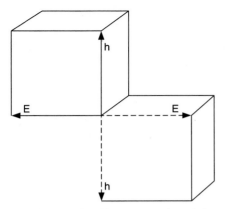

FIGURE 13 Graphic representation of formula applying to magnetic flowmeter.

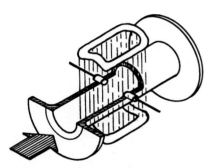

FIGURE 14 Essential components of practical magnetic flowmeter. (*Foxboro.*)

A schematic cross section of a magmeter is shown in Fig. 14. Generally a magmeter generates a voltage that is proportional to the average fluid velocity of the profile at the plane of the electrodes. This results from the fact that each element of fluid in the electrode plane develops an element of voltage that is in proportion to its instantaneous velocity. The voltage sensed at the electrodes is proportional to the sum of these elements of generated voltage and, therefore, accurately represents the average velocity of the fluid in the plane. This voltage represents the volumetric flow, provided the tube is completely filled with fluid at all times. Designs with four electrodes have been introduced to cope with the velocity profile problem more effectively.

Direct-current magmeters are somewhat smaller, have a more stable zero, and are lower in cost as compared with the earlier ac designs. However, in handling slurries with small particles, a low-frequency noise on the flow signal can result—caused by particles scraping the electrodes as they flow by. There is considerable research underway to develop special circuits that may minimize this noise problem.

Dual-frequency excitation also has been introduced into magmeter circuitry. As shown in Fig. 15, the magnetic field coils are excited by current with a compound waveform.

Microprocessors and other sophisticated electronics can provide multirange, low-liquid-flow cutoff, bidirectional flow measurement, self-diagnostics, and loop tests, among other features that may be required by system integration. The characteristics of magnetic flowmeters are summarized in Table 3.

TURBINE FLOWMETERS[2]

Turbine flowmeters consist of a rotating device, called rotor, that is positioned in the fluid path of a known cross-sectional area, the body or pipe, in such a manner that the rotational velocity of the rotor is proportional to the fluid velocity. Since the cross-sectional area of the pipe is known, fluid velocity can be converted directly to volumetric flow rate by counting the number of turbine-wheel revolutions per unit of time. The following equation relates the conversion from fluid velocity (feet per second) to volumetric flow rate (gallons per minute):

$$Q = v \times A \times C$$

where Q = volumetric flow rate
   v = fluid velocity
   A = cross-sectional area
   C = constant

[2] Bill Roeber, Great Lakes Instruments, Inc., Milwaukee, Wisconsin.

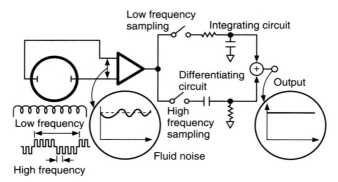

FIGURE 15 Dual-frequency excitation method. The magnetic field coils are excited by current with a compound waveform. One component has a rectangular waveform with a frequency greater than line frequency (60 Hz). This provides a signal that is immune to the low-frequency noises generated by electrochemical reactions, high viscosities, and in low-conductivity liquids. The second component is a rectangular waveform with a frequency much less than line frequency. This provides a large improvement in zero stability. The low-frequency component is integrated via a long time constant to provide a smoother, stabilized flow signal. The high-frequency component is conditioned by high-frequency sampling and processed in a differentiating circuit with the same time constant as the integrating circuit. By adding these two signals, a flow signal is obtained that is free from slurry noise, in addition to other advantages. (*Yokogawa.*)

This rate can be related to a digital pulse train generated by the turbine through the equation

$$Q = \frac{\Gamma \times f}{K}$$

where Q = volumetric flow rate, gal/min
  Γ = time constant, = 60 for flow rate per minute
  f = frequency, hertz
  K = pulse per unit volume, also referred to as the meter's K factor, pulses/gal

TABLE 3   Characteristics of Magnetic Flowmeters

Service	Electrically conductive liquids and slurries
Flow range	0.01 (0.04 L/min) to 500,000 gal/min (1900 m³/min)
Operating temperature (maximum)	360°F (182°C)
Operating pressure (maximum)	740 psig (5 MPa)
Scale	Linear
Accuracy	±0.15% (flow rate)
	±1% (full scale)
Rangeability	10:1
Advantages	Generally insensitive to viscosity changes; well suited for water, wastes; measures bidirectional flows; relatively low or no head loss; no flow obstructions
Limitations	Relatively high cost; heavy, bulky; not applicable to non-conductive fluids

K Factor

A turbine flowmeter's *K* factor is determined by the manufacturer by displacing a known volume of fluid through the meter and summing the number of pulses generated by the meter. This summation of pulses, divided by the known volume of fluid that has passed through the turbine flowmeter, is the *K* factor. The turbine flowmeter's *K* factor is specified to be linear, within a tolerance, over a 10:1 range of flow rates. Extended flow ranges (greater than 10:1) normally restrict the user to specific fluid parameters. A representative graph of common turbine flowmeter characteristics is given in Fig. 16.

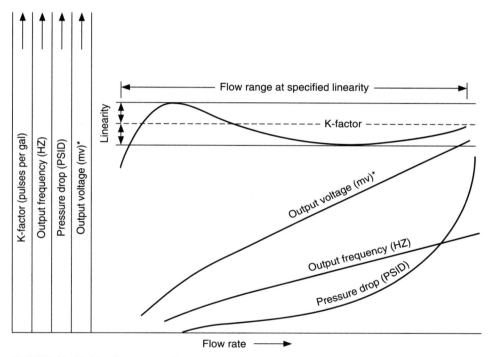

FIGURE 16   Turbine flowmeter performance characteristics. *variable reluctance pickup only. (*Great Lakes Instruments, Inc.*)

Reynolds Numbers

An important condition affecting the flow measurement is the Reynolds number of the fluid being measured. Reynolds numbers represent a unitless value which defines the ratio of a fluid's inertial forces to its drag forces. The Reynolds number equation relating these forces is

$$R = \frac{3160 \times Q \times G_t}{D \times \mu}$$

where R = Reynolds number
Q = fluid flow rate, gal/min
G_t = fluid specific gravity
D = inside pipe diameter, inches
μ = liquid viscosity, centipoise (cP)

The fluid flow rate and specific gravity represent inertial forces; the pipe diameter and viscosity are the drag forces. In a typical system the specific gravity and the pipe diameter are constants. Consequently low Reynolds numbers exist when the fluid flow rate is low or the fluid viscosity is high. The problem with measuring the flow of fluids with low Reynolds numbers using turbine flowmeters is that the fluid will have a higher velocity at the center of the pipe than along the pipe's inside diameter. This results in a K factor which is not constant over the flow rate range of the turbine flowmeter.

Turbine Flowmeter Construction

Turbine flowmeters are constructed of a permeable metal rotor, housed in a nonmagnetic body, as depicted in Fig. 17. This allows proximity sensors to sense the rotational blades of the rotor directly through the body of the meter. Rotor designs vary from manufacturer to manufacturer, but in general, rotors with tangential axes are used for flow rates smaller than 5 gal/min (19 L/m) and axial rotors for higher flow rates. The rotor is designed to be of low mass with respect to the fluid momentum. This relationship shortens the response time of the turbine flowmeter by decreasing the rotational inertia required to accelerate the rotor. Turbine flowmeter bearings are of either the ball or the sleeve type. This provides long service life while minimizing frictional losses.

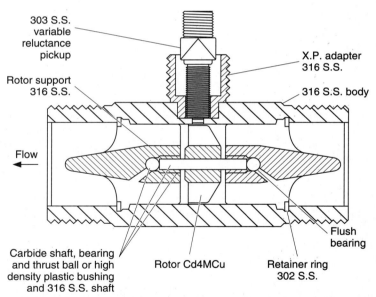

FIGURE 17   Typical axial-rotor [0.5- to 3-inch (12.7- to 76.2-mm)] turbine flowmeter construction. (*Great Lakes Instruments, Inc.*)

The mechanical losses in a turbine flowmeter can include bearing friction, viscosity shear effects, and magnetic or mechanical pickoff drag. These are illustrated in Fig. 18. These losses are most pronounced when turbine flowmeters are operated in the transitional or laminar flow region (at low fluid velocities or high fluid viscosities). Best results are obtained when turbines operate in media having Reynolds numbers in excess of 4000 or 5000. This varies with each manufacturer. Turbine flowmeter bodies have straightening vanes at the inlet and outlet to stabilize the fluid flow in the rotor area. A minimum of 10 pipe diameters of straight pipe upstream and 5 diameters downstream from the flowmeter installation is recommended for best measurement accuracy.

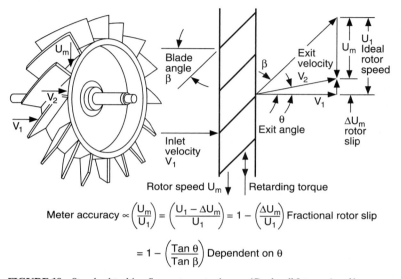

Meter accuracy $\propto \left(\dfrac{U_m}{U_1}\right) = \left(\dfrac{U_1 - \Delta U_m}{U_1}\right) = 1 - \left(\dfrac{\Delta U_m}{U_1}\right)$ Fractional rotor slip

$= 1 - \left(\dfrac{\text{Tan } \theta}{\text{Tan } \beta}\right)$ Dependent on θ

FIGURE 18   Standard turbine flowmeter rotor losses. (*Rockwell International.*)

The rotor speed is sensed through the flowmeter body using several different methods. The most common for liquid service is the use of a variable-reluctance pickup. This device generates a magnetic field through a coil and the nonmagnetic flowmeter body into the rotor area. As each rotor blade enters the magnetic field, the discontinuity excites a voltage in the coil, producing an electrical sinusoidal wave current with frequency and voltage proportional to the rotor blade velocity.

Another popular method of sensing rotor speed is with a modulated-carrier electronic pickup. This type of pickup is designed to eliminate the physical loss caused by magnetic drag with the variable-reluctance pickup. This is especially important in low-velocity and gas flowmetering applications. The modulated-carrier pickup generates a "carrier" signal of a frequency much higher than that of the rotor blade. A coil, which is integral to the high-frequency oscillator, is positioned near the rotor. As each blade passes the coil, eddy-current losses in the coil increase, causing the amplitude of the carrier signal to decrease until the blade has passed. Filtering out the high-frequency component of the carrier retains the "modulated" signal, which is proportional to the fluid velocity.

Turbine Flowmeter Applications

Turbine flowmeters accurately measure flow over a large range of operating conditions. Primary uses include flow totalizing for inventory control and custody transfer, precision automatic batching for dispensing and batch mixing, and automatic flow control for lubrication and cooling applications. Typical measurement accuracy is ±0.5 percent of reading over a

10:1 flow range (turn-down ratio) and repeatability within ±0.05 percent of rate. A degradation in performance can be expected if there are variations in fluid viscosity, swirling of fluid within the pipes, or contamination which causes premature bearing wear. Turbine flowmeters are available for pipe sizes of 0.5 to 24 inches (12.7 to 610 mm) and flow ranges to 50,000 gal/min (11,358 m³/h). Working pressures are generally determined by the inlet and outlet connections. Some designs are capable of working pressures in excess of 10,000 psig (68,950 kPa). Temperature limits are imposed by the pickup electronics. Most manufacturers specify 400°F (204°C) standard and extend to greater than 800°F (427°C) on special order. Popular end fitting connection styles include NPT, flared tube, flanged, sanitary, ACME, grooved, and wafer.

Paddle-Wheel Flowmeters

The paddle-wheel-type flowmeter, also referred to as an impeller flowmeter, is characterized in much the same way as the turbine flowmeter. The rotating portion of the flowmeter, called the rotor or impeller (Fig. 4) is positioned into the outside perimeter of the liquid path. As the liquid moves past the rotor, a moment is imposed on the blades of the rotor that are present in the stream. This moment causes the rotor to accelerate to a velocity equal to the liquid velocity. Since the cross-sectional area of the pipe is known, the volumetric flow rate can be easily obtained from the liquid velocity. The paddle-wheel flowmeter has inherently lower cost because the same sensing mechanism can be used with several pipe sizes and hardware configurations. Paddle-wheel flowmeters require a full pipe for operation and measure liquid velocities between 0 and 30 ft/s (9.1 m/s). This technology has typical linearity of +1 percent of full-scale and a repeatability of ±0.5 percent of full-scale. Operating conditions include pressures up to 2000 psig (139 kPa) and temperatures to 400°F (204°C).

Various types of mounting hardware adapt the paddle-wheel flowmeter to installation requirements. These include T-mounts, as shown in Fig. 19, insertion mounts for use with pipe saddles and weldments, spool pieces for mounting the sensor between pipe flanges, and hot tap hardware, which allows the insertion sensor to be serviced without depressurizing the system in which it is installed. The characteristics of turbine flowmeters are summarized in Table 4.

TABLE 4   Characteristics of Turbine Flowmeters*

Service	Liquids (clean), gases, and steam
Accuracy	±0.5% (reading) over 10:1 flow range; depends on specific designs; on average as quoted in literature, ±0.25% (rate) for liquids, ±1% (rate) for gases; factory calibration should include viscosity and lubricity factors for liquids
Repeatability	±0.05% (reading) over 10:1 flow range
Sizes	Generally 0.5 to 24 inches (12.7 to 610 mm), flanged or threaded connections
Rangeability	10:1 (generally); can be extended to 50:1, but with restricted flow parameters
Scale	Nominally linear, especially for Reynolds numbers higher than 10,000
Flow ranges	Extremely low flow rates claimed; depends on design; generally, for liquids, 5 gal/min (19 L/min) to 50,000 gal/min (11.400 m³/h); gases, 10 mil ft³/h (285,000 m³/h)
Pressure	Up to 3000 psig (20 MPa)
Temperature	Up to 400°F (204°C), but can be extended to 800°F (427°C) on special order
Advantages	Accuracy; operating range; useful for very low flow rates; some designs optimized for gas service; installation and maintenance relatively easy; can be used as bypass meter around pipeline orifice
Limitations	Problems may be encountered with two-phase flows; upstream straight piping required, including straighteners

* Prepared by handbook staff.

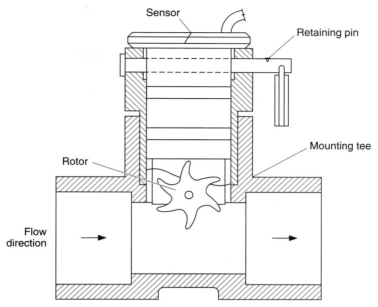

FIGURE 19 Typical paddle-wheel flowmeter construction. (*Great Lakes Instruments, Inc.*)

OSCILLATORY FLOWMETERS

As with the turbine flowmeter, the oscillatory meters sense a physical property of a moving fluid that is related to stream velocity. There are two principal meters in this classification, (1) vortex meters and (2) fluidic meters. Vortex meters are classified as vortex shedding or vortex precision types.

Vortex Flowmeters

The principle of vortex shedding has been known for many years, but only since the mid-1970s has this concept been used seriously in flowmeters. Vortex meters now are used widely, with installations numbering in the tens of thousands. One of their major impacts has been on orifice-type meters.

Vortex formation can occur when a nonstreamlined obstruction is placed in a flowing stream. The obstruction, sometimes referred to as the shedder, must be designed and shaped to produce vortices which create a differential pressure of sufficient magnitude to be detected. Considerable research has gone into the design of shedding elements (Fig. 20). In another vortex meter design, the rate of precessing vortices or whirlpools is sensed.

As shown in Fig. 21, as the fluid flows through the flowmeter, it is divided into two paths by the shedding element, which is positioned across the flowmeter body. High-velocity fluid parcels flow past the lower-velocity parcels in the vicinity of the element to form a shear layer. There is a large velocity gradient within this shear layer, thus making it inherently unstable. After some length of travel, the shear layer breaks down into well-defined vortices. Differential-pressure changes occur as the vortices are formed and shed. This

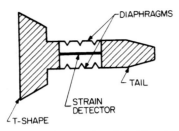

FIGURE 20 Cross section of shedding element. The force of the T-shaped end is sometimes referred to as a bluff body. The tail is very important in the control of stable vortex shedding and also provides a good location for the vortex detector because the signal-to-noise ratio at this location is high. The location also provides good protection for the detector from any objects that may be in the fluid stream. (*Foxboro.*)

pressure variation is used to actuate a sealed detector at a frequency proportional to vortex shedding.

With further reference to Fig. 20, the detector may consist of a double-faced circular diaphragm capsule filled with a liquid. A piezoelectric crystal may be located in the center of the capsule in such manner that the vortex-produced pressure changes are transmitted through the fill liquid to the crystal, whereupon the piezoelectric crystal produces a voltage output when a pressure change is detected. The alternate vortex generation reverses the differential, and a voltage of opposite polarity is generated. Thus a train of vortices generates an alternating voltage output with a frequency identical to the frequency of vortex shedding. The characteristics of vortex flowmeters are listed in Table 5.

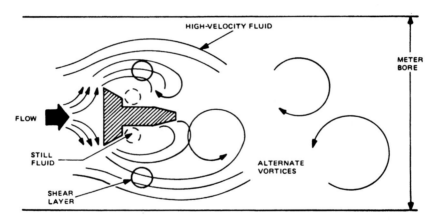

FIGURE 21 Vortex shedding phenomenon. Flow patterns are altered by presence of obstruction (shedding element).

Fluidic Flowmeters

Fluidics is the technology of sensing, controlling, and information processing with devices that use a fluid medium and the operation of which is based solely on the interaction between fluid streams. The particular function of each device, none of which has any moving parts, is dependent on its geometric shape. Fluidics received considerable attention from the military in the 1960s as a control medium that could not be jammed, as is possible with unprotected electronic devices. The wall attachment fluid phenomenon, also called the Coanda effect, was first discovered in 1926 by Henri Coanda, a Rumanian engineer. The possible industrial application of fluidics was explored in the 1960s, with the creation of amplifiers, valves, oscillators, and flowmeters, among other configurations.

Industrial fluidic flowmeters are currently available. These incorporate the principle of the fluidic oscillator shown in Fig. 22. In flowmeter applications the fluidic oscillator has the advantages of (1) linear output with the frequency proportional to the flow rate, (2) rangeability up to 30:1, (3) being unaffected by shock, vibration, or field ambient temperature

TABLE 5   Characteristics of Oscillatory Flowmeters (Averaged for Several Models and Designs)

Flow range:	
Vortex shedding	3 to 5000 gal/min (11 to 19,000 L/min) for liquids
	10 mil scfh (283,200 m^3/h) for gases
Vortex precession	1.8 to 3082 gal/min (6.8 to 11,665 L/min) for liquids
	10 mil scfh (283,200 m^3/h) for gases
Fluidic (Coanda)	1 to 1000 gal/min (4 to 4000 L/min) for liquids
Accuracy:	
Vortex shedding	±1% (rate) or better for liquids
	±2% (rate) for gases
Vortex precession	±1% (rate) or better
Rangeability:	
Vortex shedding	8:1 to 15:1
Vortex precession	8:1 to 25:1 (size- and application-dependent)
Fluidic (Coanda)	Up to 30:1
Scale:	
Vortex shedding	Linear (high Reynolds numbers)
Vortex precession	Linear (high Reynolds numbers)
Fluidic (Coanda)	Linear (high Reynolds numbers)
Operating pressure:	
Vortex shedding	Up to 3600 psig (25 MPa)
Vortex precession	Up to 1400 psig (9.5 MPa)
Fluidic (Coanda)	Up to 600 psig (4 MPa)
Operating temperature:	
Vortex shedding	Up to 750°F (399°C)
Vortex precession	−100 to 350°F (−73 to 177°C)
Fluidic (Coanda)	0 to 250°F (−18 to 121°C)
Advantages and limitations:	
Vortex shedding	Requires straight piping
	Sampling and bypass types available
	Sensitive to viscosity below minimum Reynolds number
	No moving parts
	Handles wide range of fluids, including steam
	Relatively good cost/performance ratio
Vortex precession	Operating minimum depends on flow rate and gas density
	Relatively high cost
	No moving parts
	Particularly useful for difficult and corrosive gases
Fluidic (Coanda)	Requires straight piping
	Sensitive to viscosity below minimum Reynolds number
	No moving parts
	Handles wide range of liquids
	Relatively good cost/performance ratio
	Bypass types available
Size ranges:	
Vortex shedding	0.5 to 8 in (15 to 200 mm)
Vortex precession	0.5 to 12 in (15 to 300 mm)
Fluidic (Coanda)	1 to 4 in (25 to 100 mm)

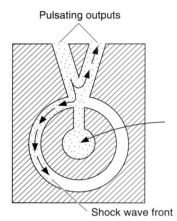

FIGURE 22   Fluidic oscillator.

changes, (4) calibration in terms of volume flow being unaffected by fluid density changes, and (5) no moving points and no impulse lines (Fig. 23).

A similar fluidic flowmeter, but without diverging sidewalls, uses momentum exchange instead of the Coanda effect and has also proved effective in metering viscous fluids. The characteristics of fluidic flowmeters are also summarized in Table 5.

MASS-FLOW MEASUREMENT

Traditionally fluid flow measurement has been made in terms of the volume of the moving fluid—even though the meter user may be more interested in the weight (mass) of the fluid. Volumetric flowmeters also are subject to ambient and process changes, such as density, which changes with temperature and pressure. Viscosity changes also may affect volumetric flow sensors.

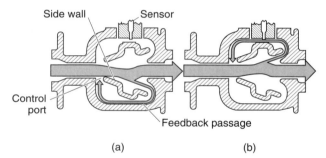

(a)   (b)

FIGURE 23   Operating principle of fluidic flowmeter. The geometric shape of the meter body is such that when flow is initiated, the flowing stream attaches to one of the side walls as a result of the Coanda effect. A portion of the main flow is diverted through a feedback passage to a control port. The feedback flow increases the size of the separation bubble. This peels the main flow stream away from the wall until it diverts and locks onto the opposite wall, where the feedback action is similar. The frequency of the self-induced oscillation is a function of the feedback flow rate, which in turn is directly proportional to the flow rate of the mainstream. (*Moore Products.*)

Thus for a number of years there has been much interest in finding ways to measure mass directly rather than to use calculating means to convert volume to mass. As of the early 1990s, there are three ways to determine mass flow: (1) the application of microprocessor technology to conventional volumetric meters, (2) the use of Coriolis flowmeters, which measure mass flow directly, and (3) the use of thermal mass flowmeters, which infer mass flow by way of measuring heat dissipation between two points in the pipeline. The latter two concepts have been investigated and attempted for several years, but were not found practical or commercially available in fairly large numbers until the mid-1980s.

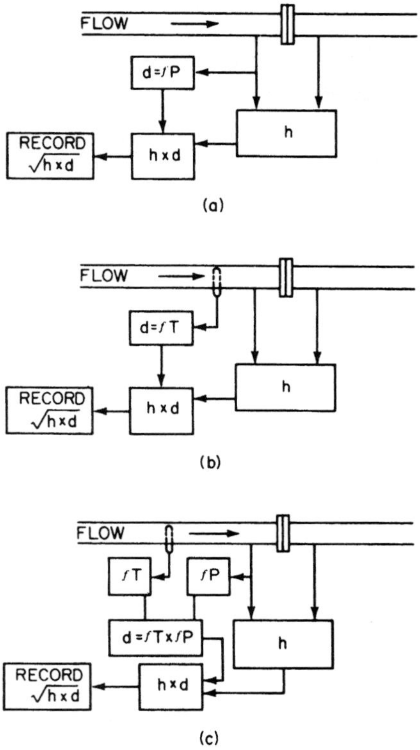

FLOW ——→

d=ʃP

h

RECORD √h×d ← h×d

(a)

FLOW ——→

d=ʃT

h

RECORD √h×d ← h×d

(b)

FLOW ——→

ʃT   ʃP

d=ʃT×ʃP   h

RECORD √h×d ← h×d

(c)

FIGURE 24 Method for compensating a volumetric flowmeter for mass flow. (*a*) Pressure-compensated meter wherein the differential pressure is measured by an appropriate sensor and the signal is fed into a combining module, along with a signal representing the pressure correction. The output from the combining module is used for display and to regulate the meter integrator. (*b*) Meter with temperature compensation. (*c*) Meter with combined pressure-temperature compensation.

Microprocessor-Based Volumetric Flowmeters

As shown in Fig. 24, with micropressors it is relatively simple to compensate a volumetric flowmeter for temperature and pressure. With reliable composition (hence density) information, this factor also can be entered into a microprocesor to obtain a mass-flow readout. However, when density changes may occur with some frequency, and particularly where the flowing fluid is of high monetary value (for example, in custody transfer), precise density compensation (to achieve mass) can be costly. For example, a gas mass flowmeter system may consist of a vortex gas velocity meter combined with a gas densitometer. The densitometer can be located upstream of the flow device and produce a pressure difference that is linearly proportional to the density of the flowing gas at line conditions. This unit will automatically correct for variations in pressure, temperature, specific gravity, and supercompressibility.

The gas sample from the pipeline passes across a constant-speed centrifugal blower and returns to the pipeline. The pressure rise across the blower varies directly with the gas density. A differential-pressure signal from the densitometer is combined with a flow-rate signal from the gas meter. The cost of such instrumentation can be severalfold that of an uncompensated meter.

The relatively high cost of this instrumentation, combined with an increasing need for reliable mass-flow data, established the opportunity for direct mass-flow instruments of the Coriolis and thermal types.

Coriolis Flowmeters

Although the Coriolis effect[3] was considered earlier, serious work in terms of its application to instrumentation and the development of a flowmeter did not occur until the mid-1970s. The formal introduction of a commercial unit occurred in the early 1980s. Since then numerous design improvements have been made, including a reduction of the effects of ambient vibration and of the size and weight of the unit. Acceptance of

---

[3] Any object moving above the earth with constant space velocity is deflected relative to the surface of the rotating earth. This deflection was first discussed by the French scientist Gaspard Coriolis about the middle of the last century and is now usually described in terms of the Coriolis acceleration or the Coriolis force. The deflection is found to be to the right in the northern hemisphere and to the left in the southern hemisphere. The Coriolis effects must be considered in a great variety of phenomena in which motion over the surface of the earth is involved, as, for example, in the following cases. (1) Rivers in the northern hemisphere should scour their right banks more severely than their left, and the effect should be more evident for rivers at high latitudes. Studies on the banks of the Mississippi and Yukon rivers indicate the predicted results. (2) The motions of air over the earth are governed to an appreciable extent by the Coriolis force. (3) A term due to the Coriolis effect must be included in the equation for exterior ballistics.

the refined unit within the last few years is attested by the estimated (1991) installations of over 80,000 units. Coriolis meters currently are available from several suppliers.

The complete Coriolis unit consists of (1) a Coriolis force sensor and (2) an electronic transmitter. The sensor comprises a tube (or tubes) assembly, which is installed in the process pipeline. As shown in Fig. 25, in one configuration a U-shaped sensor tube is vibrated at its natural frequency. The angular velocity of the vibrating tube, in combination with the mass velocity of the flowing fluid, causes the tube to twist. The amount of twist is measured with magnetic position detectors, producing a signal which is linearly proportional to the mass flow rate of every parcel and particle passing through the sensor tube. The output is essentially

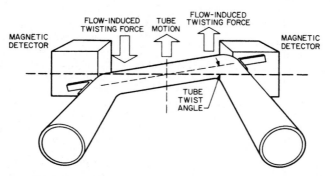

FIGURE 25   Operating principle of Coriolis mass flowmeter. (*Micro Motion.*)

unaffected by variations in fluid properties, such as viscosity, pressure, temperature, pulsations, entrained gases, and suspended solids.

There is no contact with the flowing fluid, except at the inside wall of the tube, which usually is made of stainless steel or some other corrosion- and erosion-resistant material. Two magnetic position detectors, one on each side of the U-shaped tube, generate signals which are routed to the associated electronics for processing into an output. The principal characteristics and representative applications of the Coriolis flowmeter are summarized in Table 6.

Thermal Mass Flowmeters

Like the Coriolis flowmeter, after many years of design work and limited applications, the thermal mass flowmeter did not become widely accepted until the late 1970s and early 1980s. A thermodynamic operating principle is applied. As shown in Fig. 26, a precision power supply directs heat to the midpoint of a sensor tube that carries a constant percentage of the flow. On the same tube equidistant upstream and downstream of the heat input are resistance temperature detectors (RTDs). With no flow, the heat reaching each temperature element is equal. With increasing flow the flow stream carries heat away from the upstream element T_1 and an increasing amount toward the downstream element T_2. An increasing temperature develops between the two elements, and this difference is proportional to the amount of gas flowing, or the mass flow rate. A bridge circuit interprets the temperature difference and an amplifier provides the 0- to 5-volt dc and 4- to 20-mA output signal. The principal characteristics and representative applications of the thermal flowmeter are also summarized in Table 6.

TABLE 6   Characteristics of Mass Flowmeters

Characteristic	Coriolis type	Thermal type
Power requirements	ac/dc	ac/dc
Type of measurement	Whole body	Whole body, by-pass, sampling
Installation	Meters in series should be separated by 15 pipe diameters	With inertion types, sensor head should not touch opposite pipe wall
Orientation	Gas bubbles or sediment must not collect in measuring region of meter	Installation of meter with same orientation as when calibrated is desirable in some designs
Service	Liquids and slurries; limited gas service	Gases; some designs useful for liquids
Design pressure	Up to 19 MPa (2800 psig)	Up to 99 MPa (1450 psig)
Design temperature	Up to 205°C (400°F)	Up to 65°C (150°F)
Scale	Linear	Proportional
Accuracy	±0.2% (flow rate)	±1% (full-scale)
Rangeability	10:1 (or better)	10:1 (or better)
Typical applications	Custody transfer, paint blending, food processing (peanut butter, milk, etc.), electrically conductive fluids, petro-chemicals, blood plasma, corrosive slurries	Semiconductor manufacture, flue stack monitoring, pilot plant operations, leak testing, thermal spray manufacture, gas blending control

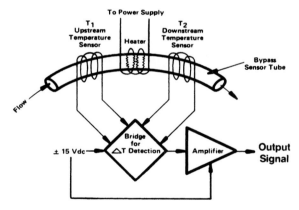

FIGURE 26   Operating principle of thermal mass flowmeter. (*Brooks Instrument Division, Emerson Electric Company.*)

ULTRASONIC FLOWMETERS

Ultrasonic or acoustic flowmeters are of two principal types: (1) Doppler-effect meters and (2) transit-time meters. In both types the flow rate is deduced from the effect of the flowing process stream on sound waves introduced into the process stream. In clamp-on designs, these meters make it possible to measure the flow rate without intruding into the stream, and thus are classified as noninvasive. But even in configurations where transducers are contained in shallow wells, the flowmeters are essentially nonintrusive.

The principles of ultrasonic flow measurement have been known for many years, but only within the last few decades have these meters made measurable penetration of the flowmeter field. This lag in acceptance has been variously explained, but the general consensus is that too many designs were introduced too soon—prematurely and without testing them against adverse plant environments—to the point that for several years ultrasonic flowmeters had somewhat of a tarnished image. Ultrasonic flowmeters have numerous innate advantages over most of the traditional metering methods: linearity, wide rangeability without an induced pressure drop or disturbance to the stream, achievable accuracy comparable to that of orifice or venturi meters, bidirectionality, ready attachment to the outside of existing pipes without shutdown, comparable if not overall lower costs, attractive features which can make a product prone to overselling. Within the last few years, ultrasonic flow measurement technology has found a firmer foundation, and most experts now agree that these meters will occupy a prominent place in the future of flowmetering.

Doppler-Effect (Frequency-Shift) Flowmeters

In 1842 Christian Doppler predicted that the frequencies of received waves were dependent on the motion of the source or observer relative to the propagating medium. His predictions were promptly checked for sound waves by placing the source or observer on one of the newly developed railroad trains. Over a century later, the concept was first considered for application in the measurement of flowing streams.

For the principle to work in a flowmeter, it is mandatory that the flowing stream contain sonically reflective materials, such as solid particles or entrained air bubbles. Without these reflectors, the Doppler system will not operate. In contrast, the transit-time ultrasonic flowmeter does not depend on the presence of reflectors.

The basic equations of a Doppler flowmeter are

$$\Delta f = 2 f_T \sin \theta \; \frac{V_E}{V_S} \tag{1}$$

and, by Snell's law,

$$\frac{\sin \theta_T}{V_T} = \frac{\sin \theta}{V_S} \tag{2}$$

Simultaneously solving Eqs. (1) and (2) gives

$$V_F = \frac{\Delta f}{f_T} \; \frac{V_T}{\sin \theta_T} = K \, \Delta f \tag{3}$$

where V_T = sonic velocity of transmitter material
  θ_T = angle of transmitter sonic beam
  K = calibration factor
  V_E = flow velocity
  ΔF = Doppler frequency change
  V_S = sonic velocity of fluid
  f_T = transmission velocity
  θ = angle of f_T entry in liquid

Doppler-effect flowmeters use a transmitter that projects a continuous ultrasonic beam at about 0.5 MHz through the pipe wall into the flowing stream. Particles in the stream reflect the ultrasonic radiation which is detected by the receiver. The frequency of the radiation reaching the receiver is shifted in proportion to the stream velocity. The frequency difference is a measure of the flow rate. The configuration shown in Fig. 27 utilizes separated dual transducers mounted on opposite sides of the pipe. Other possible configurations are illustrated in Fig. 28. In essence the Doppler-effect meter measures the beat frequency of two signals. The

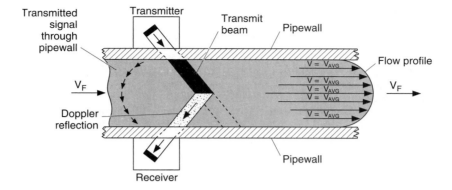

FIGURE 27   Principle of Doppler-effect ultrasonic flowmeter with separated opposite-side dual transducers.

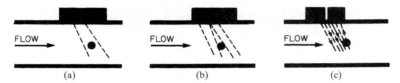

FIGURE 28   Configurations of Doppler-effect ultrasonic flowmeter. (*a*) Single transducer. (*b*) Tandem dual transducer. (*c*) Separate dual transducers installed on same side of pipe.

beat frequency is the difference frequency obtained when two different frequencies (transmitted and reflected) are combined.

When the measured fluid contains a large concentration of particles or air bubbles, it is said to be sonically opaque. The more opaque the liquid, the greater the number of reflections that originate near the pipe wall, a situation exemplified by heavy slurries. It can be noted from the flow profile of Fig. 27 that these reflectors are in the low-flow-rate region. In contrast, the preponderance of particle reflectors will occur in the center of the pipe (where the flow rate is highest) when the fluid is less sonically opaque. Where there are relatively few reflective particles in a stream, there is a tendency for the ultrasonic beam to penetrate beyond the centerline of the pipe and to detect slow-moving particles on the opposite side of the pipe. Because the sonic opacity of the fluid may be difficult to predict in advance, factory calibration is difficult.

It will be noted from Fig. 27 that the fluid velocity is greatest near the center of the pipe and lowest near the pipe wall. An average velocity occurs somewhere between these two extremes. Thus there are numerous variables, characteristic of a given fluid and of a specific piping situation, that affect the interactions between the ultrasonic energy and the flowing stream. Should a measured fluid have a relatively consistent flow profile and include an ideal concentration and distribution of particles, these qualities add to the fundamental precision of measurement and thus simplify calibration. Various designs are used to minimize interaction inconsistencies. For example, separation of transmitters and receivers makes it possible to restrict the zone in the pipe at which the principal concentration of interaction occurs. This zone usually occurs in the central profile region of the pipe less affected by variations in sonic opacity than the region near the pipe wall.

Transit-Time Ultrasonic Flowmeters

With this type of meter, air bubbles and particles in the flowing stream are undesirable because their presence (as reflectors) interferes with the transmission and receipt of the ultrasonic radiation applied. However, the fluid must be a reasonable conductor of sonic energy (Fig. 29). At a given temperature and pressure, ultrasonic energy will travel at a specific velocity through a given liquid. Since the fluid is flowing at a certain velocity (to be measured), the sound will travel faster in the direction of flow and slower against the direction of flow. By measuring the differences in arrival time of pulses traveling in a downstream direction and pulses traveling in an upstream direction, this ΔT can serve as a measure of fluid velocity. Transit-time or ΔT flowmeters transmit alternately upstream and downstream and calculate this time difference. The operation is illustrated by the following equations:

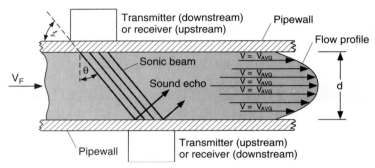

FIGURE 29   Principle of transit-time ultrasonic flowmeter, clamp-on type. Transducers alternately transmit and receive bursts of ultrasonic energy.

$$V_F = \frac{(T_U - T_D)V_s}{\sin \theta} \; \frac{V_s \cos \theta}{d} = \frac{\Delta t V_s}{\sin \theta} \; \frac{1}{T_L} \qquad (4)$$

By Snell's law,

$$\frac{V_s}{\sin \theta} = \frac{V_C}{\sin \alpha} = K \qquad (5)$$

$$V_F = \frac{K \, \Delta t}{T_L} \qquad (6)$$

where T_U = upstream transit time
T_D = downstream transit time
T_L = zero-flow transit time
V_S = liquid sonic velocity
d = pipe inside diameter
V_R = liquid flow velocity
α = angle between transducer and pipe wall
V_C = transducer sonic velocity

In the clamp-on transit-time flowmeter, where the transducers are strapped to the outside of the pipe, the sonic echo is away from the receiver and thus the device can retransmit sooner and operate faster. Clamp-on meters are installed on a normal section of piping, which is particularly attractive for retrofit applications. Other designs make use of what are called wetted transducers, which are mounted within the pipe. In both designs, of course, the transducers are installed diagonally to the flow, that is, *not* directly across from each other. Wetted transducers are usually installed in a shallow well, but because there are no projections beyond the pipeline wall, they are still considered nonintrusive. There is no significant disturbance to the general flow profile. However, slight, localized eddy currents may form in the vicinity of the wells. To avoid eddy currents, at least one manufacturer puts the transducers within the wells, forming what is termed an epoxy window. This results in completely filling the stream side of the well, making it essentially flush with the inner pipe wall.

In another design (dual-path ultrasonic flowmeter), two pairs of transducers are installed in the piping. The upstream and downstream propagation times between each pair of transducers are integrated in a microprocessor-based electronics package to determine the flow rate.

Ultrasonic flowmeters have been used successfully for certain open-channel flow measurements. The transducer is installed above the channel. Pulses emitted are reflected by the liquid surface back to the transducer in a form of sonar. The time required for the pulse to return from the liquid surface is related to the height of the liquid surface. By knowing the cross-sectional properties of the open channel, height measurements can be converted to a measurement of flow. The characteristics of ultrasonic flowmeters are summarized in Table 7.

TABLE 7   Characteristics of Ultrasonic Flowmeters

Flow velocity:	
Doppler (frequency shift)	40 ft/s (12 m/s)
Doppler (transit time)	40 ft/s (12 m/s)
Service:	
Doppler (frequency shift)	Liquids with entrained gas or suspended solids
Doppler (transit time)	Comparatively clean liquids, some gas designs
Operating temperature:	
Doppler (frequency shift)	−300 to 500°F (−184 to 260°C)
Doppler (transit time)	−300 to 500°F (−184 to 260°C)
Operating pressure:	
Doppler (frequency shift)	1000 psig (6.8 MPa) and up (wetted transducers)
Doppler (transit time)	1000 psig (6.8 MPa) and up (wetted transducers)
Accuracy:	
Doppler (frequency shift)	±5% (full-scale) or better
Doppler (transit time)	±1% (rate) to ±5% (full-scale)
Rangeability:	
Doppler (frequency shift)	10:1 (typical)
Doppler (transit time)	Up to 40:1
Size:	
Doppler (frequency shift)	¼ in (6 mm) and up
Doppler (transit time)	⅜ in (10 mm) and up
Advantages and limitations:	
Doppler (frequency shift)	Clamp-on version can be installed without process interruption; handles inorganic slurries and aeriated liquids; not applicable to clean liquids; requires straight upstream piping
Doppler (transit time)	No flow obstruction; bidirectional; useful with most clean fluids; gas models available; clamp-on version available; uniform flow profile

POSITIVE-DISPLACEMENT FLOWMETERS

Positive-displacement (PD) flowmeters measure flow directly in quantity (volume) terms instead of indirectly or inferentially as rate meters do. In several rather ingenious ways, the PD meter separates the flow to be measured into different discrete portions or volumes (not weight units) and, in essence, counts these discrete volumes to arrive at a summation of total flow. Fundamentally, PD meters do not have a time reference. Their adaptation to rate indication requires the addition of a time base. Similarly, by making appropriate corrections for changes in density resulting from pressure, temperature, and composition changes, PD meters can serve to indicate flow in terms of mass or weight.

Although inferential rate-type flowmeters appear most frequently in process control loops, departments of facilities concerned with materials accounting (receiving, shipping, inter- and intraplant transfers, distribution, and marketing) largely depend on PD flowmeters. Also, in recent years there has been growing acceptance of PD flowmeters for use in automatic liquid batching and blending systems. Utilities and their consumers are among the largest users of PD flowmeters, with millions of units in use for distributing and dispensing water, gas, gasoline, and other commodities. Where PD meters are used to dispense these commodities, they are subject to periodic testing and inspection by various governmental weights and measures agencies. Accuracy requirements vary with type of meter and service. Usually water meters for domestic service and small industrial plants require an accuracy of only ±2 percent and gas meters ±1 percent, whereas meters associated with retail gasoline and diesel fluid pumping have a tolerance (when new or after repair) of only 3.5 cm^3 in 5 gal (19.9 L). Routine tests permit a tolerance of 7 cm^3 in some states.

Nutating-Disk Meter

The nutating-disk meter is probably the most commonly encountered flowmeter found throughout the world for commercial, utility, and industrial applications. The meter is of particular importance in the measurement of commercial and domestic water. Although there are some proprietary design differences, the fundamental operation of the nutating-disk meter is shown in Fig. 30. In most water-metering situations the meter commonly incorporates a digital integrated readout of the "speedometer" style. Meters of this general type for industrial process use, such as in batching and blending, may incorporate a number of accessories. Among the common sizes are 0.5 inch (13 mm), which can deliver up to 20 gal/min (76 L/min), 0.75 inch (19 mm), 1 inch (25 mm), 1.5 inches (38 mm), and 2 inches (51 mm), which can deliver up to 160 gal/min (606 L/min).

Industrial nutating-disk meters provide accurate measurements for low-flow rates and are relatively easy to install and maintain. Depending on the construction materials used, the nutating-disk meter can handle a wide range of chemicals, including caustics. The useful temperature range is from −150 to 120°C (−238 to 248°F).

Oscillating-Piston Meter

In principle, the oscillating-piston meter is similar to the nutating-disk meter, with the important difference that mechanical motion takes place in one plane only (no wobble). In one design[4] the measuring chamber consists of five basic parts: (1) top head, (2) bottom head, (3) cylinder, (4) division plate, and (5) piston. The only moving part in the measuring chamber is the piston, which oscillates smoothly in a circular motion between the two plane sur-

---

[4] Brooks Instrument Division, Emerson Electric Company, Hatfield, Pennsylvania.

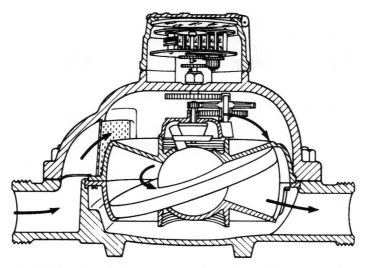

FIGURE 30 Sectional view of representative nutating-disk flowmeter. Each cycle (complete movement) of the measuring disk displaces a fixed volume of liquid. There is only one moving part in the measuring chamber, the disk. Liquid enters the inlet port and fills the space above and below the disk in a nutating motion until the liquid discharges from the outlet port. The motion of the disk is controlled by a cam which keeps the lower face in contact with the bottom of the measuring chamber on one side, while the upper face of the disk is in contact with the top of the chamber on the opposite side. Thus the measuring chamber is sealed off into separate compartments, which are successively filled and emptied, each compartment holding a definite volume. The motion is smooth and continuous with no pulsations. The liquid being measured forms a seal between the disk and the chamber wall through capillary action, thus minimizing leakage or slippage and providing accuracy even at low flow rates.

faces of the top and bottom heads. The division plate separates the inlet ports *A* and the outlet ports *B*. The piston is slotted to clear the division plate, which also guides the travel of the piston in its oscillating motion. A gear train transmits the piston motion to the register. The major components and operation of the meter are shown in Fig. 31.

The piston has a significantly smaller circumference than the chamber. This provides for maximum liquid area displacement for each oscillation. A small differential pressure across the meter produces motion of the piston within the measuring chamber. In order to obtain oscillating motion, two restrictions are placed on the movement of the piston. First, the piston is slotted vertically to match the size of a partition plate fixed to the chamber. This plate prevents the piston from spinning around its central axis and also acts as a seal between the inlet and outlet ports of the chamber. Second, the piston has a center vertical pin which is confined to a circular track that is part of the chamber. Differential pressure across the meter causes the piston to sweep the chamber wall in the direction of flow. This oscillating motion displaces liquid from the inlet to the outlet port in a continuous stream.

To further prevent unmeasured liquid from passing through the chamber, the piston has a horizontal partition or web. This web is perforated to promote balanced buoyancy of the piston within the chamber and a linear flow pattern of liquid through the meter. A drive bar and shaft are positioned through the top of the chamber so that, as the piston oscillates, the piston pin drives the bar and shaft in a circular or spinning motion. This rotating shaft is the driving link between the piston and the register or readout unit. Sizes range from 0.75 to 2 inches (19 to 51 mm), with capacity ranging from 5 to 150 gal/min (19 to 570 L/min).

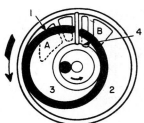

DIAGRAM 1
SPACES 1 AND 3 ARE RECEIVING
LIQUID FROM THE INLET PORT, A,
AND SPACES 2 AND 4 ARE DIS-
CHARGING THROUGH THE OUT-
LET PORT B.

DIAGRAM 2
THE PISTON HAS ADVANCED AND
SPACE 1, IN CONNECTION WITH
THE INLET PORT, HAS ENLARGED,
AND SPACE 2, IN CONNECTION
WITH THE OUTLET PORT, HAS
DECREASED, WHILE SPACES 3
AND 4, WHICH HAVE COMBINED,
ARE ABOUT TO MOVE INTO PO-
SITION TO DISCHARGE THROUGH
THE OUTLET PORT.

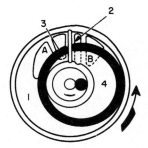

DIAGRAM 3
SPACE 1 IS STILL ADMITTING
LIQUID FROM THE INLET PORT
AND SPACE 3 IS JUST OPENING
UP AGAIN TO THE INLET PORT,
WHILE SPACES 2 AND 4 ARE DIS-
CHARGING THROUGH THE OUT-
LET PORT.

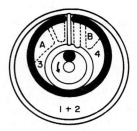

DIAGRAM 4
LIQUID IS BEING RECEIVED INTO
SPACE 3 AND DISCHARGED FROM
SPACE 4, WHILE SPACES 1 AND 2
HAVE COMBINED AND ARE ABOUT
TO BEGIN DISCHARGING AS PIS-
TON MOVES FORWARD AGAIN TO
OCCUPY POSITION AS SHOWN IN
DIAGRAM 1.

FIGURE 31   Principle of operation of oscillating-piston meter. (*Brooks Instrument Division, Emerson Electric Company.*)

Fluted Rotor Meter

Meters of this type are used in the flow measurement of crude and refined petroleum products and a variety of other commercial fluids. Frequently they are used on product loading
racks and small pipelines. Usually the meters are equipped with direct readouts and often
with ticket printers to provide authentically recorded documents for liquid-transfer transactions (Fig. 32).

Oval-Shaped Gear Flowmeters

In these meters, precision-matched oval-shaped gears are used as metering elements
(Fig. 33). Meter sizes (connections) range from ½ to 1, 1½, 2, 3, and 3 inches (~1.3 to 3.8, 5.1,
7.6, and 10.2 cm), with capacity ranges for different fluids as follows:

Cold water	0.2–1.4 to 110–705 gal/min (U.S.) 0.8–5.3 to 416.4–2668.7 L/min
Hot water	0.3–1.0 to 132–484 gal/min 1.1–3.8 to 499.7–1832.1 L/min
Liquefied petroleum gas (LPG)	0.4–1.6 to 176–837 gal/min 1.5–6.1 to 666.2–3168.4 L/min
Gasoline (0.3–0.7 cP)	0.3–1.6 to 132–837 gal/min 1.1–6.1 to 499.7–3168.4 L/min
Kerosene (0.7–1.8 cP)	0.2–1.6 to 110–837 gal/min 0.8–6.1 to 416.4–3168.4 L/min
Light oil (2–4 cP)	0.1–1.9 to 705–1010 gal/min 0.4–7.2 to 2668.7–3826.3 L/min
Heavy oil (5–300 cP)	0.04–1.9 to 44–1010 gal/min 0.2–7.2 to 166.6–3826.3 L/min

In meter application engineering, three basic viscosity classifications are taken into consideration: (1) standard-viscosity class from 0.2 to 200 cP, (2) medium-viscosity class from 300 to 500 cP, and (3) high-viscosity class above 500 cP. Tests have shown that a meter calibrated on a 1-cP product and then applied to a 100-cP product does not shift more than 1.2 percent above the initial calibration. Normally, where the viscosity is 100 cP or greater, there is no significant shift in accuracy. Oval-gear PD meters are sized for maximum flow so that the pressure drop is less than 15 psi (103 kPa).

Normal operating pressure ranges from 255 to 710 psi (1758 to 4895 kPa), depending on whether steel or stainless-steel flanges are used. The maximum operating temperature is 230°F (110°C), but special meters are available for higher temperatures. The lower limit is 0°F (−18°C). Housings, rotors, and shafts are constructed of type 316 stainless steel or type

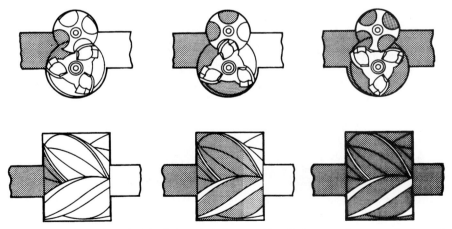

FIGURE 32 Operating principle of a Brooks *BiRotor* positive-displacement flowmeter. As the product enters the intake of the measuring unit chamber, the two rotors divide the product into precise segments of volume momentarily and then return these segments to the outlet of the measuring unit chamber. During what may be referred to as "liquid transition," the rotation of the two rotors is directly proportional to the liquid throughput. A gear train, located outside the measuring unit chamber, conveys mechanical rotation of the rotors to a mechanical or electronic register for totalization of liquid throughput. (*Brooks Instrument Division, Emerson Electric Company.*)

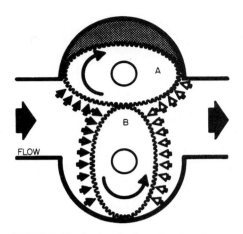

FIGURE 33 Sectional schematic of oval gear flowmeter, showing how a crescent-shaped gap captures the precise volume of liquid and carries it from inlet to outlet. (*Brooks Instrument Division, Emerson Electric Company.*)

0°F (−18°C). Housings, rotors, and shafts are constructed of type 316 stainless steel or type Alloy 20 (CN-7M stainless steel). Bushings are of hard carbon.

Meters are available with numerous accessories, including an electric-impulse contactor, which is used to open and close an electric circuit at intervals proportional to the number of units measured.

Other Positive-Displacement Meters

In the lobed-impeller meter shown in Fig. 34 two rotors revolve with a fixed relative position inside a cylindrical housing. The measuring chamber is formed by the wall of the cylinder and the surface of one-half of one rotor. When the rotor is in vertical position, a definite volume of fluid is contained in the measuring compartment. As the impeller turns, owing to the slight pressure differential between inlet and outlet ports, the measured volume is discharged through the bottom of the meter. This action takes place four times for a complete revolution, the impeller rotating in opposite directions and at a speed proportional to the volume of fluid flow. Meters of this design can handle from 8 gal/min (30 L/min) to 25,000 barrels/h. They perform well at temperatures up to 400°F (204°C) and pressures up to 1200 psi (8276 kPa). They are used for gases and a wide range of light to viscous fluids, including asphalts, and are best suited to high rates of flow.

In the sliding-vane rotary meter, vanes are moved radially as cam followers to form the measuring chamber (Fig. 35). Another version, the retracting-vane type, is shown in Fig. 36.

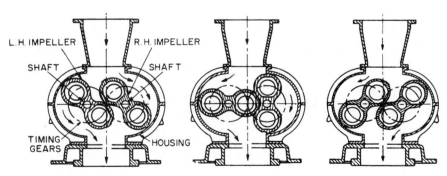

FIGURE 34 Lobed-impeller flowmeter.

OPEN-CHANNEL FLOW MEASUREMENTS

An open channel is any conduit in which a liquid, such as water or wastes, flows with a free surface. Immediately evident examples are rivers, canals, flumes, and other uncovered conduits.

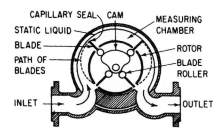

FIGURE 35   Sliding-vane rotary meter.

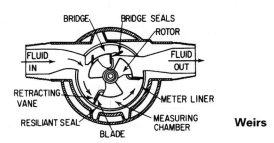

FIGURE 36   Retracting-vane rotary meter.

Certain closed conduits, such as sewers and tunnels, when flowing partially full (not under pressure), also may be classified as open channels. Two measurement units are used in measuring open-channel flows: (1) units of discharge and (2) units of volume. The discharge (or rate of flow) is defined as the volume of liquid that passes a certain reference section in a unit of time. This may be expressed, for example, as cubic feet per second, gallons per minute, or millions of gallons per day. In irrigation the units used are the acre-foot (defined as the amount of water required to cover 1 acre to a depth of 1 foot ($43{,}560 \text{ ft}^3$), or the hectare-meter ($10{,}000 \text{ m}^3$).

To determine open-channel flows, a calibrated device is inserted in the channel to relate the free-surface level of liquid directly to the discharge. These devices sometimes are called head-area meters and are used extensively for the allocation of irrigation water as well as in municipal sewage and industrial wastewater facilities.

Weirs

A weir is a barrier in an open channel over which liquid flows. The edge or surface over which the liquid flows is called the crest; the overflowing sheet of liquid is the nappe. The top of the weir has an opening, which may be of different dimensions and geometry (Fig. 37).

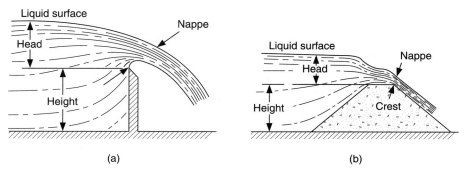

            (a)   (b)

FIGURE 37   Basic forms of weirs. (*a*) Sharp-crested weirs, which are useful only as a means of measuring flowing water. (*b*) Non-sharp-crested weirs, which are incorporated into hydraulic structures as control or regulating devices, with measurement of flow as a secondary function. When the weir has a sharp upstream edge so that the nappe springs clear of the crest, it is called a sharp-crested weir. The nappe, immediately after leaving a sharp-crested weir, suffers a contraction along the horizontal crest, called crest contraction. If the sides of the notch have sharp upstream edges, the nappe also is contracted in width and the weir is said to have end contractions. With sufficient side and bottom clearance dimensions of the notch, the nappe undergoes maximum crest and end contractions, and the weir is said to be fully contracted.

Of the various types of weirs, the three most commonly used are shown in Fig. 38. Weir calculations tend to be complex and reference to a source, such as *Stevens Hydrographic Data Book* (Leopolid & Stevens, Beaverton, Oregon—revised periodically), is suggested.

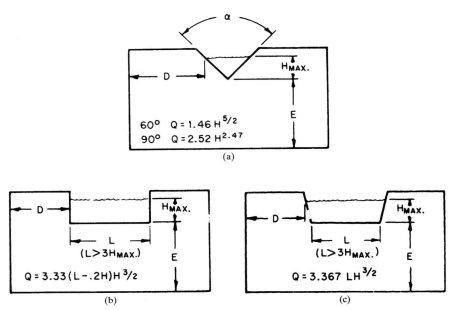

$$60° \quad Q = 1.46\,H^{5/2}$$
$$90° \quad Q = 2.52\,H^{2.47}$$
(a)

$$(L > 3H_{MAX.})$$
$$Q = 3.33\,(L - .2H)H^{3/2}$$
(b)

$$(L > 3H_{MAX.})$$
$$Q = 3.367\,LH^{3/2}$$
(c)

FIGURE 38 Weirs commonly used. (*a*) V-notch. (*b*) Rectangular. (*c*) Cipolletti. Occasionally, other weirs, such as hyperbolic (Sutro), broad-crested, and round-crested, are used.

Parshall Flumes

These devices are available for measuring open-channel flows up to 1500 million gal/day (5.7 million m³/day). The size of a flume is designated by the width W of the throat, which may range from 1 inch (2.5 cm) to 40 ft (12 meters) (Fig. 39). Generally economy of construction dictates that the smallest standard throat size be selected, provided it is consistent with the depth of the channel at maximum flow and permissible head loss. As a general rule, the throat size should be one-third to one-half the width of the channel. Prefabricated flumes often can be a cost advantage (Fig. 40).

Open-Flow Nozzle (Kennison)

The Kennison nozzle shown in Fig. 41 is an unusually simple device for measuring flows through partially filled pipes. The nozzle copes with low flows, wide flow ranges, and liquids containing suspended solids and debris. Because of its high-accuracy, nonclogging design and excellent head-versus-flow characteristics, the Kennison nozzle is well suited for the measurement of raw sewage, raw and digested sludge, final effluent, and trade wastes. Nozzle sizes range from 6 inches (15 cm) with a capacity of 90 gal/min (equivalent to 493 m³/day) up to 36 inches (91.4 cm) with a capacity of 14,000 gal/min (equivalent to 75,700 m³/day).

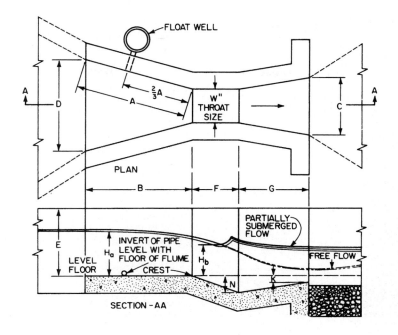

W	A	⅔A	B	C	D	E	F	P	K	N	R	T
1"	1'2 9⁄32"	9 17⁄32"	1'2"	3 21⁄32"	6 19⁄32"	1'8"	3"	2'1"	3⁄4"	1 1⁄8"	1 3⁄4"	5⁄16"
2"	1'4 3⁄16"	10 1⁄8"	1'4"	5 5⁄16"	8 13⁄32"	9"	4 1⁄2"	2'6 1⁄2"	7⁄8"	1 11⁄16"	1 3⁄4"	5⁄16"
3"	1'6 3⁄8"	1'1⁄4"	1'6"	7"	10 3⁄16"	2'0"	6"	3'0"	1"	2 1⁄4"	2 1⁄2"	5⁄16"
6"	2'7⁄16"	1'4 3⁄16"	2'0"	1'3 1⁄2"	1'3 5⁄8"	2'0"	1'0"	5'0"	3"	4 1⁄2"	2 1⁄2"	5⁄16"
9"	2'10 5⁄8"	1'11 1⁄8"	2'10"	1'3"	1'10 5⁄8"	2'6"	1'0"	5'4"	3"	4 1⁄2"	2 1⁄2"	5⁄16"
1'0"	4'6"	3'0"	4'4 7⁄8"	2'0"	2'9 1⁄4"	3'0"	2'0"	9'4 7⁄8"	3"	9"	2 1⁄2"	1⁄4"
1'6"	4'9"	3'2"	4'7 7⁄8"	2'6"	3'4 3⁄8"	3'0"	2'0"	9'7 7⁄8"	3"	9"	2 1⁄2"	1⁄4"
2'0"	5'0"	3'4"	4'10 7⁄8"	3'0"	3'11 1⁄2"	3'0"	2'0"	9'10 7⁄8"	3"	9"	2 1⁄2"	1⁄4"
3'0"	5'6"	3'8"	5'4 3⁄4"	4'0"	5'1 7⁄8"	3'0"	2'0"	10'4 3⁄4"	3"	9"	2 1⁄2"	5⁄16"
4'0"	6'0"	4'0"	5'10 5⁄8"	5'0"	6'4 1⁄4"	3'0"	2'0"	10'10 5⁄8"	3"	9"	2 1⁄2"	3⁄8"

FIGURE 39 Parshall flume.

BULK-SOLIDS FLOW MEASUREMENT

In many industries, for the control of inventory, maintaining throughput rates, and materials balances, and in batching for recipe control, it is essential to measure the flow of bulk solids carefully. Although with some very uniform solids, flow may be measured in terms of volume, and volumetric feeders are available, but with some sacrifice of accuracy, in gravimetric devices, measurements are made on the basis of weight. Here flow measurement is expressed in weight (pounds or kilograms) per unit of time (minutes or hours). Thus, in essence, a belt conveyor scale is a mass flowmeter. For many years, prior to the

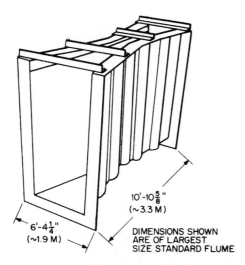

FIGURE 40   Rigid one-piece prefrabicated fiberglass Parshall flume. Advantages include smooth surface, dimensional accuracy and stability, with no external bracing needed, long life, and reduction of installation costs. Dimensions shown are for available large size.

FIGURE 41   Kennison nozzle. (*Leeds & Northrup, BIF Products.*)

introduction of strain-gage load cells and high-technology sensors, conveyor belts were coupled to mechanical weigh scales.

Belt-Conveyor Weighers

Load-weighing devices can be installed on practically any size length and shape of conveyor—from 12-inch (30.5-cm)-wide belts, handling as little as 1 lb (0.45 kg)/min at a minimum belt speed of 1 ft (30.5 cm), to 120-inch (3-meter)-wide belts, handling as much as 20,000 tons (18,145 metric tons)/hour at speeds of up to 1000 ft/min (305 m/min). The basic design of a belt conveyor weigher is shown in Fig. 42.

Wherever possible, the scale should be installed where it will not be subjected to conveyor influences that cause a "lifting effect" of the belt off the weigh-platform section. Belt tensions and belt lift have more influence on accuracy than any other factor. It is important that the material (load) travel at the same velocity as the belt when crossing the weigh platform and not be in a turbulent state, as it is when leaving a feed hopper or chute. The best test for accuracy involves carefully weighing a batch of material and passing this exact weight several times over the belt scale. Another testing method is using a chain that is very uniform and matches closely the weight of the solid material normally weighed. Some manufacturers guarantee their product to weigh within ±0.5 percent of the totalized weight (when properly installed). The National Institute of Standards and Technology (NIST) has developed a standard certification code for conveyor scales.

In proportioning various solids and liquids, belt feeders and flow controllers, including controlled-volume pumps, can be integrated into a total digitized blending system. Microprocessor technology has been available for continuous weighing for several years. The microprocessor provides accuracy, automation, and flexibility features not feasible with previous analog and digital systems. The basic difference is that the microprocessor does things sequentially, using its memory and an instruction set (software), whereas analog and digital systems accomplish these tasks continuously by way of various electrical circuits (hardware).

In addition to belt-type weighers, the basic principle can be applied to screw feeders, sometimes used for very dry materials that tend to flush and flood. So-called solids flowmeters are also available. These are not true gravimetric devices, but rather, utilize impact or centrifugal force to generate a signal that is proportional to material flow. These devices are capable of providing long-term repeatability in the 2 to 3 percent range (Fig. 43). In another gravimetric type (loss-in-weight feeder), a weigh hopper and metering screw are supported on load cells. As the metering screw delivers material from the system, the resultant load-cell signal decreases proportionately with time. The differential dG/dt produces the true feed rate F, as is shown graphically in Fig. 44.

Nuclear radiation methods also have been applied to weighing materials on a moving belt.

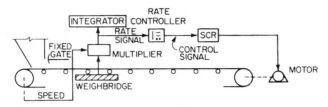

FIGURE 42 Gravimetric weigh-belt feeder with true rate control system where there are no inferred constants. Weight × speed = flow rate.

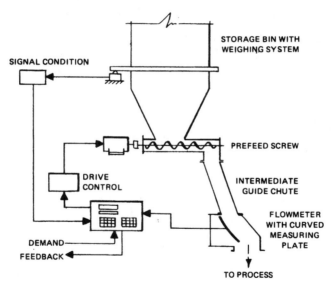

FIGURE 43 Flow-rate control system utilizing a solids flowmeter with an in-line calibration system. (*Schenck Weighing Systems.*)

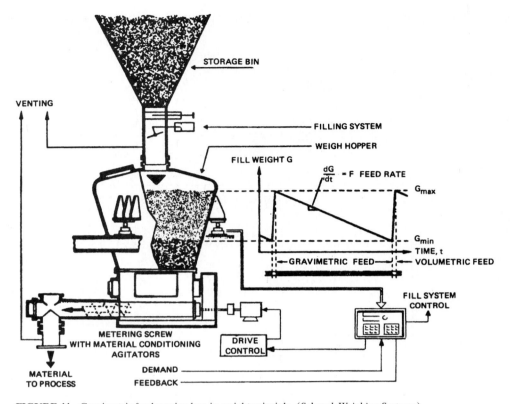

FIGURE 44   Gravimetric feeder using loss-in-weight principle. (*Schenck Weighing Systems.*)

FLUID LEVEL SYSTEMS

Measuring and, frequently, controlling the level of materials (liquids and flowable bulk solids) contained in storage and processing vessels, such as tanks, wells, reservoirs, ponds, bins, and hoppers, is one of the most common procedures of industrial instrumentation. There is a great diversity of liquids, bulk, and fluidized solids which require level measurement, ranging from pure, clean water to viscous, sticky, and corrosive and abrasive fluids and slurries, and from free-flowing, dry crystals to moist, lumpy solids. The processing environments for level sensors extend from vacuum to high-pressure service, and from subzero to elevated temperatures. With the possible exception of chemical composition measurement, the available means for measuring level exceed, in both number and diversity, those of any other process variable. This is reflected by the large number of suppliers that serve this field.

IMPORTANCE OF LEVEL MEASUREMENT AND CONTROL

In many processes involving liquids contained in vessels, such as distillation columns, reboilers, evaporators, crystallizers, and mixing tanks, the particular level of liquid in each vessel can be of great importance in process operation. A level which is too high, for example, may upset reaction equilibria, cause damage to equipment, or result in spillage of valuable or hazardous material. A level that is too low may have equally bad consequences. Combined with such basic considerations, there is the advantage in continuous processing of reducing storage capacity throughout the process. This reduces the initial cost of equipment; but less storage capacity accentuates the need for sensitive and accurate level control. Effective measurement and control of level usually can be justified in terms of economy and safety. To the operator, knowledge of this variable provides data on (1) the quantity of raw material available for processing, (2) the available storage capacity for products in process, and (3) satisfactory or unsatisfactory operation of the process.

Control of Level at One Point.   In many process applications, the level must be maintained accurately at a predetermined height, irrespective of load conditions of the process. For example, in a steam or vapor generator, such as a boiler, it is desired to maintain the level at a predetermined value in order that two sets of conditions will be present at all times, regardless of the output from the generator:

1. The quantity of liquid inventory in the vessel must be maintained in order to provide feed for the evaporation process.
2. The vapor volume space must be maintained in order to have available storage capacity for the vapor, plus a volume space that will prevent carryover of entrained liquids in the vapor.

In continuous processes a correct level head in certain equipment is of marked importance. In evaporators, for example, the heating medium may be inside a tube bundle which must at all times be covered to an optimum depth, thus requiring level control. Too low a level will uncover the heating surface, lowering the efficiency of the process, while also causing overheating and inherent tube damage. Too high a level, in contrast, will require a greater heat input as the head pressure increases, lowering efficiency while increasing the production of "off-quality" materials.

Averaging Level Control.   Continuous processes may incorporate accumulators or storage (buffer) vessels in line to provide storage (inventory). Process upsets or disturbances are absorbed in such accumulators without passage down-line, which will result in better overall control. In some process applications, control of the level at a constant height is not always desirable. It is more important that the outflow of the storage vessel does not change suddenly, allowing an upset to pass along to a subsequent process stage. Any sudden increase in input to the storage vessel should be absorbed in the vessel. To accomplish this, "averaging" liquid-level control can be used, where a wideband proportional-plus-reset mode of control can be incorporated in the level controller. With modern electronic circuitry this is easier to accomplish than with the controllers available a decade or so ago.
In principle, with this type of control, if the uncontrolled input suddenly increases, the wide proportional band permits the level to rise temporarily, with little change in the controller output regulating outflow from the vessel. If the input remains at its higher value for a period of time, the automatic reset functions to return the level to the set point, and gradually

changes the outflow rate. Conversely, if the uncontrolled input suddenly decreases, the level is permitted to drop; the outflow is not similarly affected, but is changed at a gradual rate if the decreased input continues.

The size of the vessel required becomes a function of (1) the maximum process upset to be expected in the system and to be absorbed by the accumulator, (2) the duration of the upset, and (3) the elapsed time allowed before the full value of the continued upset is to be passed along to the next stage of the process. The term "holding time" is often used for the function just described. The size of the vessel also may be limited by physical height considerations, by the change in level height permitted, and by the economics of vessel cost. All these factors become variables for design and instrument engineers in determining the size of vessels and accumulators, and such calculations make a good case for applying modern process simulation techniques.

Reduction of Vessel Size.   In simple single-capacity systems, the utilization of level control can be advantageous in keeping the capacity of processing vessels within practical limits. If level measurement and control are used, the size of a mixing or reaction vessel may be small. It is not necessary that large vessels be used to handle all available liquid to be processed, since the liquid-level controller will feed only the fluid required to keep the liquid concentration or its height at a predetermined level. Large, bulky vessels can be eliminated, with accompanying economy. This also means that at any instant only a small amount of process material is under reaction or in process, thus reducing hazards, potential losses, or spoilage.

Protection of Centrifugal Pumps.   Where it is desired to maintain a head pressure against the suction of a centrifugal pump, the level of the liquid in the storage tank must be maintained at an optimum value. If the level drops too low, flashing and cavitation will occur in the pump suction, with resultant erratic pump discharge and extreme wear on pump impellers. If the level rises too high, there may be a loss of accumulator volume in the vessel, thereby affecting the process from an operating viewpoint.

Product Quality Control.   Warp sizing in the textile industry is a good example of how close the control of liquid level directly affects product quality. The warp yarn is run through a bath of sizing solution which adds a protective coating to the yarn. The amount of size absorbed by the yarn is a function of the time during which the yarn is in contact with the size solution. As the yarn passes through the solution in a prescribed path (usually around a large cylinder rotating at a fixed number of revolutions per minute), the time of contact is a function of the level height of the size solution. A variation in solution level, therefore, will change this contact time and thus destroy warp uniformity, later causing breakage of threads on the loom.

Cost Accounting.   The flowmeter, weighing scale, and liquid-level gage are the process cost accountant's main tools for obtaining data concerning quantities of raw materials, in-process materials, and finished products. This is an important element in the overall statistical process control (SPC) procedures currently stressed so much in the literature. Many contemporary level transmitters make it possible to connect acquired data into simple as well as the most sophisticated computing and networked systems that exist in a computer-integrated manufacturing (CIM) environment.

LEVEL SYSTEM MATHEMATICS

Important to the engineer who designs and specifies liquid-level instrumentation systems are such factors as (1) the relationship of flow to level in vessels and (2) the capacity versus level height in variously shaped vessels.

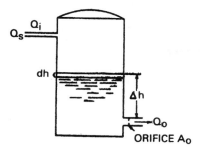

FIGURE 1 Basic flow versus level factors.

Constant Head for Steady Process Flow. Often steady process flow, such as the introduction of a raw material to a process, is maintained by holding a constant head pressure on the feed line. This can be achieved by control of the liquid level in the feed tank whose feed line exits from the bottom of the tank, as shown in Fig. 1.

As regards the theory of this arrangement, in the free flow of liquid through an opening or orifice, or across a weir plate, the quantity of liquid discharged is a function of the level height above the orifice or weir. Briefly, the flow is governed by the equation

$$Q = CA\sqrt{2gH}$$

where
Q = quantity, ft^3/s
C = orifice constant
A = area of flow, ft^2
g = acceleration of gravity, = 32.2 ft/s^2
H = height of liquid, ft

Although U.S. customary units are used, the relationship, of course, also applies for appropriate metric units.

Thus, with the area of flow A as well as the other factors constant, steady process flow Q will be obtained if the head of liquid H above the orifice or weir is accurately measured and controlled.

Relationship of Flow to Level in Process Vessels. In designing process equipment and determining the requirements of liquid-level controlling equipment, an engineer often must calculate in advance how a change in liquid inflow will affect the level in the vessel. In any given system of vessels, a stable quantity of inflow to the vessel equals the outflow. With a stable inflow, the height of the level H above the drawoff increases until the head pressure developed causes flow through orifice A by an amount Q_o which is equal to the inflow Q_i.

From the above equation and referring to Fig. 1, this relation may be expressed by

$$Q_i = Q_o = 0.897\, CA_o\sqrt{2gH}$$

where
Q_i = flow rate into vessel, gal/min
Q_o = flow rate out of vessel, gal/min
C = orifice constant or coefficient
A_o = orifice area, in^2
g = acceleration of gravity, = 32.2 ft/s^2
H = height of liquid level above top of outlet orifice, inches

Although U.S. customary units are used, the relationship, of course, also applies for appropriate metric units.

Variation in the inflow or in the orifice area, with the other variables remaining constant, causes a level change. A control valve substituted for the fixed orifice provides an easy means of varying the orifice area. If the orifice area is increased, the level will fall until, at the new area, it stabilizes at a point where its head effect on the orifice causes outflow to equal inflow.

If the orifice area is decreased, the level will rise until the product of a smaller orifice value and a larger head effect causes outflow to equal inflow. The rate at which the level rises or falls can be expressed by the following relationships. The standard flow formula is

$$Q_o = 0.897 CA_o\sqrt{2gH}$$

The height change, in units per minute, is given by

$$\frac{dH}{dt} = (Q_i - Q_o)\frac{231}{A}$$

$$= (Q_i - Q_o)\frac{231}{0.785d^2}$$

$$= \frac{294}{d^2}(Q_i - Q_o)$$

$$= \frac{294}{d^2}(Q_i - CA_o\sqrt{2gH})$$

$$= \frac{294}{d^2}(Q_i - 7.20CA_o\sqrt{H})$$

where dH = change in liquid-level height, inches
dH/dt = rate of change in liquid-level height, in/min
A = transverse tank area, in^2
d = diameter of tank (cylindrical), inches

Although U.S. customary units are used, the relationship, of course, also applies for appropriate metric units. Where U.S. customary units are used, it should be noted that 1 gal = 231 in^3 and 1 psi = 27.7 inches of water.

To find height H_2 at a time t_2, when Q_i is initially equal to Q_o and is instantaneously changing at time t_1, use the height-change equation,

$$\frac{dH}{dt} = \frac{294}{d^2}(Q_i - 7.20CA_o\sqrt{H})$$

$$dt = \frac{dH}{(204/d^2)Q_i - (294/d^2)7.20CA_o\sqrt{H}}$$

For easy manipulation, let

$$B = \frac{294}{d^2}\,7.20CA_o \qquad \text{and} \qquad D = \frac{294}{d^2}Q_i$$

Then $$dt = \frac{dH}{D - B\sqrt{H}}$$

and $$dt\int_{t_1}^{t_2} = \int_{H_1}^{H_2}\frac{dH}{D - B\sqrt{H}}$$

From a table of integrals after transformation into standard form,

$$2\left[-\frac{\sqrt{H}}{B} - \frac{D}{B^2}\log_e(D - B\sqrt{H})\right]_{H_1}^{H_2} = t_2 - t_1$$

or $$\frac{2}{B^2}\left[\sqrt{HB} + D\log_e(D - B\sqrt{H})\right]_{H_2}^{H_1} = t_2 - t_1$$

It should be noted that if there is a change in inflow Q_i, the resultant liquid-level height H_2 may be calculated when outflow Q_o again equals inflow by use of the standard flow formula given previously. By use of the last equations given, the amount of time required for the liquid level to reach any height between H_1 and H_2 can be calculated. By substituting a figure for t_2 in the equation, the change in height can also be calculated for any given change in time. Furthermore, if an uncontrolled flow is introduced into a vessel and a level controller is placed on the outflow, the rate of flow and level changes can be calculated.

Capacity Versus Level Height in Various Vessels. Many processing and storage vessels where liquid-level measurement is a factor are shaped cylindrically and mounted vertically on end. Thus the content for any level height can be simply calculated by

$$\text{Content} = \text{cross-sectional area} \times \text{level height}$$

For example, in U.S. customary units, $\text{ft}^3 = \pi r^2 h$, where r is the inside radius of the tank and h the level height, both in feet.

This relationship becomes more complex where cylindrical tanks are mounted horizontally. This is best illustrated by calculating a typical problem, the dimensions of which are in U.S. customary units and illustrated in Fig. 2.

Spherical Tanks. Useful relationships for determining the volume of a sector of a sphere are given in Fig. 3.

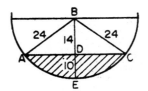

FIGURE 2 Dimensions used in calculating the capacities of horizontal cylindrical tanks. *Example:* Determine the volume in a cylindrical tank (flat ends) mounted horizontally, with the following factors known: diameter of tank 48 inches, depth of liquid 10 inches, and length of tank 120 inches. Then

$$\text{Area } ACE \text{ (shaded portion)} = \text{area } ABCE - \text{area } ABC$$

and

$$\text{Area } ABCE = \left(\frac{2\angle ABD}{360}\right) \times \text{area of circle}$$

$\angle ABD$ is found from its cosine, which is $^{14}\!\!/_{24}$. Hence,

$$\angle ABD = 54.25°$$

$$\text{Area } ABC = 14 \times 24 \times \sin ABD$$

or

$$14 \times 24 \times \sin 54.25 = 14 \times 24 \times 0.8116 = 272.7$$

$$\text{Area } ABCE = \frac{108.50}{360} \times \pi \times (24)^2 = 545.4$$

Hence,

$$\text{Area } ACE = 545.4 - 272.7 = 272.7 \text{ in}^2$$

The volume in U.S. gallons per foot of length is

$$\frac{272.7 \times 12}{231} = 14.17 \text{ gal/ft}$$

and hence,

$$\text{Total volume} = \frac{14.17 \text{ gal} \times 120}{12} = 141.7 \text{ gal}$$

FIGURE 3   Volume of partially filled spherical tank equals one-half the total volume of the sphere (if the tank is 50% full) plus the volume of the spherical sector filled with liquid above the midline of the tank. If the tank is less than half full, the volume is that of the spherical sector filled with liquid above the bottom of the tank. Volume of a sphere = $\frac{4}{3} \times \pi \times$ radius^3. With reference to the diagram, volume of spherical sector = $\frac{1}{6} \times \pi \times h \times (3a^2 + 3a_i^2 + h^2)$.

Volume and Weight Measurement from Level. Liquid-level measurement resolves itself into position measurement, namely, the position (height) of a liquid surface above a datum line. Measurement, however, need not always be expressed in terms of inches, feet, or meters above the datum line. With a knowledge of the dimensional and contour characteristics of the containing vessel, it can be conveniently interpreted (hence, calibrated) in terms of the volume of liquid contained; and further, with information concerning the specific gravity of the liquid, it can be expressed in terms of the weight of the liquid in the vessel.

Volume Determination.   If the purpose of level measurement is to determine the volume of liquid contained in the vessel, then direct measurement of level height is preferable because

$$V = A \times H$$

where V = volume of vessel
   A = area of vessel
   H = height of level

Thus the volume measured is independent of liquid density.

If measurement of the pressure due to hydrostatic head must be used because of specific requirements, volume is determined by the relation

$$V = \frac{A \times P}{D}$$

where V = volume in vessel at given level
   A = area of vessel
   P = pressure due to hydrostatic head
   D = density of liquid in vessel

In this case the volume measurement depends on the density of the liquid.

Weight Determination.   If the purpose of level measurement is to determine the weight of the liquid in the tank, then measurement of the pressure due to hydrostatic head has advantages because

$$W = H \times D \times A = A \times P$$

where W = weight of liquid in vessel
   H = height of level
   D = density of liquid
   A = area of vessel
   P = pressure due to hydrostatic head

Thus the measurement is independent of liquid density.

If direct measurement of level height is used, then

$$W = A \times D \times H$$

and the weight measurement depends on knowing the density of the liquid.

Measurement Errors in Storage Tanks. Where very accurate measurements of the liquid contents of storage tanks are needed, it is the usual practice to correct the fluid density for temperature changes and neglect the other temperature and pressure effects. Generally this is an adequate approach, but it is reassuring to know the magnitude of the uncorrected errors. The following equations allow measurements to be easily and completely corrected to any desired reference conditions.

It is assumed that the true volume of a container, empty and at its reference temperature, is known accurately. Also, it is assumed that the measuring element is without error. Under these ideal conditions, the remaining errors are:

1. The tank may not be perfectly round (if cylindrical) or geometrically true (if some other shape).
2. The tank wall is stretched, owing to hydrostatic pressure.
3. The fluid is not at its reference temperature.

Corrections for each of these unavoidable conditions are presented as relatively simple, dimensionless expressions in which any consistent data can be substituted. The results are expressed as relative errors *e*, defined as

$$e = \frac{\text{observed value} - \text{true value}}{\text{true value}}$$

Errors due to tank "out-of-round" and pressure expansion are illustrated in Figs. 4 and 5.

Some level measuring systems have designed-in corrections for temperature variations. With the availability of microprocessors, signal conditioners, and so on, a level system may be as sophisticated as one would desire. With increasing emphasis on statistical process control, the adaption of a level system to the ultimate accuracy desired is a major procurement decision.

Examples of temperature and pressure corrections that can be programmed automatically into a level system are given below.

FIGURE 4 Error due to out-of-round tank. *Example:* Assume that the tank is measured by strapping and that the area is calculated from perimeter *P*,

$$A' = \frac{P^2}{4\pi} = \text{calculated area}$$

$$A = \pi ab = \text{true area}$$

$$e = 0.298 \left(\frac{a}{b} - 1 \right)^{1.952}$$

where *a* and *b* are measured dimensions.

FIGURE 5 Error due to pressure expansion. *Example:*

$$e = \frac{\pi S}{E} \times \frac{H_1^2 - H_0^2}{H_1(H_1 - H_0)}$$

and

$$S = \frac{6PD}{t}$$

where H_0 = pressure above liquid
H_1 = pressure at bottom of liquid
S = average wall stress
P = average pressure
E = modulus of elasticity
D = tank diameter
t = wall thickness

For open tanks, $H_0 = 0$ and $e = \pi S/E$.

Example 1.   A tank 20 ft in diameter contains 30 ft of liquid of specific gravity 1.7. If the wall is ⅜-inch steel plate, what is the error due to pressure?

$$P = 30 \times 1.7 \times 0.433 = 22 \text{ psi}$$

$$S_{max} = \frac{6PD}{t} = \frac{6 \times 22 \times 20}{^3/_8} = 7040 \text{ psi}$$

$$e = \frac{\pi S}{E} = \frac{\pi \times 7040}{30,000,000} = 0.00074 = 0.074\%$$

Example 2.   A manometer reads the pressure at the bottom of a steel tank. What error is introduced if the water temperature is 80°F instead of 25°F?

In weight units of calibration,

$$e = -2gT = -2(80 - 25) \times 0.0000132 = -0.00145 = -0.145\%$$

In volumetric units of calibration,

$$e = -T(f + 2g) = -55(0.000207 + 0.000026) = -0.0128 = -1.28\%$$

Example 3.   A tank has a maximum diameter of 10 ft 5 in and a minimum diameter of 10 ft 4 in. What error is introduced by assuming it to be a circle of the same perimeter?

$$\frac{a}{b} = \frac{125 \text{ in}}{124 \text{ in}} = 1.0161$$

$$e = 0.298 \times 0.0161^{1.952} = 0.000097 = 0.01\%$$

LEVEL MEASUREMENT TECHNIQUES

Point-Contact Methods

Centuries ago, humans controlled the water level by building dams (bulwarks). A further step led to leaving an opening in the wall and filling it with logs or timbers which could be manually changed in number to control the height of water in a reservoir. Later, conveniences such as valves, sluice gates, and coffer dams were added to the system. The need for point-level measurements probably arose when liquids of value, such as oil, were stored in vessels, precipitating the need for inventory control.

In early reservoirs, a notched stick served as a crude level measuring device, the gager noting the interface line between a wetted and a dry stick. With the millions of dipsticks used in transportation vehicles and stationary engines, this unsophisticated tool appears to be the most common form of level measurement despite its very early roots.

An early refinement was the hook gage shown in Fig. 6a. The hook gage point is moved up from beneath until it just pierces the liquid surface. The scale is usually engraved on a square brass rod. Another early device was the plumb bob or plummet gage shown in Fig. 6b. This gage is lowered from above until a depression shows contact with the liquid surface. A variation of these point-contact devices is the bronze or steel tape, used with a plumb bob gage, as shown in Fig. 6c. For some applications, these remain inexpensive, reliable, and unsophisticated approaches, where convenience and accuracy are not major considerations.

Over the years, particularly since the advent of solid-state electronics a few decades ago, these basic approaches have been modernized and greatly improved in terms of convenience and absence of human error, as well as permitting centralized measurement of numerous vessels at remote locations. One system, for example, consists of a plummet, perforated tape,

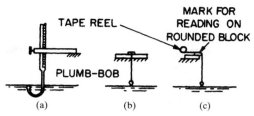

FIGURE 6 Point-contact measuring devices (*a*) Hook gage. (*b*) Plumb bob gage. (*c*) Tape and plumb bob gage.

measuring sprocket, counter, take-up reel, drive motor, and detecting and timing circuits. As the level changes, the drive motor moves the plummet to correspond to the level. The tape drive turns a measuring sprocket shaft and counter and a synchro indicating the level. The counter, reading in inches (or centimeters), tenths, and hundredths, provides a local readout for checking and calibration purposes. The synchro provides a transmission signal. Tape and plummet systems are also used to measure the depth of sludge in clarifying or settling tanks. In one such device, high-intensity infrared beams illuminate a phototransistor detector until such time that sludge blocks the probe gap. Other kinds of transducers can become a part of the plummet and cause a change in the signal when the plummet engages a liquid or solid surface.

Visual Methods

The dipstick requires visual reading of a mark on a scale. Visual methods also are represented by the gage glass. A desire for a clear indicating scale and a more convenient location for the scale inspired the introduction of a simple open-end manometer or tube, as shown in Fig. 7. The liquid height in the tube equalizes with the level in the vessel and, by means of a transparent tube material such as glass, allows the level in the vessel to be read at any outside point. Gage glasses for pressure and vacuum vessels are shown in Fig. 8. Capillary attraction in small-diameter tubing causes water, for example, to rise an extra height of about 0.046/*d,* where *d* is the internal diameter of the tube in inches; this extra height due to capillarity would be over ¼ inch (≈6.4 mm) for a ⅛-inch (≈3.2-mm) diameter tube. Larger-diameter tubes are therefore normally used in gage glasses.

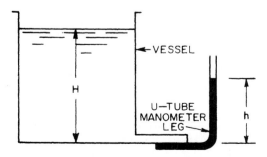

FIGURE 7 Schematic diagram of U-tube manometer for open vessels.

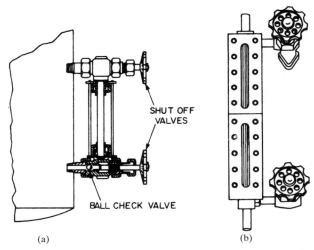

(a)   (b)

FIGURE 8   Gage glasses. (*a*) Detail of typical gage column. (*b*) Reflex type. Gage glass *a* can be considered a manometer in which the level in the glass seeks the same position as the level within the vessel. These simple instruments remain popular where local direct reading of level is satisfactory and where the application falls within their limitations.

Gage glasses are used with transparent tubes made of glass or plastic with sufficient strength to withstand the pressure in the vessel. To this end, tubular gage glasses are generally limited to 150 lb (1.04 MPa) at 400°F (204°C) service, although they are available for service at 450 lb (3.1 MPa) at 450°F (232°C). In the type illustrated, the liquid chamber is commonly of steel. Reflex gage *b* is of the prismatic or reflex type, in which the inside face of the glass is provided with 90° prisms running lengthwise, but still retaining a suitable gasket bearing surface. Rays of light normal to the face of the glass strike the prisms at an angle of 45°. If no liquid is in contact with the prisms, total reflection takes place, since the critical angle for a ray passing from glass to air is 42°; thus the gage appears silvery white. The critical angle of a ray passing from glass into water is 62°; thus the light ray passes on through the water, making it possible to see the inside of the chamber. The inside is usually painted black, so the viewer sees the liquid portion in the gage as black, though the liquid may be colorless. The reflex type does not permit observation of the color of a liquid or an interface between two different liquids. Variations of this basic design include models for (1) low-temperature liquids which require a transparent frost-preventing unit around the vision slot, (2) acids and other corrosive liquids which require a resistant lining or coating of all chamber surfaces in contact with the liquid and (3) Provision for circulating a heat-transfer medium around the liquid chamber of the gage for heating viscous liquids to enhance their flowability, and for cooling high-temperature liquids to avoid any tendency to boil. Where it is desired to note the color, characteristics, or interface of a liquid, a transparent-type gage, similar in construction to the reflex type, can be used. In this design heavy sight glasses are placed on opposite sides of the liquid chamber; neither glass has the prism on one side.

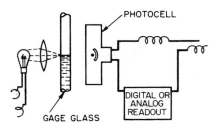

FIGURE 9 Gage glass with photocell detector.

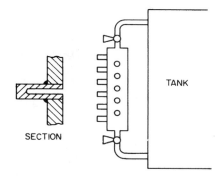

FIGURE 10 Frost-plug level gage. This method is usable for level detection where the liquid is below 32°F (0°C), such as propane and butane. With sufficient moisture in the air, frost will form on all plugs located below the surface level of the liquid.

A gage glass with a photoelectric readout is shown in Fig. 9. The scheme permits analog or digital displays. A frost-plug level detector is shown in Fig. 10.

Dip Tube. This is often used in liquefied petroleum gas (LPG) and anhydrous ammonia tanks under pressure. The dip tube is installed through a pressure-tight fitting in the vessel. The outer end of the tube has an orifice to restrict flow and a shutoff valve which, when opened, allows observation of the state of the discharged fluid, whether liquid or vapor, thereby indicating whether the end of the dip tube is submerged in liquid or vapor. A variation of this device consists of a dip tube through a packing box in the side of the vessel. The inner end of the tube is bent at an angle so that a half-rotation of the tube in the packing box will traverse the inner end of the tube through the vertical range of liquid level. The angular position of the dip tube as it starts to discharge liquid indicates the liquid level.

Buoyancy Methods

Archimedes' principle states that the resultant pressure of a fluid on a body immersed in it acts vertically upward through the center of gravity of the displaced fluid and is equal to the weight of the fluid displaced. With reference to Fig. 11, it will be noted that by measuring the difference in weight of a partially submerged element at various degrees of submergence, the level of the liquid in which the displacer element is submerged can be determined. This level measurement system has been used for decades and still creates some demand, particularly where pneumatic systems are preferred (Fig. 12).

Float Methods

These designs use the principle of a buoyant member which floats on the surface of the liquid and changes position as the liquid level varies. In a ball-float design, a physical member (hollow ball) floats on a liquid and is referred to as a datum point, provides a direct means of level measurement. A plastic or coated metal ball usually serves as the float. The ball is attached to a rod (Fig. 13). In some designs a rotary shaft operating in a bearing, trunnion, or packing gland, with a pointer and scale, may be used. Although this design has been used for about a

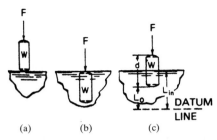

FIGURE 11   Relation between level position and displacer element. (*a*) Level position at bottom or below displacer, where

$$F = W$$

where F = force or weight to be supported
W = weight of displacer

(*b*) Level position at top of displacer, where

$$F = W - \frac{V \times G}{D}$$

where V = volume of displacer
G = specific gravity of liquid at reference temperature [usually 60°F (15.6°C)]
D = density of fluid

In the case where V is given in cubic inches, F and W are in pounds and G is in lb/in^3. For water G = 62.4 lb/ft^3 or 1/27.7 lb/in^3. The relationship, of course, also applies where appropriate metric units are used.
(*c*) Level position in intermediate position, where

$$F = W - \frac{V \times G}{D} \times \frac{L_{in} - L_0}{d}$$

where L_{in} = level position in intermediate position
L_0 = level position at bottom or below displacer
d = length of displacer

In the U.S. customary units given for c, L_{in}, L_0, and d are in inches.
With a cylindrical displacer, F varies as the level position around the displacer varies. The value of F is measured by suitable means, such as a torsion spring, pneumatic force balance, or electronic transducer. Thus this value becomes a function of the level position above the datum line.

century, a demand remains for certain kinds of industrial applications, particularly where cost is a very significant factor. [Also, it is interesting to note that even the most sophisticated facility featuring a computer integrated manufacturing (CIM) environment most likely will use more ball floats than any other level controllers, just for operating the toilets!]

Force-balance displacer methods have been used for decades and still are used for installations where cost and ruggedness may be major factors (Fig. 14).

Hydrostatic Pressure Methods

Level measurement involving the principles of hydrostatics have been available for many years. These gages have taken numerous forms, including the diaphragm-box system, the air-

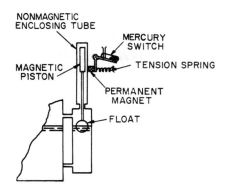

FIGURE 12 Principle of magnetically operated float switch. Switches of this general type still are used for relatively nonsophisticated applications at a reasonable cost.

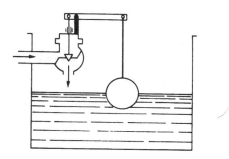

FIGURE 13 Schematic diagram of mechanically operated float valve.

bubble tube or purge system, hydrostatic differential-pressure meters, the dry-type differential-pressure meter, and the force-balance diaphragm system. Just a few years ago, for a short while, it appeared that hydrostatic level systems would gradually fade into history because of the development of so many other quite satisfactory level systems. However, with the development of the present generation of "smart" and superior hydrostatic tank gaging (HTG) transmitters, which incorporate all types of modern electronic circuitry, computing, self-diagnostics, and flexibility for connecting into data acquisition networks, the hydrostatic pressure methodology is enjoying a resurgence of demand. A 1990 survey indicates that nearly 30 percent of all level systems installed are of the modernized HTG type.

By way of background, hydrostatic head may be defined as the weight of liquid existing above a reference or datum line; it can be expressed in various units, such as pounds per square inch (psi), grams per square centimeter, and feet of meters of liquid measured. The head is a real force, due to liquid weight, and, as shown in Fig. 15, is exerted equally in all directions. It is independent of the volume of liquid involved or the shape of the containing vessel. Measurement above the datum line may be expressed by the relationship

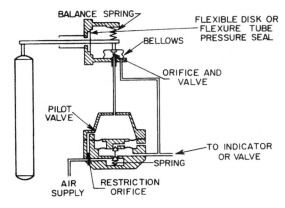

FIGURE 14 Schematic diagram of force-balance displacer unit with flexible-disk pressure seal. Once very popular, there remains a limited demand for this unit.

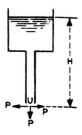

FIGURE 15   Basic elements of hydrostatic head.

$$\frac{P}{D_wG} = \frac{P \times M}{G}$$

where H = height of liquid above datum line
P = pressure due to liquid head
D = density of liquid at operating temperature
D_W = density of water at a reference temperature
G = specific gravity of liquid at operating temperature
M = multiplying factor, depending on units of measurement used. For example, if H is in inches, P is in psi, D is in lb/in^3, G is in 62.4 lb/ft^3 (0.036 lb/in^3), and the reference temperature is 60°F (15.6°C), $M = 27.70$.

From this relationship it is seen that a measurement of pressure P at the datum or reference point in a vessel provides a measure of the height of the liquid above that point, provided the density or specific gravity of the liquid is known. Also, this relationship shows that changes in the specific gravity of the liquid will affect liquid-level measurements by this method, unless corrections are made for such changes. During the past relatively few years, compensation for environmental changes which affect measurement accuracy has been automated in some systems through the use of microprocessors and sensors that continuously or intermittently detect changes in such factors as liquid temperature or density.

When a pressure greater than atmospheric is imposed on the surface of the liquid in a closed vessel, this pressure adds to the pressure due to the hydrostatic head and must be compensated for by a pressure measuring device which records the liquid level in terms of pressure.

Contemporary Hydrostatic Tank Gaging.   In the late 1980s pressure transmitters and microprocessor technology enabled the development of a "smart" or "intelligent" tank gaging system for small and large tank farms. The components of a complete system are shown in Fig. 16. An exploded view of one type of hydrostatic pressure transmitter is presented in Fig. 17; a block diagram is given in Fig. 18. The internal working of pressure sensors varies with the particular supplier. Other types are described under "Fluid Pressure Systems" earlier in this handbook section. The RTD temperature sensors are described under "Temperature Systems," also in this handbook section.

Radio-Frequency Level Measurement[1]

Radio frequency (RF) can be used to monitor the level of media in a vessel. Basic elements of an RF system include (1) the sensing probe or electrode, (2) the instrument electronics, and (3) a reference ground. An RF system can be used for point-level detection with on-off control and for continuous-level measurement in a wide variety of process medium applications. The technology is based on the simple concept of a capacitor which has an active or positive plate, a negative or ground plate, and material between the plates (Fig. 19).

The dielectric constant is a numerical value on a scale of 1:100, which relates to the ability of the dielectric (material between the plates) to store an electrostatic charge. The dielectric constant of a material is determined in an actual test cell. Values for many materials are pub-

[1] John Masters, Great Lakes Instruments, Inc., Milwaukee, Wisconsin.

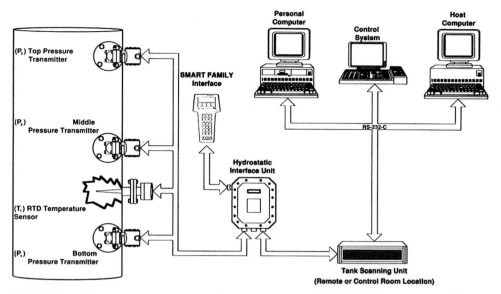

FIGURE 16 Hydrostatic tank gaging system. The hydrostatic pressure gages (HPGs) incorporate a high-accuracy free-floating sensor. The transmitters are specially characterized over a temperature range of –40 to 185°F (–40 to 85°C) and for pressures of 0 to 30 psi (0 to 207 kPa). Units are designed to conform with international intrinsic safety and explosionproof requirements. The system incorporates microprocessor-based electronics with continuous diagnostics. Accuracy performance over a short time period during product transfers is ±0.02% of upper range limit (URL). Reference accuracy is ±0.04% URL. Total performance (worst case) from –40 to 185°F (–40 to 85°C) is ±0.06% URL.

Open communication protocols are used throughout (HART™ network). This protocol is compatible with most float and servo gage transmitters. Thus HPGs and float and servo gages can coexist on the same data highway. A translator permits full integration of the Mark-Space™ network into a MODBUS™ host. Software programs are available, permitting use with some PCs and PLCs. Because the system requires no intrusive elements, it may be used with difficult products, such as asphalt and waxy chemicals. High-pressure tolerance permits use for butane, propane, ammonia, and other similar compounds. It features instrumental calculations of special parameters, such as storage level, mass, and density, or specific gravity scales. (Smart Tank™ HTG system, *Rosemount Inc.*)

lished by the U.S. National Institute of Standards and Technology (NIST). The dielectric constant of air is always 1.0, and all other media have a greater dielectric constant value. Consequently, as the level of the medium increases, the amount of capacitance generated also increases as the air is displaced by the medium. Through this relationship, the actual level in the vessel can be determined.

The unit of measure for capacitance in an RF circuit is the farad. The small amount of capacitance typically measured in an RF level application is stated in picofarads (1 picofarad $= 10^{-12}$ farad). The amount of picofarads generated is determined by three variables: the size of the capacitance plates, the distance between the capacitance plates, and the dielectric of the medium.

In actual practice, capacitance change is produced in different ways, depending on the material being measured and the electrode selection. However, the basic principle always applies. If a higher dielectric material replaces a lower one, the total capacitance output of the system will increase. If the electrode is made larger (effectively increasing the surface area), the capacitance output increases. If the distance between measuring electrode and reference decreases, the capacitance output also increases. The dielectric of the medium being

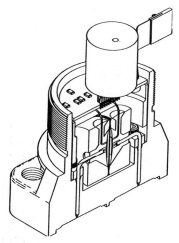

FIGURE 17 Sectional view of capacitive pressure sensor module used in system of Fig. 16. It incorporates microprocessor and digital technology. Use of application-specific integrated circuits (ASICs) and surface-mount electronics reduces transmitter size and weight. Pressure is transmitted through an isolating diaphragm and fill fluid to the sensing diaphragm in the center of the capacitance cell. Capacitor plates on both sides of the sensing diaphragm detect the diaphragm position. Transmitter electronics convert the differential capacitance between diaphragm and capacitor plates—as well as a temperature sensor measurement—into digital data for correction and linearization by the microprocessor. The proprietary cell is laser-welded and isolated mechanically, electrically, and thermally from the process medium and the external environment. These features relieve stress on the cell caused by line pressure, thus improving static pressure performance. All modules incorporate self-diagnostics. The module also stores information in nonvolatile EEPROM. Data are retained in transmitter when power is interrupted so that transmitter is immediately functional on power-up. The process variable is stored as digital data, enabling precise corrections and engineering unit conversion. Corrected data then are converted to a standard 4- to 20-mA current applied to the output loop. The operator can access the sensor reading directly as a digital signal, bypassing the digital-to-analog conversion process for higher accuracy. (*SMART FAMILY*™, *Rosemount Inc.*)

measured must remain constant. If the dielectric shifts, the picofarads generated will also shift, making it appear as though the level has actually changed.

Categories of RF Level Measurement. RF level measurement falls into three basic categories: (1) nonconductive materials, (2) conductive materials, and (3) detection of an interface between two immiscible liquids.

Nonconductive Materials. Suppose a level sensor is measuring the increasing level of a nonconductive hydrocarbon, such as gasoline, in a metal vessel (Fig. 20).

Conductive Materials. The same logic applies as for nonconductive materials, except that the conductive material, rather than the tank wall, acts as the ground plate of the capacitor.

In conductive materials it is necessary to use an electrode coated with an insulating material such as Teflon, Kynar, or polypropylene. Since the measured medium is highly conductive, it provides a direct path to ground (that is, tank wall or reference electrode). Therefore the conductive material brings the ground reference directly to the probe and the actual measured medium becomes the probe-insulating material which is immersed in the medium. In a highly conductive medium the capacitance generated by the probe is determined by the choice of insulation materials. Highly conductive materials with a dielectric greater than 60 are said to have a saturation capacitance. If a bare probe is used, the circuit will be short-circuited as soon as the conductive medium contacts the electrode, preventing the level from being measured.

Interface Detection. RF technology is an effective means of monitoring the interface between two materials with different dielectrics. For interface detection to be effective, the total vessel level must remain constant, or action must be taken to compensate for changes in total level. A common interface detection measurement is oil in water. This application typically involves monitoring the water level in an oil storage tank. The measurement is made by maintaining a constant total level in the tank and monitoring the water level on the probe. The dielectric of oil is typically between 2 and 4, and the dielectric of water is greater than 60. Consequently the probe is zeroed when completely covered with oil (very low dielectric) and spanned when it is completely immersed in water (very high dielectric). When the water level rises on the probe, the electronics easily sense the increased capacitance and provide a linear output based on the actual water level.

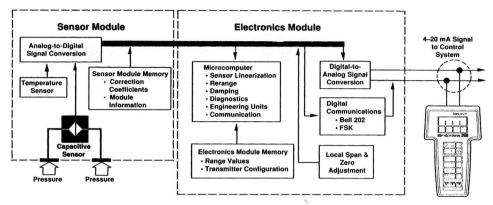

FIGURE 18   Block diagram of differential pressure transmitter of Fig. 16. (*SMART FAMILY™, Rosemount Inc.*)

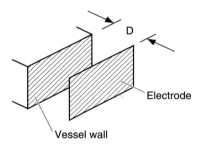

FIGURE 19 A capacitor is formed when a level-sensing electrode is installed in a vessel. The metal rod of the electrode acts as one plate of the capacitor and the tank wall (or reference electrode in a nonmetallic vessel) as the other plate. As the level rises, air or gas normally surrounding the electrode is displaced by material having a different dielectric constant. A change in the value of the capacitor takes place because the dielectric between plates has changed. Radio-frequency (RF) capacitance instruments detect this change and convert it into a relay actuation or a proportional output signal. The capacitance relationship is illustrated by the equation

$$C = 0.225K \left(\frac{A}{D} \right)$$

where C = capacitance, picofarads (pF)
K = dielectric constant of material
A = area of plates, in²
D = distance between plates, inches

Point-Level and Continuous-Level Applications.
One of the main advantages of RF technology is that it can be used in a wide variety of media and applications. Within certain limitations, RF technology can be used to measure most liquids, slurries, and bulk solids, and it can provide point-level control or continuous-level indication.

Point-Level Control.   This mode has many variations, from the simplest single-point alarm to multipoint level control. A single-point level alarm application can be accomplished with a horizontally or vertically mounted probe. As medium contacts the preselected set point on the probe, a relay closes to signify that the level has reached that trip point.

Multipoint control can be accomplished using a single vertically mounted probe with electronics designed to provide relay closures at two or more desired trip points. It also can be accomplished using multiple probes mounted horizontally at the desired trip points. Because of cost considerations, the latter is generally not used, except in cases of separate low-low or high-high alarm schemes.

Multipoint devices are ideal for pump or valve applications where differential control is required to control a liquid level between two points in a tank. The electronics are easily calibrated to accommodate single or multiple pumps and also to provide a high or low alarm.

Continuous-Level Indication.   RF technology also applies to continuous-level indication for most process applications. As stated previously, when the level rises on the probe, the capacitance increases. On a calibrated system this increase in capacitance is directly proportional to a level increase. Many

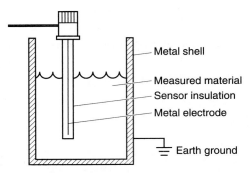

FIGURE 20   Capacitive measurement of nonconductive materials. While the actual capacitive equation is complex, it can be approximated for the gasoline-metal vessel application:

$$C = \frac{0.225(K_{air} \times A_{air})}{D_{air}} + \frac{0.225(K_{material} \times A_{material})}{D_{material}}$$

Since electrode and tank wall are fixed in place, the distance between them will not vary. Similarly, the dielectric of air and that of the measured material remain constant (air is 1, the hydrocarbon is 10). Consequently the capacitance output of the example system can be reduced to this very basic equation:

$$C = (1 \times A_{air} + (10 \times A_{material})$$

As this equation demonstrates, the more material there is in the tank, the higher the capacitance output will be. The capacitance is directly proportional to the level of the measured material.

electronic level controllers provide a digital or analog indication of the measured level as well as an analog or voltage output proportional to the level. These devices are typically offered with additional relay outputs, which can be used for simple alarm functions or for more complex pump control.

Electrode Application and Selection.   Selecting the proper level-sensing electrode and installing it in the proper location are important factors that contribute to the success of any application. The electrode is the primary measurement element (Fig. 21).

RF Electrode Location.   Mounting positions must be carefully considered. They must be clear of the inflow of material as impingement during a filling cycle can cause serious fluctuations in the capacitance generated. Side-mounted electrodes with point-level controls can be mounted at a downward angle to allow the measured material to drain or fall from the electrode surface. Electrodes mounted in nozzles contain a metal "sheath" extending a few inches past the nozzle length. The sheath renders that part of the electrode insensitive to capacitance change, and therefore ignores the material which may build up in the nozzle. Vertically mounted electrodes must be clear of agitators and other obstructions and far enough from the vessel wall to prevent "bridging" of material between the electrode and the vessel wall.

Special RF Application Factors.   These include the following.

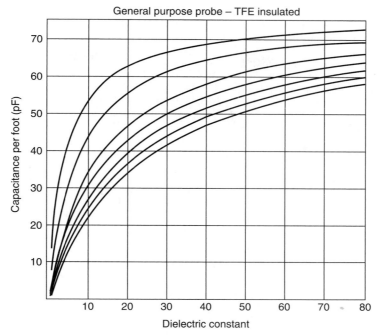

FIGURE 21   Level-sensing electrode. The electrode must be capable of producing sufficient capacitance change as it becomes submerged in the material to be measured. Several electrode types are offered, each having specific design characteristics. Capacitance (per foot of submersion) versus dielectric constant curves are published for each electrode type as installed in various-size vessels. For nonconductive materials these curves are nonlinear. Shown here is a typical set of curves. As the size of the tank becomes smaller, the capacitance per foot of submersion increases. A conductive material essentially allows the tank to serve as the electrode insulation. In this case the saturation capacitance is used. Electrodes selected for point-level detection should be capable of producing a minimum capacitance change of 3.0 pF. Continuous-level transmitter applications typically require a minimum span of 10.0 pF and a maximum span of 10,000 pF.

Temperature.   The dielectric constant of some materials varies with temperature, which affects the capacitance measured by the electrode. In general, materials with a higher dielectric constant are less affected by temperature variation.

Moisture Content.   The dielectric constant of granular materials changes with changing moisture content. This can cause significant measurement errors.

Static Charge.   Air-conveyed, nonconductive granular materials such as nylon pellets build up a static charge on the electrode which can damage the electronic components in the measuring instrument. Most instruments contain static discharge components to provide protection.

Composition.   The dielectric constant of the measured material must remain constant throughout its volume. Mixing materials with different dielectric constants in varying ratios will change the overall dielectric constant and the resultant capacitance generated. Solutions have a high dielectric constant and are less affected due to the saturation capacitance of the electrode system.

Conductivity.   Large variations in the conductivity of the measured material can introduce measurement error. The proper electrode selection can minimize this effect. A thick-wall electrode insulation is recommended in this case.

Material Buildup.   The most serious effect on the accuracy of RF capacitance measurements is caused by the buildup of conductive material on the electrode surface. Since this coating provides a path directly to ground, this can be a problem. In point-level applications this may cause an alarm condition even after the level has fallen. In continuous measurement this buildup may represent an artificially high reading. Nonconductivity buildup is not so serious since it only represents a small part of the total capacitance. Ways to handle the effects of coating in point-level detection are shown in Fig. 22; in continuous-level measurement, in Fig. 23.

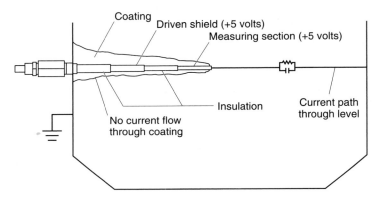

FIGURE 22   For point-level detection a special type of RF electrode, referred to as driven-shield type, is used to ignore the effect of conductive coating on the electrode surface. A second element is added to the electrode. This element is maintained electronically at the same voltage and frequency as the measuring section. Since no potential difference can exist across the two sections, no current can flow through the coating to the vessel wall. When the level reaches the bare electrode section, current passes only through the detected material to the tank wall, completing the detection circuit. This technique works even if the material dries out and becomes nonconductive—because the measurement is RF capacitive.

Other RF Considerations.   RF level technology offers many benefits over most technologies when used in the proper application. Level probes are available in both rigid-rod and flexible-cable types in many different materials to ensure compatibility with specific process media. Practical application of rigid probes generally is limited to about 12 ft (3.5 meters). Longer probes are awkward and difficult to install. Where longer lengths are required, flexible probes, both insulated and noninsulated, are available up to several hundred feet (meters). This type of probe extends the practical use of RF technology to tall tanks and even deep-well applications.

Grounding is an important consideration. RF systems require two conductive elements to form a capacitor. The metal electrode of the probe is one element. The second element normally is the wall of a metal vessel. Plastic vessel walls, however, are not conductive and thus cannot serve as the second element. Thus one must be provided for stable and reliable measurements. Grounding instructions should be obtained from the system supplier.

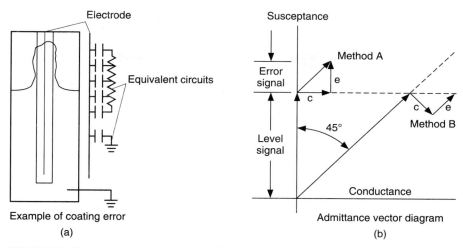

FIGURE 23   Various means are used to minimize the coating error in continuous RF level measurement. (*a*) Coating error. The submerged portion of the electrode generates nearly a pure capacitive susceptance. Since the electrode is insulated, a conductive component is virtually nonexistent. However, the upper section of the electrode, coated with conductive material, generates an error signal consisting of two components—capacitive susceptance and conductance. The result is an admittance component which is 45° out of phase with the main level signal. A study of transmission-line theory is required to prove this phenomenon. An equivalent circuit for coated section, shown as a ladder network producing the phase-shifted error signal. With reference to (b) above, one means of canceling this error signal is to measure the conductive component *c*. Since the 45° relationship exists, the capacitive error component *e* is of the same magnitude and can be subtracted from the total output signal, thereby effectively canceling the error signal.

   In another cancelation method a 45° phase shift is introduced to the entire measurement. This automatically cancels the coating error portion because the conductance component *c* still has the same magnitude as the error component *e,* resulting in the appropriate level signal. Instruments which incorporate these techniques are known as admittance types. The coating error also can be reduced by increasing the capacitive susceptance. This is accomplished by increasing the frequency of measurement or decreasing the electrode insulation wall thickness.

Ultrasonic Level Methods

   Sonic and ultrasonic technology has been applied to the problem of level measurement of liquids and bulk solids for many years. Generally, these methods have been used for point (rather than continuous) level measurement and control. There are two main categories: (1) measurement of the transit time between a sound wave transmitter and a receiver, which detects the returning sound wave pulse (echo), and (2) the absorption or attenuation of acoustic energy, such as occurs when a material (at a given height in a container) interferes with the transmission of energy from the transmitter to the receiver. In another system, a vibrating member, placed at some height in a container, displays a frequency change when it is covered by another medium. Sonic methods also have been found useful in making interface level measurements. This method is based on the alteration in the speed of sound in different media. Like probes (capacitive or conductive), sonic methods are particularly useful when the more traditional methods do not work well or at all, as exemplified by foaming liquids.

   A continuous sonic-type level measuring unit is shown in Fig. 24. The equipment includes a transmitter that periodically sends a sound pulse to the surface from the transducer, a receiver (which is included in the transducer) that amplifies the returning pulse, and a time

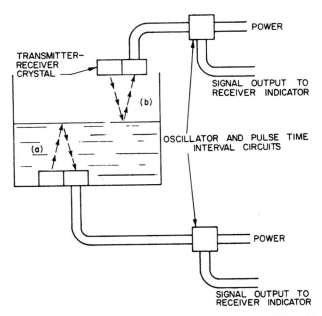

FIGURE 24   Continuous sonic-type level measuring units. (*a*) Liquid
phase. (*b*) Vapor phase.

interval counter that measures the time elapsing between the transmission of a pulse and
receipt of the corresponding pulse echo. Echo pulses are ordinarily reflected back from the
surface of the liquid; however, the line of demarcation or interface between immiscible liquids
also reflects sufficient energy to allow the system to gage these obscure interface levels. One
transducer and its single coaxial cable constitute the only installation equipment usually
required to gage an individual tank. The receiving indicator may be switched to gage as many
transducer-equipped tanks as needed. Designers of such systems must take into very serious
consideration the effect of environmental changes on the measurements. Notably these
changes are in temperature, pressure, and chemical composition—all factors which affect the
velocity of acoustic propagations and on which the measurement is fundamentally based. In
some processing operations, for example, where chemical-composition changes may be
expected, such changes can severely affect calibration unless additional electronic means are
incorporated in the system to anticipate and correct for such changes. The velocity of sound
at 0°C in air is 1087.42 ft (331.45 m)/s. In ammonia, the velocity is 1361 ft (415 m)/s; in carbon
dioxide, 1106 ft (337 m)/s; in chlorine, 674 ft (205 m)/s; in helium, 3182 ft (970 m)/s.

A representative single-sensor liquid-level indicator is shown in Fig. 25. The sensor typically is a small, hermetically sealed probe whose front face oscillates at a relatively high frequency (9 to 160 kHz). When rising liquid covers at least half the face of the sensor (if
mounted horizontally), or the entire face (if mounted vertically), the oscillating action of the
sensor will be damped out. This damping is recognized by the amplifier and causes a relay in
the amplifier to drop out and actuate either an on or an off signal (high level) or suitable control action through relays or other actuators. The sensors usually are about 1 inch (2.5 cm) in
diameter, are made of stainless steel or another corrosion-resistant alloy, and can easily be
placed in a fitting or tank. These sensors are not limited by the physical properties of various
liquids, such as pressure, conductivity, and density, and can be used with flammable liquids.

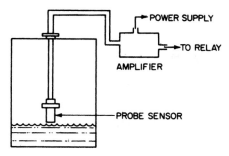

FIGURE 25 Sonic-type level indicator using top-mounted probe.

In the sonic system shown in Fig. 26, one transmitting sensor creates the sonic beam, and sound waves are picked up by a receiving sensor. This can be accomplished by a direct path or by reflective waves from the transmitter, which strike a solid surface and are reflected back to the receiving sensor. Sensors can be spaced as close as ¼ inch (6 mm) or as far apart as 10 ft (3 meters) in a direct beam path. They can pipe their beams through tubing where sensor beams cannot be direct, but generally the tubing length is limited to several feet. The sensor's sound beam is unaffected by mist, smoke, dust, or fumes, since it is interrupted only by a solid or liquid entering the beam path. This type of system is generally used for dry bulk solids.

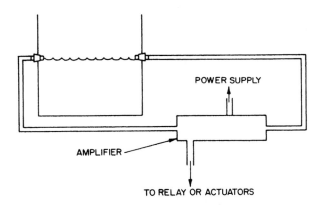

FIGURE 26   Sonic-type level detector utilizing two sonic probes.

Microwave Level Systems

There are parallels in the operating principles of sonic and ultrasonic systems with microwave radar level systems except that, of course, much higher frequencies are used in radar systems. The radar beam is not affected by density changes that may occur in the beam path as is the ultrasonic beam.

Radar level systems have been used for at least a quarter-century to measure liquefied natural gas and other hazardous hydrocarbon levels on-board seagoing tankers. Built-in electronics make it possible to compensate for any sloshing of tank contents.

In principle, a radar generator-transmitter is contained within a sealed tube. The beam is picked up by a specially designed antenna, which then beams the wave to the surface of the stored materials in a focused path. Return of the wave reflection causes a frequency shift (Doppler principle) which is picked up by a sensor. Generally the electronics associated with a radar system is much like that of other reflection-type level systems.

During recent years a growing acceptance, even at a cost premium, has been developing for the system, particularly in hazardous waste material handling and processing. More recently, acceptance has grown in the food-processing field. Sensors have been developed that withstand the high temperatures and abuse associated with clean-in-place systems required in food processing.

Nuclear Level Systems

Nuclear radiation from a selected source can be related to liquid or solids levels in a vessel. As a detector for converting nuclear gamma ray radiation into electrical quantities related to level, some systems use Geiger counters; others utilize a specially designed gas ionization cell (Fig. 27).

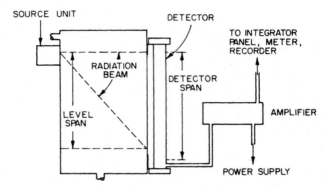

FIGURE 27 Nuclear-radiation-type level detection system. The detector converts gamma radiation received into an electrical output. Detector may be Geiger tube or ionization chamber. Radiation source is a minute quantity of an encapsulated radioactive isotope. The detector is fixed to the outside of the vessel directly across from the source location so that the material, as its level changes, will intercept the emission path and change the quantity and radiation intensity received by the detector. Some radiation sources require a license to be used. In some cases the radiation source can be located on a float.

Resistance-Tape Level Systems

Special resistance tapes are mounted on the interior of a vessel. The tapes are spirally wound against a steel tape that is mounted vertically from top to bottom on the tank. Fluid pressure created by the presence of fluid in the tank causes the tape to be short-circuited, thus changing the total resistance of the tape. Measurement of the resistance thus is directly related to the fluid level in the tank. The accuracy claimed is approximately ±⅛ inch.

Thermal Level Systems

Thermal systems have been used for several decades for determining level. Some earlier designs have long disappeared from the marketplace. Through the creation of new thermal sensors, new interest has been shown during the recent past. Point-, multipoint-, and continuous-level measuring systems are available. The principle involved relates the heat loss (lowering temperature) of a heated electrode when the measured medium rises, permitting heat dissipation from the electrode and vice versa. The criterion measured is the electrical resistance of the heated probe. See Fig. 28. The systems also are compensated for temperature changes of the measured medium, as shown in Fig. 29.

FIGURE 28 Level switch for detecting presence or nonpresence of process medium in a vessel. The switch consists of a dual-element sensing probe assembly attached to a head housing the electronics and field wiring connections. Operating details are given in Fig. 29. Suitable for corrosive and harsh environment. Liquids, gases, slurries, or foams can be handled with same sensing probe. Features total solid-state electronics. Operating temperature range −100 to 390°F (−70 to 200°C). Temperature compensation circuit eliminates false switching due to process temperature changes. Operates in electrically conducting or nonconducting media. Switching action with ±½-inch (0.8-mm) change in level. (*Delta m Corporation.*)

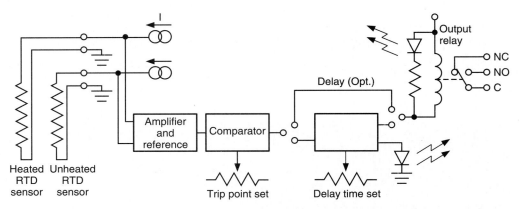

FIGURE 29   Operating principle of thermal-probe-type level sensor. Each element of the two-probe assembly contains a miniature RTD tightly encased within a stainless-steel tube. One element is internally heated to establish a temperature differential above the ambient temperature sensed by the other element. Different media have different heat transfer coefficients. Thus when the process product comes in contact with the heated sensor, the thermal differential is diminished. The unheated element provides a reference to the ambient conditions, thus providing temperature compensation over the entire operating range. Once the two-element probe senses a change in level, the temperature difference is converted into a voltage, which is compared with the adjustable set-point reference voltage. When the two values are equal, a relay is energized to open or close the alarm or control contacts. (*Delta m Corporation.*)

Magnetostriction Level System

An entirely new position-detection principle, developed as a position detector and described in Section 5, Article 1, in this handbook, also has found application for process level measurements, notably in tank gaging systems.

INDUSTRIAL WEIGHING AND DENSITY SYSTEMS

INDUSTRIAL SCALES

For centuries the underlying principle of the steelyard, that is, determining weight through the use of levers and counterweights, persisted. Modern industrial and commercial scales as recently as the midtwentieth century continued to rely on mechanical levers, pivots, bearings, and counterweights. Even without the benefits of advanced computing, electronics, and microprocessing techniques, a large truck or railroad platform scale could weigh 1 pound of butter or 1 part in 10,000. The principal design disadvantages of the former very rugged mechanical scales were their bulk, weight, and manual requirement to achieve balance. Self-balancing scales involving springs or counterbalancing pendulums essentially eliminated the need for manual operation. Invention of the strain gage and associated load cells revolutionized scale design. It should be pointed out that mechanical scales still are found in numerous isolated installations, although replacement parts are difficult to find.

Concurrent with the development of the strain gage, pneumatic and hydraulic load cells were developed and continue to be found in some installations (Fig. 1). However, the strain gage, by far, is the principal weight sensor used today. Refer to the article, "Fluid Pressure Systems" in this handbook section.

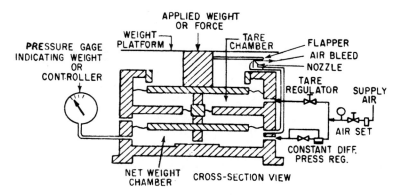

FIGURE 1   For some applications, a pneumatic load cell may be preferred over the widely used strain-gage cell.

Industrial scales may be classified by several criteria.

1. Physical configurations, influenced by application needs, include (1) ordinary platform scales as found in warehousing, receiving, or shipping departments, (2) very large highway or railroad track scales, which today frequently are designed to weigh "on the move," (3) tank scales, in which entire tanks of liquids can be weighed periodically or continuously—for inventory control (actually used as flow controllers for some difficult-to-handle materials) and for feeding batch processes, (4) bulk-weighing or hopper scales, and (5) counting scales used in inventory and production control.

2. Scales also fall into a number of capacity categories, ranging from the most sensitive laboratory-type balances (not covered in this article) to platform scales from as low as 50 lb (23 kg) through 1000, 50,000, 200,000 up to 1 million lb (453,597 kg). The upper limit on platform scales usually is about 30,000 lb (13,607 kg). Scales also are available for weighing materials which travel along a monorail, such as in a meat-packing plant. Another specialized application for weight indication is that of a crane or derrick for lifting and lowering heavy loads.

A number of installations are shown in Figs. 2 through 5.

FLUID DENSITY

Density may be defined as the mass per unit volume and usually is expressed in units of grams per cubic centimer, pounds per cubic foot, or pounds per gallon. Specific gravity is the ratio of

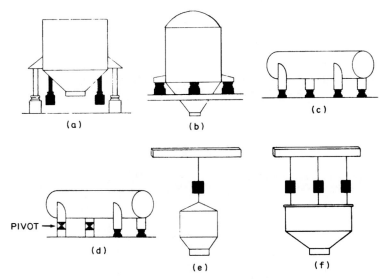

FIGURE 2 Arrangement of load cells and mounting assemblies (shown in black) for various hopper and tank configurations. (*a*) Four-load-cell compression arrangement (vertical tank built above floor) commonly used in batching and vehicle loading. (*b*) Three-cell compression arrangement (vertical tank built through floor or supported from beams) commonly used for storage of liquid or powders and for vehicle loading. (*c*) Four-cell compression arrangement (horizontal tank with pivot arrangement) for storage of liquids and powders. (*d*) Two-cell compression arrangment (horizontal tank with pivot arrangement) for storage of liquids. (*e*) Single-load-cell suspension arrangement (vertical hopper) for process weighing or batching of small loads. (*f*) Three-cell suspension arrangement (vertical hopper) for process weighing or batching of large loads.

FIGURE 3 Full-load-cell railway track scale. (*Reliance Electric.*)

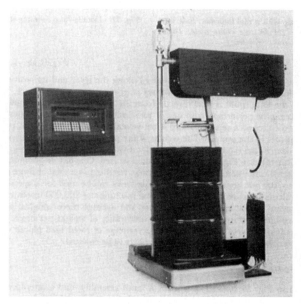

FIGURE 4   Microprocessor-based automatic drum-filling scale. (*Reliance Electric.*)

the density of the fluid to the density of water. For critical work, the reference is to double-distilled water at 4°C; in practical applications, the reference commonly used is pure water at 60°F (15.6°C).

Several specific gravity scales are used in industry.

Balling.   Used in the brewing industry to estimate the percentage of wort, but also used to indicate percent (weight) of either dissolved solids or sugar liquors. Graduated in percent (weight) at 60°F (17.5°C).

Barkometer.   Used in tanning and the tanning-extract industry, where water equals zero and each scale degree equals a change of 0.001 in specific gravity, that is, specific gravity equals $1000 \pm 0.001n$, where n is in degrees Barkometer.

Brix.   Used almost exclusively in the sugar industry, the degrees represent percent sugar (pure sucrose) by weight in solution at 60°F (17.5°C).

Quevenne.   Used in milk testing, the scale represents a convenient abbreviation of specific gravity: 20° Quevenne means a specific gravity of 1.020. One lactometer unit is approximately equivalent to 0.29° Quevenne.

Richter, Sikes, and Tralles.   Three alcoholometer scales which read directly in percent ethyl alcohol by weight in water.

Twaddle.   This scale represents an attempt to simplify the measurement of industrial liquors heavier than water. The range of specific gravity from 1 to 2 is divided into 200 equal parts, so that 1 degree Twaddle equals 0.0005 specific gravity.

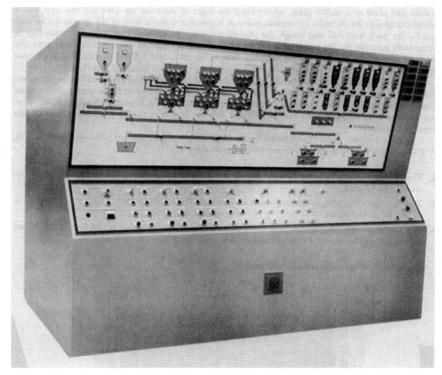

FIGURE 5 Some of the first industrial batching operations were centered about a sequence of weighing. Beyond the weight recipe handling, these batching systems need not be of the complexity and sophistication of some of the batching systems described in Section 3, Article 7, of this handbook. The batching center shown here involves several ingredient hoppers, collecting hoppers, and conveyors. (*Reliance Electric.*)

API.   Selected by the American Petroleum Institute and various standards institutions for petroleum products:

$$\text{Degrees (hydrometer scale at } 60°\text{F)} = \frac{141.5}{\text{specific gravity}} - 131.5$$

Baumé.   Widely used to measure acids and light and heavy liquids such as syrups, it was proposed by Antoine Baumé (1768), a French chemist. The scale is attractive because of the simplicity of the numbers representing liquid gravity. Two scales are used:

For light liquids
$$°\text{Be} = \frac{140}{\text{specific gravity}} - 130$$

For heavy liquids
$$°\text{Be} = 145 - \frac{145}{\text{specific gravity}}$$

The standard temperature for the formulas is 60°F (15.6°C).

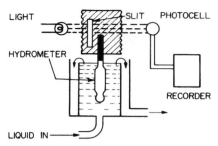

FIGURE 6 Photoelectric hydrometer. A glass hydyrometer, similar to the hand-held type, is placed in a continuous-flow vessel. The instrument stem is opaque, and as the stem rises and falls, so does the amount of light passing through a slit in the photocell. Thus photocell input is made proportional to specific gravity and can be recorded by a potentiometric instrument. The system is useful for most specific gravity recording applications not harmful to glass and is accurate to two or three decimal places.

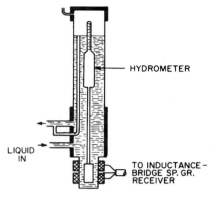

FIGURE 7 Inductance bridge hydrometer. The level of the measured liquid is held constant by an overflow tube. A glass hydrometer either rises or falls in the liquid as the specific gravity varies. The lower end of the hydrometer supports an armature in an inductance coil. Any movement of this armature is duplicated by a similar coil in a recording instrument. With this system, the temperature of the liquid usually is recorded along with the value of specific gravity, so that corrections can be made.

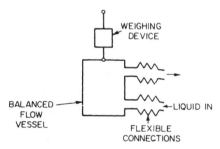

FIGURE 8 Balanced-flow vessel for measuring specific gravity and fluid density. A fixed-volume vessel is used, through which the measured liquid flows continuously. The vessel is weighed automatically by a scale, spring balance, or pneumatic force-balance transmitter. Since the weight of a definite volume of the liquid is known, the instrument can be calibrated to read directly in specific gravity or density units. Either open or closed vessels with flexible connections can be used. A high-accuracy measurement results, which is especially useful in automatic density control.

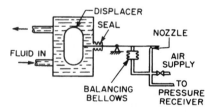

FIGURE 9 Displacement meter for measuring specific gravity and density. Liquid flows continuously through the displacer chamber. An upward force acts on the balance beam because of the volume of liquid displaced by the float. A pneumatic system, similar to the one shown, balances this upward force and transmits a signal proportional to the density of the liquid. Liquids with specific gravities 0.5 and higher can be measured with this equipment so long as suitable materials are used to prevent damage from corrosion. If the temperature of the flowing liquid changes, a thermostatic heater may be used to maintain a constant temperature.

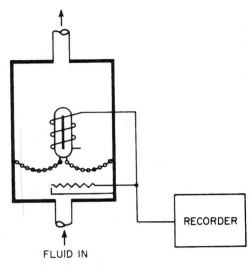

FLUID IN

FIGURE 10 Chain-balanced, float-type, density-sensitive element. This instrument uses a submerged plummet which is self-centering and operates essentially frictionless. It is not affected by minor surface contaminants. The volume of the body is fixed and remains entirely under the liquid surface. As the plummet moves up and down, the effective chain weight acting on it varies, and for each density within the range of the assembly, the plummet assumes a definite equilibrium point.

In order to transmit these changes in density, the plummet contains a metallic transformer core whose position is measured by a pickup coil. This voltage differential, as a function of plummet displacement, is a measure of specific-gravity change. To compensate for a density change due to temperature, a resistance thermometer bridge notes the temperature change and impresses a voltage across the recorder, which is equal to and opposite the voltage transmitted to it by the pickup coil as a result of the temperature-induced density change.

A number of density and specific gravity measuring principles have developed over the years. Compared with the needs for sensing temperature, pressure, flow, and so on, the requirements for density instrumentation are relatively infrequent (Figs. 6 through 10).

One of the most important developments in fluid density measurement over the last decade or so has been the Coriolis flowmeter, which measures flows in terms of their mass. See the article, "Flow Systems" in this handbook section.

Controlling the density of oil-well drilling mud is described by K. Zanker (*Sensors,* vol. 40, Oct. 1991).

HUMIDITY AND MOISTURE SYSTEMS

Humidity measurement is a special case of moisture measurement, that is, of water vapor in air (or possibly better termed "water gas"). For purposes of instrument classification, moisture in air constitutes humidity, whereas in general the term moisture refers to the water content of a solid substance, such as paper, other webbed materials, and a variety of other commercial and industrial products, including baked goods, candies, some textile products, coatings, and films.

HUMIDITY FUNDAMENTALS[1]

One of the easiest ways to put humidity in its proper perspective is through application of Dalton's law of partial pressures to the most commonly encountered gas—air.

Dalton's law states that the total pressure P_m exerted by a mixture of gases or vapors is the sum of the pressure of each gas if it were to occupy the same volume by itself. The pressure of each individual gas is called its partial pressure. The total pressure of an air–water gas mixture, containing oxygen, nitrogen, and water, is equal to the sum of the partial pressures of each gas:

$$P_m = P_{N_2} + P_{O_2} + P_{H_2O} + \cdots$$

Therefore the partial pressure of water vapor in air is directly related to the measurement of humidity. This vapor pressure varies from 1.22×10^{-3} millibar (mbar) of mercury (0.122 Pa) at the $-75°C$ (59°F) frost point of "bone-dry" arctic or industrial dry air, to 1.013×10^3 mbar of mercury (0.1013×10^6 Pa) at the 100°C (212°F) dew point of saturated hot air in a product dryer. This is a change of almost a million to one over the span of interest in industrial humidity measurement.

The ideal humidity instrument would be a linear, wide-range pressure gage, specific to water vapor and employing a primary or fundamental measuring method. Such an instrument, although physically possible, would be cumbersome. Most humidity measurements are made by a secondary instrument which is responsive to humidity-related phenomena.

Common Humidity Parameters

The humidity parameters most often encountered in scientific and industrial applications are listed in Table 1. In addition to these common parameters, numerous other formats exist for use in narrow applications or specific technologies. However, most of these are variations of the basic parameters given in Table 1.

Psychrometric Chart. The psychrometric charts in Fig. 1 provide a quick means for converting from one humidity format to another, because dew point, relative humidity, ambient temperature, and wet bulb temperature can be conveniently related to each other on a single sheet of paper. The psychrometric chart has long been the basic tool of air-conditioning engi-

---

[1] Peter R. Wiederhold, President, General Eastern Instruments Corporation, Watertown, Massachusetts.

TABLE 1 Humidity Parameters and Representative Applications

Parameter	Description	Units	Typical applications
Wet bulb temperature	Minimum temperature reached by wetted thermometer in airstream	°F or °C	High-temperature dryers, air-condition-ing, meteorology, test chambers
Percent relative humidity	Ratio of actual vapor pressure to saturation vapor pressure, with respect to water, at the prevailing dry bulb temperature	0–100%	Monitoring conditioning rooms, test chambers, pharmaceutical and food packaging
Dew point or frost point	Temperature to which the air must be cooled to achieve saturation is the dew point; if temperature is below 32°F, it is called frost point	°F or °C	Heat treating, annealing atmos-pheres, dryer control, instrument air monitoring, meteorological and environmental measurements
Volume or mass ratio	Parts per million (ppm) by volume is ratio of partial pressure of water vapor to partial pressure of dry carrier gas; ppm by weight is identical to ppm by volume, but the ratio changes according to the molecular weight of the carrier gas	ppm_v, ppm_w	Used primarily to ensure dryness of industrial process gases such as air, nitrogen, oxygen, methane, hydrogen

neers, and Fig. 1 shows the temperatures most often encountered in comfort control or product conditioning applications. Psychrometric charts are available for higher temperatures and humidities and are quite useful in dryer and condensation system design. Charts are also available for lower temperatures, but tend to be less useful because wet bulb measurements are difficult to make with any accuracy at temperatures below −10°C (14°F).

Wet Bulb–Dry Bulb Measurements

Psychrometry has long been a popular method for monitoring humidity, primarily because of its simplicity and inherent low cost. A typical industrial psychrometer consists of a pair of matched electrical thermometers, one of which is maintained in a wetted condition. Water evaporation cools the wetted thermometer, resulting in a measurable difference between it and the ambient or dry bulb measurement. When the wet bulb reaches its maximum temperature depression, the humidity is determined by comparing the wet bulb–dry bulb temperatures on a psychrometric chart. In a properly designed psychrometer, both sensors are aspirated at an airstream rate of between 4 and 10 m/s for proper cooling of the wet bulb, and both are thermally shielded to minimize errors from radiation.

A properly designed and utilized psychrometer, such as the Assman laboratory type, is capable of providing accurate data. However, since very few industrial psychrometers meet these criteria, the psychrometer is limited to applications where low cost and moderate accuracy are the underlying requirements. It does have certain inherent advantages: (1) It is capable of highest accuracy near 100 percent relative humidity (RH). From an accuracy standpoint, it is superior to most other humidity sensors near saturation. Since the dry bulb and wet bulb sensors can be connected differentially, the wet bulb depression (which approaches zero as the relative humidity approaches 100 percent) can be measured with a minimum of error. (2) Although large errors can occur if the wet bulb becomes contaminated or is improperly fitted, the simplicity of the device allows easy repair at minimum cost. (3) The psychrometer can be used at an ambient temperature above 100°C (212°F), and the wet bulb measurement is usable up to 100°C.

Major limitations of the psychrometer include: (1) As RH drops below 20 percent, the problem of cooling the wet bulb to its full depression becomes difficult. The result is impaired accuracy below 20 percent RH, and few psychrometers provide reliable data below 10 percent

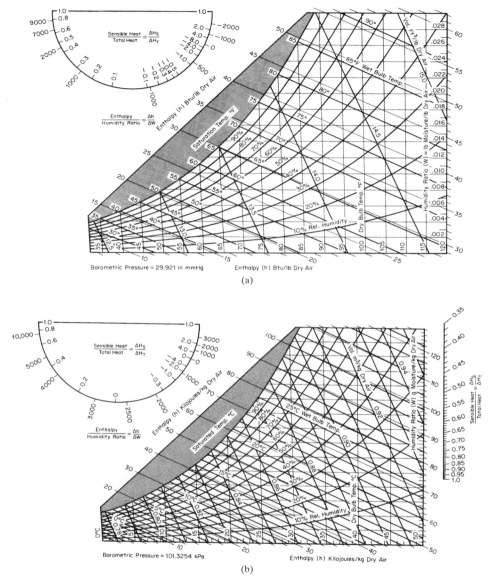

FIGURE 1 Psychrometric charts. (*a*) U.S. customary units. (*b*) SI units. Charts apply only at atmospheric pressure. (*American Society of Heating, Refrigerating, and Air Conditioning Engineers.*)

RH. (2) Wet bulb measurements at temperatures below 0°C are difficult to obtain with any high degree of confidence. Automatic water feeds are not feasible because of freezing. (3) Because a wet bulb psychrometer is a source of moisture, it can be used only in environments where added water vapor from the psychrometer exhaust is not a significant component of the total volume. (4) Generally speaking, psychrometers cannot be used in small, closed volumes.

Percent Relative Humidity

Percent relative humidity (% RH) is the best known and perhaps the most widely used method for expressing the water vapor content of air. It is defined as the ratio of the prevailing water vapor pressure e_a to the water vapor pressure if the air were saturated e_s, multiplied by 100:

$$\% \text{ RH} = \frac{e_a}{e_s} \times 100$$

The term "percent relative humidity" appears to have derived from the invention of the hair hygrometer in the seventeenth century. The hair hygrometer operates on the principle that many organic filaments, such as hair, goldbeater's skin, and even nylon, change length as a nearly linear function of the ratio of the prevailing water vapor pressure to the saturation vapor pressure.

Basically, % RH is an indicator of the water vapor saturation deficit of the gas mixture, rather than an indicator of sorption, desorption, comfort, or evaporation. A measurement of RH without a corresponding measurement of the dry bulb temperature is not of particular value, since the water vapor content cannot be determined from % RH alone.

Sensors for Measuring % RH.   Over the years, devices other than the simple hair hygrometer have evolved, which permit a direct measurement of % RH. These devices are, for the most part, electrochemical sensors, which offer a degree of ruggedness, compactness, and remote electronic readout ability not afforded by hair devices.

Contemporary electronic % RH instruments essentially are derivatives of two basic designs, which were first introduced in the 1940s and which, during the interim, have been greatly improved through the application of solid-state circuitry, microprocessors, and so on. Named after its inventor, the Dunore cell used a bifilar-wound inert wire grid on an insulative substrate coated with a lithium chloride solution of a controlled concentration. The hygroscopic nature of lithium chloride causes it to take up water vapor from the surrounding atmosphere. The ac resistance of the sensor is an indication of the prevailing % RH. Dunmore-type cells proved to be excellent RH sensors, but because of the characteristics of lithium chloride, were usually designed to cover a narrow range of interest. For example, a single sensor may cover from 40 to 60 percent RH. A cluster of narrow-range sensors can be combined in a common housing and served by an electronic matching network, but this can result in a rather bulky sensor.

The Pope cell, also named for its developer, was introduced shortly after the Dunmore cell. The Pope cell also used a bifilar conductive grid on an insulative substrate. Instead of relying on the hygroscopic nature of the substrate coating, however, a different electrochemical phenomenon was utilized. The polystyrene substrate was treated with sulfuric acid, that is, the longer-chain polystyrene molecules were sulfonated. Highly mobile sulfate radicals were created and these, in the presence of hydrogen ions (available from water molecules in vapor form), altered the surface resistivity of the substrate as a function of humidity. An ac-excited bridge is required to avoid polarization caused by dc excitation. Pope sensors had an advantage over the Dunmore cells because of a much wider useful range (15 to 99 percent RH).

Dew-Point Hygrometry

Dew-point measurements are widely used in scientific and industrial applications when precise measurement of water vapor pressure is needed. The dew point, the temperature at which water condensate begins to form on a surface, can be accurately measured from −75°C to +100°C across the entire range of humidity with a condensation (chilled-mirror) hygrometer.

Three types of instruments have received acceptance in dew-point measurement: (1) the saturated-salt dew-point sensor, (2) the condensation-type hygrometer, and (3) the aluminum oxide sensor. Many other instruments are used in specialized applications, including pressure ratio devices, dew cups, and fog chambers. The latter are operated manually.

Saturated-Salt Dew-Point Sensors.   The saturated-salt [lithium chloride (LiCl)] dew-point sensor is a widely used sensor because of its inherent simplicity, ruggedness, and low cost. Various government weather services have used this type of sensor for most official ground-based humidity measurements. However, these instruments are being converted to the more accurate condensation hygrometers.

The principle of the saturated-salt dew-point sensor is that the vapor pressure of water is reduced in the presence of a salt. When water vapor in the air condenses on a soluble salt, it forms a saturated layer on the surface of the salt. This saturated layer has a lower vapor pressure than water vapor in the surrounding air. If the salt is heated, its vapor pressure increases to a point where it matches the water vapor pressure of the surrounding air, and the evaporation-condensation process reaches equilibrium. The temperature at which equilibrium is reached is directly related to the dew point.

A saturated-salt sensor is constructed with an absorbent fabric bobbin covered with a bifilar winding of inert electrodes coated with a dilute solution of lithium chloride (Fig. 2). Lithium chloride is often used as the saturating salt because of its hygroscopic nature, which permits application in relative humidities of between 11 and 100 percent.

FIGURE 2   Lithium chloride sensor. (*General Eastern Instruments.*)

An alternating current is passed through the winding and salt solution, causing resistive heating. As the bobbin heats up, water evaporates into the surrounding air from the diluted lithium chloride solution. The rate of evaporation is determined by the vapor pressure of water in the surrounding air. When the bobbin begins to dry out, as a result of the evaporation of water, the resistance of the salt solution increases. With less current passing through the winding, because of increased resistance, the bobbin cools and water begins to condense, forming a saturated solution on the bobbin surface. Eventually, equilibrium is reached and the bobbin neither takes on nor loses any water.

Properly used, a saturated-salt sensor is accurate to ±1°C between dew-point temperatures of −12 and +38°C. Outside these limits, small errors may occur as a result of multiple hydration characteristics of lithium chloride. These may produce ambiguous results at 41, −12, and −34°C dew points. Maximum errors at these ambiguity points are 1.4, 1.6, and 3.4°C, respectively, but actual errors encountered in typical applications are usually less.

The saturated-salt sensor has certain advantages over other electrical humidity sensors, such as % RH instruments. Because the salt sensor operates as a current carrier saturated with lithium and chloride ions, the addition of contaminating ions has little effect on its behavior compared to the situation for a typical RH sensor which operates "starved" of ions and is easily contaminated. A properly designed saturated-salt sensor is not easily contaminated since, from an ionic standpoint, it can be considered precontaminated. But if a saturated-salt sensor does become contaminated, it can be washed with an ordinary sudsy ammonia solution, rinsed, and recharged with lithium chloride. It is seldom necessary to discard a saturated-salt sensor if proper maintenance procedures are observed.

Limitations of saturated-salt sensors include (1) a relatively slow response time and (2) a lower limit to the measurement range imposed by the nature of lithium chloride. The sensor cannot be used to measure dew points when the vapor pressure of water is below the saturation vapor pressure of lithium chloride, which occurs at 11 percent RH. With certain gases, ambient temperatures can be reduced, increasing the RH to above 11 percent, but the extra effort needed to cool the gas usually warrants selection of a different type of sensor. Fortunately a large number of scientific and industrial measurements fall above this limitation and are readily handled by the sensor.

Condensation-Type Hygrometers.   The condensation-type dew-point hygrometer is one of the most accurate and reliable instruments and offers the widest range of sensors for humidity measurements. These features are obtained; however, through increased complexity and cost. In the condensation-type hygrometer, a surface is cooled (thermoelectrically, mechanically, or chemically) until dew or frost begins to condense out. The condensate surface is maintained electronically in vapor pressure equilibrium with the surrounding gas, while surface condensation is detected by optical, electrical, or nuclear techniques (Figs. 3 and 4). The surface temperature is then the dew-point temperature, by definition.

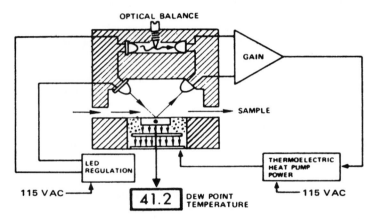

FIGURE 3   Dew point is detected in a condensation hygrometer by cooling a surface until water begins to condense. Condensation is detected optically or electronically. Signal is fed into an electronic control circuit which maintains the surface temperature at the precise dew point.

The largest source of error in a condensation hygrometer stems from the difficulty in measuring the condensate surface temperature accurately. Typical industrial versions of the instrument are accurate to ±0.2°C over very wide temperature spans, and laboratory models offer accuracies up to ±0.1°C.

A wide span and minimal errors are the two main features. A properly designed condensation hygrometer can measure from dew points of 100°C down to frost points of −75°C.

The response time of a condensation dew-point hygrometer is usually specified in terms of its cooling or heating rate, typically 1.5°C/s, making it considerably faster than a saturated-salt dew-point sensor and nearly as fast as most electrical % RH sensors. Perhaps the most significant feature of the condensation hygrometer is its fundamental measuring technique, which essentially renders the instrument self-calibrating. For calibration, it is only necessary to override the surface cooling control loop manually, causing the surface to heat, and witness that the instrument recools to the same dew point when the loop is closed. On the assumption that the surface temperature measuring system is calibrated, this is a reasonable and valid check on the instrument's performance.

FIGURE 4   Representative optical condensation (chilled-mirror) detector. (*General Eastern Instruments.*)

Because of its fundamental nature and superior accuracy and repeatability, this kind of instrument is used as a secondary standard for calibrating other lower-level humidity instruments.

The inert construction of the condensation hygrometer makes it virtually indestructible. Although the instrument can become contaminated, it is easy to wash and return to service without impairment of performance or calibration.

The condensation (chilled-mirror) hygrometer measures the dew- or frost-point temperature. Unfortunately many applications require measurement of % RH, water vapor in parts per million, or some other humidity parameter. In such cases the user must decide whether to use the fundamental high-accuracy condensation hygrometer and convert the dew- or frost-point measurement to the desired parameter, or to use lower-level instrumentation to measure these parameters directly. In recent years microprocessors have been developed which can be incorporated in the design of a condensation hygrometer, thus resulting in instrumentation which can offer accurate measurements of humidity in terms of almost any humidity parameter. Two instruments of this type are shown in Figs. 5 and 6.

Electrolytic Hygrometer.   A typical electrolytic hygrometer utilizes a cell coated with a thin film of phosphorus pentoxide (P_2O_5), which absorbs water from the sample gas (Fig. 7). The cell has a bifilar winding of inert electrodes on a fluorinated hydrocarbon capillary. Direct current applied to the electrodes dissociates the water, which is absorbed by the phosphorus pentoxide, into hydrogen and oxygen. Two electrons are required to electrolyze each water molecule, and thus the current in the cell represents the number of molecules dissociated. A further calculation, based on flow rate, temperature, and current, yields the parts-per-million concentration of water vapor.

In order to obtain accurate data, the flow rate of the sample gas through the cell must be known and constant. Since the parts-per-million calculation is partially based on flow, an error in the flow rate causes a direct error in measurement.

FIGURE 5 Chilled-mirror hygrometer with built-in microprocessor to provide humidity data in various parameters, such as ppm$_v$, ppm$_w$, dew point, and % RH. (*General Eastern Instruments.*)

FIGURE 6 Microprocessor-based humidity analyzer using a chilled-mirror sensor. (*EG&G Environmental Equipment.*)

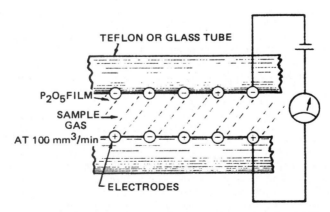

FIGURE 7 An electrolytic hygrometer dissociates water absorbed by phosphorus pentoxide into hydrogen and oxygen by electrolysis. Since two electrons are required to electrolyze a molecule of water, the amount of current used by the hygrometer relates to ppm of water vapor.

A typical sampling system for ensuring constant flow is shown in Fig. 8. Constant pressure is maintained within the cell. Sample gas enters the inlet, passes through a stainless-steel filter, and enters a stainless-steel manifold block. It is very important that all components prior to the sensor be made of an inert material, such as stainless steel, to minimize contamination. After the sample gas passes through the sensor, its pressure is controlled by a differential-pressure regulator which compares the pressure of the gas leaving the sensor with the pressure of the gas venting to the atmosphere through a preset valve and flowmeter. In this way, constant flow is maintained, even though there may be nominal pressure fluctuations at the inlet port.

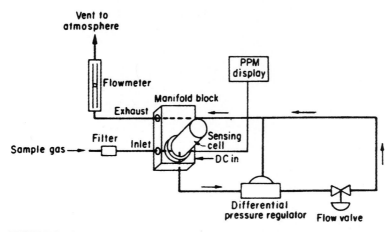

FIGURE 8   Calculation of water vapor content in an electrolytic hygrometer is dependent on precise control of the flow rate. The arrangement shown controls the sample pressure across the cell, ensuring correct flow regardless of input pressure fluctuations.

A typical electrolytic hygrometer can cover a span from 0 to 2000 ppm with an accuracy of ±5 percent of the reading, which is more than adequate for most industrial applications. The sensor is suitable for most inert elemental gases and organic and inorganic gas compounds that do not react with phosphorus pentoxide.

Electrolytic hygrometers cannot be exposed to high water vapor levels for long periods of time, because this results in a high usage rate for the phosphorus pentoxide and high cell currents.

Aluminum Oxide Humidity Sensor.   This type of sensor is a capacitor, formed by depositing a layer of porous aluminum oxide on a conductive substrate and then coating the oxide with a thin film of gold. The conductive base and the gold layer become the capacitor's electrodes. Water vapor penetrates the gold layer and is absorbed by the porous oxidation layer. The number of water molecules absorbed determines the electrical impedance of the capacity, which in turn is a measure of water vapor pressure.

Advantages of the aluminum oxide sensor are (1) small size and suitability for in situ use; (2) economical use in multiple-sensor arrangements; (3) suitability for very low dew-point levels without the need for sensor cooling (as required in condensation-type sensors); typically, dew points down to −100°C can be measured without serious difficulty; and (4) a wide measurement span.

The aluminum oxide sensor has the following limitations: (1) It is a secondary measurement device and must be periodically calibrated to accommodate aging effects, hysteresis, and contamination. (2) Sensors require separate calibration curves, which are typically nonlinear.

Aluminum oxide humidity instruments are available in a variety of types, ranging from a low-cost single-point system, including portable battery-operated models, to multipoint microprocessor-based systems with an ability to compute and display humidity information in different parameters, such as dew point, parts per million, and % RH (Fig. 9).

FIGURE 9   Microprocessor-controlled multichannel humidity analyzer using aluminum oxide sensors. (*Hygroscanner, General Eastern Instruments.*)

The aluminum oxide sensor is also used for moisture measurements in liquids (hydrocarbons), and because of its low power usage, it is suitable for use in explosionproof installations. These sensors are frequently used in petrochemical applications where low dew points must be monitored on line and where the reduced accuracies and other limitations are acceptable. The advantages of the sensor must be weighed against the fact that its accuracy is lower than that of any of the fundamental measurement sensor types. As a secondary measurement device, it can provide reliable data only if kept in calibration and if damage due to incompatible contaminants is avoided.

General Observations

Professionals stress that humidity remains one of the most difficult of the process variables to measure and that no universal methodology is on the horizon. Device selection is difficult. For each application there remain numerous tradeoffs in selecting the optimum device from the standpoints of performance and cost, including maintenance. As the need for accuracy increases, as is encountered in various kinds of research involving humidity, the selection problem becomes more difficult, and in some cases, user "in-house" engineering may be required in adapting commercially available sensors.

Some users consider a chilled-mirror device equipped with an optical control system as the most fundamental and reliable of the measurement techniques. With these devices, however, contamination may be a problem, particularly with fouling mirror surfaces. Mirror scratching must be avoided. In a problem atmosphere, mirror contamination can cause accuracy variations of a fraction to over a degree of temperature within a 24-hour period due to mirror fouling. Thus initial and maintenance costs with these devices are large factors.

Capacitive and bulk polymer technologies are considered as secondary measurements because they infer humidity from secondary (usually electrical) phenomena. With capacitive types, the effects of temperature can affect the electrical characteristics of the sensor and thus cause erroneous readings. Temperature compensation can be accomplished automatically by the instrument's electronics.

Water-soluble contaminants affect vapor pressure in capacitance systems and thus cause errors in % RH.

Prolonged exposure to humidities over 85 percent RH also can be a problem with capacitive-type sensors.

MOISTURE IN MATERIALS

With the exception of the close coupling of relative humidity measurement with moisture measurement, as in the case of some systems for moving-web moisture control, for example, the fundamental methods for measuring moisture in air and other gases are described earlier in this article. The remaining portion of this article is concerned with the measurement of moisture in solids and liquids. There are numerous industrial and commercial applications, notably in the paper, boxboard, and other moving-web industries. Frequent demands for such instrumentation occur in the food-processing industries, where moisture in food is both a positive and a negative characteristic. Unless they contain ample detectable moisture, many food products are regarded as dry and stale. In other cases, particularly with fresh produce, exposure to too much moisture over an extended period attracts molds and other spoilage microorganisms.

Laboratory Moisture Determinations

Basically there are four laboratory methods for determining moisture content: (1) the Karl Fischer technique, (2) distillation methods, (3) oven drying, and (3) equilibrium methods.

The Karl Fischer technique offers high sensitivity (5 ppm) for moisture contents ranging from 10 ppm to 100 percent. The method involves titrating a sample, using Karl Fischer reagent (a solution of iodine, sulfur dioxide, and pyridine in methanol or methyl Cellusolve). The system has been highly automated in recent years. An amperometric system controls titrant delivery to a preset end point automatically. End-point detection is determined by measuring the current through a solution, using a dual platinum electrode and a dc voltage source.

In distillation methods, the water is separated from the liquid sample by virtue of the differences in boiling points of the other liquids and water. The water fraction is collected carefully and measured volumetrically.

The oldest and still a common analytical method is that of determining the moisture content by heating the sample to ensure complete drying. Moisture is calculated on the basis of loss in weight. Other volatile materials interfere with the method.

The equilibrium moisture content of the air at the surface of a material can be representative of the moisture content within the material. A number of factors may interfere with the determination. This is particularly true of hygroscopic granular and fibrous materials. Humidity instrumentation, as described previously, is used in this method.

The Nature of Moisture Content

Moisture content has a number of synonymous terms, several of which are specific to certain materials and applications. The water content of solid, granular, or liquid materials usually is referred to as moisture content on either a wet basis or a dry basis. For industrial measurements, the wet basis is most common. Wet basis refers to the quantity of water per unit weight or volume of wet material. The textile industry uses the dry basis in measuring the moisture content of textile fibers. Often referred to as the regain moisture content, the dry basis refers to the quantity of water in a material expressed as a percentage of the weight of bone-dry material. The relationship between the wet and dry moisture content bases is shown in Fig. 10.

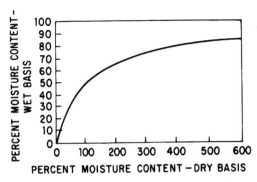

FIGURE 10 Relation between dry and wet moisture content bases.

Instruments for moisture measurement may be considered periodic or continuous. Periodic instruments generally are automated versions of conventional laboratory moisture analysis procedures—with a measurement typically requiring 2 min or longer and sometimes up to 20 minutes. Thus periodic instruments are quite impractical for automatic control systems. Moisture measuring instruments also may be classified by the operating principle involved. Instruments using electrical conductivity (dc or ac), absorption of electromagnetic energy (radio-frequency regions), electrical capacitance (dielectric constant change), and infrared energy radiation are most adaptable to continuous measurement because of the comparatively fast speed of response.

Moisture Measurement Methods

Infrared Absorption Methods. Several instruments are available that are based on the principle that the water molecule becomes resonant at certain infrared (IR) frequencies, and thus the amount of energy absorbed by the water absorption band is a measure of moisture content. One device for calculating the moisture content of sheet paper uses IR energy at two different wavelengths and measures the differential absorption (Fig. 11).

Electrical Capacitance Methods. The change in dielectric constant of a material under dry and moist conditions is the physical principle used. The dielectric constant of most vegetable organic materials is 2 to 5 when dry. Water has a dielectric constant of 80. Therefore the addition of small amounts of moisture to such materials causes a considerable increase in the dielectric constant. The material being measured forms part of a capacitance bridge which has radio-frequency (RF) power applied from an electronic oscillator as shown in Fig. 12. Electronic detectors measure bridge imbalance or frequency change, depending on the method used. Electrode design varies with the type of material under test. Parallel-plate electrodes are used for sheet materials, whereas cylindrical electrodes generally are used for liquids or powders. The moisture content measurable by electrical capacitance methods ranges from 2 to 3 percent to 15 to 20 percent, depending on the product involved (Figs. 13 and 14).

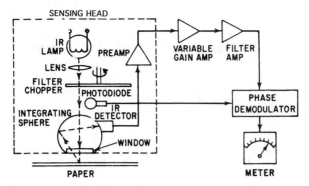

FIGURE 11 Infrared radiation is passed through a rotating disk chopper that contains two filters. One filter passes a wavelength of 1.94 μm, which covers the water-absorption band; the second filter passes a band of 1.6 μm, which is not significantly affected by moisture content. The target paper sheet is exposed to dual-frequency narrowband radiation, and the reflected radiation is collected in an integrating sphere where it activates a suitable IR detector. The signals are amplified, and the ratio of the two narrowband signals is read out directly as moisture content. The system is quite immune to the effects of other process variables, such as basis weight, coatings, density, and composition. The useful range is from 0–0.1 percent (minimum) to 95 percent (maximum). Inasmuch as electrical circuitry is used throughout, advantage can be taken of contemporary electronics, including remote digital displays and alarms. Instruments of this type find wide application in the foodstuffs, dairy, forest products, paper, tobacco, and chemical and mineral products industries.

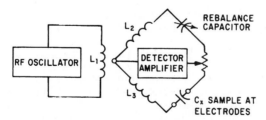

FIGURE 12 Basic bridge circuit for capacitance method of moisture content measurement.

In an effort to increase the speed of measurement over that of mechanical scanning systems, a system which uses an electronically scanned multielectrode array for determining the instantaneous moisture profile in a paper web, was developed. The system is typically used at the dry end of a paper machine or coater on most grades of paper, from newsprint to boxboard and pulp sheets. The sensor array normally consists of 8 to 16 individual electrode elements which are in light contact with the paper web. Each sensing element consists of a center electrode, and the surrounding grounded electrodes are separated by an insulator. A high-frequency sensing field is applied to the center electrode to measure the amount of field coupling through the paper web to the grounded electrode. This value is basically related to

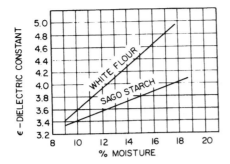

FIGURE 13 Moisture content versus dielectric constant for starch and flour.

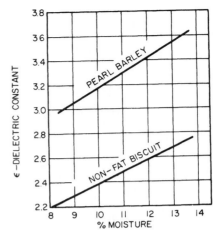

FIGURE 14 Moisture content versus dielectric constant for biscuit and barley.

the dielectric coefficient of the paper. It is claimed that the sensor can operate on materials up to 0.5 inch (13 mm) in thickness without a serious drop in sensitivity. It is reported that the sensor measures the average moisture content of multilayer sheets, such as those produced on multicylinder machines.

As reported, the system scans up to 1000 points per second. It is possible to pick out a particular wet streak and use it in the control strategy, or to delete particular wet streaks as desired.

Because of the commercial importance of moisture content in agricultural products, much effort in recent years has been directed toward grain and other moisture meters, including battery-operated portable instruments that assist not only grain elevator operators and processors but growers as well, and in some instances moisture detectors have been attached to harvesting equipment.

Similarly, much research and development activity has been directed toward soil moisture measurement. In one system the sensor is basically a liquid-crystal oscillator (30 MHz) arranged so that its frequency is sensitive to the capacitance of external electrodes which compose the walls of the case. A fringing field penetrates into the soil in the region of the insulating gap between the electrodes (Fig. 15). The measurement region sensitive to soil moisture is indicated. Oscillator output is coupled to the coaxial line which also carries the dc power supply and control pulses.

Upon digital command (advance pulse), the electrode is electrically disconnected (SW-1) from the oscillator so that the oscillator base frequency can be measured. Long-term drifts in oscillator frequency are compensated for by recording the frequency difference as the soil capacitance is switched in. Both switches are controlled by a modulus 4 (2-bit) counter, which is cleared by the reset pulse and clocked by the advance pulse.

Electrical Conductivity Methods. The relationship between dc resistance and the moisture content of such materials as wood, textiles, paper, and grain is the physical basis of measurement. Specific resistance when plotted against moisture content results in an approximate straight line up to the moisture saturation point. Beyond saturation, where all the cells and intermediate spaces in the material are saturated with free water, electrical conductivity methods are not reliable. This point varies from about 12 to 25 percent moisture content, depending on the type of material. The sample under test is applied to suitable electrodes in

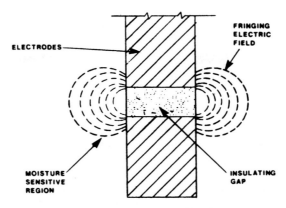

FIGURE 15 Schematic diagram of soil moisture measurement system. Only moisture in the vicinity of the sensor insulating gap [0.5 inch (13 mm)] is detected by the sensor. The sensor response corresponds to the average moisture within one or two gap diameters. For permanent installation, the hole is refilled, with the cable left in place. The sensor will operate under water.

the form of needle points for penetration into the material. Granular or fibrous materials may make use of electrodes in the form of a cup or clamp arrangement to confine the material to a fixed volume. The electrodes and sample under test comprise one arm of a Wheatstone bridge, as shown in Fig. 16.

The high sensitivity required of the detector dictates the use of electronic amplifiers. The range of resistance values corresponding to the normal moisture content varies from less than 1 to 10,000 MΩ or higher, depending on material, electrode design, and moisture content. An increasing moisture content results in decreasing resistance values. The effects of moisture on the electrical characteristics of various materials are shown in Fig. 17.

Radio-Frequency Absorption. The attenuation of electromagnetic energy when passed through a material is the physical principle used. Radio-frequency (RF) devices are operated at frequencies below 10 MHz, and for best results the products being measured must be made up of polar materials. The RF energy is passed through the polar material, and the water molecules absorb some of the energy in the form of molecular motion. A block diagram of a typical RF instrument is shown in Fig. 18. By means of suitable electrodes for the sampling process, the instrument can be applied to solid or sheet materials. Granular materials can be

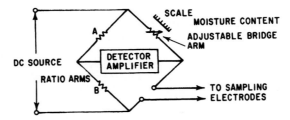

FIGURE 16 Basic bridge circuit used for conductivity method of moisture content measurement.

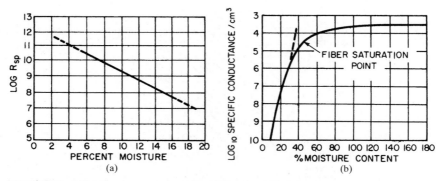

FIGURE 17 (*a*) Moisture content versus logarithm of specific resistance for spruce, basswood, fir, cedar, and hemlock woods. (*b*) Effect of moisture content on specific conductance of redwood. Note change in slope that occurs at approximately 30 percent moisture content. This point is the fiber saturation point and correlates with other testing methods. (*U. S. Forest Products Laboratory.*)

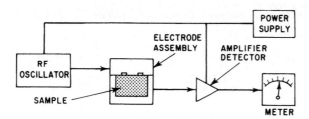

FIGURE 18   Radio-frequency absorption-type moisture indicator.

placed in cell-type electrodes. The test results will be affected by any polar material in the sample. Thus the results are nonspecific to water. However, water of hydration in a product will not be detected. Instruments of this type must be calibrated to the particular product under test. Although these instruments usually are available as periodic devices, they can be adapted for continuous measurements on some products. The moisture content range is dependent on the material and ranges from 0.1 to 60 percent. For certain materials the range is quite narrow.

Microwave Absorption.   The frequency of the electromagnetic energy used in these devices is in the region of 1000 MHz and higher. The 2.45-GHz (*S*-band), the 8.9- to 10.68-GHz (*X*-band), and the 20.3- to 22.3-GHz (*K*-band) regions all have been used. Water molecules greatly attenuate the transmitted signal with respect to other molecules in the material in the *S*- and *X*-band frequencies. In the *K*-band frequencies the water molecule produces molecular resonance. There are no other molecules that respond to this particular resonant frequency. Thus the frequency is most specific to the moisture (free water) content in paper products. The wavelengths of the *K*-band frequencies are approximately 1.35 to 1.5 cm. At these very short wavelengths, the energy may be guided or transmitted by waveguides. A block diagram of a microwave moisture instrument is shown in Fig. 19. Energy absorption can be detected by attenuation (loss) or by phase shift methods, or both measurements can be made simultaneously. The useful range of these devices is from less than 1 to 70 percent moisture and higher. At low moisture contents, *S*- and *X*-band devices require a minimum mass of

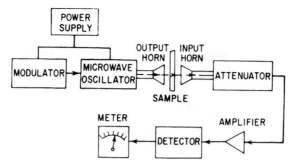

FIGURE 19   Microwave absorption-type moisture indicator.

material for practical operation. The accuracy of these instruments is within ±0.5 percent of the indicated value—up to approximately 15 percent moisture content.

Aluminum Oxide Impedance Moisture Sensor.   Aluminum oxide sensors, previously described for humidity measurements, also are effective detectors of moisture in liquids (Fig. 20). The aluminum oxide sensor consists of an aluminum strip anodized to provide a porous oxide layer. A very thin coating of gold is evaporated over this structure. The aluminum base and the gold layer form the two electrodes of what is essentially an aluminum oxide capacitor. Water vapor is rapidly transported through the gold layer and adsorbed by the pore walls of the sensor. The number of water molecules so adsorbed is proportional to the fugacity[2] of the water in the liquid. At equilibrium conditions, the fugacity of the water in the liquid is equivalent to the water vapor pressure above the liquid. In most cases it is directly proportional to the concentration of dissolved water in the organic liquid.

Since the pore openings are small in relation to the size of most organic molecules, admission into the pore cavity is limited to small molecules, such as water. The number of water molecules adsorbed determines the pore wall conductivity which influences the value of the sensor's electrical impedance. The surface of an aluminum oxide sensor can then be viewed as a semipermeable structure allowing the measurement of water vapor pressure in liquid organics in the same manner as is accomplished in a gaseous medium. In general, the aluminum oxide transducer provides identical values of electrical output, whether immersed directly in the fluid or placed in the gas space immediately above, that is, the measured vapor pressure is identical in both phases.

The measured vapor pressure can be used to determine the weight content of dissolved water within an organic liquid by application of Henry's law, which states that the mass of gas dissolved by a given volume of solvent, at constant temperature, is directly proportional to the pressure of the gas with which it is in equilibrium. The concentration of water in an organic liquid, expressed in parts per million by weight or in milligrams per liter, equals the partial pressure of water vapor times a constant:

$$C_W = KP_W$$

Vibrating Quartz Crystal.   In one system the vibrational frequency changes of a hygroscopically sensitized quartz crystal are measured. The sample gas is divided into two streams. One stream is passed directly to the measuring crystal; the other is passed through a dryer and then

_____

[2] In thermodynamics, fugacity is a measure of the tendency of a substance to escape by some chemical process from the phase in which it exists.

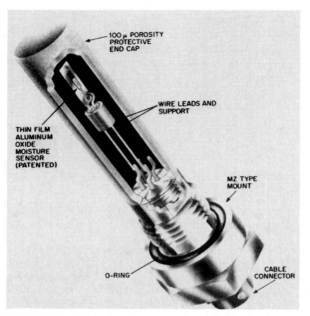

FIGURE 20   Aluminum oxide sensor with principal elements indicated. The impedance of the probe ranges from 2 MΩ to 50kΩ at 77 Hz, depending on the vapor pressure of water. (*Panametrics.*)

to the same measuring crystal, alternating with the first stream every 30 seconds. The wet and dry vibrational frequencies are electronically monitored and compared to the frequency of an uncoated sealed reference crystal. These data are further processed, and the correct moisture level is computed. The instrument is available with a programmable microprocessor (programmable read-only memory) to furnish an up-to-date display and binary-coded decimal (BCD) output compatible with computer processing. The instrument is designed for on-line monitoring of noncorrosive process gases and measures moisture levels as low as 1 ppm in such gases as hydrogen, ethylene, refrigerants, and natural gas. The standard output is in BCD form with digital readout, directly in parts per million (volume) or pounds per cubic foot (kilograms per cubic meter). Optionally, the output can be converted to a continuous analog signal of 0 to 10 mV, or 4- to 20-mA direct reading, updated every 30 seconds for display on a strip chart or used as an alarm or control signal. The control module can be located up to 2000 ft (610 m) from the sensor.

SECTION 5
GEOMETRIC AND MOTION SENSORS*

J. N. Beach

MICRO SWITCH, a Division of Honeywell Inc., Freeport, Illinois (Electromechanical Switches), (Proximity Sensors)

R. E. Gebelein

Moore Products Company, Spring House, Pennsylvania (Automatic Gaging—prior edition)

S. Longren

Longren Parks, Chanhessen, Minnesota

T. A. Morrow

Assistant Division Manager, Omron Electronics, Inc., Schaumberg, Illinois (Proximity Sensors), (Geometric and Motion Sensors)

Staff

Compumotor Division, Parker Hannifin Corporation, Rohnert Park, California (Encoders and Resolvers)

Staff

Daedal Division, Parker Hannifin Corporation, Harrison City, Pennsylvania (Metrology)

Staff

Daytronic Corporation, Miamisburg, Ohio (Geometric Transducers, Accelerometers)

Staff

Lucas Ledex Inc., Vandalia, Ohio (Metrology)

Staff

Lucas Schaevitz, Pennsauken, New Jersey (Linear and Angular Displacement Transducers)

Staff

Lucas Sensing Systems Inc., Phoenix, Arizona

---

* Persons who authored complete articles or subsections of articles, or otherwise cooperated in an outstanding manner in furnishing information and helpful counsel to the editorial staff.

Staff

MTS Systems Corporation, Minneapolis, Minnesota

Staff

Parker Digiplan, Ltd., Poole Dorset, U.K.

Staff

TSI Incorporated, St. Paul, Minnesota (Constant-Temperature Thermal Anemometer)

L. Thompson

MTS Systems Corporation, Eden Prairie, Minnesota (Magnetostrictive Linear Displacement Transducers)

Robert M. Whittier

Manager of Development Engineering, Endevco Corporation, San Juan Capistrano, California (Vibration Measurement)

METROLOGY, POSITION, DISPLACEMENT, AND THICKNESS TRANSDUCERS, SURFACE TEXTURE, QUALITY CONTROL, AND PRODUCTION GAGING[1]

Traditionally, the discrete-piece manufacturing and assembly industries, as represented by the appliance, land vehicle, aircraft, electronic components and instruments, business machine, fixture, jewelry, implement, and tool production, for example, have depended largely on the measurement and control of what may be termed the geometric variables for controlling product quality and production efficiency. Geometrically related variables of concern include such physical factors as dimension, straightness, shape, contour, displacement, position, and linear and rotary speed. Most of these variables, in essence, are related and can be derived from position determination.

[1] The cooperation of the technical and engineering staffs of the following organizations in furnishing information for this article is gratefully acknowledged: *Daytronic Corporation,* Miamisburg, Ohio; *Lucas Ledex Inc.,* Vandalia, Ohio; *Lucas Sensing Systems Inc.,* Phoenix, Arizona; *Lucas Schaevitz,* Pennsauken, New Jersey; *MTS Systems Corporation,* Minneapolis, Minnesota; *Parker Digiplan, Ltd.,* Poole Dorset, BH17 7DX (U.K.); *Parker Hannifin, Compumotor Division,* Rohnert Park, California. Special appreciation is also extended to Steve Longren, Chanhassen, Minnesota.

When these variables are integrated into a total system, in contemporary parlance, they are commonly referred to as motion control systems.

BASIC METROLOGY

From a practical standpoint, industrial metrology falls into several main categories: (1) the basic instruments, tools, and standards normally found in the well-equipped metrology laboratory responsible for establishing and maintaining manufacturing quality standards; (2) instrumental methods for assisting inspectors in manually gaging parts, workpieces, jigs, fixtures, and final assemblies; (3) automated systems for gaging parts and pieces in high-volume production; (4) dimensionally oriented motion control systems for machine tools, assembly lines, welding, painting, fabricating, and testing, among scores of geometrically sensitive operations; and (5) thickness measurement and control of metal, plastic, wood, paper, glass, films, and plated and sprayed coatings. Closely allied position-sensitive operations include registration control, as found in printing and labeling, and object-detection systems, where control actions are taken because an object occupies (even for an instant) a definite (or approximate) position (location).

DIMENSIONAL STANDARDS

Length is expressed in terms of the meter, abbreviated m. The meter is defined as 1,650,763.73 wavelengths in vacuum of the orange-red line of the spectrum of krypton 86. An interferometer is used to measure length by means of light waves.

A revised definition of the meter, based on very accurate measurement of the speed of light, was adopted by the General Conference on Weights and Measures in 1983. The meter has been defined as the length of path traveled by light in a vacuum during a time interval of 1/299,792,458 of a second.

Practical realization of this definition is obtainable through time-of-flight measurements, by frequency comparisons with laser radiations of known wavelengths, or through interferometric comparisons with radiations of stated wavelengths, such as that from krypton 86.

Other SI length-related units include:

1. *Area,* expressed in terms of the square meter (m^2). Land is often measured by the hectare (10,000 m^2, or approximately 2.5 acres).

2. *Volume,* expressed in terms of the cubic meter (m^3). The liter is a special name for the cubic decimeter (0.001 m^3).

Common equivalents and conversions are given in Table 1. Prefixes that may be applied to all SI units are listed in Table 2.

Interferometer

An interferometer is a precision instrument which uses the interference of light waves as the basis of measurement. The optics of an interferometer are designed so that the variance of known wavelengths and path lengths within the instrument permits accurate measurement of distances. The French physicist Jacques Babinet suggested in 1827 the possibility of using the wavelength of light as a standard of length. However, it was not until 1960 that the meter was officially defined by the International Bureau of Weights and Measures in terms of interferometry, as described previously, thus replacing the former prime standard of length, namely, the length of an artifact, that is, a platinum-iridium bar, exact replicas of which were maintained at various international standards institutes. Now, with adequate skills, the standard of length can be duplicated anywhere in the world with equal precision.

TABLE 1 Common Equivalents and Conversions

Approximate common equivalents		Conversions accurate to parts per million	
1 inch	= 25 millimeters	inches × 25.4*	= millimeters
1 foot	= 0.3 meter	feet × 0.3048*	= meters
1 yard	= 0.9 meter	yards × 0.9144*	= meters
1 mile	= 1.6 kilometers	miles × 1.60934	= kilometers
1 square inch	= 6.5 square centimeters	square inches × 6.4516*	= square centimeters
1 square foot	= 0.09 square meter	square feet × 0.0929030	= square meters
1 square yard	= 0.8 square meter	square yards × 0.836127	= square meters
1 acre	= 0.4 hectare	acres × 0.404686	= hectares
1 cubic inch	= 16 cubic centimeters	cubic inches × 16.3871	= cubic centimeters
1 cubic foot	= 0.03 cubic meter	cubic feet × 0.0283168	= cubic meters
1 cubic yard	= 0.8 cubic meter	cubic yards × 0.764555	= cubic meters
1 quart (liq)	= 1 liter†	quarts (liq) × 0.946353	= liters
1 gallon	= 0.004 cubic meter	gallons × 0.00378541	= cubic meters
1 ounce (avdp)	= 28 grams	ounces (avdp) × 28.3495	= grams
1 pound (avdp)	= 0.45 kilogram	pounds (avdp) × 0.453592	= kilograms
1 horsepower	= 0.75 kilowatt	horsepower × 0.745700	= kilowatts
1 millimeter	= 0.04 inch	millimeters × 0.0393701	= inches
1 meter	= 3.3 feet	meters × 3.28084	= feet
1 meter	= 1.1 yards	meters × 1.09361	= yards
1 kilometer	= 0.6 mile	kilometers × 0.621371	= miles
1 square centimeter	= 0.16 square inch	square centimeters × 0.155000	= square inches
1 square meter	= 11 square feet	square meters × 10.7639	= square feet
1 square meter	= 1.2 square yards	square meters × 1.19599	= square yards
1 hectare†	= 2.5 acres	hectares × 2.47104†	= acres
1 cubic centimeter	= 0.06 cubic inch	cubic centimeters × 0.0610237	= cubic inches
1 cubic meter	= 35 cubic feet	cubic meters × 35.3147	= cubic feet
1 cubic meter	= 1.3 cubic yards	cubic meters × 1.30795	= cubic yards
1 liter	= 1.057 quarts (liq)	liters × 1.05669	= quarts (liq)
1 cubic meter	= 250 gallons	cubic meters × 264.172	= gallons
1 gram	= 0.035 ounces (avdp)	grams × 0.0352740	= ounces (avdp)
1 kilogram	= 2.2 pounds (avdp)	kilograms × 2.20462	= pounds (avdp)
1 kilowatt	= 1.3 horsepower	kilowatts × 1.34102	= horsepower

* Exact.
† Based on the U.S. survey foot (= 0.3048006 meter).

TABLE 2 Prefixes That May Be Applied to All SI Units

Symbol	Prefix	Pronunciation	Multiples and submultiples
E	exa	ĕx'á	10^{18}
P	peta	pĕt'á	10^{15}
T	tera	tĕr'á	10^{12}
G	giga	ji'gá	10^{9}
M	mega	mĕg'á	10^{6}
k	kilo	kĭl'ō	10^{3}
h	hecto	hĕk'tō	10^{2}
da	deka	dĕk'á	10
d	deci	dĕs'ĭ	10^{-1}
c	centi	sĕn'tĭ	10^{-2}
m	milli	mĭl'ĭ	10^{-3}
μ	micro	mĭ'krō	10^{-6}
n	nano	năn'ō	10^{-9}
p	pico	pĕ'kō	10^{-12}
f	femto	fĕm'tō	10^{-15}
a	atto	ăt'tō	10^{-18}

The interferometer derives its name from the fact that it makes it possible to see light wave interference patterns as a series of bright and dark lines. If two light beams from a given source are directed optically over separate paths that may differ by as little as $\frac{1}{10}$ of a wavelength and then are recombined, detectable destructive interference will occur. Thus the interferometer may serve two roles: (1) to measure distances in terms of known wavelengths and (2) to make precise measurements of wavelengths. Currently laser interferometers are used industrially in the construction and assembly of precision machine tools where the laser can provide accuracies of better than 1.27×10^{-6} mm (20 millionths of an inch) in lengths up to 5080 mm (200 inches) (Fig. 1).

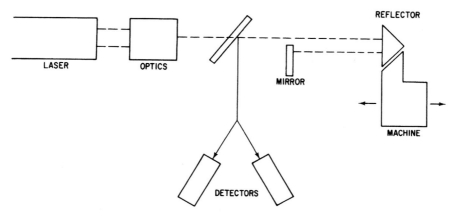

FIGURE 1   Laser interferometer capable of accuracies of better than 1.27×10^{-6} mm (20 millionths of an inch). Early designs required that the laser beam be reflected off a moving mirror that traveled along a track at least 1 meter long. This distance has been reduced to a few centimeters.

Optical Gratings

Diffraction gratings are used in optical instruments associated with precise dimensional measurements. Close, equidistant, and parallel grooves are ruled on a polished surface, commonly a glass base coated with aluminum. Gratings can be (1) reflection or (2) transmission types. The number of grooves range from several hundred to many thousand per inch (25 mm), depending on the dispersion required. The relative movement of two gratings produces optical patterns or fringes, sometimes referred to as moiré patterns (Fig. 2).

Gage Blocks

Usually furnished in sets of 81 blocks, ranging from 0.05 to 4.0 inches (1.27 to 101.6 mm), gage blocks serve as secondary standards traceable to the prime length standard. The blocks are rectangular pieces of hardened steel or carbide that are ground and polished flat with a square or oblong cross section. The length of a block is the perpendicular distance between the two opposite, polished faces. By carefully sliding the gaging surfaces of two blocks one over the other, they can be "wrung" together and thus built up to provide a useful range of standard lengths. The dimensions of individual blocks are established by interferometric methods with apparatus set up along the lines shown in Fig. 3. Class AA blocks have a tolerance of $\pm 2 \times 10^{-6}$ inch ($\pm 5.1 \times 10^{-5}$ mm) and are intended for reference purposes in temperature-controlled gage laboratories. These masters should be sent periodically to the National Institute of Standards and Technology (NIST), Gaithersburg, Maryland, for checking. Class A blocks are used in inspection departments and as masters. Blocks of class B and C quality are used throughout manufacturing for accurate measurements, tool setting, and instrument calibration.

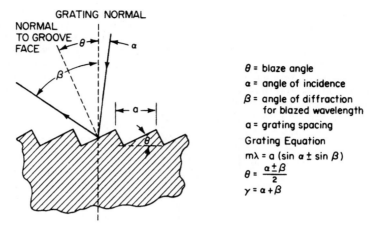

θ = blaze angle

α = angle of incidence

β = angle of diffraction for blazed wavelength

a = grating spacing

Grating Equation

$m\lambda = a\,(\sin\alpha \pm \sin\beta)$

$\theta = \dfrac{\alpha \pm \beta}{2}$

$\gamma = \alpha + \beta$

FIGURE 2   Cross section of diffraction grating showing the "angles" of a single groove, which are microscopic in size on an actual grating.

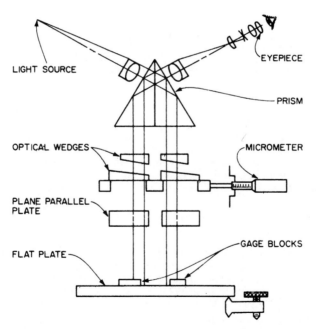

FIGURE 3   Interferometric comparator for comparing the lengths of two gage blocks. The difference in gage block lengths is measured by the horizontal distance through which optical wedges must be moved to bring the two interference patterns successively into coincidence with a cross hair in the eyepiece. Wedges and micrometer are rotated 90° for illustration.

Gage blocks are termed end standards. In addition there is need for line standards in the form of precision scales. Before the development of laser interferometric techniques, line standards were compared with master scales visually by use of a microscope. NIST now uses an automatic fringe-counting interferometer, which employs a laser light source to measure line standards directly in lengths up to 1 meter with a precision of a few parts in 100 million.

Autocollimator

An autocollimator is an optical instrument for directly measuring small angles of tilt of a reflective surface. The instrument effectively combines the functions of a collimator and a viewing telescope into a single system.

Goniometer

A goniometer is an angular measuring device essentially similar to a circular table or a dividing head, except that autocollimating telescopes are used to establish the datum values for the readings taken from the divided circle, either directly or through a micrometer eyepiece. The instrument is used for determining circular divisions and utilizes the basic principle of the divided circle.

Clinometer

Essentially a divided-circle instrument, the clinometer simplifies the transfer of angles between planes. Bubbles are used to establish the null setting principle of a precision clinometer, while less precise instruments compare a measured angle to a datum surface (Fig. 4).

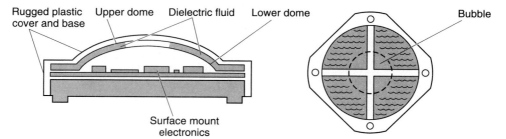

FIGURE 4   Dual-axis clinometer, which provides the function of two separate instruments. The sensor is composed of two hermetically sealed domes spaced about 3 mm (⅛ inch) apart. The lower, polyester plastic dome has four capacitance plates while an aluminum upper dome acts as a ground. A fluid with a high dielectric constant is sealed within the dome sandwich, leaving an air bubble about the size of a quarter. The bubble is centered at level position and will move from one side to the other as the device is tilted. When the sensor is tilted, the bubble (moving under the force of gravity) changes the capacitance. The resulting differential generates an output signal which reflects the relative tilt of the device in either the x or the y axis. The pin-selectable output signal provides either PWM or a dc analog voltage output. A fully electronic instrument is also available.

Clinometers are used in a number of industrial and commercial applications, including (1) level positioning of machine tools and other heavy machinery, manlifts, and cranes, and satellite and microwave antennas; (2) making camber, caster, and steering axis inclination adjustments in vehicle wheel alignment; (3) measuring pitch, roll, tilt, and angular position in construction equipment; (4) adjusting inclination of exercise equipment; (5) measuring heel angle or roll of sailboats and pitch and roll of undersea robot vehicles; and (6) leveling motor homes and RVs; automatic level control for private and commercial vehicles, among others. (*Lucas Sensing Systems Inc.*)

Theodolite

A theodolite is a surveying instrument that has been adopted for use in metrology. The instrument is comprised of a telescope in which rotation about the vertical axis is measurable. Theodolites are used for the alignment of large jigs and fixtures. Through triangulation, the instruments also can be used for length measurement.

Protractor

The familiar drafting protractor for measuring angles is now available in a digital electronic format (Fig. 5).

Optical Flats

An optical flat is a transparent disk, usually of quartz, two sides of which are parallel. One side is polished for clear vision of the surface of the gage, part, or tool on which the flat is placed. The other side of the disk is ground optically flat. Under proper conditions, interference bands are created. Observation of these bands can be used for (1) determining the flatness of a surface, that is, to assess the amount of concavity or convexity, and (2) measuring the linear difference between a reference gage and an inspection gage, or between a gage block and a highly accurate part. Any linear difference can be detected between approximately 0.051 mm (0.002 inch) as a maximum and 2.5 to 1.3×10^{-5} mm (1 to 2 millionths of an inch) as a minimum (Fig. 6).

Spherometer

The spherometer is used for determining the radius of curvature of lenses and other spherical surfaces. A widely accepted design consists of a depth-measuring device, such as a micrometer screw, mounted at the center of a tripodlike support. The concave or convex surface to be measured is centered directly under the spherometer, and the micrometer screw is rotated until its tip just touches the spherical surface. The displacement of the tip of the micrometer screw above or below the plane of the three support points is a measure of the radius of curvature of the surface. A typical spherometer reads to 0.01 mm, although much greater precision can be obtained with highly refined instruments.

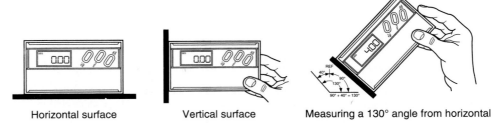

Horizontal surface   Vertical surface   Measuring a 130° angle from horizontal

FIGURE 5 Digital protractor (battery-operated, portable) measures angles in degrees, mils, percent grade, millimeters per meter, and inches per foot. The range is ±60 percent with a resolution of ±0.01° (0 to 20°); ±0.1° (20 to 60°); or ±0.05° (3 min) (0 to 19.99°), and a repeatability of ±0.05°. The instrument incorporates digital statistical process control (SPC) output for data storage and processing. Applications include level positioning of machinery tables, positioning of aircraft runway lights, aircraft flight control surfaces and prop angle, automobile drive angle, and for the measurement of innumerable manufactured parts during various stages of production as part of SPC requirements. (*Lucas Sensing Systems Inc.*)

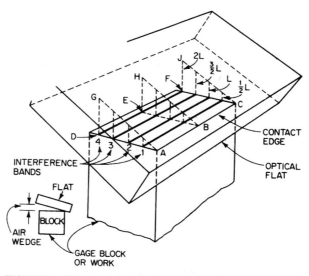

FIGURE 6 When an optical flat is manipulated so that a wedge of air is created, dark bands (interference bands) result at locations where the separation of flat and work equals a multiple of $L/2$, or one half-wavelength of the monochromatic light used.

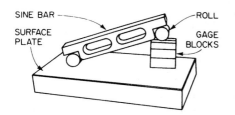

FIGURE 7 The sine bar allows work to be located in an accurate relationship to a plane or surface plate.

Sine Bar

A sine bar consists essentially of a bar serving as a straightedge and two cylindrical buttons, which may be on the side or on the undersurface. If the side-button type is used, one button rests on a gage block, and the thickness of that gage block is added to the height of the gage block stack used to set the second button. If a base-button sine bar is used, the first button rests directly on the surface plate or master flat. The second button rests on a stack of blocks equivalent to the sine of the wanted angle (Fig. 7).

Ellipsometer

The ellipsometer is used to determine the thickness of monomolecular dimensions. The instrument is basically a polarization interferometer utilizing a photometer as a readout device. A production-type ellipsometer can measure surface film thickness from a few angstroms to 1 μm and permits rapid determinations of thickness and refractive index. In one type, a laser source operates in a fully lighted room, providing a small intense spot that permits the study of very small samples. A solid-state detector is used. Automatic ellipsometers are available. In operation, a specimen is placed on the stage and the start button is depressed. A minicomputer displays and prints the film thickness in less than 4 seconds for most samples. High speed and simplicity make such an instrument ideal for monitoring rapidly changing films and for high-volume measurements in semiconductor manufacture.

Optical Comparator

Also known as a contour projector, the optical comparator is used for accurate visual inspection and measurement of a part in which a magnified image of the part or portion thereof is projected onto a viewing screen. Among their uses are the toolroom checking of cutter shapes and sizes and some delicate assembly operations. A contour projector can be conceived of as a variety of gages combined in a single unit.

Optical Bench

Although not compactly assembled in a black box, an optical bench (with accessories) may be considered a modular, highly flexible, disassembled instrument capable of convenient, customized assembly by a skilled user. An optical bench is a graduated support on which carriages for holding lenses, mirrors, and other components can be mounted and positioned for making precision optical measurements.

Measuring Machine

This is a highly instrumented device used to measure and record linear dimensions and xyz coordinates of holes and surfaces in parts and tools. Various styles operate under manual, motorized, or computer control. Direct-measuring machines with digital readout have been used widely. Some machines provide direct or absolute measurement of probe position or movement in three axes with a resolution ranging from 0.02540 to 0.00127 mm (0.001 to 0.00005 inch) (Fig. 8).

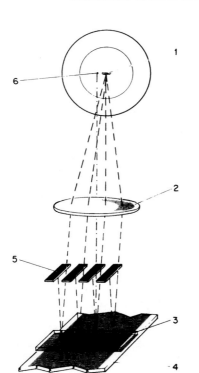

FIGURE 8   Optical system of one type of measuring machine. (1) Line filament; (2) collimating lens; (3) index grating; (4) scale grating; (5) photocell strips; (6) principal focus of lens.

Miniature Positioning Tables (Stages)

Discrete-part manufacture and assembly is not confined to comparatively large items as encountered, for example, in automotive, aircraft, appliance, and machinery production, but the instrumentation technology also applies to very small parts and assemblies as may be encountered in the construction of instruments, controllers, electronic components, circuit systems, jewelry, and watchmaking, among numerous other products. Although, in some cases, these miniaturized production and testing setups are highly automated, in cases of low volume or in product development, miniature positioning tables (or stages) are used widely. As the size of the components and assemblies increases (in recent years by orders of magnitude), the need for precision and performance repeatability obviously increases.

Specific applications of miniature and small positioning tables include (1) frequent or one-time fine adjustments, (2) pinhole micrometer positioning (piggyback on a larger workstage), (3) fiber-optics research and development and alignment, (4) quality control testing and gaging, (5) laser scribing and cutting, (6) automated assembly and component insertion, (7) positioning probes and fine gas purges, (8) axial alignment of tubes and rods, and (9) individual positioning of elements of small gas lasers, among others.

Small and miniature positioning tables (stages) are available in a wide variety of configurations, including 1–2–3 (x–y–z) axes, manual and motorized drives, linear and rotary positioning, and ball and dovetail slides, and they can be equipped with a variety of instrumentation, including computer-controlled programmable systems (Figs. 9 through 11).

Positioning Table Geometry

Several key geometric factors that apply to large and small tables and stages affect their performance characterics, such as the following.

Resolution.   Resolution is the smallest attainable increment of adjustment or positioning. With manually adjusted positioners, resolution is defined as the smallest movement achievable by controlled rotation of the adjustment screw or micrometer.

FIGURE 9a   Linear stages for controlled, precise point-to-point positioning along a linear axis. Stages are comprised of (1) a precision linear ball slide which serves as a linear bearing and guide and (2) a drive mechanism which moves and positions the slide top along the linear axis. Three drive mechanisms are available, a fine screw, a micrometer, and a differential screw. The fine screw is used for fine-resolution positioning. The micrometer is used whenever a position readout is required. The differential screw is used for applications that require extremely fine-resolution positioning. Maximum mounting surface ranges from 32 by 32 to 152 by 152 mm; maximum load capacity is 3 to 45 kg (horizontal) and 0.5 to 27.5 kg (vertical); travel is 12.5 to 50 μm; straight-line accuracy is 2 μm per 25 mm of travel; and positional repeatability is 1.3 μm. (*Daedal Division, Parker Hannifin Corporation.*)

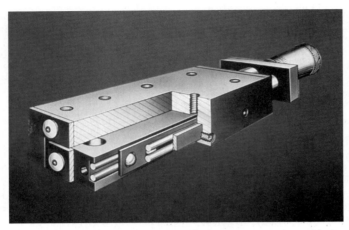

FIGURE 9b   Linear stages for controlled, precise point-to-point positioning along a linear axis. (*Daedal Division, Parker Hannifin Corporation.*)

FIGURE 10   Rotary stages and tables of varying sizes. They are used for controlled rotation and angular positioning. Each table is comprised of (1) a fixed housing (base), (2) a rotating member (shaft), (3) a bearing system, and (4) a control or drive mechanism. The bearing system rigidly supports the shaft and allows it to rotate freely within the housing. The drive mechanism controls shaft rotation, thereby converting the unit from a free rotating bearing assembly to a controllable rotary positioning device. In some cases the drive mechanism is a tangent arm drive. In other configurations a precision worm gear drive is used. This provides continuous angular positioning over a full 360° range. The stages range from 48 to 120 mm in diameter; load ranges from 4.5 to 22 kg (horizontal) and 1.5 to 9 kg (vertical); range is 360° continuous; vernier is 6 arc minutes. (*Daedal Division, Parker Hannifin Corporation.*)

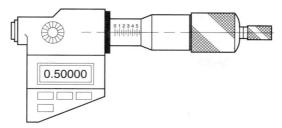

FIGURE 11 Precision electronic digital micrometer head provides a liquid crystal display (LCD) readout to 0.001-mm resolution. The micrometer features incremental or absolute positioning modes, zero set at any position, millimeter and inch readout, display hold, and automatic shutdown after 2 hours to conserve the integral battery. The micrometer travel is 25 mm. The battery will power the unit for 500 hours of use. (*Daedal Division, Parker Hannifin Corporation.*)

Repeatability.   This term defines how accurately a given position can be repeated or returned to an original location.

Positional Accuracy.   This is defined as the maximum achievable difference (error) between expected and actual travel distance. Pitch and yaw affect positional accuracy, as depicted in Fig. 12.

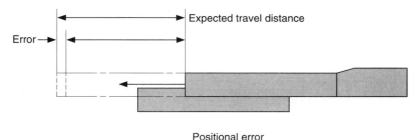

(a)

FIGURE 12a   Positional error as affected by pitch error and yaw error. (*Daedal Division, Parker Hannifin Corporation.*)

Straight-Line Accuracy.   Also called straightness and flatness of travel, it is, in theory, a linear slide or stage that moves along its axis of travel in a perfectly straight line. In reality the actual travel path deviates from the true straight line and flat line in both the horizontal and the vertical directions, respectively. Straight, flat line accuracy is defined as the maximum distance that the travel path deviates from the theoretical straight line in either plane, measured from the moving carriage surface center. Deviations result from the effects of yaw, pitch, and roll (Fig. 13).

Concentricity.   In theory, it requires that any point on the surface of a rotating table should travel along a path that forms a perfect circle. In reality, the actual path of travel will deviate from the perfect circle. Concentricity defines the maximum difference between a true circle and the actual circular path formed by the rotating point (Fig. 14).

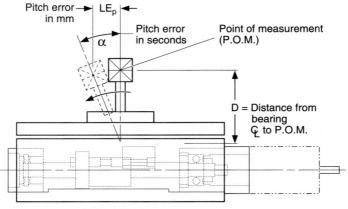

Table side view

Formula: $L_{EP} = [\text{Tan}\,(\alpha \div 3600)] \times D$
Example: $\alpha = 10$ sec; $D = 60$ mm
$L_{EP} = [\text{Tan}\,(10 \div 3600)] \times 60$ mm $= 2{,}9$ μm

Pitch error

(b)

FIGURE 12b Positional error as affected by pitch error and yaw error. (*Daedal Division, Parker Hannifin Corporation.*)

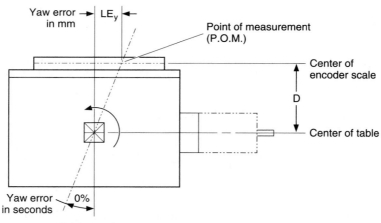

Table top view

Formula: $LE_Y = [\text{Tan}\,(\alpha \div 3600)] \times D$
Example: $\alpha = 10$ sec; $D = 85$ mm
$LE_Y = [\text{Tan}\,(10 \div 3600)] \times 85$ mm $= 4{,}1$ μm

Yaw error

(c)

FIGURE 12c Positional error as affected by pitch error and yaw error. (*Daedal Division, Parker Hannifin Corporation.*)

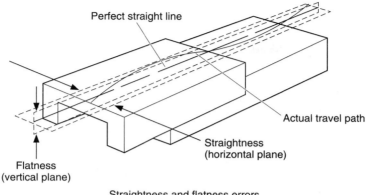

Straightness and flatness errors

(a)

FIGURE 13a Straightness of travel error as affected by yaw error and roll error. (*Daedal Division, Parker Hannifin Corporation.*)

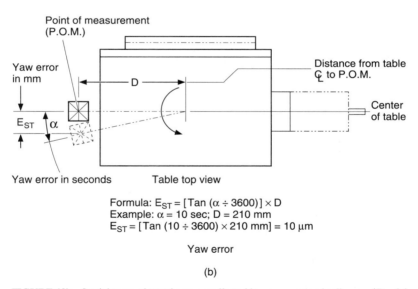

Formula: $E_{ST} = [Tan\ (\alpha \div 3600)] \times D$
Example: $\alpha = 10$ sec; $D = 210$ mm
$E_{ST} = [Tan\ (10 \div 3600) \times 210$ mm$] = 10\ \mu$m

Yaw error

(b)

FIGURE 13b Straightness of travel error as affected by yaw error and roll error. (*Daedal Division, Parker Hannifin Corporation.*)

Runout (Wobble). In theory, it requires that any point on a rotary table should remain within a perfectly flat plane that is perpendicular to the axis of rotation. Runout describes the maximum distance that point will deviate from that plane (Fig. 15).

Other factors that must be considered when specifying a positioning table include (1) horizontal load capacity, the maximum load when the device is mounted or placed on a horizontal plane; (2) vertical load capacity, the maximum (pitch) load capacity (center of gravity, a specified design distance) of a positioning device when the device is mounted in the vertical

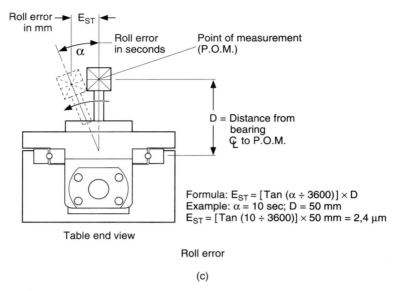

Formula: $E_{ST} = [Tan (\alpha \div 3600)] \times D$
Example: $\alpha = 10$ sec; D = 50 mm
$E_{ST} = [Tan (10 \div 3600)] \times 50$ mm = 2,4 μm

Table end view

Roll error

(c)

FIGURE 13c Straightness of travel error as affected by yaw error and roll error. (*Daedal Division, Parker Hannifin Corporation.*)

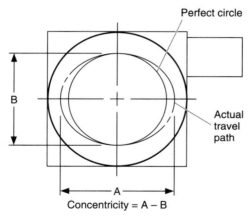

FIGURE 14 Demonstration of concentricity and deviation. (*Daedal Division, Parker Hannifin Corporation.*)

plane (vertical load capacity is a measure of the load capacity effects of table bearing capacity and screw pitch); (3) overall load capacity, the maximum weight that a positioning device can support without causing excessive wear or damage to the device.

Machine Conditions and Requirements

The degree of positioning accuracy and repeatability required by various production machines differs widely: 0 ± 0.0025 mm (±0.0001 inch) for boring machines, ±0.025 mm (±0.001 inch) for drilling and contour

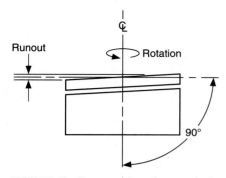

FIGURE 15 Demonstration of runout (wobble). (*Daedal Division, Parker Hannifin Corporation.*)

milling, ±0.127 mm (±0.005 inch) for armature insulator assembly, ±0.254 mm (±0.01 inch) for tube bending and frame welding, ±0.381 (±0.015 inch) for automobile seat cushion spring welding, and ±0.1 percent of full-scale reading for asphalt batching machines. In the manufacture of solid-state circuits and components in the electronics industry, involving laser and electron beam manipulation, accuracy is in terms of a fraction of a micrometer. The complexity of the control system varies with the type of machine, particularly with the number of axes that must be controlled—ranging from two up to six or more axes. The speed of response needed is related to the total cycling time of the machine and ranges from seconds and fractions of seconds up to 2 minutes or greater.

Backlash.   To position a machine member with acceptable accuracy, it is necessary to establish the extent of the backlash or dead band region for the positioning mechanism used. The measuring transducer and its attendant dead band characteristics, when used with the machine member, determine the amount of dead band or backlash to be included in the total control system loop. For all types of repeat-back devices whose mechanical input is provided by rotating a shaft, an important consideration is the means of coupling the device to the positioned machine element. Frequently, a gear train is required to reduce member travel to one revolution or less of the transducer shaft. It is necessary to determine whether any backlash in this train is comparable, when expressed as an arc of the transducer shaft, with the positional accuracy requirement of the machine member. Satisfactory results on the basis of this comparison rest on the assumption that the electrical and mechanical error factors for the transducer are small ($\frac{1}{10}$ or less) compared with the machine member positioning tolerances. Sometimes a transducer that does not require a mechanical input shaft should be considered. Where this is impractical, a separate rack and gear train, both exhibiting minimal backlash, may be used to position the mechanical input shaft of the associated transducer.

Stable Machine Base.   The controlled member may have undesired movement with respect to the machine base in a direction transverse to the controlled axis of travel. If the slide and table ways wear nonuniformly, variation in the transverse position of a point on the table of a machine may cause a variation in the air gap of a magnetic slot transducer system, for example. The same problem will result in a misalignment of optical transducer systems if the table motion becomes crablike after wear of the slides has progressed.

Vibration.   Nondata components of both a cyclic and a random nature may be superimposed on the true data because of machine-induced vibration of the transducer. Thus every effort should be made to reduce these effects and to take the residue effects fully into account when designing the total positioning system.

POSITION SENSORS AND TRANSDUCERS[2]

Position measurements fall into two broad classifications: (1) measurements concerned with rotary motions where the position of rotary machine driving members, such as shafts and

---

[2] Pneumatic sensors are discussed later in this article under "Production Gaging Systems."

screws, is translated into terms of linear motion or differences between two or more positions, each with its own coordinates; and (2) measurements concerned directly with linear motions.

ROTARY MOTION

In general terms, an encoder may be defined as a device which translates mechanical motion into electronic signals used for monitoring position or velocity. Encoders are available for use in both rotary and linear motion systems.

Encoders are of two basic types: (1) Absolute encoders provide a unique output signal for each single or multiple revolution of shaft gearing. An absolute encoder outputs a complete binary code (digital output) for each position. These devices are generally used in applications where position information rather than change in position is important. Absolute encoders have an individual digital address for each incremental move, and thus the position within a single revolution can be determined without a starting reference. By gearing two or more absolute encoders together, so that the second advances one increment for each complete revolution of the first (reminiscent of a mechanical counter), the range of absolute position can be extended. (2) Incremental encoders produce a symmetrical pulse for each incremental change in position. Pulses from the incremental encoder are counted for each incremental movement from a calibrated starting point in an up-down counter to track position.

Rotary Encoders

Over many years, a number of rotary encoders have been developed, including magnetic, contact, resistive, and optical devices. In recent years, the optical encoder has been preferred for many applications, including both absolute and incremental forms.

Rotary Incremental Optical Encoders

As shown in Fig. 16, optical encoders operate by means of a grating which moves between a light source and a detector. When light passes through the transparent areas of the grating, an output is seen from the detector. For increased resolution, the light source is collimated and a

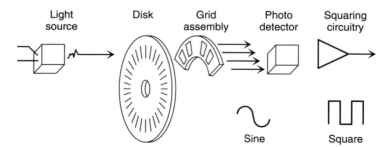

FIGURE 16   Rotary incremental optical encoder. Incremental encoders rely on a counter to determine position and a stable clock to determine velocity. These systems use a variety of techniques to separate "up" counts from "down" counts, and maximize the incoming pulse rate. Correct position information depends on accurate counting and the proper transmission of pulses. Use of a differential line drive can prevent false counting due to electrical noise developing stray electrical pulses. (*Parker Hannifin Corporation, Compumotor Division.*)

mask is placed between the grating and the detector. The grating and the mask produce a shuttering effect, so that only when their transparent sections are in alignment is light allowed to pass to the detector.

A rotary incremental optical encoder consists of five basic components: (1) a light source (LED or incandescent), (2) an encoder disk, (3) a grid assembly, (4) a photodetector, and (5) amplification electronics.

The disk shown in Fig. 17 is a key element of the encoder and is typically a glass material with imprinted marks of metal with slots precisely positioned. The number of marks or slots is equivalent to the number of pulses per turn. A typical resolution is 500 or 1000 pulses per revolution. (For example, a glass disk imprinted with 1000 marks would have moved 180° after 500 pulses. The maximum resolution of the encoder is limited by the number of marks or slots that can be physically located on the disk. Obviously, the larger the disk diameter, the greater the potential number of pulses per revolution. In general, commercially available rotary incremental encoders have an overall diameter of between 40 and 75 mm (1.5 and 3 inches).

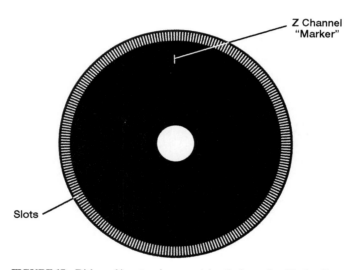

FIGURE 17   Disk used in rotary incremental optical encoder. (*Parker Hannifin Corporation, Compumotor Division.*)

Most rotary encoders also provide a single mark on the disk, called Z channel or marker. The pulse from this channel provides a reference once per revolution to detect error within a given revolution. The LED or other light source is constantly enabled. As the disk rotates, light reaches the photodetector at each slot or mark location. The detector is usually a phototransistor or, more commonly, a photovoltaic diode. This simple arrangement (Fig. 18), apart from its low output signal, has a dc offset which is temperature-dependent, making the signal difficult to use.

Thus in practice, two photodiodes are used with two masks, arranged to produce signals with 180° phase difference for each channel, the two diode outputs being subtracted so as to cancel the dc offset (Fig. 19). This quasisinusoidal output may be used unprocessed, but more often it is either amplified or used to produce a square-wave output. Incremental rotary encoders thus may have sine-wave or square-wave outputs and usually have up to three output channels.

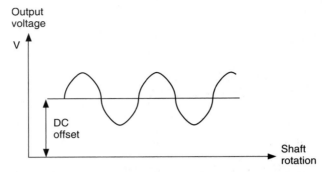

FIGURE 18   Output voltage (before conditioning) of rotary incremental optical encoder. (*Parker Hannifin Corporation, Compumotor Division.*)

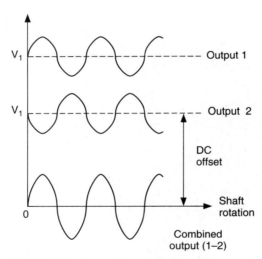

FIGURE 19   Output from dual-photodiode system used in rotary incremental optical encoder. (*Parker Hannifin Corporation, Compumotor Division.*)

A two-channel encoder, as well as giving the position of the encoder shaft, is also capable of providing information on the direction of rotation by examination of the signals to identify the leading channel. This is possible since the channels are normally arranged to be in quadrature, that is, 90° phase-shifted (Fig. 20).

For most machine tool or positioning applications, a third channel, known as the index channel or *Z* channel, is also included. This gives a single output pulse per revolution and is used when establishing the zero position.

Figure 20 also shows that for each complete square wave from channel *A,* if channel *B* output is also considered during the same period, then four pulse edges may be seen to occur. This allows the resolution of the encoder to be quadrupled by processing the *A* and *B* outputs to produce a separate pulse for each square-wave edge. For this process to be effective, however, it is important that quadrature be maintained within the necessary tolerances so that the pulses do not run into one another.

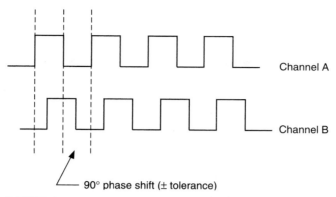

FIGURE 20 Quadrature output signals of rotary incremental optical encoder. (*Parker Hannifin Corporation, Compumotor Division.*)

Square-wave output encoders are generally available in a wide range of resolutions (up to about 5000 lines per revolution), and with a variety of different output configurations, such as the following.

Transistor-Transistor Logic. The encoder is commonly available for compatibility with transistor-transistor logic (TTL) levels and normally requires a 5-volt supply. TTL outputs are also available in an open-collector configuration, allowing the system designer a choice of pull-up resistor value.

CMOS (Complementary Metal-Oxide Semiconductor). The encoder is available for compatibility with the higher logic levels normally used with CMOS devices.

Line Driver. A low-output impedance device, it is designed for driving signals over a long distance, and usually is used with a matched receiver.

Complementary Outputs. These outputs are derived from each channel, giving a pair of signals, 180° out of phase. These are useful where maximum immunity to interference is required.

Noise of Rotary Incremental Encoder

The control system for a machine is normally screened and protected within a metal cabinet, and an encoder may be housed similarly, but unless suitable precautions are taken, the cable connecting the two can be a source of trouble due to its picking up electrical noise. This noise may result in the loss or gain of signal counts, giving rise to incorrect data input and loss of position. Noise problems are briefly addressed in Figs. 21 through 23.

Accuracy of Rotary Incremental Encoder

In addition to noise problems, there may be machine-related sources of error, such as the following.

Slew Rate (Speed). An incremental rotary encoder will have a maximum frequency at which it will operate (typically 100 kHz), and the maximum rotational speed or slew rate will be determined by this frequency. Beyond this, the output will become unreliable and accuracy will be affected. Further, if an encoder is rotated at speeds higher than its design maximum, conditions may occur which will be detrimental to the mechanical components of the assembly.

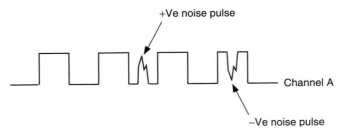

FIGURE 21   Example of effects of noise on encoder signal, showing how the introduction of two noise pulses has converted a four-pulse train into one of six pulses. Shielding of interconnecting cable may provide a solution, but inasmuch as the signals may be at low level (5 volts) and may be generated by a high-impedance source, further action may have to be taken. An effective way is to use an encoder with complementary outputs and connect this to the control system by means of shielded twisted-pair cable. (*Parker Hannifin Corporation, Compumotor Division.*)

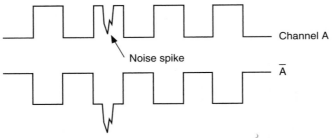

FIGURE 22   Encoder complementary output signals. The two outputs are processed by the control circuitry so that the required signal can be reconstituted without noise. (*Parker Hannifin Corporation, Compumotor Division.*)

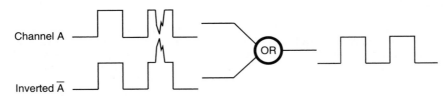

FIGURE 23   Complementary system. With reference to Fig. 22, if the A signal is inverted and is fed with the A signal into an OR gate (whose output depends on one signal or the other being present), the resultant output will be a square wave. (*Parker Hannifin Corporation, Compumotor Division.*)

Quantization Error. In common with digital systems, it is not easy to interpolate between output pulses, so that knowledge of position is accurate only to the grating width (Fig. 24).

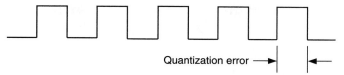

FIGURE 24 Demonstration of quantization error. (*Parker Hannifin Corporation, Compumotor Division.*)

Eccentricity Error. This error may be caused by bearing play, shaft run-out, incorrect assembly of the disk on its hub or of the hub on the shaft. Some of these error conditions include (1) amplitude modulation (Fig. 25*a*), (2) frequency modulation (as the encoder is rotated at constant speed, the frequency of the output will change at a regular rate, see Fig. 25*b*), and (3) interchannel jitter. The latter may occur if the optical detectors for the two encoder output channels are separated by an angular distance on the same radius. Then any "jitter" will appear at different times on the two channels, resulting in interchannel jitter.

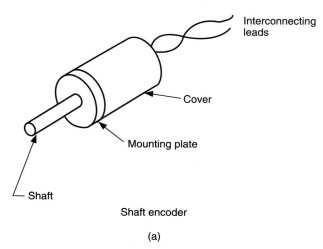

(a)

FIGURE 25a Errors caused by eccentricity problems. (*Parker Hannifin Corporation, Compumotor Division.*)

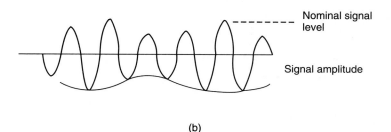

(b)

FIGURE 25b Errors caused by eccentricity problems. (*Parker Hannifin Corporation, Compumotor Division.*)

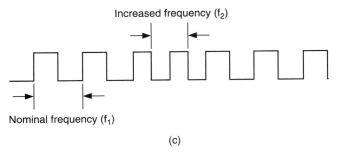

(c)

FIGURE 25c   Errors caused by eccentricity problems. (*Parker Hannifin Corporation, Compumotor Division.*)

Rotary Absolute Optical Encoders

An absolute encoder is a position verification device that provides unique position information for each shaft location. The location is independent of all other locations, whereas in the incremental encoder a count from a reference is required to determine position.

The disk used in the absolute optical encoder differs markedly from that used in the incremental encoder as previously shown in Fig. 17. In an absolute optical encoder there are several concentric tracks (unlike the incremental encoder with its single track). Each track has an independent light source (Fig. 26). As the light passes through a slot, a high state (true 1) is created. If light does not pass through the disk, a low state (false 0) is created. The position of the shaft can be identified through the pattern of 1s and 0s.

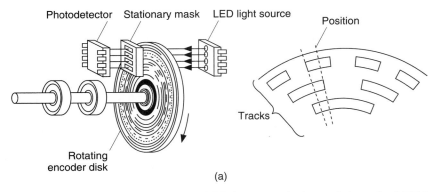

(a)

FIGURE 26a   Representative patterning of disks in a rotary absolute optical encoder. (*a*) Disk and associated elements of system and schematic representation of tracks and position. (*Lucas Ledex Inc.*)

The tracks of an absolute encoder vary in slot size, moving from smaller at the outside edge to larger toward the center. The pattern of slots is also staggered with respect to preceding and succeeding tracks. The number of tracks determines the amount of position information that can be derived from the encoder disk, that is, its resolution. For example, if the disk has 10 tracks, the resolution of the encoder usually would be 1024 positions per revolution, or 2^{10}.

For reliability it is desirable to have the disks constructed of metal rather than glass. A metal disk is not as fragile and has lower inertia.

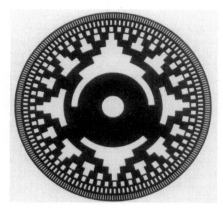

FIGURE 26b Another disk format. (*Parker Hannifin Corporation, Compumotor Division.*)

The disk pattern of an absolute encoder is in machine-readable code, usually binary, gray code, or a variety of gray. Figure 27 represents a simple binary output with 4 bits of information.

Multiturn Absolute Rotary Encoders

Gearing an additional absolute disk to the primary high-resolution disk provides for turns counting, so that unique position information is available over multiple revolutions (Fig. 28).

Advantages of Absolute Encoders

Both rotary and linear absolute encoders offer several advantages in industrial motion control and process control applications. These include the following.

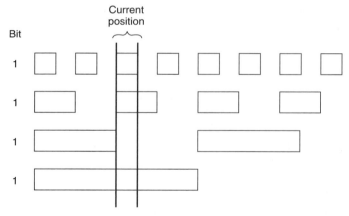

1101 = Decimal 11

FIGURE 27 Disk pattern of absolute rotary encoder is in machine-readable code. Illustration represents a simple binary output with 4 bits of information. The current location is equivalent to the decimal number 11. Moving to the right from the current position, the next decimal number is 10 (0-1-0-1 binary). Moving to the left from the current position, the next position would be 12 (0-0-1-1). (*Parker Hannifin Corporation, Compumotor Division.*)

1. *No position loss on power down or loss of power.*   An absolute encoder is not a counting device like an incremental encoder because an absolute system reads the actual shaft position. Lack of power does not cause the encoder to lose position information. Whenever power is supplied to an absolute system, it is capable of reading the current position immediately. In a facility where frequent power failures occur, an absolute encoder is necessary.

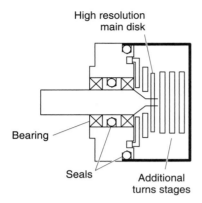

FIGURE 28   Multiturn absolute rotary optical encoder. The primary high-resolution disk has 1024 counts per revolution. A second disk with three tracks of information will be attached to the high-resolution disk geared 8:1. The absolute encoder now has eight complete turns of the shaft, or 8192 discrete positions. Adding a third disk geared 8:1 will provide for 64 turns of absolute positions. In theory, additional disks could continue to be incorporated but, in practice, do not exceed 512 turns. Encoders using this technique are called multiturn absolute encoders. This same technique can be incorporated in a rack and pinion style linear encoder, resulting in long length of discrete absolute locations. (*Parker Hannifin Corporation, Compumotor Division.*)

2. *Operates in electrically noisy environments.* Equipment, such as welders and motor starters, often generates electrical noise which can look like encoder pulses to an incremental encoder. Noise does not alter the discrete position that an absolute system reads.

3. *High-speed long-distance data transfer.* By using a serial interface, such as RS-422, absolute position data can be transmitted up to 1200 meters (4000 feet).

4. *Eliminates "go home" or referenced starting point.* An absolute system always "knows" its location. In numerous motion control applications it is difficult or impossible to find a "home" reference point. This situation may occur in multiaxis machines and on machines that cannot reverse direction. This feature can be particularly important in a "lights-out" manufacturing facility. This reduces scrap and setup time resulting from power loss.

5. *Reliable information in high-speed applications.* The counting device often limits the use of incremental encoders because a counter is limited to a maximum pulse input of 100 kHz. An absolute encoder does not require a counting device. This is not a limitation of an absolute encoder.

Encoders for Stepping and dc Servomotors

Particularly designed for application on rotary shafts where standard encoders cannot be used, and intended primarily for direct shaft and face mounting on stepping and dc servomotors is the encoder shown in Fig. 29. The unit is designed for velocity and position sensing.

Resolvers

These devices have a long history of use in machine position control systems. Because of their fundamental analog output signal, conversion is required by a synchro-to-digital (S/D) converter.

In principle, a resolver is a rotating transformer. The primary of the transformer is a winding on the shaft (rotor) much like a motor. The secondaries (stators) are wound in the case, again much like a motor. The difference in terminology of a resolver and a synchro is the number of stator windings. A resolver has two stator windings 90° apart, whereas a synchro has three windings 120° apart. In both devices, as the shaft turns, the relative positions of the rotor and stator windings change, and the root-mean-square (rms) voltage output of the stator winding varies as the sine of the angle between them. Only the ratio of the outputs is used. It should be pointed out that although a synchro is a three-wire motorlike device, it is not a three-phase device. The ac outputs are either in phase or 180° out of phase. Phase shift does not change with angle except to reverse the phase at certain angles. The resolver or synchro is coupled to the shaft to be measured, such as a pinion gear, lead screw, or robot arm, and then wired directly to an S/D converter. The latter then outputs a digital word for further processing (Figs. 30 through 32).

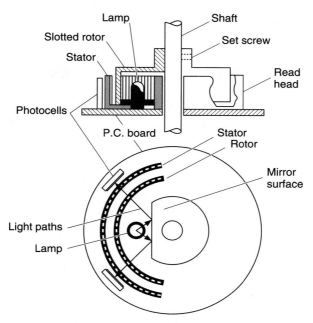

FIGURE 29 Position and velocity sensor intended primarily for direct shaft and face mounting on stepping and dc servomotors. Three types of disks are available—metal deposition on glass, photoemulsion on plastic, and etched metal. (*Lucas Ledex Inc.*)

The excitation voltage may be coupled to the rotating winding by slip rings and brushes, a disadvantage when used with a brushless motor. In such cases a brushless resolver may be used, as shown in Fig. 33.

Pancake Resolvers and Synchros

These transducers normally have large diameters and are of short length as compared with other resolvers. Diameters range from 50 to 90 mm (2 to 3.5 inches) (Fig. 34).

LINEAR MOTION

Linear Encoders

These encoders, which may be absolute or incremental, are used to make direct measurements of linear movements. As shown in Fig. 35, many linear encoders utilize rack and pinion technology. A linear encoder comprises a linear scale, which in standard models ranges from about 200 mm to 6 meters (8 inches to 20 feet), but considerably longer linear systems are constructed. Resolution is expressed in lines per unit length (normally, lines per centimeters or lines per inch). A linear absolute encoder is shown in Fig. 36. Typical mounting configurations are illustrated in Fig. 37.

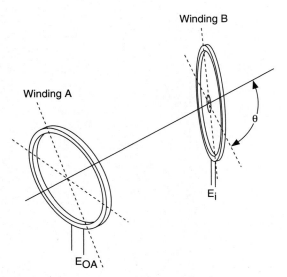

FIGURE 30   Resolver principle. Consider two windings *A* and *B*. If winding *B* is fed with a sinusoidal voltage, then a voltage will be induced into winding *A*. If winding *B* is rotated, the induced voltage will be a maximum when the planes of *A* and *B* are parallel and will be a minimum when the windings are at right angles. Also, the voltage induced into *A* will vary sinusoidally at the frequency of rotation of *B* so that $E_{OA} = E_i \sin \phi$. If a third winding *C* is positioned at right angles to winding *A,* then as *B* is rotated, a voltage will be induced into that winding, and that voltage will vary as the cosine of the angle ϕ, so that $E_{OC} = E_i \cos \phi$. (*Parker Hannifin Corporation, Compumotor Division.*)

Other Linear Position Transducers

Over the years, numerous methodologies have been developed to achieve linear positions and length measurements, some of which have been phased out while others are still in use. This limited description includes the following.

Inductive-Bridge Transducer.   This transducer type is used in production machines with restricted axis motion. Operation is based on the use of a fixed inductive member slightly longer than the axis to be measured (Fig. 38).

Inductive-Plate Transducer.   With this transducer type a slider moves across a scale, but with an air gap between the two. Since there is no physical contact, there is no apparent wear on the feedback device.

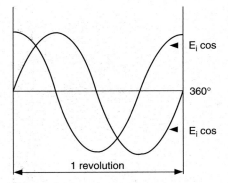

FIGURE 31   Resolver output. If the relative amplitudes of the outputs of the two windings *A* and *C* can be measured at a particular point in the cycle, these two outputs will be unique to that position. (*Parker Hannifin Corporation, Compumotor Division.*)

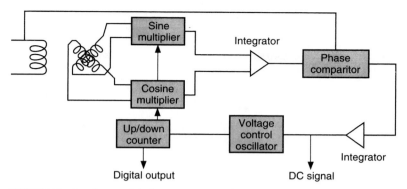

FIGURE 32 Resolver-to-digital converter. The information outputs (see Fig. 31) from the two phases are usually converted from analog to digital form for use in a digital positioning system. Resolutions up to 65,536 counts per revolution are typical. (*Parker Hannifin Corporation, Compumotor Division.*)

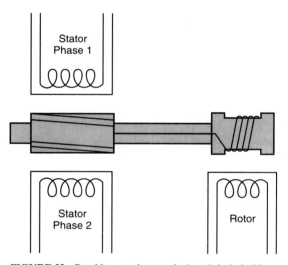

FIGURE 33 Brushless resolver used where it is desirable to avoid use of brushes and slip rings. (*Parker Hannifin Corporation, Compumotor Division.*)

The slider is attached to a movable push rod which can traverse up to 1270 cm (500 inches) per minute and attain a total of 2.5 mil travel cycles without replacing the seal. Scales are laser-checked to achieve an accuracy of ±0.0025 mm (±0.001 inch), are furnished in 60-meter (10-inch) lengths, and can be placed adjacent to one another for long travels of up to 60 meters (200 feet). Major applications have been on jig borers, horizontal boring machines, contouring machines, turning, milling, and drilling machines, as well as positioning tables, vertical turret lathes, and grinders (Fig. 39).

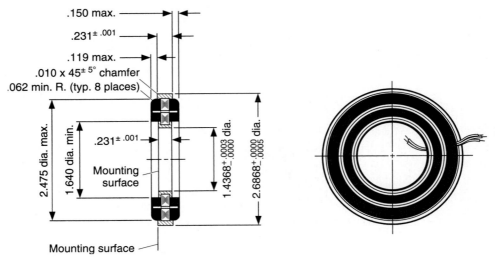

FIGURE 34 Representative pancake-type resolver, end and side views.

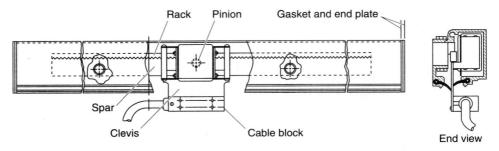

FIGURE 35 Linear encoder that utilizes rack and pinion technology. (*Parker Hannifin Corporation, Compumotor Division.*)

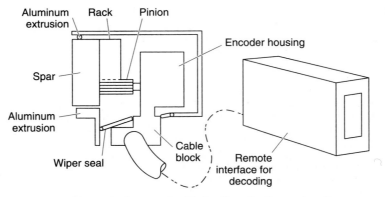

FIGURE 36 Linear absolute encoder. (*Parker Hannifin Corporation, Compumotor Division.*)

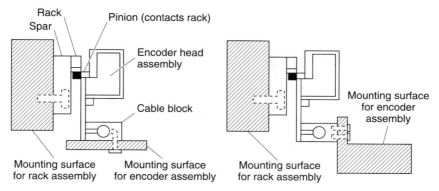

FIGURE 37 Typical mounting configurations of linear absolute encoder. (*Parker Hannifin Corporation, Compumotor Division.*)

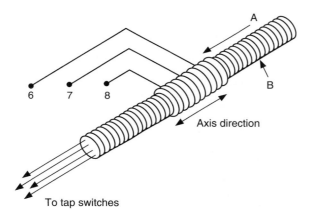

Position transducer

(a)

FIGURE 38a Inductive-bridge transducer for position measurement. Operation is based on the use of a fixed inductive member *B* slightly longer than the axis to be measured, and a movable member *A* approximately half the length of *B*. Selectable taps are placed on *B* in a successive decade with externally located inductors to provide a bridge configuration that may be externally unbalanced by placing *A* (coil) and *N* (point) across a pair of tap points; then the coil is moved until equal voltage prevails between the two ends of the coil, as evidenced by the occurrence of a small voltage at *O* and *O'*. A disadvantage of the system is the relatively large number of wires that must be taken from the device through the machine to the control system. An advantage is the high output voltage per unit of displacement, 2 mV/0.01 mm (5 mV/0.001 inch). Supply frequency usually is between 400 and 1500 Hz.

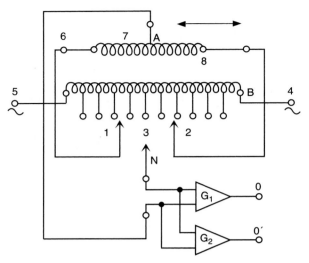

Schematic circuit diagram

(b)

FIGURE 38b   Inductive-bridge transducer for position measurement. Operation is based on the use of a fixed inductive member *B* slightly longer than the axis to be measured, and a movable member *A* approximately half the length of B. Selectable taps are placed on B in a successive decade with externally located inductors to provide a bridge configuration that may be externally unbalanced by placing *A* (coil) and *N* (point) across a pair of tap points; then the coil is moved until equal voltage prevails between the two ends of the coil, as evidenced by the occurrence of a small voltage at *O* and *O'*. A disadvantage of the system is the relatively large number of wires that must be taken from the device through the machine to the control system. An advantage is the high output voltage per unit of displacement, 2 mV/0.01 mm (5 mV/0.001 inch). Supply frequency usually is between 400 and 1500 Hz.

Magnetic Position Transducers.   A number of schemes have been developed which incorporate magnetic properties. They are shown diagrammatically in Fig. 40.

DISPLACEMENT TRANSDUCERS

Generally, displacement is thought of in terms of a motion of a few millimeters or less. Displacements as small as a wavelength of light can be measured. When displacements are quite large, the term distance may be preferred. Frequently a measurement of displacement is made to relate to some other measure, and hence displacement transducers are fundamental components of many instrumentation systems. A long established use is that of measuring the motion of the free end of a bourdon, the movement of a scale beam, or the deflection of an accelerometer. Displacement transducers are found in the measurement of thickness, among other variables.

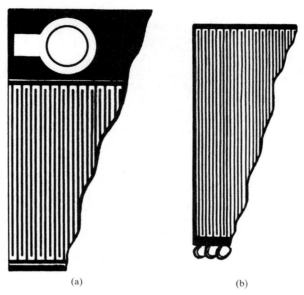

(a)   (b)

FIGURE 39 Inductive-plate position and motion transducer. (*a*)
Section of scale made up of a copper pattern bonded to heat-treated
steel. (*b*) Slider.

Displacement implies motion from one point to another; and it also implies position, that
is, a change from one position to the next. Displacement also implies the establishment of a
new position as related to a stable, normal, or reference position.

Linear Variable Differential Transformers

The linear variable differential transformer (LVDT) is an electromechanical device that produces an electrical output proportional to the displacement of a separate nonmagnetic movable core. As shown in Fig. 41, one primary and two secondary coils are arranged
symmetrically to form a hollow cylinder. A magnetic nickel-iron core, supported by a nonmagnetic push rod, moves axially within the cylinder in exact accordance with the mechanical
displacement of the probe tip.

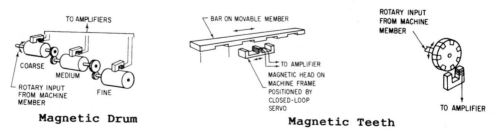

Magnetic Drum   **Magnetic Teeth**

FIGURE 40 Types of magnetic configurations used in position transducers.

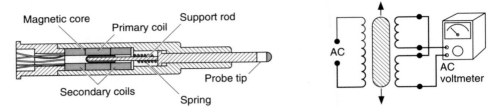

FIGURE 41   Typical linear variable differential transformer sensing probe and simplified circuit. (*Daytronic Corporation.*)

With ac excitation of the primary coil, induced voltages will appear in the secondary coils. Because of the symmetry of magnetic coupling to the primary, these secondary induced voltages are equal when the core is in the center (null, or electric zero) position. When the secondary coils are connected in series opposition, as shown in Fig. 42, the secondary voltage will cancel and, ideally, there will be no net output voltage. If, however, the core is displaced from the null position, one secondary voltage will increase, while the other decreases. Since the two voltages no longer cancel, a net output voltage will now result. If the transducer has been properly designed, this output will be exactly proportional to the magnitude of the displacement, with a phase polarity corresponding to the direction of displacement. (See also Fig. 43.)

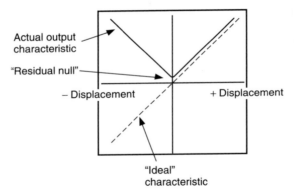

FIGURE 42   Output versus displacement in linear variable differential transformer. (*Daytronic Corporation.*)

The LVDT enjoys wide diversity in its application as a displacement sensor, and, consequently, hundreds of configurations and sizes of the device are available from several manufacturers. The device, singly or in multiples, is used for sensing displacements from millimeters (microinches) to several meters (yards) and is useful in force, pressure, thickness, and other applications where a target variable can be converted to linear displacement. The following advantages account for the wide usage of the LVDT:

1. Essentially frictionless measurement because there is no physical contact between the movable core and the surrounding core structure. This permits its use in critical measurements that can tolerate the addition of the low-mass core, but cannot tolerate friction loading, as found in dynamic deflection or vibration tests of delicate materials and tensile or creep tests on fibers or other highly elastic materials.

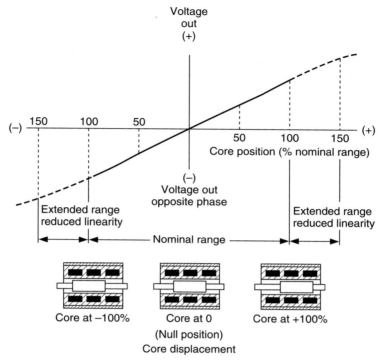

FIGURE 43   Linear variable differential transformer voltage and phase as a function of core position. (*Lucas Schaevitz.*)

2. Extremely long mechanical life as particularly required in high-reliability mechanisms and systems found in aircraft, missiles, space vehicles, as well as some industrial equipment.

3. Essentially infinite resolution, allowing the LVDT to respond to minute motion of the core.

4. Null repeatability, permitting the device to be used as a null-position indicator in high-gain closed-loop control systems.

5. Cross-axis rejection because the LVDT is sensitive to axial core motion, but relatively insensitive to radial core motion.

6. Environmental compatibility because, with the selection of proper construction materials, the LVDT can operate at cryogenic temperatures (immersed in liquid nitrogen or oxygen, for example, as well as in nuclear reactors with high radiation levels and in fluids at elevated temperatures and pressures).

The application versatility of the LVDT is illustrated in Fig. 44.
Other design variations of the LVDT include the following:

DC-LVDT.   It provides the characteristics of the AC-LVDT and the simplicity of dc operation. This design is based on the use of miniature high-performance solid-state components. Prior dc designs exhibited low sensitivity, poor stability, and output that varied with ambient temperature changes. The circuit of one contemporary dc unit is shown in Fig. 45.

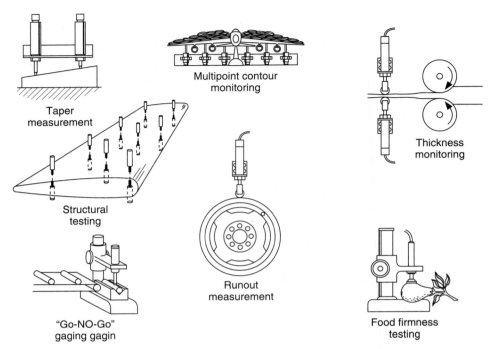

Multipoint contour
monitoring

Taper
measurement

Structural
testing

Thickness
monitoring

Runout
measurement

"Go-NO-Go"
gaging gagin

Food firmness
testing

FIGURE 44 Examples of widely diverse applications of linear variable differential transformers. (*Daytronic Corporation.*)

Rotary Variable Differential Transformer. The rotary variable differential transformer (RVDT) utilizes a specially shaped ferromagnetic rotor that simulates the linear displacement of the straight cylinder core of an LVDT. Although capable of continuous rotation, most RVDTs operate within a range of ±40°, with linearity better than ±0.5 percent of full-scale displacement. As with the LVDT, the RVDT's output voltage characteristically shifts 180° in phase around a null or zero shaft angle position.

Rotary Variable Inductance Transducer. An inductive position sensor, the rotary variable inductance transducer (RVIT) combines a noncontacting variable inductance transducer with a proprietary (*Lucas Schaevitz*) digital Autoplex decoder. Operating on 5 volts dc, the sensor can provide both digital and analog output signals.

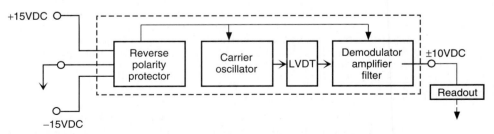

FIGURE 45 Block diagram of dc module (proprietary) for use with DC-LVDT. (*Lucas Schaevitz.*)

Linear Potentiometers

These devices for the measurement of displacement and position take numerous forms. The simplest, least costly form is a single length of wire along which a slider or other form of moving device contacts the wire. The position of the slider determines the effective length of the conductor. Hence a change in electrical resistance or a voltage drop is related to the position or displacement of the slider. This simple device is useful for laboratory demonstrations, but seldom is used industrially.

One example of a linear displacement transducer of the potentiometric type has been used in aircraft and missile production. As shown in Fig. 46, the device is comprised of two resistor elements (j), which are molded along with the slide bars (d) in the element block (s), which, in turn, is contained within an outer case (n). Wiper assemblies (h) are attached to wiper carrier (e) and are aligned to coincide with the resistance elements and contact bars immediately opposite each other. See sectional view. The wiper carrier is secured to the actuating shaft (a) in such a manner as to eliminate backlash and yet allow the shaft to rotate freely 360° when installed. O-ring sealings protect against sand, dust, and so on. Applications include aircraft control surface indication, landing gear retraction systems, missile stage separation, and various industrial uses.

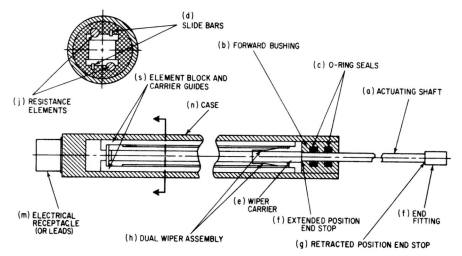

FIGURE 46   Cylindrical wire-wound potentiometric-type linear displacement transducer.

Linear Transformers

This device is a specialized synchro consisting of a salient-pole rotor and a single-phase stator, distributively wound. The winding on the stator is designed to produce an output voltage that varies linearly with the rotor position. This linear function is valid only within a restricted band about the zero position, generally ±50° or ±85°, which is known as the excursion region. Past the excursion region, the output voltage bends to become sinusoidal.

Magnetostrictive Linear Displacement Transducers

Although the principle of this transducer was discovered in 1858, it was not introduced for industrial use until the mid-1970s and since that time has undergone much refinement and has gained significant acceptance.

In this magnetostrictive transducer, a torsional strain pulse is induced in a specially designed magnetostrictive waveguide by the momentary interaction of two magnetic fields. One of these fields emanates from a rare-earth magnet which passes along the outside of the transducer tube. The other field is produced by a current pulse, which travels at over 2750 m/s (9000 ft/s) down a waveguide.

The interaction between the two fields produces a strain pulse, which is detected by a coil arrangement at the head of the device. The position of the rare-earth magnet is measured precisely by the lapsed time between (1) the launching of the electric current pulse and (2) the arrival of the strain pulse at the end of the waveguide. Thus this device differs markedly from the previously described position transducers (Fig. 47).

The transducer measures linear absolute linear position over ranges from 25 mm (1 inch) to 7.6 meters (25 feet) by attaching the magnetic ring to movable elements. The ring travels along the stationary probe without contacting it. Reported performance for this transducer is resolution from 0.1 to 0.001 mm (0.005 to 0.00005 inch), depending on electronics; repeatability to ±0.001 percent of full-scale; and hysteresis 0.02 mm (0.0008 inch) maximum.

The transducers are inherently digital, but they provide a wide range of both digital and analog outputs to match almost any interfacing need. Analog versions measure both position and velocity with either voltage or current output. Digital versions are available in pulse width, start-stop, serial encoder, and parallel binary or binary-coded decimal outputs.

THICKNESS TRANSDUCERS

Thickness is the lesser of the three dimensions that define an object. All three dimensions, of course, are of major importance in the discrete-piece manufacturing industries. In terms of control engineering, thickness measurement and control are of particular importance to what may be called the flat-goods or continuous-length manufacturing industries that are involved in producing sheeted, webbed, and extruded end products, where thickness is of the utmost

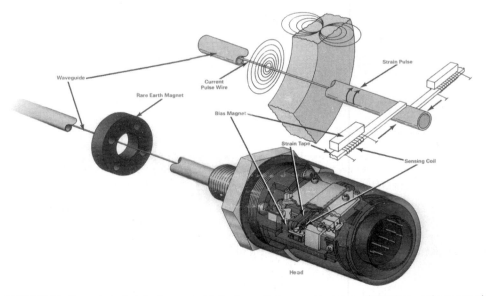

FIGURE 47   Operating principle of magnetostrictive linear displacement transducer. (*MTS Systems Corporation.*)

importance to the consumer. Examples include sheets of metal, plastic, paper, veneer, and plate glass. Thickness is also of paramount importance in the production of various films and coated or plated materials.

Most of the transducers for measuring position and displacement, as described earlier in this article, can measure thickness reliably for many applications. Figure 44, shown earlier, for example, illustrates the use of a pair of linear variable differential transformers (LVDTs) for thickness measurement. Further, some of the sensing principles of object detectors (described in the next article of this handbook section) can be adapted to thickness measurement. Covered in the following are systems that have been configured particularly for thickness measurement.

NONCONTACTING THICKNESS GAGES

Nuclear Radiation Thickness Gages

Both x-ray and nuclear-radiation principles have been applied to thickness measurement. The regulations imposed on instrumentation involving the use of ionizing radiation vary from country to country (Figs. 48 and 49).

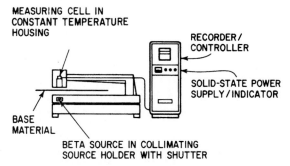

FIGURE 48   Operating principle of beta gaging system for measurement of continuous sheet materials. The gage can measure a single point or scan the sheet automatically. For paper, measurement is made in terms of basis weight (or other weight per unit area). The beta radiation source is located beneath the sheet. A detector cell is located above the sheet. Experienced precision is ±1 percent of range.

In addition to beta radiation, nuclear fluorescence has been used in thickness gages for measuring the coatings on sheet steel or aluminum. Nuclear fluorescence is produced when gamma radiation excites electrons in the metal of the coated strip. Excitation is continuous, but each occurrence is only temporary. Electrons returning to the original unexcited state produce a characteristic low-energy radiation called fluorescence.

Continuous gaging can be an integral part of a computer-operated rolling mill. The gaging system may include an interface that receives commands for thickness settings and, in return, verify that such commands have been accomplished. A system also may include a display and record of the difference in thickness between the edge and the center, or "crown," of a metal sheet or bar, for example. A gaging system may be used to sense trends of change in thickness and may use these signals to actuate screw-down, speed, and tension controls to keep the material "on gage." Such control permits rolling to close tolerances, that is, maximum "on gage" lengths are being produced per ton of material, thus reducing scrap or "out-of-gage" materials at both ends of a coil or run.

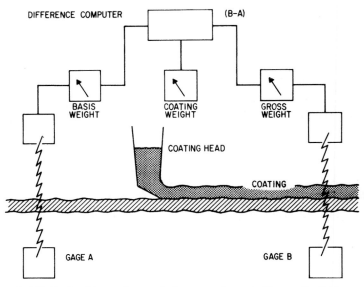

FIGURE 49   Double-gage beta-radiation system for controlling coating thickness on sheet materials.

X-Ray Thickness Gages

X-ray thickness gages can measure the thickness or density of hot or cold materials while the material is in motion. Steel, aluminum, brass, copper, glass, paper, rubber, plastic films, foils, and material coatings are amenable to such gaging. All materials absorb x rays to varying degrees, depending on thickness and density. Thickness can be determined by measuring the amount of x-ray energy absorbed by a material as it passes between an emitter and a receiver (Fig. 50). The gage is set to the desired thickness standard in mils or micrometers. A sample piece of material is used as a reference for calibration. An x-ray gaging system comprises three basic units: (1) a scanner that contains the x-ray generator and a detecting unit, (2) an operator control station, and (3) a power unit. The scanning unit generally is C- or O-frame mounted in a stationary position or on a traversing track, as on a rolling mill, process, or inspection line.

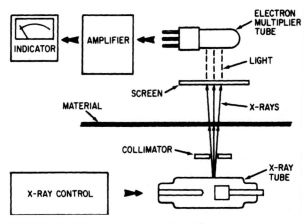

FIGURE 50   Principal elements of x-ray thickness gage.

Ultrasonic Thickness Gages

The principle of the echo-type ultrasonic gage for presence/nonpresence object detectors is described in the following article.

In resonance-type ultrasonic thickness gages a frequency-modulated continuous-wave signal is produced. This provides a corresponding swept frequency of sound waves which are introduced into the part being measured. When the thickness of the part equals one half-wavelength, or multiples of half-wavelengths, standing-wave conditions or mechanical resonances occur. The frequency of the fundamental resonance, or the difference in frequency between two harmonic resonances, is determined by the instrument electronics. The thickness is calculated by the following formula:

$$\text{Th} = \frac{V}{2F}$$

where   Th = thickness of part
V = speed of sound in material
F = frequency, Hz

In the pulse-echo method, the following formula pertains to thickness measurement:

$$\text{Th} = \frac{VT}{2}$$

where   Th = thickness of part
V = speed of sound in material
T = transit time of sound pulse through one round trip in material

A common readout for ultrasonic gages is a cathode-ray tube (CRT). Frequency indications on the display are compared with an overlay scale calibrated in direct thickness readings. The readings are instantaneous, and thickness variations can be monitored as the transducer is scanned over the parts.

Direct-reading panel meters are available. For portability, small, battery-operated packages are available, with accuracies of between 0.5 and 1 percent of full-scale. Both analog and digital versions are obtainable.

A limitation of ultrasonic gaging is the requirement for continuous coupling of the sound beam between the transducer and the part. Sound does not pass across an air-solid or air-liquid boundary. Liquid coupling, either a continuous thin film or some other type, is required. In some instances, complete immersion of the transducer and the material is feasible. In others it is possible to use a bubbler or partially contained water column to provide the continuous coupling path.

SURFACE TEXTURE MEASUREMENT

Continuous or periodic thickness measurement technology also can be applied in a customized manner for measuring the surface characteristics or texture of some materials. With increasing attention to quality control of in-process as well as finished products, this has become an important variable to measure and control. A version of thickness determination can be applied to some aspects of this problem.

For example, any surface produced by machining departs from the perfect form because of a variety of causes, such as inaccuracies in the machine tool, deformation of the work under the cutting force, and irregularities caused by vibration. Irregularities also may be caused by rupture of the material during separation of the chip. These factors, in turn, produce geometrical inaccuracies associated with errors of form, including surface texture—waviness and roughness. Roughness of surface is affected by both the size and the shape of the undulation. Wavelength spacing is just as important as height. What may seem to be a perfect surface can be changed to a very rough surface simply by changing the wavelength of the undulation. As the quantity of the undulation becomes smaller, the quality of the surface deteriorates. This deterioration becomes increasingly apparent to the eye as the wavelength becomes shorter, even though the height of the undulations remains the same.

The patterns formed on the surface by waviness often are varied. Some patterns, such as pronounced chatter marks and the coarse feed marks of a badly trued grinding wheel, can be identified at a glance. Others may require an instrument to reveal their presence. In the case of surfaces of revolution, the marks extending along the lay of the roughness often become the circumferential departures from roundness. Waviness, as seen in a profile graph, often can be appraised both as an undulation of the mean line and as an undulation of a line drawn through the more prominent crests. Thus the terms "crest line waviness" and "mean line waviness" (Fig. 51).

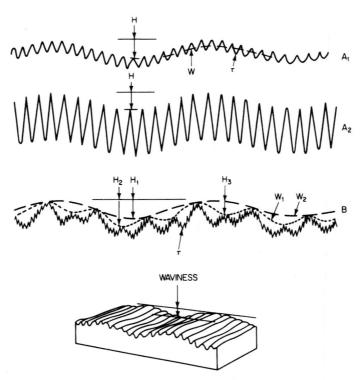

FIGURE 51   Examples of waviness. When measuring across the roughness lay, several types of waviness can be found. The profile A_1 has a wavy undulation W on which is superimposed roughness r of smaller amplitude. The profile A_2 is of the same kind, but the roughness is larger in amplitude than the waviness. In these cases, mean line and crest line waviness are substantially the same. The crest line is irregular and rarely identical with its mean line. On ground surfaces, the crest spacing of closely spaced waviness surfaces often is in the region from 0.02 to 0.1 inch (0.5 to 2.54 mm), and the height is generally less than half the overall height of the grinding texture. Surfaces also may be encountered where spacing and height are the same, yet the surfaces are quite different.

The most common specification used to control surface texture is average roughness expressed in micrometers or microinches. This value represents the actual vertical distance from a datum line of every point of the profile occurring in the length of the sampled surface. The position of the datum line or centerline is not a constant. The elevation varies with each specific length of profile being measured. By definition, the position of the centerline is such that the total areas of the profile lying above and below the line are equal. As shown in Fig. 52, surface texture includes roughness, waviness, lay, and flaws.

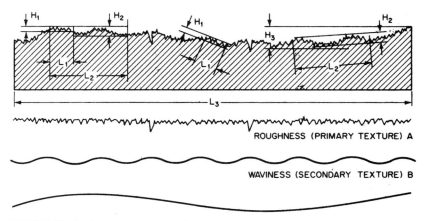

FIGURE 52   Surface texture representing the combined effects of several causes.

Since the measurement of surface roughness involves determination of the average linear deviation of the actual surface from the nominal surface, there is a direct relationship between the dimension tolerance for a part and the permissible surface roughness. It is evident that a requirement for the accurate measurement of a dimension is that the variations introduced by surface roughness not exceed the tolerance placed on a dimension. If this were not the case, measurement of the dimension would be subject to an uncertainty greater than the required tolerance.

Surface geometry has a fairly direct bearing on metrology and fits and limits. Measuring instruments ordinarily have anvils and gaging tips which, because of their size, make contact only with the highest points of surface irregularities. The intervening valleys, however, may have appreciable depth and, if this depth amounts to a large proportion of the tolerance, it may affect the size of the part. Subsequent removal of the high spots may differ from that indicated by the measuring instrument by an appreciable amount.

In order to accomplish measurements functionally and to an average numerical value, stylus-type instruments generally have been used. Noncontact methods, usually optical interference means, are effective for the interpretation of fringe values—hence interpretation of the surface. New methods employing scanning electron microscopy are now being used as a means of establishing a three-dimensional evaluation. The essential elements of a stylus-type instrument are (1) a sharply pointed stylus for tracing the profile at a cross section, (2) a means for generating a datum, and (3) a way of amplifying and indicating the stylus movement. Of the stylus-type instruments, two are in general use: (1) a carrier-modulated device in which the magnitude of a carrier current is controlled at every instant of time in accordance with the position of the stylus relative to the datum—regardless of how long the stylus remains in a given position, and (2) a device in which a current or potential is generated in accordance with the motion of the stylus as the stylus is displaced from one level to another.

Carrier-modulated instruments are useful for obtaining graphs because, in acting like simple levers, these devices faithfully reproduce every movement of the stylus relative to the datum, regardless of the spacing. These instruments are calibrated by using gage blocks or interferometrically. The instruments behave in a manner similar to a mechanical lever with magnification ratios of up to 1 million times. The generating instruments reproduce only if the stylus is rising and falling at a rate above the low-frequency limit. Therefore, widely spaced irregularities over which a stylus may be rising and falling only slowly will not be reproduced. This instrument is not desirable for a profile recording but is suitable for numerical evaluation. It is apparent that no matter how valuable and necessary the profile graph may be, some form of numerical assessment is required even if only for purposes of establishing a print value. However, a numerical value cannot be readily established until sufficient data concerning measurement of the component have been established.

Roughness width cutoffs utilized with average values are regarded as the greatest spacing of repetitive surface irregularities to be included in the measurement of average roughness height. A roughness width cutoff is rated in inches and must always be greater than the roughness width. When no value is specified, the value of 0.030 inch (0.8 mm) is assumed. The fundamental object of taking the length of the surface into consideration is based on the fact that different makes of instruments and different operators should obtain the same answer for any given surface. The profile graph on the other hand accurately defines and establishes all irregularities, giving values of height as well as width.

QUALITY CONTROL AND PRODUCTION GAGING

Statistical Quality Control

Maintaining product quality in accordance with acceptable standards has been a major role for industrial instrumentation since its inception decades ago. With the ever-growing interest in speeding up production, one becomes increasingly aware of the fact that rejects as well as acceptable products can be produced at very high rates. What constitutes product quality? Apparently, in the long run, it is that degree of excellence which the ultimate consumer demands for an affordable price. Obviously, manufacturers must compare cost versus quality acceptance in the marketplace. Based upon years of experience, the manufacturer learns what affordable quality really means in terms of market acceptance. Astute competitors also learn these basic facts.

With the foregoing knowledge, the manufacturer establishes quality standards, which then are reduced to engineering specifications for each product. But knowing in advance that there will be variations from exact specifications, acceptable variations must be established. These variations then are reduced to plus or minus tolerances.

The tolerable variations and the intolerable variations must be determined and, in an ideal situation, immediately fed back to production machines or processes. Adjustments made would again, ideally, affect every part or unit of substance being produced. But because the target of acceptance is bracketed ($\pm$), a strictly go/no-go type of sorting does not suffice. Thus at least as early as a century ago, the concept of statistical quality control (SQC) was introduced. The general intent of SQC is that of sampling units and parts being produced and essentially determining trends in deviation from production as continuously (affordable and achievable) as possible. Since the early part of this century, the literature on SQC theory has continued to grow, and because a full understanding of the concept involves rather intricate mathematics, most of the principles have been applied. But since the 1960s and continuing, the emphasis on SQC has shifted from essentially the exertion of manual controls and interpretation to the present semiautomation of data collection and interpretation. The mathematics have improved as the result of computer technology, including the development of algorithms and other shortcuts, and the interface between SQC and management has been streamlined by modern display technology, notably computer graphics, as well as by imbedded computer calculations.

Basic Assumptions of SQC

1. Variations are inherent in any process. Inherent variations are not the primary target of SQC. If there are *only* inherent variations, the process is said to be in statistical control. It is assumed that inherent variations affect all measurements and that, over a period of time, these variations will stabilize.

2. Other variations are designated as correctable, that is, they are in the realm of statistical control.

3. A key element in the SQC concept is the familiar normal distribution (bell curve), shown and explained later.

4. In some interpretations of SQC, a further objective is considered. The acceptable tolerance limits (+ and −) are not the target to seek, but those data within the + and − range should be further correlated so that the true "aim" of +0 is achieved as closely and as often as possible. For example, a manufacturer may specify that all products must fall within a given ± tolerance, but that the majority should be closer to the midrange of the tolerance, that is, as "near perfect" as possible. This differs from simply tightening the ± tolerance because it still allows less perfect units to escape full rejection.

SQC Glossary of Terms

An abridged glossary of terms used in SQC can be helpful. The following symbols are used in the next several paragraphs

where
N = number of data points
$\underline{R}$ = range
$\bar{R}$ = average range
s = standard deviation
x_i = value of specific data point
USL = upper specification limit
LSL = lower specification limit

Capability and Control (Concept of).   A process is in control if the only sources of variation are common causes. The mean spread of such a process will appear stable and predictable over time (Fig. 53).

The fact a process is in control does not imply that it will yield only good parts, that is, parts within specification limits. Control only denotes a stable process. The size variation due to the common causes may be so large that some parts are outside the specification limits. Under these conditions, the process is said to be *not* capable. Capability is the ability of the process to produce parts that conform with engineering specifications (Fig. 54).

Cp (Inherent Capability of Process).   Cp is the ratio of the tolerance to 6 sigma. The formula is

$$Cp = \frac{USL - LSL}{6\sigma}$$

The Cp ratio is used to indicate whether a process is capable.

1.33 or greater	Process is capable.
1.0 to 1.3	Process is marginally capable, should be monitored.
1.0 or less	Process is *not* capable.

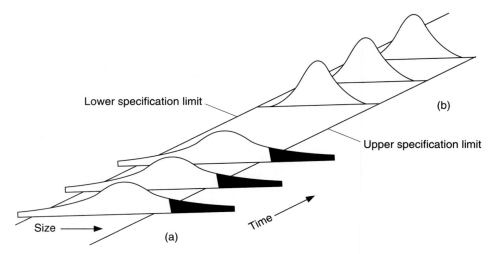

FIGURE 53   Demonstration of process variations. (*a*) In control, but not capable. (*b*) Variations from common causes are excessive. (*Moore Products.*)

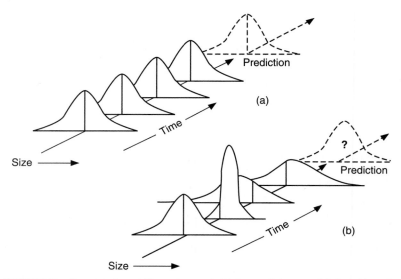

FIGURE 54   Concept of capability and control. (*a*) Process is in control. It is stable and predictable over time. (*b*) Process is neither stable nor predictable over time. (*Moore Products.*)

It should be noted that Cp does not relate the mean to the midpoint of the tolerances. If the mean is not at the midpoint, out-of-tolerance parts may still be probable, even if the process is capable (Fig. 55).

Cpk (Capability in Relation to Specification Limits).   Cpk relates the capability of a process to the specification limits, that is, Cpk equals the lesser of

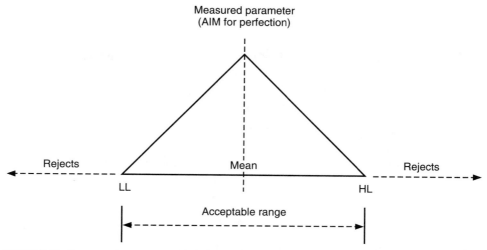

FIGURE 55 Parts or units are acceptable if they barely achieve either of the specified limits (Low, LL; High, HL). In the quality control program, the manufacturer can gear the operation such that a majority of units will lie in the middle of the acceptable range. This is the basis for the AIM mode of operation.

$$\frac{USL - mean}{3\sigma} \quad \text{or} \quad \frac{mean - LSL}{3\sigma}$$

The Cpk value is useful in determining whether a process is capable and is producing good parts based on the specification limits. Values for the Cpk index have the following meaning:

Greater than 1.0   Both of the 6σ limits fall within the specification limits. The process is both capable and producing good parts (99.73 percent or greater).

1.0   At least one of the 6σ limits falls directly on the specification limits.

Less than 0.0   The mean is outside of the specification limits.

Capability Ratio. This ratio is the inverse of Cp, that is,

$$\frac{6\sigma}{USL - LSL}$$

The value of this ratio can be thought of as the portion of the part tolerance consumed by 6σ. A common interpretation for the various values of the capability ratio is as follows:

50 or less   Desirable
51 to 70   Acceptable
71 to 90   Marginal
91 and greater   Unacceptable

Common Cause. A source of random variation that affects all of the individual measurements in a process. The distribution is stable and predictable. (See also *Special Cause* in this glossary.)

Control Limits (LCL and UCL). The upper and lower control limits are values used to determine whether or not a process is in statistical control. The values are inherent to the process and should not be confused with specification limits.

Histogram. A chart that plots individual values versus the frequency of occurrence and is used for statistical data analysis. Note that earlier, these diagrams were created manually. The diagrams now can be displayed on a CRT (computer graphics) and, of course, stored in memory (Fig. 56).

Individual. A single measurement of a particular process characteristic.

Kurtosis. An indication of whether the data in a histogram have a normal distribution. Specifically, it is a measure of the "flatness" or "peakness" of a curve. The formula for kurtosis is

$$\sum_{i=1}^{n} \frac{(X_i - \overline{X})^4}{4s^4}$$

Values for kurtosis have the following meaning:

3	Normal distribution.
Less than 3	Leptokurtic curve, that is, the curve has high peak (data are concentrated close to the mean).
Greater than 3	Platykurtic curve, that is, the curve has low peak (data are disbursed from the mean).

Another version of the foregoing equation is

$$\frac{\eta^3 \sum_{i=1}^{n} X_i^4 - 4\eta^2 \sum_{i=1}^{n} X_i \sum_{i=1}^{n} X_i^3 + 6\eta \left(\sum_{i=1}^{n} X_i \right)^2 \sum_{i=1}^{n} X_i^2 + 3 \left(\sum_{i=1}^{n} X_i \right)^7}{\left[\eta \sum_{i=1}^{n} X_i^2 - \left(\sum_{i=1}^{n} X_i \right)^2 \right]^2}$$

```
TEST CHART              OVER SCALE    0
PROC. "Q" 94.495%       +.019000      0
SKEWNESS      +.00       +.017000      0
KURTOSIS     +3.19       +.015000      0
                         +.013000      1
                         +.011000     51  ▮
HI LIMIT +.010000        +.009000    129  ▮▮
-----------------        +.007000    300  ▮▮▮▮▮▮
-----------------        +.005000    960  ▮▮▮▮▮▮▮▮▮
SAMPLE "N"   25442       +.003000   1590  ▮▮▮▮▮▮▮▮▮▮▮▮▮▮▮
      MEAN -.003000      +.001000   2840  ▮▮▮▮▮▮▮▮▮▮▮▮▮▮▮▮▮▮▮▮▮▮▮▮▮▮
STD DEV  +.004346        -.001000   4350  ▮▮▮▮▮▮▮▮▮▮▮▮▮▮▮▮▮▮▮▮▮▮▮▮▮▮▮▮▮▮▮▮▮▮▮▮
+3 SIGMA +.010040        -.003000   5000  ▮▮▮▮▮▮▮▮▮▮▮▮▮▮▮▮▮▮▮▮▮▮▮▮▮▮▮▮▮▮▮▮▮▮▮▮▮▮▮▮
-3 SIGMA -.016040        -.005000   4350  ▮▮▮▮▮▮▮▮▮▮▮▮▮▮▮▮▮▮▮▮▮▮▮▮▮▮▮▮▮▮▮▮▮▮▮
     RANGE +.032000      -.007000   2840  ▮▮▮▮▮▮▮▮▮▮▮▮▮▮▮▮▮▮▮▮▮▮▮▮▮▮
                         -.009000   1590  ▮▮▮▮▮▮▮▮▮▮▮▮▮▮▮
LO LIMIT -.010000        -.011000    960  ▮▮▮▮▮▮▮▮▮
-----------------        -.013000    300  ▮▮▮▮▮▮
-----------------        -.015000    129  ▮▮
  +999999       12       -.017000     51  ▮
  -999999       13       -.019000      1
  -??????       22       UNDER SCALE    0
TOTAL PARTS 25489
TEST ON      H                         FXT   STATS   PRINT   CAL
                         Date/Time     #1     ON     OFF    MODE
```

FIGURE 56 Facsimile of computer-generated histogram.

Mean X.   The value of the middle individual when the data are arranged in order from lowest to highest.

Mean $\overline{X}$.   The arithmetic average value of the data. The process mean is the average of all the process data. The subgroup mean averages just those values in the subgroup.

Median $\tilde{X}$.   The value of the middle individual when the data are arranged in order from lowest to highest. If the data have an even number of individuals, the median is the average of the two middle values.

Mode.   The most frequently occurring value, that is, the highest point on a histogram.

Normal Distribution.   Data often are summarized graphically as a means of better understanding and analyzing the variation. A plot of the frequency of occurrence of a particular variable is one of the common tools used for analysis. If a definite pattern emerges from the data, this plot is referred to as a distribution.

Many distributions have been identified and named as their pattern of variation is repeatable and certain mathematical characteristics can be defined for each distribution. One of the most commonly occurring patterns of distribution is the normal distribution. It describes many natural and other phenomena encountered in the field of statistics.

The mean and the standard deviation define a specific normal distribution. Knowing the values of the normal distribution, the total spread of expected outcomes can be predicted. It is important to note that 99.73 percent of the population lies between −3 standard deviations (often called −3σ) from the mean and +3 standard deviations (+3σ) from the mean. This fact is the basis for many statistically calculated indications of the status of a process, including capability and control limits.

It is equally important to note that if the distribution of a process is not normal, a number other than 99.73 percent of produced parts or units will fall within 6 standard deviations. Because of this, many of the statistical control indicators calculated from a nonnormal distribution will not describe that distribution accurately. Often a distribution is sufficiently close to being normal, however, that is, the errors are insignificant (Fig. 57).

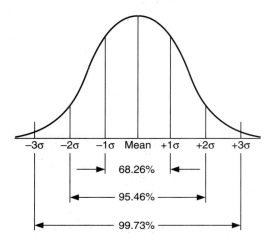

FIGURE 57 Normal distribution chart. This is basic to the formal statistical quality control (SQC) concept. The curve is useful for many other statistical purposes in other fields, such as variations in a population of people (clothing sizes, for example). The concept dates back to Laplace and Gauss and is sometimes referred to as gaussian distribution.

Pareto Chart.   This type of chart ranks events according to a specified parameter, such as rejected individuals by frequency of occurrence (Fig. 58).

Process Quality.   The process quality or process capability coefficient denotes the area under the normal curve that falls within the specification limits. It is expressed as a percentage. The calculation is a multistep procedure.

1. Calculate the distance of the mean from the specification limits in terms of standard deviations:

$$L_U = \frac{\text{USL} - \overline{X}}{s} \qquad L_L = \frac{\overline{X} - \text{LSL}}{s}$$

(For this equation to be valid, the values for L_U and L_L must be greater than or equal to 0. If either is negative, the corresponding area is calculated using $A = 1 - A'$, where A' is obtained using $L' = -L$.)

2. Determine the area outside of the specification limits, expressed as a fraction of the normal curve:

$$A_{\text{upper}} = \frac{1}{\sqrt{2}} \cdot e^{-(L_U^2/2)} \cdot [B_1 Y_U + B_2 Y_U^2 + B_3 Y_U^3 + B_4 Y_U^4 + B_5 Y_U^5]$$

$$A_{\text{lower}} = \frac{1}{\sqrt{2}} \cdot e^{-(L_L^2/2)} \cdot [B_1 Y_L + B_2 Y_L^2 + B_3 Y_L^3 + B_4 Y_L^4 + B_5 Y_L^5]$$

TOTALS

GAGED	8550		████████████████████████
ACCEPTED	7835	91.6	███████████████████████
REJECTED	715	8.4	██▌

LABELS:	OVER		UNDER		SCALE AS % OF REJECTED
N.N	400	55.9	315	44.1	████████████████████████████
P P	915	44.1	315	44.1	███████████████████████████
3 0	500	69.9			████████████████████
A A	100	14.0	390	54.5	████████████████████
G G	333	46.8	111	15.5	██████████████████
M M	201	28.1	203	28.4	████████████████
O O	190	25.2	180	25.2	███████████████
E DIA E	312	43.6			█████████████
O O	68	9.5	199	27.8	████████████
L L	180	25.2	41	5.7	██████████
I I			170	24.9	████████
H H	100	14.0	34	4.8	█████
F F	43	6.0	47	6.6	████
B B	20	2.9	24	3.4	██
J J			1	.1	
K K	1	.1			

	REJECT	XR	STATS	PRINT	TEST
TEST ON		0	ON	OFF	MODE
Date/Time					

FIGURE 58 Facsimile of computer-generated pareto diagram. Listings include total parts gaged, parts accepted, and parts rejected, with the parameters for which parts are rejected in descending order of frequency. This type of diagram stresses the relative importance of corrections to be made.

where $Y_U = \dfrac{1}{1 + R \cdot L_U}$

$Y_L = \dfrac{1}{1 + R \cdot L_L}$

$R = 0.2316419$
$B_1 = 0.31938153$
$B_2 = -0.356563782$
$B_3 = 1.781477937$
$B_4 = -1.821255978$
$B_5 = 1.330274429$

(When a tolerance limit is unused, that is, has a U prefix, the area under the normal curve that is outside that limit does not enter into the process quality calculation.)

3. Then the process quality is found from the equation:

$$\text{Process quality} = [1 - (A_{\text{upper}} + A_{\text{lower}})]\, 100\%$$

Range R.   The difference between the highest and lowest values in the group. The average of subgroup ranges is denoted by $\overline{R}$.

Run Chart (xR Chart).   A chart that displays the most recent individual measurement x and the absolute difference from the previous measure R on a consecutive basis.

Sample.   A synonym of subgroup in process control applications. However, sometimes "sample" is used to denote an individual reading within a subgroup. Because this can be confusing, "subgroup" is the preferred term.

3 Sigma. $\pm 3\sigma$ are the two specific points on a normal distribution centered about the mean. 99.73 percent of the population will fall between these values. Since this is essentially the entire population, $+3\sigma$ and -3σ represent the probable range of variation:

$$+3\sigma = X + 3s$$

$$-3\sigma = X - 3s$$

Skewness.   An indication of whether the data in a histogram have a normal distribution. Specifically, skewness is a measure of symmetry. Values for skewness have the following meaning:

0	Symmetrical distribution.
Greater than 0	Positive skewness, that is, the distribution has a "longer tail" to the positive side. The median is greater than the mode.
Less than 0	Negative skewness, that is, the distribution has a "longer tail" to the negative side. The median is less than the mode.

The formula for skewness is

$$\sum_{i=1}^{n} \frac{(X_i - \overline{X})^3}{3s^3}$$

Another version of this formula is

$$\frac{\eta^2 \sum_{i=1}^{n} X_i^3 - 3\eta \sum_{i=1}^{n} X_i \sum_{i=1}^{n} X_i^2 + 2\left(\sum_{i=1}^{n} X_i\right)^3}{\sqrt{\eta \sum_{i=1}^{n} X_i^2 - \left(\sum_{i=1}^{n} X_i\right)^2}}$$

Special Cause.   A source of nonrandom or intermittent variation in a process. The distribution is unstable and unpredictable. It is also referred to as an assignable cause. (See also *Common Cause* in this glossary.)

Specification Limits (LSL and USL).   The upper and lower specification limits (engineering-blueprint tolerances) are values that determine whether or not an individual measurement is acceptable. They are engineering tolerances established external from the process and should not be confused with control limits.

Standard Deviation.   A measure of the variation among the elements in a group. The formula for the standard deviation is

$$s = \sqrt{\sum_{i=1}^{n} \frac{(X_i - \bar{X})}{\eta - 1}}$$

Another version of this formula is

$$\sqrt{\frac{\sum_{i=1}^{n} X_i^2 - \left[\left(\sum_{i=1}^{n} X_i\right)^2 / \eta\right]}{\eta - 1}}$$

Subgroup.   A group of individual measurements, typically 2 to 10, used to analyze the performance of a process.

Variation Concept.   Variation occurs in manufacturing because the process conditions are never exactly identical for any two parts or units. If variation did not exist, quality would not be a problem. Every piece or unit would be identical—so in the case of parts and pieces, all components would assemble and test correctly. Inasmuch as differences do exist, the goal of SQC is to reduce the variation as much as possible, thereby improving product consistency and quality.

$\bar{X}R$ Chart.   A chart that displays the subgroup mean and range on a consecutive basis. Upper and lower control limits are plotted to help analyze the process (Fig. 59).

Note that SQC data are plotted in various columnar forms (Fig. 60).

System Approach to SQC

Modern instrumentation and computer technology have eliminated, or can eliminate, manually entered and manual charting by way of tremendously speeding up SQC data acquisition and the numerous computations required. SQC can be made more encompassing and penetrating. Histograms in electronic storage, and thus easily called up on graphic displays, essentially eliminate paperwork. The remaining areas, namely, the interpretation and corrective

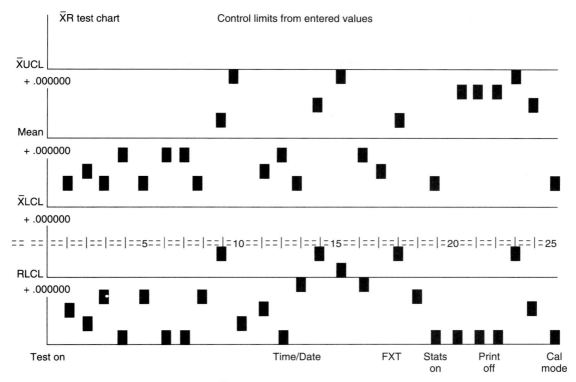

FIGURE 59 Facsimile of computer-generated $\bar{X}R$ chart.

action aspects, can be further speeded up through the application of further automation technology. Much of the hardware is available to create SQC systems of varying extent and complexity. In applying advanced instrument and computer technology to SQC systems, there will be a tendency, as occurred in the past in attacking a new area for instrumentation, to overdo in some instances. The established guidelines still apply, that is, do *not* collect more data than are required. A flood of essentially useless information taxes the hardware, the software, and the personnel who are confronted with it.

SQC is very easy to identify with controlling parts. It is somewhat more difficult to implement as the parts become subassemblies and the latter, in turn, become finished products ready for shipment. It is in the areas of subassembly and final assembly testing where much remains to be learned and achieved.

SQC in the Process Industries

The fundamentals of SQC also are applied in the continuous-process industries, but because of the marked differences in the characteristics of the end products and of the variables that are measured and controlled, there are some differences in the application of SQC. This has led to the general acceptance of the term statistical process control (SPC) in these industries. The subject is addressed elsewhere in this handbook.

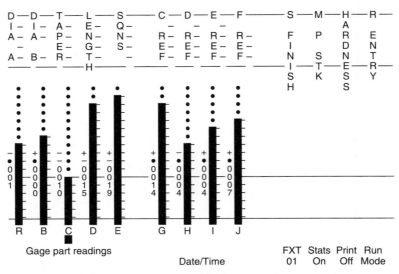

FIGURE 60 Facsimile of computer-generated SQC columnar data presentation.

PRODUCTION GAGING SYSTEMS

The wide variety of dimension and position sensors previously described in this article comprises the basis for parts gaging systems, whether they are manual or semiautomated. The range of gaging needs is wide, involving different parts materials and sizes. Also, production runs, ranging from essentially continuous to short runs, add to the complexity and costs of automating gaging operations.

Prior to the availability of digital data processing, gaging operations depended on analog technology. The centralization of data gage information was not easy and SQC analyses were

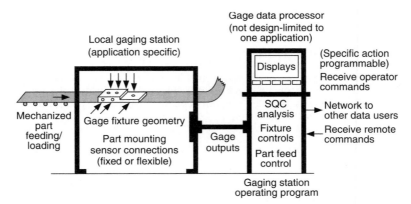

FIGURE 61 Concept of gage data processor that is programmable and need not be redesigned for each gaging station with which it will be used. Gage fixturing remains essentially very application-specific, but gaging paperwork needs do not have to be incorporated through redesign of the data processor, but rather this can be done via reprogramming.

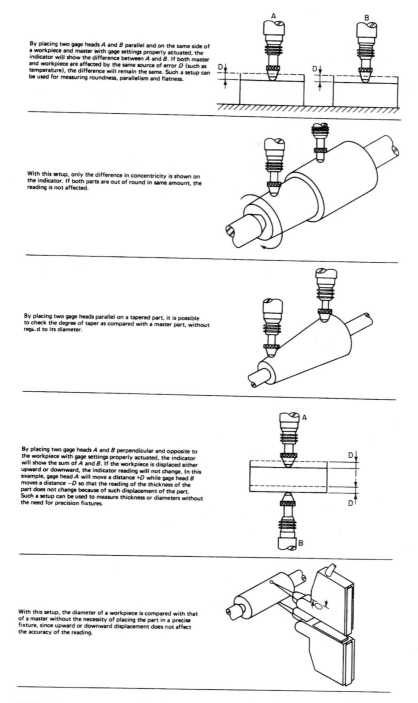

By placing two gage heads *A* and *B* parallel and on the same side of a workpiece and master with gage settings properly actuated, the indicator will show the difference between *A* and *B*. If both master and workpiece are affected by the same source of error *D* (such as temperature), the difference will remain the same. Such a setup can be used for measuring roundness, parallelism and flatness.

With this setup, only the difference in concentricity is shown on the indicator. If both parts are out of round in same amount, the reading is not affected.

By placing two gage heads parallel on a tapered part, it is possible to check the degree of taper as compared with a master part, without regard to its diameter.

By placing two gage heads *A* and *B* perpendicular and opposite to the workpiece with gage settings properly actuated, the indicator will show the sum of *A* and *B*. If the workpiece is displaced either upward or downward, the indicator reading will not change. In this example, gage head *A* will move a distance +*D* while gage head *B* moves a distance −*D* so that the reading of the thickness of the part does not change because of such displacement of the part. Such a setup can be used to measure thickness or diameters without the need for precision fixtures.

With this setup, the diameter of a workpiece is compared with that of a master without the necessity of placing the part in a precise fixture, since upward or downward displacement does not affect the accuracy of the reading.

FIGURE 62 Examples of the use of impedance-type dimension gages. The first three cases are concerned with "difference" measurements, whereas the last two cases deal with "sum" measurements. (*Brown & Sharpe.*)

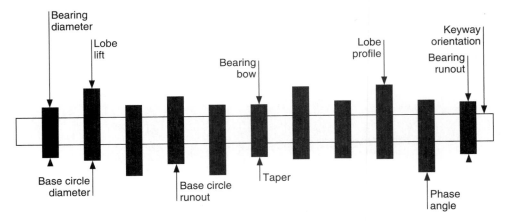

FIGURE 63 Multiple gaging of critical dimensions on the production floor. Difficult characteristics such as lobe profile no longer require sample checks in a laboratory environment. Tooling mounts the camshaft on centers. Gage tracking closely represents the actual usage in an engine.

made manually. Depending on the size of the plant, gaging applications were approached on an application-by-application basis, with gage data display engineered along with the design of the fixtures for gaging.

Much progress has been made during the past decade toward streamlining and normalizing (not standardizing) the processing of gage data so that the display, analysis, and control aspects of localized gaging stations can be packaged in a manner such that a "gage data processor" can be used with most local gaging stations once the fixture is readied and the sensors are located and connected. Depending on a user's present and planned requirements, a data gage processor can be configured with a range of capacities and functions. Not having to reengineer the display, analysis, and control aspects of gaging each time a new gaging problem presents itself, is indeed a tremendous advantage. Gage data processors also can generate for and receive from computers and data processors at a higher level. Whereas the gage fixture is essentially inflexible, the gage data processor can have unlimited flexibility (Fig. 61).

Examples of the variety of properties that must be gaged in the discrete-piece manufacturing industries are given in Figs. 62 and 63. Typical applications where data gage processors have been used are listed in Table 3.

TABLE 3 Examples of Gage Data Processor Applications

	Valve lifter body
Measurements	Pin-hole location to open end
	Pin-hole location to seat
	Parallelism of flats to pin hole
	True position of flats to pin hole
	True position of flats to OD
	Pin-hole diameter size
	Pin-hole centrality to OD
	Pin-hole squareness to OD
	Slot width
	True position of slot to OD

Location	Plant floor; gage dedicated to a single production part
Probes	Back-pressure air circuitry transduced to electric LVDTs
	Leaf flexures
	D-plug locators
Gage data processor	Microprocessor-based; connected to local area network

Wheel analyzer

Measurements	Radial runout of each bead seat with respect to pilot, hole and bolt hole pattern
	Lateral runout of each bead seat with respect to pilot hole and bolt hole pattern
	Bolt hole pattern to pilot hole concentricity
Location	Plant floor; gage is adjustable with some change part tooling; designed to handle family of about 50 wheels per hour
Probes	Roller followers
	LVDTs
	Linear encoders
	System uses harmonic analyzer to determine runouts and concentricities
Gage data processor	Microprocessor-based; connected to local area network

Tapered roller bearing inner race

Measurements	Raceway diameter
	Roundness
	Width
	Undercut presence
Location	Plant floor; adjustability and change part tooling allow gage to handle wide range of "green" roller bearing inner races
Probes	LVDTs
	All measurements are dynamic as part is rotated during gaging cycle
Gaging rate	1800 parts per hour
Gage data processor	Microprocessor-based; connected to local area network

Tapered roller bearing outer race

Measurements	Upper, center, and lower IDs
	Taper, crown, and width
	Upper and lower ODs
	Taper OD
Location	Plant floor; dedicated change part tooling allows gage to handle wide range of bearing sizes while still maintaining required accuracy and repeatability
Probes	Tapered bore plugs with air circuits
	LVDTs
Gaging rate	1800 parts per hour
Gage data processor	Microprocessor-based; connected to local area network

Piston

Measurements	Skirt diameter
	Ring groove diameters
	Ring groove width and location
	Wrist pin bore size, taper, and roundness
Location	Plant floor; gage is dedicated to single piston
Probes	LVDTs
	Bore plug with air circuits
Gage data processor	Microprocessor-based; connected to local area network

Single flank gear

Measurements	Relative angular velocities of two mating gears; frequency spectrum of driven gear is analyzed to locate source of transmission noise
Location	Laboratory
Probes	Rotary encoders
Gage data processor	Microprocessor-based; connected to local area network

Tight mesh gear	
Measurements	Functional tooth thickness
	Pitch diameter runout
	Sectional runout
	Out-of-round
	Tooth-to-tooth
	Nicks
	Average lead
	Lead variation
	Average taper
Location	Plant floor; gage is dedicated to single part number
Probes	LVDTs
	Gear is run in tight mesh with two rolling master gears. The axis of the first master gear (called center-distance rolling master) is only allowed to translate. The second rolling master (lead and taper rolling master) is mounted on a two-axis gimbal. The positions of these rolling masters are monitored as they mesh with the part gear. These signals are analyzed to determine values for the various characteristics.
Gaging rate	About 250 per hour
Gage data processor	Microprocessor-based; connected to local area network
Transmission sprocket	
Measurements	Thickness
	Tooth width
	Overall height
	Web thickness
	Sleeve diameter, size, straightness, and roundness
	Runout of spline to sleeve diameter
	Parallelism of hub to reference face
	Parallelism of thrust face to reference face
	Spline tooth—runout to sleeve diameter, pitch diameter, and nicks
Location	Plant floor; gage is dedicated to single transmission sprocket
Probes	Back-pressure air circuits transduced to electric LVDTs
Gaging rates	415 parts per hour
Gage data processor	Microprocessor-based; connected to local area network

OBJECT DETECTORS AND MACHINE VISION

Control systems concerned with position measurement tend to fall into one of two main categories:

1. Systems as exemplified by various production machines, where achieving a position, measured in exact geometric coordinates, is the goal. These systems are described in Article 1 in this handbook section.

2. Systems where control action occurs because a given object occupies (even for an instant) a specific location, that is, a definite position within a manufacturing space. This article is concerned with that domain.

Object detectors may be used to initiate motion, stop motion, and return motion, for example, in a pick-and-place robot or in the repetitive cycling of a machine. As the name implies, limit switches are used by the millions on machinery to limit a machine stroke (of vastly varying nature) in order to avoid injury to operators, machines, or materials in process. Object detectors (sometimes aided by machine vision) count and inspect products for filling level, label placement, and the absence (nonpresence) of too many or too few units in packaging, among many other applications.

Object detectors have been so commonplace in the manufacturing scene for so many years that in a way they have "disappeared into the woodwork." Nevertheless, object detectors have been refined over generations of use and generally have kept pace with technology through the incorporation of scientific advancements, as found in Hall-effect and ultrasonic detectors. After many years of considering statistical quality control as an entity apart from instrumentation, the data inputs from object detectors now are incorporated into control system electronics for the purpose of making sometimes complex mathematical calculations and displaying the results (often right on the manufacturing floor) for rapid determinations of product quality (in-process and final) for both small and vast integrated manufacturing operations.

ELECTROMECHANICAL LIMIT SWITCHES

These switches are the "workhorses" of automated systems. Because of their extensive use, they usually are readily available out of stock. Depending on specific use conditions (temperature, hazardous atmosphere, light to heavy duty, and so on) most manufacturers offer a wide line of limit switches. In general, limit switches fall into three size (dimensional) categories. The switch box proper may range in height from 50 to 150 mm (2 to 6 inches), in width from 15 to 80 mm (0.6 to 3.2 inches), and in depth from 40 to 85 mm (1.6 to 3.4 inches). The upper operating temperature range, depending on the model selected, is 121°C (250°F); the lower temperature range is –32°C (–25°F). Housing material may be zinc, aluminum, or steel treated for corrosion and weather resistance. In some designs the inner portions of the switch circuitry are prewired and potted. Switches that meet European (DIN) specifications are also available.

Specifying Details. The approximate specifying details, considering a range of models and brands, include the following.

Control Actions. Single-pole double-throw (SPDT) and double-pole double-throw (DPDT).

Switching Capacity. 10 to 15 amperes 125 volts ac (inductive load); 5 amperes 125 volts ac (resistive load).

Standard. Overtravel 60° minimum, pretravel 15° maximum, differential travel 3° (single-pole) and 7° (double-pole) maximum.

Low-Differential-Travel Design. Overtravel 68° minimum, pretravel 7° maximum, differential travel 3° (single-pole) and 4° (double-pole) maximum.

Low-Operating-Torque Design. Overtravel 60° minimum, pretravel 15° maximum, operating torque 1.7 in · lb (0.19 N · m) maximum.

Low-Torque, Low-Differential-Travel Design. Overtravel 68° minimum, operating torque 1.7 in · lb (0.19 N · m) maximum, differential travel 3° (single-pole) and 4° (double-pole).

Sequence-Action Design. Delayed action between operation of two poles, in each direction; overtravel 48° minimum.

Center-Neutral Design.   One set of contacts operates on clockwise rotation, and another set on counterclockwise rotation; overtravel 53° minimum.

Maintained-Contact Design.   Operation maintained on counterclockwise rotation, reset on clockwise rotation, and vice versa; overtravel 20° minimum.

Actuators.   Most problems of solving the geometry for matching the limit switch to the controlled machine have been solved with standard switch designs. These include (1) rotary operating heads, (2) plunger operating heads, and (3) wobble lever operating heads (Fig. 1). Gravity-return switches are also available. With these designs, the weight of the actuating lever must provide sufficient force to restore it to the free position. The very small 0.035-N · m (5-in · oz) operating torque is useful in some conveyor applications because it permits operation with small or lightweight objects.

Solid-State Limit Switches.   Conventional limit switches incorporate Hall-effect principles and solid-state circuitry. The output is computer-compatible and requires no in-between electronics. The output interfaces directly with most electronic circuits, discrete transistor circuits, microprocessors, and integrated logic circuits. This switch cannot be used in areas where extremely high magnetic fields are present.

Fiber-Optic Limit Switches.   A comparatively recent innovation in limit-switch design is the incorporation of fiber optics into the device's operation. Switching is accomplished with a cool light signal. The switch is not affected by electromagnetic radiation (EMI) or radio-frequency (RF) interference. The device is also intrinsically safe for use in hazardous locations. It is available for normally open or normally closed circuitry (in fiber-optic terminology). In the normally open mode, the light path is blocked by the shutter; in the normally closed mode, the light path is complete. Switch actuation blocks the light path.

PROXIMITY SENSORS

There are numerous instances in automated systems where an object (presence or nonpresence) must be detected, but not physically contacted (contactless). Several methods have been developed and refined over the years to achieve this task, including (1) electromagnetic inductive [radio-frequency (RF)] devices and other magnetic devices, such as variable-reluctance sensors, magnetically actuated dry-reed switches, Hall-effect switches, and Wiegand switches; (2) ultrasonic detectors; (3) photoelectric sensors; and (4) capacitive sensors. Machine vision also plays a role in some installations. Some devices are limited to ferrous materials, others, when properly calibrated, are applicable to other metals. Some proximity sensors are not affected by the target materials.

INDUCTIVE PROXIMITY SENSORS

The technology of inductive proximity sensors was established in the 1950s. Early sensors essentially were limited to the detection of iron objects, but as more was learned, their use expanded to the detection of other materials, including aluminum, brass, copper, heavily alloyed steels, and many others that make up industrial forms and parts production. It should be pointed out that sensors designed specifically for ferromagnetic materials are also available.

The key element of the detector is an oscillator which generates an RF field at the sensor face. The oscillator consists of an *LC* tank circuit (tuned circuit) and an amplifier circuit with positive feedback. The inductance and the capacitance of the *LC* network determines the oscillator frequency, which can range from 20 kHz to several megahertz (MHz).

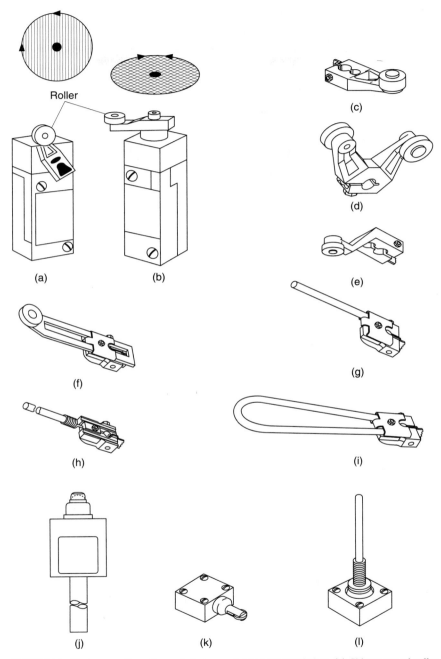

Roller

(a) (b) (c) (d) (e) (f) (g) (h) (i) (j) (k) (l)

FIGURE 1 Various configurations of electromechanical limit switches. (*a*) Side-mounted roller. (*b*) Top-mounted roller. (*c*) Standard roller lever. (*d*) Yoke roller lever. (*e*) Offset roller lever. (*f*) Adjustable-radius roller lever. (*g*) Rod lever. (*h*) Spring-rod lever. (*i*) Flexible loop lever. (*j*) Top plunger. (*k*) Side roller plunger. (*l*) Wobbler switch. (*MICRO SWITCH.*)

The inductance portion L of the tuned circuit is formed by an air-wound or ferrite-core coil. The oscillator circuit has just sufficient positive feedback to sustain oscillation. This generates a sine wave that varies in amplitude, depending on the target's distance from the sensing face.

With a metal target present in front of the oscillator coil, the RF field generates eddy currents on the target's surface. This causes the amplitude to decrease or allows the oscillation to die off. This change in amplitude provides target presence or absence information that is sent either to the output (analog) sensor or through a level detector (Schmitt trigger) that produces a digital (square-wave) output. The digital output is either on or off. The analog version of the sensor is described later.

A schematic circuit of an inductive proximity sensor is shown in Fig. 2. A linearizer between the amplifier and its output provides a more accurate position of the target relative to the sensing face. The output signal has a direct linear relationship with the gap between the sensor and the target surface.

Certain variables must be considered when applying the sensor. These include (1) the nominal distance between sensor and target, (2) the target size, (3) the target material, (4) the target approach, (5) the switching frequency, and (6) the type of signal required, namely, digital, linear, or analog.

Target Distance and Size.   These are very closely related variables. Small targets decrease the nominal sensing distance. In contrast, however, beyond a point, targets larger than the standard do not increase the sensing distance. Suppliers, for each size and configuration of sensor, publish a value of "nominal sensing distance," which is the distance between target and sensor at which a given sensor will turn on. The actual sensing distance may be affected by a number of factors, including manufacturing and temperature tolerances. Variations of as much as ±10 percent may be encountered, but they can be adjusted during calibration at the time of installation. Temperature drift tolerances also must be considered. Over a range of −25 to 85°C (−13 to 185°F), the tolerance increases by ±10 percent.

A shielded sensor will sense only the front of its face and ignores objects to the side. The presence of side materials, however, may cause a slight shift in operating characteristics.

A standard target is an object used for making comparative measurements of the operating distance. Usually a square of mild steel, 1 mm thick, is used. The length of the side of the square is equal to either the diameter of the circle inscribed on the active surface of the sensitive face of the sensor, or three times the rated operating distance, whichever is greater.

The movement pattern of the target also affects the sensing distance capability and ultimately determines at which point the sensor will switch.

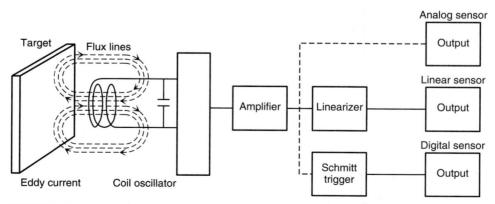

FIGURE 2   Inductive proximity sensor operating principle.

Target Material.   When a target material differs from the standard, as described previously, a material correction must be made. The correction factors (multipliers of standard distance) listed in Table 1 can be used.

TABLE 1   Correction Factors for Target Materials

Object material	Standard distance X
400 series stainless steel	1.15
Cast iron	1.10
Mild steel	1.00
Aluminum foil (0.05 mm)	0.9
300 series stainless steel	0.7
Brass MS63F38	0.4
Aluminum ALMG3F23	0.35
Copper CCUF30	0.3

Switching Frequency.   Defined as the actual number of targets to which the sensor can respond in a given time period, it is usually expressed in hertz (cycles per second).

General Sensor Characteristics.   Cylindrically shaped sensors are used widely and have face diameters as small as 4 mm (0.16 inch) upward to 76 mm (3 inches). Standard sensing distances range from 0.8 to 32 mm (0.03 to 1.3 inches). Electrically there are several options, including 3-wire ac, 4-wire ac, 2-wire dc, 3- and 4-wire dc, and 2-wire ac/dc universal voltage sensors. External configurations also include limit-switch style and vane, rectangular, and ring self-contained noncylindrical sensors. Some designs are equipped with light-emitting diodes (LEDs) to indicate power on, short circuit, and so on (Fig. 3).

Analog Proximity Sensors.   These devices provide a variable current output that is proportional to the distance between a metal target and the sensor face. The current also will vary if different metals or different numbers of small parts are placed in the sensing field. Consequently analog sensors are a convenient way of positioning, discriminating between assorted metals, and

(a)   (b)   (c)

FIGURE 3   Representative inductive proximity sensors. (*a*) Short-length cylindrical sensor in metal housing, available in shielded or unshielded models. Ac and dc incorporate reverse polarity protection. Most feature short-circuit protection. Detecting distance 0.8 to 18 mm (0.03 to 0.71 inch); diameter 4 to 5.4 mm (0.15 to 0.21 inch); length 25 to 57 mm (0.98 to 2.24 inches). Incorporates operation indicators. Nickel-plated brass body. (*b*) Short-length cylindrical sensor. Wide operating voltage and extended service temperature ranges. Rugged construction. Shielded and unshielded models. Ac or dc incorporates reverse polarity protection. Detection distance 1 to 18 mm (0.04 to 0.71 inches); diameter 12 to 30 mm (0.47 to 1.2 inches); length 80 mm (3.15 inches). (*c*) Threaded cylindrical, shielded sensor. Accepts plug-in connection. Weld field immunity to 2k gauss. RFI immune. Short-circuit protection.

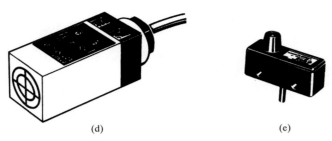

(d)   (e)

FIGURE 3 *(Continued)* *(d)* Short, limit-switch style sensor. Shielded for flush mounting in metal. Weld field immunity available. End- and side-sensing types. Detecting distance 12.5 mm (0.49 inches); height 77.7 m (3.06 inches); width 34.5 mm (1.36 inches); depth 34.5 mm (1.36 inches). *(e)* Basic switch size. Oil-tight epoxy housing. Operation indicator. useful for retrofitting mechanical positioning switches. Detection distance 2 mm (0.08 inches); height 28.7 mm (1.13 inches); width 17.5 mm (0.69 inches); depth 49.2 mm (1.94 inches). (*Omron Electronics, Inc.*)

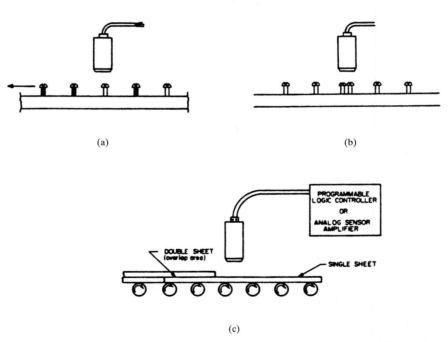

FIGURE 4 Some advantageous applications for analog inductive proximity sensor. (*a*) A stainless-steel screw mixed in with steel screws can easily be detected. The two metals yield two different current outputs, even without changing the sensing distance. (*b*) When an analog sensor detects two small parts instead of one, the change in the output is twice that of one part. The sensing distance remains constant. (*c*) Aluminum sheets slide down a conveyor, at a certain distance from the sensors, prior to a stamping operation. If two or more sheets are present, the surface passes closer to the sensor and causes the output curve to change. (*MICRO SWITCH.*)

checking parts count. Historically the cost of the added electronics was a deterrent in the use of analog sensors. With the costs dropping, analog sensors are regaining acceptance (Fig. 4).

Hall-Effect Proximity Sensors

When a semiconductor, through which a current is flowing, is placed in a magnetic field, a difference in potential (voltage) is generated between the two opposed edges of the conductor in the direction mutually perpendicular to both the field and the conductor. This effect is utilized in some proximity sensors and has been incorporated into electromechanical limit switches, as shown in Fig. 5.

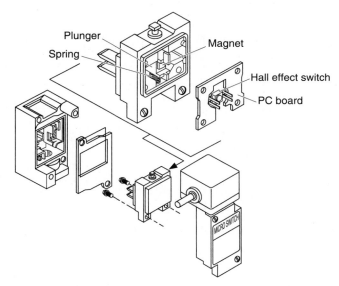

FIGURE 5   Hall effect is used in proximity detectors as well as in infra-structure of other devices, as shown here incorporated in a solid-state limit switch. (*MICRO SWITCH.*)

Wiegand-Effect Switches

A Wiegand wire is a small-diameter wire that has been selectively work-hardened so that the surface and the core of the wire differ in magnetic permeability. When subjected to a magnetic field, the wire emits a well-defined pulse that requires little signal conditioning. This pulse induces a voltage in the surrounding sensing coil. The wire is insensitive to polarity and emits a pulse whether the magnetic field is flowing from north to south or vice versa. A Wiegand proximity sensor senses the presence or absence of ferromagnetic material.

Magnetically Actuated Dry-Reed Switches

Consisting of a thin reed (wire) contained in a hermetically sealed tubelike container, this type of switch is both inexpensive and rugged. Whenever an activating magnet approaches the critical range of the switch, a contact closure is made. Life expectancy generally is in excess of 20 million operations at contact ratings of about 15 VA. These switches generally can operate loads directly, thus avoiding the cost and complexity of comparable solid-state systems. Since the actuating magnet can be installed on a rotating or reciprocating object, the switch can be used in a wide variety of applications for counting, positioning, and synchronizing. Contact closure speeds can be up to 100 per second. Mercury switches with flexible electrodes that can be attracted by the proximity of a magnet also have been used.

CAPACITIVE PROXIMITY SENSORS

Based on the fundamental phenomenon of electrical capacitance, these devices produce an oscillating electric field that is sensitive to (1) dielectric materials, such as glass, rubber, and oil, and (2) conductive materials, such as metals, salty fluids, and moist wood. The principle is shown schematically in Fig. 6.

Capacitance is a function of the size of the electrodes, the distance between the electrodes, and the dielectric constant D of the material between the electrodes. $D_{air} = 1$. The capacitance is given by

$$C = \frac{D \cdot A}{d}$$

where A = area
D = dielectric constant of material between electrodes
d = distance between electrodes

A simple capacitive sensor is shown in Fig. 7. The top electrode is the face of the sensor. A seal ring, the target, passes between it and the ground electrode (a metal conveyor belt). The sensor housing insulates the electrode from galvanic coupling to ground. The rubber seal ring has a dielectric constant D of 4.0. When it enters the electric field, the capacitance increases. The sensor detects the change in capacitance and provides an output signal.

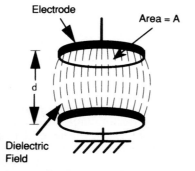

FIGURE 6 Principle of capacitive proximity sensor shown schematically. (*MICRO SWITCH.*)

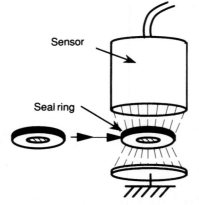

FIGURE 7   Use of capacitive proximity sensor on metal belt conveying rubber seal rings. (*MICRO SWITCH.*)

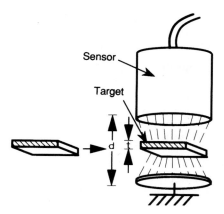

FIGURE 8 Capacitive proximity sensor detecting a metal target or some other conductive material entering the electrical field. Resulting increase in capacitance creates output signal. If the effective distance *t* is reduced, this will cause an increase in capacitance. (*MICRO SWITCH.*)

Figure 8 illustrates a metal target, or some other conductive material, entering the electric field. The resulting increase in capacitance is detected and converted to an output signal. If the effective distance between electrodes is reduced, the result is an increase in capacitance.

Unshielded Capacitive Sensor. This sensor is principally used to detect conductive materials at maximum distances. When detecting nonconductive materials, a path to ground is required. Unshielded sensors are designed to sense conductive materials through a nonconductive material, such as water in a glass or plastic container.

Conductive and nonconductive materials cause an increase in capacitance due to a dielectric change in comparison with air. The ground electrode, as shown previously in Figs. 6 through 8, is not required because a path to ground will serve equally well.

The sensor shown in Fig. 9 is generally referred to as unshielded. This type of device often is used for "looking through" a nonconductive material in order to control the level of the material. The device can also be used for controlling the level of a nonconductive material if a path to ground, such as a metal bin wall, is present.

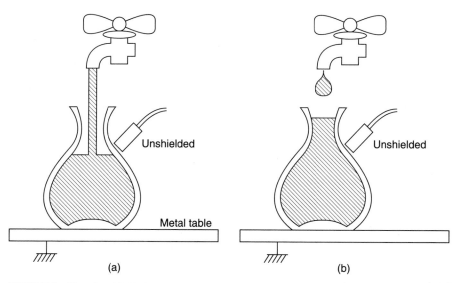

(a)   (b)

FIGURE 9 Use of unshielded capacitive sensor. (*a*) Level of conductive fluid pouring into glass bottle is below sensor. (*b*) Fluid has reached level of sensor, providing the ground electrode. This occurs even though fluid and metal table are separated by the glass of the bottle. The three materials form a capacitor. The alternating current provides a path to ground. With the ground electrode now in place, the circuit closes and a signal results. (*MICRO SWITCH.*)

Shielded (Metal-Body) Capacitive Sensor.   A general-purpose device, it is used for sensing nonconductive materials, such as wood, plastic, cardboard, and glass. A path to ground does not always exist (Fig. 10).

Silo (Plastic-Body) Touching Sensor.   This device is actuated by a material touching the sensor's detecting surface. A sensitivity adjustment is provided for tuning out certain materials as well as for setting detection for a specific mass of material.

The silo-type unit works on the same principle as the standard unshielded sensor. However, the silo is designed specifically for touching and detecting solid nonconductive materials, such as granular or powdered material in a bin or silo, and thus for determining the level of bulk materials.

Compensating for Contaminants.   When using capacitive proximity sensors, any material entering the sensing field can cause an output signal. This includes water droplets, dirt, dust, and other contaminants on the sensor face. The use of a compensation electrode in the sensor can eliminate this problem (Fig. 11).

Combined Use of Shielded and Unshielded Sensors.   Some applications can take advantage of this combination, as illustrated in Fig. 12.

Physical Configurations of Capacitive Proximity Sensors.   These sensors commonly are of a cylindrical shape, as shown in Fig. 13a. Other formats are illustrated in Fig. 13b and c.

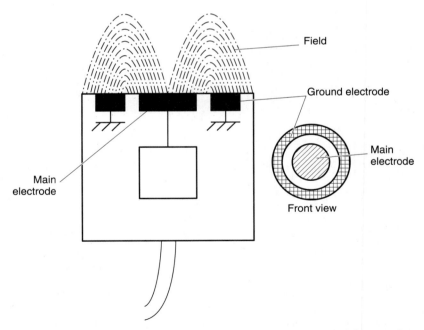

FIGURE 10   In a capacitive proximity sensor application, a path to ground does not always exist. By incorporating the ground electrode in the sensor, an electric field is created independently of any outside path to ground. The field functions the same as when the electrodes face each other. The shielded sensor shown can sense any material, grounded or ungrounded. (*MICRO SWITCH.*)

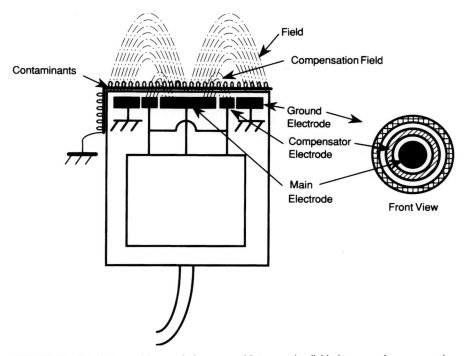

FIGURE 11 Shielded capacitive proximity sensor with two sensing fields, its own and a compensation field which the electrode creates. When contaminants lie directly on the sensor face, both fields are affected and the capacitance increases by the same ratio. The sensor does not "see" this as a change in capacitance, and an output is not produced. The compensation field is very small and does not extend very far from the sensor. When a target enters the sensing field, the compensation field is unchanged. The disproportionate change in the sensing field (with respect to the compensation field) is detected and converted to an output. (*MICRO SWITCH.*)

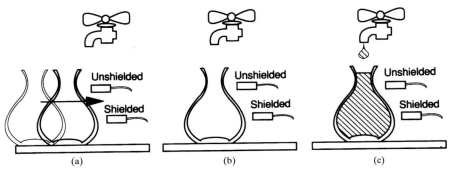

FIGURE 12 An unshielded sensor and a shielded sensor can work together. The shielded sensor locates the glass bottle so it can be filled with liquid. The unshielded sensor indicates that the fill level is reached and can be turned off. (*a*) Neither sensor switches as the bottle approaches. (*b*) The shielded sensor senses the entrance of the glass into its electric field, and it switches. (*c*) The fluid has reached the level of the unshielded sensor, and it switches. Shielded sensors may be flush-mounted in any solid material. (*MICRO SWITCH.*)

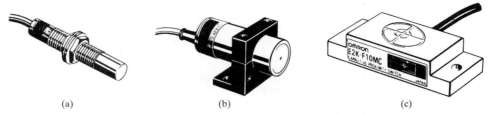

(a)   (b)   (c)

FIGURE 13   Representative configurations of capacitive proximity sensors. (*a*) Threaded cylindrical sensor. Detects glass, plastic, wood, liquids, and metallic objects. Detects materials inside nonmetallic containers. Fixed sensing distance (depending on model) 4 mm (0.16 inch), 8 mm (0.32 inch), 15 mm (0.59 inch); diameter (depending on model) 12 mm (0.47 inch), 18 mm (0.71 inch), 30 mm (1.2 inches); length 80 mm (3.15 inches). Plastic body. (*b*) Unshielded cylindrical sensor. Detects glass, plastic, wood, water, and metallic objects. Detects materials inside nonmetallic containers. Adjustable sensitivity for wide detection range. Furnished with mounting bracket. Detecting distance 3 to 25 mm (0.12 to 0.98 inch), adjustable; diameter 34 mm (1.34 inches); length 82 mm (3.23 inches). Plastic body. (*c*) Thin, flat-pack sensor. Detects metallic and nonmetallic objects. Compact size for mounting on conveyor walls or flush against metallic surfaces. Detecting distance 10 mm (0.39 inch); height 10 mm (0.39 inch); width 50 mm (1.97 inches); depth 20 mm (0.79 inch). Plastic body. (*Omron Electronics, Inc.*)

ULTRASONIC PROXIMITY SENSORS

Specific applications of ultrasonic measurement technology are associated with numerous processing and manufacturing variables and thus are mentioned in other sections of this handbook. The serious consideration of ultrasonic devices as proximity sensors is relatively recent as compared with their use in connection with other instrumental variables.

Proximity detection is one of the lesser demanding and sophisticated applications for ultrasonic technology because the application only requires the detection of presence or nonpresence, as contrasted, for example, accurately measuring the thickness of a part, the density of a fluid, or detecting exact locations of flaws in materials, among other demanding applications.

Ultrasonic gages generally fall into two categories: (1) resonance types, which produce a frequency-modulated continuous-wave signal, and (2) pulse-echo types, which operate somewhat like a sonar system. In the latter system, if a target is present, a small amount of the signal is reflected back to the transducer. Most ultrasonic sensors use a transducer which also serves as an emitter and receiver. Upon receipt of the echo, the amount of time required for the signal to travel between sensor and target is a measure of the target distance. Most sensors incorporate temperature compensation because of thermal influences on the speed of the ultrasonic wave. By definition, any sound exceeding 20,000 Hz is considered in the ultrasonic range. By using high frequencies, a device becomes less immune to environmental noise in its proximity. However, higher frequencies reduce the sensor's "seeing" distance, and thus the sensor designer must achieve a suitable compromise. Typically an industrial ultrasonic transducer will operate at 215,000 Hz.

From the standpoint of proximity sensing, ultrasonic devices have the advantage of working well in connection with nearly all materials when the devices are properly applied. Ultrasonic devices also have the advantage of detecting presence or nonpresence over longer distances than most other sensors, but some of the negative aspects include effects of temperature, surface finish (reflection), humidity, air turbulence, and inaccuracies that may arise from improper geometry between sensor and target.

The effect of distance between target and sensor is very significant. If a target inside the ultrasonic beam is positioned at 200 mm (7.9 inches) from the sensor, the received echo is 4 times stronger than if this distance were 400 mm (15.8 inches). Thus it is possible that a small object placed far from the sensor may not be detected.

Almost all materials and targets reflect sound and thus can be detected. Only sound-absorbing materials, such as textiles, weaken the echoes such that the maximum sensing distance must be greatly reduced. Thin-walled targets [below 0.005 mm (0.002 inch)] are also difficult to detect. The most effective sensor-target distance in such instances should be determined experimentally. Additional factors are illustrated and described in Figs. 14 through 17.

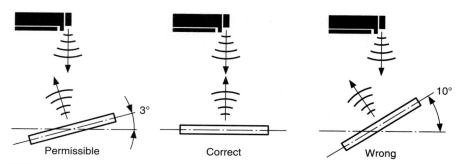

FIGURE 14   Effect of inclination on ultrasonic sensor beam. If a smooth, flat target is inclined more than ±3° to the normal of the beam axis, part of the signal is deflected away from the sensor and the sensing distance is decreased. However, for small targets located close to the sensor, the deviation from normal may be increased to ±8°. If the deflection angle exceeds ±12°, all of the signal is deflected and there is no signal response. (*MICRO SWITCH.*)

PHOTOELECTRIC PROXIMITY SENSORS

A cornerstone of early automated systems, practical applications of photoelectric devices date back several decades. They have profited over the years from advances in electronics, fiber optics, and practical application experience. Thus these devices continue to play very important roles in industrial control systems. The photoelectric effect, discovered by Hertz in 1887 and fully explained by Einstein in 1921, is the basis of numerous devices in a number of fields, including presence-nonpresence sensors.

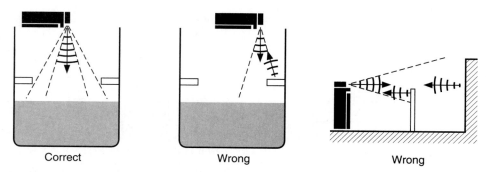

FIGURE 15   All sound-reflecting targets with sufficient surface situated inside the beam cone will be detected. To avoid false (parasitic) echoes, the interfering objects should be removed, or the ultrasonic sensor relocated. Similarly any sound-absorbing materials should not be in the beam path. (*MICRO SWITCH.*)

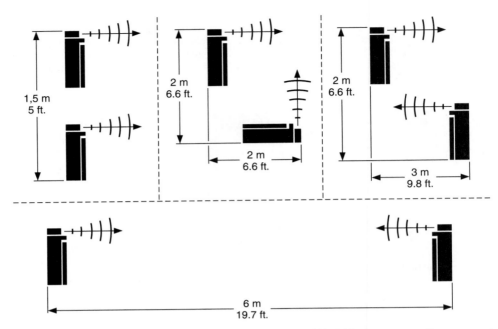

FIGURE 16 Where several ultrasonic detectors may be required within fairly close space confines, care must be taken to avoid acoustic interference between separate devices. The minimum practical sensor-to-sensor space is shown. (*MICRO SWITCH.*)

Scope of Usage. Photoelectric controls are found in many kinds of applications because they respond to the presence or absence of either opaque or translucent materials at distances from a fraction of an inch (a few millimeters) up to 100 or even 700 feet (30 to 210 meters). Photoelectric controls need no physical contact with the object to be triggered, which is very important in some cases, such as those involving delicate objects and freshly painted surfaces.

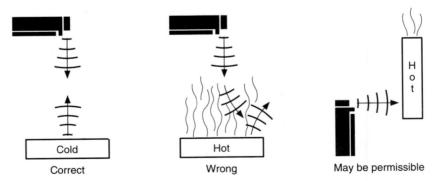

FIGURE 17 An internal temperature detector may be used within the ultrasonic device for adjusting the clock frequency of the elapsed time counter and the carrier frequency to compensate for air-temperature variations. However, large temperature fluctuations can cause dispersion and affect refraction of the ultrasonic signal adversely. If a hot object must be detected, the sensor should aim at the lower (cooler) portion of the target and thus avoid or minimize the effects of warm air currents. (*MICRO SWITCH.*)

Some of the more common applications include thread break detection, edge guidance, web break detection, registration control, parts-ejection monitoring, batch counting, sequential counting, security surveillance, elevator control, conveyor control, bin level control, feed or fill control, mail and package handling, and labeling. These applications are described later.

Photoelectric System Configurations.   A self-contained control includes a light source, a photoreceiver, and the control base function, which amplifies and imposes logic on the signal to transform it into a usable electrical output. A modular control uses a light source–photoreceiver combination or reflective scanner separate from the control base. Self-contained retroreflective controls require less wiring and are less susceptible to alignment problems, while modular controls are more flexible in allowing remote positioning of the control base from the input components and hence are more easily customized.

Photoelectric controls are further classified as nonmodulated or modulated. Nonmodulated devices respond to the intensity of visible light. Thus, for reliability, such devices should not be used where the photosensor is subject to bright ambient light, such as sunlight. Modulating controls, employing LEDs, respond only to a narrow frequency band in the infrared. Consequently they do not recognize bright, visible ambient light.

Controls typically respond to a change in light intensity above or below a certain value of threshold response. However, certain plug-in amplifier-logic circuits cause controls to respond to the rate of light change (transition response) rather than to the intensity. Thus the control responds only if the change in intensity or brightness occurs very quickly, not gradually.

Operating Mode.   Both modulated and nonmodulated controls energize an output in response to either a light signal at the photosensor when the beam is not blocked (light-operated, LO), or a dark signal at the photosensor when the beam is blocked (dark-operated, DO). Although some controls have built-in circuitry that determines a fixed operating mode, most controls accept a plug-in logic card or module with a mode selector switch that permits either light or dark operation.

In addition to a light source, light sensor, amplifier (in the case of modulated LED devices), and power supply, a complete system includes an electrical output device (in direct interface with logic-level circuitry—the output transistor of a dc-powered modulated LED device or of an amplifier-logic card).

Scanning Techniques.   There are several ways to set up the light source and photoreceiver to detect objects. The best technique is the one that yields the highest signal ratio for the particular object to be detected, subject to scanning distance and mounting restrictions. Scanning techniques fall into two broad categories: (1) thru (through) scan and (2) reflective scan.

Thru Scan.   In thru (direct) scanning the light source and photoreceiver are positioned opposite each other, so light from the source shines directly at the sensor. The object to be detected passes between the two. If the object is opaque, direct scanning will usually yield the highest signal ratio and should be the first choice (Fig. 18).

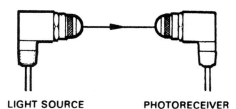

LIGHT SOURCE   PHOTORECEIVER

FIGURE 18   In direct, or thru, scan configuration, the light source is aimed directly at the photoreceiver. Sometimes the configuration is referred to as the transmitted-beam system. (*MICRO SWITCH.*)

As long as an object blocks enough light as it interrupts the light beam, it may be skewed or tipped in any manner. As a rule, the object size should be at least 50 percent of the diameter of the receiver lens. To block enough light when detecting small objects, special converging lenses for the light source and photoreceiver can be used to focus the light at a small, bright spot (where the object should be made to pass), thereby eliminating the need for the

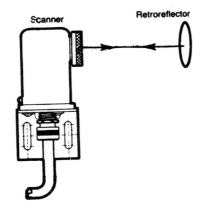

(a)

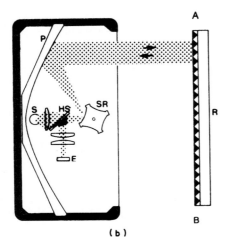

(b)

FIGURE 19 (*a*) Reflected beam (retroreflective scan) system in which light source and photoreceiver are contained in a single enclosure. This simplifies wiring and avoids critical alignment of source and sensor. (*MICRO SWITCH.*) (*b*) By adding a rotating-mirror wheel (SR), a parabolic reflector (P), and a semitransparent mirror (HS), a parallel-scanning beam can be obtained. This beam moves at high speed from *A* to *B*, thus forming a "light curtain," any interruption of which is detected and signaled by a relay. S—light source; E—photoreceiver. (*Sick Optik Elektronik.*)

object to be half the lens diameter. An alternative is to place an aperture over the photoreceiver lens in order to reduce its diameter. Detecting small objects typically requires direct scan.

Because direct scanning does not rely on the reflectiveness of the object to be detected (or a permanent reflector) for light to reach the photosensor, no light is lost at a reflecting surface. Therefore the direct scan technique permits scanning at farther distances than reflective scanning. Direct scanning, however, is not without limitations. Alignment is critical and difficult to maintain where vibration is a factor. Also, with a separate light source and photoreceiver, there is additional wiring, which may be inconvenient if the application is difficult to reach.

Reflective Scan. In reflective scanning the light source and photoreceiver are placed on the same side of the object to be detected. Limited space or mounting restrictions may prevent aiming the light source directly at the photoreceiver, so the light beam is reflected either from a permanent reflective target or surface, or from the object to be detected, back to the photoreceiver. There are three types of reflective scanning: (1) retroreflective scanning, (2) specular scanning, and (3) diffuse scanning.

Retroreflective Scanning. With retroreflective scanning the light source and photosensor occupy a common housing. The light beam is directed at a retroreflective target (acrylic disk, tape, or chalk), one that returns the light along the same path over which it was sent (Fig. 19). Perhaps the most commonly used retro target is the familiar bicycle-type reflector. A larger reflector returns more light to the photosensor and thus allows scanning at a further distance. With retro targets, alignment is not critical. The light source–photosensor can be as much as 15° to either side of the perpendicular to the target. Also, inasmuch as alignment need not be exact, retroreflective scanning is well suited to situations where vibration would otherwise be a problem.

Retroreflection from a stationary target normally provides a high signal ratio so long as the object passing between the scanner and the target is not highly reflective and passes very near the scanner. Retroreflective scanning is preferred for the detection of translucent objects and ensures a higher signal ratio than is obtainable with direct scanning. With direct scanning, the dark signal may not register very dark at the photosensor, because some light will pass through the object. With retroreflective scanning, however, any light that passes through the translucent object on the way to the reflector is diminished again as it returns from the reflector. The system is also useful where retroreflective tape or chalk coding can be placed on cartons for sorting. Retroreflective scanning normally can be used at distances up to 30 feet (9 meters) in clear air conditions. As the distance to the target increases, the retro target should be made larger so that it will intercept and return as much light as possible. Single-unit wiring and maintenance are secondary advantages of retroreflective scanning.

Specular Scanning. The specular scan technique uses a very shiny surface, such as rolled or polished metal, shiny plastic, or a mirror to reflect light to the photosensor (Fig. 20). With a shiny surface, the angle at which light strikes the reflecting surface equals the angle at which it is reflected from the surface. Positioning of the light source and photoreceiver must be precise (mounting brackets which fix the light source–photoreceiver relationship should be used), and the distance of the reflecting surface from the light source and photoreceiver must be consistently controlled. The size of the angle between the light source and photoreceiver determines the depth of the scanning field. With a narrower angle there is more depth of field; with a wider angle there is less depth of field. For a fill-level detection application, for example, this means that a wider angle between light source and photoreceiver allows detection of the fill level more precisely.

Specular scanning can provide a good signal ratio when required to distinguish between shiny and nonshiny (mat) surfaces, or when using depth of field to reflect selectively off shiny surfaces of a certain height. When monitoring a nonflat, shiny surface with high or low points that fall outside the depth of field, these points appear as dark signals to the photosensor.

Diffuse Scanning. Nonshiny (mat) surfaces, such as kraft paper, rubber, and cork, absorb most incident light and reflect only a small amount. Light is reflected or scattered nearly equally in all directions. In diffuse scanning, the light source is positioned perpendicularly to a dull surface. Emitted light is reflected back from the target to operate the photoreceiver (Fig. 21). Because the light is scattered, only a small percentage returns. Therefore the scanning distance is limited (except with some high-intensity modulated LED controls), even with

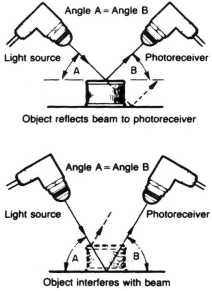

FIGURE 20 Specular scan technique uses a very shiny surface such as rolled or polished metal, shiny plastic, or a mirror to reflect light to the photosensor. (*MICRO SWITCH.*)

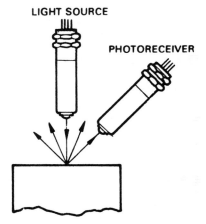

FIGURE 21 Diffuse scan is used in registration control and to detect material (corrugated metal, for example) with a slight vertical flutter—which might present a consistent signal with specular scan. Alignment is not critical in picking up diffuse reflection. (*MICRO SWITCH.*)

very bright light sources. It is often difficult to obtain a sufficient signal ratio with diffuse scanning when the surface to be detected is almost the same distance from the sensor as another surface (for instance, a nearly flat or low-profile cork liner moving along a conveyor belt). Contrasting colors can help in such situations.

Diffuse scanning is used in registration control and to detect material (corrugated metal, for example) with a slight vertical flutter—which may prevent a consistent signal with specular scanning. Alignment is not critical in picking up diffuse reflection.

Color Differentiation (Registration Control).   In distinguishing color, as in registration mark detection, contrast is the key. High contrast (dark color on light, or vice versa) provides the best signal ratio and control reliability. Therefore, if possible, bright, well-defined contrasting colors should be considered in the interest of the registration control system.

Diffuse scanning is normally used to detect color change. Table 2 gives some of the common color combinations that must be distinguished in registration control, plus the most suitable type of photosensor and scan technique.

TABLE 2   Factors That Determine Selection of the Photosensor and Scan Technique in Registration Control Applications

Background	Mark	Photosensor	Scan technique
Clear film	Black, blue, red	Any	Direct scan
White (kraft paper, metal foil)	Black, blue	Phototransistor or CdSe photocell	Diffuse scan
	Red	CdS photocell with blue-green filter	Diffuse scan
Black, blue, or other dark colors	Red	Any	Diffuse scan
Red	Black, blue	Any	Diffuse scan

When the background is clear (transparent), the best method is to detect any color mark with direct scanning. When the background is a second color, contrasts such as black against white usually ensure a sufficient signal ratio (difference between dark and light signals) to be handled routinely with diffuse scanning. Red, or a color that contains considerable red pigment (yellow, orange, brown), on a white or light background is a special case. In such instances, a photoreceiver with a cadmium sulfide cell for detecting red marks is preferred because it makes red appear dark on a light background.

A retroreflective scanner with a short-focal-length lens (but without a retro target) can be used to detect registration marks. It is placed near the mark and is actually used in the diffuse scan technique. If a retroreflective scanner is employed to detect marks on a shiny surface, the scanner should be cocked somewhat off the perpendicular to make certain that only diffuse reflection will be picked up. Otherwise the shiny surface of the mark could mirror-reflect so brightly that it would overcome the dark signal a CdS cell normally receives from red. This would mean a light signal from both the background and the mark. In detecting colors, a rule of thumb is to use diffuse (weakened), rather than specular (mirror), reflection.

Sensitivity Adjustment.   Most photoelectric controls have a sensitivity adjustment that determines the light level at which the control will respond. Conditions which may require an adjustment of sensitivity include (1) detecting translucent objects, (2) a high speed of response, (3) a high cyclic rate, (4) line voltage variation, and (5) a high electrical noise atmosphere.

Light Sources and Sensors.   Early photoelectric control systems used incandescent light sources and traditional photocells—a combination still widely used. A photocell changes its electrical resistance with the amount of light that falls on it. A number of photocells have been used over the years for different applications (photoelectric controls, copying machines, television pickup tubes, and so on). Widely used for photoelectric controls are cadmium sulfide and cadmium selenide cells. During the relatively recent past, phototransistors and photodiodes have become available as sensors, and LEDs have been used as light (infrared) sources. The advantages of the newer components in certain types of applications are described shortly.

Photocells. The sensitivity of a photocell can be defined in two ways. Static sensitivity expresses the resistance of the cell at a given light intensity; the lower the resistance, the more sensitive the photocell. So long as the resistance falls within the range of the control unit with which the photocell is used, static sensitivity is usually not an important consideration. Dynamic sensitivity is an expression of the ratio of the photocell resistance at one light level to its resistance at a different level; the greater the ratio, the higher the sensitivity. This is a much more useful expression of photocell sensitivity.

The speed of response of a photocell is the time it requires to produce a change in resistance in response to a given change in light intensity. Although all photocells are fast, they require a finite amount of time to respond to changing light. And their speed of response depends on the amount of light falling on the cell; the greater the light intensity, the faster the response.

Light history effect refers to a characteristic of a photocell that has been kept dark or light for extended periods. Such a cell overresponds to a change in light before returning to its normal response (somewhat analogous to the response of the human eye to sudden changes in light). Although in some applications it is well to know about this effect, it normally is not a significant consideration in industrial uses.

The effect of temperature on a photocell is to increase its resistance and thus decrease its current with increases in temperature for a given level of illumination. The temperature effect is smaller when the level of illumination is high than when it is low.

Photocells also respond differently to different colors of light. Photocells generally used in industrial applications have a far greater response to colors in the red and infrared range than in the blue-violet range. Except in certain applications, as described previously for registration control, the color response is not a significant factor.

Phototransistors. A phototransistor produces a collector current that is a function of both base current and light. Since the base lead of a phototransistor is usually left unconnected, only variations in light intensity produce variations in current output. There are a number of differences between the phototransistor and the photocell. (1) The current output of a phototransistor is largely independent of the voltage across it, whereas that of a photocell is not. As a result, controls designed to work with photocells will not necessarily work well with phototransistors, and vice versa. (2) The response of phototransistors is affected by changes in temperature, but in a way opposite to that of photocells; the higher the temperature, the higher the current output. (3) Phototransistors have a polarity which must be observed, while photocells do not. (4) Phototransistors respond to light much faster than photocells, but typically have a lower sensitivity.

The photodiode response is narrower than that of the phototransistor, making the diode more effective in blocking stray light from incandescent, sun, or other light sources.

Light-Emitting Diodes. The introduction of LEDs as radiation sources brought a number of advantages for certain applications. The useful life of an LED is estimated at 100,000 hours, which is about 10 times that of an incandescent lamp. However, incandescent lamps are frequently used because they have a spectrum from the ultraviolet to the visible to the infrared, allowing a wide range of colored targets to be detected. LEDs have the advantage that they can be modulated directly, whereas incandescent lamps require a mechanical chopper. Silicon phototransistors and photodiodes are excellent matches for infrared LEDs because their greatest sensitivity peaks almost match precisely at the transmitter's (LED) wavelength. The role of LEDs in other instrumentation applications (displays, communications, and so on) is described in other parts of this handbook.

Photoelectric Sensor Configurations. Contemporary sensors tend to fall into three classifications: (1) general-purpose devices that handle a wide range of industrial applications, including packaging and materials handling; (2) special-purpose designs, including miniature heavy-duty models, some of which include mark detection, transparent object detection, conveyor sensors that detect the object rather than the belt per se, belt grooved head and slotted sensors; and (3) designs that incorporate fiber optics, which are superior in long-distance applications, for detecting very small objects and shiny objects, and with an improved performance in hot environments (Fig. 22).

Fiber-Optic Sensors. Separate-type fiber-optic cables detect opaque objects that break the beam. They require mounting space for separate emitter and receiver sensing heads. Diffuse reflective fiber-optic cables reflect the light off the object to be detected. Reflective sensors deliver and receive the light in a single sensing head. The detecting distance, however, is reduced.

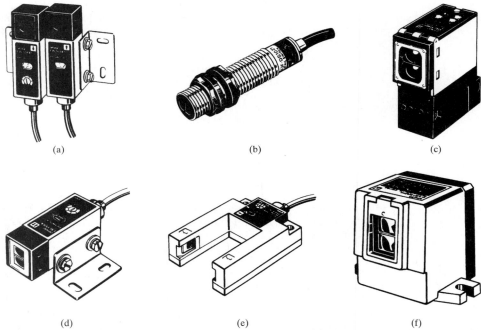

(a) (b) (c)

(d) (e) (f)

FIGURE 22 Representative configurations of photoelectric sensors. (*a*) Miniature sensor. Built-in dc amplifier offers fast response times. Horizontal or vertical mountings. Selectable light-on/dark-on operation. Adjustable sensitivity. Operation indicators. Sensing distance 2 meters (6.56 feet) or 5 meters (16.4 feet) separate type; 2 meters (6.56 feet) rectroreflective; 100 mm (3.94 inches) or 300 mm (11.8 inches) diffuse reflective. Fiber-optic versions available. (*b*) Threaded cylindrical. Built-in amplifier provides long sensing distance. Response time 30 ms (ac), 2.5 ms (dc). Polarized retroreflective model detects shiny objects. Operation indicators. Sensing distance 3 meters (9.84 feet) separate type; 0.1 to 2 meters (3.94 inches to 6.56 feet) retroreflective; 0.1 to 1.5 meters (3.94 inches to 4.92 feet) polarized retroreflective; 100 mm (3.94 inches) diffuse reflective. (*c*) Slim general-purpose. Universal power supply. Plug-in construction. Response time 30 ms (0.5 to 20 ms with built-in time delays). Operation indicators. Sensing distance 7 meters (22.97 feet) retroreflective; 5 meters (16.4 feet) polarized retroreflective; 2 meters (6.56 feet) diffuse reflective. (*d*) Color mark sensor. Self-contained dc amplifier for fast response and high-speed operations. Response time 1 ms. Horizontal or vertical mounting. Selective light-on/dark-on operation. Sensitivity adjustment. Sensing distance 12 mm (0.47 inch) or 50 mm (1.97 inches) diffuse reflectance. (*e*) Grooved head and slotted sensor for edge control, positioning, and mark detection. Response time 1 ms. Sensing distance 10 mm (0.39 inch) or 30 mm (1.18 inches). (*f*) Conventional conveyor sensor. Detects just the target, *not* the conveyor belt. Light source directed upward to avoid reflection from conveyor. Built-in time delays available. Sensing distance 10 m (32.8 feet) separate type; 0.3 to 3 meters (11.81 inches to 9.84 feet) retroreflective; 700 mm (27.56 inches) diffuse reflective. (*Omron Electronics, Inc.*)

Fiber-optic sensors in particular are preferred for hard-to-reach situations. Most plastic fiber-optic cables can be cut to custom lengths in the field from the original 2-meter (16.56-foot) length. When threaded heads are too large to reach the detection site, a cable with bendable steel tubing should be considered. The latter retains complex shapes and is well suited for multiple-sensor inspections of minute assemblies and parts. Where the sensing distance of separate-type cables is required, an optional lens kit for increasing (by 7 times) the distance between emitter and receiver should be considered.

Needle probes may be selected to detect objects as small as 0.0006 inch (0.02 mm) passing flush by the fiber-optic cable lens. Side-view accessories and "periscope"-type needle probes provide space-efficient ways to achieve right-angle detection.

The use of convergent-beam sensing can be helpful when detecting highly polished reflective surfaces. The temperature-tolerant range for fiber-optic cables is –40 to 70°C (–40 to 158°F). For higher-temperature applications, plastic-sheathed or armored glass cables may be used; the latter are useful up to 400°C (750°F).

For equipments which involve flexure, such as a robot arm, fiber-optic cables with a retractable coiled section are usually well suited. Representative fiber-optic photoelectric sensors are shown in Fig. 23.

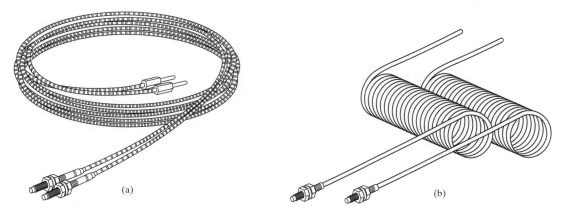

(a)   (b)

FIGURE 23   A wide selection of fiber-optic cables and accessories for photoelectric amplifiers makes it possible to solve difficult detection problems. (*a*) Armored through beam-sensing head. (*b*) Coiled cable for reciprocating/flexing equipment. (*Omron Electronics, Inc.*)

MACHINE VISION

Since its initial serious recognition in the late 1970s, machine vision (MV) has been variously defined:

MV is the process of extracting information from *visual sensors* for the purpose of enabling machines to make intelligent decisions.

or

MV is part of the larger technology of *artificial visual perception* that substitutes (partially or totally) the *human* visual capability by instruments which are backed up by electronic data processing (notably complex computations).

As will be noted from the foregoing, the real and practical objectives of MV tend to be nebulous and of far-reaching expectations.

MV became a "buzz word" of the 1980s—to the extent that MV, combined with robots, would bring about the most ambitious goals in terms of production automation. The fact is that as of the early 1990s, MV has lost much of its earlier luster. With relatively few exceptions, most contemporary MV systems are confined to sophisticated and demanding object-detection problems, with the majority of such detection problems currently being solved by the less exotic types of sensors previously described in this article. Because total MV systems generally still remain overly complex and costly, other nonvisual sensors have been greatly improved and applied more imaginatively to achieve many of the objectives that once appeared to lie within the province of MV. Cost, once again, has been the traditional motivating factor.

"Seeing" robots, once a major goal of MV, has proved disappointing from the standpoint of MV. To be true, in practice, robots, once programmed, do perform in many instances as though they could "see," but this has been brought about by other instrumental detectors (tactical, for example) than by literal "viewing" of parts, pieces, machines, and entire manufacturing scenes. For example, in the late 1980s a large automotive manufacturer originally invested rather heavily in MV systems in connection with robotic operations, but after a year or so of trials, abandoned them for possible consideration at some future date.

To be sure, MV developments will continue, and systems will be installed where other simpler, lower-cost approaches do not suffice. Coupled with trends toward reducing computation costs, MV may become more competitive.

Artificial Visual Perception

Industrial MV techniques initially stemmed from the interests of the military in a technology known as pattern recognition. This may be defined as seeing, analyzing, and interpreting patterns, as of scenery, juxtaposition, dimensional magnitudes, color, and other characteristics of the visual environment.

Human visual communication (information transfer) with the outside environment depends on the interactions (predominantly absorption, reflection, and refraction) of light (visual) radiation emanating from physical objects (in point, two, or three dimensions), thus enabling the human observer to cope with the surrounding environment in a safe and efficient manner. The vision activity (human or machine) must be complemented by some form of processor (human brain or electronic counterpart) to make a final identification of what has been seen. This process, defined by a few words, is pattern recognition.

The means that are applied to perform pattern recognition is known as the pattern processor. The human brain, in processing a pattern, does an amazing job of sorting out extraneous information in the input to quickly identify what really is present (the objects of interest that are in the pattern).

Interest in pattern recognition dates back some 70 years. The more the process is studied, the more one finds how fundamental the process really is to the human brain function. The amount of information (inputs) that the human system can observe and process is tremendous, but nevertheless remains poorly understood. This explains why MV, from a technological viewpoint, is closely associated with the study of artificial intelligence.

Elements of Pattern Recognition

Any pattern recognition system contains the same three basic elements: (1) sensing, (2) processing, and (3) implementing actions based upon input data.

The primary objective of pattern recognition is to classify a given unknown pattern as belonging to one of several classes of patterns. The applications of pattern recognition are many and varied. In the case of character recognition, for example, the patterns are easily generated and recognized by humans, and the basic goal has been to improve the human-machine communication. In other situations, the patterns are difficult for humans to recognize rapidly, as, for example, the very rapid interpretation of an electrocardiogram. The wide range of applications, as well as the relationship to diverse disciplines, including machine vision, communication and control, and the area of linguistics (word and sentence reading), have broadened the interest basis in pattern recognition.

Information Content of an Image

In terms of MV as used in industrial production situations, the information content of the image falls into three basic categories:

1. *Geometry,* which, in turn, portrays shape, position, dimension, and a number of other associated properties, including density and texture (which can be inferred from the known geometry in most cases).
2. *Color,* which is very helpful when present. There are, of course, color-blind persons and instruments that preclude its use.
3. *Movement,* which is present in two or more images of a dynamic process.

Extracting Information from Images

As with other forms of industrial instrumentation, the MV system senses (reads may be a better term) the image, but before the data obtained from the image can become meaningful, the sensed information must be compared with some form of standard, or prelearned, pattern. In more familiar terms, the combined actions of sensing and comparing constitute measurement.

As compared with the usual type of industrial sensor, such as a thermocouple, where measured data are conveniently available in the form of a ready-made electrical signal, in MV one deals not with just one or a very few points or locations of emanating data, but with thousands-plus bits of information—because what is being observed (measured) is comprised of a multitude of points (picture pixels[1]). In an artificial vision system, image data, as may be gathered by an electronic camera, must be compared electronically with information, that is, in some fashion with data that are stored in electronic memory.

Initially the application of a general-purpose computer to MV was the only choice available. The problem immediately encountered was the fact that the computer was designed to process computational data, *not* data patterns. The data from a video camera in an MV system are pattern data and, thus, are very different from computational data as exemplified by financial balance sheets or linear regression analysis. The situation is summarized as follows:

Image data quantity 484×320 pixel density

Hence, 154,880 pixels per frame

At 6 bits of gray-level data per pixel, 929,280 pixel values per frame

At a 30 times a second refresh rate, 27,878,400 bits per second

Since the early days of MV, a number of innovations have contributed to the simplification of this problem, including improvements in gray-scale systems, better algorithms, the pipelined image processing engine, the geometric arithmetic parallel processor, and associative pattern processing.

---

[1]A pixel is a picture element, a small region of a scene within which variations of brightness are ignored.

MV Sensors

Factors which usually must be considered are (1) optics and lighting, (2) field of view, (3) resolution, (4) signal-to-noise ratio, (5) time and temperature stability, and (6) cost. MV sensors which have been or are currently in use include the following.

Line Scanners.   These include solid-state arrays, flying-spot scanners, and prism, mirror, or holographically deflected laser cameras. These scanners are fast high-resolution devices, which are relatively free of geometric distortion in one dimension. In order to capture a complete two-dimensional scene, the second dimension is obtained either by motion of the object past the scanner or by mirror or prism deflectors. Mechanical motion tends to slow down data acquisition and, in some cases, produces geometric distortion.

Area-Type Scanners.   These scanners were used in early systems utilizing closed-circuit television cameras with vidicon image sensors. Advantages were comparatively low cost and relatively high resolution (300 to 500 television lines). They suffer geometric distortion, temperature instability, lag (requiring several television frames for full erasure), and sensitivity to nearby magnetic fields.

Solid-State Cameras.   These cameras represent a major advance in image-sensing technology. Scenes can be digitized onto an array of photosensitive cells. Charge-coupled devices (CCDs) or charge injection devices (CIDs) have been used in these cameras. The arrays form a pixel grid containing the data currently appearing on camera. The solid-state camera is available in a variety of pixel densities, notably, 128×128, 256×256, 512×512, 1024×1024, and so on.

The limiting factor of cameras for imaging systems is not necessarily the resolution or speed. Camera technology essentially has kept up with the ability of computationally based processes to handle information, especially when multiple cameras are used. Traditionally, in MV systems sensors can detect information in real time a lot faster and in larger quantities than processors can handle.

MV Image Processing

From the beginning, a major task of MV engineering has been that of reducing the amount of information actually needed, and thus lessening and simplifying the processing task. Shortcuts are frequently taken to reduce data volume. Some processing systems that have been or are in current use include the following.

Binary MV System.   The representation of all combinations possible in every pixel of every scene is unrealistic. Conversion of each pixel point into a binary value reduces the pixel data, but it also reduces the accuracy of the analysis. Binary systems evaluate each pixel as black or white. A threshold-adjusting capability can allow users to select what intensity of signal is to be the black-white border.

Gray-Level System.   This system can interpret each pixel's value as a specific gray tone. These systems vary in precision, with each pixel being evaluated as 16, 64, or even 256 different values.

Windowing the Scene.   This system can reduce the total data system load. Since analysis of an entire scene often is unnecessary for proper recognition, the use of a window can eliminate unneeded image data.

Segmentation.   A common means of data reduction, this technique divides the image data into areas of interest and then interpolates surrounding pixels into those areas. There are several ways to segment an image, but all detract from accuracy. Four procedural options for segmentation may be considered:

1. *Algorithms.*   In general, algorithms are a set of mathematical models that can be used to describe an image. A few years ago, the SRI algorithms were developed by the Stanford Research Institute. These consisted of about 50 different features that are extracted from a binary image, such as the size of blank areas (holes) or objects (blobs) and their *centroids* or *perimeters.*

2. *Neighborhood Processing.*   This is an averaging method, achieved by treating each pixel value as if it were part of a group or neighborhood. This gray-level data-reduction technique can change an otherwise complicated image into areas with well-defined lines. Each pixel's point value is calculated by considering the value of a neighboring pixel. The process effectively averages the scene into regions or areas.

3. *Convolution.*   If an image is represented by the rate of light change per pixel instead of light density, the image will look like a line drawing because the greatest intensity of a pixel change is at the boundaries.

4. *String Encoding.*   In run-length, string, or connectivity encoding, the values of the first and last pixel positions of each scan line are compared to see if they are equal and, therefore, belong to the same region. The tables generated by this process also consider the vertical changes in pixel state and, as may be expected, are rather short because most simple binary images contain relatively few transition borders.

5. *Associative Pattern Processing.*   Rather than increasing the capability of processing large amounts of data in a shorter time span through the route of improving computer and data-handling techniques (hardware and software) as just described, a fundamentally different approach has three functional sections (as developed by APP (pattern processing technologies): (1) An image analyzer contains all the pixel data for the current image on the camera. (2) It converts the data into a unique statistical fingerprint, which is a unique set of response memory addresses. The response memory contains the statistical fingerprints for all trained images. (3) A matching histogram compares the fingerprints of the current image with those stored in the response memory. The theory of operation is based on a statistical phenomenon that reduces an image's data into a precise fingerprint, which is a set of values for an image that differs from other images and is a minute fraction of the total data of the image. Instead of one image being trained, all the images needed are trained and no programming is involved. One simply shows an image to the camera, allows the fingerprint memory locations to drive the response memory, and loads in a label for each image.

The Recognition Process

Once the user describes the image mathematically, the second half of the process step can occur. Data processing has been mentioned previously. The programming of a general-purpose computer requires programming expertise that is costly. Using this computational approach may entail performance levels that are too slow for real-time operation, unless the application only involves very basic procedures. Gains in processing power can be quickly taken up by application complexity with its increasing data demands.

MV Applications

Some form of actuation or control is the last step in implementing MV. Reasonably standardized methods of communication usually can be used with such common interfaces as RS-232 or IEEE-488.

MV suppliers in recent years have turned to specialized products oriented to specific industries. The advent of generic software has been possible because of numerous application similarities.

MV applications essentially fall into four categories: (1) inspection, (2) location, (3) measurement, and (4) recognition.

MV can inspect to determine if parts are acceptable. Inspection tasks can be further categorized into one of four functions: (1) surface flaw detection, (2) presence or absence of certain features, (3) dimensional tolerance verification, and (4) shape verification. Three examples that use a gray scale are shown in Figs. 24 through 26.

DISCRETE-PIECE IDENTIFICATION—BAR CODING

The lack of uniformity of parts, pieces, subassemblies, and so on, that persists in the discrete-piece manufacturing industries imposes a myriad of complexities, as just alluded to in the preceding description of machine vision. Thus MV applications as generally defined must be highly customized to very specific applications. This imposes a crucial cost restraint.

If, however, pieces and parts could be "packaged" in a way to simplify their recognition and exact identification, most of the complexities of MV can be eliminated. That is, of course,

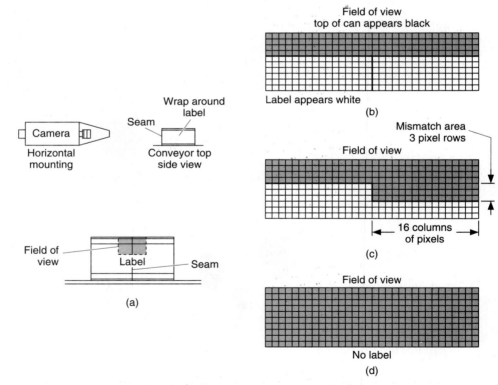

FIGURE 24   (*a*) Example of where a template or pattern match algorithm is used to check accuracy of a label placed on a can of food. The camera is located so that the field of view is horizontally across the area where the label ends touch, as shown in lower diagram. To set up the reference template, the operator adjusts the horizontal and vertical axes of the camera by viewing a display on the controller, during which time the operator also adjusts the mounted camera to obtain the kinds of views shown. (*b*) Label is properly placed. (*c*) Label is poorly placed, indicating a mismatch area of three pixels. (*d*) No label on can. (*Eaton.*)

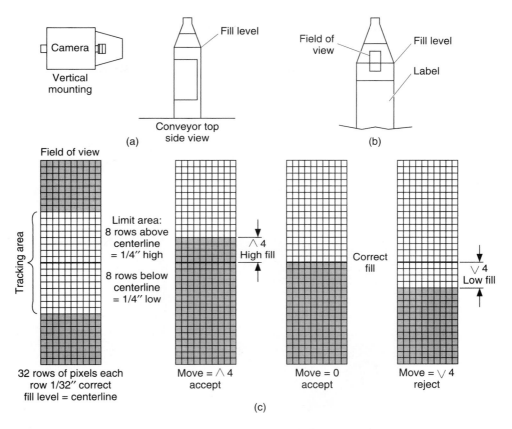

FIGURE 25 Example of use of *y*-axis algorithm to check bottle-filling operation. (*a*) Arrangement of camera and conveyor juxtapositions. (*b*) Location of field of view for given high and low levels. (*c*) Reference images showing high fill, correct fill, and low fill. (*Eaton.*)

exactly what has been done for decades in terms of a large variety of manufactured end products, such as individual boxes of nuts and bolts, screws, cases of canned goods, boxed machine parts, containerized detergents, paints, hardware, plastic-encapsulated items, and even palletized groupings of identical merchandise.

The general concept of containerization, of course, originally was developed to simplify shipping and distribution to the end user. Later these packaging concepts became the basis for vastly improved inventory control and for automated warehousing by manufacturers, distributors, wholesalers, and, indeed, the final point of purchase.

With many millions of items produced, it became obvious that an extensive codification system was needed if piece identification were ever to be automated. Early trials with color coding proved that color-based systems were grossly inadequate. Entering from another direction (commercial and financial), the concepts of magnetic and optical coding were developed and rather quickly adopted by financial institutions. However, early magnetic and optical systems required a certain closeness of the object with the detecting means, as, for example, in identifying a check or reading the address on an envelope. Specially designed alphanumeric characters have served financial and commercial needs, but it is expected that bar coding ultimately may be used.

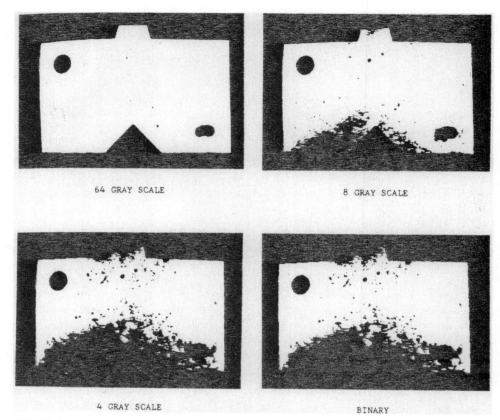

FIGURE 26 Illustration of advantages of high-intensity resolution. With 64 levels of gray, for example, part boundaries are distinguishable from background. Fine features, such as the shadow cast by the post in the lower right-hand corner of the part, are discernible. Using fewer levels of gray (8, 4, and 2, as shown) renders the scene more subject to the effects of shading. The slightly shaded portion of the part in the binary system has deteriorated, as have other subtle features. (*Analog Devices.*)

In the late 1960s the first supermarket checkout stand based on bar coding of merchandise was installed. The system did not get underway for several years, awaiting the willingness of manufacturers to include a bar code on each product. Ultimately the universal bar code, as shown in Fig. 27, was developed. During the late 1980s the bar code became a favorite of manufacturers in their finished goods operations, notably automated warehousing and inventory accounting. As the result, the bar code segment of the electronic industry has grown at an accelerated pace.

Invention of the laser scanner has made it possible to read bar codes on boxes in a warehouse as far as 10 feet (3 meters) away. Well-designed bar code readers are considered to be very close to error-free. Replacement of older style code readers with the hand-held wand has increased the acceptance and versatility of the system. Bar-coded tags for unpackaged materials, such as fabrics, have received wide acceptance.

FIGURE 27 Universal bar code.

SPEED, VELOCITY, ACCELERATION, AND VIBRATION INSTRUMENTATION

The more apparent applications for speed and velocity measurement and control relate to rotating electric motorized machines and equipment. There are numerous other applications, including the measurement of air (wind) and other effluent gas velocities, as may be needed, for example, in determining the pathways followed by air- or waterborne manufacturing pollutants for environmental detection, prevention, and enforcement purposes. The speed-related group of variables also is important in the research, design, and manufacture of a vast variety of aerodynamic and hydrodynamic machines, including vehicles. Central to the analysis and control of vibration, including its deleterious effects on equipment operation, useful life, and safety, is the measurement of acceleration. Numerous means have been developed over the years to detect the speed-related variables.

DEFINITION OF TERMS

Velocity.   Time rate of change of position. (Unless angular velocity is specified, this term generally is understood to refer to *linear* motion.) Strictly, the velocity of a moving point must specify both the speed and the direction of the motion and is, therefore, a *vector,* although the term sometimes is used more loosely as merely synonymous with speed. The velocity of a point is the time rate of the distance s from a fixed origin O, expressed as the vector derivative of s with respect to time, ds/dt (Fig. 1), while speed is the magnitude of the velocity and is not a vector. If the direction of motion is constant, so that the motion is in a straight line (but not necessarily with constant speed), and if the line of motion is clearly understood, it is convenient to treat the distance s and the velocity ds/dt as *scalars* with respect to some zero point on that line and with appropriate algebraic signs (Fig. 2). Otherwise they must be regarded as vectors. If the velocity is variable, account must be taken of the acceleration. Examples of both curved and rectilinear motion are approached from the mathematics of kinematics.

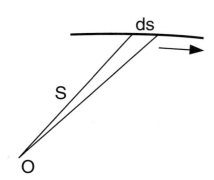

FIGURE 1   Velocity expressed as a vector derivative.

FIGURE 2   Treatment of distance and velocity as scalars.

Angular Velocity.   Quantity relating to rotational motion. While the term angular velocity may be extended to any motion of a point with respect to any axis, it is commonly applied to cases of rotation. Its instantaneous value is defined as the vector whose magnitude is the time rate of change of the angle Θ rotated through, as, for example, $d\Theta/dt$, and whose direction is arbitrarily defined as that direction of the rotation axis for which the rotation is clockwise. The usual symbol is ω or Ω.

The concept of angular velocity is most useful in the case of rigid body motion. If a rigid body rotates about a fixed axis and the position vector of any point P with respect to any point on the axis as origin is $\mathbf{r}$, the velocity $d\mathbf{r}/dt$ of P relative to this origin is $d\mathbf{r}/dt = \omega \times \mathbf{r}$, where ω is the instantaneous vector angular velocity. This indeed may serve as a definition of ω.

The average angular velocity may be defined as the ratio of the angular displacement divided by the time. In general, however, this is not a vector, since a finite angular displacement is not a vector. The instantaneous angular velocity is used more widely.

Angular velocities, like linear velocities, are vectorially added. For example, if a top is spinning about an axis which is simultaneously being tipped over toward the table, the resultant angular velocity is the vector sum of the angular velocities of spin and of tipping. This enters into the theory of precession. The derivatives of the eulerian angles are sometimes very useful in describing the angular motion of a rigid body which has components of angular velocity about all its principal axes.

Speed.   Scalar quantity equal to the magnitude of velocity. Industrially, linear speeds are frequently inferred from rotational measurements simply because of the manner in which most machines are designed—with rotating shafts, wheels, and gears to which speed transducers can be conveniently attached. At one time, analog-type sensors of speed were used almost exclusively, and they still are in demand. Commencing on a small scale in the mid-1950s, digital speed sensors were developed; they are preferred for many applications because of the ease with which they can be integrated into otherwise digital systems.

Acceleration.   Rate of change of velocity with respect to time. Acceleration is expressed mathematically by $d\mathbf{v}/dt$, the vector derivative of velocity $\mathbf{v}$ with respect to time t. If the motion is in a straight line whose position is clearly understood, it is convenient to treat the velocity v and the acceleration dv/dt as scalars with appropriate algebraic signs; otherwise they must be treated by vector methods.

Acceleration may be rectilinear or curvilinear, depending on whether the path of motion is a straight line or a curved line. A body which moves along a curved path has acceleration components at every point. One component is in the direction of the tangent to the curve and is equal to the rate of change of the speed at the point. For uniform circular motion this component is zero. The second component is normal to the tangent and is equal to the square of the tangential speed divided by the radius of curvature at the point. This normal component, which is directed toward the center of curvature, also equals the square of the angular velocity multiplied by the radius of curvature. The acceleration due to gravity is equal to an increase in the velocity of about 32.2 feet (981.5 cm)/second/second at the earth's surface and is of prime importance since it is the ratio of the weight to the mass of a body.

TACHOMETERS

In some servo-controlled motion systems, feedback is required in terms of *position,* as furnished by encoders, resolvers, linear transformers, and so on, which are described in the first article of this handbook section. In other situations, feedback must be in terms of *velocity.*

In contemporary systems, a permanent-magnet dc motor may be used as a tachometer, because when driven mechanically, it generates an output voltage which is proportional to shaft speed (Fig. 3). A principal requirement of a tachometer for such control is that the output voltage be smooth over the operating range and that the output be stabilized against temperature variations.

Small permanent-magnet dc motors frequently are used as speed-sensing devices. These usually incorporate thermistor temperature compensation. They also use silver commutator

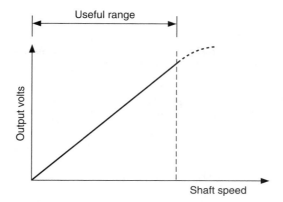

FIGURE 3   Desirable output characteristics of tachometer.

and silver loaded brushes to improve commutator reliability at low speeds and at the low currents which are typical of the application. To combine high performance and low cost, dc servomotor designs often incorporate a tachometer mounted on the motor shaft and enclosed within the motor housing, as shown in Fig. 4.

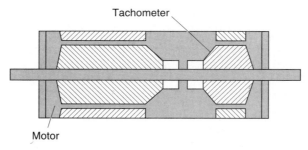

FIGURE 4   Motor with integral tachometer.

There are two fundamental types of dc tachometers, (1) brush type and (2) brushless. Brush-type dc tachometers are of two constructions, (1) iron core and (2) moving coil. Fundamentally the ac tachometer is a three-phase electric generator with a three-phase rectifier on the output. Each type has relative advantages and limitations.

DC Tachometers

As shown in Fig. 5, the dc tachometer depends on the relative perpendicular motion between a magnetic field and a conductor, which results in voltage generation in the conductor. A dc tachometer system consists of a dc generator and a dc indicator or recorder (Fig. 6). The composite characteristics of a representative dc tachometer generator are given in Table 1.

In the moving-coil brush-type dc tachometer, the winding is in the form of a shell or cup. In this construction there is a magnet on one side and an iron slug on the other. Thus the magnetic field passes through the cup-shaped winding. This overcomes much of the inertia because only the winding is rotating. The electrical inductance is also markedly reduced.

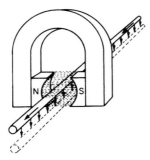

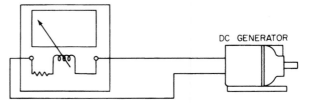

FIGURE 5 Operating principle of commutator-type dc tachometer generator. The magnitude of the voltage produced is a direct function of the strength of the magnetic field and the speed with which the conductor moves perpendicularly to it. Current will flow if the ends are connected to a load, such as an instrument. The polarity of the voltage and, therefore, the direction of current flow depend on the polarity of the field and the direction of conductor motion. This same effect can be obtained by rotating the magnet and holding the conductor still, that is, the principle of the ac or rotating-magnet tachometer.

FIGURE 6 A dc tachometer system such as that shown may be used when the top operating generator speed is at least 100 and does not exceed 5000 r/min. Special indicators may be used for top operating generator speeds as low as 100 r/min, and special recorders are available for speeds as low as 10 r/min. Generally the characteristics of the tachometer can be matched to the equipment whose speed is being determined by using suitable gearing for effecting speed reduction or multiplication. Several types of dc tachometer generators are available.

TABLE 1 Composite Characteristics of Representative DC Tachometer Generator

Voltage output at 1000 r/min	6 V ± 1%
Accuracy	±1%
EMF linearity	±0.15%
Permissible current drain	50 mA
Maximum rms value of ac ripple	2%
Allowable end play	0.005 in (0.13 mm)
Maximum operating temperature	250°F (121°C)
Internal resistance at 77°F (25°C)	20 Ω ± 2%
Composition of brushes	Palladium-silver alloy
Armature	12 bars, 12 slots
High-potential test	500 V for 1 min
Bearings	Ball
Temperature compensation	$\frac{1}{10}$% per 10°C change
Normal continuous speed	2000 r/min
Minimum top speed	100 r/min
Maximum top speed	5000 r/min
Weight	~2 lb (0.9 kg) (dust-resistant models)
	~25 lb (11.3 kg) (weatherproof and explosion-proof models)
Shaft diameter	³⁄₁₆–¾ in (5–13 mm)
Adjustable magnetic shunt range	±4%
Length/width/height (approximate)	4½–5 in/3 in/3 in (114–127 mm/76 mm/76 mm) (dust-resistant models)
	12½ in/6 in/5 in (318 mm/152 mm/127 mm) (weatherproof and explosion proof models)

Brush-type dc tachometers are usually limited to relatively clean environments. Brush life is shortened in many cases because of particulate and erosive contaminants. Some airborne contaminants may also build up in the form of films on the commutator. Sealed enclosures can be used, but these create a thermal problem because of entrapped heat. This heat is not generated within the tachometer per se because of the very low currents involved, but it can be conducted through the shaft. Magnets are sensitive to temperature (estimated to be 0.01 to 0.05 percent/°C) and therefore, if the stability of the output is critical, temperature compensation may be required.

Speed-Ratio Systems with dc Tachometers.   By using two dc tachometer generators (Fig. 7) connected to a ratio meter mechanism, measurements that are dependent on differential processing speed, such as percent stretch and ratio of draw, can be taken and controlled through additional elements in the system. The system shown for a textile application has wide use in the paper and steel industries, as well as where it is important to know the ratio between two quantities expressible in terms of rotation. In these installations the minimum generator speed must be 400 r/min because of voltage requirements of the indicator. Full-scale range limits are from 10 to 100 percent shrink. Percent stretch = output − input × 100% input. If the input generator is 100 units/min and the output generator reads 125 units/min, the percent stretch is (125 − 100)/100 = 25 percent. Through suitable switching arrangements, the outputs of several pairs of generators may be selectively fed through the ratio and production rate instruments to provide readings from various sections of multistage machines.

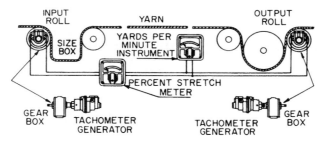

FIGURE 7   Speed-ratio tachometer system used to control percent stretch in the textile industry or ratio of draw in other materials-processing industries.

AC Tachometers

There are (1) voltage-responsive tachometer systems and (2) frequency-responsive systems.

Voltage-Responsive Tachometer Systems.   Consisting of an ac generator and a rectifier-type indicator, as shown in Fig. 8, these systems may be used in any installation where the generator speed for full-scale is not less than 500 and not greater than 5000 r/min. With adequate attention given to bearings, conventional ac generators may be used at speeds up to 10,000 r/min. The ac tachometer generator embodies a stator surrounding a rotating permanent magnet. The output of the generator for voltage-response systems is temperature-compensated and is proportional to speed.

Frequency-Responsive Tachometer Systems.   This type of system consists of a dc indicator or recorder, a frequency-responsive network which may be contained in the recorder or a separate transformer box, and an ac tachometer generator of either the conventional or the bearingless form. Several types of ac tachometer generators are available. The composite characteristics of a representative device are given in Table 2.

FIGURE 8 Highly schematic circuit of ac voltage-responsive tachometer system.

TABLE 2   Composite Characteristics of Representative AC Tachometer Generator

Voltage output at 1000 r/min	10 V ± 1% open circuit
Accuracy	±1%
Permissible current drain	150 mA
Frequency at 900 r/min	60-Hz sine wave
Allowable end play	0.005 in (0.13 mm)
Maximum operating temperature	250°F (121°C)
Internal resistance at 77°F (25°C)	100 Ω ± 1% (voltage-responsive units)
	32 Ω ± 20% (frequency-responsive units)
	8 poles
Stator	
Rotor	Alnico V, 8 poles
Bearings	Ball
Temperature coefficient	$\frac{2}{10}$% per 10°C change (voltage-responsive units)
	No temperature compensation (frequency-responsive units)
EMF linearity	Depends on load and speed
Minimum top speed	500 r/min
Maximum top speed	5000 r/min
Weight	About 3 lb (1.4 kg) (dust-resistant models)
	About 25 lb (11.3 kg) (spray-resistant and explosion-proof models)
Shaft diameter	$\frac{3}{16}$–$\frac{3}{4}$ in (5–13 mm)
Length/width/height (approximate)	4½–5 in/3 in/3–3½ in (114–127 mm/76 mm/76–89 mm) (dust-resistant models)
	12½ in/6 in/5 in (318 mm/152 mm/127 mm) (spray-resistant and explosion-proof models)

Bearingless Tachometer Generators

These devices are ac generators of the most basic form, consisting of only a permanent magnet rotor and a stator. The devices have no bearings or brushes. They are designed to be impervious to oil, grease, and relatively high temperatures and, consequently, may be installed in inaccessible areas, such as gearboxes, which permits saving of space. They have very low torque burdens of less than 1 oz·in and are capable of speeds up to 100,000 r/min.

In general, when a bearingless generator is used, the frequency-responsive approach is employed. Since the system is solely dependent on the frequency output of the generator, voltage variations caused by reductions in the magnetic strength of the rotor due to handling, poor alignment of stator and rotor, or axial travel of the rotor with respect to the stator will not affect the overall accuracy. The rotor of the generator unit should be mounted to the true center of the shaft with extreme care, particularly in high-speed installations. A few representative methods are illustrated in Fig. 9. The relatively low inertia of the rotor

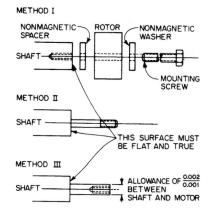

FIGURE 9 Examples of methods for mounting the rotor of a tachometer generator unit.

makes it possible to secure it to the shaft with a right-hand thread, regardless of the direction of rotation. It is recommended that a steel screw having an SAE thread be used to permit maximum tightening. The rotor should not be pressed onto the shaft because magnetic rotor material is brittle and may shatter.

Magnetic Speed Sensors

A magnetic pickup is essentially a coil wound around a permanently magnetized probe (Fig. 10). When discrete ferromagnetic objects, such as gear teeth, turbine rotor blades, slotted disks, or shafts with keyways, are passed through the probe's magnetic field, the flux density is modulated. This induces ac voltages in the coil. One complete cycle of voltage is generated for each object passed. If the objects are evenly spaced on a rotating shaft, the total number of cycles will be a measure of the total rotation, and the frequency of the ac voltage will be directly proportional to the rotational speed of the shaft. The magnetic pickup shown in Fig. 10 is used in conjunction with a 60-tooth gear to measure the revolutions per minute of a rotating shaft. Such a gear is often selected because the output frequency (in hertz) is numerically equal to revolutions per minute—a situation that allows frequency meters to be used without calibration. For very high rotational speeds, a smaller number of teeth may be better suited. This type of arrangement is used with some turbine-type flowmeters, as described in the article, "Flow Systems" in Section 4 of this handbook.

A magnetic pickup also can be used as a timing or synchronization device. Examples would include ignition timing of gasoline engines, angular positioning of rotating parts, or stroboscopic triggering of mechanical motion.

One commercially available unit is a passive or self-generating device, requiring no external excitation. When mounted in proximity to the teeth (or blades) of a conventional

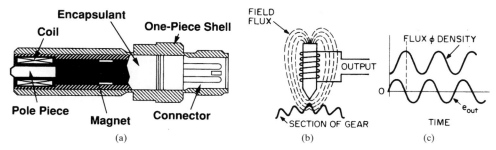

(a)   (b)   (c)

FIGURE 10 Magnetic speed sensor. (*a*) Sectional view. (*b*) Placement of probe, allowing a small air gap between pickup and gear teeth. (*c*) Output waveform, which is a function not only of rotational speed, but also of gear-tooth dimension and spacing, pole-piece diameter, and air gap. The pole-piece diameter should be less than or equal to both the gear width and the dimension of the tooth's top (flat) surface. The space between adjacent teeth should be approximately 3 times the diameter. Ideally, the air gap should be as small as possible, typically 0.13 mm (0.005 inch). A number of steel or cast-iron gears, precisely manufactured standards, are available. The standard solid gear comes with various dimensions and with 48, 60, 72, 96, and 120 teeth. (*Daytronic Corporation.*)

rotating gear (or turbine), it produces an approximately sinusoidal ac voltage signal output. The amplitude of the voltage also is generally proportional to the speed of rotation (Fig. 11).

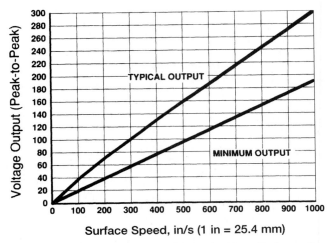

FIGURE 11   Output performance of magnetic pickup. (*Daytronic Corporation.*)

The magnetic pickup circuit shown in Fig. 12 contains its own signal-conditioning circuitry for generating a clean square-wave output pulse for each ferrous discontinuity passing the head of the pickup. The output is either on or off, depending on the presence or absence of ferrous material. The unit senses motion down to "zero" velocity and continuously produces a pulse train of constant amplitude, irrespective of the rotational speed of the gear. Magnetoresistors are flux-responsive devices that detect magnetic field changes independent of the field's rate of change. Two magnetoresistors are used to cancel the effects of temperature drift and supply voltage variations. There are some restrictive mounting requirements.

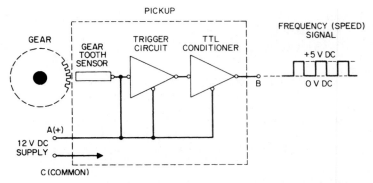

FIGURE 12   Circuit of zero-velocity magnetic pickup. (*Daytronic Corporation.*)

Adaptations of proximity switches can be used for measuring speed and frequently are called speed switches. A description of these devices is given in a prior article in this handbook section. These units are popular for slowdown indication for conveyors and other process machinery (Fig. 13). On start-up, the incoming pulses from the probe are ignored for a brief, fixed time interval (such as 5 seconds) to permit equipment to accelerate to normal operating speed. This time delay is adjustable up to one full minute. The system is rugged and can be used for heavy-duty machinery in demanding atmospheres. The elements detected by the probe are ferrous pieces mounted directly on belt (or other) drive pulleys.

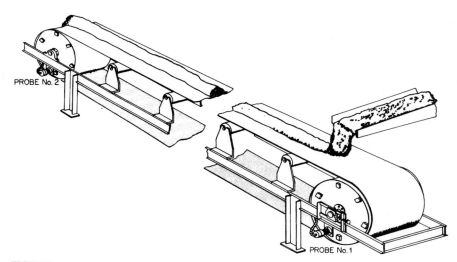

FIGURE 13 Conveyor loss-of-motion detection system that uses two probes, one on the head pulley and another on the tail pulley. By computing the plus or minus speed relation of the tail to the head pulley and comparing this value with percentage-slip set points, slip protection is ensured. A minimum percent feature of the system checks for mechanical failure between the motor and the tail pulley. Each of these features has its own time delay after start-up and one common delay to ignore nuisance alarms or shutdown. The four alarms are fed to a first-out annunciator and latch in the relay output circuit. Reset of the relay may be manual or automatic. (*Milltronics, Inc.*)

A capacitor-type or inductive proximity switch also can be used in speed measurement. A trigger cam is mounted on a rotating or reciprocating element of a machine so that it appears within the range of the proximity switch at every revolution or stroke. The distance between the cam and the switch must be no more than one-half the nominal detection range of the switch. The cam may be metallic or nonmetallic, depending on whether an inductive or a capacitive switch is used. The switch generates a pulse each time the cam appears within range. A controller measures the instantaneous rate between successive pulses and compares it with the set point. Through the use of appropriate electronics, the arrangement can be used for either overspeed or underspeed correction. Because these switches measure the time interval between two successive pulses, there is an inherent time lag in the response when the pulse rate decreases below the set point; and the size of the lag depends on the value of the set point. There is no time lag when the pulse rate increases above the set point.

These switches are described in more detail in the preceding article, dc, of this handbook.

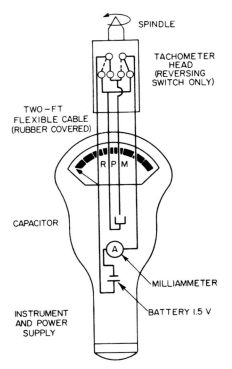

FIGURE 14   Capacitor-type impulse tachometer.

Impulse Tachometers

In the instrument shown in Fig. 14, the charging current of a capacitor is used. The pickup head usually contains a reversing switch, operated from a spindle, which reverses twice with each revolution. Thus battery potential is applied to the capacitor in each direction, and with each impulse a current is passed through the milliammeter. The indicator responds to the average value of these impulses. Therefore the indications are proportional to the rates of the pulses, which in turn are proportional to the rates of the spindle revolutions.

No current is drawn from the battery when the spindle is not revolving. The pulse current is approximately 1 mA. The spindle speed and battery voltage influence the indicator deflection. Thus it is important to check and correct the battery voltage at frequent intervals. This is accomplished by means of an adjustable resistor placed in the circuit.

The oscillating switch may be connected directly for speeds of 200 to 10,000 r/min and, with suitable gears, speeds below or above these values can be measured. The readings of a properly standardized instrument are not affected by temperature, humidity, vibration, or magnetic fields. The indicator and head may be separated up to a distance of 1000 feet (300 meters), and where suitably shielded connections are used, the distance may be increased. The indicator scale is uniform.

A high-accuracy instrument is also available wherein the capacitor and the reversing switch are connected to one leg of a bridge circuit. The pulses from the periodically charging capacitor upset the balance of the bridge and thus cause an indication on the milliammeter. Multiple ranges are obtained by using different capacitor values.

Optical Encoders

Many modern position control systems use an incremental optical encoder for position determination, as explained in the first article in this handbook section. By taking advantage of the calculating power of a microprocessor, the impulses from an optical encoder can be converted to a velocity measurement.

Optical encoders are available for handling very wide dynamic ranges, such as 10,000 to 20,000 to 1. Accuracies are claimed to be better than 0.01 percent per revolution.

Stroboscopic Tachometers

A stroboscope permits intermittent observation of a cyclically moving object in such a way as to produce an optical illusion of stopped or slowed motion. This phenomenon is readily apparent, for example, when rewinding a tape at many revolutions per minute when the tape deck is located under a 60-Hz incandescent lamp. Patterns on the reel tend to slow and then appear to stop before reversing their direction. Stroboscopic effects have been known for decades,[1] of course, one of the first scientific applications being found in very high-speed photography. Intermittency of observation can be provided by mechanical interruption of the line of sight (as with a motion picture camera) or by intermittent illumination of the object being viewed. The industrial stroboscope basically is a lamp plus the electronic circuits required to turn the lamp on and off very rapidly, at rates as high as 150,000 flashes per minute and higher.

The schematic diagram of an electronic stroboscope is shown in Fig. 15. The device includes a strobotron tube with its associated discharge capacitors, a triggering tube to fire the strobotron, an oscillator to determine the flashing rate, and a power supply. With the use of harmonic techniques, speeds up to 1 million r/min can be measured. Accuracy is nominally ±1 percent of the dial reading after calibration.

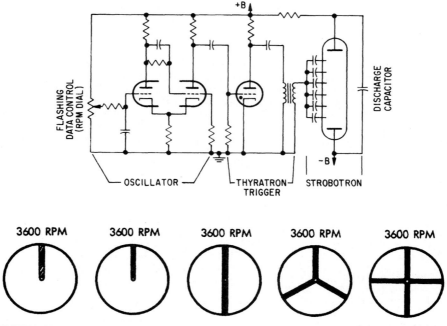

FIGURE 15 Electronic stroboscope. (*a*) Schematic representation of circuit. (*b*) Images obtained at harmonic and subharmonic flashing rates of a stroboscope. Even with an asymmetrical object, the correct fundamental image is repeated when the stroboscope is flashing at one-half, one-third, and so on, the speed of the object. The proper setting for a fundamental speed measurement is the highest setting at which a single stationary image can be achieved. This does not hold, however, if the fundamental is beyond the flashing rate of the stroboscope. There are several ways to distinguish fundamental from submultiple images. The flashing rate can be decreased until another single image appears. If this occurs at half the first reading, the first reading was the actual speed of the device. If it occurs at some other value, then the first reading was a submultiple. Or the user can double the flashing rate and check for a double image. Or the user can flip the range switch to the next higher range. Because of the 6:1 relationship between ranges, a 6:1 pattern should appear. The 6:1 relationship between ranges also makes it convenient to convert speed readings from revolutions per minute into cycles per second. One simply flips to the next lower range and divides the new reading by 10.

Because of their portability and easy setup, stroboscopes find a variety of applications, principally in machine and vehicle research, development, and testing. (*GenRad.*)

---

[1]Invented independently by Stampfer of Vienna and Plateau of Ghent in 1832. Stampfer chose the name "stroboscope," which is derived from the Greek words meaning "whirling watcher."

To serve as a tachometer, a stroboscope must have its own flashing-rate control circuits and calibrated dial. Stroboscope tachometer test disks are available. These disks can be cut out and mounted on light cardboard or metal. The center must be carefully located and fitted onto the drive shaft. Although nowadays more automatic means are available to measure belt slippage, as described previously, this was commonly accomplished by stroboscopes in earlier times.

Variable-Reluctance Tachometers

A pickup of this type produces pulses that are proportional to speed, are amplified and rectified, and control the direct current to a milliammeter. This type of instrument is rated at 10,000 to 50,000 r/min, with an accuracy of ±½ percent of full-scale reading. The pickup is rated to withstand ambient temperatures from −60 to 500°F (−51 to 260°C).

Photoelectric Tachometers

In one instrument of this type, designed to measure speeds up to 3 million r/min, the movable part subject to measurement is arranged to provide reflecting and absorbing areas. The interrupted reflected light produces, by means of a photocell, electric impulses which are applied to a frequency meter which generates a square wave from the pulse voltage and applies it to a discriminating circuit. A fixed current pulse at each half-cycle is produced. These pulses are rectified and applied to a dc milliammeter which indicates the average value. Thus the meter readings are proportional to the number of pulses per second, or the frequency.

Eddy-Current Tachometers

The eddy-current or drag-type tachometer has been widely used for certain types of speed measurements. A preponderance of these units has been employed in automobile speedometers, in which case a flexible shaft arrangement is used, but they also find industrial usage.

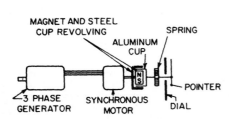

FIGURE 16   Drag-type eddy-current tachometer.

In its basic form, as shown in Fig. 16, the drag-type instrument uses a permanent magnet which is revolved by the source being measured. Close to the revolving magnet is an aluminum disk, pivoted so as to turn against a spring. A pointer attached to the pivoted disk is associated with a calibrated scale. As the permanent magnet is revolved, eddy currents are set up in the disk. The magnetic fields caused by these eddy currents produce a torque which acts in a direction to resist this action and turns the disk against the spring. The disk turns in the direction of the rotating magnetic field and turns (or is dragged) until the torque developed equals that of the spring. This torque is proportional to the speed of the rotating magnet. The instrument has a uniform scale.

The rotating field usually is produced by a permanent magnet but may be of any form which is steady. The disk may also take the form of a cup and may be of copper.

Remote indication is obtainable with one form of this tachometer. A three-phase generator is driven from the shaft whose speed is to be measured. The generator output is connected to a three-phase synchronous motor, attached to the indicator, which rotates the magnetic field. Several indicators, each with its own synchronous motor, may be connected to the three-phase generator and indicate in proportion to the speed of the generator. Since the synchronous motors keep in step with the generator frequency over a wide range, the indications are independent of the voltage developed by the three-phase generator.

Velocity Head or Hydraulic Tachometers

In devices of this type, advantage is taken of the fact that pumps or blowers produce a velocity which can be converted into a static pressure. The hydraulic tachometer incorporates a rotary pump as the transmitter and a piston as the receiver. The pump, usually contained in the indicator case, is driven by a flexible shaft and a gear train which automatically handles reversed speed, but the instrument normally is not equipped to show the direction of rotation. Pump displacement, which is positive and free of pulsations, raises or lowers a counterweight piston. The piston operates a pointer through a rack and gear. It also drives a tape-marking stylus for recording. The recorder tape is driven by the flexible shaft from the speed pickup.

The indicator may be read to 0.4 percent of full-scale value and is claimed to be accurate to within ±1 percent of full-scale reading. One application is in railroad locomotives, where the instrument accuracy is affected very little by ambient temperature changes.

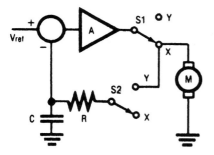

FIGURE 17   Motor-speed regulation without a tachometer. In the sample-and-hold diagram shown, the motor is switched from a free-running state, where it charges capacitor *C*, to a second state, where it is driven by an error voltage produced by the last stored sample. When *S1* is in position *Y*, the motor is disconnected from the amplifier and is free running. At the same time, *S2* connects the free-running motor to capacitor *C*.

TACHOMETERLESS REGULATION OF SERVO SPEED

Within the last few years, an interesting approach to regulating the speed of a motor without a tachometer has emerged. Basically, the arrangement consists of allowing a motor to coast for a very short interval, during which the back electromotive force (EMF) is measured. In one technique, sample and hold, the motor armature is time-shared. About 90 percent of the time it operates as a motor. During a 10 percent coasting period, the motor functions as a generator or dc tachometer. Thus it can provide an output voltage that is directly proportional to its speed. The applicable sample-and-hold block diagram is shown in Fig. 17. The motor inductance must be sufficiently small, as in the case of a printed-circuit motor, to qualify for this approach. Equations and more details are given by Geiger (1979) in the reference listed.

GOVERNORS

A governor is an automatic controller used to maintain the rotative speed of a machine at a desired value. It measures the speed, compares the measured value with the desired value, and acts to correct any error between the two values—usually by adjusting the flow of energy to the machine. Governors may be divided into two main types: (1) devices in which the speed-sensing element operates the energy metering device directly and (2) devices that use one or more stages of power amplification between the speed-sensing element and the energy control device. There is a natural distinction between these two types, arising from the fact that the first type usually gives stable control on an engine or other prime mover, whereas the second type requires the presence of some stabilizing factor to prevent continual oscillation of the speed (hunting).

AIR AND GAS VELOCITY MEASUREMENT

With accelerated interest in environmental measurements and control, there are numerous situations in which a determination of air, vapor, and effluent gas velocities must be made.

Meteorological interest in air velocity has intensified, particularly in the operation of airport facilities for aircraft takeoffs and landings.

Pitot Tube Air-Speed Indicators

For many years, in testing the air speed of aircraft and in wind tunnel testing the Pitot tube has been used widely. The Pitot tube air-speed indicator consists of two elements: (1) a dynamic tube, which points upstream and determines the dynamic pressure, and (2) a static tube, which points normal to the air stream and determines the static pressure at the same point, as shown in Fig. 18. The tubes are connected to the two sides of a manometer or inclined gage so as to obtain a reading of velocity pressure, which is the algebraic difference between total pressure and static pressure.

The relationship between air velocity and velocity pressure is

$$V = \sqrt{2GH}$$

where V = velocity
G = acceleration due to gravity
H = velocity head or pressure

The pressure differential created is quite small in relation to air velocity. At 110 ft/min (33.5 m/min) the velocity pressure is only 0.0625 inch (1.6 mm) water. Consequently the instrument is not generally used for measuring velocities less than 1000 ft/min (305 m/min). Turbulence in the air stream affects device accuracy, and the tubes are subject to clogging where dusty, unclean air is involved.

Venturi Air-Speed Indicators

Limitations of the Pitot tube led to the design of a venturi air-speed indicator in which a greater differential pressure is created. The device, shown in Fig. 19, requires individual calibration for best accuracy. Parts are not readily interchangeable.

Revolving-Vane Anemometers

This widely used device comprises a paddle wheel which is revolved by the moving air stream (Fig. 20). The wheel is attached to a counter, and by selection of the proper gear ratios and

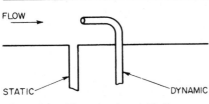

FIGURE 18 Pitot tube air-speed indicator.

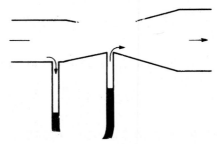

FIGURE 19 Venturi air-speed indicator.

vane pitch the counter can be calibrated to read directly in feet of air. The air velocity can be determined by measuring the time interval. The device is supplied with a curve for correction of the nonlinear relation between air velocity and rotational speed of the vanes. Air density also should be considered when high accuracy is needed. The measurement tends toward the average air speed. The range of the device usually is 300 to 3000 ft/min (91.5 to 915 m/min).

Propeller-Type Electric Anemometers

This is a version of the basic rotating-vane device. Figure 21 shows the propeller type in which the blades are fastened to the shaft of an electric generator which develops an EMF or frequency proportional to speed. The EMF or frequency signal is fed to an indicator. The generator and propeller are pivoted so that the directional vane can keep the device headed directly into the direction of airflow. The device reads average air velocity. The direction of airflow can also be indicated mechanically or electrically.

FIGURE 20   Revolving-vane anemometer.

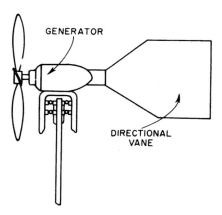

FIGURE 21   Propeller-type electric anemometer.

Revolving-Cup Electric Anemometers

One type is shown in Fig. 22. The generator is mounted on a vertical axis, and, like the propeller type, its EMF or frequency output is proportional to the speed of the revolving cups. The speed readings are average, but the device is not directional.

Constant-Temperature Thermal Anemometers

In these devices, such as that shown in Fig. 23, the sensor element contains a heated wire. An electronic control circuit maintains the sensor element at a constant temperature regardless of the air velocity. The faster the air passes by the sensor, the more power is required to maintain the preset sensor temperature. The power dissipated by the sensor is directly related to air velocity.

An uncompensated constant-temperature hot-wire anemometer will measure accurately only if the temperature of the air flowing past the sensor remains constant. The sensor illus-

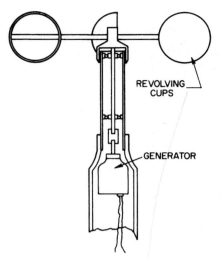

REVOLVING
CUPS

GENERATOR

FIGURE 22   Revolving-cup electric anemometer.

trated is compensated to allow accurate measurements over a wide range of air temperatures. Temperature compensation is achieved by incorporating an air temperature sensor into the control circuit. The sensors are calibrated to measure mass velocity at standard conditions [21°C (70°F) and 760 mmHg (14.7 psia)]. Built-in electronics also linearize the signal by using calibration data stored on a read-only memory (ROM). The sensors are available in three styles: (1) a general-purpose transducer, ruggedly constructed, with a protected probe tip; (2) a windowless transducer for less flow blockage, used for measurements in confined spaces, such as between circuit boards and ventilation slots; and (3) an omnidirectional transducer, also well suited to measurements in confined spaces.

Claimed accuracy for most sensors is ±1.5 percent of reading. Linear output is 0.5 volts dc or 4 to 20 mA; minimum resolution is 0.1 percent full-scale. Response time to flow is 0.2 second; response time to temperature ranges from 30 to 60 seconds. Temperature range is −45 to 93°C (−50 to 200°F); input power is 12 to 15 volts dc, 250 mA maximum. There are five standard velocity ranges, 20 to 500, 20 to 1000, 20 to 2000, 50 to 5000, and 100 to 10,000 ft/min (0.1 to 2, 0.1 to 5, 0.1 to 10, 0.2 to 20, and 0.5 to 50 m/s).

VIBRATION MEASUREMENT[2]

Vibration measurements are required during the development and construction of many kinds of machines and systems. Measurements are often made to determine levels of vibration which could be destructive or cause excessive noise. Once a system is operational, vibration monitoring is used as an important indicator of mechanical health. Excessive or increasing vibration is often an early indicator of mechanical degradation.

Vibration sensing is required over a wide range of amplitudes and frequencies. For example, vibration from rotating machinery, such as pumps, motors, compressors, and turbines, occurs from about 1 Hz to over 20,000 Hz, the principal interest being from 10 to 2000 Hz. Vibration amplitudes vary widely, depending on the equipment design. For example, a smooth-running motor may vibrate at $0.01g$ [$1g = 386$ in/s^2 (980 cm/s^2)], but a high-speed gearbox can easily vibrate at more than $100g$ at a frequency of more than 10,000 Hz.

Ideally, a vibration sensor is attached to a body in motion and provides an output signal proportional to the vibrational input from that body. Sometimes it is not practical to attach a sensor directly to the moving body. In these cases, measurement is made by attaching the sensor to another body and making a measurement relative to the motion of that body. In any event, the performance of the measurement technique should not be degraded by the location of the sensor.

[2]Robert M. Whittier, Manager of Development Engineering, Endevco Corporation, San Juan Capistrano, California.

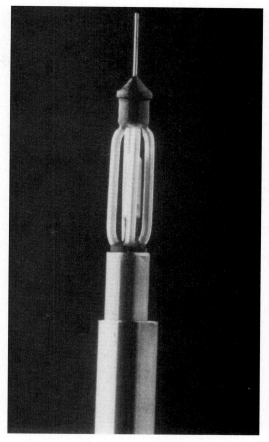

FIGURE 23   Windowless-type constant-temperature thermal anemometer. (*TSI Incorporated.*)

Inertial Motion Sensing

Inertial motion measurement is achieved by attaching a sensor to the moving structure. By doing so, the vibration is fundamentally changed because of the addition of the mass of the sensor; however, the changes are usually insignificant. The specific requirement is that the dynamic mass of the sensor be much less than the dynamic mass of the structure at the point of attachment.

Inertial motion sensors consist of a mass, a spring, a viscous damper, and a means of electrical pick-off. Figure 24 shows a simplified mechanical schematic of such a system. The response of this to vibration is well known. When the mass is small and the spring is stiff, the system can be used at frequencies below its resonant frequency, where its response is constant for acceleration inputs. Thus it is an accelerometer. When the mass is large and the spring is flexible, the device can be used to sense relative displacement at a high frequency. Damping is sometimes added to these, which limits the relative response at its resonance. However, damping does create phase shift. When damping is used, it is advisable to maintain its value at approximately seven-tenths of critical damping to achieve proportional or linear phase shift. When damping is not used in an accelerometer, the frequency range is limited to typically one-fifth of the resonant frequency.

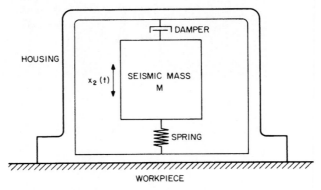

FIGURE 24   Mechanical schematic of inertial sensor.

ACCELEROMETERS

The most common type of vibration sensor is the accelerometer. It can be made small, lightweight, and rugged—all necessary attributes. Both self-generating accelerometers and those requiring electrical excitation are available. The most common is the self-generating piezo-electric device. Typical performance characteristics for accelerometers are listed in Table 3.

TABLE 3   Typical Accelerometer Performance Characteristics

Characteristic	Piezoelectric accelerometers		Piezoresistive accelerometer	Servo accelerometer
Sensitivity, pC/g, mV/g	10	100	20	250
Frequency range, Hz	4–15,000	1–5000	0–750	0–500
Resonance frequency, Hz	80,000	20,000	2500	1000
Amplitude range, g	500	1000	25	15
Shock rating, g	2000	10,000	2000	250
Temperature range, °C	−50 to +125	−50 to +260	0 to +95	−40 to +85
Total mass, g	1	29	28	80

Piezoelectric Accelerometers

These devices utilize a mass in direct contact with a piezoelectric component, or crystal. When a varying motion is applied to the accelerometer, the crystal experiences a varying force excitation ($F = ma$), causing a proportional electrical charge q to be developed across it,

$$q = d_{ij}F = d_{ij}ma$$

where d_{ij} is the material's piezoelectric strain constant.

As the equation shows, the electrical output from the piezoelectric material is dependent on its properties. Two commonly used materials are lead-zirconate titanate ceramic (PZT) and quartz. As self-generating materials, they both produce a large electric charge for their size, although the piezoelectric strain constant of PZT is about 150 times that of quartz. As a result, accelerometers using PZT are more sensitive or are much smaller. The mechanical spring constants for the piezoelectric components are high, and the inertial masses attached to them are small. Therefore these accelerometers are useful to extremely high frequencies. Damping is rarely added to these devices. Figure 25 shows a typical frequency response for such a device. Piezoelectric accelerometers have comparatively low mechanical impedance. Therefore their effect on the motion of most structures is small. They are also rugged and have outputs that are stable with time and environment.

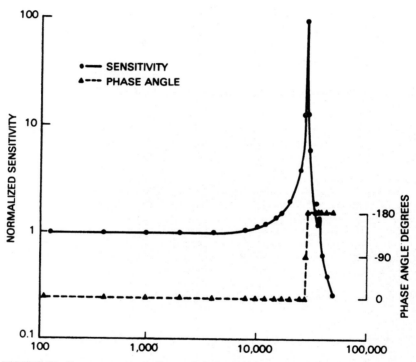

FIGURE 25   Frequency response from typical piezoelectric accelerometer.

Two principal design configurations are used for piezoelectric accelerometers. One stresses the piezoelectric material in compression, while the other stresses it in shear. Simple diagrams of these two types are shown in Fig. 26. When the accelerometer is accelerated upward, the mass is moved downward toward the bottom. Conversely, downward acceleration moves the mass element upward. With vibration motion, the resultant dynamic stress deforms the piezoelectric element. In the compression accelerometer, vibration varies the stress in the crystal which is held in compression by the preload element. In the shear accelerometer vibration simply deforms the crystal in shear. The mechanical construction of actual designs can be more complex, but the model is the same.

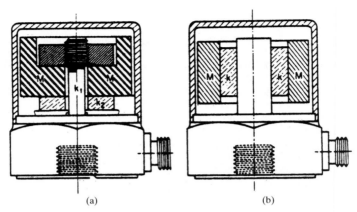

FIGURE 26   Conceptual diagram for piezoelectric accelerometers. (*a*) Compression. (*b*) Shear.

Piezoresistive Accelerometers and Strain-Gage Sensors

Piezoresistive accelerometers are strain-gage sensors which use semiconductor strain gages in order to provide much greater gage factors than are possible with metallic gages. Higher gage factors are achieved because the material resistivity changes with stress, not just its dimensions. The increased sensitivity is critical to vibration measurement in that it permits miniaturization of the accelerometer.

A typical piezoresistive accelerometer uses either two or four active gages in a Wheatstone bridge. It is more important to use multiple gages than when using metallic gages because the temperature coefficients of the semiconductor elements are greater than those of metallic gages. To control the electrical bridge balance and sensitivity variations with temperature, other resistors are used within the bridge and in series with the input.

The mechanical construction of an inertial system using piezoresistive elements is illustrated in Fig. 27, and the construction of a complete accelerometer is shown in Fig. 28. This design includes overload stops to protect the gages from high-amplitude inputs, and oil to improve damping. Such an instrument is useful for acquiring vibration information at low frequencies (for example, below 1 Hz), and the device can be used to sense static acceleration.

Servo Accelerometers

The construction and operating principle of a servo accelerometer are illustrated in Fig. 29. When subjected to acceleration, the proof mass deflects relative to the base of the accelerometer, and the pick-off changes its capacitance as a result of changes in the damping gap. As this occurs, the servo supplies current to the coil, which is located in the gaps of the permanent magnets. The resulting force restores the coil to its equilibrium position. The output signal is a measure of the coil current and is proportional to the applied acceleration.

Signal Conditioning for Accelerometers

Signal conditioners interface accelerometers to readout and processing instruments by (1) providing power to the accelerometer if it is not self-generating, (2) providing proper electrical load to the accelerometer, (3) amplifying the signal, and (4) providing an appropriate filtering and drive signal. Piezoelectric and piezoresistive transducers both require conditioners with certain characteristics, as is now discussed.

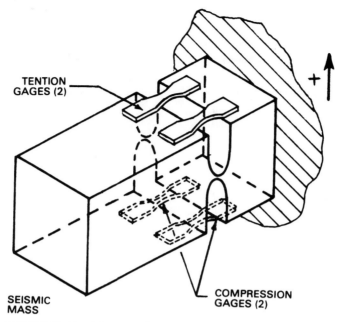

FIGURE 27   Inertial system using piezoresistive elements.

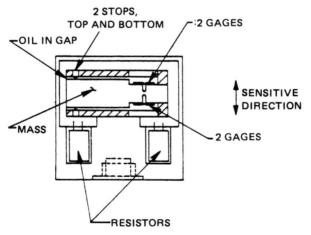

FIGURE 28   Construction of typical piezoresistive elements.

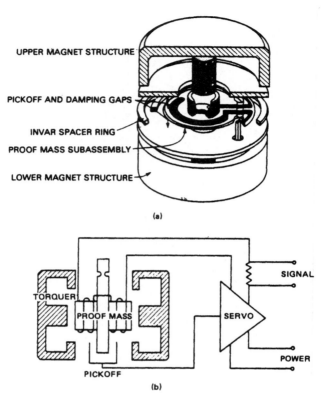

FIGURE 29   Construction of servo accelerometer.

Conditioning Piezoelectric Accelerometers

The piezoelectric accelerometer is self-generating and supplies a very small amount of energy to the signal conditioner. It presents a very high source impedance, mainly capacitive, to the conditioner. Two forms of schematic representation of a piezoelectric accelerometer are shown in Fig. 30. It may be regarded as a voltage source in series with a capacitance, or as a charge source in parallel with a capacitance. The signal conditioner determines how the transducer is treated in a given system. Both voltage and charge sensing are used. The charge amplifier is by far the most common approach. The charge amplifier is advantageous because the system gain and low-frequency response are well defined and are independent of the cable length and accelerometer capacitance.

The charge amplifier consists of a charge converter and a voltage amplifier, as shown in Fig. 31. The system does not amplify charge per se. It converts the input charge to a voltage and then amplifies the voltage.

A charge converter is essentially an operational amplifier with integrating feedback. The equivalent circuit is shown in Fig. 32. With basic operational-type feedback, the amplifier input is maintained at essentially zero volts and therefore looks like a short circuit to the input. The amplifier output is a function of the input current. Having integrating operational feedback, the output is the integral of the input current—hence the name "charge amplifier."

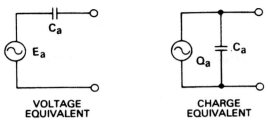

FIGURE 30   Electrical schematic representations of piezo-electric accelerometers.

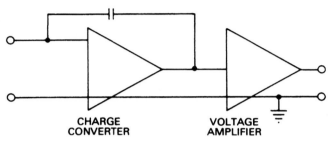

FIGURE 31   Block diagram of charge amplifier.

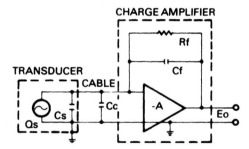

FIGURE 32   Equivalent circuit of charge converter.

Conditioning Low-Impedance Piezoelectric Accelerometers

Piezoelectric accelerometers are available with simple electronic circuits internal to their cases to provide signal amplification and low-impedance output. Some designs operate from low-current dc voltage supplies and are designed to be intrinsically safe when coupled by appropriate barrier circuits. Other designs have common power and signal lines and use coaxial cables.

The principal advantages of piezoelectric accelerometers with integral electronics are their relative immunity to cable-induced noise and spurious response, the ability to use lower-cost cable, and a lower signal conditioning cost. In the simplest case, the power supply might consist of a battery, a resistor, and a capacitor.

These advantages do not come without compromise. Because the impedance matching circuitry is built into the transducer, gain cannot be adjusted to utilize the wide dynamic range of the basic transducer. Ambient temperature is limited to that which the circuit will withstand,

and this is considerably lower than that of the piezoelectric sensor itself. In order to retain the advantages of small size, the integral electronics must be kept relatively simple. This precludes the use of multiple filtering and dynamic overload protection and thus limits their application. But when conditions are relatively benign, these accelerometers can economically provide excellent noise immunity and signal fidelity.

Conditioning Piezoresistive Transducers

Piezoresistive transducers are relatively easy to condition. They generally have high-level output, low-output impedance, and very low intrinsic noise. These transducers require an external power supply. This supply is usually direct current, but it may be alternating current, provided the carrier frequency is at least 5 to 10 times the maximum frequency of interest.

Most transducers are designed for constant voltage excitation and are used with relatively short cables. With long cables, wire resistance is not negligible. Moreover, resistance changes with temperature, and the voltage drop along the line varies as the transducer resistance or load changes. For these applications, transducers should be calibrated for constant current excitation so their output will be less dependent on external effects.

Many piezoresistive transducers are full-bridge devices. Some have four active arms to maximize sensitivity (Fig. 33). Others have two active arms and two fixed precision resistor arms to permit shunt calibration by precision calibration resistors in the signal conditioner. Miniature transducers are usually half-bridge devices, with bridge completion accomplished in the signal conditioner.

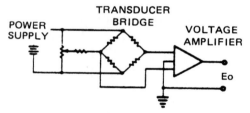

FIGURE 33 Typical system and bridge circuit for piezoresistive accelerometer.

Adjustment of the unbalanced output of an accelerometer can easily be performed in the signal conditioner. For full-bridge transducers, the balancing potentiometer R_1 is connected across the excitation terminals and a current-limiting resistor is connected between the wiper arm of the potentiometer and the bridge. This is shown in Fig. 34a. For half-bridge transducers, a small balance potentiometer (typically 100 Ω) is connected between the bridge completion arms as shown in Fig. 34b.

Environmental Effects

Temperature. Accelerometers can be used over wide temperature ranges. Piezoelectric devices are available for use from cryogenic temperatures, –270°C (−454° F), to over 650°C (1220° F). The sensitivity changes with ambient temperature, but the changes are systematic and can be calibrated. If the ambient temperature changes suddenly so that strains develop within the accelerometer and within the time response of the measurement system, further

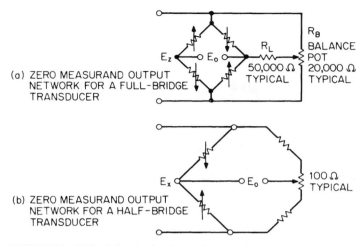

FIGURE 34   Bridge balancing for piezoresistive accelerometer.

errors can occur. These are evaluated by testing the response of accelerometers to step function changes in temperature per industry standard test procedures. Errors usually appear as a wandering signal or a low-frequency oscillation.

Cable Movement.   Cabling from the accelerometer to the signal conditioner can generate spurious signals when it is subjected to movement or dynamic forces. This is usually significant only for systems using high-impedance piezoelectric accelerometers. The major noise-generating mechanism is triboelectric noise, which is caused by charge trapping due to relative motion, or localized separation between the cable dielectric and the outer shield around the dielectric. To reduce this effect, cabling is available which is "noise-treated." These cables have a conductive coating applied to the surface of the dielectric, which prevents charge trapping. Another way to eliminate this effect is to use a sensor which includes an electronic circuit for reducing the impedance to ~100 Ω.

Dynamic Strain Inputs.   In vibration environments, some structures may dynamically flex, stretch, or bend at the mounting location of the accelerometer. Being in intimate contact with this strained area, the base of the accelerometer can also be strained. A portion of this base strain is transmitted directly to the crystal sensing element and generates error signals. In addition to strains in the structure, it is also possible to induce errors from forces or pressures onto the case of the accelerometer. Outputs from these forces vary greatly, depending on the internal design of the accelerometer. The errors from these sources are usually checked against industry standard test procedures, and the results are included in specifications.

Electrostatic and Electromagnetic Fields.   Electrostatic noise can be generated by stray capacitance coupling into the measurement system. It is important that the cabling between a high-impedance piezoelectric sensor and the signal conditioner be fully shielded. Ground loops can be avoided by grounding the system at a single point, usually at the output of the signal conditioner. Magnetically coupled noise can best be avoided by not placing signal cables in close proximity to power or high-current conductors and by avoiding electromagnetic sources when possible. Accelerometers should also be checked for their sensitivity to electromagnetic fields.

VELOCITY TRANSDUCERS

Electrodynamic velocity pickups can be used directly with readout instruments. The self-generating voltage in the transducer is proportional to the velocity of the motion being measured and is usually of sufficient amplitude so that no voltage amplification is required. The disadvantages of velocity pickups are their large size and mass and their inability to be used for measurements at frequencies below ~10 Hz. Also, the output at high frequencies, above ~1000 Hz, is quite small in most applications. Care must be taken in using these devices in strong magnetic field environments.

A typical velocity sensor consists of a seismically mounted, and usually critically damped, magnetic core suspended in a housing rigidly attached to the vibrating surface. A coil of wire attached to the housing surrounds the core. Relative motion between the magnetic core and the housing causes magnetic lines of flux to cut the coil, inducing a voltage proportional to the velocity. These sensors operate above their first natural frequency.

Relative Motion Sensing—Eddy-Current Probe

In some cases it is not practical to place a sensor in contact with the moving part. Relative motion measurement approaches are then used. The most commonly used device is the eddy-current probe. Noncontact eddy-current displacement-measuring systems have achieved general acceptance for industrial machinery protection and condition monitoring. An eddy-current displacement probe, generally about 0.30 inch (7.5 mm) in diameter, contains a small coil of fine wire at its tip which is excited by a remote RF oscillator to generate a magnetic field. As the tip of the probe is brought close to a conductive surface, such as a rotating shaft, eddy currents induced in the conductor by the probe's magnetic field oppose the field and reduce the amplitude of the carrier by an amount proportional to the change in proximity. A demodulator, usually encapsulated in the same enclosure as the oscillator, converts the change in carrier amplitude to a low-impedance, calibrated voltage output.

An eddy-current displacement sensor and its companion oscillator-demodulator therefore constitute a gap-to-voltage measuring system. The average gap, or the distance between the probe tip and the conductive surface, is represented by a dc bias or offset on which is superimposed an ac analog of the surface's dynamic motion. A typical linear amplitude range is 1 to 2 mm (~104 to 0.08 in) with a frequency response capability from static to more than 2000 Hz. The sensitivity changes for different target materials and with changes in cable length.

SECTION 6
PHYSICOCHEMICAL AND ANALYTICAL SYSTEMS*

L. Arnold
Johnson-Yokogawa Corporation, Newnan, Georgia (Zirconium Oxide Oxygen Sensor)

C. M. Albright, Jr.
Instrumentation Engineer, E. I. DuPont de Nemours & Company, Inc., Wilmington, Delaware (Classification of Analysis Instruments)

R. H. Cherry
Consultant, Huntingdon Valley, Pennsylvania (Thermal-Conductivity Gas Analyzers)

Jimmy G. Converse
Chief Chemist, Sterling Chemicals, Inc., Texas City, Texas (Chromatography; Sampling for On-Line Analyzers)

R. M. Durham
Instruments Division, Infrared Industries, Inc. (Gas and Process Analyzers)

G. F. Erk
Philadelphia, Pennsylvania (Classification of Analysis Instruments)

David M. Gray
Senior Application Specialist, Leeds & Northrup (A Unit of General Signal), North Wales, Pennsylvania (pH Measurement; Thermal Conductivity Gas Analyzers)

J. N. Harman III
Senior Chemist, Beckman Industrial Corporation, Fullerton, California (pH Measurement)

David W. Howard
Brookfield Engineering Laboratories, Inc., Stoughton, Massachusetts (Rheological Systems)

* Persons who authored complete articles or subsections of articles, or otherwise cooperated in an outstanding manner in furnishing information and helpful counsel to the editorial staff.

Fred Kohlmann
Great Lakes Instruments, Inc., Milwaukee, Wisconsin (Electrical Conductivity Measurements)

J. Kortright
Leeds & Northrup (A Unit of General Signal), North Wales, Pennsylvania (Electrical Conductivity; pH Measurement)

G. Kuebler
Great Lakes Instruments, Inc., Milwaukee, Wisconsin (Turbidity Measurement)

Donald Lex
Great Lakes Instruments, Inc., Milwaukee, Wisconsin (Turbidity Measurement)

Gregory Neeb
DeZurik (A Unit of General Signal), Sartell, Minnesota (Rheological Systems)

Eugene Norman
Consultant, Green Bay, Wisconsin (Rheological Systems)

James Overall
Georgia-Pacific Corporation, Atlanta, Georgia (Rheological Systems)

B. Pelletier
Rosemount Inc., Measurement Division, Eden Prairie, Minnesota (Electrical Conductivity; Turbidity Measurement)

J. G. Puls
DeZurik (A Unit of General Signal), Sartell, Minnesota (Rheological Systems)

R. S. Saltzman
Instrumentation Engineer, E. I. DuPont de Nemours & Company, Inc., Wilmington, Delaware (Gas and Process Analyzers)

E. Sperry
Chief Chemist, Beckman Industrial, Cedar Grove, New Jersey (Electrolytic Conductivity—prior edition)

F. T. Tipping
Instrumentation Engineer, Taylor Servomex Limited, Crowborough, Sussex, England (Oxygen Determination—prior edition)

Richard Villalobos
Engineering Department, The Foxboro Company (A Siebe Company), Foxboro, Massachusetts (Process Chromatography)

Gene Yazbak
Manager of Applied Systems Engineering, MetriCor, Inc.,
Monument Beach, Massachusetts (Refractometry)

CLASSIFICATION OF ANALYSIS INSTRUMENTS[1]

Matter is made up of complex, systematic arrangements of particles that are characterized by their mass and their electric charge. From a practical standpoint, these particles consist of neutrons, having mass, but with no charge; protons, having essentially the same mass as neutrons, but with a unit positive charge; and electrons, having negligible mass, but with a unit negative charge. Neutrons and protons comprise the nuclei of atoms, and each nucleus ordinarily is provided with sufficient orbital electrons, in a progressive shell-like arrangement of different energy levels, to neutralize the net positive charge on the nucleus. The total number of neutrons plus protons determines the atomic weight. The number of protons, which in turn fixes the number of electrons, determines the chemical and physical properties, except the mass, of the resulting atom.

---

[1] Based on an original classification by C. M. Albright, Jr., Instrumentation Engineer, E.I. DuPont de Nevours & Company, Wilmington, Delaware.

The chemical combination of atoms into molecules involves only the electrons and their energy states. Chemical reactions involving both structure and composition usually occur through loss, gain, or sharing of electrons among atoms. Every configuration of atoms in a molecule, crystal, solid, liquid, or gas can be represented by a definite system of electron energy states. Moreover, the particular physical state of the molecules, as represented by their mutual arrangement, also is reflected in these electron energy states. These energy states, which are characteristic of the composition of any particular substance under consideration, can be most readily inferred by observing the consequences of interaction between the substance and an external source of energy. This external energy source may be in any of the following basic groups:

1. Electromagnetic radiation
2. Chemical affinity or reactivity
3. Electric or magnetic fields
4. Thermal or mechanical energy

These groups differ fundamentally in their mode of interaction with matter. Moreover, the types of information which these interactions afford may vary considerably in specificity or uniqueness, a situation which sometimes can be controlled by combining techniques. Many properties can be measured or inferred by more than one type of interaction, as can be readily observed by inspection of Table 1. The philosophy of Table 1, with special regard to the definition of the four basic energy groups, is based on the considerations given in the following paragraphs.

ELECTROMAGNETIC RADIATION

Interaction of electromagnetic radiation with matter affords information of a most basic kind owing to the fact that photons of electromagnetic radiation are emitted or absorbed whenever changes occur in the quantitized energy states occupied by the electrons associated with atoms and molecules. For example, x rays, which consist of photons or electromagnetic wave packets, having relatively high energy, penetrate deeply into the electron orbits in an atom and provide, upon absorption, the large amount of energy required to excite one of the innermost electrons. The pattern of x-ray excitation or absorption is related to the identity of the atoms whose orbital electrons are excited, making the x-ray technique useful for determining the presence of atoms and elements in dense samples. However, because of their great penetrating power, x rays are not adapted to the excitation or observation of low-energy states corresponding to outer-shell or valence electrons or of interatomic bonds involving vibration or rotation. In such applications, the use of x rays would be analogous to attempting to stop a windmill with a rifle bullet. On the other hand, electromagnetic energy at longer wavelengths, in the infrared region, is made up of photons having relatively low energy, corresponding to the energy transformations involved in the vibration of atoms in a molecule due to the stretching or twisting of the interatomic bonds. Using the windmill model to represent a molecule, infrared radiation would correspond more to a breeze, which can indeed have a profound effect on windmill operation.

CHEMICAL AFFINITY OR REACTIVITY

In a very real sense, chemical reactions permit the recognition of certain substances because, by mixing, the two reactants can be brought into rather intimate contact so that the valence

TABLE 1 Generalized Relationships* between Matter and Energy That Can Be Measured to Ascertain Chemical Composition

Phenomenon to be measured	Group I interaction with electromagnetic radiation	Group II interaction with other chemicals	Group III reaction to electric and magnetic fields	Group IV interaction with thermal or mechanical energy
Definition	Measurement of the quantity and quality of electromagnetic radiation emitted, reflected, transmitted, or diffracted by the sample.	Measurement of the results of reaction with other chemicals in terms of amount of sample or reactant consumed, product formed, or thermal energy liberated, or determination of equilibrium attained.	Measurement of the current, voltage, or flux changes produced in energized electric and magnetic circuits containing the sample.	Measurement of the results of applying thermal or mechanical energy to a system in terms of energy transmission, work done, or changes in physical state.
Relations of measurements to chemical variables	Electromagnetic radiation varies in energy with radiation frequency, that of the highest frequency or shortest wavelength having the highest energy and penetration into matter. Radiation of the shortest wavelengths (gamma rays) interacts with atomic nuclei, x rays react with the inner shell electrons, visible and ultraviolet radiation with valence electrons and strong interatomic bonds, and infrared radiation and microwaves with the weaker interatomic bonds and with molecular vibrations and rotation. Most of these interactions are structurally related and completely unique. They may be used to detect and measure the elemental or molecular composition of gas, liquid, and solid substances within the limitations of the available equipment.	The selectivity inherent in the chemical affinity of one element or compound for another, together with their known stoichiometric and thermodynamic behaviors, permits positive identification and analysis under many circumstances. In a somewhat opposite sense, the apparent dissociation of substances at equilibrium in chemical solution gives rise to electrically measurable valence potentials, called oxidation-reduction potentials, whose magnitude is indicative of the concentration and composition of the substance. While individually all these effects are unique for each element or compound, many are readily masked by the presence of more reactive substances, so they can be applied only to systems of known composition limits.	The production of net electric charge on atoms or molecules by bombardment with ionizing particles or radiation or by electrolysis or dissociation in solution or by the induction of dipoles by strong fields establishes measurable relationships between these ionized or polarized substances and electric and magnetic energy. Ionized gases and vapors can be accelerated by applying electric fields, focused or deflected in magnetic fields, and collected and measured as an electric current in mass spectroscopy. Ions in solution can be transported, and deposited if desired, under the influence of various applied potentials for coulometric or polarographic analysis and for electrical conductivity measurements. Inherent and induced magnetic properties give rise to specialized techniques, such as oxygen analysis based on its paramagnetic properties and nuclear magnetic resonance, which is exceedingly precise and selective for the determination of the compounds of many elements.	The thermodynamic relationship involving the physical state and thermal energy content of any substance permits analysis and identification of mixtures of solids, liquids, and gases to be based on the determination of freezing or boiling points and on the quantitative measurement of physically separated fractions. Useful information can often be derived from thermal conductivity and viscosity measurements, involving the transmission of thermal and mechanical energy, respectively.

TABLE 1 Generalized Relationships between Matter and Energy That Can Be Measured to Ascertain Chemical Composition (*Continued*)

Phenomenon to be measured	Group I interaction with electromagnetic radiation	Group II interaction with other chemicals	Group III reaction to electric and magnetic fields	Group IV interaction with thermal or mechanical energy

	Emitted radiation
	1. Thermally excited
	  a. Optical-emission spectrochemical analysis
	  b. Flame photometry
	2. Electromagnetically excited
	  a. Fluorescence
	  b. Raman spectrophotometry
	  c. Induced radioactivity
	  d. X-ray fluorescence
	
	Transmission and reflection measurements
	1. X-ray analysis
	2. Ultraviolet spectrophotometry
	3. Conventional photometry (transmission colorimetry)
	4. Colorimetry
	5. Light scattering
	6. Optical rotation (polarimetry)
	7. Refractive index
	8. Infrared spectrophotometry, infrared process analysis
	9. Microwave spectroscopy
	10. Gamma-ray spectroscopy
	11. Nuclear quadrupole moment
		Consumption of sample or reactant
		1. Orsat analyzers
		2. Automatic titrators
		
		Measurement of reaction products
		1. Impregnated-tape devices
		2. Continuous chemical reaction types
		
		Thermal energy liberation
		1. Combustion types, combustibles, total hydrocarbons, carbon monoxide analyzers
		
		Equilibrium solution potentials (oxidation reduction)
		1. Redox potentiometry
		2. pH (hydrogen ion concentration)
		3. Metal ion equilibria
			Mass spectroscopy
			1. Quadrupole mass spectrometry
			
			Electrochemical properties
			1. Reaction product analyzers
			
			Electrical properties
			1. Electrical conductivity and resistivity measurements
			2. Dielectric constant and loss factor
			3. Oscillometry
			4. Gaseous conduction
			
			Magnetic properties
			1. Paramagnetism, oxygen analyzers
			2. Nuclear magnetic resonance
			3. Electron paramagnetic resonance
				Effects of thermal energy
				1. Thermal conductivity gas analyzers
				2. Melting and boiling point determinations
				3. Ice point
				4. Dew point
				5. Vapor pressure
				6. Fractionation
				7. Thermal expansion
				
				Effects of mechanical energy or forces
				1. Viscosity
				2. Sound velocity
				3. Density, specific gravity

* Chromatography, as a separation and analytical procedure, utilizes numerous different types of detectors.

potential or activity coefficient, or what might be called the "potential driving force toward reaction," can come into play. This situation can be characterized by a high degree of specificity and permits composition determinations in liquids, slurries, and the like, where the information afforded by electromagnetic radiation absorption would be somewhat less significant.

ELECTRIC OR MAGNETIC FIELDS

This is a powerful method for the determination of chemical composition when it is possible to rely on some inherent or conferred electrical or magnetic distinction between the sought-after components. It is employed by incorporating the sample in a suitable electric or magnetic circuit so that the distinguishing features can be sorted out and measured. The mass spectrometer, which sorts out the constituent ions in a sample—according to their mass and conferred charge—in a combination of electric and magnetic fields, can produce a complete, although empirical, chemical analysis of gas or vapor samples. Techniques have also been devised for the analysis of both liquids and solids. A more commonly encountered system for the determination of ions in solution is electric-conductivity measurement. In this case, however, no distinction is afforded between different ions.

THERMAL OR MECHANICAL ENERGY

This technique involves interactions of a gross nature compared with the other three techniques just described. For example, the distinguishing ability of some gas molecules to become highly excited during vibration, twisting, and rotation enables them to conduct larger amounts of heat away from heated bodies with which they collide than other gas molecules. The gross cooling effect on the heated body can be used to determine the quantity of the particular molecule present in a mixture. The simple and widely used thermal-conductivity analyzers that depend on this principle are indeed limited in their specificity or ability to recognize just one molecule, but where a gas of high thermal conductivity, such as H_2, occurs in a gas of lower thermal conductivity, such as N_2, the method is an excellent choice. Another example of gross energy transfer is the measurement of the viscosity of a substance by doing work on it with such devices as rotating disks or paddles or dropping a weight through it. Here again the measurement affords an insight regarding the actual intermolecular forces that must be overcome, whether the measurement is used for determining concentration, degree of polymerization, or composition. A table of conversion units is frequently helpful in evaluating chemical analysis problems (Table 2).

TABLE 2 Factors for Interconversion of Concentration Units of Gases and Vapors At 27°C, 3 in H_2O Gage Pressure

Desired units	Present units						
	Percentage by volume	Parts per million by volume	Moles per liter	Milligrams per cubic centimeter	Milligrams per liter	Milligrams per cubic meter	Milligrams per cubic foot
Percentage by volume	—	10^{-4}	2450	$24{,}500/M$	$2.45/M$	$2.45 \times 10^{-3}/M$	$0.0863/M$
Parts per million by volume	10^4	—	24.5×10^6	$24.5 \times 10^6/M$	$24{,}500/M$	$24.5/M$	$863/M$
Moles per liter	4.1×10^{-4}	4.1×10^{-8}	—	$1/M$	$10^{-3}/M$	$10^{-6}/M$	$35.3 \times 10^{-6}/M$
Milligrams per cubic centimeter	$4.1\,M \times 10^{-4}$	$4.1\,M \times 10^{-8}$	M	—	10^{-3}	10^{-6}	35.3×10^{-6}
Milligrams per liter	$0.41\,M$	$4.1\,M \times 10^{-5}$	$M \times 10^3$	10^3	—	10^{-3}	0.0353
Milligrams per cubic meter	$410\,M$	$0.041\,M$	$M \times 10^6$	10^6	10^3	—	35.3
Milligrams per cubic foot	$11.6\,M$	$1.16\,M \times 10^{-3}$	$28{,}300\,M$	$28{,}300$	28.3	0.0283	—

To use the table:
1. Locate column along top of table which gives present unit.
2. Locate row along left of table which gives desired unit.
3. Read down and across to locate multiplying factor.
4. Multiply present quantity by factor.
Example: Given 700 ppm to convert to moles per liter: $700 \times 4.1 \times 10^{-8} = 2.87 \times 10^{-5}$ mol/L.

NOTE: In table, M is molecular weight of gas or vapor.

APPLICATIONS

There are very few phases of industrial operations where chemical-composition variables are not important. Following are some of the applications of major significance:

1. Raw materials
   a. Composition analysis to check purchase specifications
   b. Detection of contamination by trace impurities
   c. Analysis check on materials priced on an active-ingredient basis
   d. Continuous analysis of materials delivered by pipeline; water analysis
2. Process control
   a. Speed up and improve control by automatization, or replacement, of control laboratory tests on "grab" samples
   b. Improve control by replacing or augmenting inferential measurements, such as temperature or pressure, with more significant composition data
   c. Permit use of continuous processes that could not be controlled except by continuous analysis instrumentation
3. Process troubleshooting
   a. Temporary use of analysis instruments for process studies aimed at overcoming occasional upsets
4. Yield improvement
   a. Continuous analysis of process streams to measure effects of variables influencing yields
   b. Analysis of overflow or purge streams, recirculated material, sumps, and the like, to determine product losses and detect buildup of undesirable by-products that affect yield
5. Inventory measurement
   a. Analytical monitoring of material flowing between process steps and plant areas to establish consumption and in-process inventory on the basis of active or essential ingredients
6. Product quality
   a. Determination of product composition
   b. Assess structurally dependent attributes, such as color, melting or boiling point, and refractive index
   c. Assist in adjustment of product to meet specifications
7. Safety
   a. Detection of leaks in equipment
   b. Survey operating areas for escape of toxic materials from leaks or spills, especially materials not readily detected by human senses
   c. Detection of flammable or explosive mixtures in atmosphere or process lines
8. Waste disposal
   a. Monitoring plant stacks for accidental discharge of toxic or nuisance gases, vapors, or smoke
   b. Analysis of waste streams for toxic or other objectionable materials
   c. Control of waste treatment or product recovery facilities
9. Research and development
   a. Continuous analysis to speed up research and optimize results
   b. Provide structural and compositional information not otherwise obtainable
   c. Produce results in a more directly usable form

In considering the application of instrumental methods for the determination of chemical composition, it is important to bear in mind that measurement is the first step toward control and that the closer the information can be brought to the process, the better will be the control. This is true whether the information is merely presented continuously to the operator or actually used to control the process itself automatically. Complete automatic control becomes more desirable as process throughput is increased and holdup is decreased in modern high-speed processing equipment. This trend also places a premium on high speed of the response in analytical instruments and their associated control equipment.

PRACTICAL CONSIDERATIONS

Any practical appraisal of the merits of chemical-composition variables for process control purposes must recognize certain inherent physical limitations in their measurement. Generally speaking, these limitations are the following

1. *Sample must be representative.*   Although this requirement may appear obvious, it is a factor that is very frequently overlooked. In the first place, the sample must be gathered or drawn off in such a fashion that it will be of the same composition as the body of the processed material. Moreover, there must be assurance that any change in conditions, such as temperature or pressure, between the sampling and measuring points cannot influence sample composition. In addition, in nearly all cases the probable composition of the sample must be known ahead of time through some independent method before an analysis technique can be selected.

2. *Physical state of sample.*   The technique must provide for interaction between the applied energy and the entire sample, as well as for observation of the total result. This can seldom be accomplished. It is for this reason that a large majority of techniques is applicable to gases, where the molecules are widely spaced and free to react in a characteristic manner, and that fewer techniques are applicable to liquids and still fewer can be applied to solids.

3. *Uniqueness of specificity of method.*   The selection of the method must be tailored to the sample composition and to the information requirements. Some methods or techniques involving atomic and molecular structure are rather universal in that they permit exact identification and measurement of every elemental or molecular constituent present in the sample. These methods are usually the most complex and costly. They are sometimes considerably less sensitive than simple methods whose only drawback is an inability to distinguish between related substances having similar gross interactions with energy. Where the related substances are known not to be present in the sample, these simpler, less specific methods should always be considered.

TRENDS IN ANALYTICAL INSTRUMENTATION[2]

The trends in specific areas of analysis are brought out in the articles which follow in this handbook section. With the accelerated development of new chemical analytical techniques in the 1950s, a trend commenced that would move the formerly isolated chemical control laboratory on-stream in terms of using continuous instrumental analyzers. A parallel trend occurred with the availability of improved nondestructive inspection techniques and semiautomated testing procedures in the discrete-piece manufacturing industries. These trends essentially started with the larger manufacturing firms and have been growing in acceptance and practicality with smaller manufacturers ever since. The driving forces for these actions have included the following:

---

[2] Parts of this summary by G. F. Erk, Philadelphia, Pennsylvania.

1. Conservation of energy has increased efforts to use on-line analytical instruments. Fuel costs have risen over the last decade to a point where on-line instrumentation has become even more cost-effective. One area immediately affected by this is the application of an oxygen and combustibles analyzer in determining optimum fuel-to-air ratios in the combustion process and the resulting heat generation. Likewise, calorimetric analysis of fuel quality is becoming even more important, as are density and specific gravity analyses.

2. Requirements for monitoring pollution generated additional needs for high-quality analytical devices with the capability to provide good records of various pollutants and particulate emissions. Stack gas monitors are widely used for pollution control and monitoring and have design features permitting long-term unattended operation within stacks or through sampling systems.

3. Pressing demands for more accurate, thorough, and rapid means of testing materials and products—from the receipt of raw materials and inspection throughout manufacturing to the completion of production, warehousing, and distribution.

Designers of contemporary analytical systems have taken advantage of all the technological amenities that have become available during the past decade or so, namely, advanced communication and networking, improved data displays, microprocessors and personal computers, self-diagnostics, and miniaturization of sensors. Reliability and precision have progressed steadily.

During the next decade, much more can be expected to come from the analytical instrument field. Scores of excellent analytical techniques still confined to the laboratory are waiting to be exploited by enterprising designers, who will convert them to configurations that will make them practical to use on the process and in factory floor environments.

SAMPLING FOR ON-LINE ANALYZERS

by Jimmy G. Converse[1]

Most often the major problems with measuring chemical composition and concentration of process streams arise from inadequate sample preparation. The solution in many cases is not customization or complex sampling system design, but rather, the effective use of already available reliable equipment. Success involves not only an understanding of the chemical sensor to be used, but equally important, a perception of the chemical and physical characteristics of the process stream.

In a majority of on-line analytical applications, the process streams are *not* clean, *nor* are they homogeneous. Consequently one cannot ideally reach into the process, grab a sample, and perform a direct analysis. On the average, a process stream is "dirty"—loaded with substances other than those targeted for analysis—and often it is of a physical nature (viscous, sticky, corrosive, sometimes very volatile, frequently hazardous, and the like) so that direct presentation of an unconditioned sample to the detector is impractical.

[1] Chief Chemist, Sterling Chemicals, Inc., Texas City, Texas.

Although over the years numerous efforts have been made to design a simple probe that will function when inserted into a "dirty" process, this approach has been notably unsuccessful to date. Other efforts in sampling system design have evolved into building essentially "miniature chemical processing units." These units are expensive to implement and difficult to maintain and, in the author's opinion, responsible for a very high percentage of analyzer-system failures. Often such complexity is really unnecessary [1].

RETHINKING OF SAMPLING PROBLEM

Adequate technology is available for proper sample preparation, but many conventional and contemporary approaches are unsatisfactory and the cause of numerous difficulties. Some basic rethinking is needed. One may ask: "Why extract a tank car of material that has to be conditioned when only a drop is required for analysis?" Extracting a minimum amount of material reduces disposal requirements and energy usage. Any more than the minimum is wasteful.

The needed sampling rate also should be questioned. The rate required to control most processes is *not* continuous. Therefore only when continuous measurement of a stream component is absolutely necessary should this practice be used. For example, chromatographs and other flow-injection analyzers (FIA) place a small, discrete sample into the measurement system on a periodic cycle (seconds or minutes). The preconditioned sample flows continuously through the injection valve, but only 2 percent of the conditioned material is used.

A comparison of continuous and discrete sample preparation is given in Table 1. Discrete sampling reduces waste, improves reliability, and provides much more flexibility in sample preparation. The sample should be extracted for a short period. A discrete quantity should be conditioned and isolated for analysis, after which the conditioning system should be back-flushed to purge and clean it before the next sample is taken. This procedure produces a low duty cycle on the cleanup system, which is the step most vulnerable to failure.

TABLE 1 Relative Advantages and Limitations of Discrete versus Continuous Sampling

Step in procedure	Discrete sampling	Continuous sampling
Sample extraction	Representative of total population	Representative of total population
Conditioning	Small quantity of residue and material for disposal; automated maintenance	Large quantity of residue and material for disposal; manual system maintenance
Component isolation	All operators useful; low maintenance	Limited operators; high maintenance
Molecular manipulation	Low reagent usage	High reagent usage
Measurement	On-line zeroing; on-line calibration; periodic updates; low duty cycle	Off-line zeroing; off-line calibration; continuous updates; high duty cycle

Avoiding Common Mistakes

Successful sampling can be achieved best if a modular approach to the task of sample preparation is used. First, the roles of the measurement system and of the sample-preparation system must be distinguished:

1. *Measurement system.* Primary objective: determine the concentration of one or more species in a mixture within the required precision and accuracy.
2. *Sample preparation system.* Primary objective: maintain continuous unattended operation of the individual parts to provide a representative sample to the detector in a suitably conditioned form within the required precision.

In assessing the success of a proposed sample-preparation system, the designer should ask a number of questions. A checklist can be convenient (Table 2).

TABLE 2   Checklist for Good Sampling System Design

The system:
____ is designed for simple and easy maintainability.
____ will not require excessive manual maintenance.
____ does not incorporate complex miniature chemical units.
____ will not generate large amounts of waste to be disposed.
____ is designed for practical use—not for ease of construction.
____ is not based on cost alone.
____ was not rushed at last minute in order to close out a project.
____ does not overemphasize accessories, such as bells and horns.
____ incorporates and adheres to the fundamental principles of chemistry.

Steps for Improving Sampling System Performance

Often taking the following actions can lead to improved performance of the sample-preparation system:

1. Prior to designing the conditioning system, establish the character of the process material by extracting through filters and scrubbers and by analyzing the vapor, liquid, and solid phases.
2. Involve someone with an understanding of analytical chemistry and sample-preparation fundamentals to evaluate appropriate alternatives.
3. Minimize the quantity of material extracted from the process for conditioning by utilizing discrete sampling.
4. Design the system for simple and easy maintenance.

BASICS OF SAMPLE PREPARATION

Sample systems are comprised of four functions:
Extraction.   Obtaining a representative portion of the total process population.
Conditioning.   Quantitative removal of undesirable phases and controlling temperature, pressure, flow rate, and other independent variables.
Component Isolation.   Separation and quantitative removal of molecular species by physical or chemical means to prevent interference with measurement.
Molecular Manipulation.   Chemical reaction, such as pyrolysis, derivatization, hydrogenation, reduction, complexation, ionization, excitation, or other conversions to make it easier to measure a molecule.
Many volumes have been published on sample extraction. The first rule is that the composition of the material extracted must be representative of the process stream population. *This does not mean, however, that it has to be identical.*
Using discrete sampling eases periodic cleanup and provides the sample-preparation system with a shorter duty cycle, as compared with continuous sampling. Also, one can introduce a reference sample into the system at the point of process-sample extraction to monitor the performance of the entire system prior to concentration measurement [2]. This can be timed such that the reference measurement does not interfere with the process measurement. These data then can be used to produce statistical quality control charts without removing the analyzer from service [3].

Of equal importance, the sample-preparation steps can be controlled to yield a constant transfer function. The concentration of the target component at the detector must have a constant proportionality to the concentration of that component in the process stream. The value of the transfer function is determined by calibration of the sample-preparation system and the detector. The detector should also be calibrated separately to isolate the sample-preparation system portion of the transfer function. The repeatability of the proportionality factor for the sample-preparation system determines the *sampling error* associated with the measurement process. Measuring the precision of the measurement device, but not the sample-preparation system, is a *common error.* One must pay close attention to both to obtain quality measurements.

MULTIPLE DIMENSIONAL SAMPLE PREPARATION

Separation Dimension

A number of operational parameters come into play in a sample-preparation system. Close attention to their use can improve performance.

The discovery of column chromatography opened new vistas of component isolation for chemical analysis. The separation devices also can be used for sample-preparation operations. Vapor-liquid equilibrium does not have to be performed on a column device. Special spargers, strippers, and separators have been used. Membranes are being developed to produce selective separations, such as aqueous-organic and ion exchange. These can provide liquid-liquid extraction as well as gas-liquid separation sample-preparation operations.

Temperature Dimension

When one is faced with separation of a sample having a wide boiling-point range, the obvious answer is temperature programming. However, in the author's experience there is a better alternative. One can use multiple isothermal temperature zones and achieve superior results due to holding the temperature constant. The concept involves passing the lower-boiling-point components through the highest temperature zone quickly, to be separated at a lower appropriate temperature (Fig. 1). A separation column of low resolution is used to split boiling-range groups. A six-port valve is used to direct the groups to their appropriate designations. The high boilers are separated and measured immediately in the highest temperature zone. They are not held on the column until the temperature is programmed to a higher value. Total cycle time can be reduced significantly. The lower-boiling-point components are sent to a lower-temperature zone by directional switching or trap-and-transfer techniques. Additional separations and transfers can be used if required. The sample, or portions thereof, travel from high to low temperature in all cases.

Flow-Rate Dimension

Pressure and flow programming has been reported, but has not been widely used until recently. Better sensors and digital controllers may improve control, although flow has less effect on separation than on temperature. Independent flow in different systems can be implemented by utilizing the trap-and-transfer technique. More use of pressure and flow-rate parameters will occur as a better understanding of them is developed. They are used more widely in liquid chromatography, especially for supercritical fluids.

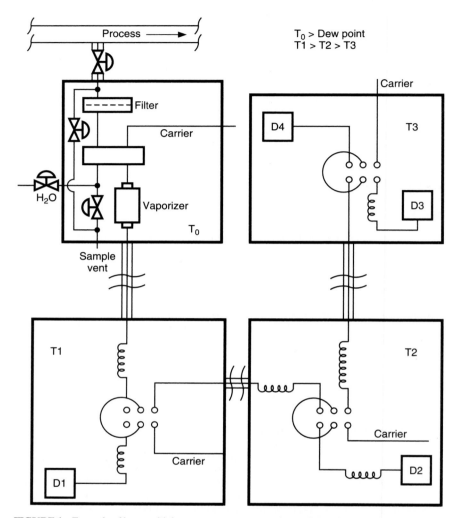

FIGURE 1 Example of how multiple temperature zones can be used. D—detector; T—temperature.

Measurement Device Dimension

Many detectors have been developed to measure chemical concentration of the isolated components. They vary in sensitivity and specificity. One application of two-dimensional detection is to place a nondestructive thermal conductivity detector (TCD) in front of a higher-sensitivity, more selective flame ionization detector (FID). Detectors can be used in parallel operation as well. Selective detectors can be used to measure specific components without completely separating nonresponding species. The choice of detector will markedly affect the sample-preparation requirements.

Carrier Fluid Dimension

The carrier fluid can be varied to affect the behavior of the separation device or the detector. This has been utilized mostly in liquid chromatography, but can aid sample preparation in other ways. The use of an acid solution on one side and a base solution on the other side of an ion exchange membrane has served to improve transfer of a selected species. Hydrophobic membranes can be utilized to reject water and concentrate the target species in an organic phase. Exchange of nitrogen for helium in a column effluent is a means of preparing the sample for improved response in an electron capture detector. The utilization of this parameter is essentially only limited by the analytical chemist's creativity and imagination.

VALVES FOR DIRECTING TRAFFIC

The main types of valves currently used in analytical systems are spool, planar sliding, rotary sliding, diaphragm, and needle designs. Primary connections are four, six, and eight ports. Valves with more ports can be considered combinations of the aforementioned three types. The valve is generally operated pneumatically, but electric solenoids can be used at low pressure. They are placed in an isothermal temperature zone and operate by direct mechanical transfer. Another technique uses mechanical valves outside the oven to produce pneumatic switching by changing the pressure differential across the separation column. Details of the internal structure and principles of operation of these valves is given in Clevett [4]. More attention to the valving configuration can pay large dividends.

A MORE FLEXIBLE SYSTEM

The discrete sample and valve-column configurations from process chromatography can be expanded and developed into a multidimensional automated sample-preparation system. One should be able to develop a limited number of configurations that will handle the preparation of most sample streams. The appropriate detector then can be attached to make the desired measurement of an isolated component. The discrete sample may be taken at a remote location by using a short sample line and a long transfer line. Work on remote discrete sampling has been reported by Converse [3] and others.

Technologies, such as foreflush, backflush, and heart cut to detector or vent, have been developed to produce component isolation. Such manipulations are actually part of sample preparation prior to concentration measurement. A portion of the sample may be trapped and stored on a section of column for later separation and measurement. An excellent example is the purge-and-trap technology used to concentrate trace volatile organic chemicals in water. The author has developed a technique, called "trap and transfer," that utilizes a six-port external volume sample valve (Fig. 2).

Once a sample has been injected into the system, any portion can be removed quantitatively and directed to a specified destination by taking a subsample. The original sample may be a vapor or a liquid which is subsequently vaporized. This technique can be utilized to capture a portion of the sample or to dispose of it. Figure 3 shows a remote sampling arrangement which disposes of the unwanted condensables quantitatively, provides for automated cleanup of the sample-conditioning devices, and assures rapid transport of the remaining components to the measurement device. Other devices used for sample preparation (sample operators) can be configured in the same manner as columns.

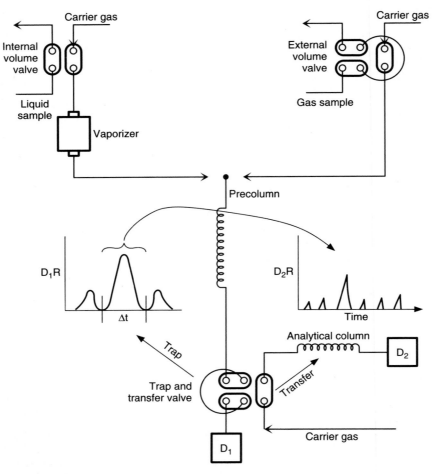

FIGURE 2   Trap-and-transfer approach. D—detector.

ADVANTAGES OF TRAP-AND-TRANSFER TECHNIQUE

Some of the major benefits of this technique include the following.

1. Any peak broadening caused by diffusion in the precolumn is eliminated when the trapped portion is injected onto the second column.

2. The second column is an "independent analytical system" with temperature and flow rate isolated from the primary separation. There are no restriction valves to adjust and balance.

3. A large sample may be used on the precolumn and most of the material discarded. A larger quantity of the key component can be isolated without loading the analytical column and detector with unwanted material. This is a concentration scheme.

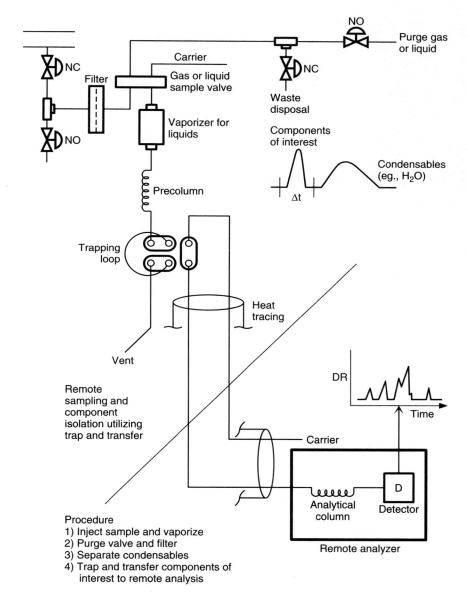

FIGURE 3   Remote sampling and component isolation using trap-and-transfer method. D—detector; NC—normally closed; NO—normally open.

4. A trace component on the tail of a major component can be trapped and transferred. The second independent system can separate the two components better because their concentration ratio has been reduced. This improves resolution.

5. An uncoated column can be used to separate components of interest from high-boiling or reactive residues which cannot be backflushed from a reactive column material.

The author believes that the trap-and-transfer technique represents a marked improvement for component isolation as compared with the older method of "directing traffic."

REFERENCES

1. Converse, J. G.: "Sample Preparation for On-Site Automated Chemical Analyzers," paper 90-458, presented at the New Orleans, Louisiana, Meeting, Instrument Society of America, Research Triangle Park, North Carolina, 1990.
2. Converse, J. G.: "Internal Reference Introduction for Performance Verification, Diagnostics, and Calibration Validation," in *Proc. Analytical Instrumentation Division Symp.,* vol. 21, Instrument Society of America, Research Triangle Park, North Carolina, 1985, p. 185.
3. Converse, J. G.: "Decisions to Change Analyzer Calibration Based on Statistical Quality Control Charts," in *Proc. Analytical Instrumentation Division Symp.,* vol. 25, Instrument Society of America, Research Triangle Park, North Carolina, 1989, p. 345.
4. Clevett, K.: *Process Analyzer Technology,* Wiley, New York, 1986.
5. Converse, J. G.: "History and Future of Remote Discrete Sampling for Automated Composition Analyzer Applications," *Analyt. Instrum. Div. Newsl.,* vol. 5, p. 5, Jan. 1984.

pH AND REDOX POTENTIAL MEASUREMENTS[1]

The pH (hydrogen ion concentration) of a water solution is a measure of the *effective* acidity, neutrality, or alkalinity of that solution. This property is of serious consequence in such diverse fields as biochemicals, pharmaceuticals, food products, water and sewage treatment, and in the manufacture of numerous chemical products which are the result of chemical reactions. Because of this importance, pH instrumentation and control are applied widely. pH is frequently used as a preventive strategy in the control of certain pollution and corrosion problems.

In modern systems pH measurement is based on the use of special selective electrodes that develop an electromotive force (EMF) which is proportional to the negative logarithm of the hydrogen ion concentration of the solution in which they are immersed. By definition,

$$pH = \log \frac{1}{\text{hydrogen ion concentration, mol/L}} \tag{1}$$

[1] Portions of this article are from the original prepared by J. N. Harman III, Senior Chemist, Beskman Industrial Corporation, Fullerton, California, and updated by the handbook staff with the cooperation of David M. Gray, Senior Application Specialist, Leeds & Northrup (A Unit of General Signal), North Wales, Pennsylvania.

All water solutions of acids and bases owe their chemical activity to their relative hydrogen and hydroxyl ion concentrations. In water, the equilibrium product of the hydrogen and hydroxyl ion concentrations is a constant, 10^{-14} at 24°C, and the pH scale is uniquely related to water at this temperature. Thus when the concentrations of H^+ and OH^- in pure water at 24°C are equal, the H^+ ion concentration must be 10^{-7}, and the pH = $\log(1/10^{-7}) = 7.0$.

The pH scale covers the range of both acid and alkaline solutions. Pure water is neither acidic nor basic (alkaline). Acid solutions increase in strength as the pH value falls below 7. Alkaline solutions increase in strength as the pH rises above 7. The scale is *not* linear with concentration. A 1-unit change in pH represents a tenfold change in the effective strength of an acid or base; for example, a solution of pH 3 is 10 times as strong an acid solution as a solution of pH 4.

It is important to recognize that pH measures only the concentration of hydrogen ions actually dissociated in a solution and *not* the total acidity or alkalinity. It is this factor that is responsible for the observed pH change in pure water with temperature. As the water temperature increases, the amount of dissociation increases and the quantity of hydrogen and hydroxyl ions increases equally. Since pH is related to the concentration of hydrogen ions alone, the pH actually decreases, although the water is still neutral. Unless the relationship between dissociation constant and temperature is known, therefore, it is not possible to predict the pH of a solution at a desired temperature from a known pH reading at some other temperature.

pH meters generally cover a range of 0 to 14 units. However, it is possible to make measurements beyond these limits in very concentrated solutions. Figure 1 shows these relationships and indicates how the pH scale encompasses measurements of solutions of any strength.

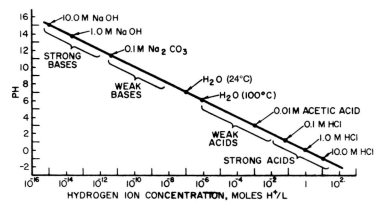

FIGURE 1 Relationship among pH, hydrogen ion concentration, and various solutions of acids and bases.

ACID-BASE THEORY

Ionization and Neutralization in Solution

Acids dissolve in water to produce conductive solutions because charged ions are formed. The equation for the ionization of hydrochloric acid may be written

$$HCl = H^+ + Cl^-$$ (2)

In the same manner, sodium hydroxide forms ions when dissolved in water,

$$NaOH = Na^+ + OH^- \tag{3}$$

When these two solutions are mixed, a neutralization reaction occurs, that is, the acidic and basic properties of the solution are lost because of the reaction of the hydrogen and hydroxyl ions to form water,

$$Na^+ + OH^- + H^+ + Cl^- = HOH + Na^+ + Cl^- \tag{4}$$

The resulting solution is still conductive because of the presence of sodium and chloride ions, but it is neither acid nor alkaline if the quantities of the original acid and base are equivalent.

Normality and Concentration

The characteristic properties of acids are due to the presence of hydrogen ions. The characteristic properties of bases are due to the presence of hydroxyl ions. A solution of acid that contains 1 g/L of ionized hydrogen is of the same strength insofar as the neutralization reaction is concerned, no matter what acid is used. Such an acid solution is said to be 1 N in acidity. Similarly a solution containing 17 g of hydroxyl ion, or 1 g/L ionic weight, is 1 N in alkalinity. This should not be confused with the more common usage of normality for solution concentrations on a gross-weight basis.

Hydrochloric acid is referred to as a strong acid and sodium hydroxide as a strong base because both are almost completely dissociated into ions in aqueous solutions. Thus the concentration of active hydrogen ions in the solution of a strong acid is approximately equal to the normality, and the same is true for the hydroxyl ion concentration of a strong base. This is not true for acids and bases which do not ionize completely; they are consequently termed weak acids and bases.

Generalized Theory

It is convenient to write HA for the general form of an acid and BOH for the general form of a base. Acidic and basic properties are related to the behavior of water, since to a limited extent water is amphoteric and acts as either an acid or a base. Both hydrogen and hydroxyl ions are produced by the dissociation of water,

$$HOH = H^+ + OH^- \tag{5}$$

The rate of ionization of water molecules is exactly equal to the rate of recombination of the ions, and the equilibrium expression can be written

$$K = [H^+][OH^-] = 10^{-7} \times 10^{-7} = 10^{-14} \text{ at } 24°C \tag{6}$$

where the brackets indicate the concentration of the ions in moles per liter. K is called the ion product constant.

Actually hydrogen ions, or protons, do not exist free in solution but are associated with solvent molecules. The ionization of water would thus more properly be written

$$HA + HOH = H_3O^+ + A^- \tag{7}$$

H_3O^+ is called the hydronium ion and, in aqueous systems, is the ion responsible for acidic properties; equations are more simply written using H^+, and this practice will be followed in this discussion. Strong acids such as hydrochloric, nitric, sulfuric, and perchloric all appear to be of the same strength in water, because each completely dissociates to form hydronium ion, which acts as the acid.

pH Theory and the pH Scale

In the foregoing discussion it has been assumed that strong acids ionize completely, producing hydrogen ion concentrations equal to the concentration of the acid present. When the concentration of hydrogen ions exceeds approximately 0.01 N, the properties begin to deviate from the ideal behavior of ions predicted by thermodynamic theory. The active hydrogen ion concentration appears to be lower than the value corresponding to complete ionization.

This apparent concentration is called the activity of the ion and is obtained by multiplying the ion concentration by the activity coefficient. The value of the activity coefficient approaches 1 in very dilute solutions. The difference in the ion concentration and the ion activity is largely accounted for by the interaction of the electric fields associated with ions when they are present at high concentrations. Hydrogen ion activity is the most useful measure of effective acid strength. For many years the general method of expressing acid strength was to give the hydrogen ion concentration in moles per liter. This was awkward because of the wide concentration range to be covered and the fact that acid strength and concentration are easily confused.

Sorenson proposed in 1909 that the expression pH be adopted for hydrogen ion concentration and defined pH as an operator standing for "power of hydrogen" in the following manner:

$$pH = -\log [H^+] \tag{8}$$

The pH of a solution may be determined by measuring the voltage of a concentration cell comprising two platinum electrodes, one immersed in a solution of known pH and the other immersed in the sample, with the two solutions separated by a salt bridge and with hydrogen gas at known pressure contacting the two electrodes. The voltage of such a cell may be expressed as

$$E_H = \frac{RT}{F} \ln \frac{a'_{H^+}}{(a'_{H_2})^{1/2}} - \frac{RT}{F} \ln \frac{a''_{H^+}}{(a''_{H_2})^{1/2}} \tag{9}$$

where a'_{H^+} and a''_{H^+} are the activities of the hydrogen ions in the known solution and in the sample, respectively, and a'_{H_2} and a''_{H_2} are the activities of the hydrogen gas. If a'_{H^+} and a'_{H_2} are unity, the reference electrode is a standard hydrogen electrode and the first term in the equation becomes zero. If the hydrogen pressure over the sample is 1 atm, then a''_{H_2} becomes unity and the voltage is dependent solely on the hydrogen ion concentration. Then

$$E_H = -\frac{RT}{F} \ln (a'_{H^+}) = -0.0591 \qquad E_H \text{ (at 25°C or 77.0°F)} = 0.0591 \text{ pH} \tag{10}$$

Thus,

$$pH = \frac{E_H}{0.0591} \text{ at 25°C or 77.0°F} \tag{11}$$

which is a useful expression of acidity.

As stated at the beginning of this discussion, the pH scale includes alkaline solutions. This is possible because of the relation

$$K_w = [H^+][OH] = 10^{-7} \times 10^{-7} = 10^{-14} \text{ at 24°C or 75.2°F} \tag{12}$$

Thus in a solution in which hydroxyl ion activity is unity,

$$a_{H^+} = \frac{10^{-14}}{1} = 10^{-14} \qquad pH = 14 \tag{13}$$

The general practice is to speak of the hydrogen ion concentration as corresponding to a given pH when actually the effective concentration or activity is meant. This practice will be followed in the remainder of this discussion.

Table 1 lists pH values and some other characteristics for a variety of acids and bases. For a more complete treatment of acid-base theory and of activity, the reader is referred to standard textbooks on physical chemistry.

TABLE 1   pH Values and Characteristics of Various Materials

Compound or material	Total normality	Effective H^+ (as normality)	pH (at 25°C)
Acids and common acidic solutions			
Acids taste sour; neutralize bases to yield water and salts; catalyze some reactions; ionize in water to yield hydrogen ions			
Hydrochloric acid	1.0	0.8	0.1
	0.1	0.083	1.08
	0.001	0.001	3.00
Sulfuric acid	1.0	0.48	0.32
	0.1	0.68	1.17
Acetic acid	1.0	0.0043	2.37
Lemon juice	—	0.01–0.0063	2.0–2.2
Acid fruits	—	10^{-3}–3×10^{-5}	3.0–4.5
Vegetables, including melons	—	10^{-5}–10^{-7}	5.0–7.0
Jellies, fruit	—	10^{-3}–3×10^{-4}	3–3.5
Fresh milk	—	3×10^{-7}–2.2×10^{-7}	6.50–6.65
Bases and basic solutions			
Bases taste bitter; neutralize acids to yield water and salts; ionize in water to yield hydroxyl ions			
Sodium hydroxide	0.1	0.057	13.73
	0.1	0.071	12.84
Ammonia (10% NH_3)	5.9	0.006	11.8
	0.1	0.0018	11.27
Lime water [$Ca(OH)_2$ sat.]	—	0.04	12.4
Trisodium phosphate $Na_3PO_4 \cdot 10H_2O$, 2%	—	0.009	11.95
Blood plasma, human	—	—	7.3–7.5
Water at various temperatures			

Temperature, °C	pH
0	7.472
22	7.00
25	6.998
50	6.631
100	6.13

Buffer Solutions

These are solutions that resist a change in pH despite the additions of acid or base. Buffer action is exhibited by any solution that contains substantial concentrations of both a weak acid and the salt of that acid. When dissolved, the salt yields a substantial concentration of anions (A^-) and the weak acid yields mostly undissociated molecules (HA). Mixtures of weak bases and their salts also act as buffers in alkaline solutions. The hydrogen ion concentration [H^+] of the solution is determined by the mass action law:

$$\frac{[H^+][A^-]}{[HA]} = K_A \tag{14}$$

where K_A is the ionization constant. The constancy of K_A over a wide range of concentrations, and thus the ratio of the free acid times the anion concentration divided by the undissociated acid concentration, forces the buffering action when the system is perturbed.

If a strong acid is added to this solution, the resulting hydrogen ion concentration will be less than if the acid were added to the same amount of pure water, because the hydrogen ion concentration cannot increase and the A⁻ and HA concentrations remain unchanged. This would be in violation of the mass action law. Conversely, if something should occur, such as dilution or neutralization, to alter the mass action relationship to decrease the hydrogen ion concentration, the HA molecules will dissociate until the law is again satisfied. In the measurement of pH, buffer solutions are used to calibrate the electrodes and the system to ensure measurement system accuracy.

pH MEASUREMENT

Over the years, pH has been measured in two ways:

1. *Colorimetric* determinations, which depend on the use of chemical substances (frequently organic dyes). These compounds markedly change their color at certain pH values. Generally this methodology is not convenient to apply in an industrial setting.

2. *Electrometric* pH measurement systems, which require a pH-responsive electrode, a reference electrode, and a potential measuring instrument.

The Hydrogen Electrode

Considered to be the standard for the electrometric measurement of pH, the hydrogen electrode was used experimentally to follow changes in acidity as long ago as 1897[2] [1]. The hydrogen electrode is employed by bubbling hydrogen gas past a wire or foil which is able to catalyze the reaction $H^+ + e \rightleftharpoons \frac{1}{2}H_2(g)$ and thus establish an equilibrium between molecular hydrogen and the hydrogen ions. The metal wire or foil usually is platinum which has been pretreated to provide a "platinum black" catalyzing surface.

Prior to the development of the glass electrode, the quinhydrone electrode and the antimony electrode were used as pH-sensitive devices. These sensors were used in fairly limited quantities prior to the appearance of the first and much more practical glass electrode in the 1940s. Modern pH electrodes are shown in Fig. 2.

Glass Electrode

The possibility that a thin glass membrane of special composition could develop a potential was described as early as 1909 by the German chemist Fritz Haber. However, little progress with this approach was made until the mid-1920s. Developments since have accelerated to the point that both laboratory and continuous-process pH measurements today are almost as common as temperature and pressure measurements. This acceptance primarily is due to development of the glass electrode together with improved means of extremely stable electronic amplification of the potential developed by the electrode.

The mechanism whereby certain types of glass develop potentials depending on the pH level is an extremely complex one. It has been observed experimentally that a glass electrode produces a predictable potential related directly to the hydrogen ion concentration of the solution in which it is immersed. The glass electrode responds in a predictable fashion

[2] W. Böttger, *Z. Phys. Chem.*, vol. 24, 1897, p. 253.

(a)

(b)

FIGURE 2 Contemporary combination pH electrodes, including glass-measuring electrode, diffusion-type reference electrode, and temperature compensator in a single ¾-inch (19-mm) NPT probe. (*a*) For in-line installation. (*b*) For submersion installation. (*Leeds & Northrup.*)

throughout the normally accepted 0 to 14 pH range, developing 59.2 mV per pH unit at 25°C (77°F), consistent with the classical Nernst equation. The recognition that actual pH electrodes seldom exhibit the theoretically predicted millivolt-pH sensitivity is a relatively recent event observed in many modern analyzers which allow slope adjustment to calibrate the analyzer for electrodes with less than the theoretical slope. Unlike the earlier types of pH electrodes, the glass electrode is not influenced by oxidants or reductants in the solution. With suitable temperature compensation, pH measurements can be made from 0 to 100°C, and to even higher temperatures under controlled conditions. The glass electrode pH measurement, by its obvious character, involves high resistances in the measuring loop. This demands that a high-impedance amplifier be used. Glass electrodes have resistances varying from a few to a thousand megohms. In recent years this has ceased to be a significant problem with the use of modern electronic circuitry.

The asymmetry potential (AP) of a glass electrode can be observed by coupling the glass electrode to a stable reference electrode of the same internal half-cell system as used in the glass electrode in a buffer and noting the voltage developed by the electrode pair. Typically, most pH glass electrode–reference electrode pairs are disposed to put out 0 mV in a pH 7 buffer at 25°C (77°F)—deviations in values are termed the AP. By definition, asymmetry potential implies that the two surfaces of the membrane may not respond in an identical fashion for various reasons. The possibility of one side of the membrane responding differently, even to the same type of solution, increases as the age and the use of the electrode increase. Most pH meters have an asymmetry potential adjustment control, which is simply a variable potentiometer, to correct for any drift which does occur.

The asymmetry potential drift, when it does occur, is not a sharp or sudden change, but almost can be regarded as a constant of the system, which requires periodic adjustment by standardization with a buffer solution. It is the author's experience that weekly standardization will keep a continuous pH measurement within ±0.05 pH unit accuracy, assuming that amplifier drifting and electrode fouling do not occur.

The most likely cause for asymmetry potential drift in continuous pH measurements is prolonged chemical or abrasive attack of the stream on the glass membrane. When either of these occurs, the outer surface becomes altered to the extent that the response of the membrane to the presence of hydrogen ions gradually alters.

Dehydration of the outer surface by prolonged exposure to alcoholic solutions also can cause asymmetry potential drift. When glass electrodes have been stored dry for long periods of time, the pH readings continue to drift for periods of up to 24 to 48 hours until a full equilibrium condition is attained. The adsorption of surface-active agents and a greasy film can cause an apparent asymmetry potential drift as the exchange capacity of the surface is upset. Exposure of the glass membrane to strong acids and caustic solutions can permanently alter the external surface of the glass. Hydrofluoric acid can quickly etch the membrane to the point of complete destruction. The term "asymmetry potential change" attempts to summarize the several effects of this usually slight decay.

Reference Electrode

In pH measurement a second electrode is required in the solution simply to complete the circuit. A good, stable reference electrode must (1) produce a predictable potential, compatible with the glass measuring electrode, (2) be linear with respect to temperature change, and (3) be simple to use.

Over time several reference electrodes have been developed, but for many years the mercury–mercurous chloride (calomel) electrode was used widely. Now, more commonly, the silver–silver chloride reference electrode is used. The calomel electrode consists of an inner glass tube packed with mercury mixed with mercurous chloride. A hole in the bottom of the tube communicates with a saturated potassium chloride solution, and this in turn is contained in a relatively large glass chamber. At the bottom of this chamber one of various types of junctions is to permit the potassium chloride to diffuse or leak into the solution being measured. To complete the circuit, a wire is inserted into the packed column and goes to the amplifier along with the shielded cable from the glass electrode. With this arrangement the potential developed at the calomel–potassium chloride interface is constant at a given temperature since the solution surrounding it remains the same.

The silver–silver chloride electrode can be nearly identical to the calomel electrode. Instead of the packed column, only a silver wire need be used if coated with a heavy coating of silver chloride. The silver–silver chloride electrode finds broader acceptance since it remains stable and reproducible at much higher temperatures than the calomel electrode. The calomel electrode may be used for sterilizing temperatures up to 130°C (260°F).

Aside from the type of internal construction employed, the other important parameter of the reference electrode is its type of junction. Different junction styles have differing advantages and limitations. Various schemes are employed, such as an asbestos fiber embedded in the glass; the use of a glass bead with a temperature coefficient different from that of the glass body surrounding it to permit a controlled crack around the bead for liquid communication; a palladium wire embedded in a near molten glass body with, again, an annular space surrounding the wire resulting after the glass cools; and a ground glass sleeve mated to a ground glass portion of the body. Reference electrodes are now available fabricated from plastic and incorporate a polymeric, ceramic, or wood diffusion junction; these junctions have found successful application in the process industries.

Temperature Compensation

Examination of the Nernst equation reveals that the magnitude of the Nernst slope factor is a function of the temperature at the measurement electrodes. Most process pH measurement systems incorporate a temperature-compensating element in the process sample to normalize the temperature-dependent slope to standard conditions by appropriate modifications of the gain of the pH amplifier. Devices used for temperature compensation are generally suitably packaged resistors, thermistors, or platinum resistance thermometers. It is important to point out that the only function of conventional temperature compensation is to correct for the change in the Nernst slope with temperature—it does *not* correct for changes in the reference electrode or pH electrode absolute potential with temperature or for a change in the pH of a test solution due to temperature changes. The latter correction is provided in some newer microprocessor-based process pH analyzers as ajustable solution temperature compensation. It is used principally in high-purity water measurements where the temperature effect on pH is large.

Developments in Measurement Hardware—Microprocessors

Significant effort has been expended in the development of novel electrode packaging configurations. Combination electrodes, incorporating the pH and reference electrode together with the temperature compensation device in one package, suitable for threading into a pipe fitting for flowing measurement applications or submersion measurement applications, are available from many manufacturers. A rather useful development is the widely available "live" insertion probe, which allows a combination electrode to be inserted into and withdrawn from a valve fitting in a tank wall or process line, allowing periodic calibration or maintenance of the probe without shutting down the process.

Ultrasonic probe cleaners have been successful in many applications where electrode life has been limited as a result of coating problems. This technique seems to work best for coating with hard crystalline precipitates, although limited success has been reported in preventing the coating of electrodes with films of grease and oil by use of the ultrasonic cleaning technique. In use, the electrodes are exposed continuously or periodically to ultrasonic excitation from a transducer mounted in the solution close to the pH and reference electrodes.

pH analyzers incorporating internal timers and relay contacts can control the flow of a cleaning solution directed to the electrodes while holding alarm and control action until normal sample measurement resumes. An extension of self-cleaning is unattended calibration, where the internal timer and relays direct buffer solution to the electrodes.

Microprocessors in pH Measurement

The microprocessor chip has had a significant impact on the features available in contemporary pH analyzers. It is possible to use the computational capability of the chip to calculate parameters of interest to the end user, such as the actual pH electrode slope (millivolts per pH unit) obtained when it has been calibrated in buffer solutions; to program in corrective factors to normalize the temperature dependence of the pH of solutions due to changes in solution composition; and to have the ability to simulate pH changes by keying in inputs to the microprocessor to verify any pH-dependent function such as alarms, operation of actuators, and recorder or computer inputs. The accuracy of temperature compensation is improved by having the microprocessor solve the Nernst equations. pH controllers have become much more flexible and self-reliant than the prior analog designs. For example, in the case of one commercially available pH controller, it is possible to program into the instrument the titration curve characteristics of a particular process to specifically tune the instrument to the specific end application. The instrument also can accept a process flow signal to provide immediate correction for flow-rate changes. Proportional control output types include current, pulse duration, and pulse frequency for electronic metering pumps (Fig. 3).

Applications for pH Measurement

There are more continuous pH analyzers in use than all other continuous analytical-type instruments combined. Installations of pH analyzers throughout the world probably are in excess of 50,000. Significant numbers are used within closed-loop automatic control systems. Numerous pH applications can be found in the pulp and paper industry, metals and metal treating fields, petroleum refining, synthetic rubber manufacturing, power generation plants, pharmaceuticals, chemical fertilizer production, and a broad spectrum of the chemical industry (Fig. 4).

When properly applied, pH measurements offer the following benefits: (1) Many chemical reactions can be controlled better, resulting in greater processing efficiency and better quality of the ultimate product, and sometimes contributing to a safer process. (2) Records are provided to assure management that optimum processing operations prevail. In the case of industrial waste discharge into public streams, lakes, or collection systems, the recorded measurement provides evidence of compliance from a regulatory standpoint. (3) In certain cases, continuous pH measurement has permitted certain processes to be converted from a batch to a continuous basis, such as certain chemical fertilizer processes and some food manufacturing processes.

REDOX POTENTIAL MEASUREMENTS

All chemical reactions involving an exchange of electrons are considered oxidation-reduction reactions. These reactions produce measurable and predictable potentials. The relative strengths of the oxidants and reductants involved can be measured by determining the oxidation-reduction potential (ORP) prevailing. This can be done by inserting an unattackable electrode, such as platinum, rhodium, or gold, into the solution together with a reference electrode and measuring the resulting EMF by means of a high-impedance potentiometer or amplifier. The EMF or voltage

FIGURE 3 Microprocessor-based pH analyzer with capabilities for automatic electrode cleaning and calibration and proportional control with pulse frequency, pulse duration, or current output types. (*Leeds & Northrup.*)

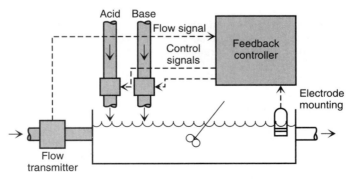

FIGURE 4 Continuous neutralization using three-mode PID (proportional-plus-integral-plus-derivative) feedback pH control and flow compensation algorithms in microprocessor-based pH analyzer-controller. (*Leeds & Northrup.*)

measured is the difference between the individual voltages developed at each electrode. If a hydrogen gas reference electrode could be used, the potential at the measuring electrode would be the true ORP of the solution, since the hydrogen gas reference electrode potential, by convention, is zero. Generally, a silver–silver chloride or calomel reference electrode is used instead of hydrogen, and the potential developed by either of these electrodes requires a correction in the ORP meter reading to relate the measuring electrode potential to the hydrogen zero potential.

To understand and properly relate the ORP values and equations it is necessary to appreciate the rather confusing sign conventions prevailing. There are two existing conventions, and they differ only by being opposite in polarity. The American convention expresses the potential with the polarity as it truly exists in the solution surrounding the noble measuring electrode. The European convention concerns itself with the opposite potential which exists on the measuring electrode. Instrument makers and many users are concerned only with the sign of the electrode since the meter is only aware of this polarity. Thus most ORP meters and recorders read out in signs followed by the European convention.

Specific oxidation-reduction values for a wide variety of reactions involving an exchange of electrons can be found in handbooks and in the literature. The values, in volts, typically are referenced to the hydrogen–hydrogen ion couple (having zero potential) at unit activity and at 25°C. These values (for the reaction being measured) can be inserted into the general form of the Nernst equation:

$$E = E_0 - \frac{0.0591}{n} \log \left(\frac{\text{oxidant}}{\text{reductant}} \right)$$

where E = measured EMF expressed in volts opposed to normal hydrogen electrode (zero potential by definition)

E_0 = standard EMF as found in literature for oxidation-reduction reaction involved under conditions specified

n = number of electrons involved in oxidation-reduction reaction

From the foregoing relationship the expected potential can be predicted, depending on the ratio between oxidant and reductant. Conversely, by reading the measured potential, the prevailing ratio of oxidant to reductant can be determined.

Limitations of Methods

Many types of chemical reactions and processes can be controlled by ORP measurement, but with the following qualifications:

1. The measuring electrode senses oxidants and reductants in a solution. For example, hexavalent chromate may be the component requiring measurement. However, other oxidants and reductants, such as iron salts and sulfides, may be present, and these can influence the net potential.

2. In many systems, variations in the pH of the solution cause variations in the net potential of the solution, since many oxidation-reduction couples involve the hydrogen ion.

3. Temperature influences the potential slightly. The necessary correction usually is less than 1 mV/°C, but cannot be predicted theoretically for most systems since mixed oxidation-reduction reactions may be present.

Calibration

If the electrodes and the amplification stage are performing properly, the measured potential is a true measurement not requiring calibration. On the other hand, the measuring electrode can become coated (and thereby insensitive) or can be "poisoned" by prior exposure to very strong oxidants or reductants and still retain a "memory" of this exposure. Thus some users find it opportune to calibrate the system periodically. This can be done by placing the electrodes first in a pH 4 buffer solution containing a few grams of quinhydrone and then placing them in a pH 7 buffer solution also containing quinhydrone. The solutions should be prepared freshly just before use. Table 2 lists the millivolt values that can be expected from such solutions at several temperatures and using various reference electrodes. The values are all plus and follow the European convention.

TABLE 2   ORP of Quinhydrone Solutions at Various Temperatures

	pH 4			pH 7		
Electrode	20°C	25°C	30°C	20°C	25°C	30°C
Calomel (mercury–mercurous chloride)	223	218	213	47	41	34
Silver–silver chloride	268	263	258	92	86	79
Hydrogen	470	462	454	295	285	275

Applications

Water treatment probably uses ORP measurement to a greater extent than any other group of processes. ORP is relied on almost universally both to monitor and to control the cyanide oxidation and chromate reduction processes so frequently required in industrial waste treatment of metal-treating wastes before discharge into sewers or streams. Regulatory agencies often monitor freshwater streams with ORP analyzers to determine the relative "health" of receiving waters and to detect unwanted dumping of strong oxidants or reductants upstream. Some sewage plants use ORP to monitor sewage influent, digester sludge, and, to a limited extent, to control chlorination for in-plant odor control. ORP is used to control the flow of chlorine or other oxidizing agents in various bleaching processes in the pulp and paper industry. A wide variety of applications exist throughout the chemical industry.

Measurement of Specific Ions

Another type of ORP measurement involves the use of a metal–metal salt as the measuring electrode instead of a noble metal electrode. As an example, a silver–silver chloride measuring electrode can measure chloride concentrations ranging from a few to over 10,000 ppm. This essentially is an oxidation-reduction potential measurement. Thus other oxidants and reductants in the solution being measured can, if in sufficient concentration, introduce a secondary potential, although most applications are not concerned with this aspect. Applications include the monitoring of streams for chlorides as an index of water quality, for salt intrusion detection, and to detect dumping of wastes of high chloride content. Other metal–metal salt combinations, such as silver–silver sulfide for measuring sulfide levels in aqueous solutions, have been used as process analyzers.

Selective ion electrodes have found wide application in laboratory determinations of many species of industrial significance. Continuous process analyzers incorporating the electrodes have been successfully applied in industrially important analyses, but significant sample pretreatment or preconditioning to afford useful operation is required.

Selective ion electrodes measure the activity of anionic species in solution by developing a potential which is related to the activity of the ion of interest and may be determined experimentally by measuring the voltage difference between the selective ion electrode and an appropriate reference electrode. The Nernst relationship expresses the potential of the selective ion electrode as a function of the ionic activity of interest:

$$E = E_0 - \frac{0.0591}{n}(\log a_{ion})$$

where E = measured EMF expressed in volts compared to a normal hydrogen electrode
         (zero potential by definition)
       E_0 = standard EMF for oxidation-reduction reaction
       n = number of electrons involved in reaction

Selective ion electrodes are usually calibrated against known standards to account for behavioral anomalies and to ensure that the system is functional for a given determination. Commercially available selective ion electrodes are classified by species in Table 3, in which the detection limit is given in parentheses after the species.

TABLE 3   Commercially Available Selective Ion Electrodes*

Glass	Solid state	Liquid membrane	Gas sensing
H^+ (10^{-14} M)	Br^- (5×10^{-6} M)	Cu^{2+} (10^{-5} M)	NH_3 (10^{-6} M)
Na^+ (10^{-7} M)	Cd^{2+} (10^{-7} M)	Cl^- (10^{-5} M)	CO_2 (10^{-7} M)
K^+ (5×10^{-5} M)	Cl^- (10^{-5} M)	NO_3^- (10^{-5} M)	NO_2 (2×10^{-5} M)
NH_4^+ (10^{-5} M)	Cu^{2+} (10^{-7} M)	ClO_4^- (10^{-5} M)	SO_2 (10^{-7} M)
Ag^+ (10^{-2} M)	Cr^- (10^{-6} M)	K^+ (10^{-5} M)	
	F^- (10^{-7} M)	$Mg^{2+} + Ca^{2+}$ (10^{-10} M)	
	I^- (1.5×10^{-10} M)		
	Pb^{2+} (10^{-7} M)		
	Ag^+ (10^{-14} M)		
	CNS^- (10^{-5} M)		

* Detection limit is given in parentheses after species.

The reader is referred to publications for information on selective ion electrode application and theory. These electrodes have found widespread use in laboratory determinations but have been less frequently used for continuous process analysis because of the problems of short life of liquid ion-exchanger-based selective ion electrodes, the problems of cross-interferences which are more difficult to deal with in the continuous analysis setting than the laboratory setting, and the lack of long-term demonstrated electrode stability.

ELECTRICAL CONDUCTIVITY MEASUREMENTS

By Fred Kohlmann[1]

All aqueous solutions conduct electricity to some degree. The measure of a solution's ability to conduct electricity is called conductance and is the reciprocal of resistivity (resistance).

[1] Great Lakes Instruments, Inc., Milwaukee, Wisconsin.

Adding electrolytes, such as salts, acids, or bases, to pure water increases conductivity (decreases resistivity).

TEMPERATURE COMPENSATION

Accurate conductivity measurement depends on temperature compensation. Since temperature coefficients of common solutions vary from 1 to 3 percent, the conductivity measurement system should include automatic temperature compensation to adjust conductivity readings to a 25°C reference (Fig. 1).

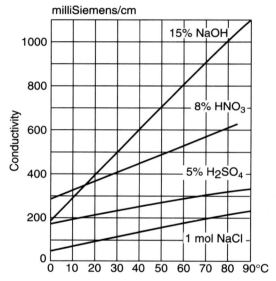

FIGURE 1 Electrical conductivity changes with temperature for several solutions. (*Great Lakes Instruments, Inc.*)

Manual temperature compensators simply are calibrated dials, usually located on the front of the instrument. These dials are set by the operator at the time the conductivity measurement is taken. They must be set to the known solution temperature. Manual temperature compensation is usually suitable for applications where only occasional readings are taken, or where little temperature variation exists. Automatic temperature compensation is preferred to manual compensation since once the instrument is in operation, it cannot be set incorrectly due to negligence or poor operating practice.

The temperature coefficients of solutions are nonlinear. They vary, depending on the actual conductivity of the solution. The latter is related to concentration, as shown in Fig. 2. For greatest accuracy, instruments should be calibrated at the actual measuring temperature.

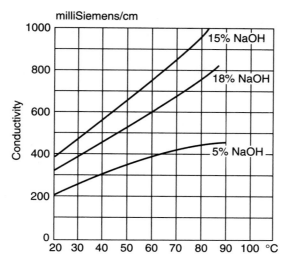

FIGURE 2   Electrical conductivity-temperature relationships as they vary with concentration. (*Great Lakes Instruments, Inc.*)

UNITS OF MEASURE

Historically, the standard unit of conductivity measurement has been mhos per centimeter (Ω^{-1}/cm) (mho is the reciprocal of ohm). A resistivity of 100 Ω · cm is equivalent to a conductivity of 1/100 Ω^{-1} · cm. This unit of measurement is being replaced throughout industry by an equal and interchangeable unit of measurement, called the Siemen per centimeter (S/cm). Conductivity usually is expressed in millionths of a Siemens, that is, microSiemens per centimeter (μS/cm). Resistivity still is expressed as megohm-centimeters for high-purity water (usually from 0.1 to 20 Ω · cm). A milliSiemens (mS) is used to represent 1/1000 of a Siemen. To convert milliSiemens to microSiemens, one multiplies by 1000; for example, 0.5 mS equals 500 μS. Since these two terms are similar, care must be taken to avoid confusing them or their abbreviations.

For a given solution, microSiemens can be related to parts per million (ppm), which is a measure of concentration. In general, 1 μS = 1.5 × TDS (total dissolved solids) ppm. Depending on solution concentration and composition, the 1.5 factor may change. If TDS in a solution is expressed in terms of sodium chloride, the microSiemens value is approximately twice the TDS (as NaCl) value. Table 1 illustrates this relationship.

TABLE 1   Relationship between Total Dissolved Solids (TDS) and Electrical Conductivity as Compared with Sodium Chloride (NaCl) Concentration

TDS, ppm	Electrical conductivity, μS/cm	NaCl, ppm
10,000	15,000	8,400
6,600	10,000	5,500
5,000	7,500	4,000
4,000	6,000	3,200
3,000	4,500	2,350

2,000	3,000	1,550
1,000	1,500	750
750	1,125	560
666	1,000	490
500	750	365
400	600	285
250	375	175
100	150	71
66	100	47
50	75	35
40	60	28
25	37.5	17.5
6.6	10	4.7

RESISTIVITY

In high-purity water, typically less than 1 µS/cm, the measurement is referred to as resistivity with units of megohm-centimeters (MΩ · cm). Pure water has a resistivity of about 18.2 MΩ · cm at 25°C. One consideration that must be made when measuring solutions is the temperature coefficient of the conductivity of the water itself. To compensate accurately, a second temperature sensor and a compensation network are used. Specific sensors and analyzers are recommended for measurement in high-purity water.

METHODS OF CONDUCTIVITY MEASUREMENT

The conductivity of a solution can be measured via two methods: (1) the contacting conductivity system and (2) the electrodeless conductivity system.

Contacting Conductivity System.   The contacting-type sensor usually consists of two electrodes, insulated from each other. The electrodes, typically of 316 stainless steel, titanium-palladium alloy, or graphite, are specifically sized and spaced to provide a known "cell constant." Theoretically a cell constant of 1.0 implies two electrodes, each 1 cm^2 in area, spaced 1 cm apart (Fig. 3).

Cell constants must be matched to the analyzer for a given range of operation. For instance, if a sensor with a cell constant of 1.0 were used in pure water with a conductivity of 1 µS/cm, the cell would have a resistance of 1 MΩ. Conversely, the same cell constant in seawater may have a resistance of 30 Ω. Since the resistances are so different, it is difficult for ordinary instruments to measure these extremes with only one cell constant.

In measuring the 1-µS/cm solution, the cell would be configured with large electrodes spaced a small distance apart. This results in a cell resistance of approximately 10 kΩ, which can be measured quite accurately. By using cells of different constants, the measuring instrument can operate over the same range of cell resistance for both ultrapure water and high-conductivity seawater.

Electron Ion Transfer.   When an alternating voltage is applied between the electrodes of an electrolytic conductivity sensor, electrons are transferred and gases are released. This electron transfer carries electrical current from one electrode to the other. As gases are released, they build up on the electrode. This buildup impairs the electrode's contact with the solution and creates a back electromotive force (EMF). This effect is called polarization. It can be

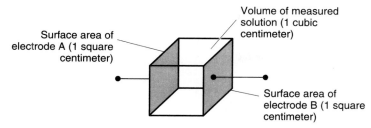

FIGURE 3   Graphic definition of theoretical cell constant of 1.0. (*Great Lakes Instruments, Inc.*)

avoided by impressing a low ac voltage on the electrodes rather than dc voltage. The duration of the ac cycle is too short and the voltage too low to cause an electrode buildup, thereby reducing polarization and associated problems.

Electrodeless Conductivity Measurement.   Conductivity measurements often must be made in solutions which will coat, foul, or plate the surface of conventional conductivity sensors. When measuring solutions over 10,000 μS/cm with electrode-type sensors, large cell constants must be used. These sensors have small electrode surface areas and, consequently, are very susceptible to fouling and polarization, which cause inaccurate measurements. Electrodeless conductivity sensors were developed to solve these problems.

OPERATION OF ELECTRODELESS SENSORS

The electrodeless sensor operates by inducing an alternating current in a closed loop of solution and measuring the magnitude of this current to determine the conductivity of the solution. As shown in Fig. 4, the conductivity analyzer's oscillator drives coil 1 (transmitter). Coil 1 then induces an alternating current in the solution to be measured. This ac signal flows in a closed loop through the sensor bore and surrounding solution. Coil 2 (receiver) senses the magnitude of the induced current, which is measured by the analyzer's electronics and processed to display the corresponding reading.

Electrodeless sensors eliminate common problems associated with contacting-type sensors which use graphite or metal electrodes. Thus oil, fouling process coating, or nonconducting electrochemical plating are no longer concerns.

APPLICATIONS FOR ELECTRODELESS CONDUCTIVITY SENSORS

Typical applications for electrodeless conductivity sensors include measuring concentrations of acid, caustic, or salt solutions. In Fig. 5 note the curve for hydrochloric acid (HCl). Both 9 percent and 34 percent concentrations of HCl are approximately 600,000 μS/cm. The conductivity of HCl increases as concentration increases up to approximately 19 percent. Then the conductivity decreases. Since two different HCl concentrations can have the same conductivity value, the measuring range must be restricted so that it does not cross the 19 percent concentration point. Wherever possible, the instrument range should be confined to a linear portion of the concentration curve. Otherwise nonlinear scales are required. When measuring solutions which have the same conductivity value for more than one concentration, the instrument can only be used to measure that specific portion of the curve between given minimum and maximum values. When working with a solution for the first time, the analyst must

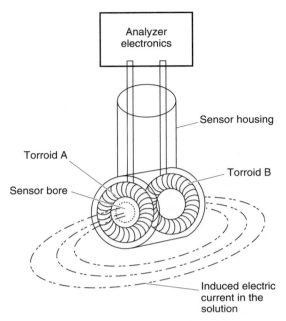

FIGURE 4 Operating principle of electrodeless sensor. (*Great Lakes Instruments, Inc.*)

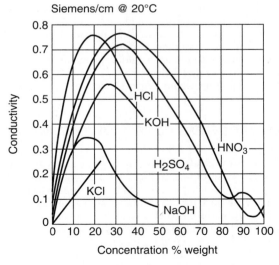

FIGURE 5 Electrical conductivity versus concentration for representative solutions. (*Great Lakes Instruments, Inc.*)

become fully aware of the conductivity "behavior" of the solution, or make inquiries to the instrument manufacturer. Representative applications for electrodeless conductivity systems are listed in Table 2.

TABLE 2 Electrodeless Conductivity System Applications

Types of liquids measured:
  Acid concentrations
  Drilling muds
  Standard aqueous solutions above 10,000 μS/cm
  Fouling, clogging, or coating materials that clog conventional sensors; electrodeless sensors have an
    opening that reduces or eliminates clogging and minimizes maintenance
  Highly corrosive solutions, electrodeless sensors typically have one material wetted by the process to
    simplify chemical resistance problems
  Mining fluids
Representative applications:
  Desalinization plants (raw water inflow)
  Dilution of caustics
  Interface determination between organic solvents and water-based solutions
  Pickling baths
  Plating baths
  Scrubbers
  Seawater contamination
  Water purity control

THERMAL CONDUCTIVITY GAS ANALYZERS[1]

For certain gases and gas mixtures, thermal conductivity has been the analytical method of choice for many years. The method is based on the fact that various gases differ considerably in their ability to conduct heat. Although this approach was first suggested as early as 1880 by Leon Somzec, it was not until 1908 that Koepsal developed a practical instrument for the continuous indication of hydrogen content in producer gas. Koepsal made use of the hot-wire method for comparing the thermal conductivities of two gases. Since that time, numerous adaptations have been made of the original hot-wire system.

   The thermal conductivity of a gas mixture depends on the nature and concentration of the constituents. Dependence of thermal conductivity on gas composition is treated extensively in the literature (Bennett and Vines; Daynes; Palmer and Weaver; Vines), but in general it is not profitable to make elaborate calculations of thermal conductivity. Furthermore, the available

---

[1] Portions of this article are from the original prepared by R. H. Cherry and updated by the handbook staff with the cooperation of David M. Gray, Senior Application Specialist, Leeds & Northrup (A Unit of General Signal), North Wales, Pennsylvania.

data on the thermal conductivity of gases generally cannot be relied on to an accuracy of better than ±5 percent. In gas analysis the only practical reason for computing the thermal conductivity of a gas mixture is to obtain a rough estimate of probable sensitivity over a limited range of composition. For this purpose, the following simple linear relation usually is adequate:

$$K_1, K_2, \ldots, K_n = (K_1 p_1 + K_2 p_2 + \cdots + K_n p_n)/100$$

where $K_1, K_2, \ldots, K_n$ = thermal conductivity values of each constituent of mixture
$p_1, p_2, \cdots, p_n$ = respective concentrations of constituents, vol %

Thermal conductivity data for a few gases of interest in industrial gas analysis are given in Table 1. Among the most common gases and mixtures measured are the following: argon in nitrogen or oxygen; carbon dioxide in air, flue gases, or nitrogen; Freon in air; helium in air or nitrogen; hydrogen in air, blast furnace top gases, carbon monoxide, carbon dioxide, nitrogen, argon, or hydrocarbons, such as reformer gases; methane in air; oxygen in argon, nitrogen, or hydrogen; and propane in air.

TABLE 1   Thermal Conductivities of Gases

Gas	K_{gas}* 0°C	K_{gas}* 100°C	K_{gas}/K_{air} 0°C	K_{gas}/K_{air} 100°C
Air	2.23	2.854	1.000	1.000
Acetone	0.906	1.558	0.406	0.546
Ammonia	2.00	3.10	0.897	1.086
Argon	1.58	2.07	0.709	0.725
Carbon dioxide	1.37	2.069	0.614	0.725
Carbon monoxide	2.15	—	0.964	—
Chlorine	0.718	—	0.322	—
Ethane	1.80	3.204	0.807	1.123
Ethylene	1.64	2.624	0.735	0.919
Ethyl alcohol	1.11	1.96	0.498	0.687
Helium	13.9	16.68	6.233	5.844
Hydrogen	15.9	20.03	7.130	7.018
Hydrogen sulfide	1.20	—	0.538	—
Methane	2.94	—	1.318	—
Methyl alcohol	1.32	2.033	0.592	0.712
Neon	4.44	5.44	1.99	1.91
Nitrogen	2.28	2.896	1.022	1.015
Nitric oxide	2.08	—	0.933	—
Nitrous oxide	1.44	2.09	0.646	0.732
Oxygen	2.33	3.006	1.045	1.053
Sulfur dioxide	0.768	—	0.344	—
Water vapor	—	2.17	—	0.760

* K is given in units of kiloergs per second per square centimeter per centimeter per degree Celsius. Multiplying the value by 41.833 the units of calories per second per square centimeter per centimeter per degree Celsius are obtained.

MEASUREMENT METHODS

A thermal conductivity gas analysis cell usually consists of an electrically conductive, elongated sensing element, mounted coaxially within a cylindrical chamber containing the gas.

The element in the cell is maintained at an elevated temperature with respect to the cell walls by passing an electric current through it. An equilibrium temperature is attained by the element when its electric power input is equalized by all thermal losses from the wire. When material having a suitable temperature coefficient of resistance is used, the sensing element can serve a dual function of heat source and sensor of equilibrium temperature. The magnitude of the temperature difference between the element and the cell walls (that is, the temperature rise of the element) at equilibrium is a function of the electric power input and the combined rate of heat loss from the wire by gaseous conduction, convection, radiation, and conduction through the solid supports of the element.

By proper design and cell geometry and by limiting the temperature rise of the heated element, it is possible to maximize heat loss due to gaseous conduction. Under these circumstances, the temperature rise of the element at constant electric power input is inversely related to the thermal conductivity of the gas confined within the cell.

A Wheatstone bridge is the usual form of network used to measure the resistance change of the sensing element. The current used to energize the bridge also serves to heat the wire. It is impractical to employ a single hot-wire cell in a bridge measurement because of the extreme sensitivity of such a measurement to changes in bridge supply voltage and ambient temperature. Instead, it is common practice to use two cells in adjacent arms of a Wheatstone bridge, one of the cells containing a reference gas and the other containing the gas to be analyzed. The bridge then responds to the difference in temperature rise (that is, the resistance) of the two sensing elements and thus is dependent on the difference in the thermal conductivities of the gases in the two cells.

A wide variety of thermal conductivity gas analyzers is available commercially (Fig. 1). These devices embody a variety of cell types, geometries, and Wheatstone bridge configurations. For example, the measuring cell may receive sample gas by diffusion, convection, or a combination of these. In some instances, as in gas chromatography, the cell cavity may be a part of the gas channel, with all the sample gas passing through the cavity. This minimizes response time but requires the capability to establish a relatively low, but reproducible and constant, gas flow rate to maintain calibration integrity and minimize signal noise.

Various thermal conductivity measuring cell configurations are used. There are (1) the diffusion exchange type, (2) the convection exchange type, and (3) the direct-flow type (Fig. 2). There are also single-pass and double-pass analyzers (Fig. 3).

APPLICATIONS AND LIMITATIONS

Thermal conductivity gas analysis is a nonspecific, nonabsolute method which depends on empirical calibration. Nevertheless, the simplicity, reliability, relative speed, and easy adaptation to continuous recording and control have made this method one of the most widely used means of industrial gas analysis. As a laboratory tool, the method has been used extensively in conjunction with analytical techniques, such as gas chromatography, where the sample gas is essentially a binary mixture for which the thermal conductivity method is ideally suited. For quantitative analysis of a binary gas mixture, provided the full-scale change in thermal conductivity is not less than about 2 percent, it is possible to obtain a useful sensitivity of better than 1 percent of full-scale and comparable zero stability. The approximate practical limits of this method, as applied to several binary mixtures, are given in Table 2. In each case the range shown is equal to a 2 percent change in thermal conductivity. The tabulated values are accurate only to approximately 10 percent because of inherent limitations in available thermal conductivity data and the assumptions made with respect to operating conditions.

In some cases the variation in the thermal conductivity of binary mixtures does not follow the simple linear law. Ammonia in air and water vapor in air are examples. Information pertaining to the application of thermal conductivity gas analysis for the determination of water

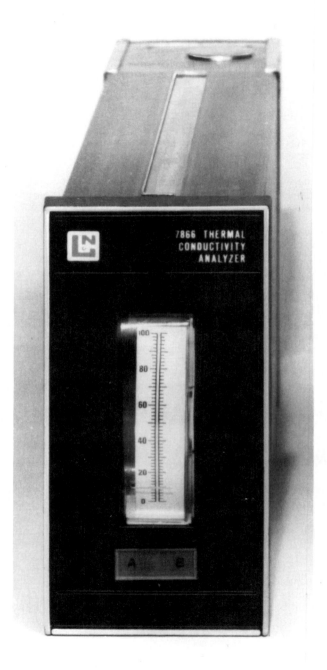

FIGURE 1 Thermal-conductivity-based control unit for monitoring hydrogen-cooled power generators. (*Leeds & Northrup.*)

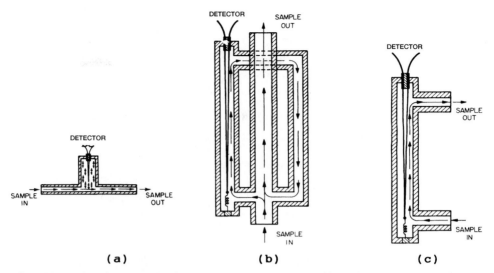

FIGURE 2   Various thermal conductivity analyzer cell configurations. (*a*) Diffusion exchange type. (*b*) Convection exchange type. (*c*) Direct-flow type.

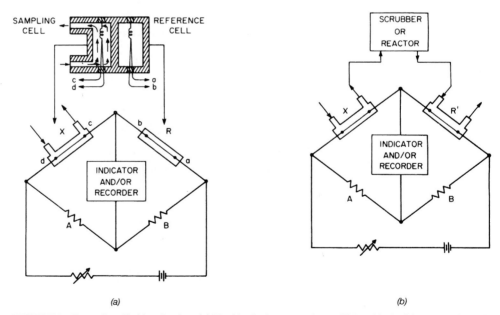

FIGURE 3   Examples of bridge circuitry. (*a*) Used in single-pass analyzer. (*b*) Used in double-pass analyzer. In a single-pass gas analyzer, the sample gas affects the resistance of only one arm of the bridge and is compared with a reference cell containing gas of constant composition in the adjacent bridge arm. In a double-pass gas analyzer, the sample gas influences the cell in one arm of the bridge and then is subjected to modification and finally passed through the cell in the adjacent arm. In this case the unmodified sample gas usually is considered the reference to which the modified sample gas is compared. Bridge circuitry is designed to compensate for small variations in bridge supply voltage through the use of resistors.

TABLE 2 Approximate Practical Range of Thermal Conductivity Method Applied to Analysis of Certain Binary Gas Mixtures

Gas mixture	Approximate practical full-scale range
Air*–carbon dioxide	0–5.3% air in CO_2
	0–7.3% CO_2 in air
Air–sulfur dioxide	0–1% air in SO_2
	0–3% SO_2 in air
Air–helium	0–2.4% air in He
	0–0.4% He in air
Nitrogen–carbon dioxide	0–5% N_2 in CO_2
	0–7% CO_2 in N_2
Nitrogen–hydrogen	0–2.3% N_2 in H_2
	0–0.3% H_2 in N_2
Nitrogen–argon	0–5% N_2 in A
	0–7% A in N_2
Carbon dioxide–oxygen	0–6.4% CO_2 in O_2
	0–4.4% O_2 in CO_2
Hydrogen–helium	0–10% H_2 in He
	0–12% He in H_2

* Where air appears in the table, the air may be regarded as a pure gas because of its substantially fixed composition.

vapor in air can be found in [1]–[3]. Information pertaining to other binary mixtures exhibiting thermal conductivity maxima and nonlinearities can be found in [4]–[6].

Complex Gas Mixtures

When a gas mixture contains three or more constituents, all of which are subject to independent variation, it is necessary to evaluate the individual effect of each constituent to determine the feasibility of applying the thermal conductivity method. There are many important industrial gas analysis problems involving such complex gas mixtures. In some of these it is known that variations in the concentration of individual constituents are interdependent. In such cases the problem may be equivalent to that of a simple binary mixture, although the matter of proper calibration may be more complex.

Selection Guidelines

Several commercial and a few special applications for thermal conductivity analysis are given in Tables 3 through 5. Table 3 lists applications for which a single-pass instrument is normally used and indicates which comparison gases are suitable. Table 4 presents applications involving pretreatment of the gas sample, followed by a single-pass measurement against a comparison gas. Table 5 lists applications usually accomplished by double-pass measurement and indicates the nature of treatment of the gas sample between passes.

A widely used application of thermal conductivity analyzers is monitoring the purity of hydrogen used for cooling power plant generator windings. Efficient and safe operation is assured by controlling the hydrogen content to greater than 95 percent in air concentration. In addition, during start-up and shutdown, the hydrogen is purged with carbon dioxide to preclude any explosive mixture from developing. Thus it is also desirable to be able to monitor carbon dioxide in air before start-up and carbon dioxide in hydrogen prior to shutdown. An analyzer that can provide all three ranges is shown in Fig. 4.

TABLE 3 Typical Applications of Thermal Conductivity
Analysis of Binary Mixtures (or Equivalent) by Direct
Comparison in Single-Pass Instrument

Gas mixture	Suitable comparison gas
H_2 in O_2	O_2, air, or H_2
H_2 in Cl_2	H_2 or Cl_2
H_2 in N_2	H_2, N_2, or air
H_2 in air	H_2 or air
H_2 in CH_4	H_2, CH_4, or $H_2 + CH_4$
H_2 in CO_2	H_2, CO_2, or $H_2 + CO_2$
H_2 in water gas ($H_2 + CO$)	H_2 or $H_2 + N_2$
He in air, N_2, or O_2	He, air, H_2 or O_2
Ne in air	Air
Cl_2 in air	Air
HCl in air	Air
Acetone in air	Air
NH_3 in air	Air
O_2 in enriched air	Air
SO_2 in air or N_2	Air or N_2
Water vapor in air, N_2, or O_2	Air, N_2, or O_2
CO_2 in air, N_2, or flue gas	Air
A in N_2, air, or O_2	N_2, air, or O_2

TABLE 4 Typical Applications of Thermal Conductivity Analysis Involving
Pretreatment of Sample Gas Followed by Direct Comparison in Single-Pass
Measurement

Gas mixture	Pretreatment (combustion)	Comparison gas
Acetone in air	$CH_3COCH_3 + 4O_2 \rightarrow 3CO_2 + 3H_2O$	Air or N_2
Benzol in air	$2C_6H_6 + 15O_2 \rightarrow 12CO_2 + 6H_2O$	Air or N_2

TABLE 5 Typical Applications of Thermal Conductivity Analysis by Double-Pass Measurement

Gas mixture	Treatment between passes
NH_3 in N_2 and H_2	Absorb NH_3
O_2 in N_2, or O_2 in flue gas	Convert O_2 to H_2O
	Add H_2 before first pass
CO in flue gas or air	Convert CO to CO_2
Water vapor in variable gas mixture	Absorb H_2O
O_2 in H_2 and CO	Convert O_2 to H_2O (under controlled conditions of combustion)

* Selection of comparison gas depends on range of composition to be covered and ability of reference cell to retain
highly diffusive gases, such as hydrogen, or to withstand corrosive action of such gases as chlorine.

Practical Considerations for Continuous Gas Measurement

For effective continuous operation, the thermal conductivity analyzer requires the following.

Conditioning of Gas Sample.   Water vapor is the most common variable component found
in industrial gases. Unless the purpose is to analyze for water vapor concentration, it is necessary either to remove the water vapor from the gas stream or in some other way to make its

FIGURE 4   Explosion-proof thermal conductivity sensor unit located on sampling-calibrating panel for monitoring hydrogen-cooled power generators. Indicating portion of this system is shown in Fig. 1. (1) Sample gas in; (2) sample gas regulator; (3) span gas in; (4) zero gas in; (5) hydrogen flow; (6) carbon dioxide, air flow; (7) sensor unit; (8) reference air in; (9) reference air regulator; (10) reference air flow; (11) bypass valve; (12) bypass flow; (13) bypass vent; (14) sample vent. (*Leeds & Northrup.*)

thermal conductivity effect negligible. In a few applications, drying the gas sample is desirable to avoid or minimize corrosion, but generally it is most convenient to fix the concentration of water vapor and eliminate it as a variable. This can be done by saturating the sample gas at a constant temperature. Several practical gas saturators requiring very little maintenance are now standard or optional equipment with commercially available analyzers. The use of saturated gas presupposes that the analyzer cells and associated parts exposed to the gas will withstand the more corrosive atmosphere caused by saturation. When the sample gas must be dried, there is a decided maintenance advantage in selecting an analyzer that will operate satisfactorily with a minimum flow rate of the sample.

Gas Sampling and Cleaning. Provision of an adequate and representative sample is mandatory. Proper location of the sampling probe or the use of multiple sampling probes must be considered in relation to the process being monitored and the desired results. This is especially important when automatic control of gas composition is involved. It is always desirable to keep sampling lines as short as possible. Particular care should be exercised to avoid low points where condensed vapors can collect and trap the line. The sampling line should have a minimum number of elbows, have a smooth inner surface, and be of a material that can withstand the thermal and corrosive action to which it will be subjected. Plastic tubing, when it meets the requirements, is the most convenient form of sampling line. Heavy-wall glass tubing and corrosion-resistant metal pipes frequently are used.

Gases from industrial processes are rarely free of suspended solids, smoke, and other particulates. The nature of the process and the type of particulate contaminants will determine the necessary measures for obtaining a satisfactory gas sample for analysis. Alundum and similar porous ceramic materials are commonly used to remove suspended solids; but these are of little value in dealing with smoke, fly ash, mists, and volatile materials such as tars, which subsequently condense in the sampling line. Water spray washers, impingers, and steam injectors are used with some success, particularly when high sampling rates are necessary.

At the high sampling rates desired in automatic control applications to reduce response lag, it is very difficult to clean the entire gas sample. Instrument maintenance, however, can be minimized by the use of secondary filters. This is frequently done by using one of the primary means mentioned for at least partially cleaning the main sample stream. A portion of the main sample is then bypassed to the analyzer cells through a secondary filter usually located in the analyzer housing. By locating the secondary filter in the analyzer housing it is possible to avoid unfavorable temperature gradients that might cause condensation of volatile contaminants. Secondary filtering materials that have proved useful include filter paper, washed long-fibered asbestos, porous sintered metal plates, and a variety of paperlike materials containing fibers of one or more types such as cellulose, glass, quartz, and asbestos. Asbestos, of course, must be considered a hazardous material.

Maintaining Sample Purity. Assuming that a representative sample of gas is obtained at the primary sampling point, it is also necessary to avoid significant changes in the composition of the sample as it passes through the sampling, cleaning, and conditioning systems. This may impose necessary limitations on the type of cleaning operation feasible, and it may be necessary to accept additional maintenance in order to deliver a chemically significant sample for analysis. Perhaps the most frequent cause of delivery of a contaminated or diluted sample is air infiltration. Since most samples are obtained by applying suction at some point in the sampling system, there is always a danger of faulty lines admitting air to the gas stream. Even systems operating at static pressures higher than atmospheric are not immune to this difficulty. It is possible, at high velocities of gas flow in such a system, to have air infiltrate by aspirator action at restricted points or turns in the sampling line. Regular inspection and maintenance of sampling lines and periodic leak tests are the best assurance of satisfactory operation.

REFERENCES

1. Cherry, R. H.: "Determination of Water Vapor by Thermal Conductivity Methods," *Anal. Chem.,* vol. 20, p. 958, 1948.
2. Cherry, R. H.: U.S. Patent 2,501,377, Mar. 21, 1950.
3. Cherry, R. H.: in *Proc. 1963 International Symp. on Humidity and Moisture,* Reinhold, New York, 1965, p. 539.
4. Angerhofer, A. W., and B. M. Dewey: *Instruments,* vol. 26, pp. 580–583, 1953.

5. Bennett, L. A., and R. G. Vines: "Thermal Conductivities of Organic Vapor Mixtures," *J. Chem. Phys.,* vol. 23, p. 1587, 1955.

6. Vines, R. G.: "The Thermal Conductivity of Organic Vapors," *Aust. J. Chem.,* vol. 6, p. 1, 1953.

PROCESS CHROMATOGRAPHY

By Jimmy G. Converse[1]

As early as 1903, the Russian botanist Mikhail Tsvet, in an effort to separate plant pigments, filled a vertical glass with an absorbent, and as the sample of the pigments was washed through the tube with a solvent, a series of colored absorption bands occurred. In essence, Tsvet had a colorgraphic display of the results. This separating process was initially called *chromatography* for the literal meaning, "color writing." The word chromatography continues in wide use today, even though its derivation no longer can be justified. However, the term is widely used in the literature and is not likely to change. A better term probably would be *chronography* for "time graph" because, as shown later, the components eluting from a column are monitored with respect to time (not color).

FUNDAMENTALS OF CHROMATOGRAPHY

Chromatography is a physical-chemical method of separating the various components of a mixture into pure fractions or bands of each component. The carrier or moving phase may be a gas, liquid, or supercritical fluid. The separation is effected by distributing the mixture between the fixed, or stationary, phase in a column and the carrier. The stationary phase may be a solid or a liquid-coated solid packed into the column, or it may be attached to the walls of a capillary. Liquid samples that can be vaporized can be separated with a gas carrier. High boiling and unstable compounds can be separated with a liquid carrier. Some materials can be handled better in a supercritical fluid and may separate faster. This latter approach, however, requires much more skill, is expensive, and thus, at present, is limited.

Components of the sample are retained in the column for different lengths of time due to adsorption-desorption, solution-dissolution, chemical affinity, size exclusion, and other mechanisms of a varying nature. Various components are continually washed from one part of the stationary phase and recaptured by another by the moving phase. Different components elute in groups from the column with respect to time from injection. Dispersion in the system causes the bands of components to emerge with a gaussian distribution or a distorted peak-shaped curve. A simplified diagram of the process is shown in Fig. 1.

All of what has been described up to this point relates *not* to analysis, but rather to separation and the preparation of a reliable sample. No matter how complex or sophisticated may be the data handling and display, computerization, or automation of the apparatus, the ultimate reliability of a chromatograph is basically and fundamentally dependent on obtaining a fully reliable sample. This subject is covered in greater detail under "Sampling for On-Line Analyzers" in this handbook section.

---

[1] Chief Chemist, Sterling Chemicals, Inc., Texas City, Texas.

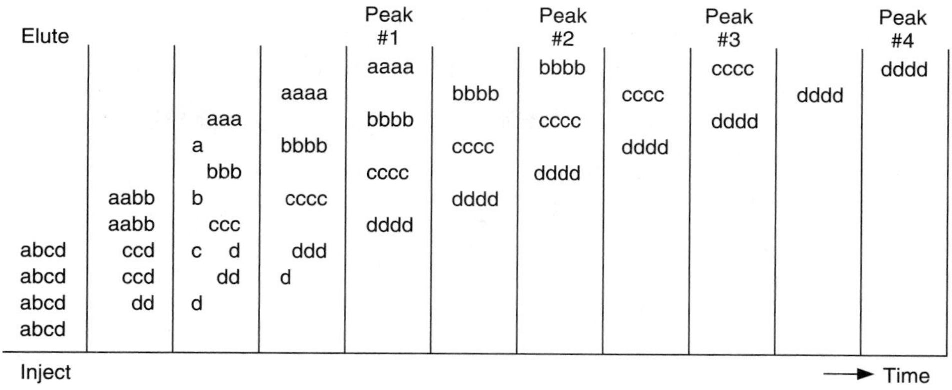

```
Elute                        Peak        Peak        Peak        Peak
                              #1          #2          #3          #4
                              aaaa        bbbb        cccc        dddd
                       aaaa   bbbb        cccc        dddd
               aaa     bbbb   cccc        dddd
               a       bbbb   cccc        dddd
               bbb     cccc   dddd
       aabb    b       cccc   dddd
       aabb    ccc     cccc
  abcd ccd   c   d     ddd
  abcd ccd    dd    d
  abcd dd    d
  abcd
Inject                                                          ———→ Time
```

FIGURE 1 Chromatographic separation process showing movement through column with time.

Flow injection techniques are utilized to prepare samples, not only for chromatography, but for numerous other analytical sensors that use thermal conductivity, flame ionization, and spectrometric principles. As important as these detectors may be, they must not overshadow the critical need to obtain a reliable sample. Flow injection analysis (FIA) is a technique for introducing a discrete sample into a flowing carrier, passing it through a conditioning "operator" system, measuring the concentration of one or more components in the modified sample, and displaying the results. Many different types of operators may be used, but component separation generally is associated with column chromatography. Flow injection is a procedure that was originally developed to automate wet lab test methods, but now has become a universal sample-preparation technique.

PROGRESS IN CHROMATOGRAPHY

When it first appeared as an industrial tool, the chromatograph was a single box with all of the components packaged together. Then it was considered an analyzer unto itself and required a separate sample conditioning system. Process engineers concerned with analysis began to develop techniques to modify the laboratory practice of using a single column. Multiple columns connected by switching valves were applied to backflush, foreflush, and cut components from the normal sequential elution scheme.

Some years later, laboratory chemists started to apply such techniques to specific industrial and process applications. Remote discrete sampling was developed in 1978. This allowed one to inject the small quantity of sample into a capillary hundreds of feet (tens of meters) from the "analyzer" and transport it to the column as a packet. The transfer line was viewed as a long oven and no longer constrained to a single box. Components of the original box now could be expanded into three-dimensional space by using several boxes with connecting capillaries. The programmer, which controls the system operation, became a main feature of the system. Multidimensional sample preparation is described under "Sampling for On-Line Analyzers."

In more recent years the capillary column, which now plays a major role in process applications, appeared. These columns produce significantly higher resolution of components and, in many cases, are faster and easier to use. Outer-surface-clad fused silica columns have been developed which are less fragile than those made of laboratory glass. Inert inner-surface metal capillary columns with cross-linked and chemically bound stationary phases also are available. The practice of technicians making their own columns has vanished into antiquity. Because there are numerous column types, details cannot be provided here. However, there are several buyers' guides available for the asking. The literature also is rich in terms of such factors as specific column selection, sample introduction techniques, analysis detectors, and overall performance qualities of a chromatographic system.

Control of chromatographic system operation has progressed from electromechanical and electrooptical devices to fully digital electronic components, which also incorporate computational capability. Microprocessors, personal computers, mini- and mainframe computers, as well as programmable logic controllers can be effectively integrated into chromatographic systems. However, for simple operations one still may use cam and digital timers.

A major improvement came with digital electronics which allows high-resolution timing (0.1 second) and exact repeatability. Precise timing of sequential events is essential to multidimensional sample preparation utilizing column separation operators and switching valves.

Electronics for controlling instrument operating parameters, such as temperature and pressure, have contributed to more stable operation. This produces better repeatability of the system for multiple cycles leading to higher precision of component isolation. The advantage of improved measurement devices can be realized once the sample preparation function is resolved. Computer hardware and software provide expanded capability for signal processing and data manipulation. Stable and precise sample preparation and measurement systems allow accurate analytical information provided that adequate maintenance and calibration practices are applied.

The wide range of measurement devices currently available makes almost any process chemical analysis possible. Column chromatographic techniques make interference-free determinations a reality.

GAS CHROMATOGRAPHY[2]

Almost any organic or inorganic compound that can be vaporized can be separated and analyzed with a gas chromatograph. As shown in Fig. 2, the minimum requirements for a system include (1) a column which contains the substrate or stationary phase, (2) a supply of inert carrier gas (moving phase) which is continually passed through the columns, (3) a means for maintaining pressure and flow constant, (4) a means of admitting or injecting the sample into the carrier gas stream, (5) a detector which senses the sample components as they elute, and (6) a recorder. The carrier gas may be any gas that does not react with the sample or adversely affect the detector. Helium, hydrogen, nitrogen, and argon are most commonly used.

Microprocessor-Based Automation

Microprocessor-based systems are now widely used to carry out functions previously performed manually by the operator. These systems may take many forms: (1) integral to and dedicated to a single chromatograph and (2) as a separate unit which controls one or several chromatographs. The latter are also available for upgrading older manually operated chromatographs to provide automated data reduction and output.

Systems are equipped with read-only memory (ROM) for the operating system and random-access memory (RAM) for storage of user-specific methods, operating instructions, and analytical data acquisition and manipulation.

Depending on user requirements, one or more of the following functions can be provided:

1. Individual temperature control and indication of column, ovens, detectors, and sample inlets
2. Dynamic control of programmed temperature profiles, starting and ending temperatures, rate of temperature increase with one or more time-controlled segments of isothermal temperature, and cooling-down cycle

---

[2] Last three sections by Richard Villalobos, Engineering Department, The Foxboro Company (A Siebe Company), Foxboro, Massachusetts.

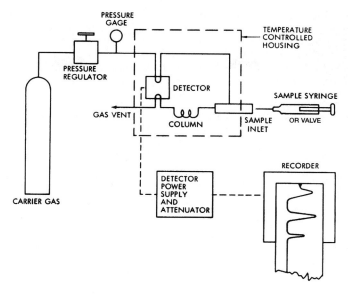

FIGURE 2   Basic elements of a gas chromatograph.

3. Automatic integration of peak areas or peak height measurements, with a choice of several calibration methods for calibration of results

4. Raw data storage with postrun calculation and manipulation of data for optimizing variables and methods development

5. Integral printer-plotter for hard copy of chromatograms and analytical results

6. Keyboard for input of analytical parameters and operator-selected variables

7. Video terminal for display of chromatograms and analytical results

8. Cassette or floppy disk extended memory for storage of files, methods, and user-specific applications programs

9. High-level language for user programming of special calculations such as physical properties and comparison with alarm levels

10. Serial communications via RS-232 ports with a central computer or other peripheral devices

11. Timed output control signals to peripheral and ancillary devices such as sample valves, column switching valves, and other external functions which are part of the automated procedure

12. On-board diagnostics to aid in troubleshooting and isolating and identifying failures

Chromatograph Data Reduction—Qualitative Analysis

For a given substrate under given conditions, each compound has a characteristic retention time which can be used for tentative identification. However, two or more compounds may have the same elution time on a particular column. In such cases the compound may be rerun on a different column with other characteristics to reduce ambiguity. Extensive compilations of individual compound retention times on different substrates are available for reference. Positive identification can be made only by collecting the compound or transferring it as it elutes directly into another apparatus for analysis by other means, such as infrared or ultraviolet spectroscopy, mass spectrometry, or nuclear magnetic resonance. Commercially available apparatus is available which combines in a single unit both a gas chromatograph and an infrared, ultraviolet, or mass spectrometer for routine separation and identification. The ancillary system may also be microprocessor-based, with an extensive memory for storing libraries of known infrared spectra or fragmentation patterns (in the case of mass spectrometers). Such systems allow microprocessor-controlled comparison and identification of detected compounds.

Quantitative Analysis

Quantitative analysis is based on the proportionality of detector response to the amount of component in the elution band. The most widely used measure of detector response is the area under the chromatogram peak. However, the peak height (amplitude of the detector signal at peak maxima) also may be used. Early methods for measuring the peak area included direct measurement on the chart with a planimeter, calculating from the height and width of the peak, and using a ball-and-disk integrator attached directly to the recorder. Electronic integrators have also been used to sense the detector signal directly, or after it is amplified, and to provide a digital printout of the peak area and time of peak maxima.

These methods of quantitation have been largely replaced by microprocessor-based data systems. The detector signal is directly coupled to a high-speed analog-to-digital converter which samples it at predetermined rates as high as 40 Hz. The digitized values are stored in memory for subsequent manipulation. Slope sensing algorithms find the beginning and end of peaks, peak maxima, and valley points between incompletely resolved peaks and can differentiate between peaks and the baseline (absence of peaks) to correct the digitized signal for baseline drift. Tangent skimming algorithms can locate and digitize small "rider" peaks on tailing edges of larger peaks.

The microprocessor sums the digitized values for a given peak to obtain the peak area. Results are then calculated by various methods:

Area Percentage, Normalized.   In this method the area of each peak is determined as a percentage of the total of the areas of all peaks. Results must add up to 100 percent. Therefore all components are detected.

Concentration by Relative Response Factors, Normalized.   Similar to area percentage, but each area is corrected by a relative response factor characteristic for a given compound.

Internal Standard.   The results are calculated relative to a standard added to the sample in a known amount. Results are independent of sample size. The internal standard must be a compound not normally in the sample and well separated from other components in the sample.

Standard Addition.   The sample is analyzed with and without the addition of a known amount of a compound which is also in the sample (spiking). The concentration is calculated from the observed increase in area.

External Standard.   The area of each component is compared to the area of a separately run standard with a known concentration of each component.

PROCESS GAS CHROMATOGRAPHY

This is a system for continuous, repetitive, and fully automatic on-line analysis of process streams which is similar to the laboratory chromatograph in all essential elements of basic technique but different in design and appearance. Factors affecting design include the need (1) to comply with the National Electrical Code for operation in hazardous atmospheres, (2) to automate the procedure, and (3) for ready adaptability to closed-loop process control and communication with process control systems and computers. The demand for maximum reliability and minimum maintenance has emphasized simplicity of hardware and methodology. Emphasis is placed on analyzing for a few rather than a number of components and on minimizing analysis time. These design targets have resulted in the extensive use of multicolumn techniques for rapid separation of selected components, with large portions of the sample being discarded. As shown in Fig. 3, the major components of a microprocessor-based process gas chromatograph system are (1) the analyzer, (2) the data processor, (3) the sample conditioner system, (4) one or more recorders, and (5) analog and serial outputs to peripheral devices or process control systems and computers.

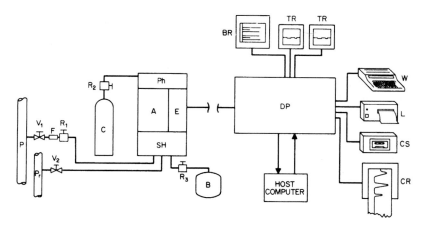

FIGURE 3 Basic elements of process gas chromatograph. The vapor sample is continuously withdrawn at a high rate from process line P, circulated through sample conditioner SH, and returned to lower pressure point P_r through shutoff valves V_1 and V_2. Particulate matter is removed by filter F, and the pressure reduced to a constant low level by regulator R_1. The sample conditioner contains flow control and other conditioning components and a valve for switching to synthetic calibration blend B through pressure regulator R_3. A sample slipstream is circulated to the sample valve in analyzer A, which also contains columns, detectors, and a temperature-control system. Carrier gas C is controlled by regulator R_2 and pneumatics control section P. A microprocessor in electronics module E stores an analytical program in RAM and controls analyzer functions and data acquisition and reduction. Analytical results are transmitted over a serial link to data processor DP, which converts results to an analog signal for presentation to bar graph recorder BR and as many as 30 to 40 trend recorders TR. Real-time constructed chromatogram is presented for maintenance on recorder CR. Serial outputs (RS-232) flow to writer or panel-mounted line printer L for data logging and to cassette recorder CS for storing applications programs. Results and alarm messages flow to host computer via serial link. An applications program is entered via data processor and downloaded to an analyzer RAM for execution in the analyzer. The processor controls several analyzers.

Analyzer

This equipment is usually located close to the sample point, enclosed in a shelter for weather protection. An analyzer is typically designed to comply with NEC Class I, Groups B, C, and D, Division 1, requirements for operation in hazardous areas by combination of explosion-proof enclosures, air purging, and intrinsically safe electric circuits. Sections of the analyzer include a controlled temperature compartment (heated air bath) for the columns, sample and column switching valves, and a detector. A pneumatics section for pressure or flow controllers for the carrier gas and other auxiliary gases (such as hydrogen and combustion air for an FID), as well as service air for the heater, electronics purge, and valve actuation. The electronics compartment contains a microprocessor with a central processing unit (CPU) and RAM and ROM for program control, data acquisition and reduction, output, and all communications functions. The RAM is battery-backed to prevent loss of applications programs due to power failure.

In its most usual form, the microprocessor performs these functions:

1. Controls all sequenced analyzer functions, such as sample injection and valve switching, by means of the applications program stored in RAM.

2. Samples and digitizes the detector signal at up to 40 Hz; performs peak area integration or peak height measurement with baseline correction and deconvolution of incompletely resolved peaks.

3. Identifies components by comparing elution times with values stored in memory.

4. Calculates the composition with a choice of several calibration and calculation methods stored in ROM.

5. Controls the sequencing of sample conditioner in multistream systems.

6. Performs automatic calibrations by analyzing calibration standards at user-selected intervals and automatically updating calibration factors.

7. Performs auxiliary calculations such as determining average molecular weight, specific gravity, heating (thermal units) value, or other properties based on the calculated sample composition.

8. Monitors electromechanical sensors in the analyzer and sample conditioner system to detect abnormal conditions: oven temperature out of limits, carrier gas flow failure, sample flow failure, etc.

9. Performs software diagnostics on the detector signal and analytical results to detect abnormal conditions: change in elution time, total peak area out of limits, excessive baseline noise or drift.

10. Communicates with data processor over a serial link to transmit analytical data and calculations and receives new or modified applications programs. A digitized form of the detector signal may also be transmitted for remote reconstruction of a real-time chromatogram.

In addition, the analyzer can accept analog signals from other field-mounted analyzers or sensors such as flowmeters and pressure transducers. The signal can be scaled, digitized, and incorporated into special calculations to determine mass flow, therms per day, reactor yields, and so on.

In alternative forms of these systems, the applications program is stored in the processor; the analyzer microprocessor digitizes only the detector signal and transmits the digitized values to the data processor. Applications program event commands are received in real time from the data processor and converted at the analyzer to electrical and pneumatic signals for sample valve actuation, column switching, sample conditioner control, and so on.

Data Processor

This unit is commonly located near or in the control room in a nonhazardous environment, and as much as 2000 to 3000 feet (610 to 914 meters) from the analyzers. The data processor also has its own microprocessor with a CPU and a complement of ROM and RAM in which the operating system and user-specific applications programs are stored. Communication with the analyzers is by a serial link.

In its most usual form, the processor is a special-purpose microprocessor and can control up to six or eight analyzers. In other forms the processor may be a microprocessor-based minicomputer and control as many as 32 analyzers.

The processor has several main functions:

1. Input of applications programs by means of a special-purpose keyboard, with an alphanumeric display and interactive dialogue. Prompting of the operator and screening of the input data ensures the input of all necessary parameters and prevents conflicting data inputs. The program is down-loaded into the analyzer memory and can be recalled for editing and modifications. In some versions, input may be accomplished through an ancillary video terminal with a keyboard and menu-driven operator communications.

2. Receives data from the analyzers and distributes them to the various output devices in analog or digital form as required.

3. Monitors the status of the analyzers and displays alarms and transmits them to peripherals.

4. Provides manual control of all analyzer functions during setup and maintenance and acts as a diagnostic center for troubleshooting and corrective maintenance. A real-time chromatogram is available.

Except for the initial setup and maintenance of the analyzer, all operations, including programming, manual operation, and calibration, take place at the processor location.

In some versions the applications program for all analyzers may be stored in the processor memory instead of the analyzer. Event commands are sent to the analyzer in real time over the serial link and converted to analog commands by the analyzer. Digitized data are received from each analyzer, with all data acquisition and reduction accomplished in the processor.

In yet other versions the processor is part of the analyzer and may be dedicated to, and integral to, a single analyzer at the analyzer location with all analyzer and processor functions described performed locally.

The simplest form of record is a bar graph. The record consists of a series of bars, one for each component, of height proportional to the component concentration. The output for each component is scaled to give a full-scale reading equivalent to a convenient concentration value. Each component may have a different full-scale range. A large number of components in one or many streams may be recorded on the same instrument. Different streams may be identified by the height of a flat-top bar preceding the series of bars for that stream.

Trend outputs consist of a continuous electrical signal [0 to 10 volts or 4 to 20 milliamperes (mA)] from the processor. As many as 30 to 40 such outputs may be available from a single processor. Each output represents the concentration of a particular component in one of the sample streams on a given analyzer scaled to some convenient range. Component identity and scale factors for each output channel are user-assigned from the processor keyboard.

The output value is held constant at the last value until a new value is determined in a subsequent analysis, causing a stepwise change in the record as the signal is updated at the end of each analytical cycle. A separate recorder pen is required for each trend output recorded. Trend outputs may also be used to input analytical data to a process control system. A separate two-wire 4- to 20-mA output line is typically required for each component input.

A real-time chromatogram for any selected analyzer may be obtained at the data processor for setup and maintenance. The chromatogram is received in digital form at a rate of 10 or more data points per second and reconstructed into analog form and scaled at the processor.

Serial output ports (RS-232) are usually provided for serial transmission of data—usually in ASCII code—to peripherals (Fig. 3). These include the following.

1. *Printer.* Provides output for logging of analytical results, data, and alarms. It also prints out, on command, a record of applications programs, analog and channel assignments, calibration data, and other user-selected parameters, with the time and data for maintenance record keeping. Some systems have two printers at different locations—one for data logging (in the control room) and the other for testing and maintenance.

2. *Panel-mounted printer.* Outputs the same data as the printer, but on narrow paper tape.

3. *Host computer.* Transmits all data, alarms, and analyzer status reports to host computer for data logging and historical archiving for subsequent input to process control systems. Special software may be required to be written for either the data processor or the host computer for compatibility of communications protocol and message structure.

4. *Magnetic tape cassette recorder.* Provides for archiving analyzer and processor applications programs on cassette tapes, usually with a conventional portable recorder. The tape may be played back to reload the program in the event of loss due to system failure.

Qualitative analysis is less important in process gas chromatography because at the outset stream composition is usually defined within narrow limits as to both compounds present and range of concentration. The need for maximum accuracy dictates a maximum separation of measured components and an accounting of all components to avoid errors due to interferences. Therefore qualitative aspects are limited to matching observed elution times to a table of expected elution time windows for identification and also for detecting and alarming elution time changes due to system malfunction.

Quantitative Analysis

Methods are in general similar to the laboratory methods described, with an emphasis on accuracy. Time slices of a detector signal are digitized and stored in RAM. Slope-sensing algorithms find the beginning and end of peaks, valleys, and rider peaks by tangent skimming. Baseline corrections are made by look-forward, look-back, or a combination of both to compensate for a drifting baseline due to column switching.

Composition is calculated by different methods, which are user-selected at the processor keyboard.

Concentration Percentage, Normalized.   Uses relative response factors and internal normalization. Does not require calibration standard.

External Standard.   Compares area or peak height to a value in memory for a calibration standard of known concentration. Only components of interest need to be measured. Normalization can be carried out if all components in the sample are measured.

Reference to Key Component.   Calibrates the system on a single key component in calibration standard and relative response factors. The key component may be a pure compound and need not be present in the sample. This method permits calibration for samples which are unstable, reactive, or hazardous by use of a safe key component.

Automatic Calibration.   Automatic introduction and analysis of a calibration standard at user-selected intervals for updating calibration factors. User input limits prevent updating if an incorrect calibration occurs as a result of a depleted or contaminated standard.

Valves

Electrically or pneumatically operated valves are used for the injection of liquid or gas samples and column switching. Rotary, spool-and-O-ring, diaphragm, and sliding-plate valves with from 4 to 12 ports are used. Vaporizing liquid sampling valves mounted through the wall of the analyzer transfer the liquid sample from the cold exterior zone to the heated vaporizing zone within the oven. The sample valve meters a fixed volume of liquid or gas into the column with a repeatability of ±0.25 percent. The sample size may vary from approximately 0.1 μL to 50 mL (with an external sample loop).

Detectors

The need for ruggedness and reliability has limited the types of detectors suitable for on-line process chromatographs. TCDs (for general use) and hydrogen FIDs (for trace organic analysis) are the most widely used detectors.

LIQUID-COLUMN CHROMATOGRAPHY

This method is particularly useful for the separation and analysis of high-molecular-weight compounds beyond the range of gas chromatography. It is generally classified according to the stationary phase or to the nature of its interaction with sample components.

1. Liquid-liquid partition chromatography, where the sample components are partitioned between a moving liquid phase and a stationary liquid phase deposited on an inert solid. The two solvent phases must be immiscible. The stationary phase may be a large molecule chemically bonded to the surface of a solid (bonded liquid phase) to prevent loss by solubility in the moving phase. This method can also be subdivided into normal-phase systems, in which the moving phase is less polar than the stationary phase, and reverse-phase systems, in which it is more polar.

2. Liquid-solid or absorption chromatography, in which the sample components are absorbed on the surface of an adsorbent such as silica gel.

3. Ion exchange, in which ionic sample components interact with functional groups on a permeable ionic resin.

4. Exclusion or gel permeation, in which compounds are separated by molecular size into a range of pore sizes in a polymeric gel. This method is useful for measuring the molecular-weight distribution of polymers.

Isocratic elution uses a solvent of constant composition throughout the analysis. Gradient elution is a modification of the technique, in which the solvent is a mixture of two solvents which differ in solvent strength. The composition or ratio of the two is changed during the analysis in accordance with a predetermined program. The change may be continuous, linear or nonlinear, stepwise, or a combination.

Apparatus for Liquid Chromatography

The need for more efficient columns and faster separations has led to the development of stationary phase packings with particles as small as 2 μm, operating at pressures as high as 10,000 psig (69 MPa), and with solvent flow rates as low as 1 mL/min or less. This has in turn led to the development of detectors with internal volumes of only a few microliters and special fittings and connectors with minimum dead volume to prevent band spreading and loss of resolution. These improvements in apparatus and technique have resulted in an ability to achieve complex separations with speeds comparable to those of gas chromatography.

The solvent is moved through the system by constant-flow or constant-pressure pumps which are driven mechanically (screw-driven syringe or reciprocating) or by gas pressure with pneumatic amplifiers. For gradient elution two pumps may be synchronized and programmed to provide a controlled, reproducible composition change.

Samples may be introduced by syringe directly into the column through a septum or by means of valves with a fixed volume which has been prefilled with the sample. Valves may be of a rotary, sliding plug, or diaphragm design and of stainless steel and fluoroplastic construction for inertness. Auto samplers are used for unattended injection of samples loaded into vials and sequentially rotated into the injection mechanism.

The differential refractive index (RI) is probably the most widely used detector; other detectors are based on photometry. Fixed-wavelength and variable-wavelength ultraviolet detectors are the most commonly used. The dielectric constant (DC) detector and the electrical conductivity (EC) detector are also used and are especially suitable for process liquid chromatography. Transport detectors transfer the eluate to a moving wire or belt through a solvent evaporator and into a pyrolyzer which fragments and transfers the sample components to a hydrogen FID.

BIBLIOGRAPHY

Annino, R., and R. Villalobos: "Process Gas Chromatography," Instrument Society of America, Durham, North Carolina, 1991.

Clevett, K: *Process Analyzer Technology,* Wiley, New York, 1986, pp. 1–84.

Converse, J. G.: "Sample Preparation for On-Site Automated Chemical Analysis," paper 90-458, presented at the New Orleans, Louisiana, Meeting, Instrument Society of America, Research Triangle Park, North Carolina, 1990.

Converse, J. G.: "Improve On-Line Chemical Concentration Measurement," *Chem. Eng. Prog.,* p. 73, Mar. 1991.

Converse, J. G.: "Discrete Sampling/Flow Injection as a Basis for On-Site Automated Sample Preparation," Anatech, Atlanta, Georgia, Apr. 1992.

Converse, J. G.: "Calibration and Maintenance Must Be a Part of Reliable Sample Preparation and Measurement System Design," presented at the Houston, Texas, Meeting, Instrument Society of America, Research Triangle Park, North Carolina, 1992.

Guiochon, G., and C. L. Guillemin: *Quantitative Gas Chromatography for Laboratory Analysis and On-Line Process Control,* Elsevier, New York, 1988.

Mowrey, R. A., Jr.: "The Potential of Flow Injection Techniques for On-Stream Process Applications," presented at the Analytical Instrumentation Division Meeting, Chicago, Illinois, Instrument Society of America, Research Triangle Park, North Carolina, 1984.

Strobel, H. A., and W. R. Heineman: *Chemical Instrumentation: A Systematic Approach in Instrumental Analysis,* 3d ed., Wiley, New York, 1989.

OXYGEN DETERMINATION

The analysis of gaseous mixtures to determine percent oxygen and of oxygen dissolved in aqueous solutions is used widely throughout industry. The importance of O_2 is its marked chemical reactivity. A common application of O_2 analysis is in combustion control systems, notably in the utility field, where various fuels react with O_2 (as a constituent of air). Excess air and unburned fuel levels can be fed back to the combustion control system, thus raising boiler efficiency and decreasing pollutants exiting out the stack. In the manufacture of certain chemicals and petrochemicals, oxygen is a direct reactant. Thus oxygen analyzers find wide application in ammonia, ethylene, air separation units, cement kilns, and metallurgical sintering and melting furnaces, among others. Because oxygen creates scaling of many materials, particularly in heating operations, O_2 detection is required in annealing, soaking pits, heat treating (often requiring a completely inert atmosphere), drying, roasting, and glass-melting furnaces.

Numerous wastes are disposed of by incineration and produce noxious gases. O_2 measurement of stack gases is an indication of the completeness of the process. Oxygen concentration monitoring is also important in certain biochemical operations, such as fermentation. Oxygen determinations often are key to the success of finding and controlling the polluting qualities of effluent streams from manufacturing and many other sources of pollution.

OXYGEN MEASUREMENT PRINCIPLES

Improvements in O_2 measurement have occurred over several decades. Dating back about 60 years, the magnetic susceptibility property of gaseous oxygen represented an early practical basis for industrial O_2 analyzers. Greatly improved, these instruments continue in use. Several other measurement principles were introduced, including electron capture, catalytic combustion, fuel cell principles, and the special properties of zirconium oxide to electrically conduct oxygen ions (O^{2-}). Instruments based on the latter principle have gained wide acceptance. For dissolved oxygen analysis, electrometric determinations, including polarographic methods, were developed and displaced the former arduous wet chemical laboratory methods.

Magnetic Susceptibility

When atoms and molecules are placed in a strong nonuniform magnetic field, either they are attracted to the strongest part of the field (in which case they are termed paramagnetic—ferromagnetism is a special case), or they tend to go to the weakest part of the field (termed diamagnetic).

The ratio of the intensity of magnetism induced in a unit volume of a substance to the magnetic intensity of the field acting on it is called the susceptibility per unit volume. The symbol K denotes volume susceptibility of gas mixture, cgs. This ratio is positive for paramagnetic substances and negative for diamagnetic materials. The susceptibility per unit mass X is equal to K/ρ, where ρ is the density of the substance. The atomic susceptibility X_A is equal to the product of X and the atomic weight, or the product of K and the atomic volume. The molar susceptibility X_M is equal to the product of X and the molecular weight, or the product of K and the molecular volume.

Diamagnetism is a universal property of matter; all substances, even though classified as paramagnetic, have a diamagnetic component. The mass susceptibility of diamagnetic substances usually is independent of temperature. The volume susceptibility of a diamagnetic substance is dependent on its density and hence at constant pressure is inversely proportional to the absolute temperature. In general, the mass susceptibility of a paramagnetic substance is inversely proportional to the absolute temperature. When volume susceptibilities are considered, paramagnetic volume susceptibility is inversely proportional to the square of the absolute temperature.

With the exception of oxygen and nitric oxide the relative volume magnetic susceptibilities of other commonly encountered industrial gases are very small or negative. Using a basis of +100 for oxygen, nitric oxide is +43. On the same normalized scale, chlorine is −0.77, methane is −0.2, carbon dioxide is −0.27, and so on (Figs. 1 through 8).

Magnetodynamic O_2 Analyzers.   Early work on the magnetic susceptibility of matter was done by Faraday, who developed a simple and sensitive method whereby the sample to be tested is suspended from the arm of a very sensitive balance in a nonuniform magnetic field at a position where the gradient of the field is maximum. The sample experiences a force which is proportional to the field strength at that position, to the gradient field, and to the difference in magnetic susceptibility between the sample and the surrounding medium. Pierre Curie used this method extensively, and later many investigators improved the technique. Havens experimented with a variety of test bodies and finally selected the dumbbell as the most suitable for his experiments in determining the magnetic susceptibility of nitrogen dioxide.

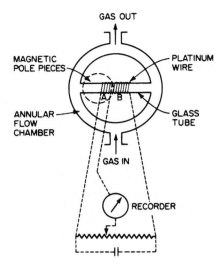

FIGURE 1 Lehrer-type oxygen analyzer. As early as 1930 Senftleben observed that the thermal conductivity of paramagnetic gases appeared to decrease under the influence of a magnetic field. Further investigators proposed that if a paramagnetic gas comes in the vicinity of a hot wire inside a nonuniform magnetic field, then the gas will be heated, reducing its magnetic susceptibility and causing it to be displaced by a cooler gas experiencing a strong magnetic field. Thus a form of magnetic wind is created, tending to cool the hot wire. In 1942 Lehrer developed an industrial O_2 analyzer based on these effects. This version is sometimes referred to as a thermomagnetic analyzer.

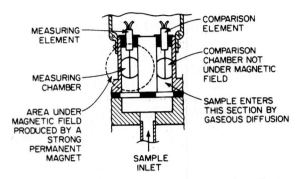

FIGURE 2 A measuring element and a comparison element form two legs of a Wheatstone bridge and are heated by bridge current. Measuring and comparison chambers are exposed to the same sample gas.

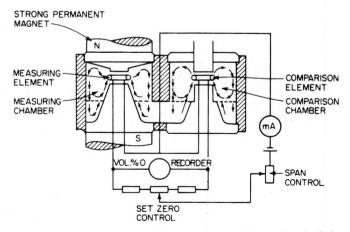

FIGURE 3 A measuring element and a comparison element form half of a Wheatstone bridge. Measuring and comparison chambers are exposed to the same sample gas.

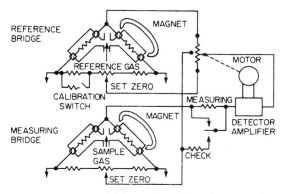

FIGURE 4 Double-bridge analyzer. Recorder output is a ratio of percent O_2 in sample to that of reference gas. System is independent of barometric pressure variations.

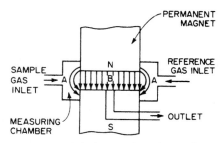

FIGURE 5 Quincke-type O_2 analyzer. When a paramagnetic gas is placed in the vicinity of a magnetic field, the gas experiences a force driving it from the nonuniform part of field A to the uniform part B. The Quincke method relies basically on the measurement of the pressure differential between the two regions. The Luft analyzer, introduced in the 1950s, relies on the Quincke method. Since then the method has been improved.

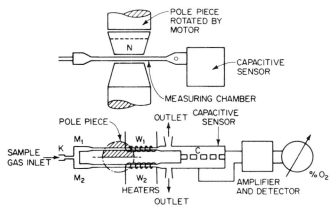

FIGURE 6 In an improved version of the Quincke O_2 analyzer, the gas sample is admitted through a capillary K into two identical measuring tubes M_1 and M_2. These tubes are located within the poles of a strong permanent magnet, which has one pole piece made of two halves. One half is formed of soft iron, producing a strong field on the tube under it. The other half is a nonmagnetic material, making the field of the second tube a weak one. The sample gas which enters the tube under the strong nonuniform field, at ambient temperature, experiences a force action on its paramagnetic component. The force is reduced at the other end by heating the gas by elements W_1 and W_2. The sample in the second tube experiences a very small force. The pressure differential between the two tubes is measured by a capacitive pressure sensor C. A motor operates the pole piece at constant speed and thus modulates the pressure sensor. The modulator signal has an amplitude proportional to the magnetic susceptibility of the gas mixture.

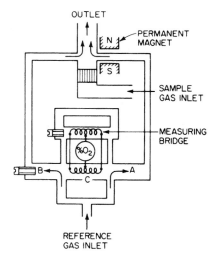

FIGURE 7 Schematic of a Maihak O_2 analyzer. A reference gas stream having a weak magnetic susceptibility is split into two identical branches A and B. These join together with a pipe through which the sample gas is introduced. There is a strong magnet at the end of one of the branches. If the sample contains components of high magnetic susceptibility, these will experience a force at the end of branch A, producing a pressure differential between A and B. This results in a reference gas flow in C. This flow is measured by a Wheatstone bridge. The bridge imbalance can be calibrated to read percent O_2.

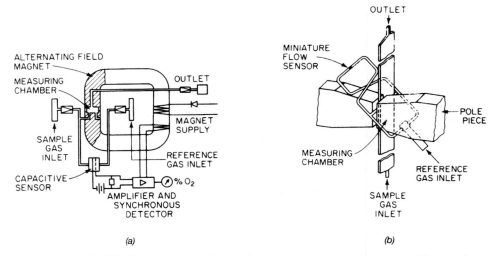

FIGURE 8 (*a*) In 1968 Hummel reported the development of a fast-response analyzer of the type shown. Sample gas and reference gas are admitted into a chamber, which is within a magnetic field. Pressure differential between the streams is due to magnetic forces acting on the high-susceptibility component of the sample gas. This is measured by a capacitive pressure sensor. The magnetic field is switched on and off, modulating the signal whose amplitude is proportional to the magnetic susceptibility of the sample. A synchronous detector is used for demodulation. (*b*) Another form of this instrument, in which the flow due to the differential pressure is measured with a miniature flow sensor.

The use of this principle in connection with determining the partial pressure of oxygen was reported by Pauling and other investigators. Some modern magnetodynamic instruments are refinements of the Pauling instrument (Figs. 9 and 10). Some improvement was achieved by the Munday cell. A null-balance form of this instrument is shown in Fig. 11.

Zirconium Oxide O₂ Sensors

Widely accepted for certain applications, the zirconium oxide sensor usually is furnished in a convenient probe format for direct in-line insertion. The sensor is based on the principle that zirconium oxide is a conductor of oxygen ions (O^{2-}). These direct in situ sensors find wide application in the measurement of excess oxygen in combustion gases over a wide range of temperatures. They do not require external sampling systems. This method does not require any cooling of gases prior to measurement. A major limitation is their unsuitability for the measurement of flammable gases (Figs. 12 and 13).

Systems generally obtainable operate between 32 and 1110°F (0 and 600°C), but high-temperature probes are available for use up to 2500°F (1400°C). Probes are designed for flange mounting. For mounting horizontally over a span of 10 feet (3 meters), support-type probes are available. A high-temperature probe for vertical mounting is shown in Fig. 14. Sensor systems are obtainable with all of the accoutrements of digital data processing and electronic displays.

Common operating ranges are 0 to 5, 0 to 10, and 0 to 25 percent O_2. Output signals are 0 to 1 volt and 4 to 20 mA. Accuracy approximates ±1 percent (reading); repeatability is approximately ±0.5 percent (full-scale); linearity is ±1 percent (full-scale). Several sensors can be connected in an averaging network.

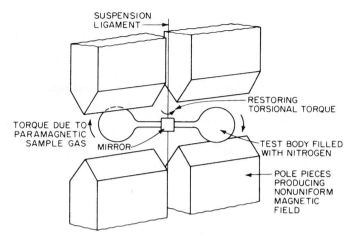

FIGURE 9 Basic measuring system of magnetodynamic O_2 analyzer. Two spheres filled with nitrogen form a dumbbell, which is suspended within a sealed cell having a sample gas inlet and outlet under a symmetrical nonuniform magnetic field produced by suitable pole pieces. The dumbbell has a resultant diamagnetic component, and when the cell is filled with a gas of low magnetic susceptibility, such as nitrogen, the dumbbell is deflected away from the intense part of the field. It assumes an equilibrium position whereby the magnetic forces are balanced by the torsional torque produced by the suspension. In some cells the suspension material is a quartz fiber.

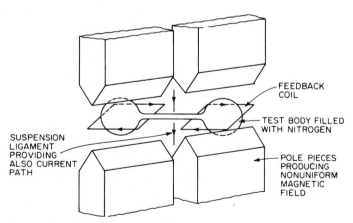

FIGURE 10 In the Munday cell the suspension utilizes platinum-iridium. When a gas is admitted to the cell, the dumbbell assumes a new position, one further out if the sample has a net paramagnetic componenete. In a simple nonlinear instrument, the position can be measured by means of a mirror fastened to the dumbbell which reflects a beam of light onto a graduated scale.

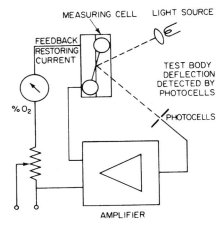

FIGURE 11 A more accurate technique involves always restoring the dumbbell to its initial zero position. This is achieved by electrostatic forces in one instrument. The dumbbell is sputter-coated with a conducting metal, and two electrodes are placed adjacent to it and held at constant potential, one above and one below ground potential. A voltage is applied to the dumbbell from a suitable source to provide the required electric field. The voltage required to restore the dumbbell to the zero position is directly proportional to the volume magnetic susceptibility of the sample. A self-nulling circuit includes two photocells for detecting the deflection of a light beam. Instruments of this type offer high sensitivity. To operate satisfactorily, the identical gas must not contain strongly parametric gases, such as nitric oxide, nitrogen dioxide, and chlorine dioxide. Contemporary instruments which compensate for atmospheric pressure changes and moisture have been designed.

Other O_2 Sensing Principles

Polarographic Principles. In the 1930s it was suggested that the current generated by an oxygen-depolarized primary battery could be used as a measure of O_2 concentration. Several experiments were made with totally immersed electrodes. Dry forms of such cells have overcome a number of the disadvantages of wet types. In these cells the sample gas, electrolyte, and cathode form a three-phase boundary where the O_2 molecules are reduced. The percentage of O_2 molecules reduced in the cell depends on the cell design. At low flow rates it is possible to approach a state where all O_2 molecules are reduced and the cell thus operates coulometrically. This improves stability and reduces the temperature coefficient. Some cells have a thin membrane fixed near the electrode surface which allows O_2 to diffuse to the electrode. In the polarographic format, a constant polarizing voltage is applied to the sensors.

Electron Capture Analyzers. Oxygen absorbs low-energy beta radiation, whereas some gases, such as nitrogen and argon, do not. The apparent reduction in the radiation of a given source is dependent on the number of oxygen molecules present. This principle can be used as a means of measuring oxygen to parts-per-million levels. Figure 15 shows a detector of this type.

The device uses a low-energy beta source (tritium). There are both inlet and outlet ports for passage of the carrier gas. Centrally placed in the cylinder is an electrode. The body of the detector is made of a stainless-steel cylinder. When a gas such as nitrogen flows through the detector cell, a cloud of electrons is formed between the electrode and the wall of the cell as a result of beta emission from the radioactive source.

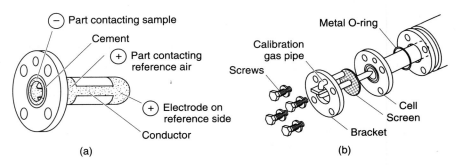

FIGURE 12 (*a*) Platinum electrodes are molecularly bonded to the zirconia element to avoid possibility of separation. Inasmuch as electrodes do not have leads, there is no risk of wire damage from corrosive gases. A special coating is used to protect electrodes from damage. (*b*) When required, the cell is easily replaced by removing four bolts. No soldering of wires is required—downtime is limited to about 10 minutes. Cell replacement does not require instrument shop work. (*Yokogawa.*)

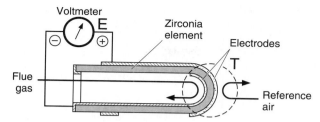

FIGURE 13   At high temperature the zirconia element is a solid electrolyte and becomes electrically conductive through the production of oxygen ions (O^{2-}). High operating temperature is produced by a cell heater. Platinum electrodes are molecularly bonded to the zirconia element exterior and interior. No connecting leads are required. In essence, the inner electrode is exposed to flue gas entering the open end of the cell. The outer electrode is exposed to a constant partial pressure of oxygen. An oxygen concentration cell is created and an electrical signal produced. (*Yokogawa.*)

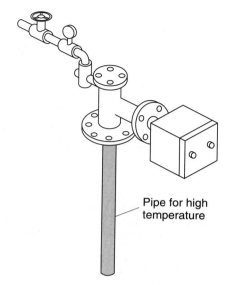

FIGURE   14   High-temperature   vertically mounted zirconium oxide probe. (*Yokogawa.*)

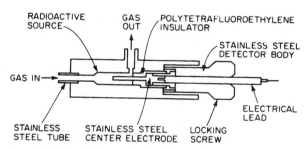

FIGURE 15   Sensing portion of electron capture analyzer.

An applied voltage between the electrode and the body gives a standing current. When an electron-capturing material, such as oxygen, mixes with the carrier gas flowing through the detector, some of the flowing electrons are captured and the standing current falls. The amount of fall is proportional to the concentration of the electron capturing material. A commercial instrument with a detector of this type and ranges from 0 to 5 ppm to 0 to 250 ppm oxygen in nitrogen is available.

Catalytic Combustion.   This method is suitable for the determination of O_2 content in noncombustible gases and saturated hydrocarbons. The method is based on the measurement of the change in temperature due to total catalytic combustion of the oxygen in the sample (or a known proportion of it) with hydrogen.

The sample gas enters the analyzer at a constant rate, after which it is mixed with a constant stream of hydrogen, heated, and admitted to the measuring cell. The measuring cell is maintained at a constant temperature suitable for catalytic combustion. The measuring cell consists of two sections, each containing a platinum sensor. The measuring sensor is exposed to the sample; the other is a compensating sensor. Only a small amount of the sample enters the compensating sensor. Combustion of the mixture takes place at both sections, raising the temperature of the measuring sensor. The sensors are connected in a bridge arrangement which drives a recorder or meter calibrated in percentage oxygen.

GAS AND PROCESS ANALYZERS

Developments in chemical analytical instrumentation during the past decade mainly have targeted the integration of modern instrumentation and control technology with long-established measurement methods, rather than the development of brand-new detecting means. To be sure, however, a few new detecting principles, such as tin dioxide (SnO_2) and other semiconducting metal-oxide sensors, have been unveiled. But by far the majority of analytical sensors in use in the early 1990s bear familiar names, such as infrared (IR), ultraviolet (UV), flame ionization, combustion, calorimetric, and chemical reaction detectors.

Rather than concentrating on the development of new sensing principles per se, suppliers have borrowed heavily from the improved processing and miniaturization techniques so successfully applied to the major process variables (temperature, pressure, flow, and so on) to reduce the size and improve the performance of established sensing principles. Microprocessors, for example, are used widely. Rather than offering clumsy "black boxes," an excellent effort has been made to cast field detectors in the form of convenient, often interchangeable transmitters. Suppliers also have exerted a major design effort in widely using digital information data-processing systems, the most advanced graphic displays, and networking technologies in offering "turnkey" analytical systems.

Although considerable work remains, advancements have been made in adapting increasing numbers of analytical detecting means to "on-line" applications, once considered almost unachievable with some detector systems.

Stemming from the demands of pollution control and the increasing stress placed on product quality and statistical product quality (SPC) systems, the demand for analytical systems has undergone a great expansion. This has enabled the success of several "vertical" firms that concentrate their efforts in this field.

Because those features which relate to instrumentation and control beyond the detector per se are described elsewhere[1] in this handbook, and in the interest of conserving space, this article will concentrate on analytical measurement means as they relate to gas analysis.

ULTRAVIOLET ABSORPTION ANALYZERS[2]

Several important classes of compounds absorb strongly in the UV radiation region. Water and the usual components of air do not absorb in this region. As a result, UV absorption ana-

---

[1] Reference to the articles, "Sampling for On-Line Analyzers," "pH and Redox Potential Measurements," "Electrical Conductivity Measurements," "Thermal Conductivity Gas Analyzers," "Oxygen Determination," and "Refractometers" is suggested.

[2] Largely based on information furnished by R. S. Saltzman, Instrumentation Engineer, E. I. DuPont de Nemours & Company, Inc., Wilmington, Delaware.

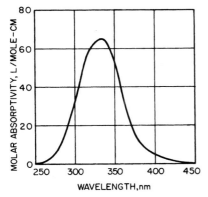

FIGURE 1 Spectral absorptivity of chlorine. The spectra of molecules generally are composed of broad bands when compared with the line spectra of atoms because each electronic state has associated with it numerous vibrational and rotational energy levels. The vibrational-rotational fine structure of the bands is not resolved except for gas-phase spectra of light, simple molecules at low pressure. For example, nitrogen dioxide and sulfur dioxide at low pressure show a fine structure when studied at high resolution. Also, the absorption spectra of simpler aromatics, such as benzene and cymene, show a distinct fine structure, even in the liquid phase. In general, however, pressure broadening and solvent interaction make the spectrum appear as an envelope of the otherwise expected fine-structure spectrum.

lyzers may be more selective and sensitive than IR and other types of analyzers used in many plant stream analysis applications. It is not uncommon to find UV analyzers used as components of a multidetector system for combustion control. The region of the electromagnetic spectrum covered by UV absorption analyzers is from 200 to 400 nm and excludes "vacuum" UV below 200 nm (Fig. 1). Absorption in this region, corresponding to photon energies between 6.12 and 3.06 eV, is due to excitation of the more loosely bound valence electrons, including the "unsaturation" (π) electrons of multiple bonds and the unpaired electrons of free radicals.

Concentration versus Absorption. UV absorption data usually are tabulated and plotted in terms of absorptivity as a function of increasing wavelength. The unit of wavelength in the UV region is the nanometer (nm) [1 nm (10^{-9} cm) = 10 Å], which is the same as the millimicron (mμ), a unit which was previously used extensively in analytical chemistry literature. A material with absorptivity of unity at a specific wavelength has an absorbance of 1.0 in a path length of 1 cm when the sample concentration is 1.0 mol/L.

UV absorption data usually are tabulated and plotted in terms of absorptivity as a function of increasing wavelength. The unit of wavelength in the UV region is the nanometer (nm) [1 nm (10^{-9} cm) equals 10 Å], which is the same as the millimicron (mμ), a unit which was previously used extensively in analytical chemistry literature. The most common unit of absorptivity is liters per mole-centimeter. A material with absorptivity of unity at a specific wavelength has an absorbance of 1.0 in a path length of 1 cm when the sample concentration is 1.0 mol/L.

The absorbance of a substance is directly proportional to the concentration of the material which causes the absorption, in accordance with the Lambert-Beer law, more commonly referred to simply as Beer's law:

$$A = abc = \log \frac{I_0}{I} = \log \frac{1}{T} \tag{1}$$

where A = absorbance
  a = molar absorptivity, L/mol · cm
  b = path length, cm
  c = concentration, mol/L
  I_0 = intensity of radiation striking detector with nonabsorbing sample in light path
  I = intensity of radiation striking detector with concentration c of absorbing sample in light path
  T = transmittance, = I/I_0

For the vapor phase,

$$A = \frac{abc'}{2450} \tag{2}$$

at 25°C and 760 torr pressure, where c' is volume percent or mole percent, or

$$A = \frac{abc'}{2450} \frac{P + 14.7}{14.7} \frac{298}{t + 273} \tag{3}$$

at any temperature or pressure, where P is pressure in psig and t is temperature in °C.

For the liquid or solid phase,

$$c = \frac{c'' \times d}{MW} \times 10 = \text{mol/L} \tag{4}$$

where
$$\begin{aligned} c'' &= \text{weight percent in liquid} \\ d &= \text{density of liquid} \\ MW &= \text{molecular weight of material to be measured} \end{aligned}$$

and hence,

$$A = \frac{10abc''d}{MW} \tag{5}$$

Basic Components of UV Analyzer.   These include a UV radiation source, optical filters, sample cell, detector, and output display system. The latter generally can be obtained with as much sophistication as the user may require. The transmittance measurement is made by calculating the ratio of the reading of the output with the sample in the cell to the reading with the cell that is devoid of UV-absorbing substances. The concentration of the measured substance can be calculated from the known absorptivity of the substance by using Beer's law [Eq. (1)], or it may be obtained by comparison with known samples.

The UV radiation source must provide the desired UV wavelength and may be a line source, such as a mercury arc or equivalent, or a continuous source, such as a hydrogen or deuterium arc or equivalent. Tungsten filament incandescent lamps are the most readily obtainable low-cost lamps used. They provide a continuous spectrum of energy that is relatively feeble in the UV range, but adequate in an efficient instrument above 350 mm. A wide variety of radiation sources is available to the designer. Some UV detectors are designed for rather specific uses, such as nitrogen oxides (NO_x), which are important in pollution control. In such instances the wide variety of system components available can be narrowed to provide the test combination for the application and minimize cost.

The analytical radiation in an UV analyzer must be as nearly monochromatic as possible in order to facilitate the calibration and to ensure high linearity, long-time stability, and sustained accuracy. Monochromatic radiation is obtained by proper selection of source, filters, and photoreceptors, each of which is selective in regard to the wavelengths that it emits, transmits, or responds to, respectively.

Early UV analyzers were of the single-beam type. Split-beam and dual-beam analyzers were developed to avoid the shortcomings of the single-beam type, including effects of drift in the light source, dirt or bubbles in the sample cell, and drift in the detector circuit. In a split-beam analyzer, radiation is partially absorbed in passing through the sample. Radiation passing through the sample is divided into two beams by a semitransparent mirror. Each beam passes through an optical filter to a phototube. The filter removes radiation at all wavelengths except the one to be measured.

Radiation at the analytical wavelength striking one phototube is absorbed strongly by the component whose concentration is being measured. Radiation at the reference wavelength, directed to the second phototube, is absorbed weakly or not at all by the component. Each phototube develops a current directly proportional to the intensity of radiation striking the phototube.

The phototube current is converted to a dc voltage directly proportional to the negative logarithm of the phototube current by a special logarithmic amplifier. The output voltages from the measuring and reference amplifiers are subtracted in the control station to produce a final output voltage. Hence the output is proportional to the difference between the logarithms of the phototube currents and proportional to the sample concentration (based on Beer's law).

This split-beam photometric analyzer has separate light source, sample, and photometer housings. The three housings of the analyzer are easily separated. Because of this, the analyzer is readily adapted for monitoring process gas and liquid streams in pipeline cells installed directly in the process line and for monitoring the thickness of film or film coatings.

A dual-beam analyzer uses a measuring and a reference beam at the same wavelength. Fundamentally this is a stable design. The addition of a short-path-length reference cell tends to compensate for dirt or optical blockage in the sample cell and the effects of aging UV sources and detectors. Instruments of this type can measure, for example, concentrations as low as 0 to 100 ppm NO, with a repeatability of better than +2 percent (span) per week.

Target Substances.   Depending on numerous factors that can interfere with a straightforward application of UV analyzers, the following substances should receive initial consideration.

Elemental Halogens.   Chlorine, bromine, and others are highly corrosive, and particular attention must be paid in selecting construction materials for exposed parts of the UV system. For example, with careful attention to design, UV analyzers have been used widely in monitoring elemental chlorine in connection with closed-loop control of chlorination processes.

Chlorine is photochemically activated by radiation of 400 nm. It reacts very rapidly with H_2 and slowly with carbon monoxide (CO) and hydrocarbons, if this radiation is not blocked from the sample. Incidentally, this phenomenon is used in monitoring hydrogen concentration in chlorine with a dual-beam analyzer by measuring the difference in the Cl_2 absorption before and after the Cl_2 photochemically reacts with any H_2 in the stream on exposure to the 405-nm radiation emitted by the source.

An example of another elemental halogen that can be measured is the monitoring of fluorine concentration in uranium recovery processes (Fig. 2).

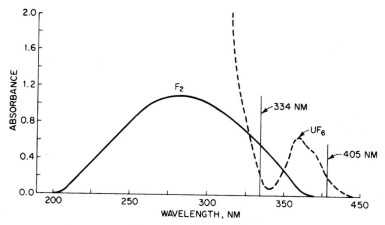

FIGURE 2 Absorption spectra (vapor phase) of fluorine (F_2) and uranium hexafluoride (UF_6), illustrating how a split-beam two-wavelength measurement can be used to compensate for interfering compounds.

Aromatic Compounds.   The benzene ring, with its conjugated double bonds, is a strong absorber of UV radiation. In general, the more substitutions of hydrogen in the ring, the greater the absorbance of the aromatic compound. Aromatic compounds with radicals containing double bonds, such as nitrogen dioxide or sulfonyl (—SO_2—) groups, are particularly strong UV absorbers, as are naphthalene, anthracenes, and related compounds. The absorption band of a para isomer of an aromatic compound usually extends to higher wavelengths than those of the ortho and meta isomers. As a result, the para isomer concentration often can be monitored selectively. Aromatics, including phenols, are monitored in wastewater streams.

Color Formers.   Water-white products, both natural and synthetic, may "yellow" with time despite elaborate processing to remove color. Impurities causing color degradation often are strong UV absorbers, and their presence can be monitored sensitively with on-line UV analyzers. Such products include paraxylene, terephthalic acid, bisphenol A, and corn syrup.

Compounds with UV Absorption Shift.   The UV absorption bands of a comparatively few, but important, chemical species shift significantly with pH changes. This phenomenon can be used to provide highly selective and sensitive analyses for such compounds. Examples include phenols (in the parts-per-billion range) in waste streams.

Oxidizing Agents.   Generally these substances have strong UV absorption bands. These include sodium hypochlorite, chlorine dioxide, hydrogen peroxide, ozone, and potassium permanganate. Bleaching operations may be controlled by monitoring the concentration of oxidizing agents.

Inorganic Compounds.   Salts of iron, nickel, manganese, copper, and other transition metals may be strong absorbers in both the UV and the visible ranges. Anions with strong UV absorption include nitrates, sulfites, and chromates. The concentration of nitric acid may be monitored with UV absorption analyzers.

Sulfur-Containing Compounds.   These compounds often strongly absorb UV radiation. In sulfur recovery plants, two UV absorption analyzers have been used to monitor concentrations of hydrogen sulfide (H_2S) for control of the H_2S/SO_2 ratio for maximum recovery efficiency.

Lower-Explosive-Range Solvents.   The high speed of response and the reliability of UV absorption analyzers qualify them for monitoring and controlling the concentration of many flammable solvents slightly below the lower explosive limits. As a result, highly efficient operation of a process can be maintained under safe conditions. Examples of solvents monitored include toluene, benzene, xylene, acetone, and dimethylformamide.

Other Compounds.   Materials not among those just described, but commonly monitored with UV analyzers, include 1,3-butadiene, furfural, caffeine, ketones, aldehydes, and proteins.

GAS ANALYSIS

Classes of compounds that essentially do not absorb UV radiation include many inorganics (argon, CO_2, CO, helium, HCl, H_2, krypton, nitrogen, O_2, H_2O, and xenon), saturated straight-chain hydrocarbons (butane, ethane, methane, propane), unsaturated straight-chain hydrocarbons (acetylene, ethylene, propylene), the lower alcohols (ethanol, methanol, butanol, propanol, ethylene glycol, isobutanol, isopropanol), acids (acetic, butyric, and largely propionic), all ethers, and esters (butyl acetate, dimethyl sulfate, ethyl acetate, 2-ethyl butyl acetate, and vinyl acetate).

INFRARED PROCESS ANALYZERS[3]

The IR portion of the electromagnetic spectrum extends from just beyond the visible to the microwave region. This is nominally from 0.75 to 1000 μm. The manner in which electromag-

---

[3] Portions of this article were furnished by R. M. Durham, Instruments Division, Infrared Industries, Inc., Santa Barbara, California.

netic radiation interacts with matter is a function of the wavelength of the radiation. IR wavelengths interact at the molecular level. It is this molecular interaction which makes the IR radiation a valuable probe in studying the properties of molecules. The IR wavelength region from 3 to 10 μm is especially valuable in detecting the presence of molecular species. IR radiation at these wavelengths has sufficient energy to excite molecular vibration. The frequency of the vibration is a function of the weights of the atomic elements bound in the molecule and the strength of the molecular bond. This characteristic vibration frequency results in an IR absorption band which is indicative of the presence of a particular molecular bond. For example, the carbon-hydrogen bond is excited by IR radiation at the wavelength of 3.4 μm, while the carbon-oxygen bond in the carbon monoxide molecule is excited by 4.7-μm radiation (Fig. 3).

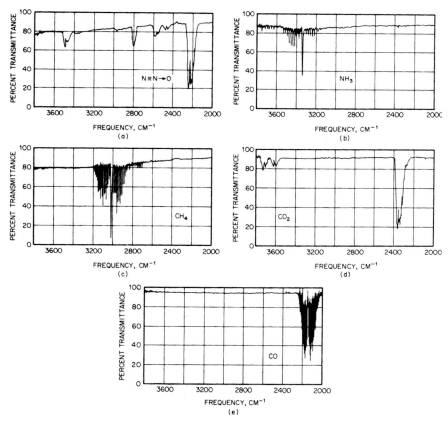

FIGURE 3   Infrared spectra. (*a*) Nitrous oxide. (*b*) Ammonia. (*c*) Methane. (*d*) Carbon dioxide. (*e*) Carbon monoxide. (*Sadtler Research Laboratories.*)

IR spectroscopy has been used for decades in scientific investigations of molecular structure. More specifically, the IR spectrophotometer is the principal laboratory instrument. These instruments are of a dispersive design, that is, a prism or grating is used to separate the spectral components in the source radiation. Contemporary instruments have a wide wavelength range from 2 to 50 μm. Frequently they are used for off-line product quality control.

Basic Components of IR Analyzer. This instrument has evolved from the laboratory spectrophotometer to satisfy the specific needs of industrial process control. While dispersive instruments continue to be used in some applications, the workhorse IR analyzers in process control are predominantly nondispersive IR (NDIR) analyzers. The NDIR analyzer can be used for either gas or liquid analysis. For simplicity, the following discussion addresses the NDIR gas analyzer, but it should be recognized that the same measurement principle applies to liquids. The use of IR radiation as a gas analysis technique is certainly aided by the fact that molecules which consist of two like elements, such as nitrogen (N_2) and oxygen (O_2), do not absorb in the IR spectrum. Since nitrogen and oxygen are the primary constituents of air, it is frequently possible to use air as a zero gas.

Many different analyzer configurations have been developed to address the diverse needs of the industrial process control industry. The basic constituents of an NDIR analyzer are (1) a source of IR radiation, (2) a means of restricting the wavelength range of the source radiation, (3) a means of detecting the IR radiation, (4) a sample chamber to hold the gas or liquid to be measured, (5) a means of modulating the source radiation, and (6) electronics to process the signal generated by the source energy falling on the detector.

Microphone Detectors. For many years IR analyzers largely depended on microphone detectors, which generally were the Veingerov single-sided microphone system or the Luft balanced-condenser microphone system. These are shown schematically and described in Figs. 4 and 5, respectively.

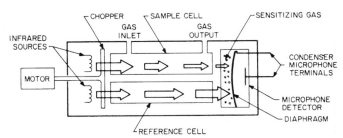

FIGURE 4 NDIR analyzer with Veingerov-type detector. The microphone detector (single-sided) uses an absorbing gas as its detecting medium. When radiation that the sensitizing gas will absorb reaches the detector, the gas heats up and expands. This causes a diaphragm to distend. The diaphragm movement varies the capacity of the condenser microphone, which is part of an electric circuit that generates an electrical output signal. Both analyzers use dual sources, which are chopped to allow energy to pass alternately through a sample cell and a reference cell. When the sample cell contains a nonabsorbing zero gas, such as nitrogen, the modulated beams reaching the detector through the two paths are of equal amplitude. In the case of the Veingerov single-sided detector, the chopper is configured so that at any given time the sum of the cross-sectional areas of the two beams as seen by the detector equals the total cross-sectional area of a single beam so that, when no absorbing sample is present, a constant signal is produced and the output is zero. When a sample is present, the sample and reference path signals become unbalanced and a signal at the chopper frequency is developed. The amplitude of this signal is a function of the concentration of the gas present in the sample cell.

Disadvantages of microphone detection have included sensitivity to vibration and high detector replacement cost. To avoid zero drift, these earlier designs required a stable, balanced output. It will be noted from the diagrams that no lenses or mirrors are used to direct the energy from the sources to the detector. These designs depend on reflection of the IR

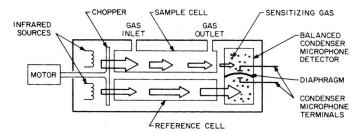

FIGURE 5 The Luft detector operates in a manner similar to the Veingerov-type detector, with the exception that it has two chambers separated by a diaphragm. The signal generated by the presence of an absorbing gas in the sample cell is at twice the chopping frequency. This arrangement has an advantage over the single-sided microphone detection system, since it is less susceptible to vibration caused by imbalance of the chopper motor. Having separate chambers does, however, allow for the possibility of a change occurring in one-half of the detector and not in the other, thus resulting in zero drift. Recently IR process analyzers have been introduced which use a Luft-type detection system but have replaced the diaphragm with flow sensors. The flow of gas from one chamber to another is sensed by the flow sensor rather than by using a capacitance detection technique. This is claimed to eliminate one of the major causes of detector failure—failure of the thin diaphragm. For a given path length, process analyzers which use a microphone detector are more effective than those which use solid-state detectors and optical filters in measuring low concentrations of gases with a lot of structure in their absorption band. This structure results from the molecular rotation spectrum being superimposed on the vibration spectrum and is easily resolved in simpler molecules, such as carbon monoxide, methane, and ammonia.

beam off the interior walls of the sample and reference cells. To achieve high reflectivity, these walls must be highly polished and are frequently lined with gold foil. Inasmuch as a microphone detector uses a gas as its detecting agent, NDIR analyzers of this configuration usually are available only for a limited number of gases. Unstable or corrosive gases cannot be used to charge the cell, although it is possible to use an alternative gas with similar IR absorption properties.

This description of one contemporary model utilizing the Luft detector may serve as an example. The model operates according to traditional non-dispersive IR measuring principles in the optical range from 2.5 to 12 microns. Since the process sample is seldom suitable to flow directly into the analyzer, the sample must first be conditioned to assure stable, long-term operation. A sample conditioning system is custom designed for each application. The optical section is readily accessible on its slide-out tray and is housed in a sealed aluminum case. IR energy is produced by a single heated filament and focused by a parabolic mirror into two parallel beams. These beams pass through a chopper wheel, which alternately strobes the two beams on and off. One beam goes through the sample being measured; the other goes through a reference gas cell. The two beams enter the two sides of a Luft detector whose measurement chambers are filled with the gas being measured. The Luft detector incorporates a unique two-layer, four-chamber design that gives extremely high sensitivity and very low zero drift while avoiding interference from other compounds that may be present in the sample.

Claimed measurement stability is within ±1 percent of the measuring range per week.

Solid-State IR Detectors. Developed in the 1980s, some analyzers, particularly for application in harsh industrial environments, use solid-state sensors for detecting IR radiation. These include thermocouples, that is, pyroelectric devices, the pioneering material being lead selenide (PbSe). These gas analyzers generally are configured as single-path, dual-beam with a reference path, or dual-channel with a reference filter (Fig. 6).

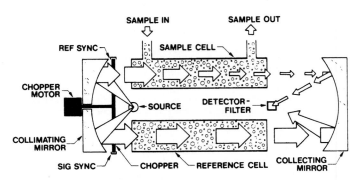

FIGURE 6 Dual-beam NDIR analyzer with a solid-state detector. Optical paths are indicated. Source energy is modulated by a chopper blade, which allows the source to pass alternately through the reference and sample paths. The reference path is always free of absorbing gas. The signals generated by the two optical paths are synchronously demodulated. The amplitude of the reference signal is used by the automatic gain-control (AGC) circuit. The use of a reference path results in stable operation with effectively reduced drift due to power-line or temperature fluctuations. A reference optical filter can be used as an alternative to the reference path. This requires that a spectral window exist where there is no interference from the sample. The 3.8- to 3.9-μm spectral region is frequently used in NDIR gas analyzers for this purpose. Both dual-beam and dual-channel (reference filter) process analyzers can be configured for measuring more than one gas.

A more recent and promising sintered n-type semiconductor consisting mostly of tin dioxide (SnO_2) is in limited production as of 1991. When heated, both reducing and oxidizing gases react with absorbed oxygen on the surface of the semiconductor material, with a resultant change in electrical conductivity. Tests of the device have indicated suitability for a number of industrial gases, including methane, propane, ammonia, H_2, and CO. The device also appears applicable to certain organic solvent vapors. Thus far the devices have been used primarily as gas leak detectors (alarms). Tin oxide sensors also may be used in any areas involving the production of fumes, including factory areas, laboratories, and automotive tunnels and garages. It has been found that the thin-film tin oxide sensors can detect hydrogen sulfide (H_2S) at very low levels (100 ppb). The device originally was developed by Wen H. Ko at Case Western Reserve University.

Fourier Transform IR Spectrometry. Under development over the past few years and presently of limited application, Fourier transform (FT) IR spectroscopy is progressively finding a position in process gas monitoring. This analytical scheme incorporates several new principles in IR measurement technology.

Prior IR analyzers are based on the absorbance of IR radiation by target fluids. In FT-IR instruments a fluid's properties to transmit, radiate, and reflect IR energy come into play—and sometimes all of these properties are taken into consideration together. Each substance has a unique combination of values (at any given wavelength) for these properties, from which one can infer the various properties of gases (species, temperature, composition, and concentration); of liquids (composition, concentration, and temperature; and of surface properties (temperature and composition). The methodoiogy also can be applied to particle populations in a gas because of the ability of a particle to scatter, reflect, and absorb energy.

Computers are required for mathematical solutions to Fourier transforms. Because the method still requires more practical testing, the complex mathematical relationships involved, because of their current limited interest, are not included here.

NEAR-INFRARED SPECTROPHOTOMETERS

The near-infrared (NIR) instrument operates in the 700- to 1300-nm wavelength range, thus differing from a conventional IR absorption device. Although NIR analysis has proved to be useful in the laboratory for a number of years, process instrumentation engineers traditionally considered it to be too complex and cumbersome for on-line applications. Trials during the early 1990s, however, have demonstrated its applicability to more rugged service. Industrial designs now are faster, more durable, simpler, and capable of incorporating all of the accoutrements and intelligence expected of contemporary instrumentation.

In the NIR spectrum multiple vibrational overtones broaden absorbance bands, thus yielding unusually clear and distinct signals. This is a dividend for practical constituent analysis. Available NIR instruments combine the modeling function with the on-line monitoring function. With appropriate software, the instrument is described as simple to operate and understand. The instrument also offers multipoint fiber-optic capability. One successful application—measuring the gasoline distillation range on line—is reported by Zhaohui and Xinlu.[4]

The NIR signature for a particular sample matrix, such as liquid, paste, solid, slurry, transparent, translucent, or opaque material, is a unique fingerprint. Minor changes in the fingerprint divulge variations in the sample's constituents with precision.

COMBUSTIBLES ANALYZERS

The measurement of combustible gases is important in controlling their concentrations within safe limits below the lower explosive limit (LEL), optimizing concentrations for efficient reactivity, and detecting process system upsets. The principal method used for combustible gas or vapor analysis employs self-heated hot-wire detectors of a material such as platinum. These detectors are mounted in a Wheatstone bridge circuit. The hot catalytic detector causes the combustibles in the sample to burn in the presence of oxygen. If the sample does not contain an excess of oxygen, then oxygen or air must be added in controlled amounts to provide an excess such that the reaction will be limited by the amount of combustibles present.

The combustion of the mixture releases heat, which is detected as a temperature rise in the catalytic detector spiral. Since the spiral's resistance is a function of its temperature, an electric unbalance occurs in the Wheatstone bridge proportional to the concentration of the combustibles. Measurement of this electric signal provides a measure of the combustibles concentration.

Instruments for measuring the concentration of total combustibles are widely used for continuously analyzing usually inert atmospheres for the presence of a wide range of combustible gases or vapors in the interest of preventing explosions. The instruments are commonly calibrated in terms of the LEL, specifically 0 to 1, 0 to 2, 0 to 5, 0 to 10, 0 to 20, 0 to 25, and 0 to 100 percent LEL. The instruments usually have high-low contacts, which actuate panel lights and can direct output relays or recorder or controller signals. These instruments find frequent application in controlling inert gas generators, solvent evaporating ovens, and heat-treating atmospheres, and in analyzing flue gas. Instruments designed for mounting in a permanent location usually are equipped for drawing a sample from a distance of up to 100 feet (30 meters). The sample gas is mixed with an equal amount of ambient air to ensure a sufficient supply of oxygen for combustion. The response time is 5 seconds for 90 percent of full-scale, and the repeatability is ±2 percent of full-scale.

---

[4] C. Zhaohui and F. Xinlu, *Hydrocarbon Processing*, 94, January 1992.

In addition to the availability of instruments for permanent mounting, semiportability, and portability (including battery-operated, hand-held meters), there are numerous models specifically designed for specific applications. Practically all of these instruments operate on the catalytic filament principle described previously. Examples include instruments designed for use in testing atmospheres that may be oxygen-enriched (more than 21 percent O_2). The rate of flame propagation of such mixtures is much higher than that of other combustibles in air. Therefore such meters are equipped with heavy-duty flashback arrestors capable of confining explosions of hydrogen or acetylene and oxygen within the combustion chamber. Instruments of this type are calibrated on hydrogen and acetylene. Special designs are available for the detection of vapors of leaded gasoline. When a hot-wire indicator is used with these vapors, the oxidation of tetraethyllead can produce a solid lead combustion product which condenses on the filament and reduces its catalytic activity, especially with respect to combustibles such as natural gas, which have high ignition temperatures. In this case the filament temperature is maintained sufficiently high to prevent the condensation of lead contaminants. In an alternative approach, an inhibitor filter may be used in the filter chamber. This filter promotes a chemical reaction with tetraethyllead vapors, which yields a more volatile combustion product and thus prevents filament contamination.

Silanes, silicones, silicates, and other compounds containing silicon in the tested atmosphere may seriously impair the response of combustibles meters because these materials can rapidly poison the filament. Where such materials are suspected, the instrument manufacturer will suggest that the instrument be checked after every fifth test.

Specially designed combustibles meters are available for use in detecting, measuring, and pinpointing leaks. Portable instruments are equipped with probe tubes or rods ranging from 3 to 4 feet (0.9 to 1.2 meters) in length, which are particularly effective for sampling from manholes or barholes and for testing tanks or other vessels which may contain liquids. The probe rod prevents the drawing of liquids into the flow system. A meter designed especially for gas utilities in routine testing for methane-in-air concentrations in manholes, sewers, curb boxes, and other street openings may be calibrated on methane in air. A similar meter for general industrial use may be calibrated on pentane in air. Pentane calibration is used because it is representative of petroleum vapors. Applications include testing tank and vessel interiors, locating leaks in pipelines and process streams, and checking confined areas in sewage disposal plants, refineries, paint factories, chemical plants, and iron and steel mills, among others. To prevent errors caused by stray magnetic fields, the instrument may use a core magnet-type meter movement. Where it is necessary to distinguish between condensable hydrocarbon vapors and natural gas, a charcoal cartridge which absorbs hydrocarbon vapors may be used. The difference in the readings with and without the charcoal cartridge indicates that vapor, gas, or a combination of the two is present in the sample.

In some instruments, two different types of filaments are used—a catalytic combustion filament for low ranges (percentage LEL) and a thermal conductivity filament for high ranges. In connection with the latter, combustibles in the sample cool this filament, decreasing its resistance and unbalancing the bridge. The imbalance, proportional to the gas concentration, is measured by the meter and read as percentage by volume.

TOTAL-HYDROCARBONS ANALYZERS

Analyzers for determining total hydrocarbons in various concentrations are widely used for pollution monitoring and control programs, particularly in connection with internal combustion engines (automobile exhausts) and fossil-fueled stationary combustion installations. They are also used for leak detection in aerosol packaging, to detect breakthroughs from carbon adsorption beds, and for leak detection in refrigerant systems, among other applications.

Some total hydrocarbon analyzers are based on the ionization of carbon atoms in a hydrogen flame. Normally a flame of pure hydrogen contains an almost negligible number of ions. The addition of organic compounds (even traces) results in a large number of ions in the flame. In one analyzer[5] the sample is mixed with a hydrogen fuel and passed through a small

---

[5] MSA total hydrocarbon analyzer.

jet. Air is supplied to an annular space around the jet to support combustion. Any hydrocarbon carried into the flame results in the formation of carbon ions. An electric potential across the flame jet and an "ion collector" electrode suspended above the flame produce an ion current proportional to the hydrocarbon count. This current is measured by an electrometer circuit whose output then provides an analog analysis signal for a direct-reading meter, graduated from 0 to 100, or for an optional potentiometric recorder. The analyzer features an electrical range attenuator, with change factors of ×1, ×3, ×10, ×30, ×100, ×300, ×1000, and ×3000. The stabilized zero setting is unaffected by range change factors. Calibration is by a span potentiometer—for greater accuracy than is provided by adjustment of the sample flow. The analyzer features sintered metal filters for the sample, air, and fuel capillary tubing. The instrument includes a flame-out alarm and an automatic fuel shutoff for maximum safety, and an optional integral sampling pump. The full-scale range of the instrument is 0 to 4 to 0 to 12,000 ppm by volume. Expressed as methane, the sensitivity is 1 percent of full-scale reading in 1 second or less, the response is 90 percent of final reading in 1 second or less, and the warm-up time is 4 hours from a cold start to stability and 1 hour after a temporary shutdown.

Another version of this instrument uses a dual-flame detector and continuously monitors ambient air for methane and nonmethane hydrocarbons. The instrument can measure (1) total hydrocarbons, (2) total hydrocarbons less methane, and (3) methane only. The full-scale range is 0 to 5 and 0 to 20 ppm. The instrument is used for pollution monitoring and frequently for checking the effectiveness of hydrocarbon control equipment.

CARBON MONOXIDE ANALYZERS

Numerous uses for carbon monoxide (CO) detection include chemical, metallurgical, and many other types of manufacturing plants, gas and utility properties, sewers, flight test centers, mines, bus terminals, and garages. The instruments are also used for testing compressed air intended for delivery to respiratory equipment and for controlling ventilating equipment (such as in a vehicular tunnel).

In addition to indication, CO meters almost always incorporate both audible and visible alarm circuits. The threshold limit value for CO adopted by the American Conference of Governmental Industrial Hygienists is 50 ppm (Fig. 7). As with the combustibles analyzers described previously, a large variety of configurations are available—from small, portable units to permanently mounted indicators and recorders, some of which are equipped to handle inputs from several detectors at various locations.

In one instrument the CO detector cell operates on the principle of catalytic oxidation. In the cell the sample passes through an inactive chemical bed and then through an active catalytic bed of hopcalite. Each bed contains a thermistor, making up part of a Wheatstone bridge circuit. Any CO present is immediately oxidized in the hopcalite half of the cell. This raises the temperature of the hopcalite thermistor and its electrical resistance, which causes a resistance imbalance in the bridge and thus a signal proportional to the CO concentration in the sample.

In another system the sensor contains an electronic interface and an electrochemical polarographic cell containing a sulfuric acid electrolyte. Air samples diffuse through a gas-porous membrane and a sintered metal disk and enter a sample area within the cell. The cell electrooxidizes CO to CO_2 in proportion to the partial pressure of CO within the sample. The oxidation generates an electrochemical signal proportional to the concentration of CO in the ambient air. The resulting electrical signal is monitored, temperature-compensated, and amplified to drive a meter. Under normal operation, the service life of the sensor assembly is usually a minimum of 6 months. Certain easily oxidized compounds in the sample atmosphere act as interferents in CO readings. These mainly include methane, ammonia, and sulfur dioxide. Hydrogen, hydrogen sulfide, nitrogen dioxide, propane, nitric oxide, ethylene, ethyl alcohol, and acetylene affect readings to a much lesser extent.

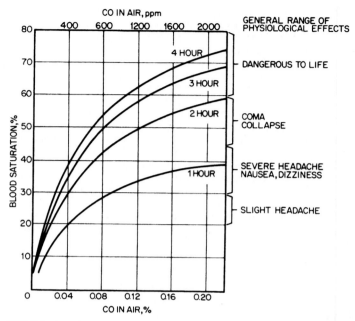

FIGURE 7   Effects of carbon monoxide on human beings. This chart can be considered only a general guide, because the percentage of CO blood saturation varies with exertion, excitement, fear, depth of respiration, anemia, and the general physical condition of the individual.

REACTION PRODUCT ANALYZERS

Some of the instruments described in this article are available for ready placement on line; others require special customizing for continuous process measurements. Most reaction product analyzers are off line, even though this method of analysis predates other analytical methodologies. Basically these methods tend to be better suited to the laboratory environment. There are a few exceptions.

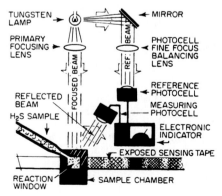

FIGURE 8   Moving reagent tape analyzer for hydrogen sulfide measurement. (*Houston Atlas.*)

Impregnated-Tape Devices.   Systems are available which utilize reagent-treated tapes to analyze for gases or suspended particles in industrial processes or atmospheres. They can be used to monitor exhaust or smoke in combustion control; dusts, aerosols, or corrosive and toxic gases in air pollution control; and to analyze and control specific components in gas mixtures over wide ranges of concentrations from fractions of a part per million to several percent (Fig. 8). The apparatus converts sulfur-containing compounds into H_2S to obtain specific measurements of the total sulfur in gas or liquid hydrocarbons. Continuous on-stream analysis for total sulfur in samples ranging from C_1 through C_{15} includes alcohols, ketones, amines, water, gasoline, kerosene, finished and intermediate products, and paraffinic and olefinic hydrocarbon gas mixtures. Direct measurement of hydrogen sulfide also is important because it is a major sulfur contaminant in fuel gas undergoing combustion. When combined and diluted with excess air and combustion products, it establishes the maximum concentration of sulfur dioxide that is permitted to leave the process and flow from the stack. Hydrogen sulfide detectors also are used for wide-area monitoring.

Titrators.   With good design, user nurturing, and frequent and reliable attention and maintenance, on-line titrators have solved some analytical problems that otherwise had gone unsolved. An on-line titrator repeats the same steps as those performed by a laboratory technician. Sample introduction, reagent addition, titrant addition, and cleanup are automated. These operations require numerous pumps, stirrers, and motors. A successful titrator application involves measuring 0 to 20 percent caustic in a gas scrubber, where the caustic is recirculated until a predetermined depletion level is reached, which is important to safety and environmental concerns. Electrical conductivity and density measurements, normally used for caustic measurements, proved unsatisfactory. Automatic titration, in this instance, were found to be the best solution. In general, however, the automation of the titration process within the plant control laboratory provides increased reliability and cost savings, and for the moment, this appears to be the best niche for this instrumental method.

Photometric Analytical Technologies.   These methodologies essentially are used best in a laboratory environment. In fluorescence analysis the amount of light emitted characteristically, under suitable excitation, is used as a measure of concentration. The method is closely related to colorimetric or spectrophotometric analysis, in which the amount of light absorbed is used to measure the concentration of dissolved species. The main advantage of fluorescence methods is their high sensitivity (1 part in 10^8) in many determinations of both inorganic and organic substances. This is two to three orders of magnitude greater than the sensitivity of many absorption methods, where sensitivity is limited by the necessity for detecting a very small fractional decrease in the light transmitted by the solutions.

Bioluminescence and Chemiluminescence.   As qualitative and quantitative indications of substances that exhibit these phenomena, these processes have grown in acceptance as important laboratory analytical tools. A differential photometer can be used to detect small amounts of nitric oxide (down to 0.05 ppm full-scale) in a gas stream (Fig. 9).

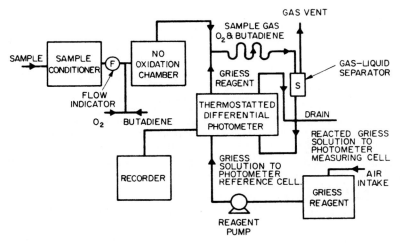

FIGURE 9 Nitric oxide analyzer using the colorimetric principle. (*Penn Airborne Products.*)

Turbidimetric Methods.   These methods are described under "Turbidity Measurement" in this handbook section.

MASS SPECTROMETRY

In the traditional mass spectrometer a sample is introduced into an evacuated area (ion source), ionized, accelerated by an electrostatic field, separated according to masses, and then collected and measured. A complete analysis is obtainable from submicrogram samples within seconds. The range extends to parts per billion. The accuracy is well within the needs of process measurement. Numerous configurations of mass spectrometers are available for laboratory control and research work. They usually are complex and quite costly.

Only recently has the quadrupole mass spectrometer been available for on-line use. An instrument of this type is diagramed and explained in Fig. 10.

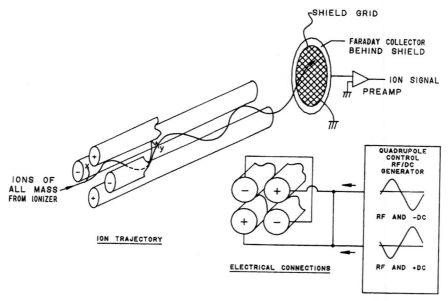

FIGURE 10   Quadrupole mass spectrometer useful for the analysis of streams containing N_2, CO, H_2O, and CO_2. In some cases a gas or liquid chromatograph is used to introduce the sample into the ionizer. When used in this manner, it is most common to scan a wide mass range [50 to 1000 atomic mass units (amu)] at rates on the order of 1000 amu/s for compound identification. Simply, a quadrupole mass spectrometer can be divided into three parts: (1) the ionizer, (2) the mass filter, and (3) the detector, all of which are contained in a vacuum chamber maintained at a low pressure. When a gaseous sample is introduced into the system's ionizer, it is bombarded with a stream of electrons, producing positively charged parent ions (ions with the same molecular weight as the neutral molecule) and fragment ions. The ionizer has a series of lenses which serve to collimate the cloud of sample molecules toward the mass filter. A quadrupole mass filter is a set of four rods disposed in parallel and symmetrically with one another, opposite rods are electrically connected. A radio frequency and dc voltage of equal potential, but opposite charge, is applied to each set of rods. By varying the absolute potential applied to the rods, it is possible to stop all ions except those of a given mass-to-charge ratio m/e. Finally, the ions flowing down the quadrupole strike a Faraday plate detector. In some cases the signal is amplified further by an electron multiplier. Thus a spectrum of signal intensity versus m/e value is obtained. Each molecule has a unique fragmentation pattern, so that a spectrum can be used as a fingerprint for compound identification. In addition, it is possible to quantitate the amount of a particular compound by comparing the sample signal intensity with the intensity produced by a known amount of the compound. (*Extranuclear Laboratories.*)

SPECTROTERMINOLOGY

The prefix *spectro* has been applied to a wide variety of analytical instruments developed over the years. For example, there are spectroscopes, spectrographs, spectrometers, and spectrophotometers, among others. The additional compounding of names as found, for example, in spectroradiometer or spectropolarimeter tend to confuse the nomenclature of this broad class of instruments. Unfortunately the nomenclature has evolved over a long period in an unplanned and often inconsistent way. Some generalizations are pointed out here, but these terms cannot be used reliably to specify an instrument simply because standardized definitions do not exist.

Spectro derives from the underlying base word, *spectrum*. Initially a spectrum referred to the rainbow's series of colors, ranging from violet through indigo, blue, green, yellow, orange, and red, which are produced when visible (white) light is split into its component colors. A simple device for accomplishing this, such as passing sunlight through a glass prism and providing a means to note the colors visually, is a form of *spectroscope*. The term spectrum no longer is confined to the visible portion of the total electromagnetic radiation span, but now is applied to all major portions of the electromagnetic spectrum, ranging from the shorter-wavelength x rays, gamma rays, and cosmic rays to the longer far-infrared waves, microwaves, and radio waves. Thus such terms for designating instrumental techniques as *infrared spectroscopy, x-ray spectroscopy, ultraviolet spectroscopy,* and *microwave spectroscopy* have evolved. Moreover, the term spectrum no longer is confined to the designation of an electromagnetic spectrum, but also is applied to nearly any orderly array of phenomena. Thus *mass spectrometers* sort atoms or radicals by their atomic weights (mass). There are spectrometers to analyze the decay "spectra" of radioactive isotopes. Some of the more important fields where the terms spectroscopy and spectrometer have been used include the following:

Electromagnetic spectrum: Gamma ray, x ray, ultraviolet, infrared, near infrared, far infrared, microwave, and radiowave

Other phenomena: Radioactive decay, Mossbauer effect, and Fourier transform

Energies of particles: Beta rays (electrons), protons, neutrons, and mass

In general, the use of the suffixes *-scope, -graph, -meter,* and *-photometer,* in combination with *spectro-,* designates the kind of detector used in the instrument. If the spectral phenomenon produced by the instrument is predominantly for reading, that is, visible to the unaided eye, the instrument will be called a *scope,* thus spectroscope. If data are recorded by film or printed, the instrument is a *spectrograph.* If a meter is used to display data detected by a bolometer, thermocouple, thermistor, or other detector, the instrument is a *spectrometer.* If the instrument provides for measuring the intensity of various portions of spectra (not necessarily confined to the range of visible light), the designation *spectrophotometer* is used.

The designation of an instrument also may describe the basic source of radiant energy whose spectrum is measured and the manner in which that energy is processed or presented by the instrument. Some instruments display a continuous spectrum of light or other radiation. Other instruments pass the energy through filters of other absorption means to eliminate or absorb certain wavelengths. An instrument of this latter type may be termed an *absorption spectrometer* or an *absorption spectrograph,* and so on. An *emission* designation generally indicates a device wherein the energy level of the specimen to be analyzed is raised by energy provided by the instrument, as by means of a spark discharge or flame heating, thus such terms as *optical emission spectrometer* or *flame photometer.*

Instrument designations can be complicated further by indicating reference to some special function. Thus an *ultraviolet-visible spectropolarimeter,* which is an instrument for measuring the rotation of the plane of polarization, according to wavelength and light intensity. A *spectrophotofluorometer* is an instrument with means both for controlling the exciting wavelength and for identifying and measuring the light output of the fluorescing sample; whereas a *fluorometer* may have a filtered light input and will measure the fluorescent light from the sample, frequently by wavelength. Also, an instrument may carry a designation that relates to a major component used, such as a prism, a diffraction grating, or a filter type. Thus a *prism spectroscope,* and *x-ray diffractometer,* and so on.

From an unscientific standpoint, suppliers tend to coin their own imaginative trade names and thus further add to the confusion of the nomenclature.

REFRACTOMETERS[1]

The refractive index of a substance as a measure of chemical identification and concentration is one of the earliest physical criteria (Snell, 1621) to serve as the basis for chemical analytical instrumentation.

Refraction refers to the change of direction which radiation (notably light or sound) takes place when a beam of light passes from one medium to another. This bending is relatively easy to measure optically and accounts for the very early use of this principle in laboratory precision instruments such as the Abbe and Pulfrich refractometers. Over the years, as industrial process control evolved, refractometers were designed specifically for such applications. Traditional refractometers were ruggedized and design attention was paid to continuous on-line measurements and to more sophisticated laboratory control instruments with increased speed and convenience. These instruments represented the practical application of Snell's law:

$$\sin i = n \sin r$$

where
i = angle of incidence
r = angle of refraction
n = refractive index

Traditional instruments use two design approaches: (1) the critical-angle method and (2) the differential method. These methods are illustrated and briefly described in Figs. 1 and 2.

In the more recent fiber-optic optoelectronic designs the refractive index is measured directly in terms of the wavefront velocity differences, that is, the incident beam and the refracted beam. In the unit shown in Fig. 3, the fluid of interest enters a cavity through capillary action and as its refractive index changes, so too does the effective path length (path length × refractive index) and thus the extent of spectral modulation. Sensors mounted directly on process vessels or piping contain a microminiature Fabry-Perot interferometer, which is connected to an optoelectronics module by fiber. The model measures the extent of light modulation which is proportional to the refractive index.

Advantages claimed for the fiber-optic sensors include intrinsic safety, immunity to electromagnetic and radio-frequency interference—hence accuracy is not impaired in electrically noisy environments and the miniature size allows minimal disturbance of the process being sensed.

---

[1] Contributed in part by Gene Yazbak, Manager of Applied Systems Engineering, MetriCor, Inc., Monument Beach, Massachusetts.

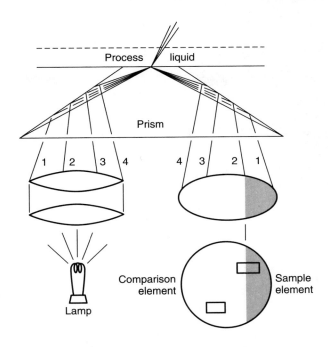

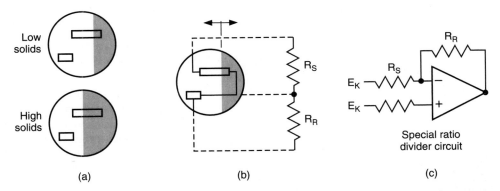

FIGURE 1 (*Top*) Light path of critical-angle refractometer. Movement of the boundary between the light (reflected) and the dark (refracted) areas is monitored by a specially designed photodetector. As the stream concentration changes, the critical angle changes proportionally and the amount of light ray reflected also changes proportionally. (*Bottom*) (*a*) The detector is positioned with one element in the constant-light area, the other at the light-to-dark edge. These are the reference cell and the sample cell, respectively. (*b*) The shadow edge on the sample moves as the process concentration varies, changing the resistance value of the sample cell. (*c*) The reference and sample cells are placed in the feedback loop of an operational amplifier. This input stage divides a stable constant voltage exactly by the ratio of the dual-element photodetector resistance, thereby converting the optical signal into a directly related linear electrical signal.

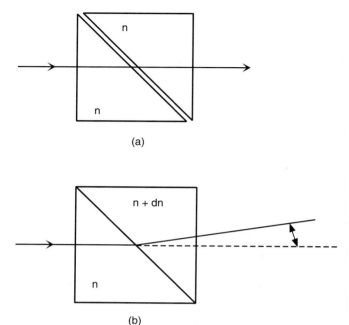

FIGURE 2   Principle of operation of a differential refractometer. (*a*) Refractive indexes of sample and reference are identical. (*b*) Refraction occurs where refractive indexes of sample and reference differ.

REFRACTOMETER APPLICATIONS

In addition to the historic uses of refractometers in the measurement of sugar solutions and the identity of many chemical substances in the process industries, other uses include the following.

1. The pulp industry for the control of strong and weak black liquors and waste by-products. They are used to monitor black liquor supply to furnaces and before and after the evaporators. A refractometer, applied on the black liquor line, can trigger alarm devices to warn of low solids, thus avoiding smelt water explosions.

2. The petroleum industry for determining the oil in wax in dewaxing plants, lube oil quality after treatment, cat cracker product quality, the composition of methylethyl ketone and toluene mixtures.

3. The chemical industry for the determination of factors in the control of resin compositions, the control of vacuum distillation product purity, the dilution of caustic, formaldehyde, sulfuric acid, and acetic acid, and for glycerine-water blending. They also find usage in the control of styrene, ethylbenzene, cellulose acetate, and acetone production, in the control of solvent polymer mixtures, as well as of soap solutions and waste streams.

4. Food-processing applications are numerous. Refractometers are used to measure the rate and extent of dissolution and the extent of agitation in operations where solids are added to liquids. In these situations, further processing usually cannot proceed until the solids are dissolved completely and mixed homogeneously. An example is the preparation of solutions to be injected into poultry. Generally, salt, sugar, and flavorings are added to water sequentially. The concentration of each component can be measured. Complete dissolution

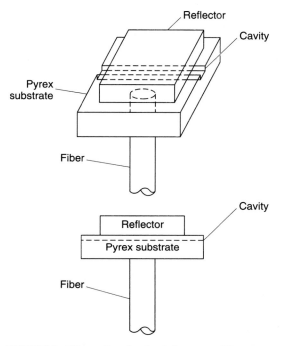

FIGURE 3 Fiber-optic refractive index sensor. The sensors are connected to an optoelectronics module, which can be located up to 3000 feet (914 meters) away. Although the probes can be made as small as 0.032 inch (0.8 mm), they usually are packaged as ⅛-inch (3.2-mm) stainless-steel probe assemblies where they are interfaced directly to process vessels or pipes, using off-the-shelf compression fittings. (*Metri-Cor, Inc.*) *Note:* Diagram of optoelectronics module can be found under "Fiberoptic Temperature Sensor" in Section 4 of this handbook.

and homogeneous mixing are indicated by a stable refractive index reading. The candy industry uses this same technique for monitoring sugar and chocolate blending and heating operations.

5. Hydrogenation operations require real-time monitoring. The extent of hydrogenation in the production of edible oil hydrogenated products, that is, the degree of saturation of the double bond traditionally is indicated by the iodine value. This has been found to correlate well with the refractive index (Fig. 4). As contrasted with several labor-intensive and time-consuming laboratory control operations, it has been determined that a fiber-optic refractive index sensor installed on processing vessels can provide real-time continuous data and eliminate the need to halt and restart reactions for analysis. The refractive index instrumentation allows the reaction to proceed until a specified degree of saturation has been attained. Toleration of high temperatures and the presence of suspended, agitated solids in a reactor had posed some difficulty for traditional refractive index instruments. Usually, extensive sample conditioning involving pumping and filtering have been sources of maintenance concerns. By contrast, the fiber-optic probe has an opening of less than 2 µm (avoiding clogging), and it is ruggedized to withstand reactor turbulence.

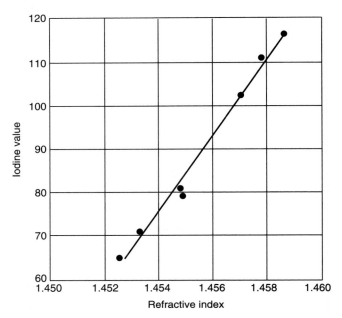

FIGURE 4   Plot of iodine values versus refractive index readings of hydrogenated vegetable oils. (*MetriCor, Inc.*)

TURBIDITY MEASUREMENT

by Donald Lex[1]

Turbidity can be defined as an "expression of the optical property that causes light to be scattered and absorbed rather than transmitted in straight lines through the sample." Simply stated, turbidity is the measure of relative sample *clarity*. The property is not to be confused with color. Turbidity is caused by the presence of undissolved material in the liquid. Undissolved matter can be defined as particles which can be filtered out of the liquid. In most cases, the use of a 0.2-μm filter is the delineation between dissolved and undissolved material. Particles smaller than 0.2 μm in size which pass through the filter can be considered dissolved. Fine, undissolved particles are sometimes known as suspended solids.

When a light beam passes through a liquid containing undissolved particles, the light is scattered in all directions. When this occurs, the transmitted light is reduced in intensity from the incident light by a ratio proportional to the amount of undissolved solids present. This is the difference between low-level turbidity and high-level turbidity measurement. Low-level turbidity detects the amount of light that the undissolved particles are scattering.

[1] Great Lakes Instruments, Inc., Milwaukee, Wisconsin.

High-level turbidity detects the change in intensity of the transmitted light due to undissolved particles.

TURBIDITY MEASUREMENT APPLICATIONS

The measurement of turbidity dates back to the wine and brew masters of the Middle Ages who would carefully check the quality of their beverage by holding a glass of the product up to the light and look for small reflectance of particles suspended in the liquid. This early use of a light to determine the quality of a beverage was the foundation upon which turbidity measurement is built. Modern applications, in addition to breweries and wineries, include water treatment for municipalities and for industrial uses, such as potable water, boiler feed water, and process water of high purity as required, for example, in pharmaceutical and electronic component manufacturing. The early use of ordinary room illumination as a light source and of the brewmaster's eye as a detector was replaced by a photomultiplier tube. With the increasing importance of turbidity measurement to various industries, measurement techniques have been greatly refined to provide more accurate and stable readings.

PARTICLE-SIZE EFFECTS

The size of specific undissolved particles will influence turbidity measurement. The relationship between the light scattered at 90° to the transmitted light and the particle size of the undissolved material is shown in Fig. 1. As the particle size increases to 1 μm, there is a steep rise in the intensity of the scattered light. As the particle increases beyond 1 μm, there is a steep decrease in the intensity of the scattered light.

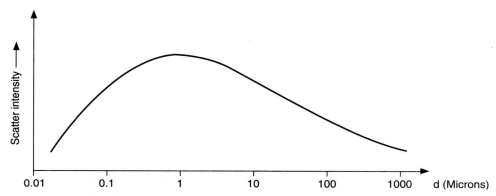

FIGURE 1   Intensity of scattered light as a function of particle size.

Similar graphs can be generated for the light scattered at angles other than 90° from the transmitted light. The combination of this information allows for the development of a light-scattering distribution pattern based on the size of the undissolved particles. There is also a difference in the intensity of the scattered light that occurs at different angles of detection. An undissolved particle that is larger than the wavelength of the incident light will have about 100 times the light scatter intensity at a 25° forward scatter than at a 90° scatter.

PARTICLE-SHAPE EFFECTS

Undissolved particles can have many shapes. The complete spectrum of shapes can include round spheres, granular cubes, flat flakes, sharp fibers, and irregular combinations of all of these shapes. As each of these different particles rotates in orientation to the incident light, the apparent size of the particle changes. This alters the relative intensity of the scattered light.

PARTICLE COLOR AND INDEX OF REFRACTION EFFECTS

Undissolved particles will differ in their color and index of refraction as well as in shape. It is theorized that undissolved particles do not actually reflect the light they scatter, but the light actually passes through the particles coming out at different angles and intensity. This is due to the difference between the refractive indexes of the material of which the particle is composed and the liquid in which it is suspended. As the index of refraction of the undissolved particles changes, the angle at which the light is scattered will be changed.

If, in addition to the particle's index of refraction, the particle has color which absorbs the specific wavelength of the incident light, the scatter intensity will be reduced in accordance with the amount of light the color absorbs. Simply put, it is easier for a turbidimeter to detect clear sand granular than black coal granular. Black coal dust will absorb the incident light, thereby providing a very weak light scattering.

OPTICAL DESIGN LIMITS

As with all measurement techniques, there is an upper and a lower level of detection accuracy. As the concentration of undissolved particles increases in a liquid, the intensity of light scatter also increases (Fig. 2). This relationship exists up to a transition point, where the rate of scatter intensity no longer increases with the increase in undissolved particles. This point is the maximum limit of the optical design of a turbidimeter.

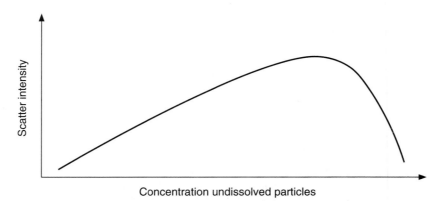

FIGURE 2   Intensity of scattered light as a function of concentration.

BASIC TYPES OF TURBIDIMETERS

There are two types of light that can illuminate the photodetector: (1) scattered light due to suspended undissolved particles and (2) light that is diverted through means other than the suspended particles and reaches the photodetector (Fig. 3). This is known as stray light and is the lower limit of the optical designs of a turbidimeter.

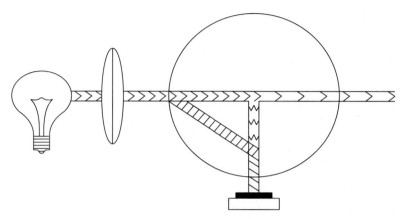

FIGURE 3   Effect of stray light on instrument design.

The fundamental components of a turbidimeter consist of a light source, a photodetector, and a sample container. The characterization and repeatability of the turbidimeter are dependent on the spectral peak and the distribution of both the light source and the photodetector. The tighter the spectral distribution, the higher the repeatability with the same type of undissolved particles. This means that with a single wavelength of light passing through the sample container, the scatter pattern and intensity will always be the same. As the spectral distribution becomes broader, the number of different wavelengths contacting the undissolved particles increases, thereby increasing the existing number of different scatter patterns and intensities. This increase in scatter patterns will greatly reduce repeatability with the same type of dissolved particles.

There are currently no single-wavelength photodetectors available. In contrast, there are several methods of producing a single-wavelength light source. This type of light is known as monochromatic light. Monochromatic light can be produced through a single-wavelength light source such as a light-emitting diode (LED) or by utilizing optical filters which allow only a single wavelength to pass through. Use of a monochromatic light source and a narrow-spectral-band photodetector is essential in developing a turbidimeter with high reproducibility of measured values.

The basic instrument design uses a single light source and a single photodetector at 90° to the transmitted light (Fig. 4). This design, which has been in existence for many years, has some inherent problems. As the light source ages, its intensity is reduced. The instrument is calibrated in terms of light intensity. Thus to offset the effects of a lowering intensity, the instrument must be calibrated frequently to establish a new value of the scattered light intensity. This design also lacks the ability to provide a stable measurement value at higher levels of turbidity. To assist with the original photodetector, additional detectors can be added at other angles, as shown in Fig. 5. A multiple-detector system increases the stability of the measured turbidity values.

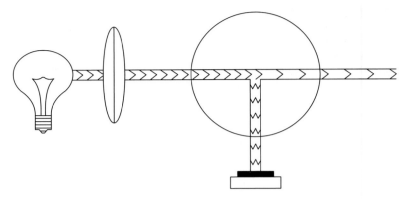

FIGURE 4   Basic turbidimeter design.

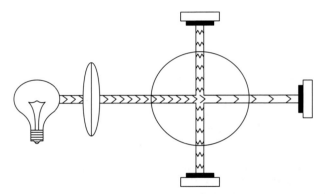

FIGURE 5   Ratio technique used in turbidimeter design.

The design of Fig. 5 also reduces the effects of light reduction caused by the color of the liquid suspension. Thus although a multiple detector using a ratio measurement technique is an improvement over the single-detector design, the problem with light source decay and frequent calibration still is present.

DUAL-BEAM TECHNIQUE

The dual-beam technique shown in Fig. 6 minimizes the aforementioned effects. In this technique the measurement is a differential value of the light intensity of a measuring beam and a reference beam. The method reduces the need for frequent calibration and, when used with a monochromatic light source, totally eliminates the need for calibration. The dual-beam technique, however, does not address the problem of instability of readings at higher turbidity values.

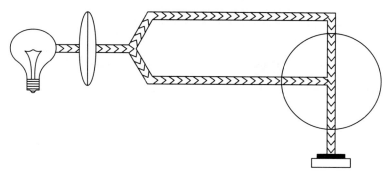

FIGURE 6   Dual-beam technique used in turbidimeter design.

RATIO FOUR-BEAM TECHNIQUE

A modulated four-beam measurement, as shown in Fig. 7, incorporates all of the advantages of the aforementioned designs. The technique of multiple detectors with a ratio measurement provides superior stability, while the multiple-beam technique with a monochromatic light source eliminates the need for frequent calibration.

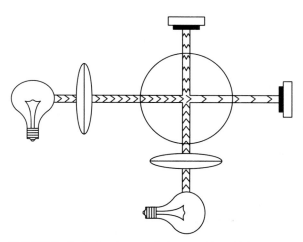

FIGURE 7   Ratio four-beam technique used in turbidimeter design.

TURBIDITY CALIBRATION STANDARDS

Each instrument, regardless of type, must be calibrated against a known standard. The most widely used unit is the formazin turbidity unit (FTU), which can be used for all instruments when formazin is the calibration material. The U.S. Environmental Protection Agency (U.S. EPA) uses formazin, but states the measuring unit in terms of nephelometric turbidity unit (NTU). The International Standards Organization refers to the unit as formazin nephelometric unit (FNU).

Formazin is a suspension produced from the polymerization of hexamethylenetetramine and hydrazine sulfate under controlled conditions. Although formazin offers advantages over sand or diatomaceous earth, it does lack the long-term stability and reproducibility needed in a universally accepted primary standard. The calibrating material is produced as a 4000-FTU suspension. The reproducibility of this suspension, using the same materials (hexamethylenetetramine and hydrazine sulfate), is ±1 percent. However, the subsequent dilutions of this suspension to lower turbidity values become increasingly less stable with greater dilution. Thus it is good practice that diluted calibration suspensions be used immediately and then discarded.

Most modern turbidimeters are equipped with a secondary calibration standard. These standards are used to standardize the instrument or check calibration to determine when calibration with formazin may be required.

STANDARD TURBIDITY SPECIFICATIONS

Two standard specifications for turbidity measurements are in use worldwide: (1) the international standard (ISO 7025 (dated 1984) and (2) the U.S. EPA standard 180.1). The specification of the ISO standard is more stringent and requires the use of a monochromatic light source. The ISO specification allows for greater reproducibility of the measured values and better agreement between other measuring instruments (Table 1).

TABLE 1 Comparison of Turbidity Measurement Standards

	U.S. EPA 180.1	ISO 7027
Wavelength	Tungsten lamp operated at color temperature between 2200 and 3000 kelvin	860 nm
Spectral bandwidth	—	60 nm no divergence (convergence 1.5°)
Measuring angle	90° ± 30°	90° ± 2.5°
Aperture angle in water sample	—	20–30°
Distance traversed by incident light and scattered light within sample	10 cm	—
Calibration standard	Formazin or AEPA-1	Formazin

NOTES: U.S. EPA = U.S. Environmental Protection Agency; ISO = International Standards Organization.

RHEOLOGICAL SYSTEMS

Rheology may be defined as the physics of form and flow of matter and embraces such material properties as elasticity, viscosity, plasticity, and consistency. The field embraces gas, simple and complex liquids, including flowable suspensions of solids in liquids, and, less frequently, substances that may be considered "flowable" solids. In terms of industrial laboratory and process and laboratory instrumentation, principal attention is given to measuring and controlling viscosity and consistency.

VISCOSITY[1]

The physical property of fluids that arises from internal friction within a fluid, *as it flows,* is viscosity. The effects of internal friction are apparent when a given "layer" of the fluid is made to move relative to another layer. The greater the amount of force required to effect this movement between layers (fluid friction) results in an increase of viscosity. This frictional property is termed shear. Shearing occurs whenever the fluid is physically moved or distributed, as in pouring, spreading, spraying, mixing, or simply transferring a fluid from one process to another.

Viscosity measurement and control are of particular importance to the chemical, food, and petroleum processing industries, as is evident in Table 1. In recent years much progress has been made in viscosity instrumentation, notably in on-line measurement automatic control. The processing of bakery products is an example of where there is a reliable inferential relationship between the viscosity of the dough and the overall characteristics of the product, such as bread, after the dough is baked. But in the case of mayonnaise and other salad dressings, viscosity per se is the target characteristic—because this translates to spreadability and the desired adherence of the dressing to the various salad ingredients. With catsup, viscosity (commonly termed thickness or flowability) is very important.

TABLE 1   Major Applications for Viscosity Measurement and Control

Low viscosity, 15 to 2 mil centipoise	Medium viscosity, 100 to 8 mil centipoise	High viscosity, 200–800 to 800–64 mil centipoise
Adhesive	Adhesives (hot melt)	Asphalt
Chemicals	Creams	Caulking
Cosmetics	Food products	Chocolate
Hot waxes	Gums	Epoxies
Inks	Inks	Gels
Latex	Organisols	Inks
Oils	Paints	Molasses
Paints and coatings	Paper coatings	Pastes
Pharmaceuticals	Plastisols	Peanut butter
Photo resist	Starches	Putty
Rubber solutions	Surface coatings	Roofing compounds
Solvents	Toothpaste	
Soups	Varnish	
Textile fibers		
Water-based systems		

Typical applications where in-line viscosity controllers are used:	
Printing	Improves print quality while reducing ink costs.
Coatings	Lacquers and coatings can be applied on difficult substrates faster and with less waste.
Power generation	Control of fuel aids combustion with lower emissions.
Pharmaceuticals	Controls quality of numerous pharmaceutical and cosmetic products during manufacture and prior to generating rejects.
Marine power	Shipboard power plants operate cleaner and more efficiently regardless of grade of fuel used.
Refining operations	Improves control of processes, blending, and product quality.
Food manufacturing	Variations in product texture and consistency can be reduced with viscosity control.
Chemical reactors	Close control of viscosity determines final product characteristics.

Source: *Brookfield Engineering Laboratories, Inc.*

Many important and sometimes unexpected relationships can be developed between viscosity and desirable product characteristics. Thus viscosity can be an important product quality control tool. For example, the viscosity of fresh orange juice differs somewhat from that of

[1] By David W. Howard, Brookfield Engineering Laboratories, Inc., Stoughton, Massachusetts.

reconstituted juice. Therefore viscosity can be a measure of freshness. Freezing ruptures some of the juice-containing sacs of the fresh fruit, releasing liquid and thus reducing the effective volume fraction of solids and consequently lowering the juice viscosity. In food processing, liquid product viscosity can frequently be related to statistics gained by human sensory evaluation panels.

Basic Relationships

In defining viscosity, Isaac Newton made reference to the model shown in Fig. 1. Two parallel planes of fluid of equal area A are separated by a distance dx and are moving in the same direction at different velocities V_1 and V_2 due to the action of force F acting on the plane moving at velocity V_2. Newton assumed that the force required to maintain this difference in speed was proportional to the difference in speed through the liquid, or the velocity gradient. This was expressed by

$$\frac{F}{A} = \eta \, \frac{dv}{dx}$$

where η is a constant for a given material and is called its viscosity.

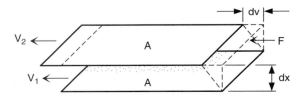

FIGURE 1 Newton's model. Two parallel planes of fluid of equal area A, separated by a distance dx, and moving in same direction at two different velocities V_1 and V_2. F is the force acting on plane A moving at velocity V_2. (*Brookfield Engineering Laboratories, Inc.*)

The velocity gradient dv/dx is a measure of the speed at which the intermediate layers move with respect to each other. It describes the shearing that the liquid experiences and is thus called shear rate. This is commonly symbolized by S, the measurement of which is called the reciprocal second (s^{-1}).

The term F/A indicates the force per unit area required to produce the shearing action. It is referred to as shear stress and is symbolized by F, the measurement unit of which is dynes per square centimeter (dyn/cm^2).

Using these simplified terms, the viscosity η may be defined mathematically by

$$\eta = \frac{F'}{S} = \frac{\text{shear stress}}{\text{shear rate}}$$

The fundamental unit of viscosity measurement is the poise. A material requiring a shear stress of 1 dyn/cm^2 to produce a shear rate of 1 s^{-1} has a viscosity of 1 poise, or 100 centipoise (cP). Viscosity measurements also are expressed in pascal-seconds (Pa · s) or millipascal-seconds (mPa · s). These are units of the international system; they sometimes are preferred over the metric designations. 1 Pa · s = 10 poise; 1 mPa · s = 1 cP.

TYPES OF FLUIDS

Newtonian Fluids

Newton assumed that all materials have, at a given temperature, a viscosity that is independent of the shear rate, that is, twice the force will move the fluid twice as fast. Fluids that behave like this are appropriately called newtonian fluids (Fig. 2).

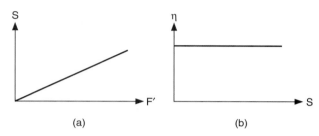

FIGURE 2 Newtonian fluid behavior. (*a*) Relationship between shear stress F' and shear rate S is a straight line. (*b*) Fluid's viscosity remains constant as shear rate is varied.

Non-Newtonian Fluids

As was learned later, however, newtonian behavior is only one of several behavioral characteristics of fluids. Others, generally termed nonnewtonian fluids, are subcategorized as pseudoplastic, dilatant, plastic, thixotropic, and rheopectic.

When the shear rate in a nonnewtonian fluid is varied, the shear stress does not vary in the same proportion (or even necessarily in the same direction). The viscosity of such fluids, therefore, will change as the shear rate is varied. Thus, although any design of viscometer can be used for measuring newtonian fluids, the detectors, such as spindles, and the speeds of these elements have their own effect on the measured viscosity. This results in what is termed apparent, or measured, viscosity, which is accurate only when explicit information pertaining to the measurement apparatus is furnished and adhered to.

Some experts in the field explain nonnewtonian behavior as the result of one or several mechanical phenomena. As nonsymmetrical objects pass by each other (as occurs during fluid flow), their size, shape, and cohesiveness determine how much force is required to move them. For example, the alignment (or sectional pattern engaging the viscosity detector) of large molecules, colloidal particles, and suspended matter (as found in clays, fibers, and crystals) will vary, and consequently, the rate of shear will vary.

Pseudoplastic Fluids.   These fluids exhibit decreasing viscosity with an increasing shear rate (Fig. 3). Materials in this class include paints, emulsions, and dispersions of many types. Another term for this characteristic is shear thinning.

Dilatant Fluids.   These fluids exhibit increasing viscosity with an increasing shear rate (Fig. 4). This behavior is frequently observed in fluids that contain high levels of deflocculated solids, including clay slurries, candy compounds, cornstarch in water, and sand-water mixtures. Another term for this characteristic is shear thickening.

Plastic Fluids.   These fluids behave as solids under static conditions. A given amount of force must be applied to the fluid before any flow is induced. This force is called the yield value. An excellent example is tomato catsup, the yield value often making it difficult to pour the contents, but when it is shaken or struck, it will gush freely. Once the yield value is exceeded and flow begins, plastic fluids may display newtonian, pseudoplastic, or dilatant flow characteristics (Fig. 5).

Thixotropic Fluids. A thixotropic fluid will undergo a decrease in viscosity with time, even when subjected to constant shearing (Fig. 6).

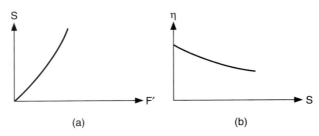

(a)   (b)

FIGURE 3 Pseudoplastic fluid behavior. (*a*) Relationship between shear stress *F′* and shear rate *S*. (*b*) Fluid's viscosity decreases with increasing shear rate.

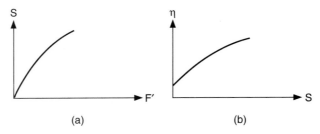

(a)   (b)

FIGURE 4 Dilatant fluid behavior. (*a*) Relationship between shear stress *F′* and shear rate *S*. (*b*) Fluid's velocity increases with increasing shear rate.

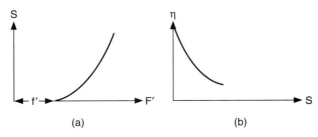

(a)   (b)

FIGURE 5 Plastic fluid behavior. (*a*) Relationship between shear stress *F′* and shear rate *S*. A given amount of force *f′* must be applied to fluid before any flow is induced. This force is called the yield value. (*b*) Fluid's viscosity decreases with increasing shear rate.

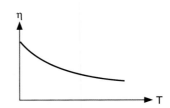

FIGURE 6 Thixotropic fluid behavior. Fluid undergoes decrease in viscosity with time, even when subjected to constant shearing.

When subjected to varying rates of shear, a thixotropic fluid may react as shown in Fig. 7. A plot of shear stress versus shear rate was made as the shear rate was increased to a certain value, then immediately decreased to the starting point. Note that the "up" and "down" curves do not coincide. This hysteresis loop is caused by the decrease in the fluid's viscosity with increasing time of shearing. Such effects may or may not be reversible. Some thixotropic fluids, if permitted to stand undisturbed for a time period, will regain their initial viscosity; others may not do so.

Rheopectic Fluids.   A rheopectic fluid will undergo an increase in viscosity with time even when subjected to constant shearing. Thus its behavior directly opposes that of a thixotropic fluid (Fig. 8).

FIGURE 7   Hysteresis curve sometimes exhibited by a thixotropic fluid. This plot of shear stress (*S*) versus shear rate *F'* was made as the shear rate was increased to a certain value, then immediately decreased to the starting point.

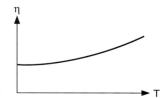

FIGURE 8   Rheopectic fluid behavior. Fluid viscosity increases with time as it is sheared at a constant rate.

Examples of Typical Fluids

Table 2 lists practical examples of fluids encountered in industry.

TABLE 2   Common Fluids Exhibiting Diverse Rheological Properties

Newtonian	Pseudoplastic	Plastic	Thixotropic	Rheopectic	Dilatant*
Water	Catsup	Chewing gum	Silica gel	Bentonite sols	Quicksand
Most mineral oils	Printer's ink	Tar	Most paints	Gypsum in water	Peanut butter
Gasoline	Paper pulp		Glue		Many candy compounds
Kerosene			Molasses		
Most salt solutions in water			Lard		
Light suspensions of dyestuffs			Fruit juice concentrates		
			Asphalts		

  * Some fluids may change from thixotropic to dilatant, or vice versa, as the temperature or concentration changes.

TYPES OF VISCOSITY

Absolute Viscosity.   Viscosity that is measured by any system geometry which is *not* under the influence of gravity for obtaining the measurement. Expressed in units of poise or centipoise.

Kinematic Viscosity.   Viscosity that is measured by any system geometry that uses gravity to obtain the measurement. Expressed in units of stokes or centistokes, where a stoke is equal to poise divided by fluid density.

Apparent Viscosity.   Viscosity of a non-Newtonian liquid. The viscosity that is measured at a single shear rate or single point. Expressed in units of poise or centipoise.

FACTORS THAT AFFECT VISCOSITY

Viscosity is more easily measured than some of the properties that affect it, making it a valuable tool for material characterization, and thus, in numerous instances, viscosity can be used as a "window" for controlling materials processing based on those characteristics inferred by viscosity. And, of course, those variables that affect viscosity, such as temperature and measurement geometry, bear heavily on the success of measuring viscosity reliably and accurately.

Temperature.   Some materials are quite sensitive to temperature. A relatively small temperature variation can produce a significant change in viscosity. The effect of cold weather on lubricating oils is an example. The viscosity of some materials, on the other hand, does not change markedly with temperature.

Shear Rate.   Non-Newtonian fluids, as described previously, tend to occur more frequently in processing and manufacturing than do the newtonian fluids. Thus an appreciation of the effects of shear rate on numerous fluids is required to avoid processing difficulties. For example, it would be disastrous to pump a dilatant fluid through a system, only to have it go solid inside the pump, bringing the whole process to an abrupt halt.

When a material may be subjected to a variety of shear rates in processing or use, it is essential to know its viscosity at the projected shear rates. If these are not known, an estimate should be made. Viscosity measurements then should be made at shear rates as close as possible to the estimated values so that these values will not fall outside the shear rate range of the viscometer.

Examples of materials that are subjected to, and affected by, wide variations in shear rate during processing and use include paints, cosmetics, liquid latex, coatings, and certain food products.

Time.   The time elapsed under conditions of shear obviously affects thixotropic and rheopectic (time-dependent) materials. But changes in the viscosity of many materials can occur over time even though the material is not being sheared. Aging phenomena must be considered when selecting and preparing samples for viscosity measurement. Many materials will undergo changes in viscosity during a chemical reaction, so that a viscosity measurement made at one point in the reaction may differ significantly from one made at another time.

Prior History of Material.   What has happened to a sample prior to a viscosity measurement can affect viscosity results significantly, especially in fluids sensitive to heat or aging. Thus storage conditions and sample preparation techniques must be designed to minimize their effect on subsequent viscosity tests. Thixotropic materials, in particular, are sensitive to prior history, as their viscosity will be affected by stirring, mixing, pouring, or any other activity which produces shear in the sample.

Composition and Additives.   Composition is a major factor in determining a material's viscosity. When the composition is altered, either by changing the proportions of the component substances, or by the addition of other materials, a change in viscosity is most likely to occur. For example, the addition of solvent to printing ink reduces the viscosity of the ink. Additives of many kinds are used objectively to control the rheological properties of paints.

Dispersions and Emulsions.   These multiphase materials consist of one or more solid phases dispersed in a liquid phase and are subject to significant alterations in their rheological characteristics. The state of aggregation of the sample material is of prime interest. Are the particles that make up the solid phase separate and distinct, or are they clumped together? How large are the clumps and how tightly are they stuck together? If the clumps (flocs) occupy a large volume in the dispersion, the viscosity will tend to be higher than if the floc volume were smaller, a condition due to the greater force needed to dissipate the solid component of the dispersion.

The attraction between particles in a dispersed phase is largely dependent on the types of material present at the interface between the dispersed phase and the liquid phase. This, in turn, affects the rheological behavior of the system. The introduction of flocculating or deflocculating agents into a system is a means of controlling its rheology.

The shape of the particles is also a significant factor. Particles suspended in a flowing medium are constantly being rotated. If the particles are essentially spherical, they will tend to rotate freely, in contrast with needle- or plate-shaped particles.

The stability of a dispersed phase is particularly critical when measuring the viscosity of a multiphase system. Any tendency for particles to settle will produce a nonhomogeneous fluid, the rheological characteristics of which will be altered.

VISCOMETER DESIGNS

Viscosity has been measured in process and product quality-control laboratories for scores of years.

Time-Based Instruments.   Initially, viscometers incorporated the dimension of time in the measuring scheme, and viscosity readings were in terms of seconds. These designs were based on measuring the time for a sample to discharge (by gravity) through an orifice or nozzle, as found in the Saybolt, Redwood, Engler, Scott, Ubbelohde, and Zahn viscometers, among others. A few time-based viscosity scales remain in common use: (1) the Saybolt scales (United States), (2) the Redwood scales (United Kingdom), and (3) the Engler scales (continental Europe). All of these scales are based on the Hagen-Poiseuille law and indicate the time of efflux under specified conditions. Kinematic viscosity (and from it, absolute viscosity) can be determined from such scales by the empirical formula

$$\mu = \frac{PR/2L \text{ (shear stress)}}{4Q/\pi R^3 \text{ (shear rate)}} = \frac{\pi P R^4}{8QL}$$

where P = pressure differential across liquid in tube
  R = inside radius of tube
  L = length of tube
  Q = volume rate of flow of liquid
  μ = viscosity

This equation is limited to conditions of laminar or viscous (streamline) flow.

Other viscometers have been based on the timed fall of a ball or rise of a bubble through a tube, where the time is proportional to absolute viscosity because the liquid moves in viscous flow through a restriction; or on the timed fall of a piston in a cylinder, the opening between the cylinder wall and the piston serving as an orifice. These approaches were not suitable for use in in-line applications.

Pressure-Drop Instruments.   In this system the sample is pumped through a friction tube in viscous flow. The pressure drop across the tube is measured by a differential-pressure transmitter. The transmitter output provides a direct solution to the Hagen-Poiseuille equation.

Torque or Viscous-Drag Instruments.   During the last several years, much research has gone into the development of rotational viscometers which measure the torque required to rotate an immersed element (spindle) in the fluid being measured. As shown in Fig. 9, the spindle is driven by a synchronous motor through a calibrated spring. The deflection of the spring is indicated by a pointer and dial (or a digital display). In some designs, by using a multiple (four to eight) speed transmission and interchangeable spindles, a variety of viscosity ranges can be measured, thus enhancing the versatility of the instrument. Pneumatic drives also are available for hazardous applications.

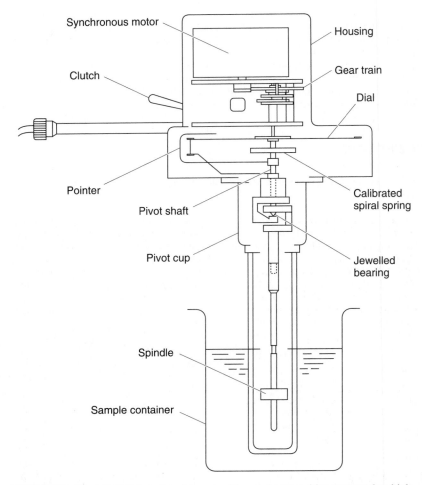

FIGURE 9 Brookfield dial-reading viscometer. The synchronous-drive motor and multiple-speed transmission are located at the top of the instrument inside the housing to which the nameplate is attached. The main case of the instrument contains a calibrated beryllium-copper spring, one end of which is attached to the pivot shaft, the other connected directly to the dial. The dial is driven by the transmission and, in turn, drives the pivot shaft through the calibrated spring. In dial-reading models the pointer is connected to the pivot shaft and indicates its angular position in relation to the dial. In digital designs the relative angular position of the pivot shaft is detected by a rotary variable displacement transducer (RVDT) and is read out on a digital display. Below the main case is the pivot cup, through which the lower end of the pivot shaft protrudes. A jewel bearing inside the pivot cup rotates with the dial or transducer. The pivot shaft is supported on this bearing by the pivot point. The lower end of the pivot shaft comprises the spindle coupling to which the viscometer's spindles are attached. (*Brookfield Engineering Laboratories, Inc.*)

For a given viscosity, the viscous drag, or resistance to flow (indicated by the degree to which the spring winds up), is proportional to the spindle's speed of rotation and is also related to the size and shape of the spindle (system geometry). The drag will increase as the spindle size or the rotational speed increases. It follows that for a given spindle geometry and speed, an increase in viscosity will be indicated by an increase in the deflection of the spring. For any particular viscometer model, the minimum range is obtained by using the largest spindle at the highest speed, whereas the maximum range will be achieved by using the smallest spindle at the slowest speed. Measurements made using the same spindle at different speeds are used to detect and evaluate the rheological properties of the test fluid.

As shown in Fig. 9, the viscometer has several major components. Standard viscometers are usually furnished to accomplish a range of viscosity measurements.

Viscometers are available for laboratory and field applications. Figure 10 illustrates a digital rotational viscometer and a close-up of the spindle; the torque sensing method is similar to the method used by the instrument diagrammed in Fig. 9.

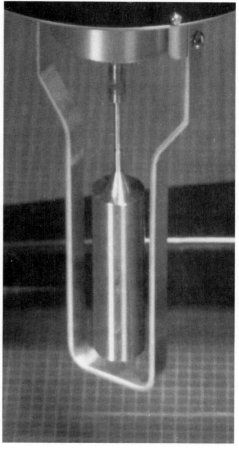

FIGURE 10   (*a*) Brookfield digital viscometer. Torque measurement is converted directly into centipoise without requiring timing or calculations. Reproducible accuracy is within 1 percent of range. Total time of test is less than 30 seconds. (*b*) Close-up of spindle arrangement. (*Brookfield Engineering Laboratories, Inc.*)

Similar instruments with digital readouts and programming capability are shown in Fig. 11. These features prove quite advantageous in laboratory situations where speed and convenience are important.

 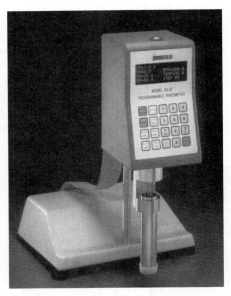

FIGURE 11   Digital instruments for laboratory measurements of rheological properties. (*a*) Brookfield digital viscometer. (*b*) Brookfield programmable rheometer. (*Brookfield Engineering Laboratories, Inc.*)

In addition to laboratory applications, viscometers and rheometers are design-engineered for in-line and field applications, as illustrated in Fig. 12.

New approaches to spindle designs also have been introduced, including (1) two coaxial cylinders, (2) disk and T-bar spindles, and (3) the cone and plate. The latter differs the most from the aforementioned spindles in terms of geometry (Fig. 13).

Over the years, a number of other physical properties, including the vibrating reed and ultrasonics, have been considered for the measurement of rheological properties.

VISCOSITY-RELATED TEST SPECIFICATIONS

Table 3 lists viscosity-related test specifications issued by the American Society for Testing and Materials (ASTM).

CONSISTENCY[2]

Substances that undergo continuous deformation when subjected to a shear stress are said to exhibit fluid behavior. The resistance offered to this deformation is termed *consistency*. This

[2] Acknowledgment is gratefully given to the following persons for their cooperation and counsel in the preparation of this summary: Eugene Norman, Consultant, Green Bay, Wisconsin; James Overall, Georgia-Pacific Corporation, Atlanta, Georgia; and Gregory Neeb and J. G. Puls, DeZurik (A Unit of General Signal), Sartell, Minnesota.

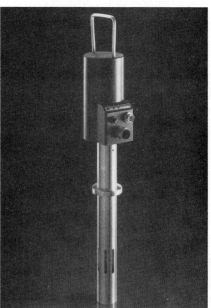

FIGURE 12 Brookfield process viscometers. (*a*) In-line flow through model. (*b*) Pneumatic model for hazardous locations. (*c*) Unit designed for tank mounting. (*d*) Portable unit. (*Brookfield Engineering Laboratories, Inc.*)

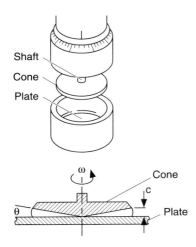

FIGURE 13 In a cone and plate detector design, the torque measuring system consists of a calibrated beryllium-copper spring which connects the drive mechanism to a rotating cone. This senses the resistance to rotation caused by the presence of sample fluid between the cone and a stationary flat plate. In essence, the stationary plate forms the bottom of a sample cup which can be removed, filled with 0.5 to 2.0 mL of sample fluid (depending on cone in use), and remounted without disturbing the calibration. The sample cup is jacketed and has tube fittings for connection to a constant-temperature circulating bath. Accuracy is within 1 percent of working range; reproducibility is within 0.2 percent; working temperature range is –10 to 100°C (–14 to 212°F). (*Brookfield Engineering Laboratories, Inc.*)

is a constant for gases and newtonian fluids when static temperature and pressure prevail. With these fluids, consistency is called viscosity.

The consistency of nonnewtonian fluids is *not* constant, but is a function of the applied stress, and in some cases, may vary with time. The term *apparent viscosity* is frequently applied with reference to the consistency of nonnewtonian fluids.

Although consistency is a real property of nonnewtonian fluids, its measurement is usually relative to arbitrary standards. Thus the measurement can be stated in no definitive units as opposed to those of absolute viscosity, temperature, and pressure.

MATERIAL EFFECTS OF CONSISTENCY

Through the years, the term consistency has acquired several connotations, largely as a result of the applications for consistency control. When considering the operation of process machinery, one is interested in the characteristics of flowing media, that is, their flowability. Here the term retains its true definition, that is, resistance to deformation. For example, consistency (and temperature) control is necessary for proper operation of a sterilizer used in the processing of canned foodstuffs. In the application of short-time high-temperature (STHT) sterilization in continuously agitated cookers, there must be no increase in the consistency of the food slurry. Careful control based on consistency measurement improves color, flavor, and texture of the final products, as has been the experience in processing cream-style corn and similar food products.

Consistency of the slurry is also an important variable in the refining of wood pulps for paper making. The objectives of refining are to brush out lumps or knots of fibers, to cut fibers to shorter lengths, and sometimes to produce more hydration. The result of each form of mechanical treatment of cellulose pulp is influenced, in both extent and kind, by the consistency (flowability) of the pulp.

Consistency also influences many other mechanical treatment operations, such as kneading of dough, mixing of oil and water and of clay and water, and the formulation of paint. In nearly all mixing operations the apparent viscosity of the mixture is an important physical variable to control.

Percent Solids and Consistency

The apparent viscosity of a suspension of certain fibrous materials is related to the percentage of fibers in suspension. This relationship provides the operating principle for many consistency-regulating devices, that is, these devices use the measurement of apparent viscosity as the base for the control of fiber concentration. Through repeated usage, the term consistency has come to mean "solids content" in some applications. Consistency is the term used

TABLE 3 ASTM Viscosity-Related Test Specifications

C 965-81	Practices for Measuring Viscosity of Glass above the Softening Point
D 115-85	Testing Varnishes Used for Electrical Insulation
D 789-86	Test Methods for Determination of Relative Viscosity, Melting Point, and Moisture Content of Polyamide (PA)
D 1076-80	Concentrated, Ammonia Preserved, Creamed and Centrifuged Natural Rubber Latex
D 1084-81	Tests for Viscosity of Adhesives
D 1417-83	Testing Rubber Latices—Synthetic
D 1439-83	Testing Sodium Carboxymethylcellulose
D 1638-74	Testing Urethane Foam Isocyanate Raw Materials
D 1824-87	Test for Apparent Viscosity of Plastisols and Organosols at Low Shear Rates by Brookfield Viscometer
D 2196-86	Test for Rheological Properties of Non-Newtonian Material by Rotational (Brookfield) Viscometer
D 2364-85	Testing Hydroxyethylcellulose
D 2393-86	Testing for Viscosity of Epoxy Resins and Related Components
D 2556-80	Test for Apparent Viscosity of Adhesives Having Shear Rate Dependent Flow Properties
D 2669-82	Test for Apparent Viscosity of Petroleum Waxes Compounded with Additives (Hot Melts)
D 2849-80	Test for Urethane Foam Polyol Raw Materials
D 2983-80	Test for Low-Temperature Viscosity of Automotive Fluid Lubricants Measured by Brookfield Viscometer
D 2994-77	Testing Rubberized Tar
D 3232-83	Method for Measurement of Consistency of Lubricating Greases at High Temperatures
D 3236-78	Apparent Viscosity of Hot Melt Adhesives and Coating Materials
D 3716-83	Testing Emulsion Polymers for Use in Floor Polishes
D 4016-81	Test Method for Viscosity of Chemical Grouts by the Brookfield Viscometer (Laboratory Method)
D 4300-83	Test Method for Effect of Mold Contamination on Permanence of Adhesive Preparations and Adhesive Films
D 4402-84	Standard Method for Viscosity Determinations of Unfilled Asphalts Using the Brookfield Thermosel Apparatus

in the paper industry, as well as in some food-processing and mining operations, to designate the concentration of a slurry. For example, in the paper industry the concentration, or weight, of air-dry pulp (that is, of total solids) in any combination of pulp and vehicle (usually water) is referred to as consistency.

The following equation finds practical use:

$$\frac{\text{Air-dry weight of solids (per unit wt)}}{\text{Weight of solids + water (per unit wt)}} \times 100 = \text{consistency, \%}$$

Papermakers sometimes use the terms "bone-dry," "oven-dry," and "moisture-free" with reference to pulp fibers or paper from which all water has been removed by evaporation. Since such drying is seldom practiced in the papermill, bone-dry is mainly used in laboratory work and in process-equipment design. Air-dry pulp is considered to contain 10 percent moisture and thus

$$\text{Bone-dry consistency, \%} = 0.9 \times \text{air-dry consistency, \%}$$

Example. For what capacity (flow rate) must a stuff pump be designed in order to handle pulp slurries at 3.0 percent air-dry consistency in a system designed to produce 100 tons/day of bone-dry paper? Assume that 1 gal of 3.0 percent consistency pulp weighs 8.34 lb.

$$\frac{100 \text{ tons/day} \times 2000 \text{ lb/ton}}{24 \text{ h/day} \times 60 \text{ min/h}} = 139 \text{ lb/min bone-dry fibers}$$

and thus

$$\text{Flow rate} = \frac{139 \text{ lb/min}}{0.03 \times 0.9 \times 8.34 \text{ lb/gal}} = 617 \text{ gal/min}$$

Other useful flow conversion formulas include:

$$\text{Gal/min} = \left(\frac{16.65}{\% \, C}\right) \text{tons/day}$$

$$\text{L/min} = \left(\frac{100}{\% \, C}\right) \text{kg/min}$$

When percent solids is the connotation, uniform consistency is also important to the efficient operation of process machinery and end-product quality control. It should be noted that this meaning of consistency is more akin to density than to the concept of viscosity. Devices that determine apparent viscosity, rather than density, are frequently used to infer solids content.

The end point in the evaporation of tomato products is determined by this type of consistency measurement. Tomato paste, for example, may be concentrated to 36 percent solids. The concentration is not indicated properly by density measurement, but may be reliably correlated with apparent viscosity. Variations in the solids content of mineral-ore slurries may be determined by observing corresponding changes in consistency.

Papermakers are vitally concerned with basis weight (weight per unit area) of the sheet because this weight is the unit on which sales are generally made. Consistency control is essential to basis-weight control because it is the means of delivering to the papermaking machine a constant volume of stock containing a uniform amount of fibers.

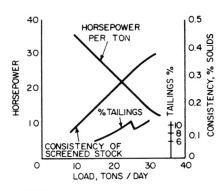

FIGURE 14 Curves showing the effect of consistency on load, power, and tailings for fine screens in paper industry.

Efficient operation of other paper processes also depends on supplying the machine with a definite and constant quantity of pulp. The operation of fine screens is a complex problem in which consistency is a major factor. Figure 14 reveals that optimum screen operation is accomplished by control of the inlet load which, at constant volume of feed, varies with consistency. The chart also shows the usual changes in the other important factors in screening. Power consumption varies with consistency, as does the percentage of tailings, or rejected material.

Moisture and Consistency

Moisture is the virtual reciprocal of consistency, and thus consistency may be considered a measure of the liquid content of the "solid" medium. This measurement is often preferred for materials of high consistencies, ranging from 12 to 99 percent. There are exceptions. For example, a paper web in the fourdrinier section of a paper machine may vary from 0.3 percent at the breast roll to 20 percent at the press roll. The dryer consistency from entrance to exit may range from 10 to 90 percent.

Freeness and Consistency

Freeness is the amount of water that leaves a sample of specific volume and consistency. The Canadian standard freeness (CSF) scale indicates the amount of bypass water out of a calibrated funnel from a 1-liter sample at 0.3 percent consistency. The freeness is read directly in cubic centimeters. The higher the number, the more readily the slurry will release the water.

Density and Consistency

Density is the mass per unit volume of a substance, whereas consistency is the mass per unit mass of a substance. As an example, 3 grams of solids in 100 grams of a mixed slurry is a 3 percent consistency.

FACTORS THAT AFFECT CONSISTENCY

Temperature.   Changes of temperature affect the shear stress of fiber suspensions in that increasing temperature provides decreasing forces on a sensor at fixed consistency. Unless compensated, the consistency detector may produce a signal indicating a lowering consistency with increasing temperature.

Velocity.   Although the velocity effect on a detector is pertinent, this is a virtually unpredictable variable. A varying velocity at the sensor alters the forces acting on the sensor at a fixed consistency. If the fluid velocity is too low, laminar flow conditions can be created in which the consistency varies across the pipe diameter. If the velocity is too high and the sensor is mounted in a long run of straight pipe, boundary layering may occur. In general it is good practice to mount the sensor in a turbulent section of the pipe (for example, directly after a pump) where a reasonably constant flow rate occurs.

Freeness.   Since freeness is a characteristic of fibrous media, a variation in freeness imposes a variation in consistency transmitter output. Increasing the freeness at a fixed consistency provides an increasing consistency signal. To compensate for the freeness effect, the freeness should be measured and included in the calibration of the sensor. Where the sensor is subjected to composite media of varying freeness, an in-line freeness test can be used to compensate the consistency signal.

pH.   As the alkalinity of a slurry increases at a fixed consistency, the sensor output increases. This is due to the effect of pH on the shear stress of the fibrous medium. Also, chemical agent additions, used for controlling the pH of the medium, may react with the fibrous mass, producing stress variations.

Inorganic Materials Content.   The presence of inorganic materials, such as filler, in a slurry does not contribute to shear stress as does the fiber. Thus an apparent reduction in consistency may be sensed. For example, a 3 percent slurry with 25 percent filler will show a fiber consistency of 2.25 percent. Where conditions are reasonably consistent, the sensor transmitter should be compensated during calibration.

CONTINUOUS CONSISTENCY MEASUREMENT

Because of stratification and nonuniformity of the fluid stream, it usually is preferred to have the consistency sensor exposed to the entire flow, not just to a sample. But because of tremendous volumes involved in some operations, sampling is a practical necessity. Sampling just after a pump is suggested.

In controlling consistency, it usually is much easier to dilute for control, rather than vice versa. Therefore the feed to the sensor should be at a consistency that will tend to be somewhat higher than the desired value. This is an important consideration in sizing the sensor and the automatic control valve.

Blade-Type Transmitter

The transmitter shown schematically in Fig. 15 operates on a shear rate principle by utilizing a blade sensor suspended in the pipeline. Consistency changes are measured by sensing the change in force required by the sensing blade *A* to shear through the flowing stock. A torque arm *B,* connected to the sensor, extends into a pneumatic torque transducer which converts torque arm movement into a pneumatic output signal proportional to consistency change. The sensor is so designed that it is not affected by changes in flow rate over a wide range of flows. The output signal is transmitted to a recording controller which records the consistency and positions an air-operated dilution valve.

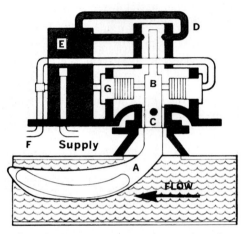

FIGURE 15   Blade sensor consistency transmitter. Consistency is determined by sensing stock fiber drag on the sensing blade (A) in the pipeline. The torque arm (B), in electronic blade transmitters, is connected to an electronic sensor. In pneumatic models the torque arm (B) connected to the sensor extends into a pneumatic torque transducer in the transmitter. As consistency changes, the sensor and torque arm pivot on a flexure (C). Torque arm movement is sensed by a linear variable differential transformer position sensor in electronic models and by a pneumatic bridge circuit (D) in pneumatic versions. Torque arm movement is converted by a signal conditioning circuit to a 4- to 20-mA output signal in electronic transmitters and by a booster pilot (E) to a 3- to 15-psi (21–103 kPa) output signal (F) in pneumatic models. The output signal varies in direct proportion to changes in consistency. The transmitter output operates a controller-recorder and controls dilution water input ahead of the transmitter. (*DeZurik, a unit of General Signal.*)

The unit has a consistency range of 1.75 to 6.00 percent, with a sensitivity capable of sensing consistency within ±0.02 percent for many applications. The unit has a 40:1 span adjustment ratio and a claimed repeatability of ±0.5 percent of chart reading. The flow velocity range is 0.75 to 5 ft (0.23 to 1.5 m)/s. The unit is available in pipe size dimensions from 4 to 30 inches (~10 to 76 cm). The 4-inch size handles a minimum of 30 gal (~114 L)/min and a maximum of 195 gal (~738 L)/min, whereas the 30-inch size handles a minimum of 1665 gal (~63 hL)/min and a maximum of 11,000 gal (~416 hL)/min. The maximum line pressure rating is 125 psi (862 kPa). The sensor, mounting neck and torque arm are constructed of type 316 stainless steel, and the body of the unit is cast aluminum. Typical horizontal and vertical installations are shown in Fig. 16.

Rotating Consistency Sensor

Operation of the instrument shown in Fig. 17 is similar to that of a rotational viscometer. The unit consists of a motor-driven sensor which rotates within the process line. The torque required to rotate the sensor at constant speed within the slurry is directly related to the slurry consistency. A change in consistency results in a change in torque, which produces a pneumatic output signal. The air signal normally is fed to a recorder-controller which then positions an air-operated dilution valve to maintain the desired consistency. These units also are available in an electronic configuration to provide output signals of milliamperes or millivolts.

These assemblies (consistency range 0.75 to 6.00 percent) are available in several sizes, from a minimum flow for small units of 15 gal and a maximum flow of 180 gal (~57 and 681 L)/min to the largest size with a minimum flow of 500 gal and a maximum flow of 10,000 gal (~19 and 379 hL)/min. It is claimed that the sensor is capable of sensing consistency within ±0.01 percent for some applications. Sensing accuracy is maintained on stock as thin as 0.75 percent or as thick as 6.0 percent, and the maximum working pressure is 125 psi (862 kPa). All parts coming in contact with stock are constructed of stainless steel.

Because consistency is a particularly difficult variable to sense, attempts to simplify the mechanics of measurement have been made. These include the use of a vibrating probe as well as a polarized-light detector.

Web Consistency Measurement

The previously mentioned methodologies and sensing techniques have referred to stock forming and conditioning processes. When the material is formed and deposited on a moving web, the consistency of the material must still be measured and controlled to characterize the final product. The method of measurement is radically different from that of the stock forming process. While contact measurement is prevalent in the former process, noncontact measurement is essential once the material has been deposited. Whereas moisture, density,

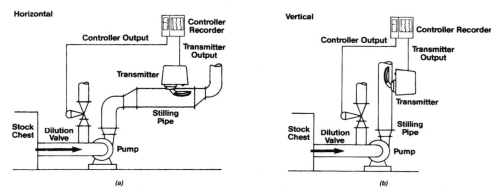

FIGURE 16   Representative horizontal and vertical installations of blade-type consistency sensor and stock control system. (*DeZurik, a unit of General Signal.*)

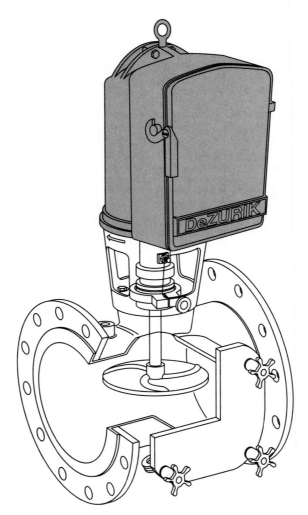

FIGURE 17 Rotating sensor consistency transmitter. Consistency is determined by sensing stock fiber drag on the sensing disk in the pipeline. To ensure accuracy, the sensing element is located in the actual pulp stock flow stream. The disk is driven at constant speed by a three-phase motor. When at constant consistency, the reactionary torque remains constant and the motor housing is stationary. As consistency changes, reactionary torque also changes, causing the motor housing to rotate on its flexure mounts. Movement of the torque arm, which is bolted solidly to the motor housing, in pneumatic models is sensed by two air gaging nozzles, and in electric models, by a linear variable differential transformer. Signals are directly proportional to stock consistency. Signal output is transmitted to a recorder and to a controller, which actuates a control dilution water input valve located ahead of the transmitter. A force-balance feedback system maintains torque arm position to allow instantaneous and continuous sensing. (*DeZurik, a unit of General Signal.*)

viscosity, and freeness are factors to be considered in the former measurement, only moisture is the consideration of the latter. Web consistency is the inverse of web moisture. This simplifies the measurement since moisture sensing is relatively easy as compared to consistency sensing. However, even with these methods, inaccuracies described as loss factors can interfere with the measurements. In direct current, audio frequencies, and up to radio frequency, electrical conductivity is responsible for this loss factor. In microwave regions, the absorption of radiation is responsible as a result of molecular rotation. In the infrared region, the loss factor is related to the vibrational energy and the binding energies of the molecules. In the visual frequency regions, the loss factor is related to the absorption of energy as the electrons change their energy levels by shifting from one orbit to another.

SECTION 7
CONTROL COMMUNICATIONS*

B. A. Loyer
Systems Engineer, Motorola, Inc., Phoenix, Arizona (Local Area Networks—prior edition)

Howard L. Skolnik
Intelligent Instrumentation, Inc. (A Burr-Brown Company), Tucson, Arizona (Data Signal Handling in Computerized Systems; Noise and Wiring in Data Signal Handling)

---

* Persons who authored complete articles or subsections of articles, or otherwise cooperated in an outstanding manner in furnishing information and helpful counsel to the editorial staff.

DATA SIGNAL HANDLING IN COMPUTERIZED SYSTEMS

by Howard L. Skolnik[1]

Prior to the advent of the digital computer, industrial instrumentation and control systems, with comparatively few exceptions, involved analog, rather than digital, signals. This was true for both the outputs from sensors (input transducers—strain gages, thermocouples, and so on) and the inputs to controlling devices (output transducers—valves, motors, and so on). In modern systems many transducers are still inherently analog. This is important because computers can only operate with digital information. Therefore a majority of contemporary systems include analog-to-digital (A/D) and digital-to-analog (D/A) converters.

An important feature of data-acquisition products is how they bring together sophisticated functions in an integrated, easy-to-use system. Given the companion software that is available, the user can take advantage of the latest technology without being intimately familiar with the internal details of the hardware. When selecting a system, however, it is useful to have a basic understanding of data-acquisition principles. This article addresses how real-world signals are converted and otherwise conditioned so that they are compatible with modern digital computers, including personal computers (PCs). The techniques suggested here are specifically aimed at PC-based measurement and control applications. These generally involve data-acquisition boards that plug directly into an expansion slot within a PC. References to specific capabilities and performance levels are intended to convey the current state of the art with respect to PC-based products (Fig. 1).

---

[1] Intelligent Instrumentation, Inc. (A Burr-Brown Company), Tucson, Arizona.

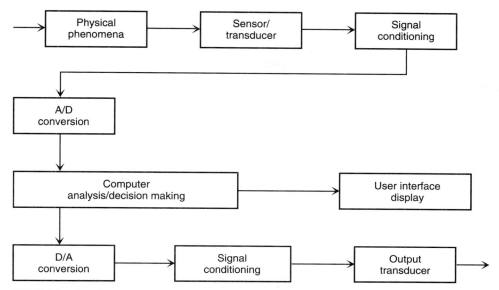

FIGURE 1   Flow diagram of modern computer-based data-acquisition and control system. The measured variable may be any physical or chemical parameter, such as temperature, pressure, flow, liquid level, chemical concentration, dimension, motion, or position. The sensor/transducer is the measuring device (primary element) that converts the measured variable into an electrical quantity. Common transducers include thermocouples, strain gages, resistance temperature devices, pH cells, and switches. The signal from a transducer can be in the form of a voltage, current, charge, resistance, and so on. Signal conditioning involves the manipulation of the raw transducer's output into a form suitable for accurate analog-to-digital (A/D) conversion. Signal conditioning can include filtering, amplification, linearization, and so on. Data conversion provides the translation between the real world (mostly analog) and the digital domain of the computer, where analysis, decision making, report generation, and user interface operations are easily accomplished. To produce analog output signals from the computer (for stimulus or control) digital-to-analog (D/A) conversion is used. Signal conditioning and output transducers provide an appropriate interface to the outside world via power amplifiers, valves, motors, and so on. (*Intelligent Instrumentation, Inc.*)

SIGNAL TYPES

Signals are often described as being either analog, digital, or pulse. They are defined by how they convey useful information (data). Attributes such as amplitude, state, frequency, pulse width, and phase can represent data. While all signals can be assumed to be changing with time, analog signals are the only ones to convey information within their incremental amplitude variations. In instrumentation and control applications most analog signals are in the range of −10 to +10 volts or 4 to 20 mA. Some of the differences between analog and digital signals are suggested in Fig. 2. Digital and pulse signals have binary amplitude values, that is, they are represented by only two possible states—low and high. While low and high states can be represented by any voltage level, transistor-transistor-logic (TTL) levels are most often used. TTL levels are approximately 0 and 5 volts. The actual allowable ranges for TTL signals are

Low level = 0 to 0.8 volt

High level = 2.0 to 5.0 volts

Thus with analog it is important how high the signal is, while with digital it matters only whether the signal is high or low (on or off, true or false). Digital signals are sometimes called

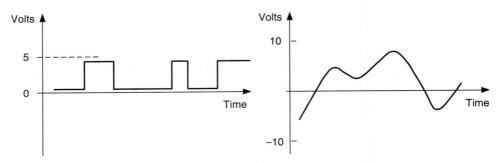

FIGURE 2   Comparison of digital signal (left) and analog signal (right). (*Intelligent Instrumentation, Inc.*)

discrete signals. While all digital signals have the potential of changing states at high speed, information is usually contained in their static state at a given point in time. Digital inputs can be used to indicate whether a door is open, a level is beyond a set point, or the power is on. Digital outputs can control (on or off) power to a motor, a lamp, an alarm, and the like. In contrast, analog inputs can indicate how high a level is or how fast a motor is turning. Analog outputs can incrementally adjust the position of a valve, the speed of a motor, the temperature of a chamber, and so on. Pulse signals are similar to digital signals in many respects. The distinction lies in their time-dependent characteristics. Information can be conveyed in the number of state transitions or in the rate at which transitions occur. Rate is referred to as frequency (pulses per second). Pulse signals can be used to measure or control speed, position, and so on.

TERMINATION PANELS

Termination panels are usually the gateway to a data-acquisition system. Screw terminals are provided to facilitate easy connection of the field wiring. Figure 3 suggests two of the many termination panel styles. Some models are intended for standard input or output functions, while others are designed to be tailored for unique customized applications. Mounting and interconnection provisions are provided for resistors, capacitors, inductors, diodes, transistors, integrated circuits, relays, isolators, filters, connectors, and the like. This supports a wide range of signal interface and conditioning capabilities. In most cases termination panels are located outside, but adjacent to, the data-acquisition system's host PC. Many mounting and enclosure options are available to suit different applications. Because the actual data-acquisition board is located inside the PC, short cables (normally shielded ribbon cables) are used to connect the termination panel's signals.

FIELD SIGNALS AND TRANSDUCERS

Whatever the phenomenon detected or the device controlled, transducers play a vital role in the data-acquisition system. It is the transducer that makes the transition between the physical and the electrical world. Data acquisition and control can involve both input and output signals. Input signals can represent force, temperature, flow, displacement, count, speed, level, pH, light intensity, and so on. Output signals can control valves, relays, lamps, horns, and motors, to name a few. The electrical equivalents produced by input transducers are most

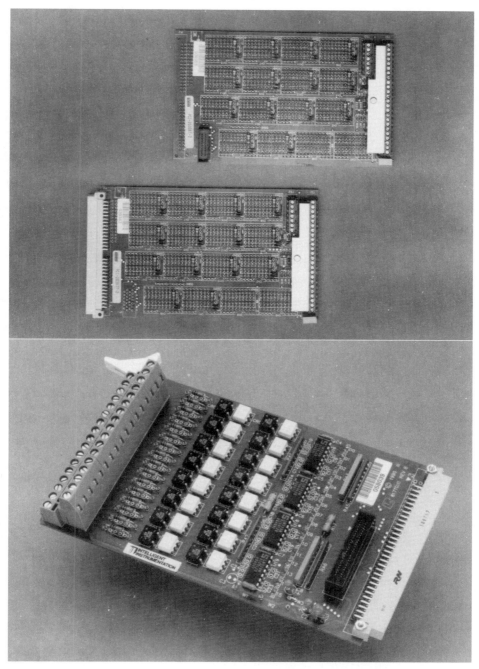

FIGURE 3   Representative termination panel styles. (*Intelligent Instrumentation, Inc.*)

commonly in the form of voltage, current, charge, resistance, or capacitance. As shown later, the process of signal conditioning will further convert these basic signals into voltage signals. This is important because the interior blocks of the data-acquisition system can only deal with voltage signals.

Thermocouples

Thermocouples are used widely to measure temperature in industry and science. Temperatures in the range of −200 to +4000°C can be detected. Physically a thermocouple is formed by joining together wires made of two dissimilar metals. The resulting junction produces a voltage across the open ends of the wires that is proportional to temperature (the Seebeck effect). The output voltage is usually in the range of −10 to +50 mV and has an average sensitivity of 10 to 50 μV/°C, depending on the metals used. However, the output voltage is very nonlinear with respect to temperature. Many different thermocouple types are in wide use. For convenience, alphabetic letter designations have been given to the most common. These include the following:

J   Iron-constantan (Fe-C)

K   Chrome-Alumel (Ch-Al)

T   Copper-constantan (Cu-C)

Tungsten, rhodium, and platinum are also useful metals, particularly at very high temperatures.

Thermocouples are low in cost and very rugged. Still, they are not without their limitations and applications problems. In general, accuracy is limited to about 1 to 3 percent due to material and manufacturing variations. Response time is generally slow. While special thermocouples are available that can respond in 1 to 10 ms, most units require several seconds. In addition to the thermocouple's nonlinear output, compensation must also be made for the unavoidable extra junctions that are formed by the measuring circuit.

As mentioned previously, a single thermocouple junction generates a voltage proportional to temperature:

$$V = k(t) \tag{1}$$

where k is the Seebeck coefficient defining a particular metal-to-metal junction, and t is in kelvins.

Unfortunately the Seebeck voltage cannot be measured directly. When the thermocouple wires are connected to the terminals of a voltmeter or data-acquisition system, new thermoelectric junctions are created. For example, consider the copper-constantan (type T) thermocouple connected to a voltmeter shown in Fig. 4. It is desired that the voltmeter read only V_1 (of J_1), but the act of connecting the voltmeter creates two more metallic junctions, J_2 and J_3.

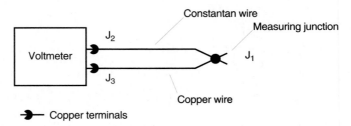

FIGURE 4   Essence of the thermocouple measurement problem. (*Intelligent Instrumentation, Inc.*)

Since J_3 is a copper-to-copper junction, it creates no thermal voltage ($V_3 = 0$), but J_2 is a copper-to-constantan junction which will add a voltage V_2 in opposition to V_1. As a result, the voltmeter reading V_v will actually be proportional to the temperature difference between J_1 and J_2. This means that determining the temperature at J_1 requires a knowledge of the temperature at J_2. This junction is referred to as the *reference junction,* or *cold junction.* Its temperature is the reference temperature t_{ref}. Note that V_2 equals V_{ref}. Therefore it follows from Eq. (1),

$$V_v = V_1 - V_{ref} = k(t_1 - t_{ref}) \tag{2}$$

It is important to remember that k is highly nonlinear with respect to temperature. However, for measurement purposes it is not necessary to know the value of k. Tables have been compiled by the U.S. National Bureau of Standards (now the National Institute of Standards and Technology, NIST) that take variations in k into account and can provide t_1 directly in terms of V_v, assuming that t_{ref} is at 0°C. Separate tables were made for each thermocouple type. J_2 (and J_3) was physically placed in an ice bath, forcing its temperature to 0°C. Note that even under these conditions [see Eq. (1)], V_{ref} is not 0 volts. The Seebeck relationship is based on the kelvin (absolute zero) scale. In computer-based applications the thermocouple tables are transformed into polynomial equations for ease of use. Depending on the thermocouple type and the accuracy (compliance with the NIST tables) desired, between fifth- and ninth-order polynomials are used.

The copper-constantan thermocouple used in this example is a special case because the copper wire is the same metal as the voltmeter terminals. It is interesting to look at a more general example using iron-constantan (type J). The iron wire increases the number of dissimilar metal junctions in the circuit as J_3 becomes a Cu-Fe thermocouple junction. However, it can be shown that if the Cu-Fe and the Cu-C junctions (at the termination panel) are at the same temperature, the resulting voltage is equivalent to a single Fe-C junction. This allows the use of Eq. (2). Again, it is very important that both parasitic junctions be held at the same (reference) temperature. This can be aided by making all connections on an isothermal (same temperature) block.

Clearly, the requirement of an ice bath is undesirable for many practical reasons. Taking the analysis to the next logical step, Eq. (2) shows that t_{ref} need not be at any special temperature. It is only required that the reference temperature be accurately known. If the temperature of the isothermal block (the reference junction) can be measured independently, this information can be used to compute the unknown temperature t_1.

Devices such as thermistors, resistive temperature detectors, and semiconductor sensors can provide a means of independently measuring the reference junction. (Semiconductor sensors are the most popular for the reasons described hereafter.) A thermocouple temperature measurement, under computer control, could proceed as follows:

1. Measure t_{ref} and use the thermocouple polynomial to compute the equivalent thermocouple voltage V_{ref} for the parasitic junctions.
2. Measure V_v and *add* V_{ref} to find V_1.
3. Compute t_1 from V_1 using the thermocouple polynomial.

Solid-State Temperature Sensors

These devices are derived from modern silicon integrated-circuit technology, and are often referred to as Si sensors. They consist of electronic circuits that exploit the temperature characteristics of active semiconductor junctions. Versions are available with either current or voltage outputs. In both cases the outputs are directly proportional to temperature. Not only is the output linear, but it is of a relatively high level, making the signal interpretation very easy. The most common type generates 1 μA/K (298 μA at 25°C). This can be externally converted to a voltage by using a known resistor. The usable temperature range is −50 to 150°C. The stability and the accuracy of these devices are good enough to provide readings within ±0.5°C. It is easy to obtain 0.1°C resolution. Si sensors are ideal reference junction monitors for thermocouple measurements.

Resistance Temperature Detectors

Resistance temperature detectors (RTDs) exhibit a changing resistance with temperature. Additional detailed information on using RTDs can be found in the section on signal conditioning for resistive devices. Several different metals can be used to produce RTDs. Platinum is perhaps the most common for general applications. Yet at very high temperatures, tungsten is a good choice. Platinum RTDs have a positive temperature coefficient of about 0.004 Ω/°C. The relationship has a small nonlinearity that can be corrected with a third-order polynomial. Many data-acquisition systems include this capability. Platinum RTDs are usually built with 100-ohm elements. These units have sensitivities of about +0.4 Ω/°C. Their useful temperature range is about −200 to about +600°C.

Most RTDs are of either wire-wound or metal-film design. The film design offers faster response time, lower cost, and higher resistance values than the wire-wound type. The more massive wire-wound designs are more stable with time. High resistance is desirable because it tends to reduce lead-wire induced errors. To convert resistance into a voltage, an excitation current is required. Care must be taken to avoid current levels that will produce errors due to internal self-heating. An estimate of the temperature rise (in °C) can be found by dividing the internal power dissipation by 80 mW. This is a general rule that applies to small RTDs in a conductive fluid such as oil or water. In air the effects of self-heating can be 10 to 100 times higher.[2]

SAMPLED-DATA SYSTEMS

Modern data-acquisition systems use sampled-data techniques to convert between the analog and digital signal domains. This implies that while data may be recorded on a regular basis, they are not collected continuously, that is, there are gaps in time between successive data points. In general there is no knowledge of the missing information, and the amplitude of missing data points cannot be predicted. Yet under special circumstances it can be assumed that missing data fall on a straight line between known data points.

Fourier analysis reveals that signals, other than pure sine waves, consist of multiple frequencies. For example, a pulse waveform contains significant frequency components far beyond its fundamental or repetition rate. Frequencies extending to approximately $0.3/t_r$ are often important, where t_r is the pulse rise time. Step functions suggest that frequency components extend to infinity.

The Nyquist theorem defines the necessary relationship between the highest frequency contained in a waveform and the *minimum* required sampling speed. Nyquist states that the sample rate must be *greater* than two times the highest frequency component contained within the input signal. The danger of undersampling (sampling below the Nyquist rate) is erroneous results. It is not simply a matter of overlooking high-frequency information, but of reaching totally wrong conclusions about the basic makeup of the signal. See Fig. 5 for an example. Note that sampling a pure sine wave (containing only the fundamental frequency) at a rate in violation of the Nyquist criterion leads to meaningless results. This example suggests the presence of a totally nonexistent frequency. This phenomenon is known as aliasing.

---

[2] Additional information on temperature sensors will be found in Section 3 of this handbook.

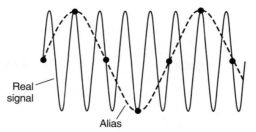

FIGURE 5   Aliasing because of insufficient sampling rate. (*Intelligent Instrumentation, Inc.*)

If it were possible to sample at an infinite rate, aliasing would not be a problem. However, there are practical limits to the maximum sampling speed, as determined by the characteristics of the particular A/D converter used. Therefore action must be taken to ensure that the input signal does not contain frequency components that cause a violation of the Nyquist criterion. This involves the use of an input low-pass filter (antialiasing filter) *prior* to the A/D converter. Its purpose is to limit the measured waveform's frequency spectrum so that no *detectable* component equals or exceeds half of the sampling rate. Detectable levels are determined by the sensitivity of the A/D converter and the attenuation of the antialiasing filter (at a given frequency).

Sequential Scanning

Systems are usually designed to collect data from more than one input channel. To reduce cost, most PC-based systems share significant components, including the A/D converter, with all of the channels. This is accomplished with a multiplexer (electronic scanning switch). However, when there is only one A/D converter, only one input channel can be acquired at a given point in time. Each channel is read sequentially, resulting in a time skew between readings. Techniques for minimizing time skew will be described.

ANALOG INPUT SYSTEMS

The fundamental function of an analog input system is to convert analog signals into a corresponding digital format. It is the A/D converter that transforms the original analog information into computer-readable data (a digital binary code). In addition to the A/D converter, several other components may be required to obtain optimum performance. These can include a sample/hold, an amplifier, a multiplexer, timing and synchronization circuits, and signal-conditioning elements.

A good starting point when choosing an A/D converter (or a system using an A/D converter) is to consider the characteristics of the input transducer. What is its *dynamic range*, maximum signal level, signal frequency content, source impedance, accuracy, and so on? A match in characteristics between the A/D converter and the transducer is usually desired. It is also important to consider possible sources of external interfering noise. This will have a bearing on the choice of A/D converter and the required signal conditioning. Some sensors have a very wide dynamic range. Dynamic range defines the span of input stimulus values that correspond to *detectable* output values. It is often expressed as the ratio of the maximum full-scale output signal to the lowest *meaningful* output signal. In this context, dynamic range and signal-to-noise ratio are the same. Caution! There is not necessarily a good correlation between sensor dynamic range and *accuracy*. Accuracy refers to how close the measured output corresponds to the *actual* input. For example, a transducer could have a dynamic range of 5000:1 and an accuracy of 1 percent. If the full-scale range is 100°C, a change as small as 0.02°C can be detected. Still, the actual temperature is only known, in absolute terms, to 1°C. 1°C is the accuracy; 0.02°C is the sensor's sensitivity. The proper choices of signal conditioning and A/D converter are essential to preserving the performance of the transducer. This example suggests the need for an A/D converter with more than 12 bits of resolution. Applications using other transducers may require 14-, 16-, or even 18-bit resolution.

Analog-to-Digital Converters

Many different types of A/D converters exist. Among these, a few stand out as the most widely used—successive-approximation, integrating, and flash (parallel) converters. Each converter has a set of unique characteristics that make it better suited for a given application. These attributes include speed, resolution, accuracy, noise immunity, cost, and the like.

Industrial and laboratory data-acquisition tasks usually require a resolution of 12 to 16 bits. 12 bits is the most common. As a general rule, increasing resolution results either in increased cost or in reduced conversion speed. Therefore it makes sense to consider the application requirements carefully before making a resolution decision.

All A/D converters accomplish their task by partitioning the full analog input range into a fixed number of discrete digital steps. This is known as digitizing or quantizing. A different digital code corresponds to each of the assigned steps (analog values). Digital codes consist of N elements, or bits. Because each bit is binary, it can have one of two possible states. Thus the total number of possible steps is 2^N. N is often referred to as the converter's *resolution* (that is, an N-bit converter). Given the number of steps S, it follows that $N = \log S/\log 2$. Caution is required when using the term "resolution" because it can also be expressed in other ways. For example, a 12-bit system divides its input into 2^{12}, or 4096, steps. Thus if the A/D converter has a 10-volt range, it has a resolution of 2.4 mV (10 volts ÷ 4096). This refers to the minimum detectable signal level. One part in 4096 can also be expressed as 0.024 percent of full-scale (FS). Thus the resolution is 0.024 percent FS. These definitions apply *only* to the ideal, internal capabilities of the A/D converter alone.

In contrast to this 12-bit example, 16 bits corresponds to one part in 65,536 (2^{16}), or approximately 0.0015 percent FS. Therefore increasing the resolution has the *potential* to improve both dynamic range and overall accuracy. On the other hand, system performance (effective dynamic range and accuracy) can be limited by other factors, including noise and errors introduced by the amplifier, the sample/hold, and the A/D converter itself.

Flash-type A/D converters can offer very high-speed operation, extending to about 100 MHz. Conversion is accomplished by a string of comparators with appropriate reference voltages, operating in parallel. To define N quantizing steps requires $2^N - 1$ comparators (255 and 4095 for 8 and 12 bits, respectively). Construction is not practical beyond about 8 or 10 bits. In contrast, most data-acquisition and control applications require more than 10 bits.

12-bit devices that run in the 5- to 15-MHz range are available using a *subranging* or half-flash topology. This uses two 6-bit flash encoders, one for coarse and the other for detailed quantizing. They work in conjunction with differencing, amplifying, and digital logic circuits to achieve 12-bit resolution. It is essential that the input signal remain constant during the course of the conversion, or very significant errors can result. This requires the use of a sample/hold circuit, as described hereafter. There is also a speed tradeoff compared to a full flash design, but the cost is much lower. Still, the size and power requirements of these flash converters usually prevent them from being used in PC-based internal plug-in products. These factors (cost, size, and power) generally limit flash converters to special applications.

Successive approximation converters are the most popular types for general applications. They are readily available in both 12- and 16-bit versions. Maximum sampling speeds in the range of 50 kHz to 1 MHz are attainable. These converters use a single comparator and are thus relatively low cost and simple to construct. An internal D/A converter (described below) systematically produces binary weighted guesses that are compared to the input signal until a "best" match is achieved. A sample/hold is required to maintain a constant input voltage during the course of the conversion.

Integrating converters can provide 12-, 16-, and even 18-bit resolution at low cost. Sampling speeds are typically in the range of 10 to 500 Hz. This converter integrates the unknown input voltage, V_x for a specific period of time T_1. The resulting output e_1 (from the integrator) is then integrated back down to zero by a known reference voltage V_{ref}. The time required T_2 is proportional to V_x:

$$V_x = V_{ref} \frac{T_2}{T_1}$$

Noise and other signal variations are effectively averaged during the integration process. This characteristic inherently smooths the input signal. The dominant noise source in many data-acquisition applications is the ac power line. By setting the T_1 integration period to a multiple of the ac line frequency, significant noise rejection is achieved. This is known as normal- or series-mode rejection. In addition, linearity and overall accuracy are generally better than with a successive approximation converter. These factors make the integrating converter an excellent choice for low-level signals such as thermocouples and strain gages.

Amplifiers

A simple analog input stage is shown in Fig. 6. This circuit can accommodate only one input channel. For multiple channels, several parallel stages can be used. However, the use of a multiplexer to share common resources can provide significant cost savings. This is suggested in Fig. 7. The amplitude of analog input signals can vary over a very wide range. Signals from common transducers are between 50 µV and 10 volts. Yet most A/D converters perform best when their inputs are in the range of 1 to 10 volts. Therefore many systems include an amplifier to boost possible low-level signals to the desired amplitude. Note that adding a fixed-gain amplifier increases sensitivity, but does not increase dynamic range. While it extends low-level sensitivity, it reduces the maximum allowable input level proportionally (the ratio remains constant). Ideally an input amplifier would have several choices of gain, all under software control. Such a device is called a programmable gain amplifier (PGA). Gains of 1, 2, 4, and 8 or 1, 10, 100, and 200 are common. Unlike a fixed amplifier, a PGA can increase the system's effective dynamic range. For example, consider the following:

- A 12-bit converter (which can resolve one part in 4096) with a full-scale range of 10 volts
- A PGA with gains of 1, 10, and 100—initially set to 1
- An input signal that follows a ramp function from 10 volts down toward zero

Under these conditions, the minimum detectable signal will be 2.4 mV (10 volts ÷ 4096). However, software can be written to detect when the signal drops below 1 volt and to reprogram the PGA automatically for $G = 10$. This extends the minimum sensitivity to 240 µV

FIGURE 6 Simple analog input stage. *Note:* Amplifier may not be required in every application. (*Intelligent Instrumentation, Inc.*)

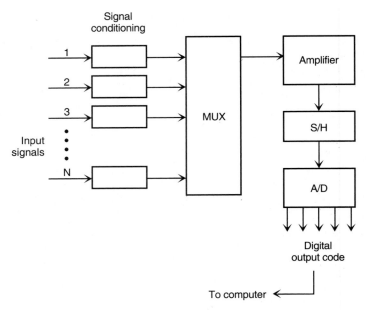

FIGURE 7   Complete analog input subsystem in which a multiplexer (MUX) is used to handle multiple signals. Software can control switches to select any one channel for processing at a given time. Since the amplifier and A/D converter are shared, the speed of acquisition is reduced. To a first approximation, the rated speed of the amplifier and A/D converter will be divided by the number of input channels serviced. Throughput rate is often defined as the per-channel speed multiplied by the number of channels. (*Intelligent Instrumentation, Inc.*)

while increasing a 0.999-volt signal to 9.99 volts. As the signal continues to drop (below 0.1 volt), the PGA can be set for $G = 100$, extending the sensitivity to 24 μV. The effect of making these corresponding changes in gain, to track signal variations, is to increase the *system's* dynamic range. Dynamically adding a gain of 100 is like adding close to 7 bits to the converter. Thus a 12-bit A/D converter can support a range of more than 18 bits. This technique is restricted to systems that have a PGA and to applications where the sample rate is slow enough to permit the time required for the auto-ranging function.

Most PGAs are differential input instrumentation amplifiers. They present a very high input impedance at both their + and − terminals. The common-mode rejection characteristic of this type of amplifier can attenuate the effects of ground loops, noise, and other error sources. Thus differential inputs are especially useful for measuring low-level signals. Most analog input systems have provisions for configuring the input multiplexer and amplifier for either single-ended or differential use.

Single-Ended versus Differential Signals

Single-ended inputs all share a common return or ground line. Only the high ends of the signals are connected through the multiplexer to the amplifier. The low ends of the signals return to the amplifier through the system ground connections, that is, both the signal sources and the input to the amplifier are referenced to ground. This arrangement works well for high-level signals when the difference in ground potential is relatively small. Problems arise when there is a large difference in ground potentials. This is usually caused by current flow (a ground loop) through the ground conductor. This is covered in further detail in the next article of this handbook section.

A differential arrangement allows both the noninverting (+) and the inverting (−) inputs of the amplifier to make direct connections to both ends of the actual signal source. In this way any ground-loop-induced voltage appears as a common-mode signal and is rejected by the differential properties of the amplifier. While differential connections can greatly reduce the effects of ground loops, they consume the equivalent of two single-ended inputs. Thus a 16-channel single-ended system can handle only eight differential inputs.

Ideally the input impedance, common-mode rejection, and bandwidth of the system's amplifiers would be infinite. In addition, the input currents and offset voltage would be zero. This would provide a measuring system that does not load or alter the signal sources. In contrast, real amplifiers are not perfect. Offset voltage V_{os} appears as an output voltage when the inputs are short-circuited (input voltage is zero). V_{os} can sometimes be compensated for in software. Input (bias) current can be more of a problem. This is the current that flows into (or out of) the amplifier's terminals. The current interacts with the signal source impedance to produce an additional V_{os} term that is not easy to correct. A resistive path must be provided to return this current to ground. It is necessary that the resistance of this path be small enough so that the resulting V_{os} (bias current × source resistance) does not degrade the system's performance significantly. In the extreme case where the inputs are left floating (no external return resistance), the amplifier is likely to reside in a nonlinear or otherwise unusable state. As a general rule, single-ended inputs do not require attention to the bias current return resistance. This is because the path is often provided by the signal source. In contrast, differential connections almost always require an external return resistor. Normally the system's termination panel will provide these resistors. Typically, values of 10 or 100 kΩ are used.

Common-Mode Rejection

The ability of a differential-input amplifier to discriminate between a differential mode (desired input signal) and a common mode (undesired signal) is its common-mode rejection ratio (CMRR), expressed in decibels (dB). For a given amplifier the CMRR is determined by measuring the change in output that results from a change in common-mode input voltage. CMRR (dB) = $20 \log(dV_{out}/dV_{CM})$. In a data-acquisition system the output signal from the input amplifier includes any error due to the finite CMRR. The A/D converter cannot discriminate between true and error portions of its input signal. Thus the relationship between the magnitude of the error and the sensitivity of the A/D converter is significant. This sensitivity is often referred to as the A/D converter's resolution, or the size of its least significant bit (LSB). If the error exceeds 1 LSB, the A/D converter responds. As suggested, CMRR is used to measure the analog output error produced by a common-mode input. This makes sense because it is the ratio of two analog signals. However, a complete data-acquisition system (including an A/D converter) has a digital output. Therefore it is more meaningful to express the system's common-mode error in terms of LSBs. This is done by dividing the common-mode error voltage dV_{out} by the sensitivity of the A/D converter (1 LSB on the given range). Amplifier gain G_{diff} must be taken into account. Sensitivity is equal to the converter's full-scale range (FSR) divided by its resolution (number of steps):

$$\text{Error}_{CM} \text{ (LSB)} = \frac{dV_{CM} \cdot G_{diff} \cdot 10^{-\text{CMRR}/20}}{\text{FSR/resolution}} \qquad (3)$$

Table 1 shows the relationship between CMRR in decibels and common-mode error in least significant bits for a hypothetical input system. A 12-bit A/D converter on a 10-volt range (0 to 10 volts or ±5 volts) is assumed in this comparison. A 10-volt common-mode sig-

TABLE 1 Relationship between CMRR Expressed in dB and Common-Mode Error in LSB for Hypothetical Input System

Signal gain	Common-mode rejection ratio, dB	Error_{CM} absolute, LSB	Error_{CM}/V_{CM}, LSB/V
1	80	0.4	0.04
10	90	1.3	0.13
100	100	4.1	0.41
1000	110	13	1.3

nal is applied to the short-circuited (connected together) input terminals of the system. Dividing the common-mode error (in LSB) by the common-mode voltage yields a direct (useful) figure of merit for the complete data-acquisition system, not just the input amplifier.

Note that independently increasing (improving) CMRR at a *given* gain improves the system's performance. However, the increase in CMRR that accompanies an *increase* in gain actually results in a decrease of the system's overall accuracy. This is because the increase in CMRR (note the log relationship) has less effect than the direct increase in common-mode error ($dV_{CM} \cdot G_{diff}$).

Sample/Hold System

There is a distinct time interval required to complete a given A/D conversion. Only the integrating converter can tolerate input amplitude changes during this period. For the other converters a detectable change will result in significant errors. In general, an analog signal can have a continuously changing amplitude. Therefore, a sample/hold (S/H) is used as a means of "freezing" or holding the input constant during the conversion period. Fundamentally the S/H consists of a charge storage device (capacitor) with an input switch. When the switch is closed, the voltage on the capacitor tracks the input signal (sample mode). Before starting an A/D conversion, the switch is opened (hold mode), leaving the last instantaneous input value stored on the capacitor. This is maintained until the conversion is complete. In all practical applications, both successive approximation and subranging converters must use an S/H at their inputs. While a full flash A/D converter does not normally require an S/H, there are applications where it will improve its spurious-free dynamic range.

Multiplexers

The multiplexer (MUX) is simply a switch arrangement that allows many input channels to be serviced by one amplifier and A/D converter (Fig. 7). Software or auxiliary hardware can control this switch to select any one channel for processing at a given time. Because the amplifier and A/D converter are shared, the channels are read sequentially, causing the overall speed of the system to be reduced. To a first approximation, the rated speed of the amplifier and A/D converter will be divided by the number of input channels serviced. The throughput rate is defined as the sample rate (per-channel speed) multiplied by the total number of channels.

The user must be careful not to be misled by the speed specifications of the individual components in the system. Conversion time defines only the speed of a single A/D conversion. Software overhead, amplifier response time, and so on, can greatly reduce a system's performance when reading multiple channels.

In an ideal system all of the input channels would be read at the same instant in time. In contrast, multiplexing inherently generates a "skew," or time difference, between channels. In some cases the system may be fast enough to make it "appear" that the channels are being read at the same time. However, some applications are very sensitive to time skew. Given the fastest A/D converters available, there are still many applications that cannot tolerate the time difference between readings resulting from multiplexing. In critical applications the technique of *simultaneous* S/H can reduce time skew by a factor of 100 to 1000 (Fig. 8).

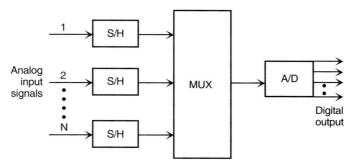

FIGURE 8   Simultaneous sample/hold system. The function of the S/H system is to grab the present value of the signal just before the beginning of an A/D conversion. This level is held constant, despite a changing input, until the A/D conversion is complete. This feature allows the accurate conversion of high-frequency signals. (*Intelligent Instrumentation, Inc.*)

The simultaneous S/H architecture is ideal when the phase and time relationships of multiple input channels are critical to a given application. For example, assume the system in Fig. 7 is sequentially scanning four analog inputs at a throughput rate of 100 kHz. The elapsed time between conversions would be 10 μs. About 40 μs would be required to digitize all four channels. If the input signals are each 10 kHz sine waves, there will be an apparent phase shift of 144° between the first and fourth channels (40 μs/100 μs · 360°). In contrast, the simultaneous S/H system in Fig. 8 can capture all four channels within a few nanoseconds of each other. This represents a phase shift of about 0.01°.

This technique is particularly useful for preserving time and phase relationships in applications where cross-correlation functions must be calculated. Prime examples include speech research, materials and structural dynamics testing, electrical power measurements, geophysical signal analysis, and automatic test equipment (ATE) on production lines.

Analog Signal Conditioning

Analog input systems, based on the components described previously, are representative of most PC-based plug-in products. These boards are usually designed to accept voltage inputs (only) in the range of perhaps ±1 mV to ±10 V. Other signal ranges and signal types generally require preprocessing to make them compatible. This task is known as signal conditioning. The type of conditioning used can greatly affect the quality of the input signal and the ultimate performance of the system. Signal conditioning can include current-to-voltage conversion, surge protection, voltage division, bridge completion, excitation, filtering, isolation, and amplification. The required components can be physically located either remotely, at the signal source (the transducer), or locally, at the data-acquisition board (the host PC). Remote applications use *transmitters* that include the required components. They generally deliver high-level conditioned signals to the data-acquisition system via a twisted-pair cable. Local transducers usually connect directly to termination panels that include the required components.

Filtering

Of all the signal-conditioning categories, filtering is the most widely needed and most widely misunderstood. Simply stated, filtering is used to separate desired signals from undesired signals. Undesired signals can include ac line frequency pickup and radio or TV station interference. All such signals are referred to here as *noise*. Filtering can be performed, prior to the A/D conversion, using "physical" devices consisting of resistors, capacitors, inductors, and amplifiers. Filtering can also be accomplished, after conversion, using mathematical algorithms that operate on the digital data within the PC. This is known as digital signal processing (DSP).

Averaging is a simple example of DSP. It is a useful method for reducing unwanted data fluctuations. By averaging a series of incoming data points, the signal-to-noise ratio can be effectively increased. Averaging will be most effective in reducing the effects of random nonperiodic noise. It is less effective in dealing with 50- or 60-Hz or other periodic noise sources. When the desired signal has lower frequency components than the error sources, a low-pass filter can be used. This includes the case where the "real" input signal frequency components can equal, or exceed, half the sampling rate. Here the filter is used to prevent sampled-data aliasing. Aliasing results in the generation of spurious signals within the frequency range of interest that cannot be distinguished from real information. Hence serious errors in the interpretation of the data can occur. Noise-filtering techniques, whether implemented in hardware or software, are designed to filter specific types of noise. In addition to low-pass filters, high-pass and notch (band-reject) filters also can be used. For example, if the frequency band of interest includes the ac line frequency, a notch filter could be used to selectively remove this one component.

Signal termination panels are available that have provisions for the user to install a variety of filters. The most common types of filters are represented by the one- and two-pole *passive* filters shown in Fig. 9. The main difference between these passive filters and the *active* filters, mentioned hereafter, is the addition of amplifiers. Figure 9*b* is an example of an effective sin-

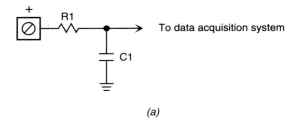

(a)

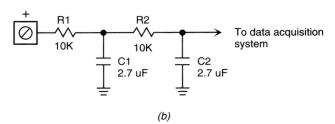

(b)

FIGURE 9 Low-pass filters. (*a*) One pole. (*b*) Two poles. (*Intelligent Instrumentation, Inc.*)

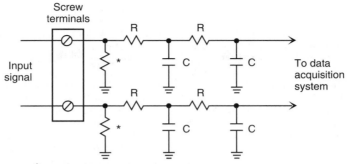

*Amplifier bias current return resistors

FIGURE 10   Two-pole differential low-pass filter. (*Intelligent Instrumentation, Inc.*)

gle-ended double-pole circuit to attenuate 50/60-Hz noise. The filter has a −6-dB cutoff at about 1 Hz while attenuating 60 Hz about 52 dB (380 times).

Figure 10 suggests a differential two-pole low-pass filter. In contrast to the circuits in Fig. 9, this can be used in balanced applications. Note that any mismatch of the attenuation in the top and bottom paths will result in the generation of a differential output signal that will degrade the system's common-mode rejection ratio. Therefore the resistors and capacitors should be matched carefully to each other. If it is given that all of the resistors and capacitors are of equal values, the pole position f_1 for this *differential* two-section filter is

$$f_1 = \frac{0.03}{R \cdot C} \tag{4}$$

and the approximate attenuation ratio ($r = V_{in}/V_{out}$), at a given frequency f_x, is

$$r = \left(\frac{f_x \cdot R \cdot C}{0.088} + 1 \right)^2$$

$$= \left(\frac{0.3 f_x}{f_1} + 1 \right)^2 \tag{5}$$

$$dB = 20 \log r$$

The equations for a single-ended single-pole filter are

$$f_1 = \frac{0.159}{R \cdot C}$$

$$r = \sqrt{\left(\frac{f_x \cdot R \cdot C}{0.159} \right)^2 + 1} \tag{6}$$

and also,

$$r = \sqrt{\left(\frac{f_x}{f_1} \right)^2 + 1}$$

The foregoing equations assume that the source impedance is much less than R and that the load impedance is much larger than R.

For filter applications, monolithic ceramic-type capacitors have been found to be very useful. They possess low leakage, have low series inductance, have very high density (small in size for a given capacitance), and are nonpolarized. Values up to 4.7 µF at 50 volts are commonly available.

In the ideal case a perfect low-pass filter could be built with infinite rejection beyond its cutoff frequency f_1. This would allow f_1 to be set just below one-half the sampling rate, providing maximum bandwidth without danger of aliasing. However, because perfect filters are not available, some unwanted frequencies will "leak" through the filter. Nyquist requires that the "2 times" rule be applied to the highest frequency that can be resolved by the A/D converter. This may not be the same as the highest frequency of interest f_1. Therefore the margin between the highest frequency of interest and the sampling rate must be adjusted. This could involve increasing the sampling rate or possibly forcing a reduction in signal bandwidth. In applications using simple passive filters, the attenuation rate might only be –20 to –40 dB per decade (one and two poles, respectively). This could require the sampling frequency to be 10 to 1000 times the filter corner frequency. The exact factor depends on the resolution of the A/D converter and the amplitude of the highest frequency component. Using high-order active filters (seven to nine poles) might only require a factor of 1.5 to 3 (relative to the original 2× Nyquist rule).

Complete ninth-order elliptic designs are available in a number of configurations. These filters have very steep rolloff (approximately –100 dB per octave), while maintaining nearly constant gain in the passband (±0.2 dB is common). In selecting elliptic filters care must be taken to choose a unit that has a stopband attenuation greater than the resolution of the system's A/D converter. For example, a 12-bit converter has a resolution of one part in 4096, which corresponds to 72 dB. The filter used should attenuate all undesired frequencies by more than 72 dB. Likewise, a 16-bit converter would require a 96-dB filter. Fixed-frequency filter modules, as well as switch- and software-programmable units, can be purchased from various manufacturers. Several of these modular filters can be installed directly on the system's input termination panels. Complete programmable filter subsystems are also available in the form of boards that plug directly into an expansion slot within the data-acquisition PC. The advantage of high-order active filters in antialiasing applications is now clear. Yet while offering excellent performance, they are physically large and expensive compared to simpler filters.

In summary, filtering is intended to attenuate *unavoidable* noise and to limit bandwidth to comply with the Nyquist sampling theorem which prevents aliasing. An antialiasing filter must be a physical analog filter. It cannot be a digital filter which operates on the data after A/D conversion. Noise suppression can often be accomplished, or assisted, with either an analog or a digital filter. Still, filters are not intended as substitutes for proper wiring and shielding techniques. Ground loops, along with capacitively or inductively coupled noise sources, require special attention.

Analog Signal Scaling

As indicated earlier, most A/D converters are designed to operate with high-level input signals. Common A/D conversion ranges include 0 to 10, ±5, and ±10 volts. When the maximum input signal is below 1 volt, accuracy is degraded. Under these circumstances it is often appropriate to amplify the signal before the A/D converter. Some A/D converter boards have amplifiers built in. If needed, external amplifiers can be added as part of the signal-conditioning circuitry on the termination panels. In addition to an input signal being too small, it is possible that it might be too large. Remember that most converters accept a maximum of 10 volts at their input. Signals could be 12, 48, or 100 volts (or more). Fortunately it is a simple matter to reduce excessive levels with a resistive voltage divider network. Figure 11 is appropriate for most analog signals. In selecting R_1 and R_2 there are practical factors to consider. Making R_1 large can introduce limitations on signal bandwidth, due to the low-pass filter pro-

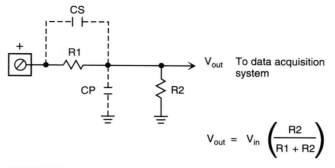

$$V_{out} = V_{in} \left(\frac{R2}{R1 + R2} \right)$$

FIGURE 11 Resistive voltage divider to reduce large analog input signals to below 10 volts. (*Intelligent Instrumentation, Inc.*)

duced by R_1 and the parasitic capacitance C_p in parallel with R_2. In some applications the network bandwidth can be extended by placing a capacitor C_s across R_1. The value should be selected to make the time constant $R_1 \cdot C_s$ equal to $R_2 \cdot C_p$. The equation assumes that the source (signal) impedance is very low compared to the series combination of R_1 and R_2, that is, $R_1 + R_2$. From this perspective, R_1 and R_2 should be as large as possible.

Input Buffering

The input characteristics of most data-acquisition boards are suitable for general applications. Yet in some cases the input resistance is too low or the bias current too high to allow accurate measurements. Input capacitance can also be an important factor. This is because some transducers, including piezoelectric and pH cells, exhibit a very high output impedance. Under these conditions, direct connection to the data-acquisition system can cause errors. These applications can be satisfied by adding a high input impedance buffer amplifier to the signal-conditioning circuitry. Figure 12 suggests one type of buffer circuit that can be used.

Resistance Signals

Resistance signals arrive at the data-acquisition system from primary sensors, such as strain gages and RTDs. Resistance is changed to a voltage by exciting it with a known current (V_{out}

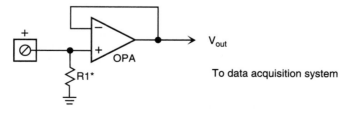

*R1 can be a very high impedance resistor up to 10^9 ohms.

FIGURE 12 High-input impedance buffer circuit. (*Intelligent Instrumentation, Inc.*)

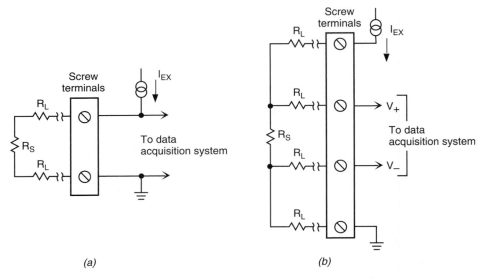

(a)   (b)

FIGURE 13   Measurement system for resistive device. R_L—lead-wire resistance. (*Intelligent Instrumentation, Inc.*)

$= IR$). Figure 13 shows the simplest way to measure resistance. As suggested in Fig. 13*a*, the parasitic (unwanted) resistance of the two connecting lead wires can introduce significant errors. This is because the excitation current flows through the signal measurement wires. Figure 13*b* uses four connecting wires and a differential input connection to the data-acquisition system to minimize the lead-wire effects. This is known as a four-terminal (or kelvin) measurement. The extra wires allow the direct sensing of the unknown resistance. In both cases the resistance of the wires going to the data-acquisition system has little effect. This is because very little current flows in these leads. However, this technique is not well suited to RTD and strain-gage applications because of the very small change in measured voltage compared to the steady-state (quiescent) voltage. The large quiescent voltage prevents the use of an amplifier to increase the measurement sensitivity.

It is usually better to measure resistive sensors as part of a Wheatstone bridge. A bridge is a symmetrical four-element circuit that enhances the system's ability to detect small changes in the sensor. In Fig. 14 the sensor occupies one arm of the bridge. The remaining arms are completed with fixed resistors equal to the nominal value of the sensor. In general, however, the sensor can occupy one, two, or four arms of the bridge, with any remaining arms being filled with fixed resistors. Note that the differential output from the bridge is zero when all of the resistors are equal. When the sensor changes, an easily amplified signal voltage is produced. Bridge-completion resistors should be of very high precision (typically 0.05 percent). However, stability is actually more important. Initial inaccuracies can be calibrated out, but instability always appears as an error.

A 100-ohm platinum RTD can be used to compare the merits of the approaches in Figs. 13 and 14. To control internal self-heating, the excitation level will be limited to 2 mA. Given that the sensitivity of this type of device is about +0.4 Ω/°C, the output will be about 0.8 mV/°C. This is indeed a small signal that will require amplification. It would be useful to multiply the signal by 100 to 1000 to make best use of the A/D converter's full-scale range (typically 5 or 10 volts). However, the quiescent voltage across the RTD is 2 mA $\cdot$ 100 Ω = 0.2 volt. In Fig. 13 this limits the maximum gain to 10. Thus in a 12-bit system, the smallest detectable

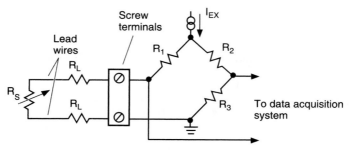

FIGURE 14   Two-wire bridge configuration. (*Intelligent Instrumentation, Inc.*)

temperature change will be about 0.5°C. In contrast, the bridge circuit in Fig. 14 balances out the fixed or quiescent voltage drop, allowing greater magnification of the difference signal. This allows the detection of changes as small as 0.005°C.

Figure 14 has the same lead-wire resistance problem that the simple circuit had. The lead-wire resistances are indistinguishable from the transducer's resistance. So while this bridge circuit has high sensitivity, it is not suitable for precision applications.

Figure 15 shows a means of correcting for lead-wire effects. While the three-wire bridge requires an additional wire to be run to the sensor, several very important advantages are gained. If the (reasonable) assumption is made that the two wires bringing current to the sensor are of the same material and length, many of the potential error terms cancel. Comparing Figs. 14 and 15 shows that the additional signal wire has moved the measurement point directly to the top of the sensor. Again, the resistance of this wire is not important because current flow to the data-acquisition system is very low. Moving the measurement point has the effect of locating one current-carrying lead resistance in the top arm of the bridge while the other remains in the lower arm. The current in each is the same, so their voltage drops tend to cancel. While the compensation is not perfect, it does offer a significant performance improvement.

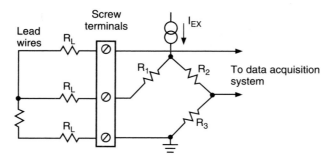

R_L is lead-wire resistance

FIGURE 15   Three-wire bridge configuration. (*Intelligent Instrumentation, Inc.*)

Transducer excitation and bridge-completion components are normally installed on the system's signal termination panels. While both voltage and current excitation can be used, current excitation is generally more desirable. This is because current excitation provides a more linear output response, making the data interpretation easier.

Current Conversion

The need to measure signals in the form of currents is quite common in a data-acquisition system. The outputs from remote transducers are often converted to 4- to 20-mA signals by two-wire transmitters. At the data-acquisition system, current is easily converted back to a voltage with a simple resistor. Figure 16 shows how this is done. For a 4- to 20-mA signal a resistor value of 250 ohms can be used to provide a voltage output of 1 to 5 volts. As a general rule, the largest resistor that does not cause an overrange condition should be used. This ensures the maximum resolution. Stability of the resistor is essential, but the exact value is not important. Most systems have software provisions for calibrating the measurement sensitivity of each channel at the time of installation. Low-cost 0.1 percent metal-film resistors are usually adequate. For larger currents it is a simple matter to scale the resistor down to yield the desired full-scale voltage ($R = V/I$, where V is the full-scale voltage range and I is the maximum current to be read).

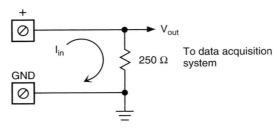

FIGURE 16   4- to 20-mA input conversion circuit (single-ended). (*Intelligent Instrumentation, Inc.*)

The technique of using only a resistor to convert from current to voltage does have limitations. If, for example, a 1-μA level is to be measured, a resistor of approximately 5 MΩ will be required. Unfortunately the use of high-value resistors leads to potentially large errors due to noise and measuring system loading. As suggested before, the data-acquisition system has a small but finite input current. This bias current (typically around 10 nA) will also flow through the conversion resistor and will be indistinguishable from the signal current. Therefore when very low currents must be measured, a different technique is used. Figure 17 suggests an active circuit that utilizes a precision field effect transistor (FET) amplifier to minimize the bias current problem. Both the simple resistor and the FET amplifier circuits require the same resistor value for a given current level. However, in the latter case the data-acquisition system's bias current is supplied by the amplifier and does not affect the measurement accuracy. A wide range of low-bias-current amplifiers are available for special applications. With the amplifier shown, currents as low as 10 pA can be read reliably.

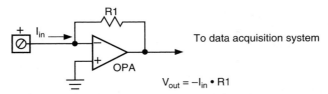

$$V_{out} = -I_{in} \cdot R1$$

FIGURE 17   Current-to-voltage converter circuit suitable for very low current levels. (*Intelligent Instrumentation, Inc.*)

Transmitters

When low-level signals (below 1 volt) are located remotely from the data-acquisition system, special precautions are suggested. Long wire runs with small signals usually result in poor performance. It is desirable to first preamplify these signals to preserve maximum signal-to-noise ratio. Two-wire transmitters provide an ideal way of packaging the desired signal-conditioning circuitry. In addition to signal amplification, transmitters can also provide filtering, isolation, linearization, cold-junction compensation, bridge completion, excitation, and conversion to a 4- to 20-mA current. Transmitters are ideal for thermocouples, RTDs, and strain gages. Current transmission allows signals to be sent up to several thousand feet (1500 meters) without significant loss of accuracy. While voltage signals are rapidly attenuated by the resistance of the connecting wires, current signals are not. In a current loop, the voltage drop due to wire resistance is compensated by the compliance of the current source, that is, the voltage across the current source automatically adjusts to maintain the desired current level. Note that power for the transmitter is conveyed from the data-acquisition system over the same two wires that are used for signal communications. No local power is required.

In addition to analog transmitters, there are also digital devices. These provide most of these features, except that the output signal is in a digital form instead of 4 to 20 mA. The output protocol is usually a serial data stream that is RS-232, RS-422, or RS-485 compatible. This is accomplished by including an A/D converter and a controller (computer) inside the transmitter. In many cases the output signal can be connected directly to a serial port on the PC without additional hardware. Two possible disadvantages of a digital transmitter are that it requires local power and, because of the added complexity, is generally more expensive.

Surge Protection

When a system can be subjected to unintentional high-voltage inputs, it is prudent to provide protection to avoid possible destruction of the equipment. High-voltage inputs can be induced from lightning, magnetic fields, static electricity, and accidental contact with power lines, among other causes.

Figure 18 suggests two different protection networks. Both circuits offer transient (short-duration) as well as steady-state protection. The circuit in Fig. 18a can tolerate continuous inputs of up to about 45 volts. When the overload disappears, the signal path automatically returns to normal. The circuit in Fig. 18b protects against continuous overloads of up to about 280 volts. In contrast, sustained overloads to this circuit will cause the fuse to open (protecting the protection circuit). A disadvantage of this network is that the fuse must be replaced before the signal path is active again. In either case, signal flow is interrupted during the overload period. The resistor (or fuse) and the metal-oxide varistor (MOV) form a voltage clamp to ensure that transients will not get to the input of the data-acquisition system. MOVs are semiconductor devices that can react very quickly to absorb high-energy spikes. The 15-volt rating shown is high enough to pass all normal signals, but low enough to protect the data-acquisition system's input. Consideration should be given to the fact that even below the MOV's threshold voltage a small leakage current flows. If the series R is too large, the leakage could appear as a significant temperature-dependent error voltage (IR).

The optional capacitor can help suppress high-frequency transients. In some applications it must be rated for high voltage. For example, transients in power stations or other noisy environments can exceed 1000 volts. The capacitance value should be as large as physically possible, and the capacitor should be positioned as close as possible to the signal entry point of the system. Capacitors with low series impedance at high frequencies should be selected.

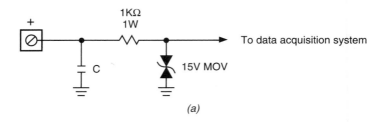

(a)

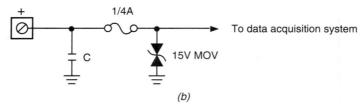

(b)

FIGURE 18   Representative input protection networks. (*Intelligent Instrumentation, Inc.*)

This requirement eliminates electrolytic-type capacitors. If the input signal can change polarity, polarized capacitors must be avoided.

Analog Isolation

Isolators can be used to protect people and equipment from contact with high voltage. They usually provide the same protection as MOVs with the addition of one very important extra feature. Isolators can block overloads (protect) while simultaneously passing a desired signal. Applications include the breaking of ground loops, patient monitoring, and the removal of large common-mode signals. For example, if a thermocouple is connected to a motor winding, it could be in contact with 240 volts ac. Yet the thermocouple output voltage might be only 30 mV. The 30 mV (the actual signal) is seen as a differential signal while the 240 volts appears as a common-mode signal. The isolator operates in a way that is similar to a differential amplifier (described earlier). Its common-mode rejection capabilities block the effects of the unwanted portion of the signal. While standard differential amplifiers are generally limited to a ±10 volt common-mode signal, isolators are available with ratings beyond 5000 volts.

A family of industry-standard 5B signal-conditioning modules is available. These complete plug-in units are designed to provide a wide range of input and output capabilities. Each module supports a single channel, allowing the flexibility to mix the various types when configuring a system. Isolation, rated at 1500 volts, provides high-voltage separation between the signals and the data-acquisition system. Input modules are available for most voltage ranges, current ranges, thermocouples, RTDs, and strain gages. All of the required conditioning functions are included: protection, filtering, linearization, cold-junction compensation, bridge completion, and excitation. Output modules support 4- to 20-mA current loops. Standard termination panels accommodate up to 16 modules.

ANALOG OUTPUTS

Digital-to-Analog Converters

Analog outputs are required in many test and industrial automation applications. For example, they can be used to generate inputs (stimuli) to a device under test and to operate valves, motors, and heaters in closed-loop feedback control systems. A D/A converter is used to transform the binary instructions from the digital computer (PC) to a variable output level. Common analog output ranges include ±5, ±10, and 0 to 10 volts and 4 to 20 mA.

A popular type of D/A converter consists internally of N binary weighted current sources. The values (the levels can be scaled to suit speed and output requirements) correspond to 1, ½, ¼, ⅛, . . . , $\frac{1}{2}^{N-1}$. N is also the number of digital input lines (bits). Each source can be turned on or off independently by the computer. By summing the outputs of the sources together, 2^N current combinations are produced. Thus a 12-bit converter can represent an analog output range with 4096 discrete steps. A current-to-voltage converter is included in voltage output models.

Faithful generation of a complex signal requires that the conversion rate (clock rate) of the D/A converter be very high compared to the repetition rate of the waveform. Ratios of 100 to 1000 points per cycle are common. This suggests that a "clean" 1-kHz output could require a 100-kHz to 1-MHz converter. This is pushing the current state of the art in PC-based data-acquisition products.

When operating in the voltage output mode, most D/A converters are limited to supplying around 5 mA of load current. This implies that most D/A converters will use some kind of signal conditioning when interfacing to real-world devices (transducers). When large loads such as positioners, valves, lamps, and motors are to be controlled, power amplifiers or current boosters need to be provided. Most data-acquisition systems do not include these high-power analog drivers internally.

Output Filtering

A D/A converter attempts to represent a continuous analog output with a series of small steps. The discontinuities inherent in a digitally produced waveform represent very high frequencies. This is seen as noise or distortion that can produce undesired effects. A low-pass filter is often used at the output of a D/A converter to attenuate high frequencies and, thus, "smooth" the steps.

DIGITAL INPUTS AND OUTPUTS

Most data-acquisition systems are able to accept and generate TTL level signals. These are binary signals that are either high or low (on or off). The low state is represented by a voltage near 0 volts (generally less than 0.8 volt), while a high state is indicated by a voltage near 5 volts (generally greater than 2 volts). Levels between 0.8 and 2 volts are not allowed. The output levels are intended to drive other "logic" circuits rather than industrial loads. As a result, drive capabilities are generally under 24 mA. Still, digital signals in many real-world applications are not TTL-compatible. It is common to encounter 24-volt, 48-volt, and 120/240-volt ac levels as digital input signals. High voltage and current outputs are often required to operate solenoids, contactors, motors, indicators, alarms, and relays.

Many types of digital signal termination panels are available to facilitate the connection of field wires to the data-acquisition system. In addition to screw terminals, the panels have provisions for signal conditioning, channel status indicators (such as light-emitting diodes), voltage dividers, and isolators. Thus the monitoring and the control of high dc levels, along with ac line voltage circuits, are readily accomplished.

Pulse and Frequency Inputs and Outputs

A variety of counting, timing, and frequency-measuring applications exists. Other applications require that devices be turned on and off for precise time periods. All of these functions can be provided by counter/timer circuits. The system's counter/timers are optimized for pulse applications, including frequency measurement and time-base generation. Counters are characterized by the number of input events that can be accumulated and by their maximum input frequency. Several independent counters are usually provided. They can be used to count events (accumulate), measure frequency, measure pulse width, or act as frequency dividers. Counting can be started from a defined initial value, and most counters can be configured to reset automatically to this value after it has been read. Software can easily interpret the counter's data as a sum or difference from an arbitrary starting point. Pulse generators (rate generators) are software programmable over a very wide range of frequencies and duty cycles. A rate generator is often used to provide the precise time base required for accurate data acquisition. Most systems use 16-bit counters that can accumulate pulses at frequencies up to 8 MHz. Up to 65,536 (2^{16}) events can be accumulated before the counter overflows. Two counters can generally be cascaded to provide 32-bit capability (more than 4 billion counts). Most counters accept only TTL level signals. Other levels require signal conditioning.

Frequency measurements using counters can be accomplished in different ways, depending on the application. When the unknown frequency is a TTL signal, it can be applied directly to the counter circuit. Analog signals can be converted to TTL levels with comparator circuits available from some manufactures. Voltage dividers using resistors, zener diodes, or optoisolators can be used to scale down high-level signals. When using any kind of signal conditioner before a counter input, consideration should be given to possible speed limitations.

Two distinct options exist for measuring high or low frequencies. The first method counts a known clock generator for the period of the unknown input signal. This provides high resolution for low-frequency signals, while minimizing the time required for the measurement. Generally this is used for frequencies below 10 Hz. The second method counts cycles of the unknown input signal for a fixed time interval. The advantage of this technique is that it allows measurements up to the limit of the counter's speed (typically 8 MHz). It is easy to implement an auto-ranging software algorithm that optimizes resolution over a very wide frequency range.

Digital Signal Scaling

For large digital signals, the circuit in Fig. 19 can be used to produce TTL-compatible levels. Most digital circuits (digital input ports and counters) require fast input level transitions to ensure reliable operation. Steps faster than 10 μs are usually adequate. Parasitic capacitance at the input to the data-acquisition system can interact with the series resistor to degrade input steps. The 10-pF capacitor in Fig. 19 is included to help correct this problem. When the

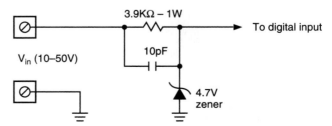

FIGURE 19   Circuit to convert large digital signals to TTL-compatible levels. (*Intelligent Instrumentation, Inc.*)

input is not fast enough, it can be made TTL-compatible with the Schmitt trigger circuit shown in Fig. 20.

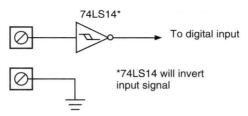

FIGURE 20   Schmitt trigger circuit to "speed up" slow input signals. Input levels must be TTL-compatible. (*Intelligent Instrumentation, Inc.*)

Digital Isolation

A family of industry standard signal-conditioning modules is available. These complete plug-in units are designed to provide a wide range of input and output capabilities. Each module supports a single channel, allowing the flexibility to mix the various types when configuring a system. Optical isolation, rated at 4000 volts, provides high-voltage separation between the signals and the data-acquisition system. This is useful for safety, equipment protection, and ground-loop interruption. Output modules use power transistors or triacs to switch high-voltage high-current ac or dc loads. Loads up to 60 volts dc or 280 volts ac at 3 amperes can be accommodated. Input modules convert digital signals between 10 and 280 volts to TTL levels. Standard termination panels accommodate up to 16 modules.

Contact Sensing

As shown in Fig. 21, contact sensing can be implemented on a signal termination panel. When interfacing to relay or switch contacts, a pull-up current must be provided. The pull-up current converts the opening and closing of the contacts to TTL level voltages. Because all metal surfaces tend to oxidize with time, poor relay contacts can result. Both level generation and contact wetting can be accomplished by connecting a resistor between the input line and the +5-volt power supply. When the switch is open, the input system sees +5 volts. When the switch is closed, the input is 0 volts. This satisfies the TTL requirements of the data-acquisition system. A value of 250 ohms for R_1 will provide 20 mA of wetting current, which is usu-

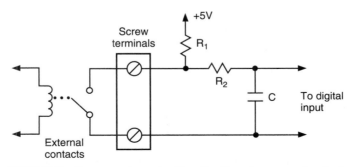

FIGURE 21   Contact sensing and wetting. (*Intelligent Instrumentation, Inc.*)

ally enough to keep most contacts free of oxide buildup. R_2 and C_1 function as a very simple debounce filter to reduce erroneous inputs due to the mechanical bouncing of the contacts. Care must be taken to avoid slowing the signal transition so much that false triggering occurs. If needed, a Schmitt trigger can be added, as shown in Fig. 20. Digital filtering techniques can also be used to eliminate the effects of contact bounce.

Relay Driving

Figure 22 shows how a TTL output from a data-acquisition board can be connected to drive an external 5-volt relay coil. The digital output must be able to switch the coil current. The specifications of digital output ports vary considerably between models. However, most can support 16 to 24 mA. When large relays, contactors, solenoids, or motors are involved, an additional driver or intermediate switching network can be used. The diode D_1 protects the internal circuitry against the inductive kickback from the relay coil. Without the diode, the resulting high-voltage spikes will damage the digital port. Note that the direction (polarity) of the diode must be as shown in the diagram. Protection diodes must be able to respond very quickly and absorb the coil's energy safely. Most standard switching diodes fill these needs.

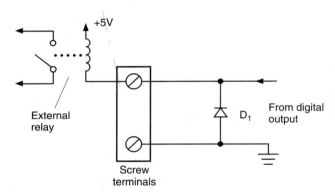

FIGURE 22 Relay driving circuit. (*Intelligent Instrumentation, Inc.*)

Motor Control

Many different types of motors are in common use today. When it comes to controlling these devices, specialized circuits are often required. Some applications, however, require only on-off operations. These can simply be driven by digital output ports, usually through optical isolators (loads of up to 3 amperes) or with various types of contactors (relays).

In general, when variable speed is desired, either analog or digital outputs from the data-acquisition system are used to manipulate the motor through an external controller. A wide range of both ac and dc controllers is available. Motor controls are discussed in more detail in Section 9 of this handbook.

Stepper-type motors are of particular interest in robotics, process control, instrumentation, and manufacturing. They allow precise control of rotation, angular position, speed, and direction. While several different types of stepping motors exist, the permanent-magnet design is perhaps the most common. The permanent magnets are attached to the rotor of the motor. Four separate windings are arranged around the stator. By pulsing direct current into the windings in a particular sequence, forces are generated to produce rotation. To continue rotation, current is switched to successive windings. When no coils are energized, the shaft is held in its last position by the magnets. In some applications these motors can be driven directly (via opto relays) by one of the data-acquisition system's digital output ports. The user provides the required software to produce the desired pulses in proper sequence. The software burden can be reduced by driving the motor with a specially designed interface device. These units accept a few digital input commands representing the desired speed, rotation, direction, and acceleration. A full range of motors and interfaces is available. Stepper motors are discussed in more detail in Section 9 of this handbook.

NOISE AND WIRING IN DATA SIGNAL HANDLING

by Howard L. Skolnik[1]

Signals entering a data-acquisition and control system include unwanted noise. Whether this noise is troublesome depends on the signal-to-noise ratio and the specific application. In general it is desirable to minimize noise to achieve high accuracy. Digital signals are relatively immune to noise because of their discrete (and high-level) nature. In contrast, analog signals are directly influenced by relatively low-level disturbances. The major noise-transfer mechanisms include conductive, inductive (magnetic), and capacitive coupling. Examples include the following:

- Switching of high-current loads in nearby wiring can induce noise signals by magnetic coupling (transformer action).
- Signal wires running close to ac power cables can pick up 50- or 60-Hz noise by capacitive coupling.
- Allowing more than one power or signal return path can produce ground loops that inject errors by conduction.

Conductance involves current flowing through ohmic paths (direct contact), as opposed to inductance or capacitance.

Interference via capacitive or magnetic mechanisms usually requires that the disturbing source be close to the affected circuit. At high frequencies, however, radiated emissions (electromagnetic signals) can be propagated over long distances.

In all cases, the induced noise level will depend on several user-influenced factors:

- Signal source output impedance
- Signal source load impedance (input impedance to the data-acquisition system)
- Lead-wire length, shielding, and grounding
- Proximity to noise source or sources
- Signal and noise amplitude

[1] Intelligent Instrumentation, Inc. (A Burr-Brown Company), Tucson, Arizona.

Transducers that can be modeled by a current source are inherently less sensitive to magnetically induced noise pickup than are voltage-driven devices. An error voltage coupled magnetically into the connecting wires appears in series with the signal source. This has the effect of modulating the voltage across the transducer. However, if the transducer approaches ideal current-source characteristics, no significant change in the signal current will result. When the transducer appears as a voltage source (regardless of impedance), the magnetically induced errors add directly to the signal source without attenuation.

Errors also are caused by capacitive coupling in both current and voltage transducer circuits. With capacitive coupling, a voltage divider is formed by the coupling capacitor and the load impedance. The error signal induced is proportional to $2\pi fRC$, where R is the load resistor, C is the coupling capacitance, and f is the interfering frequency. Clearly, the smaller the capacitance (or frequency), the smaller is the induced error voltage. However, reducing the resistance only improves voltage-type transducer circuits.

Example.   Assume that the interfering signal is a 110 volt ac 60-Hz power line, the equivalent coupling capacitance is 100 pF, and the terminating resistance is 250 ohms (typical for a 4- to 20-mA current loop). The resulting induced error voltage will be about 1 mV, which is less than 1 LSB in a 12-bit 10-volt system.

If the load impedance were 100 kΩ, as it could be in a voltage input application, the induced error could be much larger. The equivalent R seen by the interfering source depends on not only the load impedance but also the source impedance and the distributed nature of the connecting wires. Under worst-case conditions, where the wire inductance separates the load and source impedances, the induced error could be as large as 0.4 volt. This represents about an 8 percent full-scale error.

Even though current-type signals are usually converted to a voltage at the input to the data-acquisition system, with a low-value resistor this does not improve noise performance. This is because both the noise and the transducer signals are proportional to the same load impedance.

It should be pointed out that this example does not take advantage of, or benefit from, shielding, grounding, and filtering techniques.

GROUNDING AND SHIELDING PRINCIPLES

Most noise problems can be solved by giving close attention to a few grounding and shielding principles:

- Do not confuse the definitions of ground and return paths. Ground = safety; return = current-carrying.
- Minimize wiring inductance.
- Limit antennas.
- Maintain balanced networks wherever possible.

The foregoing directions appear simple, but what really is involved?

For a beginning, one should redefine some common terms. A ground is *not* a signal or power supply return path. A ground wire connects equipment to earth for safety reasons—to prevent accidental contact with dangerous voltages. Ground lines do not normally carry current. Return lines are an active part of a circuit—carrying power or signal currents (Fig. 1). Care should be taken to distinguish between grounds and returns and to avoid more than one connection between the two.

To be effective, return paths should have the lowest possible impedance. Someone once said that the shortest distance between two points is a straight line. But in geography this is not true, and it is not generally true in electronics either. Current does not take the shortest

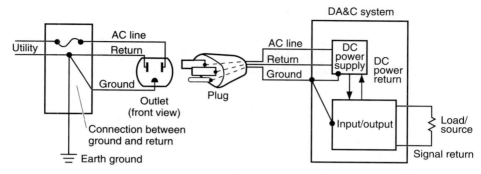

FIGURE 1 Differences between ground and return conductors. (*Intelligent Instrumentation, Inc.*)

path; rather it takes the path of least resistance (really, of least impedance). Return imped-ance is usually dominated by the path inductance. Wiring inductance is proportional to the area inside the loop formed by the current-carrying path. Therefore impedance is minimized by providing a return path that matches or overlaps the forward signal path. Note that this may not be the shortest or most direct route. This concept is fundamental to ensuring proper system interconnections.

Three different grounding and connection techniques are suggested in Figs. 2, 3, and 4. The circuit in Fig. 2 allows the signal return line to be grounded at each chassis. This may look like a good idea from a safety standpoint. However, if a difference in potential exists between the two grounds, a ground current must flow. This current, multiplied by the wire impedance, results in an error voltage e_e. Thus the voltage applied to the amplifier is not V_1, but $V_1 + e_e$. This may be acceptable in those applications where the signal voltage is much greater than the difference in the ground potentials.

When the signal level is small and a significant difference in ground potentials exists, the connection in Fig. 3 is more desirable. Note that the return wire is not grounded at the ampli-fier and ground current cannot flow in the signal wires. Any difference in ground potential appears, to the amplifier, as a common-mode voltage. In most circuits the effects of common-mode voltage are very small, as long as the sum of signal voltage plus common-mode voltage is less than 10 volts. (Ten volts is the linear range for most amplifiers.) Additional information about common-mode rejection and single-ended versus differential amplifiers can be found in a prior article in this handbook.

If cost is not a limitation, Fig. 4 offers the highest performance under all conditions. Inject-ing an isolator into the signal path faithfully conveys V_1 to the amplifier while interrupting all direct paths. In this configuration multiple ground connections can be tolerated along with

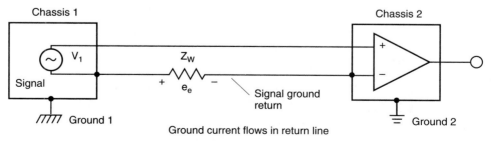

FIGURE 2 Single-ended connection. (*Intelligent Instrumentation, Inc.*)

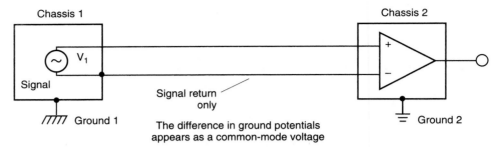

FIGURE 3 Differential connection. (*Intelligent Instrumentation, Inc.*)

several hundred volts between input and output circuits. Additional information on both analog and digital isolators can be found in a prior article in this handbook.

Cable Types

What kind of wire should be used to interconnect a system? First, it must be emphasized that a single piece of wire is not generally useful. Circuits consist of complete paths, so pairs of wires are referred to in this discussion. Basically four kinds of wire are fundamental: (1) plain pair, (2) shielded pair, (3) twisted pair, and (4) coaxial cable. All but the coaxial wires are said to be balanced. Coaxial cable differs from the others in that the return line surrounds the central conductor.

Technically, the outer conductor should not be called a shield because it carries signal current. It is significant that the forward and return path conductors do not have exactly the same characteristics. In contrast, a shielded pair is surrounded by a separate conductor (properly called a shield) that does not carry signal current.

Figure 5 suggests a simple model for a differential signal connection. The attributes of the signal source have been split to model the influence of a common-mode voltage. Let us focus on the effect of forward and return path symmetry in the cable. Assuming that the amplifier is perfect, it will respond only to the difference between V_A and V_B. The technique of superposition allows us to analyze each half of the cable model separately and then to add the results. Z_1 is usually dominated by series inductance, while Z_2 is dominated by parallel capacitance. In any case, Z_1 and Z_2 form a voltage divider. If the dividers in both legs of the cable are identical, $V_A - V_B$ will not be influenced by common-mode voltage. If, however, the capacitance represented by Z_2 is different in the two paths, a differential voltage will result and the amplifier will be unable to distinguish the resulting common-mode error from a change in V_S.

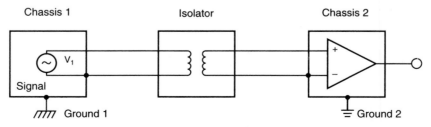

FIGURE 4 Isolated connection. (*Intelligent Instrumentation, Inc.*)

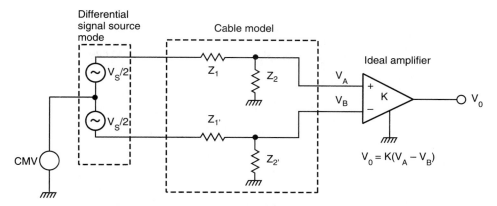

FIGURE 5   Influence of cable connectors on common-mode signal performance. (*Intelligent Instrumentation, Inc.*)

Coaxial cable offers a very different capacitance between each of its conductors and ground. Not only does the outer conductor surround the inner, it is also connected to ground. Thus coaxial cable is intended for single-ended applications only. Note that even perfectly balanced cables can still attenuate differential signals.

Sometimes even a single-ended source is best measured with a differential amplifier. Refer again to Fig. 3. To maintain a high rejection of any ground difference potential, balanced cables are required.

TROUBLESHOOTING GUIDE FOR NOISE

One method of reducing errors due to capacitive coupling is to use a shield. Generally there is little that can be done to reduce the actual capacitance. (Wire length and physical location are factors, however.) Nevertheless, placing a conductive material (at ground potential) between the signal wires and the interference source is very useful. The shield blocks the interfering current and directs it to the ground. Depending on how complete the shield is, attenuations of more than 60 dB are attainable. When using shielded wire, it is very important to connect only one end of the shield to ground. The connection should be made at the data-acquisition system end of the cable (such as input amplifier). Connecting both ends of the shield can generate significant error by inducing ground-loop currents.

A shield can work in three different ways:

- Bypassing capacitively coupled electric fields
- Absorbing magnetic fields
- Reflecting radiated electromagnetic fields

Another approach is to use twisted pairs. Twisted-pair cables offer several advantages. Twisting of the wires ensures a homogeneous distribution of capacitances. Capacitances both to ground and to extraneous sources are balanced. This is effective in reducing capacitive coupling while maintaining high common-mode rejection. From the perspective of both capacitive and magnetic interference, errors are induced equally into both wires. The result is a significant error cancellation.

The use of shielded or twisted-pair wire is suggested whenever low-level signals are involved. With low-impedance sensors the largest gage-connecting wires that are practical should be used to reduce lead-wire resistance effects. On the other hand, large connecting wires that are physically near thermal sensing elements tend to carry heat away from the source, generating measurement errors. This is known as thermal shunting, and it can be very significant in some applications.

The previous discussion concentrated on cables making single interconnections. Multiconductor cables, for connecting several circuits, are available in similar forms (such as twisted pairs and shielded pairs). Both round and flat (ribbon) cables are used widely. Because of the close proximity of the different pairs in a multiconductor cable, they are more susceptible to crosstalk. Crosstalk is interference caused by the inadvertent coupling of internal signals via capacitive or inductive means.

Again, twisted pairs are very effective. Other methods include connecting alternate wires as return lines, running a ground plane under the conductors, or using a full shield around the cable.

Still another noise source, not yet mentioned, is that of triboelectric induction. This refers to the generation of noise voltage due to friction. All commonly used insulators can produce a static discharge when moved across a dissimilar material. Fortunately the effect is very slight in most cases. However, it should not be ignored as a possible source of noise when motion of the cables or vibration of the system is involved. Special low-noise cables are available that use graphite lubricants between the inner surfaces to reduce friction.

The key to designing low-noise circuits is recognizing potential interference sources and taking appropriate preventive measures. Table 1 can be useful when troubleshooting an existing system.

After proper wiring, shielding, and grounding techniques have been applied, input filtering can be used to further improve the signal-to-noise ratio. However, filtering should never be relied upon as a fix for improper wiring or installation.

Cable-Length Guidelines

What is the maximum allowable cable length? There is no direct answer to this question. The number of factors that relate to this subject is overwhelming. Signal source type, signal level, cable type, noise source types, noise intensity, distance between cable and noise source, noise frequency, signal frequency range, and required accuracy are just some of the variables to consider. However, experience can yield some "feel" for what often works, as per the following examples:

Analog Current Source Signals.   Given 4- to 20-mA signal, shielded wire, bandwidth limited to 10 Hz, required accuracy 0.5 percent, and average industrial noise levels. Cable lengths of 1000 to 5000 feet (300 to 1500 meters) have been used successfully.

Analog Voltage Source Signals.   Given ±1- to ±10-volt signal, shielded wire, bandwidth limited to 10 Hz, required accuracy 0.5 percent, and average industrial noise levels. Cable lengths of 50 to 300 feet (15 to 90 meters) have been used successfully.

Analog Voltage Source Signals.   Given 10-mV to 1-volt signal, shielded wire, bandwidth limited to 10 Hz, required accuracy 0.5 percent, and average industrial noise levels. Cable lengths of 5 to 100 feet (1.5 to 30 meters) have been used successfully.

Digital TTL Signals.   Given ground-plane-type cable and average industrial noise levels. Cable lengths of 10 to 100 feet (3 to 30 meters) have been used successfully.

Ground-plane cable reduces signal reflections, ringing, and RFI (radio frequency interference). Special termination networks may be required to maintain signal integrity and minimize RFI. If squaring circuits (e.g., Schmitt triggers) are used to restore the attenuated high-frequency signals, improved performance can be realized.

This information is given only as a typical example of what might be encountered. The actual length allowed in a particular application could be quite different.

TABLE 1 Troubleshooting Guide for Noise

Observation	Subject	Possible solution	Notes
Noise a function of cable location	Capacitive coupling	Use shielded or twisted pair.	a
	Inductive coupling	Reduce loop area; use twisted pair or metal shield.	b
Average value of noise:			
Not zero	Conductive paths or ground loops	Faulty cable or other leakage. Eliminate multiple ground connections.	c
Zero	Capacitive coupling	Use shielded or twisted pair.	a
Shield inserted:			
Ground significant	Capacitive coupling	Use shielded or twisted pair.	a
Ground insignificant	Inductive coupling	Reduce loop area; use twisted pair or metal shield.	b
Increasing load:			
Reduces error	Capacitive coupling	Use shielded or twisted pair.	a
Increases error	Inductive coupling	Reduce loop area; use twisted pair or metal shield.	b
Dominant feature:			
Low frequency	60-Hz ac line, motor, etc.	1. Use shielded or twisted pair. 2. Reduce loop area; use twisted pair or metal shield. 3. Faulty cable or other leakage; eliminate multiple connections.	
High frequency		Complete shield.	d
Noise a function of cable movement	Triboelectric effect	Rigid or lubricated cable.	
Noise is white or $1/f$	Electronic amplifier, etc.	Not a cable problem.	

a. Complete shield to noise-return point and check for floating shields.
b. Nonferrous shields are good only at high frequencies. Use MuMetal shields at low frequencies.
c. Could be capacitive coupling with parasitic rectification, such as nonlinear effects.
d. Look for circuit element whose size is on the order of the noise wavelength (antennas). Openings or cracks in chassis or shields with a dimension bigger than the noise wavelength/20 should be eliminated.
Source: Intelligent Instrumentation, Inc.

The following relationships are offered as an aid to visualizing the influence of the most significant factors determining cable length. These relationships show how the various parameters affect cable length. These relationships are *not equations,* and will not allow the calculation of cable length.

For Current Source Signals:
Allowable length is proportional to

$$\frac{I_s \, D_n \, C_f}{f_n \, A \, N_i}$$

For Voltage Source Signals:
Allowable length is proportional to

$$\frac{V_s \, D_n \, C_f}{f_n \, A \, N_i \, R_L}$$

where I_s, V_s = signal level
C_f = coupling factor, which is inversely proportional to effectiveness of any shielding or twisting of wires
D_n = distance to noise source
f_n = noise frequency
A = required accuracy
N_i = noise source intensity
R_L = equivalent resistance to ground at signal input

INDUSTRIAL CONTROL NETWORKS

Commencing in the late 1960s and continuing through the present (1993) to the 2000s, industrial data and control networks will capture the ingenuity of control engineers, computer scientists, and, of course, communications specialists.[1] The well justified thirst for information from top management down the industrial hierarchy will continue undiminished. To date, communication system designs have survived past difficult periods in an effort to find the best network systems for given applications, always with an eye toward finding the "universal" answer. Open systems hold some promise along these lines. Millions of hours of effort have gone into defining optimal protocols, improved communication media, practical and cost-effective bus configurations, and the reconfiguration of controls, such as programmable logic controllers (PLCs), distributed control systems, and personal computers (PCs), in an effort to make them increasingly "network friendly." There are other objectives and there have been tough decisions and roadblocks, but the leading technical societies have mustered strength through special committees in defining terms and establishing standards. This work will continue apace.

---

[1] Industrial control networks also are discussed in other portions of this handbook. See, in particular, the articles, "Distributed Control Systems," "Programmable Controllers," and "Distributed Numerical Control and Networking" in Section 3.

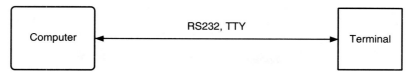

FIGURE 1   Point-to-point communication.

EARLY NETWORKING CONCEPTS

Early communication needs were served with point-to-point data links (Fig. 1). Very early standards, such as TTY (teletypewriter) current loops and RS-232, which allow different equipment to interface with one another, appeared and were accepted. From that, the star topology (Fig. 2) was developed so that multiple computers could communicate. The central, or master, node uses a communications port with multiple drops, as shown in Fig. 3. In this system the master is required to handle traffic from all the nodes attached, poll the other nodes for status, and, if necessary, accept data from one node to be routed to another. The heavy software burden on the master is also shared to a lesser degree among all the attached nodes. In addition, star topologies are inflexible as to the number of nodes that can be attached. Either one pays for unused connections (for future expansion), or a system results that cannot grow with demands.

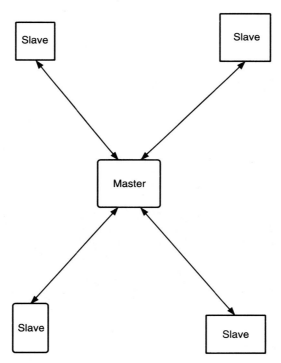

FIGURE 2   Star topology.

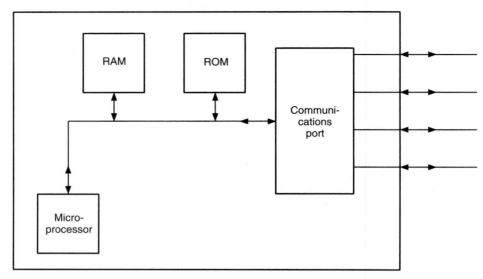

FIGURE 3   Master node for star topology.

To overcome some of these shortfalls, multidrop protocols were established and standardized. Data loop, such as SDIC (synchronous data link control), were developed as well as other topologies, including buses and rings (Fig. 4). The topology of these standards makes it easy to add (or subtract) nodes on the network. The wiring is also easier because a single wire is routed to all nodes. In the case of the ring and loop, the wire also is returned to the master. Inasmuch as wiring and maintenance are major costs of data communications, these topologies virtually replaced star networks. These systems, however, have a common weakness—one node is the master, with the task of determining which station may transmit at any given time. As the number of nodes increases, throughput becomes a problem because (1) a great deal of "overhead" activity may be required to determine which may transmit and (2) entire messages may have to be repeated because some protocols allow only master-slave communications, that is, a slave-to-slave message must be sent first to the master and then repeated by the master to the intended slave receiver. Reliability is another problem. If the master dies, communications come to a halt.

The need for multinode networks without these kinds of problems and restraints led to the development of the initial local area networks (LANs) using peer-to-peer communications. Here no one node is in charge; all nodes have an equal opportunity to transmit. An early LAN concept is shown schematically in Fig. 5.

In designing LAN architecture, due consideration had to be given to the harsh environment of some manufacturing and processing areas. Design objectives included the following.

Noise.   Inasmuch as a LAN will have long cables running throughout the manufacturing space, the amount of noise pickup can be large. Thus the LAN must be capable of working satisfactorily in an electrically noisy area. The physical interface must be defined to provide a significant degree of noise rejection, and the protocol must be robust to allow easy recovery from data errors. (See preceding article in this handbook section, "Noise and Wiring in Data Signal Handling.")

Response.   The LAN in an industrial situation should have an assured maximum response time, that is, the network must be able to transmit an urgent message within a specified time frame. The real-time aspect of industrial control demands this.

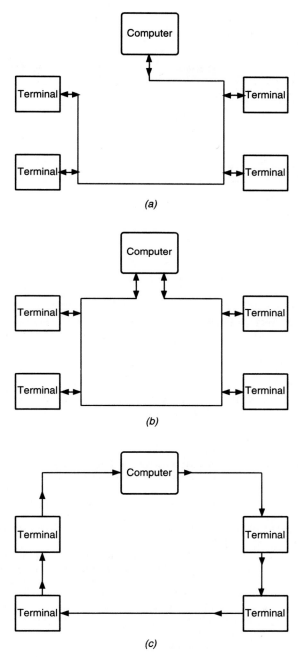

FIGURE 4 Basic communication standards. (*a*) Data-loop topology. (*b*) Bus topology. (*c*) Ring topology.

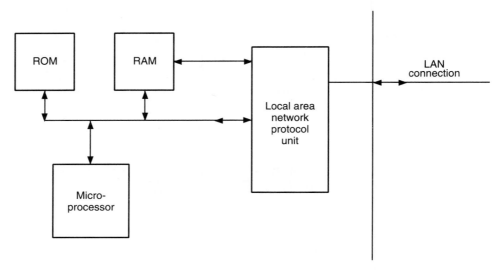

FIGURE 5   Early LAN concept shown schematically.

Priority Message.   On the factory floor, both control and status, when carried over the same network, should recognize the higher priority of the control message.

Early Data Highways

In 1972 the very first serial data communications highways were introduced. At that time the only purpose of the data highway was to allow host computers to adjust set points or, in some cases, perform direct digital control (DDC), while providing measurement data to the host computer. With such radically altered control concepts, in designing a data highway, great emphasis was placed on proposed data highways and their ability to operate at sufficiently high rates. There was concern that, during process upsets, many alarm conditions could suddenly change and these had to be reported to the entire system quickly so that remedial action could be taken. There also was major concern over start-up and shutdown procedures that cause heavy communication loads. Security also was a major concern.

In 1973 the distributed control system (DCS) appeared. It represented a major departure in control system architecture and impacted on the configuration of the data highway.

Ethernet.   The first LAN, developed by Xerox Corporation, has enjoyed years of application. The network uses CSMA/CD (carrier sense multiple access with collision) and is a baseband system with a bus architecture (Fig. 4*b*). Baseband is a term used to describe a system where the information being sent over the wire is not modulated.

DECnet.   DEC (Digital Equipment Corporation) computers, operating on the factory floor in the early 1970s, were linked by DECnet. This was a token passing technique, described later under "Network Protocols."

DECnet/Ethernet.   In 1980, with an aim to support high-speed local LANs, DECnet and Ethernet were used together to form DECnet/Ethernet, a network that has been used widely over the years. One of the major advantages of combining the two concepts is that Ethernet's delay in one node's response to another's request is much shorter than that of a token passing protocol. Users generally found that these networks provide good real-time performance. Ethernet is inherently appropriate for transmitting short, frequent messages and it effectively handles the irregular data transfers typical of interactive terminal communications (Figs. 6 and 7).

CATV Cable. In 1979 a CATV (community antenna television) cable system was announced. This system also had a central point of control and a multimaster protocol, but it used CATV cable at 1 Mbit/s. At these data rates, even in baseband, the transceiver design was based on radio-frequency (RF) technology. The topology of the network used a local star cluster with the CATV interconnecting all clusters. The data communication within the cluster was bit serial, byte parallel.

Later networks offered dual redundant mechanisms so that if one data highway failed, a second data highway would take over. The second highway was unused except for integrity diagnostics. To make certain that these data highways were in good order, elaborate mechanisms were implemented apart from the data communications to ensure cable and station integrity. CATV is mentioned later under MAP protocol.

NETWORK PROTOCOLS

A data communication protocol may be defined as "a set of conventions governing the format and relative timing of message exchange between two (or more) communications termi-

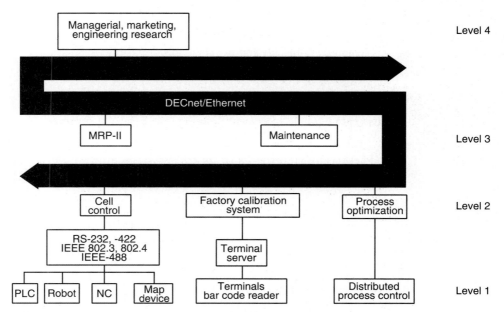

FIGURE 6 LAN for manufacturing applications combines features of DECnet and Ethernet. It may be defined as a multilevel functional model in which distributed computers can communicate over a DECnet/Ethernet backbone. Commencing in the 1980s, large numbers of these networks have been installed.

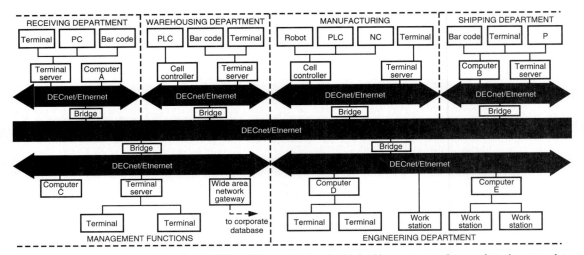

FIGURE 7   Schematic diagram shows how a DECnet/Ethernet baseband cable backbone can extend a manufacturing network to numerous plant areas, including a corporate database.

nals," or, restated, "the means used to control the orderly communication of information between stations on a data link."

The communication protocol is vital to equipment design and must be defined accurately if all elements in a system are to work together in harmony in the absence of a lot of local "fixes." Thus much international effort has been made over several years by various technical society committees to accurately identify, define, and refine protocols. Because of the heavy investment involved, a protocol usually is debated for at least a few years before the various committees officially approve standards.

CSMA/CD Protocol (IEEE 802.3)

"Carrier sense multiple access with collision" is a baseband system with a bus architecture. Normally only one station transmits at any one time. All other stations hear and record the message. The receiving stations then compare the destination address of the message with their address. The one station with a "match" will pass the message to its upper layers, while the others will throw it away. Obviously, if the message is affected by noise (detected by the frame check sequence), all stations will throw the message away.

The CSMA/CD protocol requires that a station listen before it can transmit data. If the station hears another station already transmitting (carrier sense), the station wanting to transmit must wait. When a station does not hear anyone else transmitting (no carrier sense), it can start transmitting. Since more than one station can be waiting, it is possible for multiple stations to start transmitting at the same time. This causes the messages from both stations to become garbled (called a collision). A collision is not a freak accident, but a normal way of operation for networks using CSMA/CD. The chances of collision are increased by the fact that signals take a finite time to travel from one end of the cable to the other. If a station on one end of the cable starts transmitting, a station on the other end will "think" that no other station is transmitting during this travel time interval and that transmission can be resumed. After a station has started transmitting, it must detect when another station is also transmitting. If this happens (collision detection), the station must stop transmitting. Before quitting transmitting, however, the station must make sure that every other station is aware that the last frame is in error and must be ignored. To do this, the station sends out a "jam," which is simply an invalid signal. This jam guarantees that the other colliding station also detects the collision and quits transmitting. Each station that was transmitting must then wait before trying again. To make sure that the two (or more) stations that just collided do not collide again, each station picks a random time to wait. The first station to time out will look for silence on the cable and retransmit its message again.

Token Bus Protocol (IEEE 802.4)

This standard was developed with the joint cooperation of many firms on the IEEE 802 Committee. Since becoming a standard, it was selected by General Motors (GM) for use in its manufacturing automation protocol (MAP) as a local area network to interconnect GM factories.

This is also a bus topology, but differs in two major ways from the CSMA/CD protocol: (1) The right to talk is controlled by passing a "token," and (2) data on the bus are always carrier-modulated.

In the token bus system, one station is said to have an imaginary token. This station is the only one on the network that is allowed to transmit data. When this station has no more data to transmit (or it has held the token beyond the specific maximum limit), it "passes" the token to another station. This token pass is accomplished by sending a special message to the next station. After this second station has used the token, it passes it to the next station. After all the other stations have used the token, the original station receives the token again.

A station (for example, A) will normally receive the token from one station (B) and pass the token to the third station (C). The token ends up being passed around in a logical token ring (A to C to B to A to C to B . . .). The exception to this is when a station wakes up or dies. For example, if a fourth station, D, gets in the logical token ring between stations A and C, A would then pass the token to D so that the token would go A to D to C to B to A to D. . . . Only the station with the token can transmit, so that every station gets a turn to talk without interfering with anyone else. The protocol also has provisions that allow stations to enter and leave the logical token ring.

The second difference between token bus and CSMA/CD, previously mentioned, is that with the token bus, data are always modulated before being sent out. The data are not sent out as a level, but as a frequency. There are three different modulation schemes allowed. Two are single-channel and one is broadband. Single-channel modulation permits only the token bus data on the cable. The broadband method is similar to CATV and allows many different signals to exist on the same cable, including video and voice, in addition to the token bus data. The single-channel techniques are simpler, less costly, and easier to implement and install than broadband. Broadband is a higher-performance network, permitting much longer distances and, very important, satisfying the present and future communications needs by allowing as many channels as needed (within the bandwidth of the cable).

This protocol still is evolving. Considerations have been made to incorporate the standard of an earlier standards group, known as the PROWAY (process data highway), which was discussed in the mid-1970s by several groups worldwide, including the International Electrotechnical Committee (IEC), the International Purdue Workshop on Industrial Computer Systems (IPWICS), and the Instrument Society of America (ISA). It was also at about this time that the Institute of Electrical and Electronics Engineers (IEEE) became very active with its 802 Committee. PROWAY also is a token bus concept.

Technical committee standards efforts progress continuously, but often quite slowly, so that sometimes users with immediate requirements must proceed with approaches that have not been "officially" standardized. Association and society committees meet at regular intervals (monthly, quarterly, and annually). Thus the individual who desires to keep fully up to date must obtain reports and discussions directly from the organizations involved, or attend meetings and consult those periodicals which assiduously report on such matters.

Benefits to the token bus protocol gained from PROWAY are several.[2]

Token Ring Protocol (IEEE 802.5)

Originally token ring and token bus used the same protocol with different topologies. IBM suggested a different token ring protocol to IEEE. The proposal was accepted and became the basis for the token ring protocol, thereby forming two token protocols.

The topology is that of a ring (Fig. 4c). Any one node receives data only from the "upstream" node and sends data only to the "downstream" node. All communication is done on a baseband point-to-point basis. The "right to talk" for this network also is an imaginary token. Token ring has simplicity in that the station with the token simply sends it to the next downstream station. This station either uses the token or lets it go on to the next station.

A general summary of the three aforementioned protocols is given in Table 1.

COMMUNICATION MODELS AND LAYERS

Before describing more recent network protocols, such as MAP and MMS, it may be in order to mention tools and models that have been developed to assist in defining the various "layers" of data communications required from the factory floor or process to top-level corporate management. Computer-integrated manufacturing (CIM), for example, requires excellent communication at all levels before the promises of the concept can be achieved.

Experts have placed communication systems in levels ranging from three to seven in number. Examples of a three-level and a seven-level model are given.

Three-Level Concept.   This represented the state of the art prior to the development of the OSI reference model in the early 1970s.

Lowest Level.   Links groups of people and machines. The link involves the flow of information among different workstations at a department level. These local networks must have a fast payback because it is in this area where most manufacturing and processing alterations are made.

Middle Level.   Facilitywide networks that permit all departments within a facility to share data. Examples may include (1) obtaining employee vacation data at a moment's notice so that shift assignments can be made without delay, (2) tracing the history of a quality control problem immediately when a failure is noted, and (3) permitting a service supervisor to check the current readiness of production equipment.

---

[2] PROWAY advantages include:

1. The immediate acknowledgment of a frame. If station *A* has the token and uses it to send a message to station *B*, station *B* can be requested to send an acknowledge message back to station *A*. Station *B* does not wait until it gets the token, but instead, station *B* "uses" station *A*'s token for this one message. Station *B*, in essence, is telling station *A* that it received the message without any errors. This idea is allowed in 802.4.

2. Capability to initialize and control a station by sending a message over the network. (Every token bus protocol handler would have an input to control the rest of the node.)

3. Every station has some predefined data that are available upon demand. Any other station can request the data, and this station would send the information out immediately. (The station sending the response is using the other station's token.)

4. Access time of the network is deterministic and on upper bound must be predictable for any given system. This means that if a station on a network wants to send a message, one should be able to predict the maximum possible delay before it is sent out. (The original 802.4 specified a maximum access time per node, but no upper bound on the number of nodes.)

5. It provides a method of monitoring membership in the logical token ring. If a station died, every other station would be aware of it. Every station in the logical ring would have a list of all stations that are in the ring.

6. It provides for the accumulation of network performance statistics.

TABLE 1   General Characteristics of Basic Protocols

CSMA/CD protocol	Designed for a lot of short messages.
	Works well if there are not a lot of collisions.
	With heavy traffic there is a lot of overhead because of the increased number of collisions.
	Probabilistic in nature—every station has a finite chance of hitting some other station every time it tries to send. Thus there is no guarantee that a message may not be held up forever. This could be catastrophic.
	Baseband configuration means that digital data are represented as discrete levels on the cable, thus reducing noise effects and allowing longer cable lengths.
	Probably the most cost-effective for most applications.
Token bus protocol	Robust, able to recover from errors easily.
	Cable length is limited only by attenuation of signal in cable.
	Allows three different modulation schemes. Can carry multiple voice and video channels at the same time as data.
Token ring protocol	Deterministic access and priorities on messages.
	Can use fiber-optic cables. Maximum physical length of cable can be large.
	Requires more complex wiring than a bus because the last station must be connected to the first station to form a ring.
	Redundant cabling may be needed for fault tolerance because a single break or down station can stop all data transfer.

Highest Level.   Corporatewide communications where all multifactories and departments can exchange information and report to a central information processing site. Manufacturing automation, such as CAD/CAM, materials management, and automated machine tools, can be linked to management information processing, including such functions as financial data, sales reporting, statistical quality control (SQC), statistical process control (SPC), and corporate planning, among many other functions. This arrangement enables managers to check the flow of production and product lead times as only one of many examples that could be given. This level of networking is more complex than the local area networks (LANs) or the facilitywide networks (WANs), in part because of the heavy information-exchange load.

OSI Reference Model.   The OSI (open system interconnections) model was developed in the 1970s as a joint effort of several groups, including the International Standards Organization (ISO), the American National Standards Institute (ANSI), the Computer and Business Equipment Manufacturers Association (CBEMA), and the National Institute of Standards and Technology (NIST), formerly the National Bureau of Standards (NBS). It has been estimated that this work represents tens of thousands of hours of effort by experts worldwide.

Initially the OSI reference model was oriented to telephony systems. Some of its achievements have included MAP (manufacturing automation protocol) and MMS (manufacturing message service). The seven distinct layers of OSI are shown in Table 2.

Manufacturing Automation Protocol

In the early 1970s the management of several industrial firms in the discrete-piece manufacturing industries realized that industrial computers and the networks that serve them were the key tools for achieving production automation on a grand scale as contrasted with the low-key efforts of the past, such as "isolated," or islands of, robotics and computerized numerical control of machines. Particularly in the United States the need to automate and pursue the ambitious goals of new production concepts, such as CIM, MRP I and II (materials requirement planning), FMS (flexible manufacturing systems), and just in time (delivery), was pre-

TABLE 2 Open-System Interconnections (OSI) Model

Layer	Name	Uses and applications
1	Physical	Electrical, mechanical, and packaging specifications of circuits. Functional control of data circuits.
2	Link	Transmission of data in local network—message framing, maintain and release data links, error and flow control.
3	Network	Routing, switching, sequencing, blocking, error recovery, flow control. System addressing and wide-area routing and relaying.
4	Transport	Transparent data transfer, end-to-end control, multiplexing, mapping. Provides functions for actual movement of data among network elements.
5	Session*	Communications and transaction management. Dialog coordination and synchronization. Administration and control of sessions between two entities.
6	Presentation†	Transformation of various types of information, such as file transfers; data interpretation, format, and code transformation.
7	Application‡	Common application service elements (CASE); manufacturing message service (MMS); file transfer and management (FTAM); network management; directory service.

* The session layer provides functions and services that may be used to establish and maintain connections among elements of the session, to maintain a dialog of requests and responses between the elements of a session, and to terminate the session.

† The presentation layer provides the functions, procedures, services, and protocol selected by the application layer. Functions may include data definition and control of data entry, data exchange, and data display. This layer comprises: CASE (common application services), SAS (specific application services), and management protocols required to coordinate the management of OSI networks in conjunction with management capabilities that are embedded within each of the OSI layer protocols.

‡ The application layer is directly accessible to, visible to, and usually explicitly defined by users. This layer provides all of the functions and services needed to execute user programs, processes, and data exchanges. For the most part, the user interacts with the application layer, which comprises the languages, tools (such as program development aids, file managers, and personal productivity tools), database management systems, and concurrent multiuser applications. These functions rely on the lower layers to perform the details of communications and network management. Traditionally, network vendors have provided a proprietary operating system for handling functions in the upper layers of the OSI model. These unique features have been the source of interconnection difficulties.

cipitated by severely threatening competition from abroad. An excellent example of this was the decision by the management of General Motors (GM) to develop a program that would hasten the achievement of CIM, recognizing the pitfalls that existed then in attempting to utilize the control products of numerous manufacturers, that is, products which could not easily be connected and orchestrated in a practical operating network. Not necessarily the first, but certainly the most illuminated effort was the recommendation and demand of GM for simplification and implementation of control and communication components. Thus the MAP program was initiated. The U.S. government, also at about this time, created a special section of NBS (later NIST) to assist industry in the development of improved automation techniques, including networking. Again, the motive was a serious concern over the country's diminishing leadership in manufacturing. These actions placed great emphasis on LANs and the general concept of interconnectability.

GM, prior to the formation of the MAP plan, had for a number of years used a network of CATV cable (described previously) for closed-circuit television. To this had been added several channels of low-speed serial data communications, which involved RF modems to operate in the television broadband spectrum. To get a plantwide high-speed information network standard established for GM, the firm formed an internal committee called MAP. This well-publicized committee invited proposals from many vendors and circulated several papers on GM requirements.

The primary purpose of the plantwide ("backbone") network was not process control, but rather to allow the two-way flow of high-level production data. Otherwise the whole bandwidth of the backbone network would easily be consumed by local traffic. GM had defined a true hierarchical network environment and had clearly endorsed token bus and CATV, but not for process control. Also, during this period, a lot of interest was shown in TOP (technical office protocol).

The MAP concept developed at a good rate for several years, progressing to MAP 3.0. Scores of MAP networks have been installed in the United States and abroad, with the largest number located in large discrete-manufacturing firms, including, of course, GM, but also Ford Motor and Boeing. In June 1987 a 6-year freeze was imposed on the MAP 3.0 specification. This intended to allow manufacturers to build, and users to install, MAP networks without concern that the specifications continue to change over relatively short intervals of time. A leader in the MAP field indicated that the freeze would not affect the addition of functionality, compatible backward and forward, but that technology would not be added that would make obsolete what has been installed already.

Acceptance of MAP probably peaked in 1990. A large base, including Ethernet, DECnet, ARCNET, and others, remains in place. Thus, unfortunately, evaluations and guides to system selection are well beyond the province of a "permanent" handbook reference. However, a few general suggestions may be in order:

1. Are there severe noise conditions?

   - YES—Seriously consider fiber-optic cables.
   - NO—Hard-wired.

2. Are there time-critical throughputs—monitoring, data collection, control?

   - YES—Provide equal access by all nodes.
   - NO—If transmission distance is under 1 mile (1.9 km), consider twisted pair. If over this distance, consider telephone.

3. Is equal access by all nodes a requirement?

   - YES—Use single high speed.
   - NO—Consider combination of high-speed access and twisted pair.

4. Are there plans for future expansion?

   - YES—Use standards-based network.
   - NO—Use proprietary or standards-based network.

The general characteristics of PLC-based LANs are listed in Table 3 for various network types.

Open Systems

Although not necessarily fully accredited to the developers of the OSI model described previously, that group placed early emphasis on the concept of open-system architecture. The footnotes included with Table 2, which describes the OSI model, aptly define the presentation and application layers in open-system architecture. Also, the importance of MMS is listed under the applications layer of the OSI model. MMS is accepted internationally as a standard communications protocol for integrating mixtures of unlike devices that are used on the factory floor or by processing areas. The MMS reduces development costs for adding new devices to a network and diminishes the requirement for custom software and hardware for diverse device interfacing. Within the last few years it is estimated that some 40 major suppliers have recognized the MMS protocol.[3]

As of 1993 the topic of open-system architecture remains quite fluid. Some major firms are refining their most recent offerings of "open" networks that use the term as part of a proprietary trade name.

---

[3] It is interesting to note that 10 of these firms participated in an unusual demonstration at the ISA 1991 Exhibit in Anaheim, California, for the purpose of showing how MMS can handle data transfer between PLCs, PCs, NC, robots, and others on the plant floor.

TABLE 3 General Characteristics of PLC-Based LANs*

Very Good	Good	Fair	Poor
Noisy environments			
RHF FO	HC	TP	RTRL
Speed			
HC FO RHF		TP	RTRL
Throughput			
	HC FO RHF	TP	RTRL
Purchase price			
	TP RTRL	HC FO	RHF
Lifetime cost			
	RHF	FO	HC TP RTRL
Expandability			
HC FO RTRL RHF		TP	

* FO—fiber-optic; HC—hard-wired coaxial; RHF = redundant hard-wired/fiber; RTRL—remote telephone/radio link; TP = twisted pair.

Field Bus

Since electronic measurement and control systems essentially replaced pneumatic systems several decades ago, industry has depended heavily on the 4- to 20-mA transmission standard, that is, until the *near* future! The ever-increasing use of microprocessor technology in sensors, transmitters, and control devices has created the need for a digital replacement of the 4- to 20-mA standard. Notably, this applies to "smart" transmitters.

A new field bus standard has been in preparation since 1985, sponsored essentially by the same society and institutional groups that have done admirable work in preparing other standards, network models, and so on. As of early 1992, completed portions of the new (SP 50) standard for both process control and factory automation include the physical layer and function block requirements. Field tests are under way in the United States and internationally. A low-speed version of the physical layer (31.25 kbaud) is under test at a New Jersey refinery.

One portion of the standard (H1) specifies a low-speed powered link as a digital replacement for 4- to 20-mA transmission. When implemented, microprocessors embedded in smart transmitters will be able to communicate directly with digital control systems. Another portion (H2) specifies a high-speed unpowered link to operate at 1 Mbaud.

Other tests completed to date have included what have been described as the "worst of conditions," such as inserting a bad message, a missing terminator, and an open spur. Still other tests included adding and removing a device on line, adding crosstalk, adding a wide frequency range of white noise, and placing a walkie-talkie within 2 feet (30 cm) from the open cable for RF interference tests.

It can be safely forecast that the proposed field bus will be the subject of innumerable papers and discussions over the next few years.

FIBER-OPTIC CABLES AND NETWORKS

Fiber-optic technology has been used for well over a decade in telephony. The first large-scale demonstration was made by AT&T in 1980 in connection with the Olympic Games held at Lake Placid, New York. This test installation was only 4 km (2.5 miles) long, but tested out very successfully. The first actual commercial installations were made between Washington and New York and New York and Boston. As early as 1982 Leeds & Northrup offered an earlier version of the network, as shown in Fig. 8. This is a redundant electrical-optical highway which has been used in hundreds of installations worldwide.

Fiber-optics offers many advantages and relatively few limitations as a networking medium. Some of the advantages include the following:

- Not affected by electromagnetic radiation (RMI). For example, the cable can be installed in existing high-voltage wireways, near RF sources, and near large motors and generators. No shielding is required.
- With experience, fiber cable is easy to install and at less labor cost. Only a few special tools are needed.
- Not affected by lightning surges.
- Resists corrosion.
- Intrinsically safe.
- Compatibility makes fiber cables easy to integrate with existing platforms. Fiber cable is inherently suited to open systems.

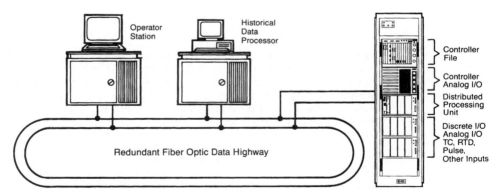

FIGURE 8   Redundant fiber-optic data highway for advanced data acquisition and plantwide control. This is an updated version of fiber-optic system first offered to industry in 1982, with ensuing hundreds of installations. Data highway consists of two redundant fiber-optic loops with repeaters to provide digital data communications among large numbers of multiloop controllers, operator stations, and computers. The optical data highway loops from one cluster to another, eventually returning to a control room with operator stations. One loop transmits digital data in a clockwise direction, while the other transmits counterclockwise. This ensures that communication between any two stations will be maintained no matter where a fault occurs. (*Leeds & Northrup.*)

- Increased security because of immunity to "bugging."
- Higher data rates. The next generation of fiber-optic network protocol (FDDI) can transfer data at a 100-Mbit/s rate on the same cable fiber that now offers a 10-Mbit/s rate.

For reasons of caution and lack of better understanding of fiber technology, coupled with a continuing (but decreasing) cost differential with other media, optic cables still are in the lower portion of their growth curve. Predictions for increased use are very optimistic.

Characteristics of Optical Fibers and Cables

As in electrical transmission systems, the transmission sequence of a light-wave system begins with an electric signal. This signal is converted to a light signal by a light source, such as a light-emitting diode (LED) or a laser. The source couples the light into a glass fiber for transmission. Periodically, along the fiber, the light signal may be renewed or regenerated by a light-wave repeater unit. At its destination, the light is sensed by a special receiver and converted back to an electric signal. Then it is processed like any signal that has been transmitted in electrical form.

The system is comprised of transmitter circuitry that modulates or pulses in code the light from a light source. An optical fiber waveguide conducts the light signal over the prescribed distance, selected because of its particularly good transmission capability at the wavelength of the light source. The terminal end of the waveguide is attached to a detector, which may be a *pn* junction semiconductor diode or an avalanche photodiode, to accept the light and change the signal into an electromagnetic form for the receiver circuitry. The latter decodes the signal, making it available as useful electronic analog or digital output. When two-way communication is needed, the system is fully duplexed and two circuit links are needed.

Optical Fibers.   Glasses of many compositions can be used for optical fibers, but for intermediate- and low-loss applications the options become increasingly limited. Multicomponent glasses containing a number of oxides are adequately suited for all but very low-loss fibers, which are usually made from pure fused silica doped with other minor constituents. Multicomponent glasses are prepared by fairly standard optical melting procedures, with special attention given to details for increasing transmission and controlling defects resulting from later fiber drawing steps. In contrast, doped fused silica glasses are produced by very special techniques that place them almost directly in a form from which fibers may be drawn.

Digital Light-Wave Systems.   Much research has been directed toward light-wave systems that are digital. In a digital system, the light source emits pulses of light of equal intensity, rather than a continuous beam of varying intensity (analog approach). Each second is divided into millions of slices of time. The light source inserts 1 bit of information into each time slot, which flashes on briefly or remains off. The receiver looks for 1 bit in each slot. If the receiver senses a pulse, it registers a 1; if the absence of a pulse, a 0. Eight such bits of information make up a digital word. From a series of such words, other elements of the transmission system can reconstruct the original signal.

The capacity of a digital light-wave system is the maximum rate at which pulses can be sent and received. The maximum pulse rate is limited by how much the signal is distorted by dispersion as it travels along the fiber. *Dispersion* means that a pulse is spread out in time, so that some of the pulses arrive in the wrong time slot. If enough is lost from the proper slot, the receiver may not sense a pulse that was sent. If enough is received in an adjoining slot, the receiver may sense a pulse when none was sent. The greater the dispersion, the longer the time slots must be for the receiver to sense accurately.

Basic Fiber Types.   Dispersion is of two kinds: (1) modal and (2) chromatic. Modal dispersion is the spreading of light as it traverses a length of fiber along different paths or modes (Fig. 9). Each path is a different length, and thus light takes a different time to travel through

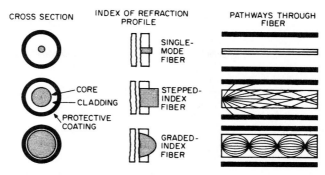

CROSS SECTION   INDEX OF REFRACTION PROFILE   PATHWAYS THROUGH FIBER

SINGLE-MODE FIBER

CORE
CLADDING
PROTECTIVE COATING

STEPPED-INDEX FIBER

GRADED-INDEX FIBER

FIGURE 9 Schematic sectional views of fiber-optic cable. Not all layers are shown. Structure of the fiber determines whether and how the light signal is affected by modal dispersion. A single-mode fiber permits light to travel along only one path. Therefore there is no modal dispersion. In contrast, a step-index fiber provides a number of pathways of different lengths, but only one index of refraction boundary between layers, which bends the light back toward the center. Here modal dispersion is high. A graded-index fiber has many layers. The resulting series of graded boundaries bends the various possible light rays along paths of nominally equal delays, thus reducing modal dispersion.

each. The highest-capacity fiber has only a single mode, so it has no modal dispersion. However, such fibers are much smaller, more difficult to couple light into, and harder to splice and connect with other types of fibers.

The more common type of fiber is multimode, either step index or graded index. These fibers have wider-diameter cores than single-mode fibers and accept light at a variety of angles. As light enters at these different angles, it travels through the fiber along different paths. A light beam passing through a step index fiber travels through its central glass core and in the process ricochets off the interface of the cladding adhering to and surrounding the core. The core-cladding interface acts as a cylindrical mirror that turns light back into the core by a process known as total internal reflection. To ensure that total internal reflection occurs, fibers are usually made from two glasses: core glass, which has a relatively higher refractive index, and clad glass, or possibly a plastic layer surrounding the core, which has a somewhat lower refractive index. When the seal interface between core glass and clad glass is essentially free of imperfections and the relative refractive indexes of the glasses used are correct, many millions of internal reflections are possible and light can travel through many kilometers of fiber and emerge from the far end with only a modest loss in brightness or intensity. A step-index fiber has just a single composition inside the cladding. Light must travel to this boundary before it is bent toward the center. The paths in this type of fiber disperse the pulse more than in a graded index fiber.

In a graded-index fiber, light is guided through it by means of refraction or bending, which refocuses it about the center axis of the fiber core. Here each layer of glass from the center of the fiber to the outside has a *slightly* decreased refractive index compared to that of the layer preceding it. This type of fiber construction causes the light ray to move through it in the form of a sinusoidal curve rather than in the zigzag fashion of the step-index variety. With this type of fiber, when the physical design is correct and the glass flaws are limited, light can also be conducted over very long distances without severe loss because it is trapped inside and guided in an efficient manner. The fiber core is the portion of an optical fiber that conducts the light from one end of the fiber to the other. Fiber core diameters range from 6 to ~250 μm.

Fiber Cladding.   To help retain the light being conducted within the core, a layer surrounding the core of an optical fiber is required. Glass is the preferred material for the cladding, although plastic-clad silica fibers are common in less demanding applications. The cladding thickness may vary from 10 to ~150 μm, depending on the particular design.

Index of Refraction.   This is the ratio of the velocity of light passing through a transparent material to the velocity of light passing through a vacuum using light at the sodium D line as a reference. The higher the refractive index of a material, the lower the velocity of the light passing through the material and the more the ray of light is bent on entering it from an air medium.

Numerical Aperture.   For an optical fiber this is a measure of the light capture angle and describes the maximum core angle of light rays reflected down the fiber by total reflection. The formula from Snell's law governing the numerical aperture (NA) for a filter is

$$\text{NA} = \sin \theta = \sqrt{n_1^2 - n_2^2}$$

where n_1 is the refractive index of the core and n_2 the refractive index of the clad glass.

Most optical fibers have NAs between 0.15 and 0.4, and these correspond to light acceptance half-angles of about 8 and 23°. Typically, fibers having high NAs exhibit greater light losses and lower bandwidth capabilities.

Light Loss or Attenuation through a Fiber.   This is expressed in decibels per kilometer. It is a relative power unit according to the formula

$$\text{dB} = 10 \log \frac{I}{I_0}$$

where I/I_0 is the ratio of the light intensity at the source to that at the extremity of the fiber. A comparison of light transmission with light loss in decibels through 1 km of fiber is as follows:

$$80\% \text{ transmission per kilometer} \simeq \text{loss of} \sim 1 \text{ dB/km}$$

$$10\% \text{ transmission per kilometer} \simeq \text{loss of} \sim 10 \text{ dB/km}$$

$$1\% \text{ transmission per kilometer} \simeq \text{loss of} \sim 20 \text{ dB/km}$$

Bandwidth.   This is a rating of the information-carrying capacity of an optical fiber and is given either as pulse dispersion in nanoseconds per kilometer or as bandwidth length in megahertz-kilometers. Light pulses spread or broaden as they pass through a fiber, depending on the material used and its design. These factors limit the rate at which light carrier pulses can be transmitted and decoded without error at the terminal end of the optical fiber. In general, a large bandwidth and low losses favor optical fibers with a small core diameter and a low NA.

The longer the fiber, the more the dispersion. Thus modal dispersion limits the product of the pulse rate and distance. A step index fiber can transmit a maximum of 20 Mbit of information per second for 1 km, and a graded-index fiber, more than 1000 Mbit. The process for making very low-loss fibers is essentially the same whether the fiber is step or graded index. Consequently nearly all multimode fibers presently used or contemplated for high-quality systems are of the higher-capacity graded-index type. Possibly for very high-capacity installations of the future, single-mode fibers may be attractive.

Cabling and Connections.   Although optical fibers are very strong, having a tensile strength in excess of 500,000 psi (3450 MPa), a fiber with a diameter of 0.005 inch (0.1 mm), including the light-guide core and cladding, has a maximum tensile strength of only ~10 psi (0.07 MPa). Unlike metallic conductors, which serve as their own strength members, fiber cables must contain added strength members to withstand the required forces. Also, pulling forces on unprotected fibers may increase their losses, as the result of bending or being under tension. Sometimes this is called microbending loss.

Imaging Requirements.   In imaging, both ends of the group of fibers must maintain the exact same orientation one to another so that a coherent image is transmitted from the source to the receiver. Flexible coherent bundles of optical fibers having only their terminal ends secured in coherent arrays are used primarily in endoscopes to examine the inside of cavities with limited access, such as stomach, bowel, and urinary tract.

Rigid fiber bundles fused together tightly along their entire length can be made to form a solid glass block of parallel fibers. Slices from the block with polished surfaces are sometimes used as fiber-optic faceplates to transmit an image from inside a vacuum to the atmosphere. A typical application is the cathode-ray tube (CRT) used for photorecording. The requirement for this type of application is for both image coherence and vacuum integrity, so that when the fiber-optic array is sealed to the tube, the vacuum required for the tube's operation is maintained. However, any image formed electronically by phosphor films on the inside surface of the fiber-optic face is clearly transmitted to the outside surface of the tube's face. High-resolution CRT images can easily be captured on photographic film through fiber-optic faceplates.

Light Sources and Detectors

LEDs produce a relatively broad range of wavelengths, and in the 0.8-μm-wavelength range, this limits present systems to ~140 Mbit/s for a 1-km path. Semiconductor lasers emit light with a much narrower range of wavelengths. Chromatic dispersion is comparatively low, so ~2500 Mbit/s may be transmitted for a 1-km path.

Another factor affecting system capacity is the response time of the sources and detectors. In general it is possible to build sources and detectors with sufficiently short response times that the fiber, rather than the devices, becomes the capacity limiting factor. With single-mode fibers, lasers, and high-speed detectors, transmission rates of more than 10^9 bit/s have been achieved experimentally. This corresponds to more than 15,000 digital voice channels. Although it is interesting to learn how fast a rate can be achieved, in practice the system designer must balance other technical, operational, and economic constraints in deciding how much capacity to require of an individual fiber.

Present semiconductor lasers can couple ~1 mW of optical power into a fiber. On the decibel scale, this is expressed as 0 dBm, meaning 0 dB above a reference power of 1 mW. Although some increase in power is possible, the small size and the temperature sensitivity of these lasers make them inherently low-power devices. LEDs can be made that emit as much power as lasers, but since they project light over a wide angle, much of it is lost just coupling it into the fiber. This loss is typically ~10 to 20 dB. Lasers are more complex and require more control circuitry than LEDs, but they are the light source of choice when repeaters must be far apart and the desired capacity is high.

Light-wave receivers contain photodiodes which convert incoming light to an electric current. The receivers used in telecommunications system are avalanche photodiodes (APDs) made of silicon. They are called avalanche devices because the electric current is amplified inside the diode. This results in a more sensitive receiver than photodiodes without internal amplification. Again, this improved performance is achieved at the expense of added complexity. APDs require high-voltage power supplies, but they are the detectors of choice when high performance is desired.

Even with APDs, light-wave receivers are less sensitive than the best electrical ones; they require a larger minimum received power. This is a consequence of the random fluctuations in optical signal intensity known as shot noise. Light-wave systems can compensate for this. They can carry a much wider bandwidth than electrical systems, and bandwidth can be used to offset noise.

SECTION 8
OPERATOR INTERFACE*

J. N. Beach
*MICRO SWITCH Division, Honeywell Inc., Freeport, Illinois
(Operator Interface—Design Rationale)*

W. A. Bruenger
Standish Industries, Inc., Lake Mills, Wisconsin (Liquid Crystal Displays)

R. W. Choronzy
Panalarm Division, AMETEK, Inc., Skokie, Illinois (Annunciators and Hazard Management)

R. Eberts
Purdue University, School of Industrial Engineering, West Lafayette, Indiana (Adaptation of Manuscript on "Cognitive Skills and Process Control)

James A. Odom
Corporate Industrial Design, Honeywell Inc., Minneapolis, Minnesota (Operator Interface—Design Rationale)

Staff
Moore Products Company, Spring House, Pennsylvania (CRT-Based Graphic Display Diagrams)

OPERATOR INTERFACE—DESIGN RATIONALE	8.3
HUMAN FACTORS	8.3
Importance of Initial Planning	8.3
Operator Variables	8.3
Habit Patterns	8.4
Operator-Interface Geometry	8.5
Alphanumeric Displays	8.5
Keyboards	8.9
Voice Recognition	8.11
ENVIRONMENTAL FACTORS	8.11
Local Environment	8.11
Ambient Light Conditions	8.11
Temperature	8.12
AESTHETIC FACTORS	8.12
Component Selection	8.14
Component Arrangement	8.15

---

* Persons who authored complete articles or subsections of articles, or otherwise cooperated in an outstanding manner in furnishing information and helpful counsel to the editorial staff.

OPERATOR INTERFACE—DESIGN RATIONALE[1]

The interface between a process or machine and the operator is the primary means for providing dialogue and for introducing human judgment into an otherwise automatic system. Although signals arrive at the interface and, once an operator judgment is made, leave the interface at electronic speeds, the operator responds at a much slower communication rate. Thus the operator, because of human limitations, is a major bottleneck in the overall system. The interface, whether it takes the form of a console, a workstation, or other configuration, must be designed with the principal objective of shortening operator response time. The interface must be customized to the operator, and through a serious training program, the operator must become accustomized to the interface. The interface designer not only considers the hardware interface, but the software interface as well.

Inadequate interface design usually is a result of (1) considering the interface late in the overall system design process, (2) giving short shrift to human factors, and (3) procuring off-the-shelf interface configurations that have been compromised for generalized application rather than for specific needs. Cost cutting at the interface level of a system carries large risks of later problems and dissatisfaction.

HUMAN FACTORS

Interface design falls within the realm of human factors, which is sometimes referred to as human engineering or ergonomics—all of which pertain to the very specialized technology of designing products for efficient use by people. Human factors is concerned with *everything* from specific characteristics of interface components to the total working environment of the operator.

Importance of Initial Planning

An excellent starting point for the interface designer is that of providing a functional description of the interface and then a job description for the operator. These two descriptions should dovetail precisely. A pro forma questionnaire can be helpful (Fig. 1).

Operator Variables

The principal characteristics of the operator in designing an interface are (1) physical parameters, (2) experience, including trainability, and (3) long-established habit patterns. The physical aspects will be described shortly. In terms of experience, the amount of instruction required to efficiently use the interface as intended obviously depends on the complexity of the process or machine under control and of the interface per se. There are tremendous differences, ranging from the simplicity of operating an average copy machine to a complex

---

[1] Technical information furnished by *MICRO SWITCH,* a Division of Honeywell Inc., and James A. Odom, Corporate Industrial Design, Honeywell Inc., Minneapolis, Minnesota.

Initial Planning Sheet—Interface Design

---

1. Functionality of interface:
   Variables to be indicated, recorded, adjusted, automatically alarmed ....
   Provide instructions, ancillary and historical information—Printed or computer-stored ....
   Extent of interactive graphics desired ....
   Special tasks of interface ....
   Will interface satisfy only functional needs, or will it be used also as a "show piece" for VIP
   visitations? ....
2. What are cost constraints? space limitations? ....
3. Are there special environmental considerations?—Installed indoors (air-conditioned space?),
   outdoors, proximity to excessive noise, near moving machinery, vibration and shock, electrical
   interference, corrosive or explosive vapors ....
   Will design require guards, barriers or protective shields for components and user safety? ....
   What degree of "ruggedness" is required? ....
4. Is there a requirement for information confidentiality? ....
   Will interface be used constantly (three shifts) or be out of action parts of a 24-hour period or
   weekends? ....
5. Will operator instructions be simple or complex? ....
   What will be the extent of operator training? ....
6. Who will install equipment and maintain it?—User technicians, supplier, outside service
   contractor? ....
   How can installation and maintenance tasks be simplified? ....
7. What are the choices in terms of overall physical configuration?—Console, workstation,
   pedestal, desk? ....
8. What are hardware options?—Visual displays, audio alarms, voice reception and activation ....
   Operator's physical contact with interface, switches, touch screens, etc.? ....
9. Are there any special procurement regulations, as in military installations? ....
10. Must the interface fit well into facility expansion plans? ....
11. What are the physical and mental profiles of your present very successful operators? ....
12. What have been common operator complaints in the past? ....
13. Will complexity justify preparing a life-size mockup for achieving an interface that combines
    human factors with aesthetics? ....
14. Functionality of operators—Complete job description, operator qualifications, physical size,
    agility, alertness, other factors (from past experience)..
    Is interface for a process or machine that could cause loss of life or equipment if permitted to go
    out of control? ....

FIGURE 1 Pro forma questionnaire for use by engineers and designers when initially considering new or revised operator interface.

machine tool or assembly line to a complex chemical process that incorporates many hundreds of control loops. When forecasting the amount of instruction that an operator will require for a given interface, the designer should establish the specific content of a training program as part of the overall interface design task.

Habit Patterns

People, as the result of past exposure, "expect" controls to move in certain ways. These expectations sometimes are called population stereotypes because they are so universally accepted. Where possible, component selection for an industrial control interface should be an extension of these stereotypes or habit patterns. For example, the wall-mounted toggle switch found in houses has established a habit pattern for turning on the lights. The upward flipping motion is associated with "on" and can be utilized with other instrumentation-type toggle-paddle switches for a natural transfer of a previously learned habit.

The clockwise motion of a rotary knob is frequently used to turn on a domestic appliance (television, range or oven, mixer). This same familiar action may be adapted to any control panel for an extension of a normal habit pattern. The scale of a slide switch or potentiometer should show an increase as the switch is moved upward or to the right. These control actions require the least amount of conscious effort to learn and are well established in our daily lives (Fig. 2).

When controls or control and display arrangements take advantage of these habit patterns, generally the following results can be expected:

1. Reaction time is reduced.
2. The first control movement by an operator is usually correct.
3. An operator can perform faster and can make adjustments with greater precision.
4. An operator can learn control procedures faster.

Operator-Interface Geometry

Extensive use of computer data terminals has generated a wide variety of interface architectures. However, there are certain common denominators. Typically, an operator will use a keyboard, a display unit, and a document station (a work holder for printed material). These elements require careful attention to their positioning with respect to the operator. A thorough ergonomic analysis must be a part of interface design to ensure that lines of sight, reach, lighting, access, and angular relationships are addressed properly (Fig. 3).

The line of sight from the operator to the center of a display should be as near to perpendicular as possible. The top of the display should be at or below eye level.

Height, tilt, and swivel adjustment of seat, keyboard, display, supports, and footrests must be provided, since they are all parts of an interactive workplace system. Adjustments for operator posture should be easy to make and without requiring special tools, skills, or strength. If an operator cannot maintain a comfortable posture, there will be complaints of fatigue related to the eyes, neck, shoulders, arms, hands, trunk, legs, and feet.

Alphanumeric Displays

Data display terminals that use cathode-ray tubes (CRTs) are commonly referred to as video display terminals (VDTs) or video display units (VDUs). Data display systems also may use gas discharge, light-emitting diodes (LEDs), and liquid-crystal displays (LCDs). Although non-CRT systems are often attractive because the interface screen can be relatively thin [2.5 to 5 cm (1 to 2 inches) front to back], the CRT enjoys wide usage because of its versatility in terms of graphics and use of color, and because of cost.

Although it is possible to display alphanumeric data on a single or abbreviated line display, as offered by some electronic typewriters, the full-page business-letter format usually is more desirable. If reference material is in the same format as the displayed information, the visual interaction between the two is compatible and operator perception problems are minimized. Many word-processing display screens offered today do not have full vertical page capacity, but they move type upward (scrolling) when the lower line limit has been reached.

Alphanumeric displays on VDTs are typically dot-matrix construction because of the discrete addressing mode which they use. Both 5 by 7 and 7 by 9 rectangular dot groupings are used, with the larger sides vertical (Fig. 4). A practical working height for these characters is 2.5 mm (0.1 inch) minimum. With adequate spacing between characters and lines, a typical

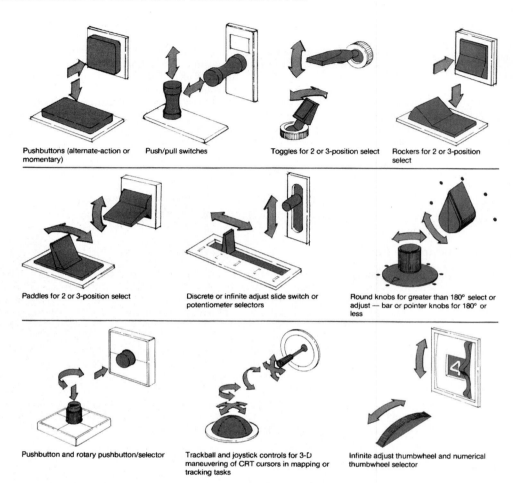

FIGURE 2   Switch response control movements in accordance with habit patterns. (*MICRO SWITCH.*)

display screen [244 mm wide by 183 mm high (9.6 by 7.2 inches)] can reasonably accommodate 47 lines of 95 characters.

Although light characters on a dark background are most common in VDTs, testing has verified that there is improved legibility with dark characters on light backgrounds. By having the same contrast format for both reference document and display, there is considerably less eye strain as an operator shifts back and forth between two surfaces. It is good practice to have CRT displays include the capability of image reversal from positive to negative in full or selected areas. Dark characters on a CRT light background display (positive image) present less contrast than equivalent light characters on a dark background (negative display). In effect, backlighting of the positive image "washes" around the dark characters to make them appear narrower. To counteract this phenomenon, the stroke width of the positive image should be approximately 20 percent heavier than that of the equivalent negative image.

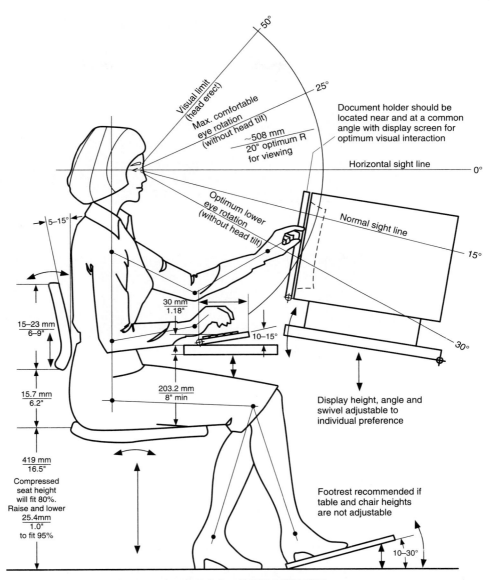

FIGURE 3 Geometry of operator-interface design. (*MICRO SWITCH.*)

Glare on the face of CRTs is one of the most frequently cited problems associated with VDT operation (Fig. 5). Glare from uncontrolled ambient light reflects into the operator's eyes, making it difficult or impossible to distinguish images on the screen, as well as producing eye strain and fatigue. High ambient light also reduces the contrast between background and displayed characters, since it adds its energy to both. Antireflective coatings, tinted windows, louvered screens, and various other filter media can be used to minimize this problem.

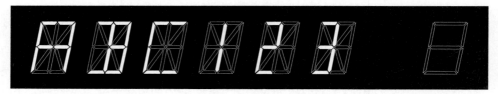

FIGURE 4 Types of alphanumeric matrix displays. (*Top*) 5 by 7 dot matrix. (*Middle*) 7 by 9 dot matrix. (*Bottom*) Bar matrix. (*MICRO SWITCH.*)

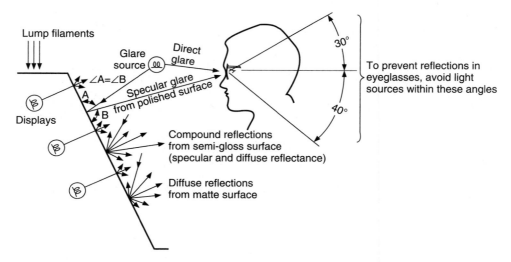

FIGURE 5 Glare from reflected ambient lighting can mask signals. (*MICRO SWITCH.*)

Some of these glare control solutions reduce ambient light bounce at the expense of contrast and image sharpness. The effectiveness of these various means should be evaluated carefully prior to making a commitment to use them. Further, the location and diffusion of ambient light sources should be investigated.

A wide range of image color is available on VDTs, but most use green, white, orange, or yellow contrasted with black. Any color image will satisfy the visual requirements of the display, so long as it does not fall at either end of the spectrum. Multicolor displays at these extreme wavelengths are also poorly perceived by the eye and should be avoided.

Keyboards

Current trends in programmable electronic data storage and retrieval systems have made the traditional typewriter keyboard all the more relevant as a critical interface between user and equipment (Fig. 6).

Key groupings for special functions and numeric entry have taken their place alongside the basic keyboard format. Full-function intelligent keyboards with integral microcomputer-based electronics may be encoded to meet user requirements.

The physical effort required should be consistent with the speed and efficiency of the operator and the duration of the operation. Without attention to these needs, the operator's performance may suffer. Poorly conceived equipment standards have sometimes been established casually, or without meaningful studies of operator needs. With years of inertia, they become so ingrained in the user population that there is virtually no way of changing or even modifying their effect. The standard QWERTY keyboard (named by the arrangement of the first six keys on the upper row) is a prime example.

An alternative arrangement, the simplified keyboard (Fig. 7) was developed by Dvorak in 1936. It distributes the keys according to the comparative strength of the fingers and the frequency of letter occurrence in common English usage. This improved design has more than doubled typing speed and it is said to reduce the average typist's daily keyboard finger travel from 12 to 1 mile (19.3 to 1.5 km). Adopting of the system has been and most likely will continue to be difficult because of the number of QWERTY machines in existence and by the extensive retraining of operators that would be required.

Volumes have been written about the optimum keyboard angle for efficient use. There have been keyboard layouts designed to operate at every angle between horizontal and vertical. However, there is a general consensus that somewhere between 10 and 20° is a comfortable workplane. Most contemporary keyboards fall within this angular reference. Recent studies have shown that by providing a keyboard with an adjustable setting for use between 10 and 25°, the needs of individual operators can be better accommodated.

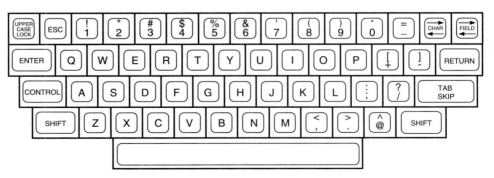

FIGURE 6   Keyboard with standard QWERTY layout.

FIGURE 7 Dvorak simplified keyboard input.

The current European standard (Fig. 8) requires that the height of the keyboard measured at the middle key row should not exceed 30 mm (1.2 inches) above the counter top. Although there are a number of low-profile keyboards thin enough to allow meeting this requirement, the combination of the keyboard enclosure dimensions and the 30-mm DIN height standard may restrict the keyboard angle to 10° or less, and as stated previously, the preferred angle is between 10 and 20°.

DIN standards also recommend that if a keyboard is mounted at a height of more than 30 mm (1.2 inches), then a palm rest must be provided to reduce static muscle fatigue. This dictum, however, may be subject to question. A palm rest may be desirable for some low-speed data entry, but a high-speed typist does not rest palms on anything. The use of a palm rest could only promote inefficient and error-prone typing. In addition, a palm rest will require more desk space, which is usually at a premium.

For rapid keyboard entry, a 19.05-mm (0.75-inch) horizontal spacing between key centers is considered comfortable. For rapid entry it is desirable for traveling keys to have a displacement of between 1.3 and 6.4 mm (0.05 and 0.25 inch) along with an associated force of 25.5 and 150.3 grams (0.9 and 5.3 ounces). Traveling keys are preferred to low-travel touch panels, which are better suited for low-speed random entry situations. When used with a data display terminal, it helps to have the keyboard enclosure separable from the display, so the user can position it for maximum comfort.

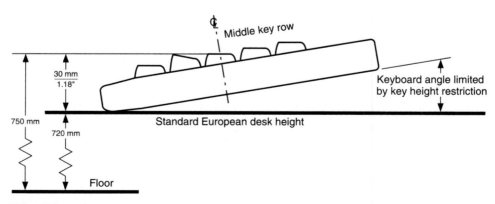

FIGURE 8 Basic DIN standard dimensions and angular restrictions. (*MICRO SWITCH.*)

Some of the more ambitious human factors efforts have created keyboards with a slight concavity, running from front to back on the surface of the combined key faces. This reflects the natural radii followed by the fingers moving up and down the key rows.

Other experimental keyboards have separated the left- and right-hand keys into two groups spaced slightly apart. The angle of each group parallels the natural angle of the hand and arm positioned in line and directly in front of a user. This minimizes the angular offset between hand and arm normally required for keyboard entry. These separated keyboards are not commercially available yet, but their invention indicates a continuing interest in improving the performance of keyboard operators.

Voice Recognition

Most voice recognition systems in current use are programmed to respond only to specific voices and accept a relatively limited command vocabulary. As speech processing technology continues to develop and as memory capabilities are expanded, these systems will be able to accept instructions from anyone. As the technology matures, voice recognition systems are expected to be able to recognize and identify both specific and nonspecific users. This process could eventually be used for authorized access to databases, and could supersede magnetic-card readers for clearance in security situations. More on voice recognition technology is given in a separate article in this handbook section.

ENVIRONMENTAL FACTORS

Numerous factors can affect the short- and long-term operating performance of the process or machine interface. Two of the most important of these are (1) the comparative gentleness or hostility of the local environment and (2) ambient lighting.

Local Environment

Industrial environments frequently pose a threat to the life and reliability of interface components. Often they are subjected to daily routines of abuse. Controls and displays may be splashed by water, oil, or solvents. They may be powdered with a layer of soot and dust, including metal particles, sticky vapors, and various granulated, gritty substances. However, even under these harsh circumstances, a resistant yet still attractive interface can be designed that will take advantage of oiltight manual controls, protective membrane touch panels, and ruggedized keyboards.

Ambient Light Conditions

External interface lighting almost always will either enhance or downgrade display visibility. As the ambient light level decreases, the visibility of self-illuminated displays will increase. For example, low-output lamps and projected color or "dead front" displays require low ambient light. Conversely, a brighter display is needed for recognition in high ambient light. By way of illustration, full indicator brilliance may be called for in the bubble of an aircraft cockpit exposed to direct sunlight, whereas a minimum glow from a display will be most appropriate at night. A situation like this will call for a brightness control for the indicator. Wherever possible, the designer should customize display brilliance to accommodate a wide range of brightness in any similar circumstances.

Glare becomes a particularly detrimental factor when the light comes from undiffused or concentrated sources, as in locations exposed to sunlight or from poorly located overhead lights that spot rather than diffuse the light source. A minor change in the panel angle can assist in reducing this problem, as previously indicated in Fig. 5. If the ambient lighting cannot be planned in advance, as may be the case of a separate integrated control room, then a survey of the existing lighting should be made and, if practical, altered to favor the control interface to be installed. Also, this will enable the interface designer to plan for whatever light hoods, shields, baffles, or diffusers may be needed and thus included with the projected costs. Control surfaces that are matted rather than polished also will help to alleviate the ambient light problem.

Temperature

Even if not installed in an air-conditioned space, interface components are designed to operate at temperatures in excess of those that can be tolerated by an operator. However, where surges of high temperature may be possible, then all-metal manual controls, such as toggles, may be specified instead of devices that incorporate plastic materials. Low temperatures are more apt to be encountered, as in the case of an outdoors installation or in an unheated or purposely cooled building. Under these circumstances, the operator may have to wear gloves or mittens and will be attending the interface under less than optimal conditions. In such cases, switches and controls equipped with large buttons, knobs, or levers should be considered. Also, an extra measure of positive tactile feedback will help the operator "feel" when a switch operation has occurred. Low temperatures also may affect the operating performance of manual switches. Seals or lubricants may stiffen, causing a switch to stick or to stay open or closed. These matters should be made known to the components maker well in advance of procurement.

AESTHETIC FACTORS

Although designing an effective control interface is a specialized example of industrial design, the basics of good industrial product design apply. These principles, in particular, affect the aesthetics of an interface design and, as will be shown, are also human factors.

Several decades ago, a pioneer of American industrial design, Henry Dreyfuss, developed a "performance creed" that applies equally well today:

We bear in mind that the object we are working on is going to be ridden in, sat upon, looked at, talked into, activated, operated or in some other way used by people.

If the point of contact between the product and the people becomes a point of friction, the industrial designer has failed.

On the other hand, if people are made safer, more efficient, more comfortable or just plain happier—by contact with the product, then the designer has succeeded.

As pointed out by James A. Odom, this creed applies to engineers as well as to industrial designers, and it exemplifies the goals of any conscientious approach to design. It is interesting to note that before anyone even touches a product, an opinion is formed based solely upon appearance. Thus this should persuade the design engineer to exploit this initial judgment by enhancing the visual qualities. This is done with a sensitive use of form, color, and comparison of components. First impressions are important. By creating a good visual impression, potential users are more likely to form a positive image, that is, of a friendly device rather than of a threatening device. For equipment manufacturers as well as ultimate users, this is very important, not only for interfaces, workstations, and the like, but for ancillary equipment that must be fitted into the control room or mounted on the manufacturing floor.

Along these lines, manual controls that are clean and simple in appearance should be preferred over those components that are ornate, decorative, or intricate. These qualities generally detract from functionality as well as appearance. Where practical, it is good practice to limit the number of components on any given panel. There is a point of diminishing returns when, by sheer weight of numbers, panel elements begin to confuse the visual definition of functional groups. This confusion can cause hesitation and error.

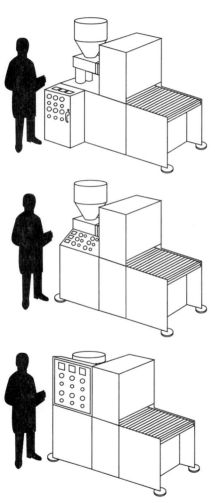

As an example, it is of value to note the differences in panel density and appearance between a strip of ganged interlocked push-button switches as contrasted with a single rotary selector switch that performs the same function. A multifunction illuminated rotary push-button control may be the solution when panel space is at a premium. The designer should review numerous alternatives prior to freezing a design. The "heavy-handed" approach that uses visible mounting screws or other practical but unattractive solutions that may downgrade panel appearance should be avoided. A statement of quality is made by the astute selection and use of appropriate well-designed interface components.

An effective interface can only be designed with a proper balance between human factors and aesthetics. To simply "plug in" human factors without regard to aesthetics (also a human factor) often can markedly detract from appearance. In contrast, a good-looking device that is difficult to use is unacceptable. With the large variety of all manner of components available today, an excellent acceptable compromise can be made.

Interface components should be considered early in the design process from both human factors and appearance standpoints. Components added to a device after it has been defined as a three-dimensional form may look like an afterthought—and they may not be arranged for maximum operator efficiency. Too often a machine, for example, may be fully designed with only the functional process in mind. Then a standard, off-the-shelf electrical enclosure will be added without first having been integrated into the basic form. In such instances, even if the control and display components are outstanding in appearance and human factors relationships, the total design will suffer, as shown in Fig. 9.

Inasmuch as processes and machines frequently will be altered some time after their initial start-up, the interface designer, whenever possible, should give some thought as to how such changes can be made without resulting in a "patched up" monstrosity.

FIGURE 9 When control interfaces are located directly on operating equipment, they should be placed to maximize operator convenience and not appear as afterthoughts. (*Top*) Poor design. (*Middle, bottom*) Satisfactory designs. (*MICRO SWITCH.*)

Component Selection

The interface designer should always keep simplicity uppermost in mind. Stress should be given to those controls that do or may require immediate attention and reaction of the operator. Oversized, infrequently used maintenance controls and exposed hardware (screws, latches, handles) or access knobs that resemble rotary controls should be avoided.

Guards or barriers (Fig. 10) should be considered only for controls where inadvertent operation could produce an irrevocable adverse effect. Guards should not be used as decora-

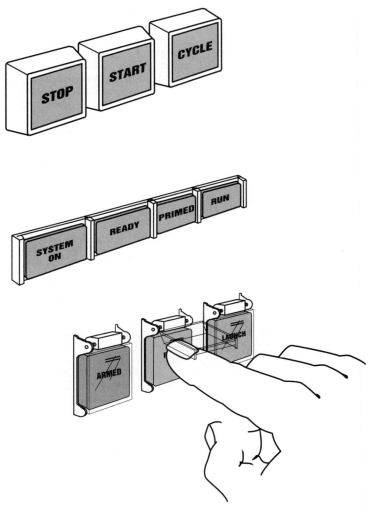

FIGURE 10   Switch guards and barriers. Strip barriers between switches help prevent inadvertent operation. Full barriers surround push buttons where more switch protection is desired. Ringed guards over push buttons in high-risk control stations. Guards also may be locked for additional security. (*MICRO SWITCH.*)

tive frames. Unnecessary guards simply contribute to visual clutter. Any "extras" of this type tend to distract the operator from the mainstream of more critical data. Equipment designers and vendors should preserve operator familiarity by not making frequent styling changes simply for the sake of change.

Component Arrangement

Interface panels become overly complex for two principal reasons: (1) the number and types of components and (2) the failure of the designer to explore alternative arrangements.

Panel arrangement can be assisted through the use of cardboard cutouts made to actual dimensions, which can be repositioned a number of times prior to making a final panel diagram. The control panel should be approached as a unified concept. The sequence of operations that the operator will follow should be studied and observed with the intent of minimizing (and eliminating) unnecessary visual "jumps" and "rebounds" across intermediate panel elements. James Odom makes the following suggestions:

1. Frequently used components should be the most accessible. For manually operated controls, somewhere between elbow and shoulder height is the best location.

2. Controls and displays should be arranged for a conventional sequence of operation, from left to right and top to bottom, as one normally reads.

3. Functional groups should be defined on the panel by allowing some space between them. Where practical, common centerlines and common edge lines throughout. Elements to be avoided include borders, separate color patches, and brackets extending from group titles—except in cases of extreme component density.

4. Emergency controls and displays should be located prominently on panel faces to ensure easy viewing and access by the operator.

5. Where large panel layouts are required, the workload should be distributed between both of the operator's hands for ease of operation and increased operator efficiency.

6. Displays should be located above (preferably) or to the left of the corresponding manual controls to prevent visual interference while the manual controls are being operated. When manual controls are at the extreme left of the panel, the associated displays should be located above the controls.

Where there is opportunity to upgrade an existing interface, the "before" and "after" experience can be gratifying. As an example, see Fig. 11.

PANEL GRAPHICS

Graphics should not overwhelm the operator in their size because they normally are viewed at no more than arm's length. Legibility is reinforced when the color of the graphics contrasts strongly with the background. As mentioned earlier, letters and numbers are more legible when they are dark against a light background.

Titles

Titles on the panel face may designate functional properties and their subgroups, specific operational requirements, and so on. James Odom suggests that the following actions be taken before laying out a graphic panel:

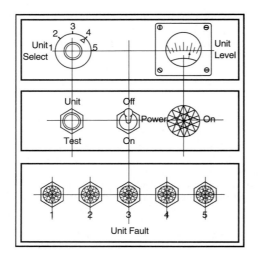 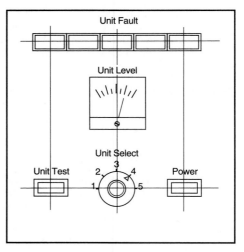

Poor   Preferred

FIGURE 11   Alternative panel layouts. Before and after views illustrate how an existing design may be upgraded through layout revision and component substitution. Both function and appearance are improved. The left-hand panel's outline frames serve only as a decorative feature, unnecessarily separating the related functions and contributing to a more crowded appearance. They are fully eliminated in the right-hand panel as the components themselves define their functional space without the added background frames. The use of square and rectangular panel elements further simplifies and unifies the appearance. UNIT FAULT indicators are given more prominence at the top of the right-hand panel. An operator's hands will not cover the indicator lights or the meter when the rotary control of UNIT TEST button is being used. The POWER switch has been combined with an integral indicator, thereby eliminating the separate POWER ON light. All graphics in the preferred scheme are consistent in their relationships with the panel elements. The legends appear above their respective components, rather than in the random arrangement used in the left-hand version. (*MICRO SWITCH.*)

1. Titles should normally appear over the control to prevent visual interference while the control is in use. An exception would be when panel components must be placed at a height that would block the operator's line of sight to the title (Fig. 12).

2. When components of different sizes are used in a horizontal array, it is good practice to pick a common baseline for all their associated titles to avoid a stepped, disorderly look. But if the grouped components are substantially different in size, some titles and components may be so far apart that the association between them is lost. In that case it is well to follow the primary rule of placing the titles above the components at a distance in keeping with the scale of all the elements involved.

3. All titles should be composed of a simple, sans serif typeface for optimum clarity (Fig. 13). A standard, normally proportioned typeface should be used, except where limited space may require condensed (narrower) lettering to avoid crowding. If vertical panel space is restricted, additional emphasis may be needed to separate primary and secondary titles. In such a case a boldface (heavy-weight) type may be used for the major titles but of the same height as the standard type. Inconsistent use of type styles, sizes, or weights should be avoided.

4. Title abbreviations should be avoided wherever possible. The entire word should be spelled out. If horizontal panel space is tight, a condensed typeface can be used. However, that typeface should be used consistently rather than mixing it with standard-width type.

5. All titles should be lettered horizontally, never vertically.

6. Type sizes should correspond to the functional priorities of the control panel components. Requirements for individual applications may vary, but grossly oversized letters should be avoided. Typically letter heights may decrease in size in accordance with their relative importance, as illustrated in Fig. 14.

Symbols

In addition to alphanumerics there is a growing use of pictorial graphics for control designation and direction at the interface. Semiology or semiotics refers to the discipline that deals with signs or sign language. The need for readily understood and universally recog-

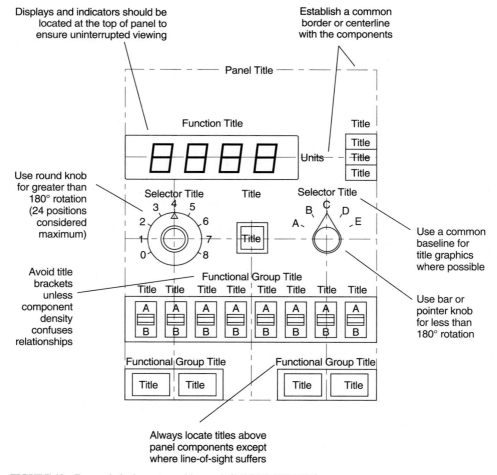

FIGURE 12   Factors in laying out graphic panel. (*MICRO SWITCH.*)

Helvetica Medium (This is the preferred type proportion and weight for most titles.)

ABCDEFGHIJKLMNOPQRSTUVWXYZ
1234567890

Helvetica Medium Condensed

ABCDEFGHIJKLMNOPQRSTUVWXYZ
1234567890

Helvetica Bold

**ABCDEFGHIJKLMNOPQRSTUVWXYZ
1234567890**

FIGURE 13   Examples of typefaces that are acceptable for interface panels. These typefaces are available in various sizes.

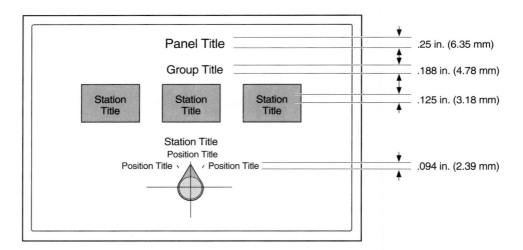

FIGURE 14   Use of various type sizes to differentiate the relative importance of various components. (*MICRO SWITCH.*)

nized signs to identify control functions has led to "picture messages" that are seen daily on all manner of control panels used with tools, automobiles, appliances, farm equipment, and audio electronics, among other examples. Some of these are innately understood, requiring no explanation. However, some must be learned, and whenever they are in doubt, they should be spelled out. Many symbols are now recognized by standards organizations (Fig. 15).

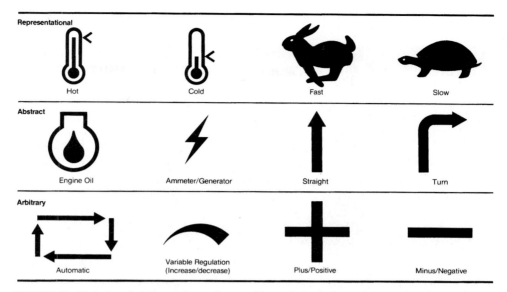

FIGURE 15   Typical symbols that can be used on graphic panels depending on the sophistication of the installation.

VISUAL DISPLAYS

Display device selection depends on the type, amount, and accuracy required of the information to be transmitted. It *should not* present more information than is necessary. Information should be displayed in its most meaningful form to allow quick and accurate reading and thus avoid operator misunderstanding, delay, and error.

Lighted Displays

These displays include indicators, manual controls, programmable display push buttons, meters, cathode-ray tubes, incandescent lamps, light-emitting diodes, plasma, thin-film electroluminescence, vacuum fluorescent displays, and liquid-crystal displays. These are described briefly here. More detail can be found in other articles in this handbook section.

Indicators.   A simple type of display that uses an illumination source, color, and associated legends to convey information. In highly critical monitoring situations, lamp redundancy (or some secondary backup system) may be required for safety reasons. Lighted indicators are ideal for signaling discrete condition changes or general situational status.

Lighted Manual Controls.   Combine display and input functions. Lighted controls provide space saving by accomplishing two functions in the space of one (Fig. 16).

Programmable Display Push Buttons.   Appear like light-emitting diode (LED) dot-matrix displays, but function as lighted push buttons. They can provide a step-through sequence (checklist or menu), diagnostics, and start-up and shutdown functions. Each press of the button produces new instructions. They can display variables, be preselected, or be preprogrammed in the form of alphanumerics or symbols. Programmable display push buttons can reduce both the size and the complexity of the control interface by handling more than one function (Fig. 17). An example is the step-through sequence used for preflight checklists

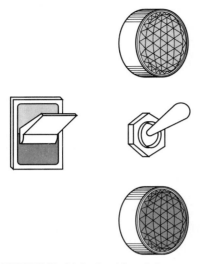

FIGURE 16 Lighted paddle switch versus separate unlighted toggle switch and lighted indicators. (*MICRO SWITCH.*)

in aircraft cockpits. Typically, long checks and the acknowledgment of numerous system statuses can be boring and error prone.

Meters. Of the analog type, they have moving pointers which provide full-value range viewing so that proportional relationships and directional trends are apparent. Parallax between pointer and scale can reduce the precision of readability. Digital meters with alphanumeric (lighted or rotary drum) scales and bar-graph scales provide high levels of accuracy and effortless reading. They also offer greater viewing distance for a given panel area. There are no parallax errors. Digital meters should be mounted as close to the interface plane as possible (Fig. 18).

Cathode-Ray Tubes. CRTs are suited for large-screen multi-image, dynamic, high-fidelity, alphanumeric, line and field, and variable-color displays. CRTs are very versatile, but the large-volume glass envelope requires considerable space behind a panel face. Often they are desk- or table-top-mounted. CRT displays can be enhanced functionally by using a form of touch-switch system. This can consist of a transparent, flexible membrane switch overlay that is made specifically to fit the curvature of a given CRT. Clear, electrically conductive coatings create a matrix grid system which translates a touch within a designated graphic area into a visual response on the CRT screen. Such a system provides more direct interaction between an operator and electronic equipment (Fig. 19). The CRT is described in greater detail in a later article in this handbook section.

Self-Illuminated Displays

Incandescent Lamps. Provide a variety of color and light levels for point source and full-face lens-cap illumination, as well as for edge lighting and hidden message, "dead front" panel configurations. These lamps are prone to vibration failures and significant heat generation, although they do produce more light than some alternatives.

Light-Emitting Diodes. LEDs produce small concentrated bright areas of spot color. They may be used alone or in bar or matrix formats for alphanumeric and linear bar-graph displays. They are available in red, green, and amber colors. Compared with incandescent sources, LEDs offer long life, low heat generation, and low power consumption. They do tend to "wash out" in direct sunlight.

Plasma (Gas Discharge, Neon) Devices. These devices are available in various forms, ranging from individual indicators to composite alphanumerics, bar graphs, and symbols. Plasma devices offer a display solution for higher (line)-voltage applications. Devices are available in basic neon orange, or in green colors that can be altered and enhanced through the use of color filters to produce red and brown colors. The light is characteristically bright, but has a soft glowing quality that is easy on the eyes for long- or short-term viewing. The physical package is relatively thin, thus requiring a minimum of front-panel depth for installation.

Thin-Film Electroluminescence. TFEL devices offer a thin, lightweight, flexible, inherently uniform source of "cold" light surface-illumination graphics. These devices are available in a wide choice of colors, based upon single or blended phosphor characteristics, and phosphors in conjunction with photoluminescent dyes or overlays. TFELs are excellent for

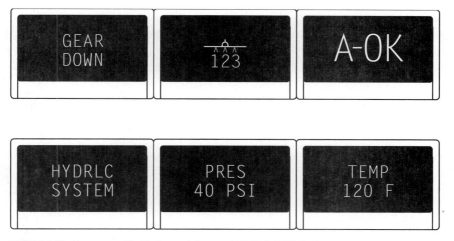

FIGURE 17   Programmable display push button. (*MICRO SWITCH.*)

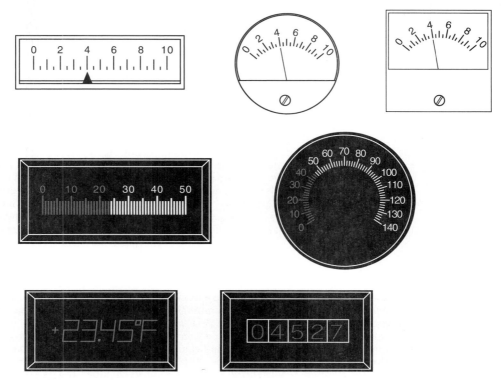

FIGURE 18   Variety of indicating meters. (*Top*) Analog meters having moving pointers with fixed scales. (*Middle*) Analog bar-graph meters with lighted portion indicating measured value. (*Bottom*) Digital meters with electronic display or rotary drum scales. (*MICRO SWITCH.*)

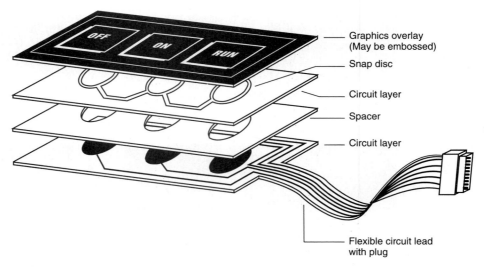

FIGURE 19   Membrane switch panel construction. (*MICRO SWITCH.*)

low ambient light conditions and backlit panel graphics. These lamps do not change color when voltage is reduced, but some colors can shift at different frequencies.

Vacuum Fluorescent Displays.   VFDs use a low-voltage phosphor light system to produce a bright, uniform illumination for the graphic display of alphanumerics, bar graphs, or symbols. Phosphor, filter, and electronic techniques permit a wide variety of color (over seven basics) in addition to the more conventional green color. The glass envelope can be as thin as 0.25 inch (6.4 mm) and requires a minimum of front-panel depth for mounting.

Liquid-Crystal Displays.   LCDs do not meet the full criterion of self-illumination because they depend on an independent light source for visibility in darkness or under low ambient lighting conditions. They usually are classified with self-illuminated devices because of their similar behavior. The matrix, segmented, or diagramatic graphic display changes visually in the same manner as LEDs and VFDs. LCDs are suitable for thin-profile, low-power, and low-cost needs. Because of their low cost, they are sometimes used in situations without properly considering ambient lighting conditions. LCDs can provide alphanumeric, bar-graph, and symbol displays in positive or negative formats. The devices are available in black, red, blue, and green, but they must be used with sufficient ambient light or provided with their own supplemental light source.

Unlighted Displays

Steady or flashing lights naturally have a tendency to draw visual attention, but in some cases may not be necessary. Sometimes the observable physical characteristics of the controls per se may provide the required information (Fig. 20).

Mechanical Flag Indicators.   Depend on ambient light and utilize fluorescent colors or high-contrast legends to do so. These indicators may be used for status and on-off signals, but should not be used for warnings.

Actuator Position.   A useful display indication technique. Toggle or paddle switches, alternate-action (two-level) push buttons, push-pull switches, rotary knobs, and slide switches can provide a quick status reference by their physical positions. The offset of a single bat in a row of toggles is immediately evident, as is the angular set of a bar knob. Any of these switch conditions can be additionally defined by graphic designations adjacent to their positions.

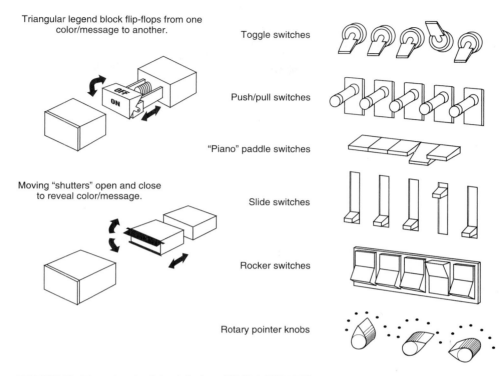

Triangular legend block flip-flops from one color/message to another.

Moving "shutters" open and close to reveal color/message.

Toggle switches

Push/pull switches

"Piano" paddle switches

Slide switches

Rocker switches

Rotary pointer knobs

FIGURE 20   Examples of unlighted displays. (*MICRO SWITCH.*)

Augmentation of Visual Displays

Visible displays can be reinforced through the use of one of the following.

Tactile Feedback.   It imparts a snap force or detent action to a manually operated device. Feeling (or hearing) the click of a switch, for example, provides additional assurance to the operator.

Auditory Feedback.   As provided by bells, buzzers, tone generators, horns, and electronic speech synthesizers, it has its principal advantage in communicating information to the operator without need for the operator to face the information source.

Auditory signals may be considered when the visual tasks of the operator are overloaded, where ambient lighting may cause viewing to be unreliable, and where continuous visual monitoring is needed, while, at the same time, additional information must be provided.

Auditory signals may be detrimental in some situations. Conflicting sounds can be confusing. Existing ambient noise levels already may approach or exceed discomfort levels. The presence of nearby noise from machines and equipment may be of such quality as to compete with the purposely generated signal sounds.

COLOR IN INTERFACE DISPLAYS

As pointed out by James Odom, color is a psychological experience, and therefore color preference is almost completely subjective. Everyone has an opinion about preferred color combinations. Which choice is best can be argued endlessly to no avail. For this reason, the selection of color for a control interface and the control components themselves should be made by an individual well versed in color and response relationships, such as an industrial designer. Color choices affect the appearance, visibility, and cleanliness of the equipment, as well as the safety, speed, efficiency, and morale of the operator. For permanent or long-term installations, color schemes should be carefully coordinated with the surroundings for maximum benefit. Compatible color relationships will prevent garish, nerve-jangling combinations that are uncomfortable for the user to work with. Certain paired colors (commonly blues and reds) are perceived to vibrate or "dance" when viewed together.

Selective reflectance and absorption of light determine the color of an object. In general, lighter colors are easier to see because they do not absorb as much ambient light as the darker colors. Controlling the surface texture can enhance the degree of reflectivity and color fidelity and minimize glare. Color may change dramatically in appearance when it has been overcoated with a matting agent. Sample runs from vendors are usually the only way to resolve the variations that result with a material's new color and texture combination.

There are no absolutes in color use. Most conscientious industries employ elaborate color control systems to ensure faithful, consistent product color results. In the final analysis, the human eye can distinguish an enormous number of different hues and is the best tool for resolving questions of color fidelity.

Color Coding

Color coding of interface components should be consistent with accepted human factors symbols. Some colors have become associated with certain relationships (red for *stop,* green for *go,* and so on). This tends to reinforce their meaning in functional applications (Table 1).

Color Transmission and Projection

Transmitted Color.   Refers to the use of colored lenses in applications where color must be apparent when the display is unlighted (Fig. 21).

TABLE 1   Generalized Meanings and Uses of Color for Coding*

Red	Incompatible or dangerous condition. Corrective action needed. *No-go, error, failure, stop, malfunction, warning, hazard, danger*
Yellow	Marginal condition exists. Proceed with caution–double-check. *Pressure: below normal, check hopper level, caution, inspection port open*
Green	Conditions are within tolerance. Operations are normal. *Continue course, situation is at ready, safe, power on*[†]
White	Conditions and situations that do not have a right or wrong implication. Indicated information does not imply a success or failure. *Boiler* #1 *on line* (alternative function) or *reservoir cycling* (transitory condition)
Blue	Limited value as a coding element. Positioned at the lower end of the spectrum, blue generally has a lower intensity than other colors and is less stimulating.

  * It is estimated that 4 percent of healthy adults (mostly males) are color deficient and may confuse red with yellow, blue with green, and so on. Vision of operators should be checked for this possible disability.

  [†] A notable exception is the power industry where "power on" is indicated by *red,* which is associated with a "hot" electrical condition.

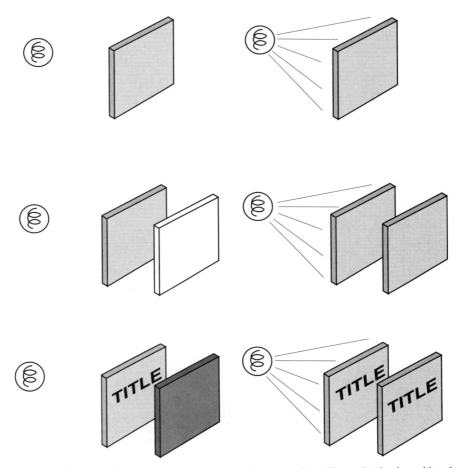

FIGURE 21 Transmitted, projected, and dead-front displays. (*Top*) Transmitted color achieved with colored lens. (Color is visible even when display is unlighted.) (*Middle*) Projected color achieved with colored filter behind white lens. (Color not visible until lamp is lighted.) (*Bottom*) Hidden legend and hidden color (dead front). Dark lens hides color message until display is lighted. (*MICRO SWITCH.*)

Projected Color.   Achieved with a white lens and color insert (or color filters over the lamps). When the lamps are off, the display is white; it becomes colored when illuminated. Projected color displays are effective in dimly lit or dark rooms, but they have the disadvantage of diluting or washing out in high ambient light. This occurs because the display signal is just one color, and the white lens reflects a large amount of incident ambient light which, in turn, tends to weaken the color signal.

Dead Front, Hidden Legend/Hidden Color Displays.   These displays generally use a transparent smoky gray lens with a legend on a color insert. The display appears black (dead front) when the lamps are off. When illuminated, color and legend appear. A dead-front display enables so-called control by exception, that is, the display can contain a great deal of potential information bits, but few legends are displayed simultaneously. When unlighted, dead-front displays are concealed. This technique is used when it is desirable to limit the signals given to the operator until some specific conditions, which should be brought to the operator's attention, appear on the screen. This tends to reduce operator error.

INTERFACE STANDARDS AND REGULATIONS

Several organizations publish standards and specifications for interfaces in the interest of accenting safe and sensible designs. Some of these organizations are active on a national basis, others are international. Examples include the following:

UL	Underwriters Laboratories, Inc. UL's Technical Assistance to Exporters (TATE) program tests U.S. made appliances to international standards.
CSA	Canadian Standards Association.
CEE	International Commission of Rules for the Approval of Electrical Equipment.
MIL-STD	U.S. Department of Defense human engineering approval body for military systems, equipment, and facilities.
IEC	International Electrotechnical Commission (electrical characteristics).
ISO	International Standards Organization (mechanical characteristics).

MAINTAINABILITY

The basic rules that apply to other complex electrical and electronic equipment apply equally to the user interface. The designer must determine how easily (or with how much difficulty) components can be installed, replaced, repaired, adjusted, modified, relamped, or relabeled. It is desirable that minor maintenance, such as relamping, be accomplished easily without need for special tools. Service matters, such as these, should be checked out well in advance of procurement. Built-in test or diagnostic features are highly desirable because they can reduce downtime. But if not available, the equipment should be designed to accommodate standard test equipment.

MINIATURIZATION

The trend in recent years has been toward greater miniaturization of components, circuits, and the like. However, at a point human factors enter into design. For example, push buttons in random entry application must be sufficiently large to be manipulated by human fingers. Thus a horizontal center spacing of 0.5 inch (12.7 mm) and a vertical spacing of 0.4 inch (10.16 mm) are the minimum dimensions for accommodating operators. In addition to finger clearance around each manual control, there must be adequate space for graphics. Many of the recent components are too small to apply titles directly to their surface, but nevertheless panel space must be reserved for this purpose. Designers must be aware of the possibility of overloading a panel with smaller components.

PROCESS OPERATOR TASK ANALYSIS AND TRAINING

Since the early 1970s a large segment of human factors research in the instrumentation and control field has been directed toward developing a better understanding of how human operators supervise a complex industrial control system. Many of the answers have come from specialists who have an excellent grasp of applied psychology, the behavioral sciences, mathematics, and statistics. Thus it is not surprising that much of the body of theory has been developed at the academic level. The application of these findings, however, rests essentially with the users and manufacturers of control components and systems. Human factors engineering has been comparatively slow in developing and by no means can it be regarded today as a mature technology.

MOTIVATIONS FOR HUMAN FACTORS RESEARCH

The needs for top process operator performance over the years have become self-evident. Error-free operations ensure maximum throughput, enhance safety, and minimize downtime and maintenance, among other important factors. Added to these drives is the important need to bring along the next generations of highly qualified operators.

Among more specific reasons for accelerating the study of process-operator-centered developments, one would include:

How can the performance of experienced operators become even better in light of the continuous streams of improved control system hardware and software?

What equipment design changes can be made that will improve the operator's ability to obtain information from the system and to communicate decisions to the system more reliably and faster?

How can operator experience be embedded in the equipment to transfer some of the corrective actions normally made by the operator to the automatic domain, relying on the operator for backup?

How can the operator overcome problems, such as embedded inner loops, without compromising performance in an emergency situation?

FIVE MAJOR OPERATOR TASKS

There is a general conclusion among human factors researchers that the process operator normally must accomplish five principal tasks. Four of these tasks are typed as *cognitive*, that is, they require some form of decision making. The tasks are:

1. *Monitoring.* The continuous scanning of signals of various kinds and degrees of importance and at various interface locations. Human factors engineers classify this as a *perceptual motor task,* requiring alertness, agility, and immunity to fatigue and boredom. (As pointed out later in this article, although not a decision-making task, this task is one of the weakest in terms of human skills. As delineated in the prior article, much can and has been done by way of panel, screen, console, and workstation design, as well as with component design and placement to enhance the operator's monitoring performance.

2. *Control tracking.* The continuous checking of process variable values, mentally noting slightly off-normal readings and other incoming information that may hint to variations from the norm. These actions, when persistent, lead to the next task.

3. *Interpreting.* Quantifying, estimating, weighing, sorting, and filtering incoming information for the purpose of separating random fluctuations from definite trends. Identifying true signals from noise, among other factors, that do not immediately square well with normal process operation.

4. *Planning.* Applying historical data (heuristics) for establishing goals and adjusting control strategies that will favor plant safety, for efficient use of resources, and for sequencing tasks in terms of priorities, among other factors, that possibly may indicate trouble and some unsafe condition ahead.

5. *Diagnosing.* The prompt identification of a problem source when a fault occurs and appropriately developing effective remedial actions, bypassing the trouble if possible—all coupled with a keen sense of timing.

Obviously the foregoing task descriptions are encapsulated. Particularly, in connection with tasks 3, 4, and 5, one can envision scores of combinations and permutations that the operator must evaluate prior to taking action. These are the factors which essentially differentiate the trained and experienced operator from the novice. The nature of these tasks also makes it apparent why some operators are better than others (like racing car drivers) and why some persons are not trainable.

APPROACHES TO HUMAN FACTORS KNOWLEDGE

Most research on human factors to date has been accomplished via two paths: (1) quantitative modeling and (2) developing expert systems.

Quantitative Modeling

In this methodology, attempts are made to reduce operator subtasks to mathematical descriptions of human performance. They also borrow some of the techniques developed from the field of artificial intelligence (AI). To date these descriptions do not approach the complexities seen in actual human performance. The models are for an *ideal* observer or performer. Many of the models developed are applications of physical system models applied to the description of human performance. Quantitative models are better for describing the simple cognitive tasks rather than the more complex. This is unfortunate because one is much more interested in understanding the complex tasks rather than the simple. Quantitative systems tend to fail when confronted with novel situations and in capturing abilities, such as flexible problem solving, that are uniquely human.

Expert Systems

This methodology has proved more fruitful in shedding light on operator performance. An expert system generally comprises an intelligent problem-solving computer program that

---

[1] Researchers have shown conclusively that hours of experience count heavily in radiology. For example, an expert radiologist will see, on average, 67 x rays per day, or up to 10,000 per year. At least up to this point the experienced radiologist continues to improve performance.

employs knowledge and inference procedures that are required for a high level of operator performance. Most attention to date has been directed to those domains that are considered to be difficult and that require much specialized knowledge and skill.

A common form of representing knowledge in an expert system is to use production rules, which can be of the form:

$$\text{IF } (premise), \text{THEN } (action)$$

The expert system methodology requires capturing the knowledge (from experience) of one or more operators who have been performing given tasks over a comparatively long time span. Researchers do not all agree on the conditions that qualify an expert. Some researchers suggest that a true expert is someone who has practiced the germane tasks for at least 5000 hours.

Support for this high figure stems from experience in another field, namely, that of radiology.[1] However, as is pointed out later, with some individual experts, the law of diminishing returns may take effect much sooner.

The extent of knowledge required for an expert system ranges widely from one field to the next. For example, researchers estimate that around 31,000 basis or primitive piece configurations are required by master chess players. Some researchers estimate that upward of 10,000 production rules, as defined previously, may be required for a complex process control system. A well-known and accepted medical program (MYCIN) has fewer than 1000 rules.

INHERENT MENTAL ATTRIBUTES OF OPERATORS

Before exploring the process of knowledge engineering used in the development of an expert system, it is in order to consider those inherent mental attributes which a top-performing process operator must have.

Automatization. With practice, a particular task requires less and less attentional resources of the operator so that the task can become automatized. Automatization can occur on several levels. With reference to the earlier list of five major operator tasks, monitoring, controlling, and interpreting tasks would most likely become automatized with practice, while planning and diagnosing would not become automatized. (Automatization here refers to the operator and is not to be confused with process and machine automation.)

From the standpoint of neuroscience, operator automatizing is akin to rote, as exemplified by memorizing the multiplication tables. Automatizing less simple tasks, such as controlling and interpreting, although difficult, should be the goal, thus allowing more available operator resources for attention to the important tasks of planning and diagnosing.

Because of their inherent ability to automatize certain tasks, expert operators develop an ability to timeshare tasks and to automatically process parts of the tasks. One method which has been used to study this ability is to use dual-task methodology, in which subjects in an experiment are required to perform two tasks at once. The experienced operator will learn from experience where dual tasking is practical. The seasoned operator is quite aware that each task requires a certain span (attentional resource) and these resources are limited. It is important in operator training to stress how dual tasking can conserve time and to practice "hands on" dual tasking for given situations. This is sometimes referred to as operator timesharing.

To date comparatively little research has been directed on the timesharing and automatization of process operator tasks. The research that has been done indicates that operator performance can be improved in some instances simply by making operators aware of these inherent abilities that most humans possess.

Operator's Mental Model of System. Process operators develop what human engineering researchers term the "internal" or "mental" model of the system—process and controls. This is an inherently mental attribute possessed by all normal persons, but research indicates that this ability is particularly important for process operators. The mental model is difficult to explain in engineering terms.

Digressing for a moment, the average person can visualize his or her home, places (buildings, scenery, and so on) visited in the past, or the faces of friends—even though time has elapsed from seeing the real things—and with practice, such "scenes" may be filled with considerable detail.

Working with a process and its control systems for the majority of days in a year, the operator develops this internal image almost effortlessly. The operator can project this image in his or her mind instantly on command. This is the image in which the operator has confidence. True, an average control room will display a total process and parts of it in more detail, but these, in terms of operator performance, are not one-to-one substitutes for the built-in internal mental image, and recall time is much longer because such displays must be "read."

From depth interviews with experts, the importance of the internal mental model is undeniable in terms of operator actions and confidence. Research has shown a close coupling between the internal model and the operator's ability to predict changes and abnormal operations. Researchers have also shown that the operator who has an accurate and effective internal model will be able to predict how the system will function. The events as they happen can then be compared to this prediction to see whether anything unusual has occurred. Some authorities characterize this ability as a "routine model" that the operator uses when the process is performing in normal situations. It has been found that expert process control operators continually update (in their heads) the current state of the system and where the system is moving. Without expectancies, or without the internal model, the operator would have to refer to lookup tables and other support data, requiring time to retrieve and absorb, when evaluating the current operation of the system.

Operator's Spatial Representation of System. Somewhat parallel to the preceding discussion, operators inherently have available to them a spatial representation of the system. A plausible alternative to a spatial representation would be a propositionally based rule system similar to that used in expert systems.

To differentiate these two kinds of representation, consider a process in which the operator knows that the water level in a tank is dropping. If the operator stored a set or rules, he or she would mentally step through the rules to find one that fits, such as, if the water level is dropping, an output valve is open. For a spatially based model the operator would represent the task spatially as a physical system with locations and movements between the locations. To find the source of a problem such as this, the operator could mentally "run" a simulation of the process until a match is found.

As an example, the operator could picture the water flowing out of the tank through an open valve. When expert steam plant operators were interviewed, they indicated that they did mentally "run" such simulations to solve system problems. Another researcher found a similar phenomenon in expert process control operators and characterized this kind of activity in fault diagnosis and troubleshooting as a *topographical search*. Spatial representation of process control tasks, especially fault diagnosis, was also found.

A spatial representation of a task is an efficient way to store information. By spatially representing a process control task, the operator must represent physical locations and must know how the systems can interact with each other. Then well-learned and versatile reasoning and problem-solving strategies can be used to make inferences about the process. As a common example of how such a representation scheme could work, if someone inquires about the number of windows in your house, this number can be arrived at by mentally picturing yourself moving through the rooms of the house counting as you go. Propositionally based rules can then be generated from this spatial information as a secondary process. In other words, rules may be a by-product of the processing rather than the direct process. Storing and retrieving 100,000 rules from memory, as an example, would be difficult; a spatial representation of the information is more efficient. A computer model of this kind of expertise, however, is difficult to generate. To do so, one would have to know how to incorporate the spatial information in computer code, how to model the reasoning that occurs on this spatial information, and how the particular strategies (such as picturing yourself walking through rooms) are chosen from the vast amount of strategies, many fruitless, which must be available. Research along these lines is continuing.

TRAINING FOR IMPROVING COGNITIVE SKILLS

Although on-the-job training remains the essential ingredient of a training program, computer-assisted instruction (CAI) can be effective toward saving both time and money. CAI is a type of computer-based training that is often used in conjunction with other instructional systems. For simulators, CAI can be used to guide the instruction. CAI also can be used as a component of a computer-managed instruction (CMI) system, which manages and prescribes the mode of instruction to be used. CAI is one of the available modes. Intelligent CAI (ICAI) is a class of instructional programs that is made more intelligent through the use of artificial intelligence (AI) techniques. The amount of computer control over the complete instructional process varies, of course, with the application. Finally, embedded training, a particular kind of CAI, can be used for tasks that use computerized equipment. This is appropriate for many process control tasks. For this training, the operating system of the computerized equipment is taken over by special software that emulates the working of the actual system, but can enhance that system with special computerized instructional techniques. Embedded training provides both hands-on training and the instructional techniques of CAI.

In terms of providing the experience needed for acquiring expertise, CAI has many advantages over conventional instruction or on-the-job experience. A main advantage is that CAI allows the instruction to be individualized because a single person is interacting with a computer instead of an instructor interacting with the whole class. In a traditional classroom situation, training time for the whole group is driven by the slowest learners. With CAI, training time is driven by the individual's capabilities and motivations. Researchers have found that instruction time due to individualization ranges from 14 to 67 percent. This means, then, that a given individual can receive much more practice in the same amount of classroom study, thus accelerating the transition from novice to expert.

Another technique used is that of time compression. For example, in a simulator a tank can be filled within a few seconds, a marked advantage as compared with the same action occurring in a process per se. For some tasks, time compression in training actually makes the spatial patterns easier to detect than if they were done in real time.

More research is required to determine how practical it may be to improve the automatization of complex tasks, such as the control task. One experiment has shown that this task could be automatized with practice so that little attentional resource consumption will occur. In this experiment, operators were required to control a second-order system which is very difficult for operators because of the difficulty in seeing the relationships between the control input and the system output. For an acceleration system like this, the operators perceive that the input-output relationships are inconsistent, or that the system will behave as a first-order velocity system.

The hypothesis of the experiment was that performance on the control task could become automatized if the operator were better able to see the relationship between the control input and the system output.

In the experiment, operators were trained on one condition with a display which had an augmented cue. This cue made the consistencies between input and output more salient to the operators. The augmented cue used was a parabola, because this is the geometric representation of the path of the acceleration system. In this condition each input position had a unique parabola associated with it and the parabola function as a predictor of where the system would be in the future. This condition was compared with a condition whereby an augmented cue was also provided, a point cue, but this did not provide a unique cue with each input position and, thus, did not make the input-output consistencies more salient. When the attentional resources required for controlling the task were measured in a dual-task situation, the group which received the augmented cue (making the input-output consistencies more salient) had much lower attentional resource consumption than the other test group. The implications of the study were that process control tasks can become automatized through training—if the operators are allowed to see the consistent relationships between control input and system output. Highlighting the consistencies can be accomplished by using appropriate augmenting cues.

Training also can be used for development of the operator's internal model as mentioned previously. One way to know that an accurate internal model has been developed is to train operators on one aspect of the system and then transfer these persons to another related task in the system in order to determine if their knowledge from the internal model is accurate in the transfer task. The important concept is that the operators must come to know the consistent relationships between control input and system output.

Operators were trained in one of three conditions, using a consistent augmenting cue, an inconsistent augmenting cue, and no augmenting cue. When transferring knowledge from the internal model to other related tasks, the consistent augmenting cue group outperformed the others. The inconsistent augmenting cue group and the no-augmenting cue group showed very serious inaccuracies in the internal models which were developed. It may be possible that these latter two groups could have developed accurate internal models with more practice. To increase training effectiveness, however, augmenting cues which make the consistencies between input and output more salient are important.

Providing a predictor of the future state of the system is a way to increase cognitive abilities. It will be recalled that an integral part of the operator's internal model is the ability to predict where the system should be, so that this prediction can be compared with the current state of the system. In a series of experiments, after training, predictors on an augmented display were demonstrated to increase the cognitive skills of operators, even if they were not shown. In this case the operators had learned to internalize the predictor so that performance would be high, even without the predictor on the screen. This is particularly advantageous if a breakdown occurs in the computerized aid, or if novel situations occur. If augmentation is used during training, some trials should be presented without the predictor so that the operator does not form a dependence on this assistance.

Another training tool for developing cognitive skills is that of providing the operators with spatial information. Recall that expert operators seem to conceptually represent a process control task spatially and that to enhance this ability through training can be useful. This was attempted as early as 1983 when the "steamer" program was developed. This was a computer-assisted instruction program developed for the U.S. Navy for training steam plant operators. For a large ship there typically are some 1000 valves, 100 pumps, and various turbines, switches, gages, dams, and indicators. The operational procedures for these kinds of plants require several volumes.

"Steamer" incorporated sophisticated graphics, an intelligent tutor, and animation on a low-cost computer. All subsystems in a steam plant could be accessed easily from the terminal by using a mouse to move through a hierarchy display. Although graphics have been used in this and other training situations, experimental validation against other methods has been inadequate.

EFFECTS OF PLANT AUTOMATION

Increased automation of plants and processes, on the one hand, has decreased human errors. On the other hand, this may result in relocating human error to another level. Errors still can occur in setup, in manufacturing and maintenance, and in programming. And, to date, automation still is not very good for performing cognitive-based tasks. The human operator still is needed at some point in the system as another check and to provide needed flexibility for unexpected events. Automation causes the human operator to be further removed from the control loop.

As more tasks become automated, the human operator is needed to monitor the automation. A basic problem is that humans have not proved to be very good monitors. This was first noticed as early as 1948, when researchers studied radar operations and found that the longer an operator continued at a task, the more targets were missed. This became known as the *vigilance decrement*. Alternatives for this problem are to automate monitoring, provide more special training, or provide effective feedback to the operators. Automating the monitoring is not a good solution because someone then must monitor the automatic monitoring. Special training can be effective, and feedback, such as unexpected drills, can and always should be provided.

Another problem with automation in regard to cognitive skills is out-of-the-loop familiarity. As more inner-loop functions become automated, the operator is required to function more at the outer-loop stages. A problem occurs when an inner-loop function fails and the operator is required to find the failure. From another field, airlines have found that pilots have poor transfer from highly automated planes to smaller planes that require inner-loop control. Also, because much of the training of an operator is acquired on the job, automation can make learning more difficult.

In a research experiment it was found that operators who developed an internal model of the system from monitoring the system were not very good at detecting system faults. However, operators who developed an internal model from controlling the system were much better at detecting system faults. This implies that skill maintenance of the operators may be required. Having the operator control the process in a simulator may be an important way to develop and maintain cognitive skills that otherwise may be lost due to automation.

Still another problem arising from automation, especially with the incorporation of expert systems, is that the operator may think that the system is more intelligent than it actually is. Again, there are several examples from the aviation field. In one situation two large passenger aircraft were on a collision course, as noted by the air traffic controller, but the controller assumed that if a collision became imminent, the system would provide an alert. When the controller took a break, a second controller saw the collision course and managed to contact the aircraft just in time to avert a crash. In another example the pilot of a military aircraft put too much faith in the autopilot. When a fault occurred in the plane, the pilot put the plane on autopilot, but the autopilot did not hold and the plane crashed.

Although many factors can contribute to accidents, it appears that an important factor in these instances was an overreliance on automation. The solution to the problems may not be less automation, but rather increased awareness training of operators that automation also has its limits and imperfections.

A similar situation was simulated. This involved the interaction of an operator with an expert-system controlled nuclear power plant. The simulation demonstrated that the operator assumed that the expert system "knew" more than it actually did. The operator assumed that the system should have "known" when a failure in the cooling system occurred. Other research has shown that novices often attribute humanlike characters to the computer and cannot understand when the computer makes nonhumanlike mistakes. When an expert system is incorporated into process control, the human operator should be thoroughly trained on what the system does and does not "know."

DISTRIBUTED DISPLAY ARCHITECTURE

Since the early 1970s instrument displays have been undergoing numerous changes that reflect and incorporate the marked progress that has occurred throughout instrumentation and control technology. These would include development of the early minicomputers and the later introduction of the personal computer (PC); the advent of the microprocessor; advances in cathode-ray tube (CRT) technology and other display methodologies, such as liquid-crystal displays (LCDs), vacuum fluorescent displays (VFDs), and plasma (gas discharge); the decentralization of control centers by way of distributed control; the development of advanced networking concepts; a better understanding of human factors as they relate to the operation of machines and processes; and the engineering of interactive graphics among numerous other advancements that have seriously affected display practices, the general result of which is a severely altered interface. The contrast is dramatically apparent by comparing Figs. 1 and 2.

Contemporary systems are quite software-intensive. Installation and maintenance require different skills than in the case of prior-generation systems. Fewer electricians and hardware technicians are needed, whereas more expertise is required in the system engineering and software areas.

THE GRAPHICS INTERFACE

"Pictures" help immeasurably to improve an operator's comprehension of what is going on in a controlled process or machine. This, of course, is wisdom known since antiquity. It was first applied in industrial instrumentation in the 1940s, when elaborate process or manufacturing diagrams were put on permanent panels, with illuminated indicators placed directly on the diagrams, and where attempts were made to locate single-loop controllers to reflect the relative geometry of where the measurement sensors were located in the process. Such panels became plant "showpieces" and often were constructed of costly enameled steel. Unfortunately changes were inevitably made in the process or instrumentation, and the beautiful initial panel soon was marked over with tape. A cluttered appearance progressively emerged, which hardly could be considered an inspiration to the operator's confidence in the system. Use of a CRT as the panel face essentially solved this problem of panel obsolescence. Later embellished through the use of color, it offered immense flexibility, not only at one location, but at several locations along a network. Currently these properties are being exploited further through the use of the X-window and X-network concepts. But the CRT has had some limitations too, notably the comparatively small space in which an entire control panel must be projected. Even though through menu techniques a process may be shown in detailed scenes, some loss of comprehension occurs, brought about by crowding and by the complexities introduced when expanding a scene into subscenes ad infinitum. Larger basic presentation techniques (such as flat screens) are constantly being refined.

Graphic displays are formed by placing a variety of symbolic forms, originally created by an artist and later reproduced electronically, on the display surface where they can be viewed by the user. These may enter the system by way of certain display standard protocols, such as GKS/PHIGS (graphical kernel system/programmer's hierarchical interactive graphics system), or otherwise software created. In some systems the objects are formed by using the basic output primitives, such as text, lines, markers, filled areas, and cell arrays.

FIGURE 1 Long panelboard containing indicators and controllers used in large processing plant. Generally this type of panel was used in the early 1970s. The very beginnings of the cathode-ray tube (CRT) displays for annunciators is indicated by the CRTs in the upper right-hand portion of the view.

The display software must provide additional higher-level graphic objects formed with the basic primitives. For graphic displays on process systems, the fundamental types of objects available should include background text, bar graphs, trends, pipes, and symbols. It is desirable to allow each of the fundamental types to be freely scaled and positioned. It is also desirable for the trends to be displayable as either point plots or histograms, and for both text and trend fields to be scrollable together if there are more data available than can be reasonably displayed in the space allocated for the object. The system should provide libraries of symbols and facility for the user to edit.

Well-designed graphics displays can show large amounts of dynamic information against a static background without overwhelming the user. The objects used to form the static portion of the display may be as simple as textual headings for a table of dynamic textual data, or a graphic representation of a process flow diagram for a section of a process, with dynamic information positioned corresponding to the location of field sensors. In a GKS system the static portion of a display usually is stored in a device-independent form in a metafile; in PHIGS, an archive file may be used. From there it can be readily displayed on any graphics device in the system, including various terminals and printers.

Display software also can provide the ability to define "pickable" objects and a specific action to take if an object is picked. Picking an object may call up a different display, show some additional information on the same display, or initiate an interaction sequence with the user.

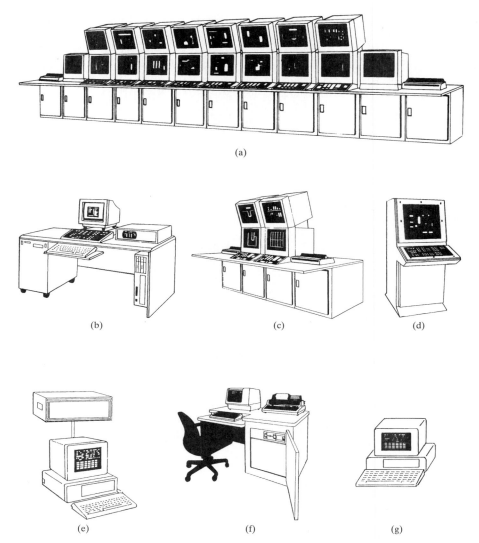

FIGURE 2   Examples of how the flexibility of CRT-based panels contributes to various interface configurations and distributed display architecture. (*a*) Central configurable CRT station with distributed historian. (*b*) Engineer's workstation. (*c*) Area CRT station. (*d*) Field-hardened console. (*e*) Independent computer interface (serial-interface–personal-computer). (*f*) Batch management station. (*g*) General-purpose computer.

The graphic display software should provide the ability to update dynamic data objects at specifiable time intervals, upon significant changes, or on demand. The update specification should be on a per object rather than a per display basis. The updating of each object should involve the execution of arithmetic expressions with functions available to access the database of the process information system. Besides changing the values of the data to be represented, the updating could result in changes to the attributes of the primitives used to form the objects.

For some users, especially process operators, color can add an important dimension to graphic displays, expanding the amount of information communicated. Color graphics are used widely today. (One must make certain that there are no color-blind operators.) Making hard copies of color displays has improved quite a bit during recent years.

USER INTERACTIONS

The most important aspect of any interface is the means by which the operator can interact with the system. This is the mechanism that the user may employ when accessing, manipulating, and entering information in the system. It is the key point in any interface design and in specifying requirements prior to procurement.

It has been established that best acceptance occurs when an operator becomes confident that there is no harm in using an interactive graphics system. To achieve this, software should be provided and used that makes certain that all interactions are syntactically consistent and that all data entered will be checked for validity. The system should reject invalid inputs and provide a message that the user can readily understand without external reference material.

The user interface must be responsive. The goal should be that all system responses to user actions occur faster than it is physically possible for the operator to perform a subsequent action. Some situations require more responsive interfaces than others. In general, the number of interactions and the complexity of each interaction should be minimized, even though this may require additional sophisticated hardware. Most interactions with a process computer system can be performed in an easier fashion with an interactive graphics system than with a textual dialogue system. However, even with interactive graphics, some textual interaction sometimes may be necessary.

"CONVERSATIONAL" INTERACTIONS

Historically systems have interacted by having the user carry out a textual dialogue in a dedicated portion of the display screen, sometimes called a dialogue area. The flow of the dialogue is similar to a vocal conversation—with the system asking for information with a prompt that the user reads in the dialogue area, and the user providing answers by entering alphanumeric text strings or pressing dedicated function keys on a keyboard. The system provides feedback to each user keystroke by echoing the user's input to an appropriate position in the dialogue area, importantly of a different color than the system's prompt. This is a natural form of interaction for the operator. The tools used to carry out such interactions are easy to construct.

The person who has programmed the system and the user should fully understand the dialogue language that is being exchanged. This can be troublesome because natural language communication can be ambiguous, considering the multiple meanings of some words that arise from the level of education and sometimes the geographic background of the user. Programmers sometimes have difficulty in creating meaningful prompts that are short and easily accommodated in the dialogue area. Live voice systems also can be effective, but have not been accepted on a wide scale.

In process control systems the dialogue area frequently appears at the bottom of the display surface, allowing the user to view the displays while the interaction is progressing. In this way the user will not, in error, attempt to use old data left on the screen during an extended or interrupted interaction. If display updating is halted during interactions, the system should either obscure the data by performing the interaction in a dialogue box in the center of the screen, or in some way make it obvious that the displayed data are not reliable.

Dialogues should be designed to minimize the number of questions required and the length of user inputs needed. Prompts should be simple and easy to understand. They should list menus of possible alternatives when appropriate. Whenever possible, default values should be provided to facilitate the rapid execution of functions.

DATA INPUTS

An interactive graphic user's interface may consist of a bit-map graphic display device (usually with a dedicated processor), a keyboard, and one or more graphics input devices. Hence the electronic interface is essentially the same as the traditional interface, except that the video display unit has some additional capabilities and a graphics input device has been added. Normally, interactive graphics capabilities will be in addition to those of the conventional panel.

Graphics input devices are a combination of hardware and software that allows the positioning of a graphic pointer or cursor to a desired location on the screen and "triggering" an event to perform some action based on the selected pointer position. The event normally is triggered by pressing or releasing a button. Examples of physical input devices that have been used include the light pen, joystick, track ball, mouse, graphics tablet, touchpad, touchscreen, and keyboards. Once a device is selected, one should concentrate on how it can be used effectively. Among the most commonly used are the keyboard, a mouse, or a touchscreen.

Keyboard.   A keyboard should be available with an alphanumeric section for entering text and an array of function buttons for certain specialized functions. In contrast to nongraphic interactions, the alphanumeric section will be used infrequently. On process systems, the function button section of the keyboard is important even in a graphic environment, because buttons can provide immediate generation of commands or random access to important information. Sometimes there is a need for keyboards with in excess of 100 function buttons on a single workstation. Also, there may be a need for lamps associated with keys that can be used to prompt the user. Keys should provide some kind of feedback when actuated. Tactile feedback is desirable, but not necessary. Many operators are prone to favor tactile feedback. Visual feedback from every keystroke is essential.

Touchscreen.   This is probably the simplest of the graphic input devices to use because it requires no associated button for triggering events. This can be done by sensing a change in the presence of a finger or some other object. A touchscreen occupies no additional desk or console space and appears to the operator as an integral part of the screen. The touchscreen is useful to naive operators for selecting one of several widely spaced items that are displayed on the screen. However, because of calibration problems, the curvature of the CRT screen, and its coarse resolution (that is, between screen and human finger), implementations using touchscreens must be considered carefully in advance—probably involving prior experience to get the "feel" of the system. From a human factors standpoint, it is important that the system be designed to trigger the event upon the user's removal of a finger from the screen rather than upon the initial contact of the finger. This allows the user to correct the initial (frequently inaccurate) position of the cursor by moving the finger until the graphic pointer is properly positioned over the designated object on the screen, at which point the finger can be removed, thus triggering the selection of the object.

Touchscreens operate on several principles, but most require an overlay on the monitor. Touchscreen technologies include the following.

Resistive.   Two conductive sheets separated by an array of clear, tiny elastic dots. The first substrate layer may be glass or plastic. This has a conductive coating (such as indium or tin oxide) which possesses uniform and long-term stability. The second layer (cover sheet) may be of polycarbonate or oriented polyester (Mylar) plastic, which has a conductive inner coating on the surface facing the substrate conductive coating. A voltage measurement indicates where the circuit is completed. Resistive screens indicate only a single point to the computer at the center of the area of contact. In discrete resistive touchscreens, one of the layers includes equally spaced horizontal rows, which are etched into the surface; the others are in the form of vertical lines (columns). Thus the two sheets form a grid. Discrete screens usually are application-specific. Analog versions allow total software control for the touch zones. Characteristics of resistive touchscreens are 4000×4000 resolution, low cost, very rapid response, and no limitation on stylus used.

Capacitive.   These operate on the principle that high-frequency alternating current couples better between conductors than direct current. The screen system is controlled by a radio-frequency source (approximately 10 MHz), the outputs of which go to each corner of the screen. When the screen is touched, current passes from the screen through the operator's finger. The amount of current emanating from each corner of the screen relates to the location of the touch. Characteristics of capacitive touchscreens are good resolution (1000×1000 points per screen side), fairly high cost, very rapid response, and some limitations on stylus used. The screen must be shielded from internal electronics noise and external electromagnetic interference. Very good optical properties.

Surface Acoustic Wave.   This is a relatively new technology. When surface acoustic waves (SAWs) propagate across a rigid material, such as glass, they are efficient conductors of acoustic (sound) energy—at precise speeds and in straight lines. Special transducer-generated SAW signals travel along the outside edges of the screen, where they encounter reflective arrays on the screen. Each array reflects a small portion of the wave. Through complex logic circuitry, the SAWs determine the coordinates of any pliable device (including a finger) that may press the screen surface to form a transient detent. Some drawbacks solved by this technology include a combination of transparency, ruggedness, and ease of installation. Characteristics of SAW screens are a modest resolution (15–100 points per inch), high cost, and very rapid response.

Piezoelectric.   Pressure-sensitive electronics detect and determine touch location. A touch on the screen causes the screen glass to flex slightly, applying pressure to each of the stress transducers. This pressure is inversely proportional to the distance from the touch point. Each transducer generates a signal proportional to the amount of pressure applied. Signals are processed into X and Y coordinates for transmission. Sometimes vibrations in the environment may adversely trigger one or more transducers. Resolution is limited (80 to 90 touch points); response is very rapid.

Infrared.   The principle used is interruption of a beam of infrared (IR) light. The system consists of a series of IR light-emitting diodes (LEDs) and phototransistors arranged in an optomatrix frame. The frame is attached to the front of the display. A special bezel is used to block out ambient light. An IR controller scans the matrix, turning on LEDs and the corresponding phototransistors. This is done about 25 times per second. Characteristics of IR systems include reasonable cost, any type of stylus may be used, and response is very rapid.

Mouse and Track-Ball Input Devices.

These devices are used to drive a cursor on the graphics screen. To move a cursor, the mouse is moved across a hard surface, with its direction and speed of movement proportionally related to the cursor's movement on the display screen. A track ball operates on the same principle, but the movement of a ball mounted on bearings within the track ball positions the cursor. It can be thought of as an upside-down mouse. Both the mouse and the track ball have one or more buttons that are used for control. One button is used to indicate when the cursor has reached the proper location. One button is used to activate or deactivate the device. The screen cursor can be used to make menu selections from the display or to place graphic elements.

Digitizer. This has been a common input device, especially for computer-aided design (CAD) systems. Three main components of a digitizing tablet are pad, tablet, or surface; the positioning device that moves over the surface; and electronic circuitry for converting points into x–y coordinates and transmitting them to the computer. The device can be used to place a graphic element on the screen, make a menu selection from the screen, or make a menu or symbol selection from the digitizing tablet itself.

Joystick. Identical in function to the joystick supplied with home computer games, these devices are sometimes preferred when working with three-dimensional drawings because of their free range of motion. The joystick works on the same general principles as the mouse and track ball.

Light Pen. One of the very early means for interacting a narrow beam of light with photoreceptors behind the screen, it is similar in principle to some of the other systems described.

VISUAL DISPLAY DEVICES

Cathode-Ray Tube. The operating principles of this ubiquitous display device were well developed and the device was thoroughly seasoned by the time of its first use in industrial instrumentation and control systems. Of course, for many years it has been used in the form of the oscilloscope, where it enjoys a leading role in test instrumentation. In terms of process control and factory applications, it made its first appearance in CATV (closed-circuit television) applications, where it was used as a "seeing eye" for monitoring remote operations. Then it was later used in some of the earliest attempts to achieve several basic advantages of computer-integrated manufacturing.

The immense flexibility of the CRT is demonstrated by Figs. 3 and 4. Also as evident from other fields of CRT usage, such as animation and CAD applications, the CRT seems almost unlimited in its potential when in the hands of ingenious programmers and software engineers. See the articles, "Distributed Control Systems" in Section 3 and "Industrial Control Networks" in Section 7 of this handbook.

Liquid-Crystal Displays. Introduced several years ago, liquid-crystal displays (LCDs) are achieving a measurable degree of design maturity, particularly after the addition of numerous colors, as indicated by the many thousands of units that are installed and ordered each year. LCDs are nonemissive and therefore require good ambient light at the proper angle for good viewing. The physical chemistry of LCDs is a bit too complex for inclusion here. However, a cross-sectional view of a typical LCD unit is shown in Fig. 5.

Electroluminescent Panels. These displays use a thin film of solids that emit light with an applied electric field. Because they are monochromatic, small, and expensive, they usually are applied when applications require a highly readable, lightweight, low-power-consuming visual display unit with very rapid update times. Design research in this field is continuing.

Vacuum Fluorescent Panels. These display units create thin images by accelerating electrons in a glass envelope to strike a phosphor-coated surface. They are bright, available in multiple colors, and have fast update times. Usually they are small in size and, therefore, find applications that require low-information displays, such as reprogrammable control panels or video-control stations, where graphics are simple and space is a premium consideration.

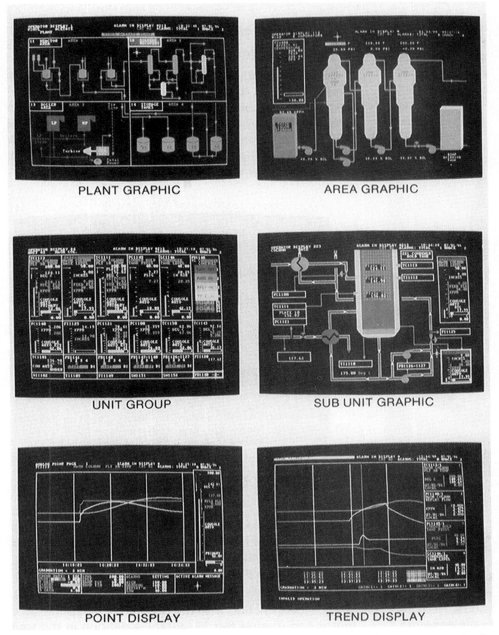

FIGURE 3 Depending on the operator's needs and preferences, the available area on a CRT screen can be used in total for a given "scene," or, more commonly, the area will be divided into windows. These arrangements are accomplished through the software program. Via menu-driven techniques the entire operation can be brought into view (plant graphic), followed by close-ups of areas, units, and subunits. Further, point and trend displays may be brought to the screen. Interactive graphics can be planned into the program wherever desired.

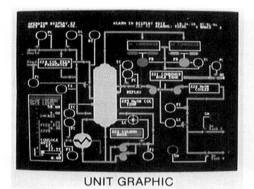

UNIT GRAPHIC

SUB UNIT GROUP

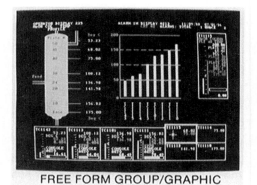

FREE FORM GROUP/GRAPHIC

FIGURE 3 *(Continued)* Depending on the operator's needs and preferences, the available area on a CRT screen can be used in total for a given "scene," or, more commonly, the area will be divided into windows. These arrangements are accomplished through the software program. Via menu-driven techniques the entire operation can be brought into view (plant graphic), followed by close-ups of areas, units, and subunits. Further, point and trend displays may be brought to the screen. Interactive graphics can be planned into the program wherever desired.

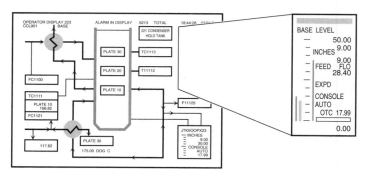

FIGURE 4 By way of software, a portion of a CRT display, namely, a window, can be singled out for detailed viewing. In the X-Window scheme, one or more windows could be designated as "dedicated" for use on an X-Network, making it possible for several stations along the network to obtain instant information that may be of interest to several operators in a manufacturing or processing network.

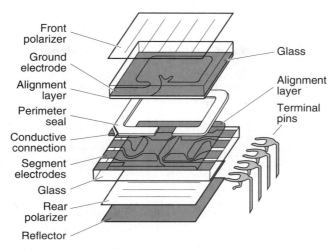

FIGURE 5   Cross-sectional view of typical TN (twisted nematic) liquid-crystal display with transparent indium–tin oxide electrodes patterned on glass substrates using photolithography techniques. (*Standish Industries, Inc.*)

INDICATORS, RECORDERS, AND LOGGERS

Information from a data-acquisition system (DAS) may be categorized as follows:

1. Dynamic data, of an immediate, essentially instantaneous value, are used as the basis for very short-term attention that pertains to the fully automatic or semiautomatic operation of a machine or process, that is, to determine whether operator intervention is required. Indicators of various kinds serve this need.

2. Historical data, of a longer-term value, are used as the basis for analytical and statistical objectives. Recorders and loggers fulfill these requirements.

During recent years, based on the advent of microprocessors and other solid-state components, the definitions of indicators, recorders, and loggers have been compromised by a markedly competitive field of suppliers who have created instruments and systems that commingle the aforementioned DASs in various commutations and permutations, to the extent that common architectures are difficult to categorize. This observation, however, should not detract from the fact that hybrid combinations can and do serve a useful purpose for application-specific needs. But it is more difficult to select the most applicable equipment today.

INDICATING MEANS[1]

The two fundamental indicating means used are analog and digital. Because so many process variables are analog, the analog indicator predates the digital by several decades. There were, however, a few exceptions, such as counters, where the input was directly digital. With advances in electronic circuitry and display techniques made available since the 1970s, there has been a strong trend toward digital indicators.

Analog Indicators. These require a calibrated scale and a pointer which moves horizontally or vertically with reference to a stationary scale (Fig. 1), or the scale may move with reference to a stationary pointer. The pointer need not be a "pointer" in the literal sense, but may take the form of a moving ribbon or an overlay that provides the indication in the form of a bar graph (Fig. 2). Analog indicators traditionally have been somewhat difficult to read and interpret, especially between graduations. Thus a few instruments, such as the hand-held micrometer, incorporate a vernier scale. Sometimes magnification is provided. In process indicators there is commonly an adjustable pointer beside the variable indicator to indicate set point and alarm points. Analog indicators do not require analog-to-digital conversion.

Digital Indicators. With the availability of solid-state electronics and improved display techniques, a strong preference developed for digital indicators. Very early digital indicators utilized curved elements of an incandescent type lamp to make figures and letters. These were soon replaced by light-emitting diodes (LEDs), liquid-crystal displays (LCDs), and, of course, cathode-ray tubes (CRTs) if they were part of a video screen.[2] A large variety of stand-alone digital indicators is available today. They are exceptionally attractive for desktop and laptop applications and for their portability. The display is all-electronic, having no sensitive moving parts per se. Such indicators can be connected easily to a number of different process inputs, or they may be incorporated into an indicator dedicated to a single variable. Digital indicators sometimes are likened to "illuminated signs." They can be easily switched from current variable reading to set-point values, alarm values, and so on (Fig. 3).

RECORDING MEANS

Contemporary recorders have several configurations and are obtainable with numerous options. Thus they are not easy to categorize.

Physical Configuration. Traditionally two broad classes have been recognized: (1) strip-chart recorders and (2) circular (round)-chart recorders. Strip-chart recorders plot the measured variable (horizontally) and time (vertically) on a strip of especially marked graph paper that is selected for the type of marking medium used. The rectilinear coordinates used are easy to read, but sometimes difficult to store because of their bulk. A strip-chart recorder can be found for almost every application. Circular (round)-chart recorders plot the measured variable along the arc of a curve and time on circular lines that emanate from the center (hub) of the instrument. By far the most common marking medium is an inked pen. Round-chart instruments still are used, but are less popular for many advanced applications because of their lack of flexibility. When used, they are usually associated with recording process variables rather than other phenomena. Contrast the foregoing observations by referring to Fig. 4.

---

[1] For those handbook users who are experienced in a computer-integrated manufacturing (CIM) environment, some of the instruments shown in this article may appear "old fashioned," and indeed these designs have been in service for many years. This, however does not detract from the fact that there are many more non-CIM settings than there are the ultramodern ones. Demand statistics and continued listings in catalogs prove this.

[2] Display methodologies are described in a prior article in this handbook section.

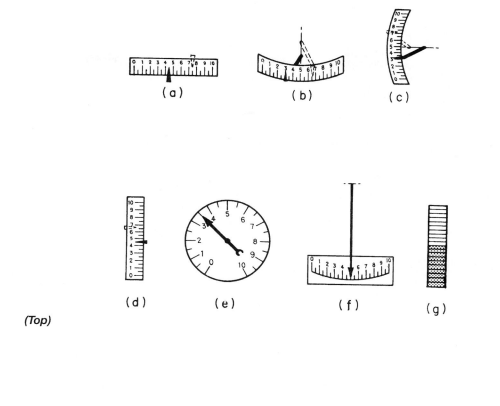

(Top)

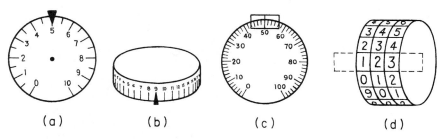

(Bottom)

FIGURE 1   Pointer and scale configurations used in analog indicators. (*Top*) Moving-pointer indicators. (*a*) Straight scale (horizontal). (*b*) Arc scale (horizontal). (*c*) Arc scale (vertical). (*d*) Straight scale (Vertical). (*e*) Circular scale. (*f*) Segmental scale. (*g*) Bar graph where, instead of a moving pointer, there is a moving semi-transparent ribbon (usually colored), the top of which indicates the present value, or several individually illuminated, vertically arranged solid-state display elements may be used. (*Bottom*) Moving-scale indicators. (*a*) Moving dial. (*b*) Moving drum. (*c*) Precision indicator with moving dial. (*d*) Drum counter.

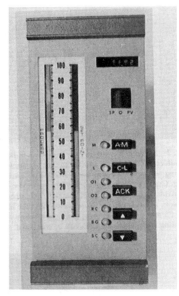

FIGURE 2 Indicating controller in which moving-ribbon bar-graph displays are used for process variable and set-point indications. (*Robertshaw.*)

Recorder Application. There are two principal fields for which recording instruments are used: (1) process measurements, including temperature, pressure, flow, and so on, and (2) scientific and engineering applications, statistical analysis, and high-speed testing and other operations where a graphic record is desired in addition to a digital record.

PROCESS RECORDERS

In modern CIM-type environment stand-alone recorders have largely been replaced by CRT displays or loggers with recording facilities. There are, however, applications in small or isolated processing units, where there may be requirements for documenting sustained temperatures, for example, in cold warehousing of perishables and in the pasteurization of dairy products. There also are users who still prefer to instrument their fairly noncomplex processes with a group of stand-alone instruments. See the article, "Stand-Alone Controllers" in Section 3 of this handbook. Thus a demand exists for panel-mounted process recorders of the types shown in Figs. 5 and 6.

FIGURE 3 A wide variety of digital indicating instruments, ranging from small hand-held indicators to other portable and bench-top indicators, are available for accepting inputs from thermocouples, RTDs, strain gages, and other process transducers. Commonly used display elements include liquid-crystal displays (LCDs) and light-emitting diodes (LEDs). (*Emerson Electric.*)

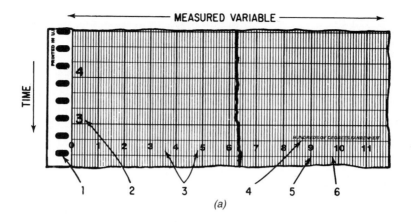

(a)

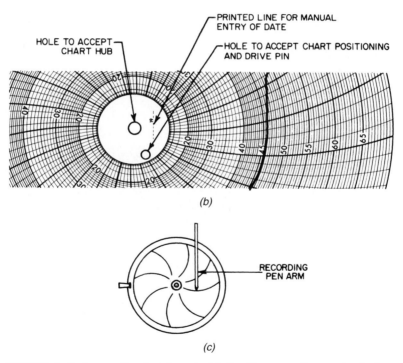

(b)

(c)

FIGURE 4 Basic recording-chart formats. (*a*) Rectilinear coordinates of strip-chart recorder. (1) Humidity-compensating holes; (2) major time markings; (3) minor graduations; (4) identifying calibration; (5) major graduations; (6) minor time markings. (*b*) Curvilinear coordinates of circular (round)-chart recorder. (*c*) Section (reduced) of a 12-inch (305-mm) chart; right—pen-arm travel.

FIGURE 5 Modern version of circular-chart recorder-controller. The instrument features microprocessor-based technology to provide improved accuracy, reliability, simple operation, and reduced maintenance. The instrument may be used as a stand-alone instrument or networked via serial data interface to a supervisory computer. Accepts inputs from a wide variety of process transducers, including thermocouples, RTDs, current-loop frequency, and pulse inputs, among others. The three output control channels can be configured to record only with alarms, perform on-off, or full PID control. Flow compensation and totalization are selectable on flow-input versions. Available in wall-, panel-, or post-mounted versions. The instrument features simplified menu prompts and front-panel membrane key operation. Overall accuracy is ±0.25 percent (operational span) with measurement resolution better than ±0.1 percent (span) for zero-based ranges. (*ABB Power Automation Inc.*)

RECORDERS FOR OTHER NEEDS

xy Recorders. These instruments are used where it is desired to plot the relationship between two variables, $y = f(x)$, instead of plotting each variable separately as a function of time. An *xy* recorder closely resembles and functions like a single-pen recorder, except that the chart (*y* axis) is moved in response to changes in a variable instead of at a uniform time rate. The chart may be driven by a separate servo-actuated measuring element similar to

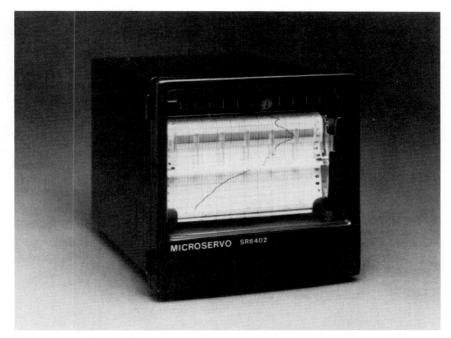

FIGURE 6 Compact industrial strip-chart recorder for panel mounting or desktop use. Conforms to DIN case standards. Records continuously for 30 days. Chart speeds range from 1 mm/s to 30 mm/h. Uses a dispensable fiber-tip pen. Accepts a single range from 1 mV through 100 mV, 1 volt, 10 volts, 1 to 5 volts, and 4 to 20 mA. Uses a Z-fold chart paper; record width 100 mm with 50 graduations; pen response speed 0.5 s (full-scale); linearity ±0.3 percent (full-scale); resolution ±0.2 percent (full-scale); utilizes a dc servo. (*Western Graphtec, Inc.*)

that used in positioning the recorder pen, or by a self-synchronous motor in a remote control system. In most applications for the xy recorder a square-shaped graph sheet is used. Thus the full-scale chart travel (y axis) is made equal to the full-scale pen travel (x axis). The addition of a second servo-actuated measuring element and recording pen to an xy recorder yields an xx_1y or an xyy_1 recorder, in which the relationship among three associated variables can be plotted.

For laboratory use of an xy recorder it may be desirable to have the horizontal or x axis driven as a function of time. Operated in this way, an xy recorder may be thought of as a low-frequency direct-writing oscillograph (Fig. 8).

Recorders for Process and Product Testing and Research. Although recorders are quite versatile in their application, some designs are more customized for laptop and desktop usage for investigating various phenomena that may arise in the scientific and industrial laboratory (Figs. 9 and 10).

CHART-MARKING MEANS

Pen and Ink. This recording methodology was developed in the early part of this century to replace the stylus-scratched records on a smoked surface, as once used in seismographic recorders. A large investment has gone into improving pens, inks, and filling systems. Refilling

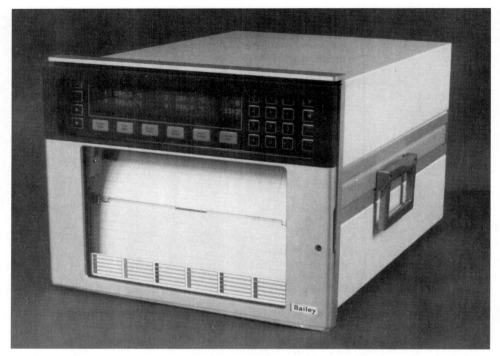

FIGURE 7   Multipoint strip-chart process recorder. The instrument features a high-speed forward and reverse chart drive with chart speed and printing rate selection. The instrument prints in six colors by means of a fast nine-needle print head. Data are displayed on the recorder mainframe for four channels at one time. The instrument scans and prints up to 30 channels per second, making it suitable for flow and pressure loops. Features RS-232C/RS-422 serial communications port. Math pack is available for derived variables computations. Sixty alarms are freely assignable and operate on rate of change as well as absolute value and deviation to provide full process monitoring and protection. (*Bailey Controls.*)

today has been made much easier through the use of ink cartridges and replacement pen assemblies. Almost every type of pen-point design has been tried. Popular contemporary designs are needle and fiber-tipped pens. Thus pen-and-ink recording remains widely accepted and, indeed, is a mature technology. Much effort also has gone into designing improved papers.

Impact Printing.   This technique also has existed for many years. Print heads or print wheels are particularly adapted to multipoint records where symbols are used rather than penned lines for each variable. The technique has been widely used in data loggers.

Thermal Writing.   This technique, although much more recent, has gained several years of experience and is used in some contemporary recorders. In this method thermally sensitive paper and a heated stylus are used (Fig. 11).

Thermal Array Recording.   Recorders of this type have no moving parts in the writing mechanism, thus eliminating problems of inertia and hysteresis. Thin-film resistors in the thermal head conduct heat generated by energizing specific resistors (styli) to the thermosensitive paper contacting the head. The chart paper is coated with a white thermosensitive layer (a few micrometers thick) that is made up of a colorless paint and a colorizing agent. Upon application of heat to the paper, the colorizing agent dissolves instantly. This

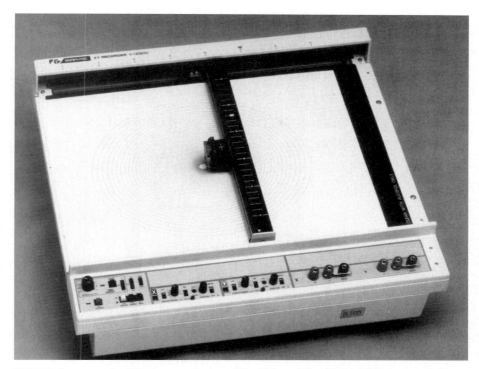

FIGURE 8   *xy* recorder. Effective recording area 381 × 254 mm (15 × 10 inches). This is controlled by an electronic limiter; dc servo; fiber-tip recording pen; chart paper available in separate sheets or rolls; electrostatic (spots of light) used to align paper. Pen response: slewing speed $X = 1500$ mm/s, $y = 1500$ mm/s; maximum acceleration $X = 3G$, $Y = 4G$. Pen lift 10 Hz; remote controls for pen (up/down); time sweep (start/reset); servo (on/off); paper (hold/free); input of x, y axes and x, y zero setting; dimensions 483 mm (19 inches) wide, 440 mm (17.3 inches) deep, 162 mm (6.4 inches) high. (*Western Graphtec, Inc.*)

causes a chemical reaction that colors the paint. The technique requires that the head be controlled to adjust rapidly (almost instantaneously) the amount of energy delivered to each element in the array. This can be solved by a technique known as "dot skipping," or controlling head temperature by "micropulsing." The instrument previously shown in Fig. 9 features thermal array recording.

Electrostatic Recording.   This method utilizes an array device in the imaging head and, like the thermal-array techniques, this eliminates inertia, hysteresis, and overshoot. Frequency response up to 35 kHz is achievable. Copper bar shoes are placed on either side of the imaging head. As pointed out by D. MacLennan (*Gould Inc.*), "When a positive charge is applied to specific shoes, and a negative charge is applied to certain array points that lie between the shoes, a resultant negative point charge is applied to the electrostatic paper. A toning head allows positively charged toner to flow over this area of the paper, depositing toner where the paper was charged. A vacuum knife then removes any excess toner, leaving the charged-particle image."

Ultraviolet Writing.   Mainly for oscillographic use, the system utilizes a halogen or ultraviolet light beam. This beam is reflected off a moving mirror onto photosensitive paper. The mirror, mounted on a galvanometer, has much lower mass than a pen motor, and thus the frequency response of the system can reach 25 kHz.

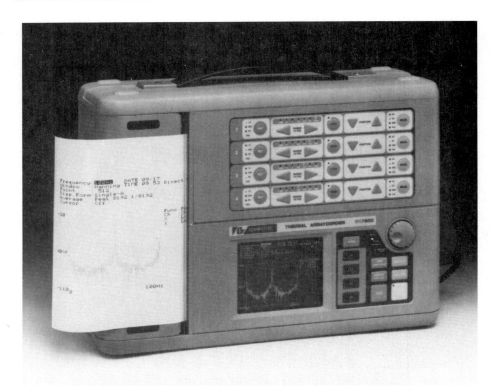

FIGURE 9 Portable recorder for on-site analysis. Incorporates built-in fast Fourier transform function to aid data analysis. Four channels, analog-voltage input types. Thermal printing. Maximum writing width 205 mm (1640 dots); display screen electroluminescent, amber color (95.9 × 76.7 mm; 320 × 256 dots). Chart speeds 1, 2, 5, 10, 20, 25, 50 mm/min; 32-kbyte and 256-kbyte memory cards for filing of setup parameters and storage of measurement data; 14-bit analog-to-digital conversion for highly accurate waveform recording and analysis. Features thermal array recording. (*Western Graphtec, Inc.*)

> ***Fiber-Optic Recorders.*** Available for several years, this method essentially is used in oscillographs (Fig. 12).

HYBRID RECORDERS AND DATA LOGGERS

Not too many years ago a logger was an instrument that presented measurement data in a digital form on a paper tape that advanced after each line was printed. A log report may be defined as a printout (or tearout) of numerical values for various channels to which the instrument is connected. Special functions or situations, such as alarms, were also printed out. This type of instrumentation augmented a recording instrument and was particularly appealing to accountants, engineers, and scientists who prefer to deal with precise digital values rather than line graphs.

Within the last few years traditional recorders have moved toward data loggers in functionality and are termed hybrid recorders because they combine analog trend representation with digital information on the same chart paper without disruption of trend printing. Likewise, traditional loggers have incorporated recorders in their overall systems, as well as con-

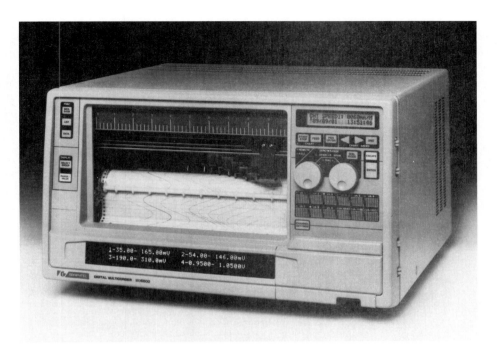

FIGURE 10 Reputedly (1991) the world's smallest 12-channel pen recorder. Features 6, 8, 12 channels of 250-mm (9.8-inch) width. Programmable inputs: 17 voltage, 11 thermocouple, and 1 RTD; digital servo drive; twin dials and backlit liquid-crystal display (LCD) for interactive programming; memory cards for filing of setup parameters and storage of measurement data; interface for data transmission and computer control; one-touch pen exchange; external alarm output. (*Western Graphtec, Inc.*)

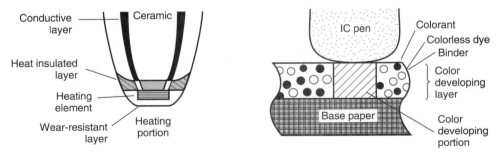

FIGURE 11 Thermal writing entails the use of a hot stylus tip and a special thermally sensitive paper. Details of the stylus are shown at left. Cross section of paper with stylus in contact is shown at right. The stylus has low thermal inertia, small contact area, and hardness much greater than steel. It supplies sufficient thermal energy to burst the microscopic capsules in which dye is embedded in the paper. To ensure long life of the specially constructed stylus (pen), a high-density ceramic is used to protect the thick-film heating element. (*Western Graphtec, Inc.*)

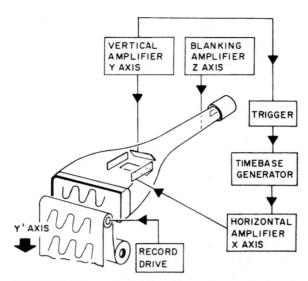

FIGURE 12   Fiber-optic-type recording oscilloscope (*Honeywell*) can record over 30 channels. Records that are relatively insensitive to normal room light are produced by ultraviolet light and directed through a special fiber-optic faceplate onto photosensitive direct print paper. Records are produced immediately without the delays associated with wet processing or toner application. A fiber-optic CRT faceplate consists of millions of glass fibers, each only a few micrometers in diameter, fused together within a 5- by 200-mm (0.2- by 8-inch) area. These fibers provide efficient transfer of ultraviolet light from the phosphor inside the faceplate to the photosensitive paper, enhancing writing speed and trace resolution. Inasmuch as the recording paper is in intimate contact with the faceplate, additional optical elements are not required.

nections to PCs and data highways. Perhaps "hybrid logger" would be an apt term to use for them. Contemporary loggers have borrowed heavily from the elements of instrumentation and electronics technology, emphasizing network connectability, friendly relations with minicomputers and PCs, and with recorders. Modern data loggers are heavily laced with visual displays, including CRTs, and perform complex mathematical operations on incoming data. Ample software is available today to make a logger a workhorse of data acquisition, analysis, and manipulation.

ANNUNCIATORS AND HAZARD MANAGEMENT

by Technical Staff*

Annunciators are industrial alarm systems which can monitor field (trouble or signal) contacts or analog signals directly. An annunciator is dedicated to a specific function, that of furnishing alarm capability should something go wrong in an industrial process.

PERSPECTIVE

Alarms and annunciators have been necessary parts of industrial instrument and control systems history. Because of the magnitude and large scale of processes and manufacturing plants, including the numerous variables involved, the warning of dangerous conditions or trends is brought to the attention of operators via the plant's instrument and control system.

Two types of annunciator systems are commonly used today: (1) separate, dedicated annunciators; and (2) alarm-condition displays that are integrated with other operational information on an operator's screen, usually a cathode-ray tube (CRT).

In critical safety situations requiring alarm recognition and reliability, both of the foregoing approaches may be used to provide visual alarms via redundancy. This article focuses mainly on the dedicated annunciator, with some later discussion of CRT screens under "Computer Alarms."

TYPES OF ALARMS

Five types of alarms are suggested for inclusion in an optimized plant-monitoring system:

1. *Most critical alarm.* Alarms that require prompt operator action in order to maintain the unit on line (conditions resulting in a complete loss of load), for protection of major equipment, or for safety of personnel.
2. *Less critical alarm.* Alarms that require prompt operator action in order to maintain the unit load (conditions resulting in a partial loss of load) or to protect equipment.
3. *Noncritical alarm.* Alarms that require corrective action, but not directly by the operator.
4. *Status information.* Displays that indicate the status of events, and which do not require corrective action.
5. *Trip analysis information.* Conditions directly related to, or that can lead up to, tripping of the unit. This category also includes information from special pretrip or postrip logs in order to determine the actual unit condition more readily before a restart.

It is further suggested that the foregoing different categories be displayed on different devices. Most critical alarms (red) and less critical alarms (white) would be displayed on

---

* Panalarm Division, Ametek, Inc., Skokie, Illinois.

visual annunciators; noncritical information would possibly be displayed on a CRT. Status information should be available as hard copy or possibly on a separate CRT. Trip analysis would be hard copy.

Further observations on an optimized alarm system include the following.

1. Visual flashing lights accompanied by an audible warning may be the best way to convey the urgency of critical alarms to operating personnel.

2. Plant operators require many more data on the events that indicate the status of the plant than can be, or should be, displayed with critical alarms.

3. Alarm and event data should be permanently logged along with the exact time and sequence for an operations event analysis and safety review.

The foregoing is discussed later in this article.

SEQUENCE OF OPERATIONS

Annunciators are usually made up of a lamp display cabinet with nameplates incorporating engraved messages, an audible device, manually operated push buttons, and sequence logic circuits (Figs. 1 and 2). These circuits are used to coordinate the response of lights, audible device, and push-button operation to the action of the alarm circuits being monitored. This is called the "sequence."

Typically, an annunciator sequence may proceed as follows: During normal, all visual and audible devices are quiescent. Upon an abnormality (off-normal or alarm condition), an audi-

FIGURE 1 Contemporary standard unit that combines I²L custom logic with microprocessor technology. Five capabilities include standard annunciation, integral or remote architecture, input-output functions for user customizing, meter set (digital readout and point selection), analog monitoring, and blind set analog monitoring. Reflects the status of conditions at the time this photo was taken. Some lightboxes are white; others are red; and others are yellow-orange. (*Panalarm Division, Ametek, Inc.*)

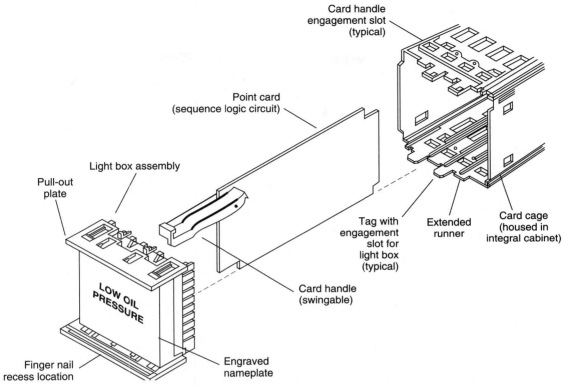

FIGURE 2   Partial sectional view of annunciator unit of Fig. 1. (*Panalarm Division, Ametek, Inc.*)

ble device, such as a horn, will sound. The horn thus advises an attendant or operator that an alert condition exists. The nameplates that flash direct the attendant to the specific points which are in the alarm stage. (Each alarm point is synonymous with the circuit it is monitoring and the associated nameplate with its engraved message—describing the function being monitored.)

Attendant response to the foregoing events involves pressing an acknowledgment push button. This results in silencing the horn as well as changing the flashing lights to a steady on state. The latter will remain illuminated as long as the point remains off-normal. If new points are alarmed, the horn will sound again and the back-lighted windows associated with those alarms will flash. Note that the flashing mode distinguishes newly alarmed points from those off-normal points acknowledged previously and whose lights remain steady on. Upon acknowledgment, once again the audible device is silenced and all points which remain off-normal have steady on lights. An operational (full-function) test can be accomplished by pressing a test push button.

The foregoing is a simplified description of a typical annunciator sequence, in this example, sequence A, as designated by ISA-S18.1 [1]. Sequence A is the most widely used sequence and is considered basic. Its features include lock-in of momentary alarm until acknowledgment as well as automatic reset of acknowledged alarm indications when the condition returns to normal (Fig. 3).

Numerous annunciator sequences are available from equipment manufacturers. Other sequences frequently used are *manual reset, ringback,* and *first out.*

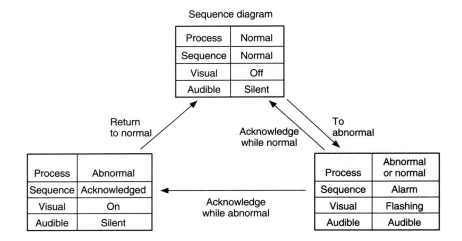

Sequence diagram

Process	Normal
Sequence	Normal
Visual	Off
Audible	Silent

Return to normal

Acknowledge while normal

To abnormal

Process	Abnormal
Sequence	Acknowledged
Visual	On
Audible	Silent

Acknowledge while abnormal

Process	Abnormal or normal
Sequence	Alarm
Visual	Flashing
Audible	Audible

Sequence table

Line	Process condition	Pushbutton operation	Sequence state	Visual display	Alarm audible device	Remarks
1	Normal	—	Normal	Off	Silent	
2	Abnormal	—	Alarm	Flashing	Audible	Lock-in
3A	Abnormal	Acknowledge	Acknowledged	On	Silent	Maintained alarm
3B	Normal		To line 4			Momentary alarm
4	Normal	—	Normal	Off	Silent	Automatic reset

Sequence features

1 — Acknowledge and test pushbuttons.
2 — Alarm audible device.
3 — Lock-in of momentary alarms until acknowledged.
4 — The audible device is silenced and flashing stops when acknowledged.
5 — Automatic reset of acknowledged alarm indications when process conditions return to normal.
6 — Operational test.

FIGURE 3   Sequence A (automatic reset) diagram and sequence table. (*Instrument Society of America.*)

Manual reset requires a reset push button. As its name implies, the off-normal condition (light steady on) persists until the point has returned to normal *and* the reset push button is pressed.

The ringback sequence provides a distinct visual and audible indication when the condition returns to normal. This sequence also requires a reset push button.

First-out sequences are probably the second most popular option, after the basic sequence A. The first-out type indicates which one in a group of alarm points is operated first. To do this, the visual display indication for the point that alarmed first is distinct from the visual display indications for subsequent alarm points—for example, an intermittent flash (first point to alert) as opposed to the fast flash (subsequent point to alarm) (Fig. 4).

Usually the sequence dictates the choice of the alarm module, which is a plug-in assembly containing the sequence logic circuit. However, many manufacturers offer multifunction

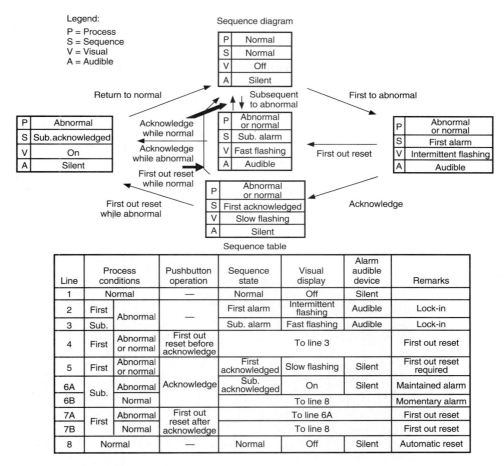

FIGURE 4 Sequence F3A (automatic reset first out with first out flashing and reset push button) diagram and sequence table. (*Instrument Society of America.*)

alarm modules which are capable of giving numerous variations of the basic sequence. These may include a signal lock-in option, lamp test in lieu of operational test, and normally open or normally closed signal contact options.

Some manufacturers offer sophisticated annunciators with built-in option capability of accepting analog signals directly, obviating the need for an interposing switch contact of the device sensing the process condition. This capability includes acceptance of thermocouple, resistance temperature detector (RTD), and thermistor outputs, or 4- to 20-mA or 1.5 volt instrument signals.

Annunciators are available in various sizes and configurations. A compact version can be obtainable in a given size with audible device, power supply, logic, visual display, and control push buttons in one common housing. Alternatively, all of the components may be separate such that the system is identified as remote type.

The evolution of modern conventional annunciators has progressed from the electrical relay type to solid-state to microprocessor-based logic systems. Packaging has improved. Cabinets can accommodate future expansion by changeover of nameplate light boxes and logic alarm modules to double, triple, and even quadruple the initial number of points specified.

Many graphic displays (sometimes identified as mimic style map boards) can incorporate the annunciator function powered by remote-mounted logic boards. Figure 5 illustrates a typical mosaic graphic display for a water treatment plant.

FIGURE 5   Typical mosaic graphic display for water treatment plant. (*Panalarm Division, Ametek, Inc.*)

An event recorder in earlier days was referred to as a recording annunciator since its main function is to provide hard copy of time-tagged events, which usually are an alarm or a return to normal. The hard copy presents an exact history of the events leading to the trip or shutdown, tagged to the nearest millisecond. Frequently a shutdown occurs, the cause of which cannot be determined by plant management, engineers, or maintenance personnel. The shutdown may not necessarily be of long duration, but the disruption could lead to product loss or plant or personnel hazardous conditions. Human error can enter into the picture. Many shutdowns are caused by incorrect operating procedures. Restart by the operator can obfuscate the cause.

Compressor shutdowns are a typical example. Such shutdowns during alarm conditions are necessary because of their interactive role in the plant process to preclude dangerous con-

FIGURE 6 Sequential-event recorder incorporating internal printer and various control push buttons for operator use. (*Panalarm Division, Ametek, Inc.*)

ditions and to prevent hazardous situations contributing to product spill, environmental hazard, or safety of plant personnel. While many compressors have their own monitoring display panels which can indicate shutdown, in many installations these are insufficient since the alarms may be cleared prior to accurate diagnosis. In a case of this kind, the event recorder can be a diagnostic tool for determining the exact sequence of events, some of which may have occurred upstream, that is, ahead of the final action. Figure 6 shows a typical sequential-event recorder that incorporates an internal printer and includes various control push buttons for operator use. Application-specific alarm systems, usually self-contained, are available (Fig. 7).

FIGURE 7   Eight-point analog alarm monitor, which can monitor temperature or instrument signals directly with dual-set-point capability and liquid-crystal display (LCD) with two lines, 16 characters per line, for description of channel and trip set points versus actual field reading. (*Panalarm Division, Ametek, Inc.*)

COMPUTER ALARMS

In past years it was not uncommon to use conventional annunciators with hundreds or even thousands of dedicated alarm windows for large installations. As the use of computers increased, annunciators were downsized to a few critical points since the alarm functions were simply incorporated into the computer via software. The alarms were displayed on a CRT screen.

Alarm recognition of costly plant mishaps remains crucial in any plant, process, or utility, so the choice between conventional light-box annunciators and CRT-type alarms must frequently be made, particularly as the price of computer power trends downward. Plant users faced with the choice between alarm window and CRT alarm may opt for a combination of both.

Dedicated annunciators have certain advantages over CRT alarm screens and vice versa. Conventional annunciators have undergone their own evolution through the use of micro-pressor techniques, adding input capability while retaining some key features, such as being direct-wired to the sensors, fixed configuration familiarity, and singular dedication to the alarm function. CRT-type alarms can be custom-designed and provide a tremendous amount of useful information on the screen. However, the CRT types first must be programmed, at some cost, and CRT alarms have been known to cause operator confusion and delays in reaction or recognition.

In either case, window or CRT, sensors must be hard-wired to the inputs, whether direct or serially strung. Present conventional annunciators allow expansion within a given overall panel size if the architecture is selected correctly at the initial design. Further, a properly chosen annunciator with RS-422/RS-232 communication port options, plus appropriate software from the manufacturer, can be coupled to a computer to provide a powerful combination.

The trend toward combining a computer system with a dedicated annunciator system is supported by a recent report summary issued by the Electric Power Research Institute [2], which states: "Conventional lightbox systems allowed operators to obtain information from

TABLE 1 Factors in Selecting Annunciator Systems

Characteristic	Annunciator	CRT alarm	Combined system*
System is independent	✓	Not independent	✓
Dedicated alarm system	✓	A secondary consideration; depends on computer objective.	✓
Intentional redundancy (overlapping reliability)	Depends on particular design.	—	✓
Alarm recognition	✓	Possible; depends on particular design.	✓
Cost-effectiveness	Depends on particular application, size, and location.	Depends on cost of software.	✓
Alarm priority designation	✓	Depends on program.	✓
Possibilities for expansion	Varies with manufacturer.	May require additional hardware and change in software.	✓
Graphics	Varies; mosaic tile graphic displays available.	✓	✓
Configuration is familiar (visual organization)	✓	Possible; depends on several factors—software, number of screens, speed of call-up.	✓
Communication with computer	Depends on manufacturer.	✓	✓
Direct-wired	✓	Requires input hardware.	Replaces input hardware.

* Combined system assumes that annunciator has communication port and software availability.

the alarm system more quickly and accurately than alarm lists on CRT screens," and "combined lightbox and CRT systems demonstrated that list of highlighted alarms on the CRT screen provided a beneficial adjunct to the main lightbox displays."

Table 1 can be helpful as a checklist. A simple "yes" or "no" answer may not be applicable to a particular plant application. The instrument and control system designer must make a judgment call concerning the type of system or combination that will be most beneficial.

REFERENCES

1. ISA-S18.1, "Annunciator Sequences and Specifications," Instrument Society of America, Philadelphia, Pennsylvania.
2. Electric Power Research Institute, Palo Alto, California.

SECTION 9
VALVES, SERVOS, MOTORS, AND ROBOTS*

Mark Adams
Power Industry Marketing Manager, Fisher Controls International, Inc., Marshalltown, Iowa (Process Control Valves)

Len Auer
Product Marketing Manager, Rosemount Inc., Eden Prairie, Minnesota (Current-to-Pressure Transducers for Control-Valve Actuation)

Allen C. Fagerlund
Senior Research Specialist, Fisher Controls International, Inc., Marshalltown, Iowa (Control Valve Noise)

D. Kaiser
Design Engineer, Compumotor Division, Parker Hannifin Corporation, Rohnert Park, California (Servomotor Technology in Motion Control Systems)

J. J. Kester
Bodine Electric Company, Chicago, Illinois (Stepper and Other Servomotors)

S. Longren
Longren Parks, Chanhassan, Minnesota (Servomotors)

Richard H. Osman
Engineering Manager, AC Drives, Robicon Corporation, Pittsburgh, Pennsylvania (Solid-State Variable-Speed Drives)

C. Powell
GMFanuc Robotics Corporation, Auburn Hills, Michigan (Robots)

---

* Persons who authored complete articles or subsections of articles, or otherwise cooperated in an outstanding manner in furnishing information and helpful counsel to the editorial staff.

Marc L. Riveland
Senior Engineering Specialist, Fisher Controls International, Inc. Marshalltown, Iowa (Control-Valve Cavitation)

A. J. Statz
Compumotor Division, Parker Hannifin Corporation, Rohnert Park, California (Servomotor Technology in Motion Control Systems)

PROCESS CONTROL VALVES

by Mark Adams[1]

The control valve, or final control element, is the last device in the control loop. It takes a signal from the process instruments and acts directly to control the process fluid. Control valves main-

---

[1] Power Industry Marketing Manager, Fisher Controls International, Inc., Marshalltown, Iowa.

tain process parameters such as pressure, flow, temperature, or level at their desired values, despite changes in process dynamics and load. Control valves are intimate to the process fluid and must be designed to accommodate its needs and characteristics. Likewise, the control valve is bonded to the process control system and must react to the protocol and needs of the controlling devices. The evolution of control valves to their current state has been a reaction to the combined forces of the processes in which valves are installed and the systems that control them. Evidence of these factors exists in the design of valve bodies, actuators, and interface accessories.

CHRONOLOGY

The modern history of control valves is rooted in the industrial revolution and began with the invention of the industrial steam engine. Steam to power these engines built up pressures, which had to be contained and regulated. Instantly valves, which had existed since the middle ages, took on new significance. The process control systems at that time were simple—the human being. Things changed quickly, however. The invention of the flyball governor heralded a new era of "feedback control" and permanent linkage of process to control valves.

The next leap of technology took place in 1875, when William Fisher invented the self-contained automatic pump governor. This device used pump output pressure to control valves throttling steam flow to the engines of the pump. It was the first process control device to achieve a set point by offsetting process pressure acting on a piston with the force of a mechanical spring. This couple, attached to a valve body, was the basis on which control-valve actuators and control valves later evolved.

For the next 50 years control valves consisted of a variety of self-contained governors (now called regulators), float valves, and lever valves. The most common method of valve actuation was by means of a spring-opposed piston or diaphragm motor (an actuator), which operated directly from the process fluid.

In the mid-1930s pneumatic pressure control instruments began to emerge. The instrument companies coaxed the valve companies to make valve actuators which reacted to standardized pneumatic signals rather than process pressure. The new control instrumentation improved the fidelity of process control dramatically and also necessitated an upgrade of control-valve components. Characterized valve plugs were developed, and the valve positioner made its debut.

As process pressures increased, high-pressure valves were developed. As flow rates increased, larger-capacity valves were developed. Valves changed to cope with process changes. The late 1970s witnessed a wholesale move to centralized electronic control, and control valves were modified to accept analog electronic signals. Today control valves are being modified to live in an environment of distributed digital control. The evolution continues. Control valves are adapting to change in the processes they control and to the instruments which control them.

This article covers the subject of control valves in their major subsegments. First it deals with valve bodies and the criteria for their application, sizing, and selection. These factors are oriented strongly to the needs and specifications of the process. The next segment covers actuation. Included are discussions of actuator styles, distinctions, and selection. This discussion centers on the needs of the control-valve body and the needs of the process and power source. Finally, the article covers actuator accessories and their use to interface valves and actuators to controlling instrumentation.

CONTROL-VALVE BODIES

General Categories of Control Valves

Control valve here means any power-operated valve, whether used for throttling or for on-off control. Varieties from which to select, as listed in Table 1, include sliding stem valves (globe or angle configuration) and rotary valves—ball and butterfly. Other varieties such as motorized gate valves, louvers, pinch valves, plug valves, and self-operated regulators are not considered here. These major types, sliding-stem and rotary, are further divided into a total of 10 subcategories according to relative performance and cost. Despite variations found within each category—such as cage-guiding versus stem-guiding—all valves within a given subcategory may be considered very much alike in the early stages of the selection process. Selecting a valve involves narrowing down to one of these subcategories and then comparing specific valves in that group (Table 2).

Sliding Stem Valves

The most versatile of the control valves are the sliding stems. Globe, angle, and Y-pattern valves can be purchased in sizes ranging from 0.5 to 20 inches (15 to 500 mm) (Figs. 1 to 4). More choices of materials, end connections, and control characteristics are included here than in any other product family. Globe valves are available in cage-guided, post-guided, and stem-guided designs, with flanged, screwed, or weld ends. Economical cast iron as well as carbon steel, stainless steel, and other high-performance body materials are available. Pressure ratings range to American National Standards Institute (ANSI) Class 2500 and beyond. Their precise throttling capabilities, overall performance, and general sturdiness make sliding stem valves a bargain, despite their slight cost premium. The buyer gets a rugged, dependable unit intended for long, trouble-free service. Sliding stem valves are built ruggedly to handle field conditions such as piping stress, vibration, and temperature changes. In sizes through 3 inches (80 mm) incremental costs over rotary valves are low in contrast to the increments in benefits received.

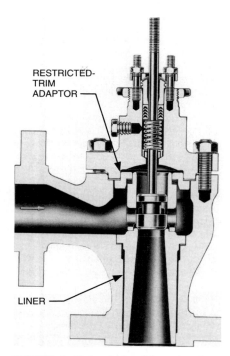

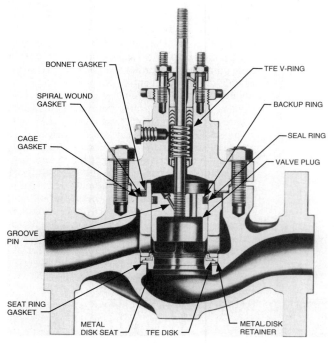

FIGURE 1   Reduced-trim globe valve shows capability for trim reduction in a globe-style valve. In addition, the product features an outlet liner for resistance to erosion. The unbalanced plug provides tighter shutoff, but requires larger actuators than new balanced designs. (*Fisher Controls.*)

FIGURE 2   Standard globe sliding stem valves are available in a broad range of sizes, materials, and end connections. The balanced plug reduces unbalance force and allows use of smaller actuators. A soft seat affords tight shutoff. Valves such as this are the first choice for applications less than 3 inches (80 mm) in size. (*Fisher Controls.*)

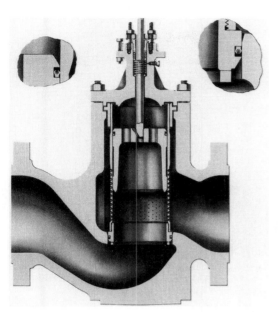

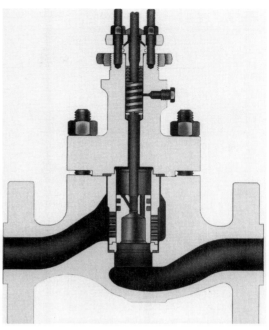

FIGURE 3 Severe service capability in globe valves is demonstrated by this large valve. It features a drilled hole cage to provide attenuation of flow noise by splitting the flow into multiple passages. Hole spacing is controlled carefully to eliminate jet interaction and high resultant noise levels. (*Fisher Controls.*)

FIGURE 4 High-pressure globe valves are typically available in sizes 1 to 20 inches (15 to 500 mm) and ANSI classes 900, 1500, and 2500. These products provide throttling control of high-pressure steam and other fluids. Antinoise or anticavitation trim is often used to handle high-pressure drop induced flow-related problems. (*Fisher Controls.*)

For many extreme applications, sliding stem valves are the only suitable choice. This includes valves for high pressure and temperature, antinoise valves, and anticavitation valves. Due to process demands, these products require the rugged construction design of sliding stem products. An excellent example of a specialty sliding stem valve is a steam conditioning valve. These combination products provide steam pressure and flow control along with desuperheating—all in one unit. They give performance improvement over separate units and require fewer piping and installation restrictions. These and other specialty products are presented in Figs. 5 to 8.

Bar stock valves are small, economical sliding-stem valves whose bodies are machined from bar stock (Fig. 9). Body sizes range from fractions of an inch up to 3 inches (80 mm); flow capacities generally are lower than those of general-purpose valves. End connections usually are flangeless (for mounting between piping flanges) or screwed. The main advantage of this type of valve is that far more materials are readily available in bar stock form than in cast form. Consequently these valves are often used where there are special corrosion considerations. However, their compactness and general high-quality construction make them attractive for flow rates below the range of the regular sliding stem subcategory, even when corrosion is not a consideration. Overall, they are an economical choice when they can be used.

The third subcategory, the lowest-cost products among sliding-stem valves, are called economy bodies. These valves are used for low-pressure steam, air, and water applications that are not demanding (Figs. 10 and 11). Available sizes range from 0.5 up to 4 inches (15 to 100 mm). Body materials include bronze, cast iron, steel, and stainless steel (SST). Pressure classes generally stop at ANSI 300. Compared to regular sliding-stem valves, these units are very simple, their actuators are smaller, and their cost is three-quarters to one-half as much. Severe service trims for noise and cavitation service are not available in these products.

TABLE 1 Principal Selection Criteria and Availability of Generic Valve Styles

Valve style	Main characteristics	Available size ranges, inches (mm)	Typical standard body materials	Typical standard end connection	Typical pressure range	Relative flow capacity
Regular sliding stem	Heavy duty, versatile	½–16 (15–400)	Cast iron, carbon steel, alloy steel, stainless steel	Flanged, welded, screwed	To ANSI 2500	Moderate
Bar stock	Compactness	½–3 (15–80)	Stainless steel, nickel alloys	Flangeless, screwed	To ANSI 600	Low
Economy sliding stem	Light duty, inexpensive	½–4 (15–100)	Bronze, cast iron, carbon steel	Screwed, flanged	To ANSI 300	Low
Through-bore ball	On-off service	1–24 (25–600)	Carbon steel, stainless steel	Flangeless	To ANSI 900	High
Partial ball	Characterized for throttling	1–24 (25–600)	Carbon steel, stainless steel	Flangeless, flanged	To ANSI 600	High
Eccentric plug	Erosion resistance	1–12 (25–300)	Carbon steel, stainless steel	Flangeless, flanged	To ANSI 600	Moderate
Swing-through butterfly	No seal	2–36 (50–900)	Carbon steel, cast iron, stainless steel	Flangeless, lugged, welded	To ANSI 2500	High
Lined butterfly	Elastomer or TFE liner	2–24 (50–600)	Carbon steel, cast iron, stainless steel	Flangeless, lugged	To ANSI 150	High
High performance butterfly	Offset disk, flexible seals	2–72 (50–1800)	Carbon steel, stainless steel	Flangeless, lugged	To ANSI 600	High
Special sliding stem	Custom to application	2–24 (50–600)	Carbon steel, alloy steel, stainless steel	Flanged, welded	To ANSI 4500	Moderate

Ball Valves

There are two subcategories of ball valves. The through-bore or full-ball type illustrated in Fig. 12 is generally used for high-pressure drop throttling and on-off applications in sizes to 24 inches (600 mm). Full-port designs exhibit high flow capacity and low susceptibility to wear by erosive streams. However, sluggish flow throttling response in the first 20 percent of ball travel makes full-bore ball valves unsuitable for throttling applications. Newer designs in full-ball, reduced-bore valves provide better response. Pressure ratings up to ANSI class 900 are available, as are a variety of end connections and body materials. Another popular kind of ball valve is the partial-ball style (Fig. 13). This subcategory is very much like the reduced-bore group, except that the edge of the ball segment has a contoured notch shape for better throttling control and higher rangeability. Intended primarily for modulating service, not merely for on-off control, partial-ball valves are generally higher in overall control performance than full-ball products. They are engineered to eliminate lost motion, which is detrimental to performance. The use of flexible or movable metallic and fluoroplastic sealing elements allows tight shutoff and wide temperature and fluid applicability. Their straight-through flow design achieves high capacity. Sizes range through 24 inches (600 mm). Pressure ratings go to ANSI class 600. Price is generally lower than that of globe valves.

Eccentric Plug Valves

This class of valves combines many features of sliding stem and rotary products. Eccentric plug valves feature rotary actuation. But unlike most rotary valves, this product utilizes mas-

Relative shutoff capability	Noise or cavitation trim option	Available control characteristic	Flow rangeability	Application temperature range	Pressure drop capability	Best economic size range, inches (mm)
Excellent	Yes	Equal percentage, linear, quick opening, special	Moderate	Quite low to very high	High	1–4 (25–50)
Excellent	No	Equal percentage, linear	Moderate	Moderate	Moderate	½–1 (12–25)
Good	Yes	Equal percentage, linear	Moderate	Moderate	Moderate	1–2 (25–50)
Excellent	Yes	Equal percentage	Low	Moderate	Moderate	4–8 (100–200)
Excellent	Yes	Equal percentage	High	Quite low to quite high	Moderate	4–8 (100–200)
Excellent	No	Linear	Moderate	Quite low to quite high	High	4–8 (100–200)
Poor	No	Equal percentage	Moderate	Very low to quite high	Moderate	6–36 (150–900)
Good	No	Equal percentage	Low	Moderate	Low	6–24 (150–600)
Excellent	No	Linear	Low	Very low to quite high	Moderate	6–72 (250–1800)
Excellent	Yes	Custom	Moderate to high	Very low to quite high	High to very high	—

sive, rigid seat design. They exhibit excellent throttling capability and resistance to erosion. Thus many of the good aspects of both rotary and sliding stem designs are combined. Sizes generally range through 8 inches (200 mm) in pressure ratings to ANSI 600. Both flanged and flangeless body styles can be obtained in a variety of materials (Fig. 14).

Butterfly Valves

Butterfly valves are divided into three subcategories, (1) swing-through, (2) lined, and (3) high performance. The most rudimentary is the swing-through design (Fig. 15). Rather like a stovepipe damper, but considerably more sophisticated, this kind of valve has no seals—the disk swings close to, but clear of, the body's inner wall. Such a valve is used for throttling applications that do not require shutoff tighter than about 1 percent of full flow. Sizes range from 2 to 96 inches (50 to 2400 mm). Body materials are cast iron, carbon steel, or stainless steel. Mounting is flangeless, lugged, or welded. Body pressure ratings up to ANSI class 2500 are common, and wide temperature ranges are also available. While a very broad range of designs is available in these products, they are handicapped by lack of tight shutoff.

Need for no or low leakage gave rise to new designs such as the lined and high performance butterfly valves. Lined butterfly valves feature an elastomer or polytetrafluoroethylene (PTFE) lining that contacts the disk to provide tight shutoff (Fig. 16). Since this seal depends on interference between disk and liner, these designs are more limited in pressure drop. Temperature ranges are also restricted considerably because of the use of elastomeric materials. A benefit, however, is that because of the liner, the process fluid never touches the metallic body. Thus these products can be used in many corrosive situations. Elastomer-lined butterfly valves are generally the lowest-priced products available as control valves in medium to large sizes.

TABLE 2 Control-Valve-Characteristic Recommendations for Liquid-Level, Pressure, and Flow Control*

Liquid-level systems

Control valve pressure drop	Best inherent characteristic
Constant ΔP	Linear
Decreasing ΔP with increasing load, ΔP at maximum load > 20% of minimum-load ΔP	Linear
Decreasing ΔP with increasing load, ΔP at maximum load < 20% of minimum-load ΔP	Equal percentage
Increasing ΔP with increasing load, ΔP at maximum load < 200% of minimum-load ΔP	Linear
Increasing ΔP with increasing load, ΔP at maximum load > 200% of minimum-load ΔP	Quick opening

Pressure control systems

Application	Best inherent characteristic
Liquid process	Equal percentage
Gas process, small volume, less than 10 ft of pipe between control valve and load valve	Equal percentage
Gas process, large volume (process has receiver, distribution system, or transmission line exceeding 100 ft of nominal pipe volume), decreasing ΔP with increasing load, ΔP at maximum load > 20% of minimum-load ΔP	Linear
Gas process, large volume, decreasing ΔP with increasing load, ΔP at maximum load < 20% of minimum load ΔP	Equal percentage

Flow control processes

Flow measurement signal to controller	Location of control valve in relation to measuring element	Best inherent characteristic — Wide range of flow set point	Small range of flow but large ΔP change at valve with increasing load
Proportional to flow	In series	Linear	Equal percentage
	In bypass†	Linear	Equal percentage
Proportional to flow squared	In series	Linear	Equal percentage
	In bypass†	Equal percentage	Equal percentage

* Based on a combination of applied control theory and actual experience. (*Fisher Controls.*)
† When control valve closes, flow rate increases in measuring element.

High performance butterfly valves such as the one shown in Fig. 17 are characterized by heavy shafts and disks, full pressure rating bodies, and sophisticated seals which provide tight shutoff. These valves provide an excellent combination of performance features, light weight, and very reasonable pricing. Eccentric shaft mounting allows the disk to swing clear of the seal to minimize wear and torque. The offset disks used allow uninterrupted sealing and a seal ring that can be replaced without removing the disk. High performance butterfly valves come in sizes from 2 to 72 inches (50 to 1800 mm) with flangeless or lugged connections, carbon-steel

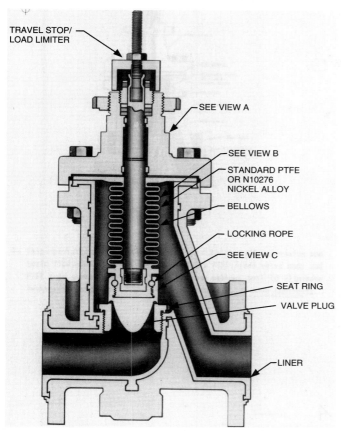

TRAVEL STOP/
LOAD LIMITER

SEE VIEW A

SEE VIEW B

STANDARD PTFE
OR N10276
NICKEL ALLOY

BELLOWS

LOCKING ROPE

SEE VIEW C

SEAT RING

VALVE PLUG

LINER

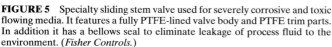

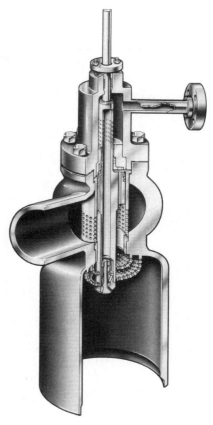

FIGURE 5   Specialty sliding stem valve used for severely corrosive and toxic flowing media. It features a fully PTFE-lined valve body and PTFE trim parts. In addition it has a bellows seal to eliminate leakage of process fluid to the environment. (*Fisher Controls.*)

FIGURE 6   All-in-one steam conditioning valve, which can be used for turbine bypass service or simply to provide control of steam pressure and temperature within one valve body. Design features forged construction and feedforward control of desuperheating spray water flow. (*Fisher Controls.*)

or stainless-steel bodies, and pressure ratings up to ANSI class 600. With their very tight shutoff and heavy-duty construction, these valves are suitable for as many process applications. Advanced metal-to-metal seals provide tight shutoff in applications that are too hot for elastomer-lined valves to handle.

VALVE SELECTION

Picking a control valve for a particular application used to be simple. Usually only one general type of valve was considered—sliding stem valves. Each manufacturer offered a product suitable for the job, and the choice depended on obvious matters such as cost, delivery, vendor relationships, and user preference. Today matters seem considerably more complicated—especially for engineers with limited experience or for those who have not kept up with

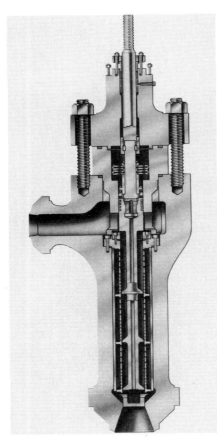

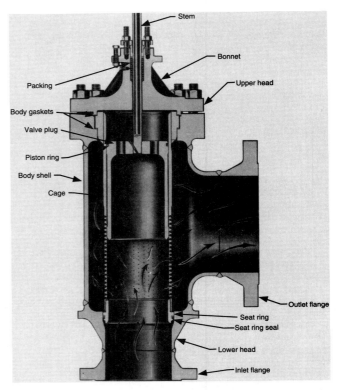

FIGURE 8   Extremely high-flow-rate globe valve applications can be handled by fabricated designs such as the one shown. The design is custom-made to match the required flow rates and piping configuration. It features a drilled hole noise-reduction trim. (*Fisher Controls.*)

FIGURE 7   Special sliding-stem valve used for superheater bypass service in power plants. Application calls for tight shutoff and flows which range from cold cavitating water to flashing water to superheated steam. (*Fisher Controls.*)

changes in the control valve industry. For many applications, an assortment of sliding-stem ball and butterfly valves is available. Some are touted as "universal" valves for almost any size and service, while others are claimed to be optimum solutions for narrowly defined needs. Like most decisions, the selection of a control valve involves a great number of variables. Presented here is an overview of the selection process.

General Selection Criteria

Most of the considerations that guide the selection of the valve type are rather basic. However, there are some matters that may be overlooked by users whose familiarity is limited. A checklist would include the following:

FIGURE 9 Bar stock valves provide economical solutions to small flow requirements. Pressures to 1500 psi (104 bar) and temperatures to 450°F (232°C) can be handled. Compact spring and diaphragm actuators complement these small valve bodies. (*Fisher Controls.*)

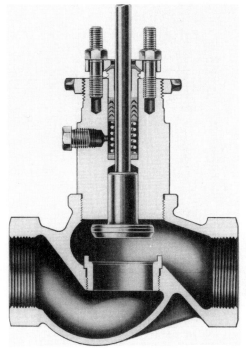

FIGURE 10 Screwed-end bronze body capable of handling many utility applications. It is complemented by a wide variety of orifice sizes and control characteristics. (*Fisher Controls.*)

1. Body pressure rating and limits
2. Size and flow capacity
3. Flow characteristics and rangeability
4. Temperature limits
5. Shutoff leakage
6. Pressure drop (shutoff and flowing)
7. End connection requirements
8. Material compatibility and durability
9. Life-cycle cost

FIGURE 11 Valve style typical of low-cost general-service products. Availability generally extends to 4 inches (100 mm) in size and ANSI class 300. These products feature compact, reversible diaphragm actuators and inexpensive positioners. (*Fisher Controls.*)

Pressure Ratings

Body pressure ratings ordinarily are considered according to ANSI pressure classes. The most common ones for steel and stainless steel are ANSI classes 150, 300, and 600 [1], [2]. For a given body material, each ANSI class prescribes a profile of maximum pressure that decreases with temperature according to the strength of the material. Each material also has minimum and maximum service temperatures based on loss of ductility or loss of strength. For most applications, the required pressure rating is dictated by the application. However, since not all products are available for all ANSI classes, it is an important consideration for selection.

Operating Temperature

Required temperature capabilities are usually also a foregone conclusion, but one which is likely to further narrow the range of selections. Considerations here include the strength or ductility of the body material as well as the relative thermal expansion of the valve internal parts. Temperature limits may also be imposed due to disintegration of soft parts at high temperatures or loss of resiliency at low temperatures. The soft materials under consideration include various elastomers, plastics, and TFE. They may be found in parts such as seat rings, seal or piston rings, packing, rotary shaft bearings, and butterfly valve liners. Typical upper temperature limits for elastomers are in the 200 to 350°F (93 to 177°C) range, and the general limit for PTFE is 450°F (232°C).

Temperature affects valve selection by excluding certain valves that do not have high- or low-

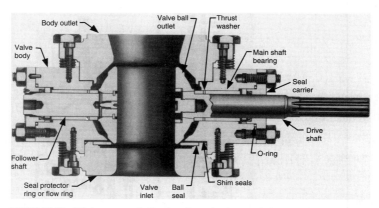

FIGURE 12 High-pressure ball valves featuring heavy shafts and full-ball designs. Design shown is suitable for pressure drops to 2220 psi (152 bar). Class 600 and 900 bodies are available; sizes range to 24 inches (600 mm). (*Fisher Controls.*)

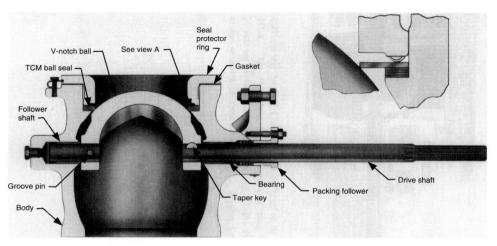

FIGURE 13 Applications to ANSI class 600 can be handled by this segmented ball valve. The flangeless body incorporates many features to improve throttling performance and rangeability. Tight shutoff is achieved by either metallic or composition seals. (*Fisher Controls.*)

temperature options, such as lined butterfly valves. It also may have some effect on the valve's performance. For instance, going from PTFE to metal seals for high temperatures generally increases the shutoff leakage flow. Similarly, high-temperature metal bearing sleeves in rotary valves impose more friction load on the shaft than PTFE bearings do, so that the shaft cannot withstand as high a pressure-drop load at shutoff.

Selection of valve packing is done largely based on the service temperature. Two packing types, PTFE V-rings and graphite, meet most packing requirements. These materials have proven reliable, inexpensive, and effective.

PTFE V-ring packing is composed of solid rings of molded PTFE. Generally, in a given packing set, there are two or more packing rings with a V cross section, a male adaptor, and a female adaptor. The packing can be used over a temperature range of –40 to 450°F (–40 to 232°C) and for nearly all chemicals. PTFE packing can be used with a spring (live loaded) or as jam-type packing. Stem friction is low. This packing is the preferred packing for most applications.

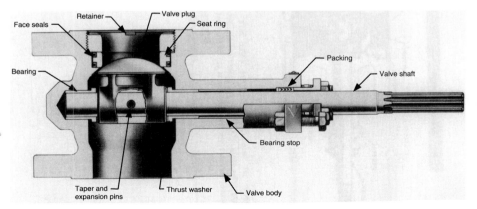

FIGURE 14 Eccentric plug valve specially designed for severe rotary applications. It features tight shutoff with globe valve style seating and excellent resistance to abrasive wear and flashing-induced erosion. (*Fisher Controls.*)

FIGURE 15   Swing-through butterfly valve providing an economical solution to high-flow-rate throttling applications. Leakage is higher than for other designs as no sealing mechanism is used. (*Fisher Controls.*)

FIGURE 16   Lined butterfly valves offering tight shutoff, but limited to low-temperature applications. Liner material keeps process away from metallic body, eliminating many corrosion problems. (*Fisher Controls.*)

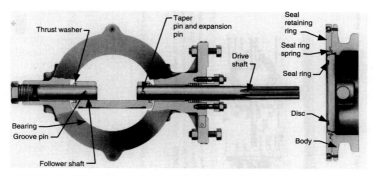

FIGURE 17   High performance butterfly valve providing excellent performance and value. High-pressure capability, tight shutoff, and excellent control are featured as standard. Designs are available in ANSI classes 150, 300, and 600 and size ranges through 72 inches (1800 mm). (*Fisher Controls.*)

Graphite packing systems are used mainly for temperatures above 450°F (232°C). They are composed of graphite ribbon rings, graphite filament rings, and sacrificial zinc washers. The graphite rings perform the sealing function while the zinc washers protect the valve stem from galvanic corrosion.

MATERIAL SELECTION

Material compatibility and durability are complex considerations. The issue may be corrosion by the process fluid, erosion by abrasive material, flashing, cavitation, or simply a matter of process pressure and temperature. The material used for piping is a good predictor of control-valve body material. However, since the velocity is higher in valves, other factors must also be considered. When these items are included, often valve and piping materials will differ. Trim materials are usually a function of the body material, temperature range, and qualities of the fluid. When a body material other than carbon, alloy, or stainless steel is required, the use of alternate valve types, such as lined or bar stock, should be considered.

Control valves are required to function with precision in some very extreme environments. A number of factors must be considered to ensure that a material will perform properly in service. These factors fall primarily into two categories: (1) the material's suitability to function mechanically and (2) the material's compatibility with the process environment. These constraints conflict in many instances, making it difficult to satisfy all considerations with a single material. In these cases, the best compromise must be identified.

Carbon-Steel Bodies and Bonnets

The most standard material for control-valve bodies is ASME SA216 grades WCB or WCC. Carbon steel is easily cast, welded, and machined. It is used for a large majority of process applications due to its low cost and reliable performance. Its use is strongly recommended over any other material if possible because of its standard availability and low cost.

Alloy-Steel Bodies and Bonnets

When higher temperatures or pressures are involved, alloy steels are often specified. Most are steels with chromium or molybdenum added to enhance their resistance to tempering and graphitization at elevated temperatures. The chromium and molybdenum additions also increase their resistance to erosion in flashing applications. Among the more popular materials are ASME SA217 grades WC9 and WC6.

Stainless-Steel Bodies and Bonnets

The most common stainless steel used for bodies and bonnets is CF8M, which is the cast version of S31600. With its nominal 19½% Cr, 10½% Ni, 2½% Mo composition, CF8M is a relatively low-cost material with good high-temperature properties and excellent resistance to corrosion.

Selection of Materials

Comparing pressure-temperature (P–T) ratings in ANSI B16.34 is much simpler when the ratings are presented in graphic form. The first discovery that is made is that the class 150 ratings for WCB, WCC, WC9, and C5 are identical over their common temperature ranges, and that CF8M is only rated slightly lower at temperatures below 550°F (288°C). The second discovery is that the rating plots for these materials in all other classes have the same shapes, and all that changes is the y-axis scale for the allowable pressures. Figure 18 is a plot of the pressure-temperature ratings, where the allowable pressure has been normalized to 100 percent. This curve is representative of the relative pressure-temperature ratings of the materials for ANSI classes 300 through 4500. If the material providing the maximum allowable pressure at any temperature is determined from the plot, three material regimes can be established. From ambient temperature to 750°F (399°C), WCC has the highest P–T ratings of this group of

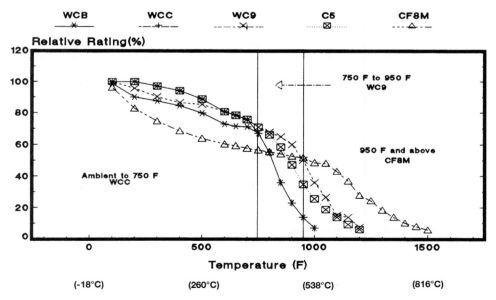

FIGURE 18 Control-valve pressure ratings depend on temperature and material. This chart compares normalized ratings as a function of these factors. These are relative ANSI B16.34 pressure-temperature ratings (ratings versus WC9 at room temperature). (*Fisher Controls.*)

materials. From 750 to 950°F (399 to 510°C), WC9 has the highest ratings, and from 950°F (510°C) up, CF8M has the highest ratings. These are the materials of choice.

Trim Parts

Valve trim components have much different material requirements than valve bodies and bonnets. They are not pressure-retaining, so they are not directly safety-related. However, since the trim components provide the flow control, they are very important with respect to the overall performance of the valve. In general, trim materials must have excellent resistance to corrosion by the process fluid in order to maintain adequate flow control and mechanical stability. Each individual component must possess certain other characteristics, depending on valve design, process fluid, and application.

FLOW CHARACTERISTIC

The next selection criterion—inherent flow characteristic—refers to the pattern in which the flow at constant pressure drop changes according to valve position. Typical characteristics are quick opening, linear, and equal percentage. The choice of characteristic has a strong influence on the stability or controllability of the process, since it represents the change of valve gain relative to travel. Most control valves are carefully "characterized" to exhibit a certain flow characteristic by means of contours on a plug, cage, or ball element. Some valves are available in a variety of characteristics to suit the application, while others offer little or no choice.

To determine the best flow characteristic for a given application quantitatively, a dynamic analysis of the control loop can be performed. In most cases, however, this is unnecessary; reference to established rules of thumb will suffice. Figure 19 illustrates typical flow characteris-

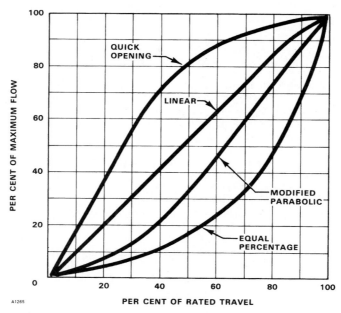

FIGURE 19   Many control valves offer a choice of control characteristics. Selection to match process requirements is guided by simple rules. Adherence to these guidelines will help assure stable process operation. (*Fisher Controls.*)

tic curves. The quick-opening flow characteristic provides for maximum change in flow rate at low valve travels with a fairly linear relationship. Additional increases in valve travel give sharply reduced changes in flow rate, and when the valve plug nears the wide open position, the change in flow rate approaches zero. In a control valve, the quick-opening valve plug is used primarily for on-off service; but it is also suitable for many applications where a linear valve plug would normally be specified.

The linear flow characteristic curve shows that the flow rate is directly proportional to the valve travel. This proportional relationship produces a characteristic with a constant slope so that with constant pressure drop, the valve gain will be the same at all flows. The linear valve plug is commonly specified for liquid-level control and for certain flow control applications requiring constant gain.

In the equal-percentage flow characteristic, equal increments of valve travel produce equal percentage changes in the existing flow. The change in flow rate is always proportional to the flow rate just before the change in valve plug, disk, or ball position is made. When the valve plug, disk, or ball is near its seat and the flow is small, the change in flow rate will be small; with a large flow, the change in flow rate will be large. Valves with an equal-percentage flow characteristic are generally used on pressure control applications, and on other applications where a large percentage of the pressure drop is normally absorbed by the system itself, with only a relatively small percentage available at the control valve. Valves with an equal-percentage characteristic should also be considered where highly varying pressure drop conditions can be expected. Table 2 lists characteristic recommendations by process type.

Rangeability

Part and parcel of a valve's flow characteristic is its rangeability, which is the ratio of its maximum and minimum controllable flow rates. Exceptionally wide rangeability may be required for certain applications to handle wide load swings or a combination of start-up, normal, and maximum working conditions. Generally speaking, rotary valves, especially partial ball valves, have greater rangeability than sliding-stem varieties.

Pressure Drop

The maximum pressure drop the valve can tolerate at shutoff and when partly or fully open is an important selection criterion. Sliding-stem valves are generally superior in both regards because of the rugged, well-supported nature of their moving parts. Unlike most sliding-stem valves, many rotary valves are limited to pressure drops well below the body pressure rating, especially under flowing conditions, due to dynamic stresses imposed on the disk or ball segment by high-velocity flow.

Noise and cavitation are two considerations which, while unrelated, are often grouped together because they both usually accompany high pressure drops and flow rates. They are handled by special modifications of more or less standard valves. Cavitation is the noisy and potentially damaging implosion of bubbles formed when the pressure of a liquid momentarily dips below its vapor pressure through a constriction at high velocity. In controlling gases and vapors, noise results from the turbulence associated with high-velocity streams. When cavitation or noise is judged likely to be a problem, its severity must be predicted from the valve's specifications according to well-known techniques, and valves with better specifications must be sought if necessary. Cavitation-control and noise-control trims for various degrees of severity are widely available in regular sliding-stem valves—at a progressive penalty in terms of cost and flow capacity. Rotary valves have more limited noise- and cavitation-control options and are also much more susceptible to cavitation and noise at a given pressure drop. Please refer to subsequent articles in this handbook section concerning control-valve noise and cavitation.

End Connections

At some point in the selection process the valve's end connections must be considered. The question to be answered is simply whether the desired connection style is available in the valve style being considered. In some situations, end connections can quickly limit the selection or dramatically affect the price. For instance, if a piping specification calls for welded connections only, the choice may be limited to sliding-stem valves. The few weld-end butterfly and ball valves that are available are rather expensive.

Shutoff Capability

Some consideration usually must be given to a valve's shutoff capability, which ordinarily is rated in terms of classes specified in ANSI/FCI 70-2-1976 (R1982) [3]. In actual service, shutoff leakage depends on many factors, including pressure drop, temperature, the condition of the sealing surfaces, and—very importantly for sliding-stem valves—the force load on the seat. Since shutoff ratings are based on standard test conditions (Table 3), which may be very different from service conditions, service leakage cannot be predicted very well. However, the ANSI shutoff classes provide a good basis for comparisons among valves of similar configuration.

Tight shutoff is particularly important in high-pressure valves since leakage can cause seat damage, leading to ultimate destruction of the trim. Special precautions in seat materials, seat preparation, and seat load are necessary to ensure success. Valve users tend to overspecify shutoff requirements, incurring unnecessary cost. Actually, very few throttling valves really need to perform double duty as tight block valves. Since tight shutoff valves generally cost more initially and to maintain, serious consideration is warranted.

Flow Capacity

The criterion of capacity or size can be an overriding constraint on selection. For very large lines, sliding-stem valves are much more expensive than rotary types. On the other hand, for very small flows, a suitable rotary valve may not be available. If the same valve is desired to handle a significantly larger flow at a future time, a sliding-stem valve with replaceable, restricted trim may be indicated. Rotaries generally have much higher maximum capacity than sliding-stem valves for a given body size. This fact makes rotaries attractive in applications where the pressure drop available is rather small. But it is of little or no advantage in high pressure drop applications such as pressure regulation or letdown.

At the risk of overgeneralizing, we may simplify the process of selection roughly as follows. For most general applications it makes sense, both economically and technically, to use sliding-stem valves for the lower ranges, ball valves for intermediate capacities, and high performance butterfly valves for the very largest sizes. For the very least demanding services, in which price is the dominant consideration, one might consider economy sliding-stem valves for the small-size applications and butterfly valves for the largest.

For sizes of 0.5 to 3 inches (15 to 80 mm), general-purpose sliding-stem valves provide an exceptional value. For a minimal price premium over rotary products, they offer unparalleled performance, flexibility, and service life. The premium for these devices over rotary products is warranted. For severe service applications, the most frequently used, and often the only available product is the sliding-stem valve.

Applications ranging from 4 to 6 inches (100 to 150 mm) are generally best served by such transitional valve styles as the eccentric plug valve or the ball valve. These products have excellent performance and lower cost. They also offer higher capacity levels than globe designs.

In sizes 8 inches (200 mm) and larger, pressures and pressure drops are generally much lower. This gives rise to the possibility of using high-performance butterfly valves for most sit-

TABLE 3 Maximum Leakage and Test Conditions for Control-Valve Leakage Classes

ANSI B16.104-1976*	Maximum leakage*			Test medium	Pressure and temperature
Class II	0.5% valve capacity at full travel			Air	Service ΔP or 50 psid (3.4-bar differential), whichever is lower, at 50 to 125°F (10 to 52°C)
Class III	0.1% valve capacity at full travel			Air	Service ΔP or 50 psid (3.4-bar differential), whichever is lower, at 50 to 125°F (10 to 52°C)
Class IV	0.01% valve capacity at full travel			Air	Service ΔP or 50 psid (3.4-bar differential), whichever is lower, at 50 to 125°F (10 to 52°C)
Class V	5×10^{-4} mL/min/psid/in port dia (5×10^{-12} m³/s/bar differential/mm port dia)			Water	Service ΔP at 50 to 125 F (10 to 52°C)
Class VI	Nominal port diameter			Air	Service ΔP or 50 psid (3.4-bar differential), whichever is lower, at 50 to 125°F (10 to 52°C)
	inches	mm	Bubbles/min	mL/min	
	1	25	1	0.15	
	1½	38	2	0.30	
	2	51	3	0.45	
	2½	64	4	0.60	
	3	76	6	0.90	
	4	102	11	1.70	
	6	152	27	4.00	
	8	203	45	6.75	

* Copyright 1976 Fluid Controls Institute, Inc. Reprinted with permission.

uations. These valves are economical, offer tight shutoff, and provide good control capability. They provide cost and capacity benefits well beyond those of globe and ball valves.

Special considerations require special valve solutions. There are valve designs and special trims available to handle high-noise applications, cavitation, high pressure, high temperature, and combinations of these conditions.

The obvious point here is this: different types of valves are appropriate for use in different size ranges, because they provide the most cost-effective solution in each given instance. If one sticks with the same type of valve over a wide size range, one sacrifices either performance at the low end or economy at the high end, or both.

After going through all the other criteria for a given application, a specifier often finds that he or she can use several types of valves. From there on, selection is a matter of price versus capability as discussed here—coupled with the inevitable personal and institutional preferences. Since no single control-valve package is cost-effective over the full range of applications that are normally encountered, it is important to keep an open mind for alternative choices.

VALVE SIZING

It used to be common practice in the industry to select valve size strictly as a function of pipe size. Soon it became apparent that this practice contributed to very poor control and resulting process problems. The wide range of flow, pressure, and fluid conditions required a more in-depth selection methodology. With time, methods were developed and the days of selecting a valve based on pipe size are gone forever.

Selecting the correct valve size for a given application requires a knowledge of the flow and process conditions the valve will actually see in service as well as information on valve function and style. Sizing valves is based on a combination of theory and empirical data. The results are predictable, accurate, and consistent.

Early efforts in the development of valve sizing centered around liquid flow. Daniel Bernoulli was one of the early experimenters who applied theory to liquid flow. Subsequent experimental modifications to this theory produced a useful liquid-flow equation,

$$Q = C_v \sqrt{\frac{P_1 - P_2}{G}}$$

where
Q = flow rate
C_v = valve sizing coefficient, determined by testing
P_1 = upstream pressure
P_2 = downstream pressure
G = liquid specific gravity

This equation rapidly became widely accepted for sizing valves on liquid service, and manufacturers of valves began testing and publishing C_v data in their catalogs.

It was inevitable that the good results obtained from the C_v equation would strongly tempt its use to predict the flow of gas. The results, however, were inaccurate. Modifications were made to the equations over time, with consequent improvement of results. Various companies used techniques they developed, but there was no common formulation until the Instrument Society of America (ISA) put forth its standardized guidelines.

In order to assure uniformity and accuracy, the procedures for measuring flow parameters and for valve sizing are addressed by ISA standards. Measurement of C_v and related flow parameters is covered extensively in ANSI/ISA S75.02, 1981 [4]. The basic test system and hardware installation are outlined so that coefficients can be tested to an accuracy of ±5 percent. Water is circulated through the test valve at specified pressure differentials and inlet pressures. Flow rate, fluid temperature, inlet and differential pressure, valve travel, and barometric pressure are all measured and recorded. This yields sufficient information to calculate necessary sizing parameters. Numerous tests must be performed to arrive at the values published by the valve manufacturer for use in sizing. It is important, also, that these factors be based on tests, not estimates, since the results are not always predictable.

Basic Sizing Procedure

The procedure by which valves are sized for liquid flow is straightforward. Again, to ensure uniformity and consistency, a standard exists which delineates the equations and correction factors to be used for a given application (ANSI/ISA S75.01-1985 [5]).

The simplest case of liquid-flow application involves the basic equation developed earlier. Rearranging the equation so that all of the fluid and process-related variables are on the right-hand side, we arrive at an expression for the valve C_v required for the particular application,

$$C_v = \frac{Q}{\sqrt{(P_1 - P_2)/G}}$$

Based on a given flow rate and pressure drop, a required C_v value can be calculated. This C_v can then be compared to C_v values for a particular valve size and valve design. Generally, the required C_v should fall in a range of between 70 and 90 percent of the selected valve's C_v capability. Allowance for minimum and maximum flow pressure conditions should also be considered.

Once a valve has been selected and C_v is known, the flow rate for a given pressure drop, or the pressure drop for a given flow rate, can be predicted by substituting and solving for the appropriate quantities in the equation.

This basic liquid equation covers conditions governed by the test assumptions. Unfortunately many applications fall outside the bounds of these standards and therefore outside of the basic liquid-flow equation. Rather than develop special flow equations for all of the possible deviations, it is possible to account for different behavior with the use of simple correction factors. These factors, when incorporated, change the form of the equation to the following:

$$C_v = \frac{Q}{NF_p F_R \sqrt{P_1 - P_2/G}}$$

where N = numerical coefficient for unit conversion
F_p, F_R = correction factors

Choked Flow

A plot of the basic equation (Fig. 20) implies that flow can be increased continually by simply increasing the pressure differential across the valve. In reality the relationship given by this equation holds for only a limited range. As the pressure differential is increased, a point is reached where the realized flow increase is less than expected. This phenomenon continues until no additional flow increase occurs in spite of increasing the pressure differential. This condition of limited maximum flow is known as choked flow. This phenomenon occurs on both liquids and gases. It is necessary to account for the occurrence of choked flow during the sizing process to ensure against undersizing a valve. Predictions must be made using a valve recovery coefficient F_L for liquids and X_T for gases.

Viscous Flow

One of the assumptions implicit in the sizing procedures presented up to this point is that of fully developed, turbulent flow. In laminar flow, all fluid particles move parallel to one another in an

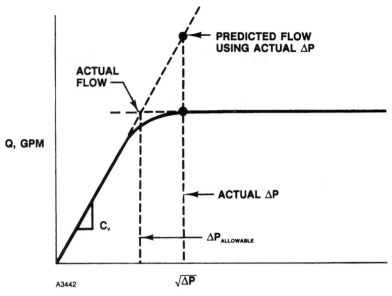

FIGURE 20   Sizing equation suggests that as ΔP is increased, flow will increase proportionately—forever. In reality this relationship holds only for certain conditions. As pressure drop is increased, choked flow caused by formation of vapor bubble in flow stream imposes a limit on liquid flow. A similar limitation on flow of gases is realized when velocity at the valve's vena contracta reaches sonic. These choked-flow conditions must be considered in valve sizing. (*Fisher Controls.*)

orderly fashion with no mixing of the fluid. Conversely, turbulent flow is highly random in terms of local velocity direction and magnitude. While there is certainly net flow in a particular direction, instantaneous velocity components in all directions exist within this net flow. Significant fluid mixing occurs in turbulent flow. The factor F_R is a function of the Reynolds number and describes the degree of turbulent flow. It can be determined by a simple nomograph procedure.

Piping Considerations

When a valve is installed in a field piping configuration which is different than the standard test section, it is necessary to account for the effect of the altered piping on flow through the valve. Recall that the standard test section consists of a prescribed length of straight pipe up- and downstream of the valve. Field installation may require elbows, reducers, and tees, which will induce additional pressure losses adjacent to the valve. To correct for this, the factor F_p is introduced.

Gas and Steam Sizing

While most comments so far pertain to liquid sizing, they closely parallel the procedures used for air, gas, and steam valve sizing. The only additional steps involve correction for the physical properties of the particular gas and pressure ratio factors which determine the degree of compression and predict choked flow. The general form of the sizing equation for compressible fluids is

$$C_v = \frac{Q}{NF_pP_1Y\sqrt{X/G\,T_1Z}}$$

where Y = expansion factor
X = $\Delta P/P_1$
T_1 = temperature
Z = compressibility factor

For additional information on valve sizing, consult the referenced ISA publications or the manufacturer's literature. Computer software sizing aids are available, which alleviate the need to solve complex equations manually and which provide exceptional accuracy.

ACTUATORS

Actuators are the distinguishing elements between just valves and control valves. The actuator industry has evolved to answer a wide variety of process needs and user desires. Actuators are available with many designs, power sources, and capabilities. Proper selection involves process knowledge, valve knowledge, and actuator knowledge. A control valve can perform its function only as well as the actuator can handle the static and dynamic loads placed on it by the valve. Proper selection and sizing are, therefore, very important. The actuator represents a significant portion of the total control-valve package price, and careful selection can minimize costs.

The range of actuator types and sizes on the market today is so great that it seems the selection process might be highly complex. It is not. With a few rules in mind and knowledge of your fundamental needs, the selection process can be very simple.

The following parameters must be known at the beginning of the selection process. They are key as they quickly narrow the selection process.

1. Power source availability

2. Fail-safe requirements

3. Torque or thrust requirements (actuator capability)

4. Control functions

5. Economics

Power Source

The power source available at the location of a valve can often point directly to what type of actuator to choose. Typically, valve actuators are powered either by compressed air or by electricity. However, in some cases water pressure, hydraulic fluid, or even pipeline pressure can be used. The majority of actuators sold today use compressed air for operation. They operate at supply pressures from as low as 15 psi (1 bar) to maxima of about 150 psi (10.4 bar).

Since most plants have ready availability of both electricity and compressed air, the selection depends on the ease and cost of furnishing either power source to the actuator location. One must also consider reliability and maintenance requirements of the power system and their effect on subsequent valve operation. Consideration should be given to providing backup operating power to critical plant loops.

Fail-Safe Characteristics

The overall reliability of power sources is quite high. However, many loops demand specific valve action should the power source ever fail. Desired action on signal failure may be required for safety reasons or for the protection of equipment. Fail-safe systems store energy, either mechanically in springs or pneumatically in volume tanks or hydraulic accumulators. When power fails, the fail-safe systems are triggered to drive the valves to the required position and then maintain this position until resumption of normal operation. In many cases the process pressure is used to ensure or enhance this action.

Actuator designs are available which allow a choice of failure mode between failing open, failing closed, or holding in the last position. Many actuator systems incorporate failure modes at no extra cost. Spring and diaphragm types are inherently fail open or closed. Electric operators nearly always hold in their last position.

Actuator Capability

An actuator must have sufficient thrust or torque for the application. In some cases this requirement can dictate actuator type as well as power-supply requirements. For instance, large valves requiring a high thrust may be limited to only electric or electrohydraulic actuators due to a lack of pneumatic actuators with sufficient torque capability. Conversely, electrohydraulic actuators would be a poor choice for valves with very low thrust requirements. The matching of actuator capability with valve-body requirement is best left to the control valve manufacturer, as there is considerable variation in frictional and fluid forces from valve to valve.

Control Functions

Knowledge of the required actuator functions will most clearly define the options available for selection. These functions include the actuator signal (such as pneumatic, electric, analog, frequency), signal range, ambient temperatures, vibration levels, operating speed, cycle frequency, and quality of control required.

Generally, signal types are grouped as being either two-position (on-off) or analog (throttling). On-off actuators are controlled by two-position electric, electropneumatic, or pneumatic switches. This is the simplest type of automatic control and the least restrictive in terms of selection.

Throttling actuators have considerably higher demands put on them from both a compatibility and a performance standpoint. A throttling actuator receives its input from an electronic or pneumatic instrument that measures the controlled process variable. The actuator must then move the final control element in response to the instrument signal in an accurate and timely fashion to ensure effective control. The two primary additional requirements for throttling actuators are (1) compatibility with instrument signal and (2) better static and dynamic performance to ensure loop stability.

Compatibility with instrument signals is inherent in many actuator types, or it can be obtained with add-on equipment. But the high-performance characteristics required of a good throttling actuator cannot be bolted on. Low hysteresis and minimal dead band must be designed into actuators.

Stroking speed, vibration, and temperature resistance must also be considered if critical to the application. Stroking speed is generally not critical; however, flexibility to adjust it is desirable. On liquid loops, fast stroking speeds can be detrimental due to the possibility of water hammer.

Vibration or mounting position can be potential problems as the actuator weight, combined with the weight of the valve, may necessitate bracing. If extremes of temperature or humidity are to be experienced by the control valve, this information is essential to the selection process. Many actuators contain either elastomeric or electronic components, which may be subject to degradation by high humidity or temperature.

Economics

Evaluation of the economics in actuator selection involves combining first cost, maintenance, and reliability factors. A simple actuator, such as a spring and diaphragm, has few moving parts, is easy to service, and will generally cause fewer problems. Initial cost is low as well. Maintenance personnel understand and are comfortable working with them. An actuator made specifically for a control valve eliminates the chance for a costly performance mismatch. An actuator manufactured by the valve vendor and shipped with the valve will eliminate separate mounting charges and ensure easier coordination of spare parts procurement. Interchangeable parts among varied actuators are also important to minimize spare-parts inventory.

ACTUATOR DESIGNS

There are many types of actuators on the market. They fall into four major categories:

1. Spring and diaphragm (Figs. 21 and 22)
2. Pneumatic piston (Figs. 23 through 26)
3. Electric motor (Fig. 27)
4. Electrohydraulic (Figs. 28 and 29)

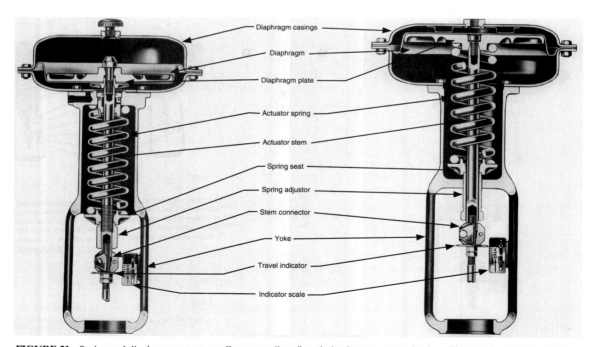

FIGURE 21   Spring and diaphragm actuators offer an excellent first choice for most control valves. They are inexpensive, simple, and have ever-present, reliable spring fail-safe action. Shown are two styles. Left—air opens, spring closed design. Right—actuator utilizes air pressure to close valve and spring to open it. (*Fisher Controls.*)

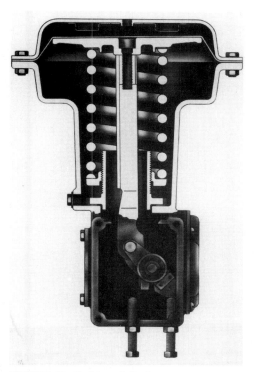

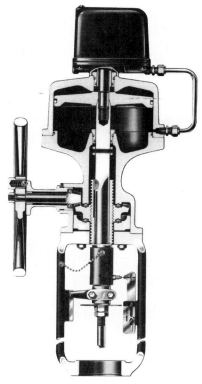

FIGURE 22 Spring and diaphragm actuators contain many features which provide precise control. The splined actuator connection, clamped lever, and single-joint linkage all contribute to low lost motion. (*Fisher Controls.*)

FIGURE 23 Double-acting piston actuators are a good choice when thrust exceeds capability of diaphragm actuators. They require higher supply pressure, but have benefits such as high stiffness and more compact size. (*Fisher Controls.*)

Each actuator has weaknesses, strong points, and optimum uses. Most actuator designs are available for either sliding stem or rotary valve bodies. They differ only by linkage or motion translators. The basic power sources are identical (Table 4).

Rotary Actuators

Most rotary actuators (Figs. 22, 25, and 26) use linkages, gears, or crank arms to convert direct linear motion of a diaphragm or piston into the 90° output rotation required by rotary valves. The most important consideration for control valve actuators is the requirement for a design that limits the amount of lost motion in the internal linkage and valve coupling. Rotary actuators are now available which use tilting pistons or diaphragms. These designs eliminate most linkage points (and the resultant lost motion) and provide a safe, accurate, and enclosed package.

When considering an actuator design, it is also necessary to consider the method by which it is coupled to the drive shaft of the control valve. On rotary valves, slotted connections mated to milled shaft flats generally are not satisfactory if throttling is required. Pinned connections, if constructed solidly, are suitable for nominal torque applications. The best connectors are clamped, splined shapes. This type of connection eliminates all lost motion, is easy to disassemble, and is capable of high torques.

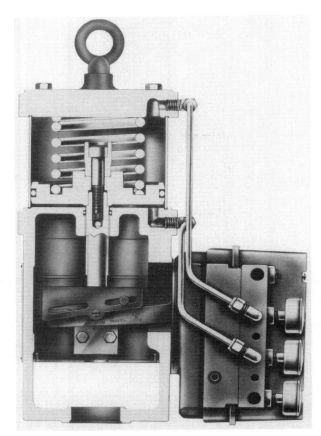

FIGURE 24 Spring fail-safe is present in this spring bias piston actuator. Process pressure can aid fail-safe action or actuator can be configured for full spring fail closure. (*Fisher Controls.*)

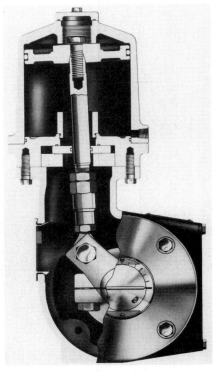

FIGURE 25 Piston actuator, adapted for control of rotary valves, provides features to eliminate lost motion and provide improved throttling accuracy. (*Fisher Controls.*)

Sliding stem actuators are rigidly fixed to valve stems by threaded and clamped connections. Sliding stem actuators are very simple in design. Since they do not have any linkage points and their connections are rigid, they exhibit no lost motion and excellent inherent control characteristics.

Since rotary and sliding-stem actuators are so similar in concept and characteristics, they will not be further differentiated in this section unless necessary.

Diaphragm Actuators

The most popular and widely used control-valve actuator is the pneumatic spring and diaphragm style (Figs. 21 and 22). Diaphragm actuators are extremely simple and offer low cost and high reliability. Diaphragm actuators normally operate over the standard signal ranges of 3 to 15 psi (0.2 to 1 bar) or 6 to 30 psi (0.4 to 2 bar). Therefore they are often suitable for throttling service using instrument signals directly. Many designs offer either adjustable springs or wide spring selections to allow the actuator to be tailored to the partic-

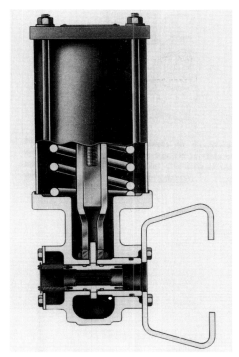

FIGURE 26 For on-off service, requirements for accuracy and minimal lost motion are unnecessary. Cost savings can be achieved by simplifying the design for these applications. The actuator shown incorporates spring return capability. (*Fisher Controls.*)

ular application. Since diaphragm actuators have few moving parts which may contribute to failure, they are extremely reliable. Should they ever fail, maintenance is extremely simple. Improved designs include mechanisms to control the release of spring compression, eliminating possible injury to personnel during actuator disassembly.

The overwhelming advantage of the spring and diaphragm actuator is the ever-present provision for fail-safe action. As air is loaded on the actuator casing, the diaphragm moves the valve and compresses the spring. The stored energy in the spring acts to move the valve back to its original position as air is released from the casing. Should there be a loss of signal pressure to the instrument or the actuator, the spring can move the valve to its initial (fail-safe) position. Actuators are available for either fail-open or fail-closed action.

The only real drawback to the spring and diaphragm actuator is a relatively limited capability. Much of the thrust created by the diaphragm is taken up by the spring and thus does not result in output to the valve. Therefore the spring and diaphragm actuator is seldom used for high force requirements. It is not economical to build and use very large diaphragm actuators because the size, weight, and cost grow out of proportion to capability. This handicap is mitigated, however, by the fact that most valves are small and have low force requirements.

Piston Actuators

Piston actuators, such as those shown in Figs. 23 through 25, are the second most popular control-valve actuator style. They are generally more compact and provide higher torque or force outputs than spring and diaphragm actuators. Piston styles normally work with supply pressures of between 50 and 150 psi (3.5 and 10.4 bar). Although piston actuators can be equipped with spring returns, this construction has limits similar to those of the spring and diaphragm style.

Piston actuators used for throttling service must be furnished with double-acting positioners, which simultaneously load and unload opposite sides of the piston. The pressure differential created across the piston causes travel toward the lower pressure side. The positioner senses the motion of the output, and when the required position is reached, the positioner equalizes the pressure on both sides of the piston.

The pneumatic piston actuator is an excellent choice when a compact high-power unit is required. It is also easily adapted to services where high ambient temperatures are involved.

The main disadvantages of piston actuators are the high supply pressures required, the requirement for positioners when used for throttling service, and the lack of inherent fail-safe systems. Two types of spring return piston actuators are available. The variations are subtle, but significant. It is possible to add a spring to a piston actuator and operate it much like a spring and diaphragm. These designs use a single acting positioner, which loads the piston chamber to move the actuator and compress the spring. As air is unloaded, the spring moves the piston back. These designs use large high-output springs, which are capable of overcoming the fluid forces in the valve.

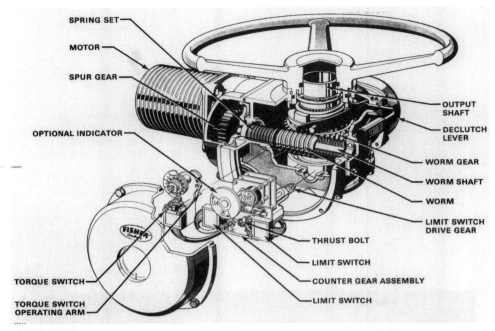

FIGURE 27 Technical improvements in recent years have made electric actuators viable for control purposes. They offer high thrust and stiffness, but are handicapped by a lack of fail-safe action and relatively high cost. (*Fisher Controls.*)

The alternative design uses a much smaller spring and relies on valve fluid forces to help provide the fail-safe action. In normal operation they act like a double-action piston. In a fail-safe situation the spring initiates movement and is helped by unbalance forces on the plug.

The only fail-safe alternative to springs are pressurized air volume tank pneumatic trip systems to move the piston actuator to its fail-safe position. While these systems are quite reliable, they add to overall system complexity, maintenance difficulty, and cost. Therefore for any fail-safe requirement prime consideration should be given to spring return operators if they are feasible.

Special care should be exercised during the selection of throttling piston actuators to get one which has minimal hysteresis and dead band. As the number of linkage points in the actuator increases, so does the dead band. As the number of sliding parts increases, so does the hysteresis. An actuator with high hysteresis and dead band can be quite suitable for on-off service. However, caution is necessary when attempting to adapt this actuator to throttling service by simply bolting on a positioner.

The cost of a diaphragm actuator is generally less than that of a comparable-quality piston actuator. Part of this cost savings lies in the ability to use instrument output air directly, thereby eliminating the need for a positioner. The inherent provision for fail-safe action in the diaphragm actuator is also a consideration.

Electric Actuators

Electric actuators can be successfully applied in many situations. Most electric operators consist of motors and gear trains (Fig. 27). They are available in a wide range of torque outputs, travels, and capabilities. They are suited for remote mounting where no other power source is avail-

FIGURE 28 Self-contained electrohydraulic actuator. This single unit contains hydraulic pump, reservoir, hydraulic positioner, and actuator cylinder. (*Fisher Controls.*)

FIGURE 29 Electrohydraulic actuators provide the ultimate in thrust, speed, frequency of response, and stiffness. The type shown operates from an external hydraulics power supply. (*Fisher Controls.*)

able or for use where there are specialized thrust or stiffness requirements. Electric operators are economical, compared with pneumatic ones, for applications in small size ranges only. Larger units operate slowly, weigh considerably more than pneumatic equivalents, and are more costly. At this time there are no electric actuators economically available with fail action other than lock in last position. And precision throttling versions of electric motor actuators are quite limited in availability. One very important consideration in choosing an electric actuator is its capability for continuous closed-loop control. In applications where frequent changes are made in control valve position, the electric actuator must have a suitable duty cycle.

While having many disadvantages, the electric actuator will generally provide the highest output available within a given package size. In addition electric actuators are very stiff, that is, resistant to valve forces. This makes them an excellent choice for good throttling control of large high-pressure valves.

Electrohydraulic Actuators

Electrohydraulic actuators, like those in Figs. 28 and 29, are electric actuators in which motors pump oil at high pressure to a piston, which in turn creates the output force. The

TABLE 4   Comparison of Valve Actuator Features

Advantages	Disadvantages
Spring and diaphragm	
Lowest cost	Limited output capability
Ability to throttle without positioner	Large size and weight
Simplicity	
Inherent fail-safe action	
Low supply-pressure requirement	
Adjustability to varying conditions	
Ease of maintenance	
Pneumatic piston	
High torque capability	Fail-safe requires accessories or
Compact	addition of spring
Lightweight	Positioner required for
Adaptable to high ambient temperatures	throttling
Fast stroking speed	Higher cost
Relatively high actuator stiffness	High supply-pressure requirement
Electric motor	
Compactness	High cost
Very high stiffness	Lack of fail-safe action
High output capability	Limited duty cycle
Slow stroking speed	
Electrohydraulic	
High output capability	High cost
High actuator stiffness	Complexity and maintenance difficulty
Excellent throttling ability	Fail-safe action only with accessories
Fast stroking speed	

electrohydraulic actuator is an excellent choice for throttling due to its high stiffness, compatibility with analog signals, excellent frequency response, and positioning accuracy. Most electrohydraulic actuators are capable of very high outputs, but they are handicapped by high initial cost, complexity, and difficult maintenance. Fail-safe action on electrohydraulic actuators can be accomplished by the use of springs or hydraulic accumulators and shutdown systems.

Actuator Sizing

The last step in the selection process is the specification of the actuator size. Fundamentally, the process of sizing is to match the actuator capabilities as closely as possible to the valve requirements. In practice, the mating of actuator and valve requires the consideration of many factors. Valve forces must be evaluated at the critical positions of valve travel (usually open and closed) and compared to actuator output. Valve force calculation varies considerably between valve styles and manufacturers. In most cases it is necessary to consider a complex summation of forces, including the following:

Static fluid forces
Dynamic fluid forces and force gradients
Friction of seals, bearings, and packing
Seat loading

Although actuator sizing is not difficult, the great variety of designs on the market and the ready availability of vendor expertise (normally at no cost) make detailed knowledge of the procedures unnecessary.

Summary of Actuator Selection Factors

In choosing an actuator type, the fundamental requirement is to know your application. Control signal, operating mode, power source available, torque required, and fail-safe position can make many decisions for you. Keep in mind simplicity, maintainability, and lifetime costs. Safety is another consideration which must never be overlooked. Enclosed linkages and controlled compression springs available in some designs are very important for safety reasons. The pros and cons of the various actuator styles are listed under "Summary Checklist."

The spring and diaphragm actuator is the most popular, versatile, and economical type. Try it first. If the limitations of available diaphragm actuators eliminate them, consider pistons or electric actuators, bearing in mind the capabilities and limitations of each.

CONTROL-VALVE ACCESSORIES

No study of control valves would be complete without a look at accessory devices which augment the valve function and interface it to control systems. Included in this category are well-known devices such as valve positioners, electropneumatic transducers, limit switches, and manual actuator overrides. These devices assure controllability, provide information about valve operation, and also allow for operation or shutdown in emergency situations.

Valve Positioners

Positioners are instruments which help improve control by accurately positioning a control valve actuator in response to a control signal. Positioners receive an input signal either pneumatically or electronically and provide output power, generally pneumatically, to an actuator to assure valve positioning. A feedback linkage between valve stem and positioner is established so that the stem position can be noted by the instrument and compared with the position dictated by the controller signal (Fig. 30).

Use of positioners is generally desirable to linearize the control valve plug position with a control signal. Generally speaking, positioners will improve the performance of control valve systems. There are situations, however, where process dynamics eliminate the use of positioners. On very fast loops it has been found that the use of positioners will degrade performance since the response of the positioner may not be able to keep up with the system in which it is installed.

Electropneumatic Transducers

Electropneumatic transducers (Fig. 31) are devices which proportionally convert an electronic input signal into a pneumatic output signal. Electropneumatic transducers are used in electronic control loops to help operate pneumatic control valves. Most transducers on the market will convert a standard 4- to 20-mA analog electronic signal into a 3- to 15-psi (0.2- to 1-bar) pneumatic output. Devices are also available that can respond to digital signals and nonstandard analog inputs. The transducer function is sometimes included with the valve positioner, in which case the device is known as an electropneumatic valve positioner. In this case the input is an electronic signal and the output is position.

FIGURE 30 Electromagnetic positioner, combining the functions of a current-to-pressure transducer with those of a positioner. It receives an input signal from the controller and assures valve position by adjusting output pressure. (*Fisher Controls.*)

FIGURE 31 Electropneumatic transducers are common actuator accessories. They take a standard analog electronic signal and produce a proportional pneumatic output. The best transducers are compact, accurate, and consume little supply air. (*Fisher Controls.*)

Volume Booster

The volume booster is normally used in control-valve actuators to increase the stroking speed. These pneumatic devices have a separate supply pressure and deliver a higher-volume output signal to move actuators rapidly to their desired positions. Special booster designs (Fig. 32) are also available for use with positioners. These devices incorporate a dead-band feature to adjust their response and eliminate instabilities. This booster, therefore, permits high actuator stroking speeds without degrading the steady-state accuracy provided by positioners in the loop.

Trip Valves

Pressure-sensing trip valves are available for control applications where a specific actuator action is required when supply pressure fails or falls below a specific point. When supply pres-

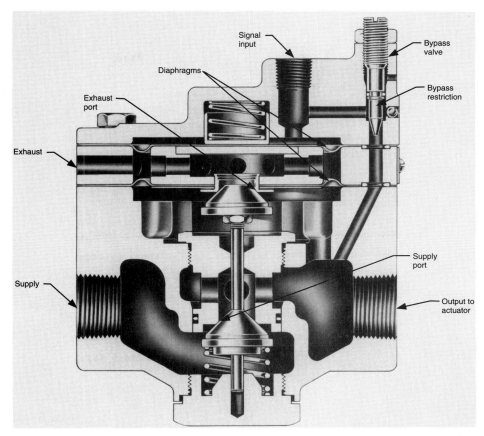

FIGURE 32 Volume booster delivering the added air volume needed to provide rapid actuator stroking. The booster shown is specifically made for positioners. It incorporates a bypass function to allow small output changes to pass through yet allow the unit to deliver high-volume output pressure for rapid stroking when signal changes exceed preset dead-band limits. (*Fisher Controls.*)

sure falls below the preadjusted trip point, the trip valve causes the actuator to fail up, lock in last position, or fail down. When supply pressure rises above the trip point, the valve automatically resets, allowing the system to return to normal operation. Auxiliary power to provide for actuator action in case of trip is provided by pneumatic volume tanks. Figure 33 shows a system as installed on a valve actuator.

Limit Switches

Electrical position switches are often incorporated on control valves to provide the operation of alarms, signal lights, relays, or solenoid valves when the control-valve position reaches a predetermined point. These switches can be either integrated, fully adjustable units with multiple switches or stand-alone switches and trip equipment. Special care should be exercised in the selection of limit switches for harsh environments to assure functionality over time (Figs. 34 and 35).

FIGURE 33   Fail-safe action on piston actuators can be accomplished by use of pneumatic trip systems. A switching valve transfers stored pressure from volume tanks to piston to initiate travel and to maintain predetermined failure position. (*Fisher Controls.*)

Solenoid Valves

Small, solenoid-operated electric valves are often used in a variety of on-off or switching applications with control valves. They provide equipment override, fail-safe interlock of two valves, or switching from one instrument line to another. A typical application involves a normally opened solenoid valve, which allows positioner output to pass directly to the actuator. Upon loss of electric power, the solenoid valve will close the port to the valve positioner and bleed pressure from the diaphragm case to the control valve, allowing it to achieve its fail-safe position.

Position Transmitters

Electronic position transmitters are available which send either analog or digital electronic output signals to control-room devices. The instrument senses the position of the valve and provides a discrete or proportional output signal. Electrical position switches are often included in these transmitters (Fig. 36).

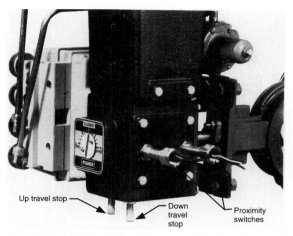

Up travel stop — Down travel stop — Proximity switches

FIGURE 34   Actuator featuring externally adjustable travel stops and integrally mounted cam-operated proximity-style limit switches. All linkages for positioner switches and stops are fully enclosed within actuator housing. (*Fisher Controls.*)

FIGURE 35   Limit switches are common actuator accessories. Unit shown can accommodate up to six switches and have trip points adjustable to any point in travel. (*Fisher Controls.*)

Manual Handwheels

A variety of actuator accessories are available which allow for manual override in the event of signal failure or lack of signal previous to start-up. Nearly all actuator styles have available either gear-style or screw-style manual override wheels. In many cases, in addition to providing override capability, these handwheels can be used as adjustable position or travel stops. Figure 37 shows the installation of a manual handwheel on a spring and diaphragm actuator.

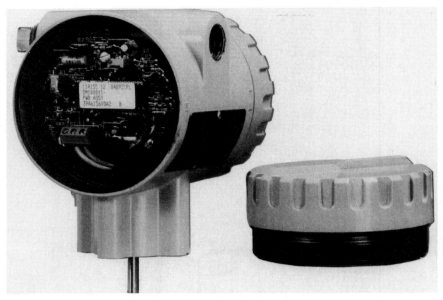

FIGURE 36   Stem position transmitters provide discrete or analog outputs of valve position for use by control-room instrumentation. (*Fisher Controls.*)

SUMMARY CHECKLIST

The subject of control valves is complex and ever evolving. Valve styles are changing to meet changing process conditions and accessories and instrumentation continue to evolve to meet the requirements of the control systems. The key to the selection process is to understand both ends of the requirement spectrum, the needs of the process, and the needs of the controlling instrumentation. Key tips for valve selection, sizing, and actuator selection follow.

1. Valve Body Selection
   a. Sliding-stem valves provide the widest variety and best capability in the industry. Their performance and versatility make them very popular. In large sizes they may be expensive, but for sizes 3 inches (80 mm) and less they are a first choice.
   b. Rotary-ball and eccentric-plug valves provide excellent control and are especially good values in sizes 4 to 6 inches (100 to 150 mm). Erosion-resistant designs and trims are available to extend their life in many difficult applications.
   c. Butterfly and high-performance butterfly valves are most popular and economical in sizes above 6 inches (150 mm). In many large-size cases they are the only available choice.
   d. Special requirements necessitate special valve solutions. Valve designs and special trims are available to handle high noise, cavitation, high pressure, high temperature, and combinations of these.
2. Sizing of Valve
   a. The liquid sizing equation is simple to use and based on empirically determined sizing coefficients.
   b. A valve size should be selected which gives the required application C_v at 70 to 90 percent of travel.

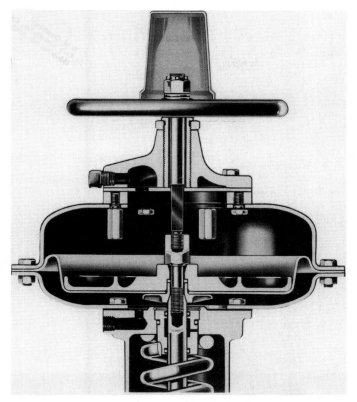

FIGURE 37   Handwheel on this actuator can act as a travel stop or a means of emergency operation. Cutaway view shows details of actuator mounted on air-to-open, spring-closed actuator. (*Fisher Controls.*)

    c. Sizing and trim selection are influenced by choked flow and the presence of cavitation. These phenomena limit flow and may cause significant damage.
    d. Viscosity and piping corrections must be made in many sizing situations. Piping considerations are especially important when high-recovery valves are specified.
    e. Sizing valves for gas flow involves physical principles similar to liquid flow. However, effects of compressibility and critical flow factors must be considered.
3. Actuator selection
    a. Actuator selection must be based on a balance of process requirements, valve requirements, and cost.
    b. Spring and diaphragm actuators are simpler, less expensive, and easier to maintain. Consider them first in most situations.
    c. Piston actuators offer many of the advantages of pneumatic actuators, with higher thrust capability than diaphragm styles. They are especially useful where compactness is desired or long travel is required.
    d. Electric and electrohydraulic actuators provide excellent performance. They are, however, much more complex and difficult to maintain.
    e. Actuator sizing is not difficult. But the wide variety of actuators and valves makes this difficult to master. Vendor expertise is widely available.

REFERENCES

1. ANSI B16.34-19, "Steel Valves," American National Standards Institute, New York.
2. ANSI B16.1-19, "Cast Iron Pipe Flanges and Flanged Fittings," American National Standards Institute, New York.
3. ANSI/FCI 70-2-1976 (R1982), "Quality Control Standard for Control Valve Seat Leakage," Fluid Controls Institute.
4. ANSI/ISA S75.02-1988, "Control Valve Capacity Test Procedure," Instrument Society of America, Research Triangle Park, North Carolina.
5. ANSI/ISA S75.01-1985, "Flow Equations for Sizing Control Valves," Instrument Society of America, Research Triangle Park, North Carolina.
6. ANSI/ISA S75.05-1983, "Control Valve Terminology," Instrument Society of America, Research Triangle Park, North Carolina.
7. ANSI/ISA S75.11-1985, "Inherent Flow Characteristic and Rangeability of Control Valves," Instrument Society of America, Research Triangle Park, North Carolina.
8. *Control Valve Sourcebook—Power and Severe Service,* Fisher Controls International, Inc., Marshalltown, Iowa.

CONTROL-VALVE CAVITATION—
AN OVERVIEW

by Marc L. Riveland[1]

Cavitation is a hydrodynamic event of significant concern to the process control industry. It can be the source of unacceptable noise, vibration, material damage, and a decrease in the efficiency of hydraulic devices. Left unchecked, cavitation can shorten the operating life of critical and expensive hardware, upset process control, and create hazardous or unsuitable work environments for plant personnel.

Theoretically cavitation can occur in any process element or fitting which induces pressure changes. However, control valves, by virtue of their process function, are inherently problematic in this regard. Understanding the basic nature of cavitation and becoming familiar with available cavitation control products and techniques are the most effective means to avoid these negative consequences in practice.

CAVITATION FUNDAMENTALS

At the most fundamental level, cavitation concerns the growth and collapse of cavities in a liquid. This growth and collapse results from fluid pressure dynamics which, in the case of control valves, form when the local pressure of the fluid drops to the vapor pressure of the liquid. Conversely, if the local pressure then rises to a value above the vapor pressure, the cavity collapses. High local pressures and velocities associated with the collapse phases are the

---

[1] Senior Engineering Specialist, Fisher Controls International, Inc., Marshalltown, Iowa.

source of most problems. This type of cavitation is sometimes called vaporous cavitation and usually is considered the most detrimental form.

Pressure dynamics conducive to cavitation arise from several sources in a typical control valve. The traditional viewpoint primarily considers the classic mean pressure profile shown in Fig. 1. The shape of this general curve is a consequence of fluid continuity and conservation of energy, that is, pressure changes result from velocity changes and available energy dissipation. The mean fluid pressure is seen to decrease from the inlet value to some minimum value and to recover partially to the outlet value. Cavitation theoretically occurs when the minimum pressure is equal to the vapor pressure and the outlet pressure is above the vapor pressure.

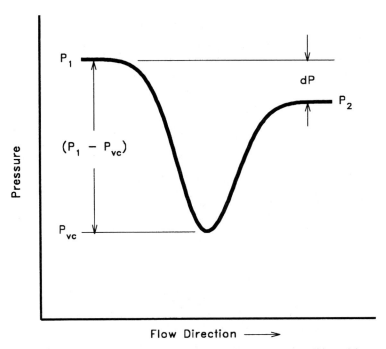

FIGURE 1   Idealized mean pressure profile through a control valve. This model represents mean pressure changes due to velocity and area changes, but does not account for local pressure fluctuations. (*Fisher Controls.*)

This model is suitable for conceptualizing the cavitation process, but it does not account for local pressure dynamics attributable to other sources. It is a particularly poor model for predicting the incipient cavitation condition. Flow through the complex geometry of control valves gives rise to boundary-layer separation, free shear zones, stagnation regions, vortices, eddies, and reentrant zones. These phenomena can produce local pressures significantly higher or lower than the mean pressure. These and the pressure fluctuations associated with fluid turbulence can be sufficient to initiate cavitation in very localized regions. Typically, then, cavitation begins well before the minimum mean pressure is reduced to the vapor pressure.

Vaporous cavitation is typically implied by the term cavitation. However, this term has also been used to describe a variety of related phenomena that involve bubble or cavity formation. Examples include effervescence or outgassing, the boiling of a liquid, and flashing. These different processes are distinguished from vaporous cavitation on the basis of the cavity contents or the controlling dynamics. A common characteristic of all of these different phenomena is the absence of the collapse of the cavity. Whereas most problems are associated with this phase, the absence of cavity collapse markedly changes the nature of the negative side effects. Each of these phenomena may have attendant problems, but they are different from cavitation (as defined above) in origin, manifestation, and treatment.

CAVITY BEHAVIOR AND NEGATIVE EFFECTS OF CAVITATION

Cavitation is actually a more complex phenomenon than depicted by the preceding simple model. Four distinct events, namely, nucleation, growth, collapse, and rebound, are identifiable in the cavity cycle. (The following discussion is based on very thorough coverage of the subject by Hammitt [1].)

The onset of cavitation is known as incipient cavitation. In order for a liquid to cavitate at or near the vapor pressure of the liquid it is necessary to have stabilized voids or inclusions such as an entrained noncondensable gas bubble or a solid particulate, available in the liquid. Known as nuclei, these provide a weakness in the liquid continuum from which vaporization and cavity formation can begin.

The forming cavity experiences an initial period of stable growth whenever one of these nuclei enters a region of reduced pressure. This growth is controlled primarily by the cavity gas pressure and the liquid surface tension. If the ambient fluid pressure falls to the vapor pressure of the liquid, the fluid is not thermodynamically stable as a liquid and begins to vaporize at the free surface. This vaporization contributes to cavity growth.

The cavity will continue to grow in this manner (assuming it remains in a low-pressure region) until it attains a critical radius. At this point, growth is no longer stable nor controlled; growth now proceeds explosively with substantial vaporization of the liquid. The ultimate degree and extent of cavity growth will be determined by the number and size of entrained nuclei, the properties of the liquid, and the extent and nature of the low-pressure region.

Cavity growth ceases and the collapse process begins when the fluid pressure increases. Unlike the comparatively symmetrical growth of the cavity, the collapse is very rapid and highly asymmetrical. This results from the inertial forces acting on the cavity as it moves into a lower-velocity (higher-pressure) region. Experimental studies [2] have revealed the presence of a very small high-velocity jet formed during such asymmetrical collapse.

Several additional growth-collapse cycles may follow the initial cycle in a phenomenon known as rebound. If the rate of cavity collapse exceeds the condensation rate, or if there are substantial noncondensable gases present, the cavity contents are compressed by the liquid rather than condensed to the liquid state. Mechanical energy is stored in the compressed gases and can be released to initiate another cycle. The total cavity volume decreases on each successive cycle until the process ceases. The collapse of a rebound cavity is generally more symmetrical than the initial collapse and is marked by the absence of the high-velocity "microjet." However, the rapid movement of the liquid surrounding the cavity induces a shock wave which propagates away from the cavity.

As mentioned previously, there are four recognized side effects of cavitation: excessive noise, excessive vibration, material damage, and a deterioration of the hydraulic effectiveness of the control valve. The first three effects stem from the collapse stage of cavity dynamics while the latter effect is attributable to the compressibility of the vapor-liquid mixture near the throat of the valve.

The collapsing cavity serves as the primary source of hydrodynamic noise and vibration. In fact, hydrodynamic noise is usually attributed entirely to cavitation; noise levels at subcavitating flows are not typically troublesome. The general subject of hydrodynamic noise and related prediction methods is very involved and in general not considered to be a mature technology. In practice, treatment of cavitation (the source) brings resultant hydrodynamic noise levels to within an acceptable level. A more detailed discussion is outside the immediate scope of this article.

Physical damage to the valve is probably the most frequent concern because of the associated cost, inconvenience, and unpredictable nature. Cavity collapse, again the primary source of the problem, initiates an attack on adjacent material surfaces. This attack, followed by the response or reaction of the material to the attack, determines the extent of total damage to the material (damage in this context is defined as any permanent deformation or loss of material).

It currently appears that the attack on a material does not consist of a single mechanism, but rather involves several forms which interact in a positive reinforcing manner. First there is evidence which suggests the presence of a mechanical component of attack in all instances of material damage. This mechanical attack can occur in either of two forms: high-velocity microjet impingement or shock-wave impingement on the material surface.

Mechanical attack must originate from a cavity collapsing near the material surface in order to impart damage to that surface regardless of which of the two forms is involved. If the small high-velocity jet established during the asymmetrical cavity collapse is close to the surface and impinges directly on the surface, a damaging attack will occur; otherwise no adverse material effect results. These microjets exhibit a preferred orientation toward rigid surfaces. Presumably the fluid resistance near the wall reduces the "supply" of fluid to that side of the cavity. Flow to the outboard side of the cavity is relatively unimpeded, so that the jet orientation is in the direction of the wall. The coupling dynamics of a highly compliant surface (such as an elastomeric material) effect the opposite behavior, that is, the orientation is *away* from the surface.

Similarly, the intensity of the shock waves dissipates rapidly with the propagation distance so that shock waves originating from cavities far removed from the surface (that is, more than one bubble diameter or so) have insufficient strength to impart significant damage.

Corrosion can play a significant role in the damage process in that it interacts with the mechanical forms of attack in a synergistic manner. Protective coatings can be removed by the mechanical attack, allowing chemical attack to occur. The chemical attack in turn deteriorates the material, making it more susceptible to mechanical attack. The process continues, resulting in a more aggressive damage process than that associated with either of the constituent forms individually.

Repeated attack by millions of microjets and shock waves results in the characteristic appearance of cavitation—a very rough pitted surface. An example of cavitation-damaged material is shown in Fig. 2, along with material damaged by an erosion process.

A number of factors can potentially affect the intensity of cavitation or, more importantly, the level of the associated negative effects of cavitation. Sometimes called scale effects or influences, these factors can either intensify or diminish cavitation-related problems relative

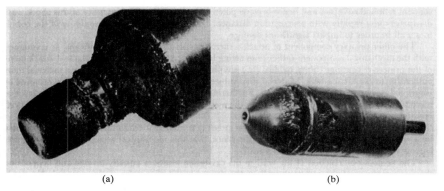

(a)   (b)

FIGURE 2   Typical appearance of damaged material. (*a*) Cavitation damage. (*b*) Flashing damage. (*Fisher Controls.*)

to equivalent installations under hydrodynamically similar operating conditions. The foremost effects of concern to control-valve applications are pressure- and velocity-related effects, size effects, and air-content effects. Other effects such as viscosity, surface tension, and various thermal property effects have also been investigated. However, in most industrial applications these are either of little significance or not sufficiently quantified to be able to adequately account for them.

The trend in general is that as the pressure, velocity, or size increases, the associated cavitation problems get worse. Numerous investigations (as documented by Hammitt) have borne out the fact that the degree of damage resulting from cavitation is very sensitive to the fluid velocity; that, in fact, the total damage imparted is an exponential function of the fluid velocity. The range of values reported for the exponent is very broad, usually between three and ten. However, there is some agreement that six is a representative number. Tullis [3] and Mousson [4] both provide data showing an increase in the damage rate as the upstream pressure increases. Investigations by Tullis [3] report an increase in the severity of the negative side effects of cavitation associated with an increase in the nominal size of the device.

Likewise, there are effects associated with the change in backpressure applied to a valve. For a monotonically decreasing outlet pressure at constant inlet conditions, two opposite trends can be rationalized: an intensifying effect (increasing vapor volume) and a diminishing effect (decreasing collapse intensity). The issue becomes a matter of determining which effect dominates at any given outlet pressure. Mousson's data [4] supply a partial answer to this by showing a maximum damage level existing roughly midway between the two extremes. Field experience with control valves is consistent with this. In some cases unacceptable levels of noise and vibration existing at flow conditions well below choked flow have been observed to diminish to satisfactory levels at choked flow conditions.

The presence of dissolved or entrained air (or any other noncondensable gas) has multiple effects on the cavitation process and associated problems. Increasing the amount of such gases has the effect of providing additional nucleation sites in the fluid. This contributes to an increase in the overall amount of cavitation and consequently the level of problems associated with cavitation. However, continued increase in the amount of air reduces the collapse velocities and disrupts the microjet and shock-wave attack mechanisms. This results in an overall attenuation of the negative effects of cavitation, even though cavitation is still occurring. Mousson reports a significant reduction in the damage levels with only a few percent of air (by volume) entrained in water under otherwise constant conditions.

SYSTEM DESIGN CONSIDERATIONS

While the need for controlling the problems created by cavitation is obvious, there is no universal method for accomplishing this. A number of techniques are available, each with inherent advantages and shortcomings. Familiarity with these practices provides greater opportunity to implement the most technically satisfactory and cost-effective solution for a given application.

A primary and preferred strategy is to consider potentially problematic conditions at the time a process system is designed. Awareness and avoidance of conditions conducive to control-valve cavitation are a highly effective means of reducing the risk of cavitation-related problems. Occasionally cavitation-related problems can be averted by simply locating the valve in a region of high overall pressure. The flow rate and pressure differential across the valve remain the same regardless of location, but overall fluid pressure within the valve body is increased proportionately, as depicted in Fig. 3. This results in a greater margin between the minimum pressures throughout the valve and the vapor pressure and thus decreases the likelihood that the fluid pressure will fall below the vapor pressure of the liquid.

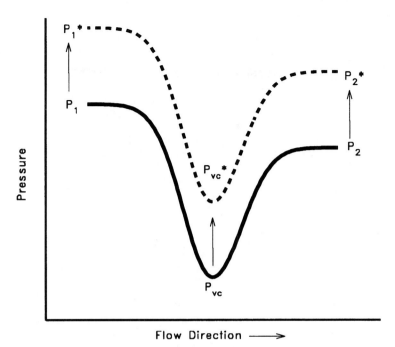

FIGURE 3   The effect of locating a valve in a region of overall higher pressure is to increase the mean pressure within the valve. (*Fisher Controls.*)

When placement of the valve within the system is not flexible, fluid pressures within the control valve may be increased artificially by introducing additional resistance to flow downstream of the valve. By placing a restriction such as an orifice plate or a second valve downstream of the valve, the back pressure is increased by the amount of the pressure differential across the restriction. Usually the pressure drop across the valve is decreased by this amount, that is, the inlet pressure remains constant but the back pressure increases, as shown in Fig. 4. Consequently the valve will realize less pressure drop for the same flow rate and the required valve coefficient must increase accordingly. In addition to increasing the fluid pressures within the valve, the fluid velocities will generally be reduced since the valve will operate at a larger opening. (This was not true of simply relocating the valve.) As discussed earlier, reduced velocity will have a pronounced mitigating effect on cavitation-related problems. The combined effects of increased fluid pressure and reduced velocity can be very effective in controlling cavitation. A word of caution is needed, however. Cavitation within the overall system may not always be eliminated or controlled by this method, but rather merely displaced from the valve to another location within the system. It is important to account for the possibility and consequence of any resulting cavitation at the downstream restriction.

The fixed-restriction alternative is best suited to on-off service since the device can only be optimized for a single flow rate. If the restriction is sized for high load and the system is operating at a lower load, the pressure drop will be very low due to decreased fluid velocity through the restriction. Consequently the valve will realize an increased pressure drop and again be at risk of cavitation-related problems. At higher flow rates than designed for, the orifice may well become the primary restriction and in turn limit or choke the flow at a lower flow rate than desired. Use of a second control valve in series affords a greater effective range, but is usually a more expensive solution and requires a more sophisticated control scheme for optimum performance.

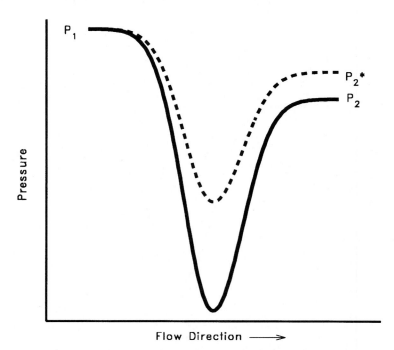

FIGURE 4 The effect of placing a fixed restriction downstream of a valve is to increase overall mean pressure within the valve (although the pressure drop usually is decreased). (*Fisher Controls.*)

Another method of controlling the damage, noise, and vibration resulting from cavitation (but not totally eliminating cavitation) is through noncondensable gas injection. This method, while effective, is very selective since not all processes will tolerate the introduction of gases. A gas which will not condense under the prevailing downstream conditions is injected (or aspirated) into the flowing fluid near the vena contracta. The continued presence of the gas phase during pressure recovery disrupts the cavity collapse process and limits the negative effects associated with cavity collapse. Caution must be exercised to introduce the gas at or downstream of the throat of the valve in order to avoid a reduction in flow from a two-phase mixture at the vena contracta. Individual valve suppliers should be consulted as to the viability of this method for the valve style being considered as well as the minimum amount of gas required and the exact location of introduction to the flow stream.

CONTROL-VALVE CONSIDERATIONS

If it is not possible to avoid cavitating conditions in the process system, it is necessary to contend with it at the control-valve level. Several alternative approaches can be considered.

First consideration should be given to material selection. If physical damage is of primary concern, it is sometimes possible to create a cavitation-tolerant environment by selecting materials more resistant to cavitation attack. Standard trims constructed of materials suited to the process fluid and service conditions often provide a cost-effective solution.

Proper material selection is not a black-and-white issue, nor is there one "best" material. The characterization of a material's resistance to cavitation lacks rigorous quantification. Currently, qualitative force ranking of material resistance to cavitation attack in combination with empirical "rules of thumb" governs selection. As a general rule, a material's hardness and resistance to corrosion are the foremost properties considered. Other properties which have shown a correlation to cavitation damage resistance to varying degrees include the ultimate resilience and strain energy to failure. However, no single property offers a consistent numerical correlation.

It is important to base material selection on the total attack, that is, considering both the mechanical and the chemical components. A notable exception to the "harder is better" rule is the widespread use of cobalt alloy 6 in cavitating service. Its combined hardness and corrosion resistance make it a preferred choice to harder materials currently available. However, even this material is not universally superior. It provides very poor protection in applications of boiler feed water treated with hydrazine. Even though the material is extremely hard, the amines attack the material and render it structurally inferior. Other materials which are chemically more resistant to the amines, such as S44004 (440C), are a preferred choice. Other popular materials frequently used in cavitating liquid service include other alloy steels, tool steels, certain stainless steels (such as the 300 series), and precipitation-hardened materials.

Ceramics is an emerging material category which shows promise of good cavitation damage resistance. Ceramics of practical interest to the control-valve industry consist of metals or metalloids combined with oxygen, carbon, nitrogen, or boron. Examples include aluminum oxide, zirconium dioxide, silicon carbide, and silicon nitride.

Other nonmetallic materials, such as elastomers and compliant materials in general, exhibit an ability to withstand levels of cavitation attack greater than standard structural indicators would suggest. This paradox apparently results from a dynamic interaction between the surface and the cavity, which orients the microjet away from the surface, thus eliminating the mechanical attack. While such behavior is appealing from a damage-control standpoint, Sanderson [5] points out that bonding difficulties, as well as potential pressure and temperature limitations, have curbed widespread use of such materials in industry.

It should be emphasized that all materials are vulnerable to cavitation attack. The rate of damage is a complicated function of the intensity of the cavitation attack, the total time of exposure, and the material characteristics. Material selection by itself can prolong the life of a component, but will not completely eliminate the possibility of damage and therefore is best utilized in conjunction with other abatement strategies.

If the protection offered by material selection is deemed inadequate by itself, or if noise and vibration are also of concern, it may be necessary to use special trim designs. A number of proprietary products are available from different valve manufacturers. These products and trims come in a wide variety of configurations, but are all based on one or more fundamental operational strategies with different tradeoffs between cost and performance.

The most common design concepts embraced by different valve manufacturers parallel many of the techniques used on a larger scale in the context of system strategies. The foremost objective of good cavitation control product design is to control energy conversions within the valve. Pressure recovery characteristics and trim velocities are favorably affected by strategically introducing resistance into the flow path. In general, overall fluid pressure recovery is reduced in such trims, as depicted in Fig. 5. Reduced-pressure recovery effectively reduces the tendency of the valve to cavitate by raising overall pressures in the valve compared to those in a high-recovery valve under the same conditions. Further benefit is realized in that, if cavitation does occur, the pressure differential driving cavity collapse $(P_1 - P_v)$ is reduced, which in turn tends to reduce cavitation intensity. The added resistance also reduces trim velocities, which in turn can reduce the negative effects of cavitation.

This objective is commonly achieved in practice by forcing fluid flow through successive stages or tortuous paths. When the pressure drop across a valve is staged, a portion of the total pressure drop is taken across each of a series of restrictions, or stages. This creates a much less efficient hydrodynamic path than an equivalent single restriction and results in

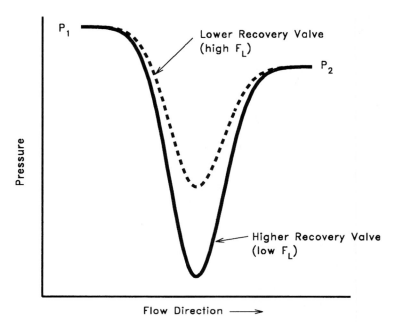

FIGURE 5 Lower recovery valves tend to have higher overall mean pressures at the same pressure drop as a higher-recovery valve. However, local pressure fluctuations may offset this effect. (*Fisher Controls.*)

lower pressure recovery. Furthermore, the decreased flow efficiency requires comparatively larger flow passages, hence lower velocities, under similar flow conditions. Tortuous-path treatment, on the other hand, utilizes a labyrinthal flow path to induce irreversible energy conversions, which in turn have the same impact on pressure recovery and velocity that staging does.

Another fundamental design strategy consists of dividing the flow stream into multiple parallel flow paths. Whereas many of the cavitation-related problems tend to scale with the physical size of the flow stream, the reduced size of individual flow paths helps to reduce overall cavitation-related problems, particularly noise and vibration. To avoid potential plugging problems associated with restrictive flow passages, a compromise between degree of cavitation control and passage clearance must be reached.

Finally it is possible to redirect the flow stream to avoid direct impingement of cavitating flows on critical control surfaces. As pointed out earlier, a cavity must collapse in close proximity to a surface in order to damage that surface. By removing collapsing cavities from the surface the threat of damage is minimized, even though the cavitation has not necessarily been eliminated or reduced.

Design of any control valve to minimize the effects of cavitation involves a tradeoff between higher comparative costs, lower relative capacity, the degree of protection required by a particular application, and the ability of the valve to tolerate dirty fluids. Highly optimized designs actually prevent the formation of any significant cavitation, whereas standard trim would cavitate heavily. This degree of protection is not warranted by all process control applications. Therefore a variety of trims generally designed to a reasonably specific set of conditions are available to meet the variety of process needs.

CONTROL-VALVE EVALUATION, SIZING, AND SELECTION

A detailed discussion of the process of sizing and selecting control valves for use in cavitating service would be quite involved and not practical in the context of this article. The following is intended to serve as an overview of the fundamental elements of this process.

Several different control-valve application guidelines currently exist in the control-valve industry, with each utilizing parameters and coefficients unique to that approach. Confusion results when identical parameters are used differently, or when different definitions are associated with the same nomenclature.

Control-valve service is generally characterized by variant forms of the pressure coefficient or pressure ratios. Several forms have been adopted over time, the exact choice of which varies between valve manufacturers.

$$\sigma_1 = \frac{P_1 - P_v}{P_1 - P_2}$$

$$\sigma_2 = \frac{P_2 - P_v}{P_1 - P_2}$$

$$\sigma_3 = \frac{P_1 - P_v}{\rho V^2/2g_c}$$

$$\sigma_4 = \frac{P_2 - P_v}{P_1 - P_2 + \rho V^2/2g_c}$$

$$K = \frac{P_1 - P_2}{P_1 - P_v}$$

Mathematical relationships can be developed which relate these different parameters in terms of each other and the indicated pressure and velocity terms. Categorically the advantages of these types of parameters are that they are functionally simple and that the required terms are generally readily available from the known service conditions. However, these parameters are not complete similarity parameters and do not, without modification, account for the numerous scale effects previously identified (such as pressure, size).

The behavior of specific control-valve hardware under different conditions must likewise be characterized in a format compatible with the service parameters. Generally, a limiting value of the aforementioned pressure ratios is established as a benchmark for decision making in the sizing and selection process. This limiting value can then be used to determine an acceptable pressure drop for the given service conditions.

Determining meaningful benchmarks from objective laboratory testing is obviously very desirable, but in practice it is difficult to achieve. One complicating factor is simply the fact that different levels of tolerance exist for side effects of noise, vibration, and material damage. In other words, the "acceptable" level of cavitation varies from application to application.

Another major problem in evaluating control-valve cavitation is that it is very difficult to directly observe and measure cavitation. No "scientifically pure," or completely objective, laboratory method exists for evaluating the cavitation which occurs in a control valve. The usual approach is to monitor the effect of cavitation on characteristics such as noise levels, vibration levels, damage rate, or flow efficiency, and infer information about the behavior of other cavitation effects under field conditions. The most valid information gleaned from such an approach is obviously that which relates directly to the actual side effect observed. For instance, quantifying noise and vibration does not account for all of the factors which affect the rate of material damage. As Robertson [6] points out, side effects exhibit different trends with respect to increasing cavitation intensity.

Various tests do exist whereby a particular cavitation attribute is measured and various conclusions are inferred from the results. One such example is determination of the onset of cavitation, or incipient cavitation. Reasonably objective tests are available to evaluate this parameter for any given valve [7, 8]. This parameter can be used in conjunction with the particular service conditions to determine the pressure drop at which cavitation will just begin. However, for the vast majority of cases this point is far too conservative for sizing purposes, and specific published information is generally lacking. Furthermore, this point is also subject to scale effects associated with the fluid properties or service conditions.

Another parameter well established by test is the pressure recovery coefficient F_L,

$$F_L = \frac{P_1 - P_2}{P_1 - P_{vc}}$$

The measurement of this parameter is covered in [9]. Universally, F_L has only one quantitative use, and that is to determine the choked flow rate through a given valve under a given set of conditions. However, because it is a pressure recovery term, it is qualitatively related to a valve's ability to accommodate cavitating flow. Reflecting on the previous discussions regarding pressure recovery, it is apparent that if the same pressure differential is applied to both a high-recovery device and a low-recovery device, the high-recovery device will have the lower vena contracta pressure (Fig. 5). Whereas low-pressure recovery devices are characterized by large values of the pressure recovery coefficient, a valve with a high value of F_L is a better candidate for cavitating service than one with a low value insofar as overall fluid pressures within the valve are greater. Used in conjunction with other parameters and additional information, F_L can play a role in sizing valves for cavitation service. Unfortunately, over time, it has been interpreted as "the" cavitation index and this is not correct. Many cavitation control hardware features are not reflected in the value of the pressure recovery coefficient.

Other tests exist whereby control-valve cavitation is characterized by the behavior of a particular side effect of cavitation. Figure 6 depicts pipe-wall vibration as a function of the desired flow parameter (in this case σ_2). Various inflection points are evident and can serve as benchmarks. However, these are generalizations in terms of effects other than vibration (or noise), and the suitability of any of these points for sizing purposes depends on the anticipated service and specific cavitation concern (see Ball and Tullis [8]). The significance of these measurements to actual installed behavior is still largely inferred. Consequently only a limited number of such benchmarks are currently available by laboratory test

While a number of the proposed methods hold promise of future development, no single universal method is currently endorsed by all valve manufacturers. In moving toward a practical industry technology, codes and standards groups are beginning to address this issue. As of this writing a subcommittee of the Instrument Society of America is in the process of preparing recommended practices concerning these issues. Formal documents should be forthcoming. However, it is still necessary to consult with the individual valve supplier for guidance as to the exact performance limits for each product, especially when it comes to proprietary valve trims. These recommendations will be based on an understanding of cavitation fundamental and scale effects, knowledge of specific product design, valve service (continuous, throttling, on-off), and experience with that product in actual service.

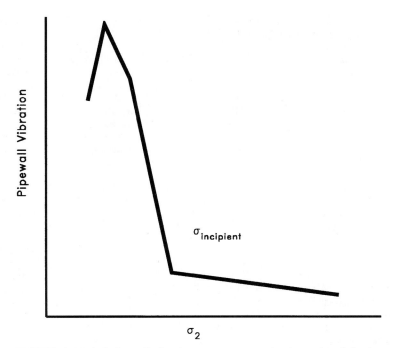

FIGURE 6 Typical pipe-wall vibration versus σ_2 curve showing various inflection points. This type of test can be used to determine the point of incipient cavitation [7, 8]. (*Fisher Controls.*)

REFERENCES

1. Hammitt, F. G.: *Cavitation and Multiphase Flow Phenomena,* McGraw-Hill, New York, 1980.

2. Knapp, R. T., and A. Hollander: "Laboratory Investigations of the Mechanism of Cavitation," *Trans. ASME,* vol. 70 (1948).

3. Tullis, J. P.: "Cavitation Scale Effects for Valves," *J. Hydraulics Div., ASCE,* vol. 99, p. 1109 (1973).

4. Mousson, J. M.: "Pitting Resistance of Metals under Cavitation Conditions," *Trans. ASME,* vol. 59, pp. 399–408 (1937).

5. Sanderson, R. L.: "Elastomers for Cavitation Damage Resistance, paper C.I.82-908, presented at the ISA Int. Conf. and Exhibit, Philadelphia, Pennsylvania, Oct. 1982.

6. Robertson, J. M.: "Cavitation Today—An Introduction," in *Cavitation State of Knowledge,* American Society of Mechanical Engineers, New York, June 1969.

7. Riveland, M. L.: "The Industrial Detection and Evaluation of Control Valve Cavitation," *ISA Trans.,* vol. 22, no. 3 (1983).

8. Ball, J. W., and P. P. Tullis: "Cavitation in Butterfly Valves," *J. Hydraulics Div, ASCE,* vol. 99, p. 1303 (1973).

9. ISA S75.01, "Control Valve Capacity Test Procedure," Instrument Society of America, Research Triangle Park, North Carolina, 1985.

CONTROL-VALVE NOISE

EDITOR'S NOTE

The principal element of control-valve noise is aerodynamic in nature. Research on this topic has been going on for several years, but important findings are not expected to be finalized until 1994–1995. Some handbook users may elect to refer to a more detailed discussion of the topic in the 3rd edition of this handbook. Other sources of valve noise are updated and delineated briefly in this current article (1993).

by Allen C. Fagerlund[1]

Fluid and transmission systems are major sources of industrial noise. Elements within the systems that contribute to the noise are control valves, abrupt expansions of high-velocity flow streams, compressors, and pumps. Control-valve noise is a result of the turbulence introduced into the flow stream in producing the permanent head loss required to fulfill the basic function of the valve.

NOISE TERMINOLOGY

Noise is commonly defined as unwanted or annoying sound. Noise is frequently described or specified by the physical characteristics of sound. The definitive properties of sound are the magnitude of sound pressure and the frequency of pressure fluctuation, as illustrated in Fig. 1.

Sound-pressure P_s measurements are normally root-mean-square (rms) values of sound pressure expressed in microbars. Because the range of sound pressure of interest in noise measurements is $\sim 10^8$ to 1, it is customary to deal with sound pressure level (L_P) instead of sound pressure. L_P is a logarithmic function of the relative amplitude of sound pressure and is expressed mathematically as

$$L_P = 20 \log_{10} \frac{P_s}{0.0002 \ \mu\text{bar}} \quad \text{dB}$$

The arbitrarily selected reference sound pressure of 0.0002 µbar is approximately the sound pressure required at 1000 Hz to produce the faintest sound that the average young person with normal hearing can detect. The characteristic of the L_P scale is such that each change of 6 dB in level represents a change in the amplitude of sound pressure by a factor of 2.

The apparent loudness of a sound varies not only with the amplitude of sound pressure but also as a function of frequency. The human ear responds to sounds in the frequency range between 20,000 and 18,000 Hz. The normal ear is most sensitive to pressure fluctuations in the neighborhood of 3000 to 4000 Hz. Therefore the degree of annoyance created by a specific sound is a function of both sound pressure and frequency.

---

[1] Senior Research Specialist, Fisher Controls International, Inc., Marshalltown, Iowa.

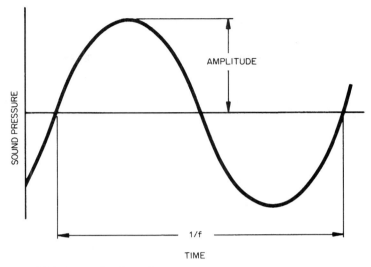

FIGURE 1 Properties of sound.

L_P measurements are often weighted to adjust the frequency response. Weighting that attenuates the lower frequencies to approximate the response of the human ear is called A-weighting. Figure 2 shows L_P correction as a function of frequency for A-weighted octave-band analysis.

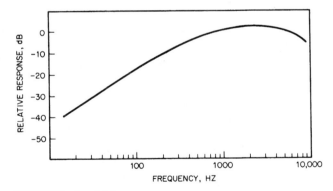

FIGURE 2 A-weighting curve.

Sound intensity I is defined as the acoustic sound power transmitted per unit area, perpendicular to a specified direction. The common unit of measurement for sound intensity is watts per square centimeter. Sound intensity for a plane wave is given by the relation

$$I = \frac{P_s^2}{\rho c} \tag{1}$$

where the product of the density ρ and sonic velocity c of the transmitting medium represents the characteristic impedance.

For measurement and comparison of sound intensities, it is more convenient to deal with sound intensity levels than with absolute values of sound intensity. Sound intensity level is commonly defined by the following relation:

$$\text{Sound intensity level} = 10 \log_{10} \frac{I}{10^{-16}} \quad \text{dB} \qquad (2)$$

The reference sound intensity is selected as 10^{-16} W/cm^2. This is approximately the minimum intensity audible to the average human ear at 1000 Hz/s for standard air. Table 1 presents the approximate overall sound levels of some familiar sound environments.

TABLE 1   Approximate Sound Levels of Familiar Sounds

Sound	Sound level, dB
Pneumatic rock drill	130
Jet takeoff [at 200 ft (61 m)]	120
Boiler factory	110
Electric furnace (area)	100
Heavy street traffic	90
Tabulation machine room	80
Vacuum cleaner [at 10 ft (3 m)]	70
Conversation	60
Quiet residence	50
Electric clock	20

SOURCES OF VALVE NOISE

The major sources of control valve noise are (1) Mechanical vibration of valve components and (2) Fluid-generated noise, namely hydrodynamic noise and aerodynamic noise.

Mechanical Noise

The vibration of valve components is a result of random pressure fluctuations within the valve body or fluid impingement on movable or flexible parts. The most prevalent source of noise resulting from mechanical vibration is the lateral movement of the valve plug relative to the guide surfaces. Sound produced by this type of vibration normally has a frequency of less than 1500 Hz and is often described as a metallic rattling. The physical damage incurred by the valve plug or associated surfaces is generally of more concern than the noise emitted.

A second source of mechanical noise is a valve component resonating at its natural frequency. Resonant vibration of valve components produces a single pitched tone, normally having a frequency between 3000 and 7000 Hz. This type of vibration produces high levels of stress that may ultimately produce fatigue failure of the vibrating part.

Noise resulting from mechanical vibration has, for the most part, been eliminated by improved valve design and is generally considered a structural problem rather than a noise problem.

Hydrodynamic Noise

Control valves handling liquid flow streams can be a substantial source of noise. The flow noise produced is referred to as hydrodynamic noise and may be categorized with respect to the specific flow classification or characteristic from which it is generated. Liquid flow can be divided into three general classifications: (1) noncavitating, (2) cavitating, and (3) flashing.

Noncavitating liquid flow generally results in very low ambient noise levels. It is generally accepted that the mechanism by which the noise is generated is a function of the turbulent velocity fluctuations of the fluid stream, which occur as a result of rapid deceleration of the fluid downstream of the vena contracta as the result of an abrupt area change.

The major source of hydrodynamic noise is cavitation. This noise is caused by the implosion of vapor bubbles formed in the cavitation process. Cavitation occurs in valves controlling liquids when the service conditions are such that the static pressure downstream of the valve is greater than the vapor pressure and at some point within the valve the local static pressure because of either high velocity or intense turbulence is less than or equal to the liquid vapor pressure.

Figure 3 depicts the pressure profile of a cavitating flow stream as a function of distance along the stream. Vapor bubbles are found in the region of minimum static pressure and subsequently are collapsed or imploded as they pass downstream into an area of higher static pressure. Noise produced by cavitation has a broad frequency range and is frequently described as a rattling sound similar to that which would be anticipated if gravel were in the fluid stream. Since cavitation may produce severe damage to the solid boundary surfaces that confine the cavitating fluid, noise produced by cavitation is of secondary concern.

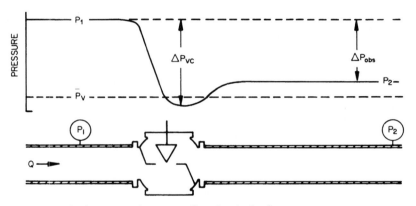

FIGURE 3   Static pressure along stream line of cavitating flow.

Pertaining to the design of quiet valves for liquid application, the problem resolves itself into one of designing to reduce cavitation. Service conditions that will produce cavitation can readily be calculated. The use of staged or series reductions provides a viable solution to cavitation and hence hydrodynamic noise.

Flashing is a phenomenon that occurs in liquid flow when the differential pressure across a restriction is greater than the differential between the absolute static and vapor pressures at the inlet to the restriction, that is, $\Delta P > P_1 - P_v$. The resulting flow stream is a mixture of the liquid and gas phases of the fluid. Noise resulting from a valve handling a flashing fluid is a result of the deceleration and expansion of the two-phase flow stream.

Test results supported by field experience indicate that noise levels in noncavitating liquid applications are quite low and generally are not considered a noise problem.

Aerodynamic Noise

Aerodynamic noise is created by turbulence in a flow stream as a result of deceleration or impingement. The principal area of noise generation in a control valve is the recovery region immediately downstream of the vena contracta, where the flow field is characterized by intense turbulence and mixing.

NOISE CONTROL

Either one or both of the following basic approaches can be applied for noise control:

1. *Source treatment.* Prevention or attenuation of the acoustic power at the source (quiet valves)
2. *Path treatment.* Reduction of noise transmitted from a source to a receiver.

Quiet Valves

Based on the preceding discussion, the parameters that determine the level of noise generated by compressible flow through a control valve for a given application are the geometry of the restrictions exposed to the flow stream, the total valve flow coefficient, the differential pressure across the valve, and the ratio of the differential pressure to the absolute inlet pressure.

It is conceivable that a valve could be designed that utilizes viscous losses to produce the permanent head loss required. Such an approach would require valve trim with a very high equivalent length, which becomes impractical from the standpoint of both economics and physical size.

The noise characteristic or noise potential of a regulator increases as a function of the differential pressure ΔP and the ratio of the differential pressure to the absolute static pressure at the inlet $\Delta P/P_1$. Thus for high-pressure-ratio applications ($\Delta P/P_1 > 0.7$) an appreciable reduction in noise can be effected by staging the pressure loss through a series of restrictions to produce the total pressure head loss required.

Generally in control valves, noise generation is reduced by dividing the flow area into a multiplicity of smaller restrictions. This is readily accomplished with a cage-style trim, as shown in Fig. 4.

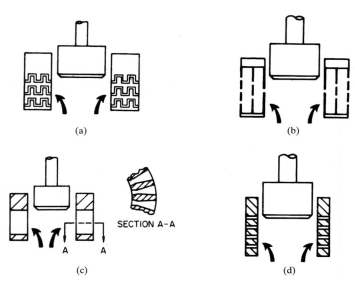

(a) (b)

SECTION A-A

(c) (d)

FIGURE 4   Valve trim designed for noise attenuation.

Critical to the total noise reduction that can be derived from the utilization of many small restrictions versus a single or a few large restrictions is the proper size and spacing of restrictions such that the noise generated by jet interaction is not greater than the summation of the noise generated by the jets individually. It has been found that optimum size and spacing are very sensitive to the pressure ratio $\Delta P/P_1$.

It should be noted that, when multiple small-hole restrictions are used in series within the valve to distribute the pressure drop, they can act as strainers and are very susceptible to plugging as a result of either solid particles in the gas stream (dirty gas) or hydrate formation prior to the last stage.

However, for control-valve applications operating at high-pressure ratios ($\Delta P/P_1 \gtrsim 0.7$) the series restriction approach, splitting the total pressure drop between the control valve and a fixed restriction (diffuser) downstream of the valve, can be very effective in minimizing the noise. In order to optimize the effectiveness of a diffuser, it must be designed (special shape and sizing) for each given installation so that the noise levels generated by the valve and diffuser are equal. Figure 5 depicts a typical valve-plus-diffuser installation.

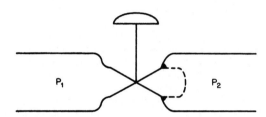

FIGURE 5   Two-stage pressure reduction with diffuser.

Path Treatment

A second approach to noise control is path treatment. Sound is transmitted through the medium that separates the source from the receiver. The speed and efficiency of sound transmission is dependent on the properties of the medium through which it is propagated. Path treatment consists of regulating the impedance of the transmission path to reduce the acoustic energy communicated to the receiver.

In any path treatment approach to control valve noise abatement, consideration must be given to the amplitude of noise radiated by both the upstream and the downstream piping. Since, when all else is equal, an increase in static pressure reduces the noise transmitted through a pipe, the upstream noise levels are always less than those downstream. Also, the fluid propagation path is less efficient moving back through the valve.

Dissipation of acoustic energy by the use of acoustical absorbent materials is one of the most effective methods of path treatment. Whenever possible, the acoustical material should be located in the flow stream either at or immediately downstream of the noise source. This approach to abatement of aerodynamic noise is accommodated by in-line silencers. In-line silencers effectively dissipate the noise within the fluid stream and attenuate the noise level transmitted to the solid boundaries. Where high mass-flow rates or high-pressure ratios across the valve exist, in-line silencers are often the most realistic and economical approach to noise control. The use of absorption-type in-line silencers can provide almost any degree of attenuation desired. However, economic considerations generally limit the insertion loss to ~25 dB.

Noise that cannot be eliminated within the boundaries of the flow stream must be eliminated by external treatment or isolation. This approach to the abatement of control-valve noise includes the use of heavy-walled piping, acoustical insulation of the exposed solid boundaries of the fluid stream, and the use of insulated boxes, rooms, and buildings to isolate the noise source.

In closed systems (not vented to the atmosphere) any noise produced in the process becomes airborne only by transmission through the solid boundaries that contain the flow stream. The sound field in the contained flow stream forces the solid boundaries to vibrate, which in turn causes pressure disturbances in the ambient atmosphere that are propagated as sound to the receiver. Because of the relative mass of most valve bodies, the primary surface of noise radiation to the atmosphere is the piping adjacent to the valve. An understanding of the relative noise transmission loss as a function of the physical properties of the solid boundaries of the flow stream can be beneficial in noise control for fluid transmission systems.

A detailed analysis of noise transmission loss is beyond the scope of this article. However, it should be recognized that the spectrum of the noise radiated by the pipe has been shaped by the transmission loss characteristic of the pipe and is not that of the noise field within the confined flow stream. For a comprehensive analysis of pipe transmission loss, see [1].

Acoustic insulation of the exposed solid boundaries of a fluid stream is an effective means of noise abatement for localized areas. Test results indicate that ambient noise levels can be attenuated 10 dB per inch of insulation thickness.

Path treatment such as the use of heavy-walled pipe or external acoustical insulation can be a very economical and effective technique for localized noise abatement. However, it should be pointed out that noise is propagated for long distances via a fluid stream and that the effectiveness of heavy-walled pipe or external insulation terminates where the treatment is terminated.

A simple sound survey of a given area will establish compliance or noncompliance with the governing noise criterion, but not necessarily either identify the primary source of noise or quantify the contribution of individual sources. Frequently piping systems are installed in environments where the background noise due to highly reflective surfaces and other sources of noise in the area make it impossible to use a sound survey to measure the contribution a single source makes to the overall ambient noise level.

A study of sound transmission loss through the walls of commercial piping indicated the feasibility of converting pipe-wall vibrations to sound levels. Further study resulted in a valid conversion technique as developed in [2].

The vibration levels may be measured on the piping downstream of a control valve or other potential noise source. Sound pressure levels expected are then calculated based on the characteristics of the piping. Judgment can then be made as to the relative contribution of each source to the total sound field as measured with a microphone. The use of vibration measurements effectively isolates a source from its environment.

REFERENCES

1. Fagerlund, A. C.: "Sound Transmission through a Cylindrical Pipe Wall," *J. Eng. Ind., Trans. ASME,* vol. 103, pp. 355–360, 1981.

2. Fagerlund, A. C.: "Conversion of Vibration Measurements to Sound Pressure Levels," Publ. TM-33, Fisher Controls International, Inc., Marshalltown, Iowa.

SERVOMOTOR TECHNOLOGY IN MOTION CONTROL SYSTEMS[1]

Motion control embraces a very wide range of applications. The motion-related variables of position, speed, and velocity are key variables in those industries which manufacture and assemble discrete metal, plastic, wood, and other solid materials (as contrasted with the fluids and bulk solids handled in the process industries). Also included are products made in long, continuous lengths, such as tubing, piping, rails, extrusions, paper sheets, printed materials, textile fabrics, films, and coatings among numerous others. Motion control is important in conveying and warehousing storage and retrieval, and of paramount significance wherever robots are used to manipulate and transfer materials. Allied applications include the transportation of materials and people over long distances over land, by air, and by sea, thus embracing navigation. Other areas, such as personal vehicles and construction, earth-moving, and mining equipment, are much less automated, but are depending more and more on the guidance received from motion instrumentation.

This article addresses motion control as it pertains to industrial manufacturing. In this area of electronic motion control, systems fall within a relatively limited power range, typically up to about 10 hp (7 kW), and require varying degrees of precision. Applications embrace three principal manufacturing objectives:

1. The positioning of materials with reference to production machines, typically metal, plastic, ceramic, and woodworking equipment, as used in contouring, shaping, drilling, cutting, drawing, and many other frequently encountered production operations. These would include the extremely precise operations encountered in microelectronic components manufacture as well as edge control and registration control for labeling, painting, and printing of parts in various stages of manufacture.

2. The positioning of two or more discrete pieces that must be matched geometrically during assembly operations.

3. Quality control inspections of in-process and completed products.
   Sensors are described in another section of this handbook.

ELECTRIC SERVOS

Although hydraulic and pneumatic servos are used in motion control systems, a majority of servomotors is electric. They take a variety of design formats and exhibit performance characteristics that range widely in terms of power, precision, size, and adaptability to specific applications. Extensive research and development has gone into electric motor design and construction, particularly during the past decade or so, in keeping with the improvements in production machinery and materials transfer, and the refinement of robotics. Notable has been the integration of servo systems into computerized control at the machine level, the cell level, and the total factory level.

[1] The cooperation of the technical and engineering staff of Parker-Hannifin, Compumotor Division, Rohnert Park, California, in the preparation of this article is gratefully acknowledged. The editors are indebted to the staff of Bodine Electric Company, Chicago, Illinois, for inputs pertaining to step motor drivers.

The basic elements of a motion control system are shown in Fig. 1. Some terms frequently used are given in the following.

Absolute Positioning.   Refers to a motion control system using position feedback devices (absolute encoders or resolvers) to maintain a given mechanical location.

Absolute Programming.   Positioning coordinate reference where all positions are specified relative to some reference or zero position. This differs from incremental programming, where distances are specified relative to the current position.

Acceleration.   Change in velocity as a function of time. Acceleration usually refers to increasing velocity and deceleration describes decreasing velocity.

Accuracy.   Measure of the difference between expected position and actual position of a motor or mechanical system. Motor accuracy is usually specified as an angle representing the maximum deviation from expected position.

Ac Servo.   General term referring to a motor drive that generates sinusoidally shaped motor currents in a brushless motor wound as to generate sinusoidal back EMF.

Ambient Temperature.   Temperature of the cooling medium, usually air, immediately surrounding the motor or other device.

ASCII (American Standard Code for Information Interchange).   Code that assigns a number series of electrical signals to each numeral and letter of the alphabet. In this manner, information can be transmitted between machines as a series of binary numbers.

Bandwidth.   Measure of system response. This is the frequency range that a control system can follow.

BCD (Binary coded decimal).   Encoding technique used to describe the numbers 0 through 9 with four digital (on or off) signal lines. Popular in machine tool equipment. BCD interfaces are being replaced by interfaces that require fewer wires, such as RS-232C.

BEMF.   Reverse bias created in a dc permanent-magnet motor as the rotor (a conductor) is rotated so as to cross the magnetic field created by the stator's energized winding; $= K_E W$.

Bode Plot.   Graph of system gain and phase versus input frequency which illustrates the steady-state characteristics of the system graphically.

Break Frequency.   Frequency or frequencies at which the gain changes slope on a Bode plot. (Break frequencies correspond to the poles and zeros of the system.)

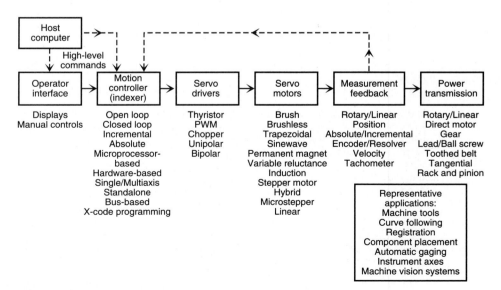

FIGURE 1   Basic elements of motion control system.

Brushless dc Servo.   General term referring to a motor drive that generates trapezoidally shaped motor currents in a motor wound as to generate trapezoidal BEMF.

Closed Loop.   Broadly applied term relating to any system where the output is measured and compared with the input. The output is then adjusted to reach the desired condition. In motion control, the term is used to describe a system wherein a velocity or position (or both) transducer is used to generate correction signals by comparison with desired parameters.

Commutation.   Switching sequence of drive voltage into motor phase windings necessary to assure continuous motor rotation. A brushed motor relies on brush or bar contact to switch the windings mechanically. A brushless motor requires a device which senses motor rotational position and feeds that information to a drive, which determines the next switching sequence.

Critical Damping.   A system is critically damped when the response to a step change in desired velocity or position is achieved in the minimum possible time with little or no overshoot.

Crossover Frequency.   Frequency at which the gain intercepts the 0-dB point on a Bode plot. Used in reference to the open-loop gain plot.

Daisy Chain.   Term used to describe the linking of several RS-232C devices in sequence such that a single data stream flows through one device and on to the next. Daisy-chained devices usually are distinguished by device addresses, which serve to indicate the desired destination for data in the stream.

Damping.   Indication of the rate of decay of a signal to its steady-state value. Related to settling time.

Damping Ratio.   Ratio of actual damping to critical damping. A value less than 1 denotes an underdamped system; greater than 1 an overdamped system.

Dead Band.   Range of input signals for which there is no system response.

Decibel.   Logarithmic measurement of gain. If G is a system's gain (ratio of output to input), then $20 \log G$ = gain in decibels (dB).

Detent Torque.   Minimal torque present in an unenergized motor. The detent torque of a step motor is typically about 1 percent of its static energized torque.

Direct-Drive Servo.   High-torque, low-speed servomotor with a high-resolution encoder or resolver intended for direct connection to the load without going through a gearbox.

Duty Cycle.   For a repetitive cycle, ratio of on time to total cycle time.

Efficiency.   Ratio of power output to power input.

Electrical Time Constant (dc Motor).   Ratio of armature inductance to armature resistance.

Encoder.   Device that translates mechanical motion into electronic signals, used for monitoring position or velocity.

Form Factor.   Ratio of the rms value of a harmonic signal to its average value in one half-wave.

Friction.   Resistance to motion. Friction can be constant with varying speed (coulomb) or proportional to speed (viscous).

Gain.   Ratio of system output signal to system input signal.

Holding Torque.   Sometimes called static torque, it refers to the maximum external force (or torque) that can be applied to a stopped, energized motor without causing the motor to rotate continuously.

Home.   Reference position in a motion control system derived from a mechanical datum or switch. Often designated as zero position.

Hybrid Servo.   Brushless servomotor based on a conventional hybrid stepper. It may use either a resolver or an encoder for commutation feedback.

Hysteresis.   Difference in response of a system to an increasing or a decreasing input signal.

IEEE-488.   Digital data communications standard popular in instrumentation electronics. This parallel interface is also known as general-purpose interface bus (GPIB).

Incremental Motion.   Term used to describe a device that produces one step of motion for each step command (usually a pulse) received.

Incremental Programming.   Coordinate system whose positions or distances are specified relative to the current position.

Indexer or PMC.   Programmable motion controller (PMC) primarily designed for single- or multiaxis motion control with input/output (I/O) as an auxiliary function.

Inertia.   Measure of an object's resistance to a change in velocity. The larger an object's inertia, the larger the torque that is required to accelerate or decelerate it. Inertia is a function of an object's mass (and shape).

Inertial Match.   For most efficient operation, the system coupling ratio should be selected so that the reflected inertia of the load is equal to the rotor inertia of the motor.

Lead Compensation Algorithm.   Mathematical equation implemented by a computer to decrease the delay between the input and the output of a system.

Limits.   Sensors, in a properly designed motion control system, that alert the control electronics that the physical end of travel is being approached and motion should stop.

Logic Ground.   Electric potential to which all control signals in a particular system are referenced.

Mechanical Time Constant.   Time for an energized dc motor to reach two-thirds of its set velocity. Based on a fixed voltage applied to the windings.

Microstepping.   Electronic control technique that proportions the current in a step motor's windings to provide additional intermediate positions between poles. Produces smooth rotation over a wide speed range and high positional resolution.

Midrange Instability.   Condition resulting from energizing a motor at a multiple of its natural frequency (usually the third order condition). Torque loss and oscillation can occur in underdamped open-loop systems.

Open Collector.   Term used to describe a signal output that is performed with a transistor. An open collector output acts like a switch closure with one end of the switch at ground potential and the other end accessible.

Open Loop.   Refers to a motion control system where no external sensors are used to provide position or velocity correction signals.

Optoisolated.   Method of sending a signal from one piece of equipment to another without the usual requirement of common ground potentials. The signal is transmitted optically with a light source (usually a light-emitting diode, LED) and a light sensor (usually a photosensitive transistor). These optical components provide electrical isolation.

Parallel.   Refers to a data communication format wherein many signal lines are used to communicate more than one piece of data at the same time.

Phase Angle.   Angle at which the steady-state input signal to a system leads the output signal.

Phase Margin.   Difference between 180° and the phase angle of a system at its crossover frequency.

PLC (Programmable Logic Controller).   Machine controller which actuates relays and other input-output units from a stored program. Additional modules support motion control and other functions.

Pole.   Frequency at which the transfer function of a system goes to infinity.

Pulse Rate.   Frequency of the step pulses applied to a motor driver. The pulse rate, multiplied by the resolution of the motor-drive combination (in steps per revolution) yields the rotational speed in revolutions per second.

Pulse-Width Modulation (PMW).   Method of controlling the average current in a motor's phase winding by varying the on time (duty cycle) of transistor switches.

Ramping.   Acceleration and deceleration of a motor. Term may also refer to the change in frequency of the applied step pulse train.

Rated Torque.   Torque-producing capacity of a motor at a rated speed. This is the maximum continuous torque the motor can deliver to a load and is usually specified with a torque-versus-speed curve.

Regeneration.   Usually refers to a circuit in a drive amplifier which accepts and either dissipates or stores energy produced by a rotating motor during either deceleration or free-wheel shutdown.

Registration Move.   Changing the predefined move profile, which is being executed, to a different predefined move profile following receipt of an input or interrupt.

Repeatability. Degree to which the positioning accuracy for a given move, when performed repetitively, can be duplicated.

Resolution. Smallest positioning increment that can be achieved by a given system. Frequently defined as the number of steps required for a motor shaft to rotate one complete revolution.

Resolver. Feedback device with a construction similar to a motor's construction (stator or rotor). Provides velocity and position information to a drive's microprocessor or digital signal processor (DSP) to commutate the motor electronically.

Resonance. Condition resulting from energizing a motor at a frequency at or close to the motor's natural frequency. Lower-resolution open-loop systems will exhibit large oscillations from minimal input.

Ringing. Oscillation of a system following a sudden change in state.

Rms (Root-Mean-Square) Torque. For an intermittent duty cycle application, the rms torque is equal to the steady-state torque which would produce the same amount of motor heating over long periods of time,

$$T_{\text{rms}} = \sqrt{\frac{\Sigma(T_i^2\, t_i)}{\Sigma t_i}}$$

where T_i is the torque during interval i and t_i the time of interval i.

RS-232C. Data communications standard that encodes a string of information on a single line in a time-sequential format. This standard specifies the proper voltage and the requirements so that different manufacturers' devices are compatible.

Servo. System, consisting of several devices, which continuously monitors actual information (position, velocity), compares these values to desired outcome, and makes necessary corrections to minimize that difference.

Slew. In motion control, the portion of a move made at a constant nonzero velocity.

Static Torque. Maximum torque available at zero speed.

Step Angle. Angle the shaft rotates upon receipt of a single-step command.

Stiffness. Ability to resist movement induced by an applied torque, often specified as a torque displacement curve, indicating the amount a motor shaft will rotate upon application of a known external force when stopped.

Synchronism. Term referring to a motor rotating at a speed correctly corresponding to the applied step pulse frequency. Load torques in excess of motor's capacity (rated torque) will cause a loss of synchronism. The condition is not damaging to a step motor.

Torque. Force tending to produce rotation.

Torque Constant. K_T, the torque generated in a dc motor per unit current applied to its windings. Simplified for brushless motors at 90° commutation angle,

$$K_T = \frac{T\,[\text{oz} \cdot \text{in (N} \cdot \text{m)}]}{A\ (\text{amperes})}$$

Torque Ripple. Cyclical variation of generated torque at frequency given by the product of motor angular velocity and number of commutator segments or magnetic poles.

Torque-to-Inertia Ratio. Motor holding torque divided by inertia of its rotor. The higher the ratio, the higher a motor's maximum acceleration capability will be.

Transfer Function. Mathematical means of expressing the output to input relationships of a system. Expressed as a function of frequency.

Trigger. Inputs on a controller which initiate, or "trigger," the next step in the program.

Voltage Constant. K_E, the back EMF generated by a dc motor at a defined speed. Usually quoted in volts per 1000 r/min.

Zero. Frequency at which the transfer function of a system goes to zero.*

---

*Source: Parker Hannifin Corporation, Compumotor Division.

Factors in Selecting a Servomotor

The diagram of Fig. 2 may be helpful in selecting the best motor for a given application. Factors of particular importance include the following.

Cost.   Because of its general cost advantage, a step motor often should be given an initial examination. This is particularly true where the dynamic requirements are not severe.

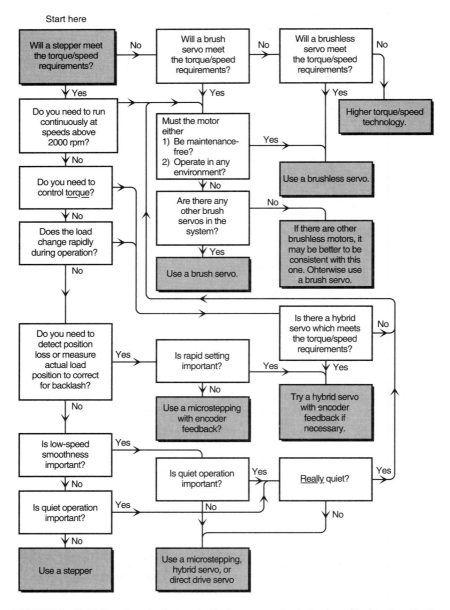

FIGURE 2   Guidelines for selecting most effective servomotor technology. (*Parker Hannifin Corporation, Compumotor Division.*)

Torque and Speed.   In high-torque, low-speed applications steppers are usually appropriate. At low speeds the stepper is very efficient in terms of torque output relative to both size and input power. Microstepping can be used to improve low-speed applications, such as a metering pump drive for accurate flow control.

In high-torque, high-speed continuous-duty applications the servomotor is well suited. In fact, the stepper should be avoided in such applications because the high-speed losses can cause excessive motor heating. A dc motor can deliver greater continuous shaft power at high speeds than a stepper of the same frame size.

Short, Rapid, Repetitive Moves.   This is the "natural" domain of the stepper or hybrid servo due to its high torque at low speeds, good torque-to-inertia ratio, and lack of commutation problems. The brushes of the dc and brush motor can limit its potential for frequent starts, stops, and direction changes.

Low-Friction, Mainly Inertial Loads.   These loads can be handled efficiently by the dc servo provided the start-stop duty requirements are not excessive. This type of load requires a high ratio of peak to continuous torque, and in this respect the servomotor excels.

Arduous, Difficult Applications.   These uses with a high dynamic duty cycle or requiring very high speeds may require a brushless motor. This solution also may be dictated when maintenance-free operation is required.

Low-Speed, High-Smoothness Applications.   Most appropriate for these applications are direct-drive servos, microsteppers, or servos with tachometer feedback.

Hazardous Environments, Including Vacuum.   In these applications a brush motor may not perform well. For example, heat dissipation may be a problem in vacuum when the loads are excessive. Either a stepper or a brushless motor is needed, depending on the demands of the load.

DIRECT-CURRENT MOTORS

The origin of the dc motor dates back to Faraday who, in the early 1800s, formulated the fundamental concepts of electromagnetism and who did extensive work with disk-type machines (Fig. 3). This design was improved upon over the years, and by the 1890s the design principles of dc motors had become well established.

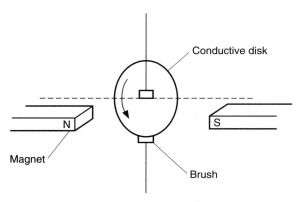

FIGURE 3   Operating principle of dc brush motor. (*Electro-Craft Ltd.*)

With the establishment of ac power supplies for general use, the acceptance of the dc motor declined in favor of the less costly ac induction motor. However, in recent years the particular

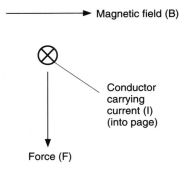

Force on conductor F = B × I

FIGURE 4 Force on conductor in a magnetic field. (*Electro-Craft Ltd.*)

characteristics of dc motors, notably their high starting torque and controllability, have led to their application in a wide range of systems that require accurate control of speed and position. Some dc motors fit well with some modern drive and computer-control systems.

It is well established that when a current-carrying conductor is placed in a magnetic field, it experiences a force (Fig. 4). The force acting on the conductor is given by $F = BI$, where B is the magnetic flux density and I the current. If this single conductor is replaced by a large number of conductors (that is, a length of wire is wound into a coil), then the force per unit length is increased by the number of turns in the coil, and this becomes the basis of a dc motor.

In order to achieve maximum performance from the motor, the maximum number of conductors must be placed in the magnetic field, so as to obtain the greatest possible force. In practice, this results in what is, in effect, a cylinder of wire, with the windings running parallel to the axis of the cylinder. A shaft is placed down the axis to act as a pivot. The shaft is the motor armature (Fig. 5). As the armature rotates, so does the resultant magnetic field, and the armature will come to rest with its resultant field aligned with that of the stator field, unless some provision is made to constantly change the direction of the current in the individual armature coils.

The force bringing about rotation of the motor armature is the result of the interaction of two magnetic fields (the stator field and the armature field). In order to produce a constant torque from the motor, it is required that these two fields remain constant in magnitude and in relative operation. This is achieved by constructing the armature as a series of small sections connected in sequence to the segments of a commutator (Fig. 6). Electrical connection is made to the commutator by means of two brushes. It can be observed that if the armature rotates through one-sixth of a revolution clockwise, the current in coils 3 and 6 will have changed direction. As successive commutator segments pass the brushes, so the current in the coils connected to those segments changes direction.

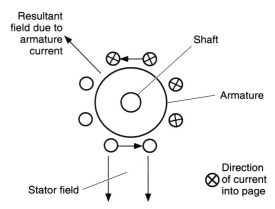

FIGURE 5 Dc motor armature. (*Electro-Craft Ltd.*)

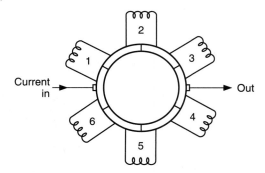

FIGURE 6 Electrical arrangement of armature in dc motor. (*Electro-Craft Ltd.*)

This commutation, or switching, effect results in a current flow in the armature which occupies a fixed position in space, independent of the armature rotation, and allows the armature to be regarded as a wound core with an axis of magnetization fixed in space. This gives rise to the production of a constant-torque output from the motor shaft.

The axis of magnetization is determined by the position of the brushes, and in order for the motor to have similar characteristics in both directions of rotation, the brush axis must be positioned such as to produce an axis of magnetization which is at 90° to the stator field.

Iron-Cored dc Motors

The motor design shown in Fig. 7 has been to date the most common type of motor used in dc servo systems. It is made up of two main parts, (1) a housing containing the field magnets and (2) a rotor made up of coils of wire wound in slots in an iron core and connected to a commutator. Brushes in contact with the commutator carry current to the coils.

Moving-Coil Motors

There are two principal forms of these motors, (1) the "printed" motor (Fig. 8), which uses a disk armature, and (2) the shell-armature motor (Fig. 9). Inasmuch as these motors do not have any moving iron in the magnetic field, they do not suffer from iron losses, and consequently, higher rotational speed with low power inputs is achievable.

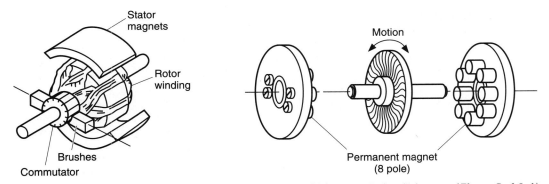

FIGURE 7   Iron-cored motor. (*Electro-Craft Ltd.*)

FIGURE 8   Disk-armature "printed" dc motor. (*Electro-Craft Ltd.*)

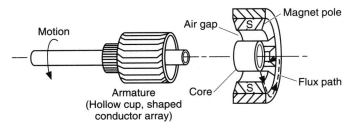

FIGURE 9   Shell-armature dc motor. (*Electro-Craft Ltd.*)

Losses in dc Brush Motors

All input power to a dc motor is not converted into mechanical power due to the electrical resistance of the armature and other rotational losses. These losses give rise to heat generation within the motor. Motor losses can be divided into two areas, (1) those that depend on the load and (2) those that depend on speed (Fig. 10).

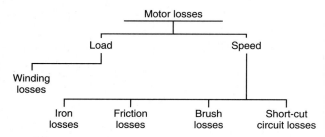

FIGURE 10   Types of losses in dc motor. (*Parker-Hannifin Corporation, Compumotor Division.*)

Winding Losses.   Caused by the electrical resistance of the motor windings, winding losses are equal to I^2R, where I is the armature current and R the armature resistance.

As the torque output of the motor is increased, so I is increased, giving rise to additional losses. Consideration of winding losses is very important because heating of the armature winding will cause an increase in R, which, in turn, will result in further losses and heating. The process can lead to self-destruction of the motor if the maximum current is unlimited. To compound this situation, at higher temperatures the current requirement becomes greater.

Brush Contact Losses.   These losses are fairly complex to analyze since they depend on a number of factors, which will vary with motor operation, but in general, brush contact resistance may represent a high proportion of the terminal resistance of the motor. The result of this resistance will be increased heating due to I^2R losses in the brushes and contact area.

Iron Losses.   There are two kinds of iron losses. (1) Eddy current losses are common in all conductive cored components experiencing a changing magnetic field. Eddy currents are induced into the motor armature as it undergoes changes in magnetization. These currents are speed-dependent and have a significant heating effect at high speeds. In practice, eddy currents are reduced by producing the armature core as a series of thin, insulated sections or laminations, stacked to produce the required core length. (2) Hysteresis losses are created by the resistance of the core material to constant changes of magnetic orientation, giving rise to additional heat generation, which increases with speed.

Friction Losses.   These losses are associated with the mechanical characteristics of the motor and arise from brush and bearing friction and air resistance. These areas generate heat and require additional armature current to overcome them.

Short-Circuit Currents.   Short-circuit currents may be generated as illustrated in Fig. 11.

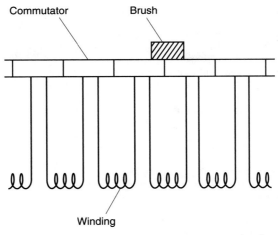

FIGURE 11 Generation of short-circuit currents. (*Parker-Hannifin Corporation, Compumotor Division.*)

Other Limitations of dc Brush Motors

Torque Ripple.   The requirement for constant torque output of a dc motor is that the magnetic fields due to the stator and the armature be constant in magnitude and relative orientation, but this ideal is not achieved in practice. As the armature rotates, the relative orientation of the fields will change slightly, and this will result in small changes in torque called torque ripple (Fig. 12). This usually does not cause problems at high speeds because the inertia of the motor and load tend to smooth out the effects, but problems may arise at low speeds. Motors can be designed to minimize the effects of torque ripple by increasing the number of windings or the number of motor poles, or by skewing the armature windings.

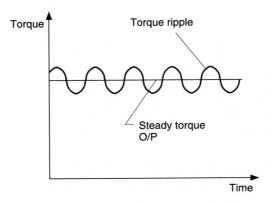

FIGURE 12 Torque ripple components in dc motor. (*Parker-Hannifin Corporation, Compumotor Division.*)

Demagnetization. The permanent magnets of a dc motor will tend to become demagnetized whenever a current flows in the motor armature. This effect is known as armature reaction and will have negligible effect in normal use. Under high-load conditions, however, when motor current may be high, the effect will cause a reduction in the torque constant of the motor and a consequent reduction in torque output. Above a certain level of armature current, the field magnets will become permanently demagnetized, so that it is important not to exceed the maximum pulse current rating for the motor.

Mechanical Resonances and Backlash. It is normal to assume that a motor and its load, including a tachometer or position encoder, are all rigidly connected together. However, this may not always be true. It is important for a bidirectional drive or positioning system to be mechanically tight and true in order to avoid backlash, which can destroy system accuracy.

Torsional Oscillation. In high-performance systems with high accelerations, interconnecting shafts and couplings may deflect under the applied torque, such that the various parts of the system may have different instantaneous velocities which may be in opposite directions. Under certain conditions a shaft may go into torsional resonance (Fig. 13).

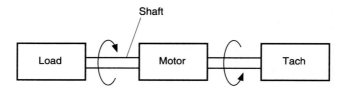

FIGURE 13 Torsional oscillation. (*Parker-Hannifin Corporation, Compumotor Division.*)

Back EMF. A permanent-magnet dc motor will operate as a generator. As the shaft is rotated, a voltage (back EMF) will appear across the brush terminals. This voltage is generated even when the motor is driven by an applied voltage. The output voltage is essentially linear with motor speed and has a slope which is defined as the motor voltage constant K_E, which typically is quoted in volts per 1000 r/min (Fig. 14).

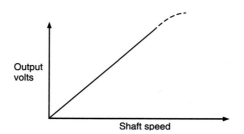

FIGURE 14 Back EMF characteristic of dc motor.

DC Brush Motor Equations

Unlike a stepper motor, the dc brush motor exhibits simple relationships between current, voltage, torque, and speed. The application of a constant voltage to the terminals of a motor will result in its accelerating to attain a steady final speed n. Under these conditions the voltage V applied to the motor is opposed by the back EMF nK_E, and the resultant voltage drives the motor current I through the motor armature and brush resistance R_s. The equivalent circuit of a dc motor is shown in Fig. 15.

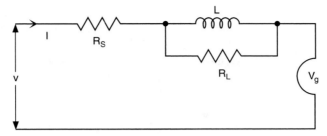

FIGURE 15   Equivalent circuit of dc motor.

If a voltage V is applied to the motor and a current I flows, then

$$V = IR + V$$

But $V = nK_e$, so

$$V = IR + nK_e \tag{1}$$

This is the electrical equation of the motor.

If K is the torque constant of the motor (typically in ounce-inches per ampere, then the torque generated by the motor is given by

$$T = IK_T \tag{2}$$

and the opposing torque due to friction T_F and viscous damping K_D is given by

$$T_M = T_F + nK_D$$

If the motor is coupled to a load T_L, then at constant speed,

$$T = T_L + T_F + nK_D \tag{3}$$

Equations (1), (2), and (3) allow calculation of the required current and drive voltage to meet given torque and speed requirements. The values of K_T, K_E, and so on, are given by the manufacturer.

Brushless Motors

Before describing brushless dc motors, some clarification in terminology is in order. The term "brushless" has become accepted as referring to a particular variety of servomotor. Clearly, a stepper motor is a brushless device, as is an ac induction motor. In fact, the stepper motor, described later, can form the basis of a brushless servomotor, often called a hybrid servo, which is also discussed later. The brushless motor described here has been designed to have a similar performance to the dc brush servo, but without the limitations imposed by a mechanical commutator.

There are two basic types of brushless motors in the dc motor category as just defined, (1) the trapezoidal motor and (2) the sine-wave motor. As is described later, the trapezoidal motor is really a brushless dc servo, whereas the sine-wave motor bears a close resemblance to the ac synchronous motor.

To convert the conventional dc brush motor, as previously shown in Fig. 7, to a brushless design, the windings on the rotor must be eliminated. This can be achieved by "turning the motor inside out," that is, making the permanent magnet the rotating part and putting the windings on the stator poles. However, some means of reversing the current automatically is needed. This can be accomplished by using a cam-operated reversing switch, as shown in Fig. 16. Obviously, such an arrangement with a mechanical switch is not very satisfactory, but the switching capability of noncontacting devices tends to be very limited. However, in a servo application some form of electronic amplifier or drive can be used, and thus this can do the commutation in response to low-level signals from an optical or Hall-effect sensor (Fig. 17). This component is called a commutation encoder. Thus unlike the dc brush motor, the brushless version cannot be driven simply by connecting it to a source of direct current. The current in the external circuit must be reversed at a defined rotor position, so that the motor is, in fact, being driven by an alternating current.

Returning to the conventional brush dc motor, it will be recalled that a rotor consisting of only one coil will exhibit a large torque variation as it rotates. In fact, the characteristic will be sinusoidal, with maximum torque produced when the rotor field is at right angles to the stator field and zero torque at the commutation point (Fig. 18). A practical dc motor has a large number of coils on the rotor, each one connected not only to its own pair of commutator segments, but to the other coils as well. In this way the chief contribution to torque is made by a coil operating close to its peak-torque position. There is also an averaging effect produced by current flowing in all the other coils, so the resulting torque ripple is very small.

The designer would like to reproduce a similar situation in the brushless motor. This would require, however, a large number of coils distributed around the stator. In itself, this could be feasible, but each coil would then require its own individual drive circuit. This is clearly prohibitive, so in practice a compromise is made and a typical brushless motor has either two or three sets of coils, or "phases," as shown in Fig. 18. The motor illustrated is a two-pole three-phase design. Usually the rotor has four or six rotor poles, with a corresponding increase in the number of stator poles. This does not increase the number of phases—each phase has its turns distributed between several stator poles.

The torque characteristic (Fig. 19) indicates that maximum torque is produced when the rotor and stator fields are at 90° to each other. Therefore to generate constant torque, it is necessary to keep the stator field a constant 90° ahead of the rotor. Limiting the number of phases to three means that advances of the stator field can only be made in increments of 60° of shaft rotation (Fig. 20). Thus a constant 90° torque angle cannot be maintained, but an average of 90° *can* be maintained by working between 60° and 120°. Figure 21 shows the rotor position at a

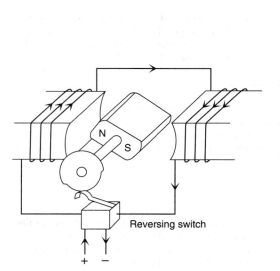

FIGURE 16   Reversing switch concept for brushless dc motor.

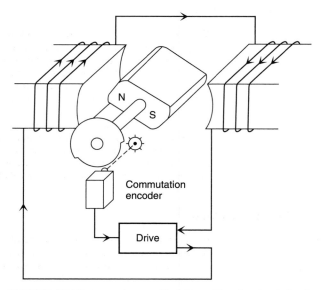

FIGURE 17   Commutation encoder that utilizes low-level signals from optical or Hall-effect sensor.

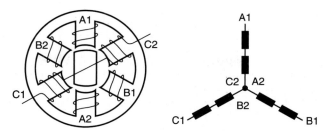

FIGURE 18   Three-phase brushless motor.

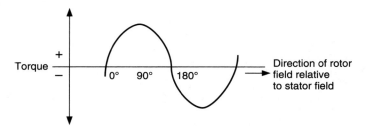

FIGURE 19   Position-torque characteristic. Maximum torque is produced when rotor and stator fields are at 90° to each other.

commutation point. When the torque angle has fallen to 60°, the stator field is advanced from 1 to 2 so that the angle now increases to 120°, and it remains there during the next 60° of rotation.

Trapezoidal Motor

With a fixed current level in the windings, the use of this extended portion of the sinusoidal torque characteristic gives rise to a large degree of torque ripple. This effect can be minimized by manipulating the motor design in order to "flatten out" the characteristics, that is, to make it trapezoidal (Fig. 22). In practice this is not easy to do, and thus some degree of nonlinearity will remain. The effect of this tends to be a slight "kick" at the commutation points, which can be noticeable when the motor is running at a very slow rate.

Torque ripple resulting from nonlinearity will obviously tend to produce a velocity modulation in the load. However, in a system using velocity feedback the velocity loop will generally have a high gain. Thus very small increases in velocity will generate a large error signal, reducing the torque demand to correct the velocity change. Thus, in practice, the output current from the amplifiers tends to mirror the torque characteristic, and therefore the resulting velocity modulation is extremely small (Fig. 23).

Sine-Wave Motor

In the sine-wave motor, sometimes called an ac brushless servo, no attempt is made to modify the basic sinusoidal torque characteristics. Such a motor can be driven in the same way as an ac synchronous motor by applying sinusoidal currents to the motor windings. These currents must have the appropriate phase displacement, 120° in the case of the three-phase motor.

However, to achieve smooth rotation at low speeds, a much higher-resolution device is needed to control the commutation. So instead of the simple commutation encoder that gen-

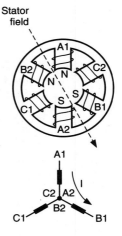

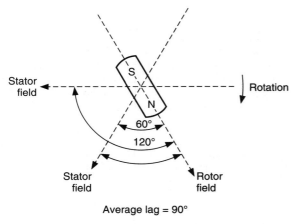

FIGURE 20 Stator field positions for different phase currents.

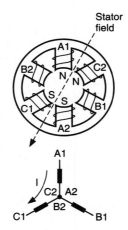

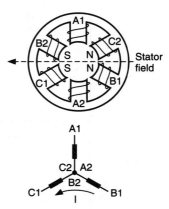

FIGURE 21 Position of rotor at commutation point. (*Parker-Hannifin Corporation, Compumotor Division.*)

erates a number of switching points, something like a resolver or high-resolution encoder is used. The drive needs to generate three currents that are in the correct relationship to each other at every rotor position. In this way it is possible to maintain a 90° torque angle very accurately, resulting in very smooth low-speed rotation and negligible torque ripple.

Clearly the drive for a sine-wave motor is more complex than for the trapezoidal version. A look-up table from which to generate the sinusoidal currents is required, and these must be multiplied by the torque demand signal to determine their absolute amplitude. With a star-connected three-phase motor it is sufficient to determine the currents in two of the windings. This will automatically determine what happens in the third. Although the sine-wave motor requires a high-resolution feedback device, this also can serve to provide position and velocity information for the controller. Thus, in practice, this is not regarded as a penalty.

HYBRID SERVOS

In terms of their basic operation, the stepper motor and the brushless servo are nearly identical. They each have a rotating magnet with slightly different flux paths and a wound stator. The primary difference is that one has more poles than the other, typically two or three pole pairs in the brushless servo and 50 in the stepper. A brushless servo could be used as a stepper, although with limited performance inasmuch as the step angle would be too large. Conversely, a stepper can be used as a brushless servo by fitting a feedback device to perform the

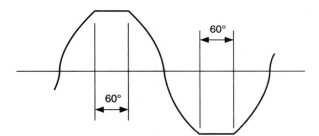

FIGURE 22 Trapezoidal motor characteristic. (*Parker-Hannifin Corporation, Compumotor Division.*)

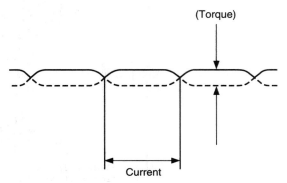

FIGURE 23 Current profile in velocity-controlled servo. (*Parker-Hannifin Corporation, Compumotor Division.*)

commutation. Hence the hybrid servo is so named because it is based on a hybrid stepper motor. The designs also have been called stepping servos or closed-loop servos. Hybrid servo, however, is the much better term to use because the design does not exhibit operating characteristics of a stepper motor.

The hybrid servo is driven in the same fashion as the brushless motor. A two-phase drive provides sine and cosine current waveforms in response to signals from the feedback device. This device may be an optical encoder or a resolver. Since the motor has 50 pole pairs, there will be 50 electrical cycles per revolution. This conveniently permits a 50-cycle resolver to be constructed from the same rotor and stator laminations as the motor itself.

A hybrid servo generates approximately the same torque output as the equivalent stepper motor, assuming the same drive current and supply voltage. However, the full torque capability of the motor can be utilized inasmuch as the system is operating in a closed loop. (With an open-loop stepper it is always necessary to allow an adequate torque margin.) The hybrid servo system will be more costly than the equivalent stepper, but less costly than a brushless servo. As with the stepper, continuous operation at high speed is not recommended since the high pole count results in greater iron losses at speed. A hybrid servo also tends to run quieter and cooler than the stepper counterpart. Since it is a true servo, power is only consumed when torque is required, and normally no current will flow at standstill. Low-speed smoothness is much improved over the open-loop full stepper.

The hybrid servo is entirely different from the open-loop stepper operated in the closed-loop or position-tracking mode. In the position-tracking mode an encoder is used to measure the load movement, and final positioning is determined by encoder feedback. While this technique can provide high position accuracy and eliminates undetected position loss, it does not allow full torque utilization, improve smoothness, or reduce motor heating.

STEPPER MOTORS

The advantages of stepper motors are low cost, ruggedness, construction simplicity, high reliability, no maintenance, wide acceptance, no "tweaking" to stabilize, and no feedback components are needed; they are inherently fail-safe and tolerate most environments. Steppers are simple to drive and control in an open-loop configuration. They provide excellent torque at low speed, up to 5 times the continuous torque of a brush motor of the same frame size or double the torque of the equivalent brushless motor. Frequently a gearbox can be eliminated. A stepper-driven system is inherently stiff, with known limits to the dynamic position error.

There are three main types of stepper motors.

Permanent-Magnet Motors. The tin-can or "canstack" motor shown in Fig. 24 is perhaps the most widely used type in commercial, nonindustrial applications. It is essentially a low-cost, low-torque, low-speed device ideally suited for use in computer peripherals, for example. The motor construction results in relatively large step angles, but the overall simplicity favors high-volume, low-cost production. The axial-air gap or disk motor is a variant of the permanent-magnet design, which achieves higher performance mainly because of its very low rotor inertia. This does restrict the applications of the motor to those situations involving little inertia, such as positioning the print wheel in a daisy-wheel printer.

Disadvantages of the stepper motor include resonance effects, relatively long settling times, and rough performance unless a microstepper is used. Undetected position loss may result in open-loop systems. Steppers consume current regardless of load conditions, and they tend to run hot. Steppers tend to be noisy, especially when operated at high speeds. Some of the foregoing limitations can be overcome by use of a closed-loop system.

Variable-Reluctance Motors. There is no permanent magnet in a variable-reluctance motor. Thus the rotor spins freely without detent torque. Torque output for a given frame size

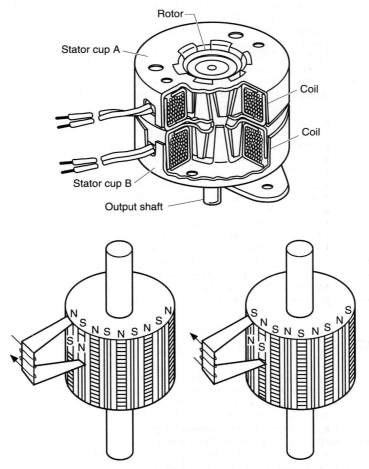

FIGURE 24   Canstack or permanent-magnet stepper motor. (*Top*) sectional view; (*bottom*) as used for positioning the print wheel in a daisy-wheel printer. (*Airpax Corporation, U.S.*)

is restricted, although the torque-to-inertia ratio is good. This type of motor is frequently used in small sizes for applications such as micropositioning tables. Variable-reluctance motors are seldom used in industrial applications. Having no permanent magnet, these motors are not sensitive to current polarity and thus require a different driving arrangement compared to other types (Fig. 25).

Hybrid Stepper Motors.   The hybrid motor is the most widely used stepper motor in industrial applications. Most hybrid motors are two-phase, although five-phase designs are available. A recent development is the enhanced hybrid, which uses flux-focusing magnets to give a significant improvement in performance, but at extra cost.

The rotor of the "model" hybrid stepper illustrated in Fig. 26 consists of two pole pieces with three teeth on each. Between the pole pieces is a permanent magnet which is magnetized along the axis of the motor, making one end a north pole, the other a south pole. The teeth are offset at the north and south ends, as shown in the diagram.

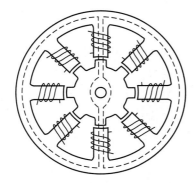

FIGURE 25   Variable-reluctance motor.

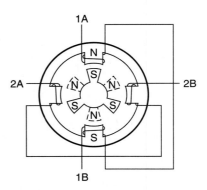

FIGURE 26 Simple 12-step-per-revolution hybrid motor.

The stator consists of a shell having four teeth which run the full length of the rotor. Coils are wound on the stator teeth and are connected together in pairs.

With no current flowing in any of the motor windings, the rotor will tend to take up one of the five positions shown in the diagram. This is because the permanent magnet in the rotor attempts to minimize the reluctance, or magnetic resistance, of the flux path from one end to the other. This occurs when a pair of north- and south-pole rotor teeth are aligned with two of the stator poles. The torque, tending to hold the motor in one of these positions, is usually small and called the detent torque. The motor shown has 12 possible detent positions.

If current is passed through one pair of stator windings, as shown in Fig. 27a, the resulting north and south stator poles will attract teeth of the opposite polarity on each end of the rotor. Thus there are only three stable positions for the rotor, the same as the number of rotor teeth. The torque required to deflect the rotor from its stable position is thus much greater, and is referred to as the holding torque.

By changing the current flow from the first to the second set of stator windings (Fig. 27b), the stator field rotates through 90° and attracts a new pair of rotor poles. This results in the rotor turning through 30°, corresponding to one full step. Reverting to the first set of stator windings, but energizing them in the opposite direction, the stator field will be rotated through another 90° and the rotor takes another 30° step, as shown in Fig. 27c. Finally, the second set of windings is energized in the opposite direction (Fig. 27d) to give a third step position. Returning to the first condition (Fig. 27a), and after these four steps, the rotor will have moved through one tooth pitch. This simple motor, therefore, performs 12 steps per revolution. Obviously, if the coils are energized in the reverse sequence, the motor will change direction.

If the two coils are energized simultaneously (Fig. 28), the rotor takes up an intermediate position since it is equally attracted to two stator poles. Greater torque is produced under these conditions because all the stator poles are influencing the motor. The motor can be made to take a full step simply by reversing the current in one set of windings. This causes a 90° rotation of the stator field, as before. In fact, this would be the normal way of driving the motor in the full-step mode, always keeping two windings energized and reversing the current in each winding alternately.

By alternately energizing one winding and then two (Fig. 29), the rotor moves through only 15° at each stage, and the number of steps per revolution will be doubled. This is called half-stepping. Most industrial applications make use of this stepping mode. Although sometimes there is a slight loss of torque, this mode results in much better smoothness at low speeds, and less overshoot and ringing occur at the end of each step.

Current Patterns in Motor Winding.   When the motor is driven in its full-step mode, energizing two windings, or phases, at a time (Fig. 30), the torque available on each step will be the

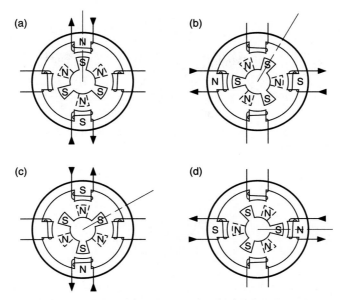

FIGURE 27   Full stepping, one phase on. (*Parker-Hannifin Corporation, Compumotor Division.*)

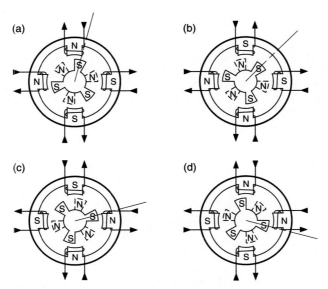

FIGURE 28   Full stepping, two phases on. (*Parker-Hannifin Corporation, Compumotor Division.*)

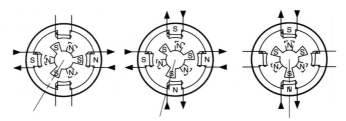

FIGURE 29   Half-stepping. (*Parker-Hannifin Corporation, Compumotor Division.*)

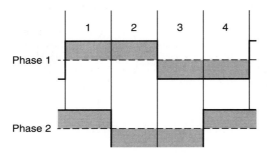

FIGURE 30   Full step current, two phases on. (*Parker-Hannifin Corporation, Compumotor Division.*)

same (subject to very small variations in the motor and drive characteristics). In the half-step mode two phases are alternately energized and then only one, as shown in Fig. 31. Assuming the drive delivers the same winding current in each case, this will cause greater torque to be produced when there are two windings energized, that is, alternate steps will be strong and weak. Although the available torque obviously is limited by the weaker step, there is a significant improvement in low-speed smoothness over the full-step mode.

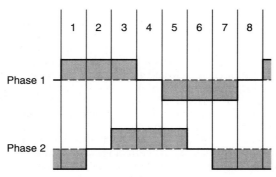

FIGURE 31   Half-step current. (*Parker-Hannifin Corporation, Compumotor Division.*)

The motor designer would like to produce approximately equal torque on every step and to have this torque be at the level of the stronger step. This goal can be achieved by using a higher current level when there is only one winding energized. This does not overly dissipate the motor because the manufacturer's current rating will assume two phases to be energized. (The current rating is based on the allowable case temperature.) With only one phase energized, the same total power will be dissipated if the current is increased by 40 percent. Using the higher current in the one-phase-on state produces approximately equal torque on alternate steps, as indicated in Fig. 32.

Microstepping.   It will be noted from the prior discussion that energizing both phases with equal currents produces an intermediate step position halfway between the one-phase-on positions. If the two phase currents are unequal, the rotor position will be shifted toward the stronger pole. This effect is utilized in the microstepping drive, which subdivides the basic motor step by proportioning the current in the two windings. In this way the step size is reduced and the low-speed smoothness is improved dramatically. High-resolution microstep drives divide the full motor step into as many as 500 microsteps, giving 100,000 steps per revolution. In this situation the current pattern in the windings closely resembles two sine waves with a 90° phase shift between them (Fig. 33). Thus the motor is driven very much as though it were a conventional ac synchronous motor. In fact, the stepper motor can be driven in this manner from a 60-Hz (U.S.) or 50-Hz (Europe) sine-wave source by including a capacitor in series with one phase. It will rotate at 60 r/min.

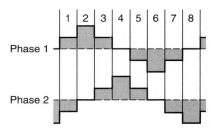

FIGURE 32 Half-step current, profiled. (*Parker-Hannifin Corporation, Compumotor Division.*)

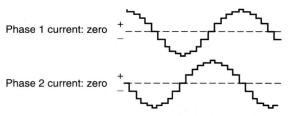

FIGURE 33 Phase currents in microstep mode. (*Parker-Hannifin Corporation, Compumotor Division.*)

Standard 200-Step Hybrid Motor.   The standard stepper motor operates in the manner just described as a model, but it has a greater number of teeth on the rotor and stator, giving a smaller basic step size. The rotor is in two sections, as described previously, but has 50 teeth on each section. The half-tooth displacement between the two sections is retained. The stator has eight poles, each with five teeth, making a total of 40 teeth (Fig. 34).

Visualize that a tooth is placed in each of the gaps between the stator poles, in which case there would be a total of 48 teeth, two less than the number of rotor teeth. If rotor and stator teeth were aligned at 12 o'clock, they would also be aligned at 6 o'clock. But at 3 and 9 o'clock the teeth would be misaligned. However, due to the displacement between the sets of rotor teeth, alignment will occur at 3 o'clock and 9 o'clock at the other end of the rotor.

In practice, the windings are arranged in sets of four, and wound such that diametrically opposite poles are the same. Thus referring to Fig. 34, the north poles at 12 and 6 o'clock attract the south-pole teeth at the front of the rotor and the south poles at 3 and 9 o'clock attract the north-pole teeth at the back. By switching current to the second set of coils, the stator field pattern rotates through 45°, but to align with this new field, the rotor only has to turn through 1.8°. This is equivalent to one-quarter of a tooth pitch on the rotor, giving 200 full steps per revolution.

Note that there are as many detent positions as there are full steps per revolution, namely, 200. The detent positions correspond with rotor teeth being fully aligned with stator teeth. When power is applied to a stepper drive, it is usual for it to energize in the zero-phase state in which there is current in both sets of windings. The resulting rotor position does not correspond with a natural detent position, so an unloaded motor will always move by at least one-half step at power on. Of course, if the system were turned off other than in the zero-phase state, or the motor is moved in the meantime, a greater movement may be seen at power-up.

For a given current pattern in the windings there are as many stable positions as there are rotor teeth (50 for a 200-step motor). If a motor is desynchronized, the resulting position error will always be a whole number of rotor teeth, or a multiple of 7.2°. A motor cannot "miss" individual steps. Position errors of one or two steps may be due to noise, spurious step pulses, or a controller fault.

Bifilar Windings.   Most motors are described as being bifilar wound, which means there are two identical sets of windings on each pole. Two lengths of wire are wound together as though they were a single coil. This produces two windings which are electrically and magnetically almost identical. If one coil were wound on top of the other, even with the same number of turns, the magnetic characteristics would be different.

The origin of the bifilar winding goes back to the unipolar drive. Rather than reversing the current in one winding, the field may be reversed by transferring current to a second coil wound in the opposite direction. (Although the two coils are wound the same way, interchanging the ends has the same effect.) Thus with a bifilar-wound motor, the drive can be kept simple. However, this requirement has now largely disappeared with the widespread availability of the more efficient bipolar drive. Nevertheless, the two sets of windings do provide additional flexibility.

If all the coils in a bifilar-wound motor are brought out separately, there will be a total of eight leads (Fig. 35). This is becoming the most common configuration since it gives the greatest flexibility. However, there are still a number of motors produced with only six leads, one lead serving as a common connection to each winding in a bifilar pair. This arrangement limits the range of applications of the motor since the windings cannot be connected in parallel.

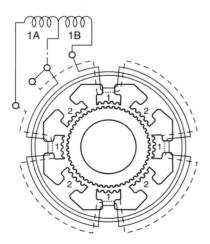

Phase 1 wind shown
Phase 2 windings on intermediate poles

FIGURE 34   200-step hybrid motor. (*Parker-Hannifin Corporation, Compumotor Division.*)

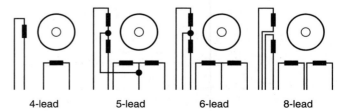

4-lead   5-lead   6-lead   8-lead

FIGURE 35   Motor lead configurations.

Some motors are made with only four leads. These are not bifilar-wound and cannot be used with a unipolar drive. There is obviously no alternative connection method with a four-lead motor, but in many applications this is not a drawback and the problem of insulating unused leads is avoided. Occasionally a five-lead motor may be encountered. These should be avoided inasmuch as they cannot be used with conventional bipolar drives requiring electrical isolation between the phases.

LINEAR STEPPERS

The linear stepper is essentially a conventional rotary stepper which has been "unwrapped" so that it operates in a straight line. The moving component is referred to as the forcer, and it travels along a fixed element, or platen. For operational purposes the platen is equivalent to the rotor in a normal stepper, although it is an entirely passive device and has no permanent magnet. The magnet is incorporated in the moving forcer together with the coils (Fig. 36).

The forcer is equipped with four pole pieces, each having three teeth. The teeth are staggered in pitch with respect to those on the platens so that switching the current in the coils will bring the next set of teeth into alignment. A complete switching cycle (four full steps) is equivalent to one tooth pitch on the platen. Like the rotary stepper, the linear motor can be driven from a microstep drive. In this case a typical linear resolution will be 12,500 steps per inch (4921 steps/cm).

The linear motor finds favor in applications involving a low mass to be moved at very high speed. In a lead-screw-driven system the predominant inertia usually is the lead screw rather than the load to be moved. Hence most of the motor torque goes to accelerate the lead screw, and this problem becomes more severe the longer the travel required. In using a linear motor, all the developed force is applied directly to the load and the performance achieved is independent of the length of the move. A screw-driven system can develop greater linear force and better stiffness. However, the maximum speed may be as much as 10 times higher with the equivalent linear motor.

With further reference to Fig. 36, the forcer consists of two electromagnets A and B and a strong rare-earth permanent magnet. The two pole faces of each electromagnet are toothed to concentrate the magnetic flux. Four sets of teeth on the forcer are spaced in quadrature so that only one set at a time can be aligned with the platen teeth.

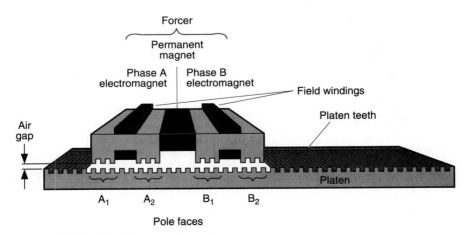

FIGURE 36   Principle of linear stepping motor.

The magnetic flux passing between the forcer and the platen gives rise to a very strong force of attraction between the two pieces. The attractive force can be up to 10 times the peak holding force of the motor, requiring a bearing arrangement to maintain precise clearance between the pole faces and the platen teeth. Either mechanical roller bearings or air bearings are used to maintain the required clearance.

When current is established in a field winding, the resulting magnetic field tends to reinforce permanent magnetic flux at one pole face and cancel it at the other. By reversing the current, the reinforcement and cancellation are exchanged. Removing current divides the permanent magnetic flux equally between the pole faces. By selectively applying current to phases *A* and *B* it is possible to concentrate the flux at any of the forcer's four pole faces. The face receiving the highest flux concentration will attempt to align its teeth with the platen. Figure 37 shows the four p.rimary states or full steps of the forcer. The four steps result in motion of one tooth interval to the right. Reversing the sequence moves the forcer to the left.

Repeating the sequence in the example will cause the forcer to continue its movement. When the sequence is stopped, the forcer stops with the appropriate tooth set aligned. At rest, the forcer develops a holding force which opposes any attempt to displace it. As the resting motor is displaced from equilibrium, the restoring force increases until the displacement reaches one-quarter of a tooth interval (Fig. 38). Beyond this point the restoring force drops. If the motor is pushed over the crest of its holding force, it slips or jumps rather sharply and comes to rest at an integral number of tooth intervals away from its original location. If this occurs while the forcer is traveling along the platen, it is referred to as a stall condition.

Linear Step Motor Characteristics. These include velocity ripple, platen mounting, environment, life expectancy, yaw (plus pitch and roll), and accuracy. To summarize, the worst-case accuracy of a linear step motor can be given by

$$\text{Accuracy} = A + B + C + D + E + F$$

where A = cyclic error due to motor magnetics, which recurs once every pole pitch as measured on motor body
 B = unidirectional repeatability—error measured by repeated moves to the same point from different distances in the same direction
 C = hysteresis—backlash of motor when changing direction due to magnetic nonlinearity and mechanical friction
 D = cumulative platen error—linear error of platen as measured on motor body
 E = random platen error—nonlinear errors remaining in platen after linear error is disregarded
 F = thermal expansion error—error caused by change in temperature, expanding or contracting the platen

Common Queries Concering Step Motors. A summary of questions frequently asked concerning the application of step motors is given in Table 1.

DRIVES (POWER TRANSMISSION)

Direct-Drive Motors

In a direct-drive situation the load of the system is coupled directly to the motor without the use of belts or gears. Most servomotors, brushed or brushless, often lack adequate torque or resolution to satisfy application needs. Therefore mechanical means, such as gear reduction, are required. The gearing and mechanics needed may be considered a separate engineering package, or the motor and mechanical functions may be integrated and marketed as a self-contained package (Fig. 39). Use of direct drives is quite common, as in rotary indexers (Fig.

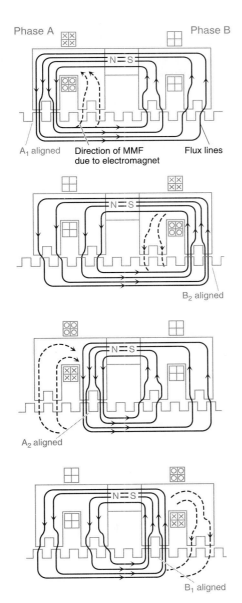

Phase A   Phase B

A_1 aligned   Direction of MMF   Flux lines
due to electromagnet

B_2 aligned

A_2 aligned

B_1 aligned

FIGURE 37 Four cardinal states of full steps of the force. (*Parker-Hannifin Corporation, Compumotor Division.*)

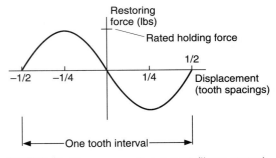

Restoring
force (lbs)

Rated holding force

$-1/2$ $-1/4$ $1/4$ $1/2$

Displacement
(tooth spacings)

One tooth interval

FIGURE 38 Force versus displacement (linear stepper). (*Parker-Hannifin Corporation, Compumotor Division.*)

TABLE 1   Common Queries Concerning Step Motors

Why do step motors run hot?
(1) Full current flow through motor windings is at standstill; (2) PWM drives tend to make motor run hotter. Motor construction, such as lamination material and riveted motors, also affects heating.

What are safe operating temperatures?
Motors have class B insulation rated at 130°C. Motor case temperatures of 90°C will not cause thermal breakdown. Motors should be mounted where operators cannot come in contact with motor case.

How can motor heating be reduced?
Many drives feature a "reduce current at standstill" command or jumper. This reduces current when the motor is at rest without position loss.

What does absolute accuracy specification mean?
This refers to inaccuracies, noncumulative, encountered in machining the motor.

How can the repeatability specification be better than that of accuracy?
Repeatability indicates how precisely a previous position can be reobtained. There are no inaccuracies in the system that affect a given position. Returning to that position, the same inaccuracy is encountered.

Will motor accuracy increase proportionately with the resolution?
No. The basic absolute accuracy and hysteresis of the motor remain unchanged.

Can a small motor be used on a large load if the torque requirement is low?
Yes. However, if the load inertia is more than 10 times the rotor inertia, cogging and extended ringing at the end of the move will be experienced.

How can end-of-move "ringing" be reduced?
Friction in the system will help dampen this oscillation. Acceleration and deceleration rates could be increased. If start-stop velocities are used, lowering, or eliminating them will help.

Why does motor stall during "no-load" testing?
The motor needs inertia roughly equal to its own inertia to accelerate properly. Any resonances developed in the motor are at their worst in a no-load condition.

Why is motor sizing important? Why not simply go with a larger motor?
If the motor's rotor inertia is the majority of the load, any resonances may become more pronounced. Also, productivity would suffer as excessive time would be required to accelerate the larger rotor inertia. Smaller may be better.

What are options for eliminating resonance?
Most often resonance will occur with full-step systems. Adding inertia would lower resonant frequency. Friction would tend to dampen the modulation. Start-stop velocities higher than the resonant point could be used. Changing to half-step operation would greatly help. Ministepping and microstepping also greatly minimize any resonant vibrations. Viscous inertial dampers also may help.

Why does motor jump at times when it is turned on?
This is due to the rotor having 200 natural detent positions. Movement can then be ±3.6°, either direction.

Do the rotor and stator teeth actually mesh?
No. While some designs used this type of harmonic drive, in most an air gap is very carefully maintained between the rotor and the stator.

Does the motor itself change if a microstepping drive is used?
The motor is still the standard 1.8° stepper. Microstepping is accomplished by proportioning currents in the drive. Higher resolutions result. The motor's inductance must be compatible.

When a move is made in one direction and then the motor is commanded to move the same distance, but in the opposite direction, why does the move end up short sometimes?
Two factors could be influencing the results: (1) The motor has magnetic hysteresis, which is seen on direction changes. This is in the range of 0.03°. (2) Any mechanical backlash in the system to which the motor is coupled could also cause loss of motion.

TABLE 1   (Continued)

Why are some motors constructed as eight-lead motors?
This allows greater flexibility. The motor can be run as a six-lead motor with unipolar drives. With bipolar drives, the
   windings can then be connected in either series or parallel.

What advantages do series or parallel connection windings give?
With series windings, low-speed torques are maximized. But this also gives the most inductance. So performance at
   higher speeds is lower than if the windings were connected in parallel.

Can a flat be machined on the motor shaft?
Yes. But care must be taken not to damage the bearings. The motor must not be disassembled. Suppliers do not warrant
   user's work.

How long can the motor leads be?
For bipolar drives, 50 meters (100 feet); for unipolar designs, 15 meters (50 feet). Shielded twisted-pair cables are
   required.

What are the options if an explosion-proof motor is required?
Installation of the motor in a purged box should be considered.

   Source: After Parker-Hannifin Corporation, Compumotor Division.

40), an optical scanner (Fig. 41), and other applications, such as spectroscopy, flight simulators, engine test stands, hydraulic actuators, fluid control, metering pumps, conveyors, printing and registration machines, fiber-optics lathes, and film transports.

Gear Drives

Traditional gear drives are more commonly used with step motors. The fine resolution of a
microstepping motor, however, can make gearing unnecessary in many applications. Gears
generally have undesirable efficiency, wear characteristics, backlash, and can be noisy. Gears
are useful, however, when very large inertias must be moved because the inertia of the load
reflected back to the motor through the gearing is divided by the square of the gear ratio. In
this manner, large inertial loads can be moved while maintaining a good load-inertia-to-rotor-
inertia ratio (less than 10:1) (Fig. 42).
      Gear drives are frequently used with antenna positioners, articulated arms, and conveyors.

Tangential Drives

Tangential drives consist of a pulley or pinion which, when rotated, exerts a force on a belt or
racks a linear load. Common tangential drives include pulleys and cables, gears and toothed
belts, and racks and pinions. Tangential drives permit a lot of flexibility in the design of drive
mechanics, and can be very accurate, with little backlash. Metal chains generally should be
avoided, as they provide little or no motor damping (Fig 43).

Lead or Ball Screw Drives

These drives convert rotary motion to linear motion and are available in a wide variety of
configurations. Screws are available with different lengths, diameters, and thread pitches.
Nuts range from the simple plastic variety to precision-ground versions with recirculating ball
bearings that can achieve very high accuracy.

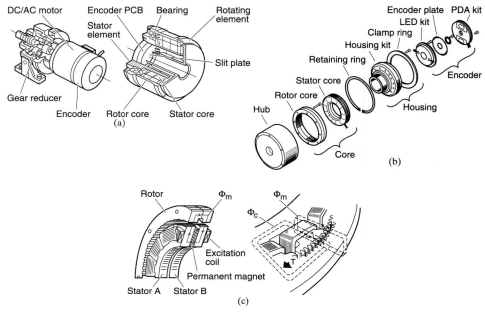

FIGURE 39   Comparison of (*a*) a conventional motor with a gear reducer and (*b*) a self-contained unit which incorporates precision bearings, magnetic components, and integral feedback in a compact package. The motor is an outer rotor type, providing direct motion of the outside housing of the motor and thus the load. The cross roller bearings which support the rotor have high stiffness to allow the motor to be connected directly to the load. In most cases it is not required to use additional bearings or connecting shafts. The torque is proportional to the square of the sum of the magnetic flux ϕ_m of the permanent magnet rotor and the magnetic flux ϕ_c of the stator windings. (*c*) High torque is generated due to the following factors: (1) Motor diameter is large. The tangential forces between rotor and stator act at a large radius, resulting in higher torque. (2) A large number of small rotor and stator teeth create many magnetic cycles per motor revolution. More working cycles means increased torque. (*Parker-Hannifin, Compumotor Division.*)

The combination of microstepping and a quality lead screw provides excellent positioning resolution for many applications. A typical 10-pitch [10 threads per inch (approximately 4 threads/cm)], when attached to a 25,000-step-per-revolution motor, provides a linear resolution of 0.000004 inch (~0.1 μm) per step.

A flexible coupling should be used between the lead screw and the motor to provide some damping. The coupling will also prevent excessive motor bearing loading due to any misalignment (Figs. 44 and 45).

Lead screws find wide application in machine tools, *xy* plotters, facsimile transmission, tool-bit positioning, cut-to-length machinery, back gaging, microscope drives, coil winders, pick and place machines, and articulated arms.

STEPPER MOTOR DRIVERS[2]

Unlike conventional motors, stepper motors are only one part of a sophisticated motion control system. They cannot simply be connected to a power source and energized to deliver

---

[2] See also the following article in this handbook section, concerning drivers for other types of motors.

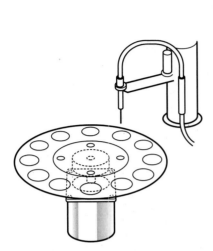

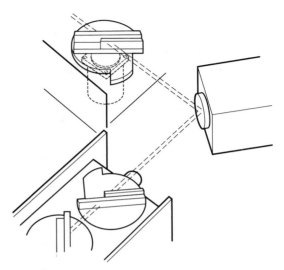

FIGURE 40 Example of directly driven load to replace an old-style Geneva mechanism. In vial-filling machine, the indexing table holds 12 vials (30° apart). Each must index in 0.5 second and dwell for 1 second. Motion must be smooth to prevent spillage. A microstepping motor answers these requirements. Operation is controlled by a low-cost microprocessor-based pulse-generation indexer. (*Parker-Hannifin Corporation, Compumotor Division.*)

FIGURE 41 A dye-laser designer needs to rotate a diffraction grating precisely under computer control to tune the frequency of the laser. The grating must be positioned to an angular accuracy of 0.05°. The microstepping motor with its high resolution and its freedom from "hunting" or other unwanted motion when stopped is an excellent choice. The inertia of the grating is found to be equal to 2 percent of the proposed motor's rotor inertia and thus is ignored. Space limitations in the cavity require a small motor with sufficient torque. An indexer with an IEEE-488 interface is selected and mounted in the rack with the computer. It is controlled with a simple program written in BASIC, which instructs the indexer to interrupt the computer at the completion of each index. (*Parker-Hannifin Corporation, Compumotor Division.*)

$$J_{Load} = \frac{W_{Load}}{2} R^2_{Load} \left(\frac{N_{Gear\,2}}{N_{Gear\,1}}\right)^2$$

or

$$J_{Load} = \frac{\pi L_{Load} \rho_{Load}}{2} R^4_{Load} \left(\frac{N_{Gear\,2}}{N_{Gear\,1}}\right)^2$$

$$J_{Gear\,1} = \frac{W_{Gear\,1}}{2} R^2_{Gear\,1} \left(\frac{N_{Gear\,2}}{N_{Gear\,1}}\right)^2$$

$$J_{Gear\,2} = \frac{W_{Gear\,2}}{2} R^2_{Gear\,2}$$

$$T_{Total} = \frac{1}{g} \left(J_{Load} + J_{Gear\,1} + J_{Gear\,2} + J_{Motor}\right)\frac{w}{t}$$

*Most gearbox manufacturers provide the reflected inertia of the gears in the user guide. This will help eliminate the need to calculate the J gear.

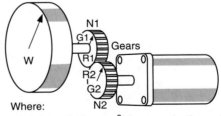

Where:
*J = inertia, oz-in (gm-cm²) "as seen by the motor"
T = torque, oz-in (gm-cm)
W = weight, oz (gm)
R = radius, in. (cm)
N = number of gear teeth (constant)
L = length, in. (cm)
ρ = density, oz/in³ (gm/cm³)
w = angular velocity, radians/sec
t = time, seconds
g = gravity constant, 386 in/sec²

FIGURE 42 Gear-drive formulas. (*Parker-Hannifin Corporation, Compumotor Division.*)

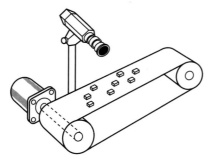

FIGURE 43 Application of tangential drive to conveyor that is part of a machine vision system for automatically inspecting small parts for defects. The conveyor is started and stopped under computer control. A flat timing belt is driven by a pulley. (*Parker-Hannifin Corporation, Compumotor Division.*)

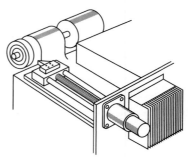

FIGURE 44 Application of lead screw on precision grinder. The objective—to replace former equipment that had a two-stage bearing grinding arrangement, where one motor and gearbox provided a rough cut and a second motor with a higher-ratio gearbox performed the finishing cut. In the modernized version the grinder is controlled by a PLC (programmable logic controller). An indexer provides the necessary velocities and accelerations. The speed change in the middle of the grinding operation is signaled to the PLC with a limit switch, whereupon the PLC programs the new velocity to the indexer. An optical encoder mounted on the back of the motor alerts the PLC if the mechanics become "stuck." (*Parker-Hannifin Corporation, Compumotor Division.*)

The torque required to drive load W using a lead screw with pitch, p, and efficiency, e has the following components.

$$T_{Total} = T_{Friction} + T_{Acceleration}$$

$$T_{Friction} = \frac{F}{2\pi pe}$$

Where:
 F = frictional force in ounces
 p = pitch in revs/in
 e = lead screw efficiency (%)

$F = \mu_s W$ for horizontal surfaces where μ_s = coefficient of static friction and W is the weight of the load.

$$T_{Accel} = \frac{1}{g}\left(J_{Load} + J_{Lead\ screw} + J_{Motor}\right)\frac{w}{t}$$

$$w = 2\pi pv$$

$$J_{Load} = \frac{W}{(2\pi\rho)^2} \quad ; J_{Lead\ screw} = \frac{\pi L\rho R^4}{2}$$

Where:
 T = torque, oz-in
 w = angular velocity, radians/sec
 t = time, seconds
 v = linear velocity, in/sec
 L = length, inches
 R = radius, inches
 ρ = density, ounces/in
 g = gravity constant, 386 in/sec²

The formula for load inertia converts linear inertia into the rotational equivalent as reflected to the motor shaft by the lead screw.

FIGURE 45 Lead-screw formulas. (*Parker-Hannifin Corporation, Compumotor Division.*)

rotary motion. Instead, steppers are dependent on external control circuitry to communicate how many steps to take, in which direction, when to start, and when to stop. Although this may seem unnecessarily complicated, stepper systems can carry out extremely varied patterns of precise movements. Position is determined by the number of steps taken in either direction, and velocity by the step rate. To produce the same sequence by other means might require more costly and error-prone apparatus, such as resolvers and tachometer generators, which require more system maintenance.

Common Driver Types

The heart of any stepper system is the driver (Fig. 46). To prevent motor overheating, each power driver circuit uses a different method to limit current beyond the specified maximum for the motor. The differences in system performance are reflected primarily in the time required for each driver type to bring the step motor up to full current, and the shape of their phase current versus time curves. In the simplest, least expensive stepper driver, resistors are connected in series with the motor windings, as shown in Fig. 47. These resistors limit the maximum winding current to a safe operating level by adding to the divisor in the formula

$$I_{max} = \frac{V}{R_{ser} + R_{wind}}$$

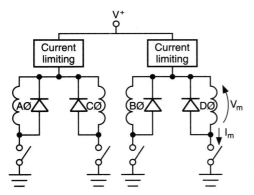

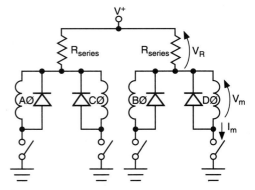

FIGURE 46 Simplified representation of stepper drive scheme. Power transistors are shown as switches. (*Bodine Electric Company.*)

FIGURE 47 Simplified representation of series R drive scheme. Power transistors are shown as switches. (*Bodine Electric Company.*)

The driver is commonly referred to as the $R(L/R)$, or series R, driver circuit. To obtain adequate high-speed performance, the winding current must rise and decay quickly. This is accomplished by using high-resistance series resistors to minimize the time constant, and correspondingly high power-supply voltages to attain adequate levels of current. Inasmuch as a significant amount of energy is dissipated as heat in the resistors, series R drivers are limited to applications which can tolerate additional heat and relatively low system efficiency. Advantages of the series R are low initial cost, system size, and general simplicity.

Choppers. With chopper drivers, external resistors are not used to limit the maximum flow of current. Limited only by the relatively small winding resistance, current would tend to rise to an unsafe level. To prevent this, the chopper driver will turn off the voltage across the windings when current reaches a preset maximum (Fig. 48). The driver then monitors current decay until it reaches a minimum level, at which it reapplies the voltage to the windings.

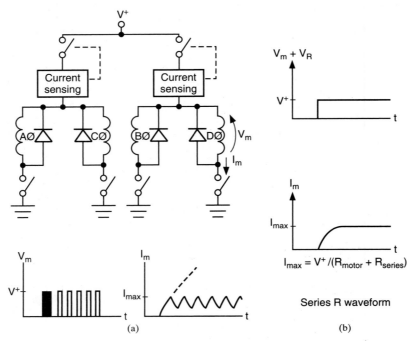

FIGURE 48   Simplified representation of chopper drive scheme. Current waveform shown at bottom of left-hand diagram. Compare with current waveform of series R drive scheme shown at right. (*Bodine Electric Company.*)

Instead of a purely exponential curve, choppers produce a sawtooth-shaped current waveform like that shown in Fig. 48. Since chopper drives do not dissipate energy through series resistors, it is practical to increase voltage for much higher horsepower output. The on-and-off "chopping" action maintains current at safe operating levels. The absence of series resistors allows chopper-type drivers to reach maximum currents much faster than series R drivers.

An operational characteristic inherent in chopper circuits that may cause problems in some uses is the tendency for oscillating current to produce line resonance at certain frequencies or motor speeds. Vertical dips on the speed-torque curve represent narrow speed ranges in which torque dips unexpectedly. Line resonance effects can be diminished and sometimes eliminated by using electronic antiresonance circuitry, such as a proprietary resonance compensation circuit.

Bilevel Drivers.   Rather than chopping current at a prescribed maximum, bilevel drivers switch between two separate input voltage levels (Fig. 49). To bring the motor windings rapidly up to a maximum current, a relatively high voltage (typically above 24 volts) is initially applied. Once the desired operating level has been reached, the driver quickly switches to a much lower "maintenance" voltage (typically under 10 volts). This dual-voltage approach provides the rapid acceleration not possible with a series R design, while eliminating torque variations common to chopper drivers. The principal disadvantages encountered with bilevel drivers are the added expense for switches or transistors and the dual power supply needed to deliver the two voltages used in the scheme.

Unipolar versus Bipolar Circuits.   There are three common drive circuits used in 1.8° stepper motor applications: unipolar, bipolar series, and bipolar parallel (Fig. 50).

Each method has its advantages in terms of cost and performance. In the unipolar mode two of the four windings are energized at any given instant, and current flows in only one direction through each winding. The sequence in which the windings are energized determines the direction of shaft rotation. In bipolar operation, all windings are on simultaneously. Rotation is produced by changing the direction of the phase current in the windings. No matter which method is used, the rotor senses the same changes in the direction of magnetic flux in the motor stator.

The real differences between the two aforementioned schemes occur in the areas of performance and system cost. Unipolar driver circuits are generally much simpler and therefore more reliable and less expensive, requiring only four drive transistors and a single power supply. Although they deliver somewhat lower torque for a given power input at low speeds, unipolar drivers usually produce higher torques at higher speeds.

With bipolar circuits as many as eight power transistors, or four power transistors and a dual power supply, are needed. This adds cost and vulnerability to system failure. But when high torque at very low speeds is a requirement and there are constraints on motor size, a bipolar driver may be the most desirable alternative. Since all four phases are energized at any given instant in bipolar circuits, a stronger magnetic field is created, making more torque available to do work.

Under static or low-speed conditions, bipolar drivers can increase torque output by 20 to 40 percent. When connected in parallel, effective phase resistance and inductance is reduced by half—so current per phase can be increased to 140 percent of the two-phase-on unipolar rating. When connected in series, the effective number of winding turns is increased, so the series bipolar circuit makes more efficient use of the windings. Voltage across the windings can be increased, while keeping current low (70 percent of two-phase-on unipolar rating), and in some cases this permits less expensive power supplies to be used.

EMERGENCY STOPS

For safety reasons, steppers and other servomotor systems should be capable of stopping *quickly.* Several means are available.

Full-Torque Controlled Stop—Servo Amplifier.   Applying zero velocity command to a servo amplifier will cause it to decelerate hard to zero speed within current limit, that is, using the maximum available torque. This will create the fastest possible deceleration to rest. In the case of a digital servo with step and direction inputs, cutting off the step pulses will produce the same effect.

Full-Torque Controlled Stop—Stepper.   The situation is rather different in the case of a stepper drive. The step pulse train should be decelerated to zero speed in order to utilize the available torque. Simply cutting off the step pulses at speeds above the start-stop rate will desynchronize the motor and the full decelerating torque will no longer be available. Thus the controller should be able to generate a rapid deceleration rate *independent* of the normal programmed rate, to be used only for overtravel limit and emergency stop functions.

Motor Disconnection.   Although this is safe in most cases, it has little to commend it as a quick-stop measure. The time taken to stop is indeterminate since it depends on load inertia

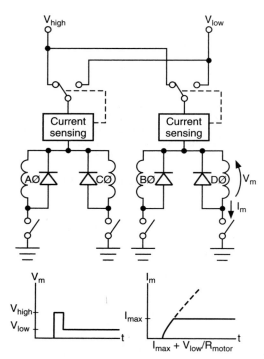

FIGURE 49 Simplified representation of bilevel drive scheme. (*Bodine Electric Company.*)

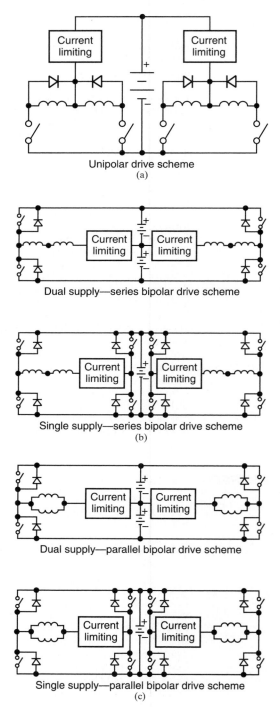

Unipolar drive scheme
(a)

Dual supply—series bipolar drive scheme

Single supply—series bipolar drive scheme
(b)

Dual supply—parallel bipolar drive scheme

Single supply—parallel bipolar drive scheme
(c)

FIGURE 50 Simplified representation of drive schemes. (*a*) Unipolar. (*b*) Bipolar series. (*c*) Bipolar parallel. (*Bodine Electric Company.*)

and friction. In high-performance systems the friction usually is kept to a minimum. Also, certain types of drive may be damaged by disconnecting the motor under power. This method is particularly unsatisfactory in the case of a vertical axis since the load may fall under gravity.

Removal of AC Input Power from Drive.   On drives which incorporate a power-dump circuit, a degree of dynamic braking is usually produced when power is removed. Therefore this is a better solution than disconnecting the motor, although the power supply capacitors may take some time to decay and this will extend the stopping distance.

Dynamic Braking.   A motor with permanent magnets will act as a generator when driven mechanically. By applying a resistive load to the motor, a braking effect is produced which is speed-dependent. Deceleration is thus rapid at high speeds, but falls off as the motor slows down.

A changeover contactor can be arranged to switch the motor connections from the drive to the resistive load. This can be made fail-safe by ensuring that braking occurs if the power supply fails. The optimum resistor value depends on the motor, but typically will be in the range of 1 to 3 ohms. It must be chosen to avoid the risk of demagnetization at maximum speed as well as possible mechanical damage through excessive torque.

Mechanical Brake.   It is often possible to fit a mechanical brake either directly on the motor or on some other part of the mechanism. However, such brakes usually are intended to prevent movement at power down and seldom are adequate to bring the system to a rapid halt, particularly if the drive is delivering full current at the time. Brakes can introduce friction even when released, and also add inertia to the system. Both effects will increase the drive power requirements.

MACHINE CONTROL SYSTEMS

Frequently a designer will be concerned with controlling an entire process. Motion control is one important and influential aspect of complete machine control. With reference to Fig. 51, the primary elements of machine control include the following:

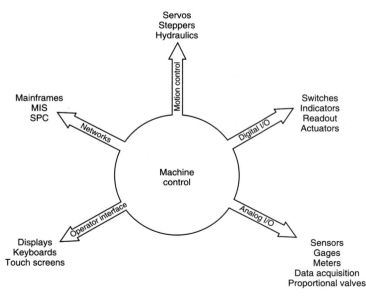

FIGURE 51   Primary elements of machine control. (*Parker-Hannifin Corporation, Compumotor Division.*)

Motion Control.   For precise programmable load movement using servomotor, stepper motor, or hydraulic actuators, feedback elements are often used.

Analog and Digital Input and Output.   For actuation of external process devices, such as solenoids, cutters, heaters, and valves.

Operator Interface.   For flexible interaction with the machine process for both setup and on-line variations. Touchscreens, data pads, and thumbwheels are examples.

Communications Support.   For process monitoring, diagnostics, and data interaction with peripheral systems.

System Architecture.   There are many different machine control architectures that integrate the foregoing elements. Each provides varying levels of complexity and integration of both the motion and nonmotion elements. Programmable logic-controller-based, bus-based, and integrated solutions are commercially available.

Controller

The controller is an essential part of any motion control system. It determines such factors as speed, direction, distance, and acceleration rate, in fact all the parameters that the motor performs. The output from the controller is connected to the input of the drive, either in the form of an analog voltage or as step and direction signals. In addition to controlling one or more motors, many controllers have additional inputs and outputs which allow them to look after other functions on a machine. Controllers can take a wide variety of forms.

Stand-Alone Controller.   A controller that can operate without the need for data or other control signals from elsewhere. Usually a stand-alone unit incorporates a keypad for data entry as well as a display, and frequently it includes a main power supply. It will also include some form of nonvolatile memory to allow it to store a sequence of operations. Many controllers that must be programmed from a terminal or computer, can also, once programmed, operate in a stand-alone mode.

Bus-Based Controllers.   This controller is designed to accept data from a host computer using a standard communications bus. Typical bus systems include STD, VME, and IBM-PC bus. The controller usually will take the form of a plug-in card, which conforms to the standards for the corresponding bus system. For example, a controller operating on the IBM-PC bus resides within the PC, plugging into an expansion slot and functioning as an intelligent peripheral.

PLC-Based Controller.   A PLC-based indexer is designed to accept data from a PLC in the form of input-output (I/O) communication. Typically, the I/O information is BCD format. The BCD information may select a program to execute a distance to move, a time delay, or any other parameter requiring a number. The PLC is well suited to I/O actuation, but marginally suited to perform complex operations such as math and complex decision making. The motion control functions are separated from the PLC's processor and thus do not burden its scan time.

X-Code Programming.   X-code has been designed to allow motion control equipment to be programmed by users with little or no computer experience. Although it includes upward of 150 commands, depending on the product, it is only necessary to learn a small percentage of these in order to write simple programs.

Most command codes use the initial letter of the function name, which makes them easy to remember. Some examples include:

V   Velocity, in revolutions per second

D   Distance, in steps

A   Acceleration rate, in revolutions per second squared

G   Go; start the move

T   Time, in seconds

A typical command string may appear as follows:

V10 A50 D4000 G T2 G

This would set the velocity to 10 r/s, acceleration to 50 r/s^2, and distance to 4000 steps. The 4000-step move would be performed twice with a 2-second wait between moves.

Single- and Multiaxis Controllers.   A single-axis controller, as the name implies, can only control one motor. The controller in an integrated indexer-drive falls into this category. However, such units are frequently used in systems with more than one motor, where the operations do not involve tight synchronization between axes.

A multiaxis controller is designed to control more than one motor and can frequently perform complex operations, such as linear or circular interpolation. These operations require accurate synchronization between axes. This generally is easier to achieve with a central controller.

A variant of the multiaxis controller is the multiplexed unit, which can control several motors on a time-shared basis. A printing machine having the machine settings controlled by stepper motors can conveniently use this type of controller when the motors do not have to be moved simultaneously.

Hardware-Based Controller.   Control systems designed without the use of a microprocessor have been used for many years and can be quite cost-effective in simple applications. They tend to lack flexibility and thus are inappropriate where the move parameters are continually changing. A majority of applications are now based on a microprocessor.

Processor-Based Controller.   The flexibility offered by a microprocessor system makes it a natural choice for motion control. Figure 52 shows the elements of a typical step and direction controller which can operate either in conjunction with a host computer or as a stand-alone unit. All control functions are handled by the microprocessor whose operating program is stored in read-only memory (ROM). This program will include an interpreter to the command language which, for example, may be X-code.

X-code commands are received from the host computer or terminal via the RS-232C communications interface. These commands are in the form of simple statements giving the required speed, distance, acceleration rate, and the like. The processor interprets these commands and uses the information to control the programmable pulse generator. This, in turn, produces the step and direction signals, which will control a stepper or servo drive.

The processor also can switch outputs and interrogate inputs via the I/O interface. Outputs can be used to initiate other machine functions, such as punching or cutting, or simply driving panel indicators to show the program status. Inputs may come from sources, such as operator push buttons or directional limit switches.

When the controller is used in a stand-alone mode, the required motion sequences are programmed from the host and stored in nonvolatile memory [normally battery-backed random-access memory (RAM)]. These sequences then may be selected and executed from switches via the I/O interface or from a separate machine controller, such as a PLC.

REPRESENTATIVE MOTION CONTROL SYSTEMS

These systems number in the multithousands. A comparatively few are illustrated and described briefly in Figs. 53 through 57. Special motion systems, such as line-following, registration, and machine-vision applications, are described elsewhere in this handbook.

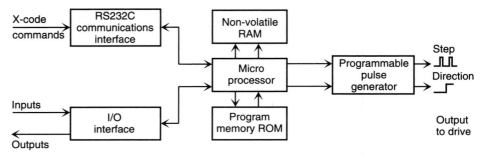

FIGURE 52   Processor-based controller. (*Parker-Hannifin Corporation, Compumotor Division.*)

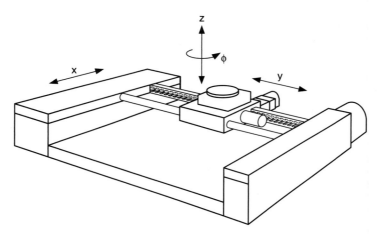

FIGURE 53   Automatic printed-circuit (PC) board component placement machine requires positioning a placement head at 75 cm (30 inches) per second with a resolution of 0.025 mm (0.001 inch) or better. Control of *X, Y,* and *Z* axes, component alignment, and gripper are required from computer programming. A belt-driven gantry is controlled by an indexer, and two servomotor drives are used for *X–Y* positioning. *Z* motion and rotational alignment are controlled by a computer microstepping drive. Joystick inputs are used to move the head manually and to teach positions to the computer. (*Parker-Hannifin Corporation, Compumotor Division.*)

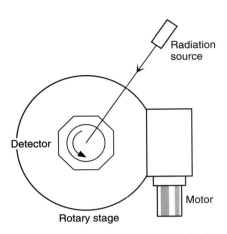

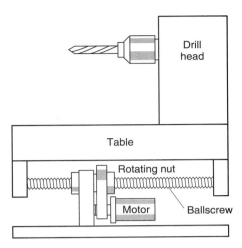

FIGURE 54 A system is required to plot the response of a sensitive detector, which must receive equally from all directions. Detector is mounted on a rotary table which requires to be indexed in 3.6° steps, completing each index within 1 second. (For setting up purposes, the table can be positioned manually at 5 r/min. The table incorporates a 90:1 worm drive.) The maximum required shaft speed (450 r/min) is well within the capacity of a stepper, which is an ideal choice in simple indexing applications. Operating at a motor resolution of 400 steps per revolution, the resolution at the table is a convenient 36,000 steps per revolution. In this case it is important that electrical noise be minimized to avoid interference with the detector. Two possible solutions are to use a low-EMI linear drive or to shut down the drive after each index. (With a stepper driving a 90:1 worm gear there is no risk of position loss during shutdown periods.) (*Parker-Hannifin Corporation, Compumotor Division.*)

FIGURE 55 A stage of a transfer machine is required to drill a number of holes in a casting using a multihead drill. The motor has to drive the drill head at high speed to within 2.5 mm (0.1 inch) of the workpiece and then proceed at cutting speed to the required depth. Drill is now withdrawn at an intermediate speed until clear of the work and then fast retracted, ready for the next cycle. Complete drilling cycle takes 2.2 seconds, with a 0.6-second delay before the next cycle. Due to proximity of other equipment, the length in the direction of travel is very restricted. An additional requirement is to monitor the machine for drill wear and breakage. The combined requirements of high speed, high duty cycle and of monitoring the drill wear all point to use of a servomotor. By checking torque load on the motor (achieved by monitoring drive current), one can watch for increased load during the drilling phase, pointing to a broken drill. Application will require a ball-screw drive to achieve high stiffness together with high speed. One way of minimizing the length of the mechanism is to attach the ball screw to the moving stage and then rotate the nut, allowing the motor to be buried underneath the table. Since access for maintenance will then be difficult, a brushless motor is suggested. (*Parker-Hannifin Corporation, Compumotor Division.*)

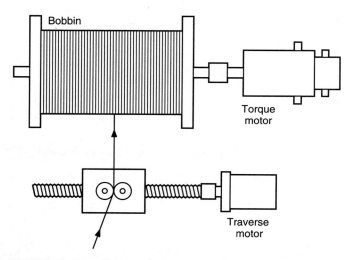

FIGURE 56 Monofilament nylon is made by an extrusion process which results in an output of filament at a constant rate. Product is wound onto a bobbin rotating at a maximum speed of 2000 r/min. Tension in filament must be held between 0.1 and 0.3 kg (0.2 and 0.6 lb) to avoid stretching. Winding diameter varies from 5 to 10 cm (2 to 4 inches). Prime requirement is to provide a controlled tension, which means operating in a torque mode rather than a velocity mode. If the motor produces a constant torque, the tension in the filament will be inversely proportional to the winding diameter. Since the winding diameter varies by 2:1, the tension will fall by 50 percent from start to finish. A 3:1 variation in tension is acceptable, so constant-torque operation is acceptable. Requirement leads to use of a servo operating in the torque mode. (Need for constant-speed operation at 2000 r/min also makes a stepper unsuitable.) Rapid acceleration is not needed, so a brush servo would be adequate. In practice, this suggests a servo in velocity mode, but with an overriding torque limit. The programmed velocity would be a little over 2000 r/min. In this way the servo will normally operate as a constant-torque drive, but if the filament breaks, the velocity would be limited to a programmed value. The traversing arm can be adequately driven by a stepper. However, the required speed will be very close to resonance, so a microstepping system would be preferable. An alternative would be to use a half-step drive in conjunction with a toothed-belt reduction of about 3:1. A ball-screw drive can be used to achieve high stiffness together with high speed. (*Parker-Hannifin Corporation, Compumotor Division.*)

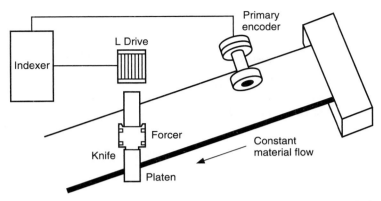

FIGURE 57   Plastic sheet cutting. Process produces a continuous flow of sheeted plastic to be cut into prescribed lengths before it is fully cured. Material is cut as it exits a machine and cannot be stopped. Depending on ambient conditions, the speed can vary. Clean angle cuts are required. In system shown, an encoder is mounted to a friction wheel driven by the plastic material. This speed signal is an input to a self-contained indexer (controller), which references all linear (cutoff-knife) velocity and position commands to the encoder, allowing precise synchronization of the web. Placing the knife at an angle to the flowing material allows for precise, straight cuts while material is moving. This is an excellent application for a linear motor. (*Parker-Hannifin Corporation, Compumotor Division.*)

SOLID-STATE VARIABLE-SPEED DRIVES

by Richard H. Osman[1]

The past two decades have seen rapid growth in the availability and usage of solid-state variable-speed drives. As of the early 1990s there is a profusion of types which are suitable for vitually every type of electrical machine, from the subfractional to the multithousand-horsepower rating (Fig. 1). Although they are quite diverse, there are two properties that are common to these drives:

1. All accept commonly available ac input power of fixed voltage and frequency and, through switching power conversion, create an output of suitable characteristics to operate a particular type of electric machine, that is, they are machine-specific.
2. All are based on semiconductor switching devices. Even though many of the principles of power conversion have been known as long as 50 years, when they were initially developed using mercury-arc rectifiers, it was not until the invention of the thyristor in 1957 that variable-speed drives became practical.

---

[1] Engineering Manager, AC Drives, Robicon Corporation, Pittsburgh, Pennsylvania.

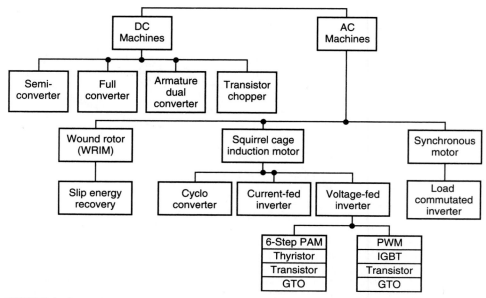

FIGURE 1 General-purpose solid-state variable-speed drives. Abbreviations are defined in text. (*Robicon Corporation.*)

Thyristor

The thyristor, or silicon-controlled rectifier (SCR), is a four-layer semiconductor device which has many of the properties of an ideal switch. It has low leakage current in the off state, a small voltage drop in the on state, and takes only a small signal to initiate conduction. (Power gains of over 10^6 are common.) When applied properly, the thyristor will last indefinitely. After introduction of the SCR, the current and voltage ratings increased rapidly. Presently it has substantially higher power capability than any other solid-state device. It dominates power conversion in the medium and higher power ranges. The major drawback of the thyristor is that it cannot be turned off by a gate signal, but the anode current must be interrupted in order for it to regain the blocking state. The inconvenience of having to commutate the thyristor in its anode circuit at a rather high energy level has encouraged the development of other related devices as power switches.

Development of Power-Switching Devices

Transistors predate thyristors, but their use as high-power switches was relatively restricted (as compared with thyristors) until the ratings reached 50 amperes and 600 volts in the same device in the early 1980s. These devices are three-layer semiconductors, which exhibit linear behavior, but which are used only in saturation or cutoff. In order to reduce the base drive requirements, most transistors used in variable-speed drives are Darlington types, where two or more transistors are cascaded in a single package. Even so, they have higher conduction losses and greater drive power requirements than thyristors. Nevertheless, because they can be turned on or off quickly via base signals, they are attractive candidates for drives within the scope of the transistor ratings, particularly for pulse-width-modulated (PWM) inverters. As of the early 1990s power transistors are available up to 1000-ampere 1200-volt ratings. Typical switching times are in the 2- to 3-μs range, excluding storage time.

Gate-Turnoff Thyristors

More recently, successful attempts to modify thyristors to enable them to be turned off by a gate signal have been made. These devices are four-layer semiconductors and are called gate-turnoff thyristors, or simply GTOs. Power GTOs have been available at least since 1965, but not until the early 1980s have these devices rated more than a few tens of amperes become available. Present GTOs have a lower forward voltage drop than a Darlington transistor, but significantly more than that of a conventional thyristor. GTOs require a much more powerful gate drive, particularly for turnoff, but the lack of external commutation circuit requirements makes them desirable for inverter use where space and weight are at a premium. GTOs are available at higher voltage and current ratings than power transistors. Unlike transistors, once a GTO has been turned on or off with a gate pulse, it is not necessary to continue the gate signal due to the internal positive-feedback mechanism inherent in four-layer devices. It is considerably more difficult to manufacture large GTOs than conventional thyristors. Their costs are accordingly much higher for a comparable rating.

Insulated-Gate Bipolar Transistors

The insulated-gate bipolar transistor (IGBT) is an integrated combination of a power bipolar transistor and a MOSFET (metal-oxide semiconductor field-effect transistor), which combines the best properties of both devices. This was developed during the 1980s in order to overcome the transistor's drawback of high base drive-power requirements. IGBTs have a very high input impedance, which permits them to be driven directly from relatively lower power logic sources, thus reducing the cost of the driver circuit. Another improvement is that the IGBT switches much more quickly (on the order of 200 to 300 ns) than transistors. The higher switching speed permits better PWM waveforms and reduced acoustic noise from the motor. Much progress has been made in increasing the voltage and current ratings of IGBTs. They are now available at very nearly the same ratings as power Darlington transistors. It remains to be seen if the cost of IGBTs will drop sufficiently to fully displace Darlington transistors.

The Technology Base

The four semiconductor devices just described (thyristor, transistor, GTO, and IGBT) form the technological base upon which the solid-state variable-speed drive industry rests in the early 1990s. There are other devices in various stages of development. These may or may not become significant, depending on their cost and availability in large current (>50 amperes) and high voltage (1000 volts) ratings. These prospects include the MOSFET and the static induction thyristor.

To date it has not been possible to construct power MOSFETs which have acceptably low on resistance while still having the 1000 volt rating necessary for reliable power conversion at the 500-volt ac level. Therefore their use has been limited to very small drives (<10 hp). Power MOSFETs are the fastest power switching devices (100 ns) of the group, and they also have very high gate impedance, thus greatly reducing the cost of drive circuits. Recent advances in IGBTs may have achieved the objectives nearly as well as the MOSFETs may in the future.

Static induction thyristors (SITs) are claimed to have the voltage and current ratings of GTOs, but with a much higher gate impedance (to reduce driver requirements). The validity of these claims has not been proved in commercial use because SITs are just emerging from the development laboratory.

Variable-Speed-Drive Hardware Development

Parallel to the development of power switching devices, there have been important advances in hardware for controlling variable-speed drives. These controls are a mixture of analog and digital signal processing.

Developments during the 1965–1975 period (integrated-circuit operational amplifiers and logic families) made possible dramatic reductions in size and cost of the drive control, while permitting more sophisticated and complex control algorithms—all without a reliability penalty. Further consolidation of the control circuits occurred somewhat later when large-scale integrated (LSI) circuits became available. Prior to LSI circuits the PWM control technique was not practical because of the immense amount of combinational logic required. The most significant trend in the early and mid-1980s was the introduction of microprocessors into drive control circuits. The resulting performance enhancements include the following:

1. More elaborate and detailed diagnostics because of the ability to store data relating to drive internal variables, such as current, speed, and firing angle, among other factors
2. Ability to communicate both ways with the user's central computers concerning drive status
3. Ability to make drive tuning adjustments by keypad entry to parameters, such as loop gains, ramp rates, and current limit stored in memory rather than as potentiometer settings
4. Self-tuning drive controls
5. More sophisticated techniques for overcoming power circuit nonlinearities

The foregoing control improvements have not increased control circuit costs significantly.

More recently the development of more capable microprocessors, particularly digital signal processors, has made possible real-time calculation of PWM waveforms and on-line solution of machine differential equations. The availability of large-capacity programmable read-only memory (ROM) is another advance. The latter has made possible the storage of huge numbers of precalculated switching angles for elaborate PWM strategies. The development of field-oriented controls has benefited markedly from these hardware and software developments. Consequently ac drives now can offer control performance equal to or better than dc drives and hence are displacing dc drives from many motion-control applications. The consensus is that numerous other improvements in ac drive performance will occur in the future.

DIRECT-CURRENT (DC) DRIVES

The introduction of the thyristor had the most immediate impact on dc drive technology. Motor-generator variable-speed drives were quickly supplanted by thyristor dc drives of the type shown in Fig. 2 for reasons of lower first cost, higher efficiency, and lower maintenance cost. This type of power circuit with three thyristors and three diodes in a three-phase bridge is generally referred to as a semiconverter. By phase control of the thyristors, the system behaves as a programmable voltage source. Therefore speed variation is obtained by adjustment of the armature voltage of the dc machine. Because the phase control is fast and precise, critical features such as current limit are easily obtained. Almost all thyristor dc drives today are configured as current regulators with a speed or voltage outer loop. The semiconverter is suitable for one-quadrant drives because it produces only one direction of current and output voltage. The input power factor is dependent on speed.

Six-Thyristor Full Converter

As the cost differential between thyristors and diodes has narrowed, the semiconverter has been displaced by the six-thyristor full converter, as shown in Fig. 3. This circuit arrangement (Graetz circuit) has become the "workhorse" of the electrical variable-speed drive industry.

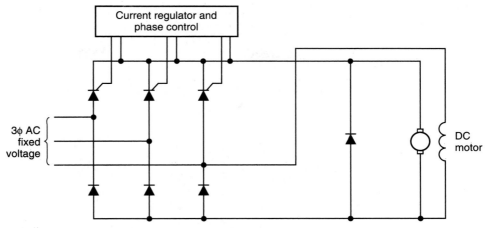

FIGURE 2 Thyristor dc drive; three-phase semiconverter. (*Robicon Corporation.*)

The control techniques are very similar to those of the semiconverter. The full converter, however, offers lower output ripple and the ability to regenerate (or return) energy to the ac line. The system can be made into a four-quadrant drive by the addition of a bidirectional field controller. Torque direction is determined by field current direction. Due to the large field inductance, torque reversals are relatively slow (100 to 500 ms), but adequate for many applications.

Dual-Armature Converter

For the best response speed of large thyristor dc drives the dual-armature converter of Fig. 4 is preferred. This is simply two converters (as shown previously in Fig. 3) connected back to back. Torque direction is determined by the direction of the armature current. Inasmuch as

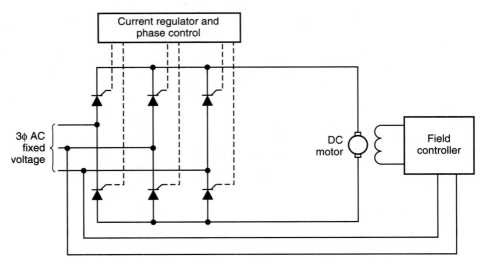

FIGURE 3 Thyristor dc drive; three-phase full converter. (*Robicon Corporation.*)

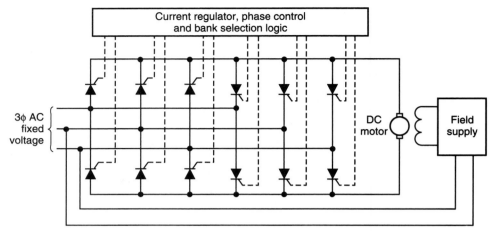

FIGURE 4   Thyristor dc drive; dual-armature converter. (*Robicon Corporation.*)

this is a low-inductance circuit, reversal can be accomplished in less than 10 ms. Obviously only one converter is conducting at one time, with the other group of six thyristors *not* being gated. This scheme is called bank selection.

Thyristor dc Drives—Summary

The three types of thyristor dc drives just described all share a common property in that the devices are turned off by the natural polarity of the input line. This is called natural or line commutation. Thus the inability to turn off a thyristor from the gate is no practical drawback in these circuits. Consequently they are simple and very efficient (typically 98.5 percent) because the device forward drop is small as compared with the operating voltage. These drives can be manufactured to match a dc machine of any voltage (commonly 500 volts) or horsepower (typically 0.5 to 2500 hp).

For certain types of applications, typically machine-tool axis drives and tape transport drives, the response of phase-controlled thyristor drives is not as fast as needed. A special class of dc drives has been developed for this reason (Fig. 5). A fixed dc bus is developed from the line via a rectifier and capacitor filter. This voltage source is applied to the armature through power transistors. The voltage is modulated by duty-cycle (or pulse-width) control. The devices usually operate from 120 volts or 240 volts ac and rarely exceed 10 hp. Frequently they are applied with permanent-magnetic-field dc machines.

ALTERNATING CURRENT (AC) DRIVES

The impacet of the new solid-state switching devices was considerably greater on ac variable-speed drives than on dc drives, but the impact occurred somewhat later. Ac drives are machine-specific and more complex than dc drives. Solid-state variable-speed drives have been developed and applied for (1) wound-rotor induction motors (WRIMs), (2) synchronous motors with wound field and permanent-magnet field, and (3) cage-type induction motors.

Historically WRIM-based variable-speed drives were commonly used long before the advent of solid-state electronics. These drives operate on the principle of deliberately creating high-slip conditions in the machine and then disposing of the large rotor power which results. This is done, for example, by using liquid rheostats to vary the resistance seen by the rotor windings.

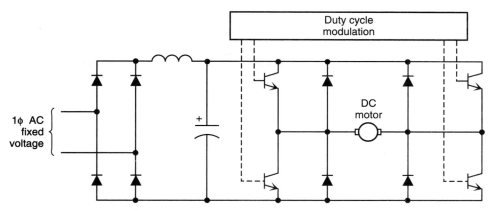

FIGURE 5 Chopper dc drive; transistor bridge type. (*Robicon Corporation.*)

Slip-Energy Recovery Systems

A more modern WRIM drive is shown in Fig. 6. This is called the slip-energy recovery system or static Kramer drive. The output of the rotor is rectified, and this dc voltage is coupled to the line via a thyristor converter. The line-commutated converter is current-regulated, which effectively controls torque. Efficient, stepless speed control results. Very large (>1000-hp) drives can be built because the stator may be wound for medium voltage, while the rotor operates at less than 600 volts to match the power converter rating. The power conversion equipment may be downsized if a narrow range (say, 100 to 70 percent) of control is adequate (for example, as may be encountered in fan drives). The performance drawbacks are (1) a poor system power factor if not corrected and (2) no above-synchronous-speed operation is available.

The WRIM is the most expensive ac machine. Thus they are noncompetitive as compared to cage induction motor (IM) drives or load-commutated inverters using synchronous machines. Consequently few WRIM systems are manufactured in the United States today.

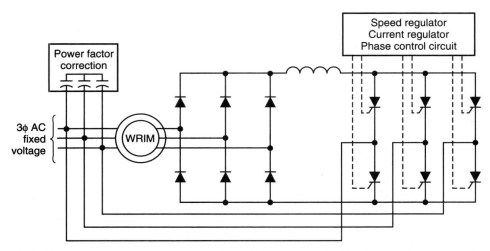

FIGURE 6 Slip-energy recovery system; wound-rotor induction motor (WRIM) drive. (*Robicon Corporation.*)

Load-Commutated Inverter

As shown in Fig. 7, the load-commutated inverter (LCI) is based on a synchronous machine. It uses two thyristor bridges, one on the line side and another on the machine side. All devices are naturally commutated because the back EMF of the machine commutates the load-side converter. This requires the machine to operate with a leading power factor, and therefore it requires substantially more field excitation (a special exciter as compared with a normally applied synchronous motor).

Another result is a reduction in the torque for a given current. The machine-side devices are fired in exact synchronization with the rotation of the machine. This maintains constant torque angle and constant commutation margin. It is accomplished either by rotor position feedback or by phase-control circuits driven by the machine terminal voltage. The line-side converter is current-regulated to control torque. A choke is used between converters to smooth the link current. LCIs were first used commercially in about 1980 and are used mainly on very large drives (1000 to 50,000 hp). At these power levels, multiple series devices are used (typically 2.4- to 4-kV input), and conversion takes place directly at 2.4 or 4 kV or higher. The efficiency is excellent and reliability has been very good. Although they are capable of regeneration, LCIs are rarely used in four-quadrant applications because of the difficulty in commutating at very low speeds, where the machine voltage is negligible. Operation above line frequency is straightforward.

A lower power version of the LCI has found widespread application in recent years. This is called the brushless dc motor, but it really is a synchronous ac machine (often with permanent-magnet rotor). It is equipped with a transistor inverter whose switching is synchronized with the rotor position by means of an encoder. The transistors are controlled to produce a sinusoidal or trapezoidal current waveform. Excellent response time is achieved.

INDUCTION MOTOR VARIABLE-SPEED DRIVES

The greatest diversity of power circuits is available with induction motor variable-speed drives. Because the squirrel-cage induction motor is the least expensive, least complex, and most rugged electric machine, much effort has gone into drive development in order to exploit the machine's superior qualities. Because of its simplicity, however, it is the least amenable of drives for variable-speed operation. Since the squirrel-cage motor has only one electrical input port, the drive must control flux and torque simultaneously through this single input. As there is no access to the rotor, the power dissipation there raises its temperature and thus very low-slip operation is essential to avoid overheating.

Cycloconverter

One approach used with an induction or synchronous motor drive is that of "synthesizing" an ac voltage waveform from sections of the input voltage. This can be done with three dual converters, the circuit of which is called a cycloconverter (Fig. 8).

The output voltage is rich in harmonics, but of sufficient quality for motor drives so long as the output frequency does not exceed one-third to one-half of the input. The thyristors are line-commutated, but there are 36 required. (Sometimes half-wave circuits are used, which need only 18 devices, but more harmonics result.) The cycloconverter usually is built in very large power ratings. The device is capable of heavy overloads and four-quadrant operation, but it has a limited output frequency and poor input power factor. For special low-speed high-horsepower (>1000-hp) applications, such as cement kiln drives, the cycloconverter has been used successfully.

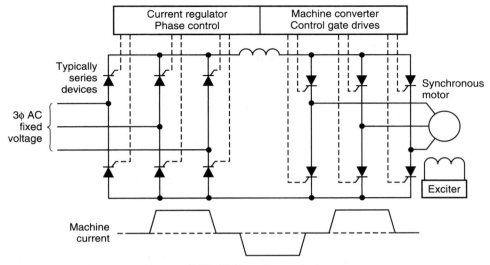

FIGURE 7   Load-commutated inverter (LCI). (*Robicon Corporation.*)

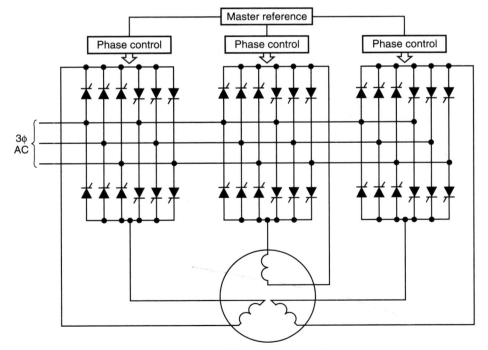

Induction motor

FIGURE 8   Cycloconverter induction motor drive. (*Robicon Corporation.*)

Autosequentially Commutated Current-Fed Inverter

This represents another approach to induction motor drive, wherein a smooth dc current is generated and fed into different combinations of windings of the machine so as to create a discretely rotating magnetomotive force (MMF). This type of inverter is called the autosequentially commutated current-fed inverter (ASCI) (Fig. 9).

This circuit was invented later than other inverters and is much more popular in Europe and Japan than in the United States. The input stage is a three-phase thyristor bridge which is current-regulated. A link choke smooths the current going to the output stage. There a thyristor bridge distributes the current into the motor winding with the same switching function as the input bridge, except at variable frequency. (Notice the similarity to the LCI.) The current waveform is a quasi-square wave whose frequency is set by the output switching rate and whose amplitude is controlled by the current regulator. The capacitors and rectifiers are used to store energy to commutate the thyristors, since the induction motor cannot provide this energy and remain magnetized (in contrast to the synchronous motor). This type of drive has simplicity, good efficiency (96 percent), excellent reliability, and four-quadrant operation up to about 120 Hz. Harmonics in the output current are reasonably low, giving a form factor of about 1.03 (the same as the LCI).

Moreover, harmonic currents are not machine-dependent and decrease at light load. The input power factor is load- and speed-dependent, but much better than for the cycloconverter. Above 100 hp the ASCI is very cost-effective. Because the units are constructed with SCRs, they have recently (1984) become available at 2.4- and 4-kV direct conversion for very large (>1000-hp) units. It is nearly always possible to retrofit an existing motor with this type of drive. Due to the controlled current properties, this drive is virtually immune to damage from ground faults, load short-circuits, and commutation failures.

Since MMF (current) is directly controlled and the drive is regenerative, ASCIs can be readily equipped with field-oriented controls. The control technique keeps track of the flux and MMF vectors inside the machine in order to provide a fast and precise torque response to an external reference. In addition, they are four-quadrant drives, capable of producing either direction of torque in either direction of rotation. This type of ac drive has been used successfully in the most demanding applications, such as traction drives and test-stand drives.

Voltage-Source Inverter

A third approach to induction motor drives is to generate a smooth dc voltage and apply that to different combinations of the machine windings so as to create a rotating flux. This circuit, called a voltage-source inverter (VSI), is illustrated in Fig. 10. It is interesting to note that this circuit was one of the first applications of thyristors to induction motor drives, and it is shown here for historical interest because other devices are now used in the circuit.

The input is a three-phase thyristor bridge which feeds a capacitor filter bank forming a controlled low-impedance voltage source. The output stage consists of six main thyristors (1–6) in a bridge with antiparallel diodes. There are six auxiliary or commutating thyristors (11–16) which, together with the L–C circuits, impulse-commutate the main devices. The output waveform is a quasi-square wave of voltage whose amplitude is set by the dc link voltage, sometimes referred to as pulse amplitude modulation (PAM). Here the output frequency is determined by the output switching rate. The output voltage is set by the voltage regulator on the input converter. In order to reduce the size of the commutating L–C, special thyristors with fast turnoff times are required (in contrast to the ordinary phase-control types used in dc drives, LCIs, and ASCIs). Even though complex, these drives have had a reasonably good reliability record. They have good efficiency and can operate at very high frequencies (180 Hz and up). Regeneration to the line is not possible. Many of these units are in service and are available from 50 to 500 hp, typically at 460 volts ac. They are not available at over 600 volts.

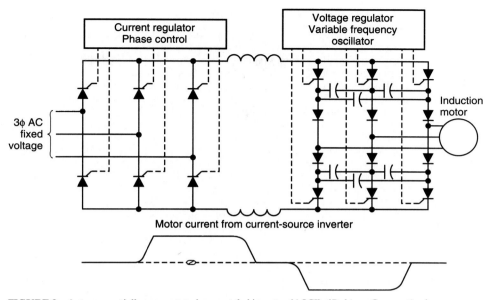

FIGURE 9   Autosequentially commutated current-fed inverter (ASCI). (*Robicon Corporation.*)

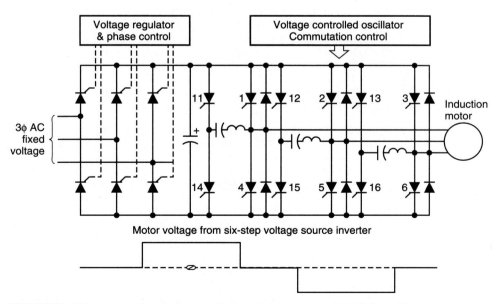

FIGURE 10   Voltage-source inverter; six-step thyristor, impulse-commutated. (*Robicon Corporation.*)

Transistors and GTOs in VSIs

In order to reduce the cost and complexity of the VSI (Fig. 10), the thyristors and their commutation circuits have been replaced with transistors or with GTOs. The resulting circuit is shown in Fig. 11. The performance features are better than the thyristor version because of the faster switching of the devices and substantial reduction in size and weight. Although the conduction losses are higher due to the higher on voltages, commutation losses are reduced substantially. Efficiencies remain above 95 percent. The transistor version has been available since about 1982 at 460-volt ac input; 230-volt units have been available since the mid-1970s. At present 100-hp transistor drives with single-output devices are available, and up to 500-hp units with parallel-output devices can be obtained.

Since GTOs have somewhat higher ratings, drives from 50 to 500 hp at 460 volts ac are available (early 1990s). For a time it was unclear which of the aforementioned devices would be most successful, but transistors and IGBTs are clearly gaining wide acceptance in the power range below 500 hp. There appears to be a three-way competition among GTO, transistor, and thyristor drives in the range above 500 hp to capture highest acceptance in that range. However, GTO and transistor costs must be reduced substantially to match thyristor designs at the upper end of this power range, although there are performance advantages offered by the GTO and transistor designs.

Pulse-Width-Modulated Drives

The induction motor drives described thus far are all similar in that the amplitude of the output is controlled by the input converter. Another category of voltage source inverters controls both the frequency and the amplitude by the output switches alone. A representative circuit based on transistors is shown in Fig. 12. Note that the input converter is replaced with a diode bridge so that the dc link operates at a fixed, unregulated voltage. The diode front end gives virtually unity power factor, independent of load and speed. This type of drive is called pulse-width-modulated. Due to the simplicity of the power circuit and the absence of magnetic components, this design has become the least expensive, particularly now that power transistors have matured.

An output voltage waveform is synthesized from constant-amplitude, variable-width pulses at a high (500-Hz to 10-kHz) frequency so that a sinusoidal output is simulated. The lower harmonics (5, 7, 11, 13, 17, 19, . . .) in six-step waveforms are not present in currently available PWM drives. One advantage is smooth torque, low output harmonic currents, and no cogging. Although the PWM approach eliminates the phase control requirements and cuts the front end losses somewhat, there are offsetting drawbacks. Since every switching causes an energy loss in the output devices and their suppressors, the total losses at high speed go up considerably over six steps (six switches per cycle) if the same devices are used.

To overcome this limitation, faster transistors and IGBTs are used. The output devices are stressed much more severely than in six-step designs. Some PWM designs do not have a voltage regulator. At any given output frequency they deliver a preset fraction of the input voltage. If the input fluctuates, so does the output. Finally the high-frequency switching may cause objectionable acoustic noises in the motor. Even with these drawbacks, the PWM drive has become the most widely applied design below 100 hp.

There are both transistor and GTO PWM units on the market today in the range of 1 to 3000 hp at 460 volts ac. As with all voltage source inverters, regeneration to the line is not inherent.

Views of representative industrial solid-state variable-speed drives are given in Figs. 13, 14, and 15.

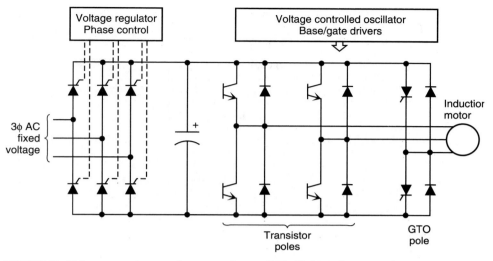

FIGURE 11   Voltage source inverter; six-step transistor or GTO. (*Robicon Corporation.*)

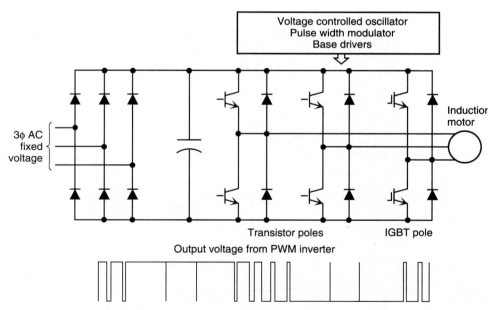

FIGURE 12   Pulse-width-modulated inverter; transistor or IGBT implementation. (*Robicon Corporation.*)

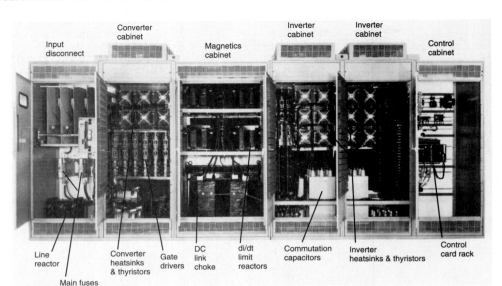

FIGURE 13   200-hp, 2300-volt ac current-fed drive. (*Robicon Corporation.*)

FIGURE 14   30-hp, 460-volt PWM drive; transistor output stage. (*Robicon Corporation.*)

FIGURE 15   600-hp, 460-volt current-fed inverter. (*Robicon Corporation.*)

ROBOTS

Early industr⋅ ⋅. robots date back several decades. Initially they were used mainly to assist or take over dangerous or difficult handling operations, essentially to protect manual laborers from undue exposure to harmful substances, temperature, radiation, and so on. Research and development expense in such situations was relatively easy to justify, but all along the ultimate objective was to design robots that could perform manual tasks better, more cheaply, and more quickly than people. As with numerous other specialized equipment technologies, the robot, in concept, was far ahead of the components needed to enhance its performance, as have later emanated from advancements in solid-state electronics, computer controls, and communications. Very large strides in robot development have been made since the mid-1970s, particularly as the result of some piece- and parts-handling industries (automotive being a major example) to improve their competitive position in terms of increased productivity and product quality.

BASIC FORMAT OF ROBOT

It is not always easy to distinguish a mechanized handling machine from what is generally considered to be a robot. For example, a modern, complex conveyor system would meet some of the general descriptive criteria of a robot, but in professional parlance, a conveyor by itself would seldom be considered a robot. However, in terms of total robotic technology, one or many conveyors could be involved. A definition, coined several years ago by the Robot Institute of America, still provides a good definitive foundation, even though some of the words used are rather general and perhaps superconclusive. The definition is:

A robot is a *reprogrammable, multifunctional* manipulator designed to *move* materials, parts, tools, or specialized devices through variable programmed *motions* for the performance of a variety of tasks.

In terms of their classification, robots may be considered from a number of viewpoints:

1. Axes of motion, including type of motion, number of axes, and the parameters of axis travel
2. Load capacity and power required, namely, weight of load, and electrically, pneumatically, or hydraulically operated
3. Dynamic properties
4. End-effectors or grippers used
5. Programming and control system
6. General-purpose or special task

AXES OF MOTION

A robot may be movable from one factory location to another, as may be required by factory layout changes or by major alterations in job assignment. However, for any given task that will be repeated over and over for long periods, a robot will be firmly fastened to the operating floor (sometimes the ceiling). The firm location establishes a fixed geometrical location of reference, an unchangeable position that will geometrically relate precisely with an associated machine, or in the case of a work cell, involving several other machines and often other robots.

For relatively moderate changes in the robot's working envelope, the average "stock" robot will incorporate considerable flexibility within its design so that changes can be made without altering the location of reference. Sometimes, in the case of a "smart" robot, final very small changes in the positioning of an arm can be made by outputs from a machine vision or tactile system.

Less frequently, a robot will be intentionally designed for movability so that it can be transferred to the worksite, rather than grouping one or more robots about specific locations, as will be mentioned later.

Degrees of Freedom

Designed or built-in axes of motion essentially define the robot's ability to move parts and materials, sometimes referred to as degrees of freedom. The axis of motion refers to the separate motion a robot has in its manipulator, wrist, and base. The designer usually will select from one of four different geometric coordinates for any given robot.

Revolute (Jointed-Arm) Coordinates.   In this system the robot arm is constructed of several rigid members, which are connected by rotary joints. Three independent motions are permitted (Fig. 1). These members are analogous to the human upper arm, forearm, and hand, while the joints are equivalent to the human shoulder, elbow, and wrist, respectively. The arm incorporates a wrist assembly for orienting the end-effector, in accordance with the demands of the workpiece (Fig. 2). These three articulations are pitch (bend), yaw (swing), and roll (swivel). In some applications, fewer than six articulations may suffice, depending on the geometry of the workpiece and the machine which the robot is serving.

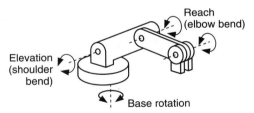

FIGURE 1   Jointed-arm manipulator, incorporating revolute coordinates.

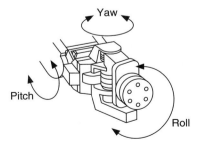

FIGURE 2   Wrist assembly on robot arm for orienting end-effector in accordance with requirements of workpiece.

Cartesian Coordinates. In this system all robot motions travel in right-angle lines to each other. There are no radial motions. Consequently the profile of a cartesian-based robot will have a rectangularly shaped work envelope (Fig. 3).

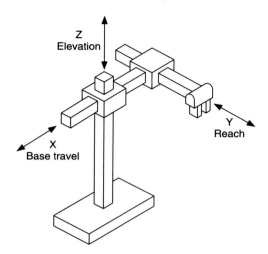

FIGURE 3 Manipulator incorporating cartesian coordinates.

Some systems utilize rotary actuators to control end-effector orientation. Robots of this type generally are limited to special applications. A robot may incorporate rectilinear cartesian coordinates as, for example, a continuous-path extended-reach robot gains much versatility through a bridge and trolley construction, which enables the robot to have a relatively larger rectangular work envelope. When ceiling-mounted, this system may service many stations with several functions, thus leaving the floor clear. X and Y motions are performed by the bridge and trolley; the vertical motions are accomplished by using telescoping tubes.

In a cartesian coordinate system the location of the center of the coordinate system is the center of the junction of the first two joints. Except for literally moving the robot to another factory location, the center does not move. In effect, it is tied to the "world" as if anchored in concrete. If the X measurement line points toward a column in the work area where the robot is placed, the X line will always point toward that same column, no matter what way the robot turns while performing its programs. These are known as the world coordinates for a given robot installation (Fig. 4).

In the operation of a robot, having an origin for a measurement reference is not sufficient. One also needs to know the point to which measurements will be made. This measurement is made from the origin of the coordinate system to a point that is exactly in the center of the circle, on which the tool (end-effector) is to be mounted. This system moves with the tool and is aptly called the tool coordinate system. In the tool coordinate system the X and Y lines lie at right angles flat on the tool-mounting surface. The Z line is the same as the axis of rotation for the point, that is, it points directly through the tool in one direction and through the wrist in the other direction. The system is *not* tied to the world. Instead it stays in position on the tool-mounting surface and moves wherever the tool moves. While the origin of the system is thus allowed to move around, the destination (where it measures to) is left to the discretion of the user. Sometimes the tool coordinate system is actually used to measure where the tip of the tool lies relative to where it is mounted; sometimes it is used to measure where one position in space lies relative to some other point in space (Fig. 5).

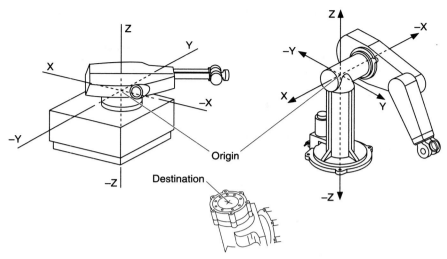

FIGURE 4   World coordinate system of robot using cartesian coordinates.

Cylindrical Coordinates.   Robots designed with this system have a horizontal shaft that goes in and out and rides up and down on a vertical shaft. The latter rotates about the base (Fig. 6). Additional rotary axes are sometimes used to allow for end-effector orientation. Cylindrical-coordinate robots are often well suited for tasks to be performed on machines to be serviced that are located radially from the robot and where no obstructions are present. A robot that incorporates cylindrical coordinates has a working area or envelope that is a portion of a cylinder.

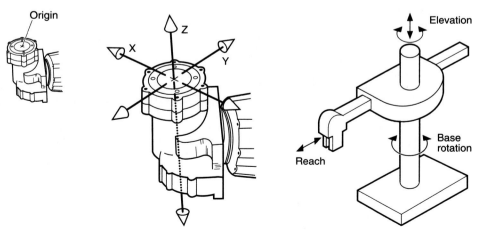

FIGURE 5   Tool coordinate system of robot using cartesian coordinates.

FIGURE 6   Manipulator incorporating cylindrical coordinates.

Spherical (Polar) Coordinates.   Robots using this system may be likened to a tank turret, that is, they comprise a rotary base, an elevation, and a telescoping extend-and-reach boom axis. Up to three rotary wrist axes (pitch, yaw, and roll) may be used to control the orientation of the end effector (Fig. 7).

Work Envelope

The area in space that a robot can touch with the mounting plate on the end of its arm is known as its work envelope (Fig. 8).

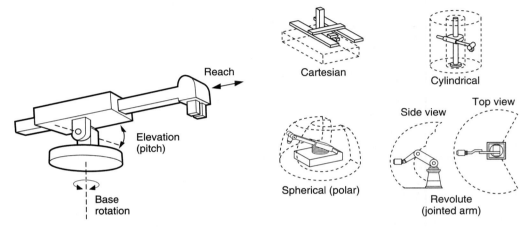

FIGURE 7 Spherical-coordinate manipulator, the operation of which is comparable to a tank turret.

FIGURE 8 Work envelope of a robot is that area in space which the robot can touch with the mounting plate on the end of its arm.

LOAD CAPACITY AND POWER REQUIREMENTS

With need and proper design, robots can be designed to handle miniature (tiny) pieces that weigh a few ounces or grams, as found, for example, in electronics manufacture, up to heavy industrial loads ranging from 135 to 1045 kg (300 to 2300 lb) and even much greater loads where robotic equipment is used, for example, in earth-moving situations, as may be found in earthquake debris removal. A recent survey of user demand shows that robots lie within the range of 9 kg (20 lb) on the low side to 136 kg (300 lb) on the high side for the majority of applications. A majority of robots are electrically actuated by servomotors, particularly stepping and permanent-magnet dc motors, as described previously in detail in this handbook section. Less frequently used are pneumatic and hydraulic actuators.

DYNAMIC PROPERTIES OF ROBOTS

Important dynamic properties of robots include (1) stability, (2) resolution, (3) repeatability, and (4) compliance. Considering these factors, the design of a robot is innately complex because of the manner in which these properties interrelate. This also contributes to the difficulties of optimizing a design. Figures 9 through 12 illustrate specific examples of dynamic problems.

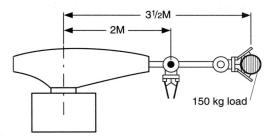

FIGURE 9   Consider a robot arm that has a retracted hand position of 2 meters and an extended hand position of 3.5 meters. Consider also that this arm might carry a load of 150 kg, and that the arm should go from position to position, with or without load, at any extension and without overshoot. For the configuration shown, the variation in moment of inertia is from 70 kgMsec^2 when tucked in and unloaded to 230 kgMsec^2 when fully extended and loaded. To achieve a critically damped servo with position repeatability of 0.5 mm under all operating conditions is difficult. Note that 0.5-mm resolution for an arm with 300° of rotation requires position encoding to an accuracy of 1 part in 33,000, or 2^{15}. The foregoing deals only with a major robot arm articulation. In a full arm the interactions among the various articulations complicate both dynamic performance and accuracy. For example, a robot arm designed to achieve an individual-articulation natural frequency of 50 Hz degenerates to an overall 17 Hz in a six-articulation arm. (*Westinghouse.*)

Stability

This characteristic is associated with oscillation in the motion of the tool. The fewer the oscillations that are present, obviously the more stable is the operation of the robot. Negative aspects of oscillations include the following. (1) Extra wear is imposed on the mechanical, hydraulic, and other parts of the robot arm. (2) The tool will follow different paths in space during successive repetitions of the same movement, thus requiring more distance between the intended trajectory and the surrounding parts. (3) The time required for the tool to stop at a precision position will be increased. (4) The tool may overshoot the intended stopping position, possibly causing a collision with some object in the system.

Oscillations may be damped or undamped. Damped (transient) oscillations will degrade and cease with time; undamped oscillations may persist or may grow in magnitude (runaway oscillation) and are the most serious because of the potential damage they may cause to the surroundings.

Variations of internal and gravitational loads on the individual joint servos (as the arm's posture changes) make the operation of the robot prone to oscillation.

In one approach to solving oscillation problems, the joint servos operate continuously. Some sophisticated servo designs (the result of experience from numerical control of machine tools) prevent oscillation from starting, regardless of the load carried. In another approach the robot controller locks each joint independently the first time it reaches its set point. Special circuitry also decelerates the joint after it comes within a prescribed distance of that position. When the joints are all locked (total coincidence), the arm is stationary, and it can then begin to move to the next position. If the position is held for more than a few seconds, the tool slowly creeps away from its programmed position.

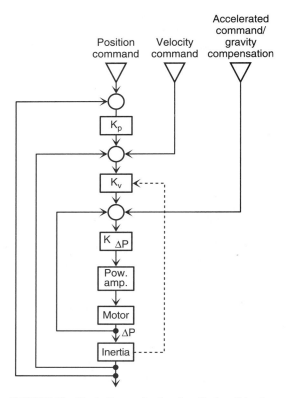

FIGURE 10 Block diagram for functionally describing key elements of a single-articulation servo system, including velocity and acceleration feedback and interarticulation bias signals. In estimating the time to complete a task (without actually simulating the entire process), the interface with the workplace complicates the process. Paths to avoid obstacles add program steps. Some steps must be very precise, calling for closing out to zero error before the program advances. Other steps may be the corners in a motion path, which can be passed through "on the fly" so to speak. The use of interlock switches may introduce transport lags. Simple programs often permit using a rule of thumb. For example, if one allows 0.8 second for each motion taught, short steps as well as long, a time for program completion can be estimated quite closely. However, if a program is complex, as in spot welding a car body, there are too many variables to permit the use of such methods. Other factors that must be included are weld-gun inertia, weld-gun operating time, metal thickness, proximity of spots to one another, among other critical variables.

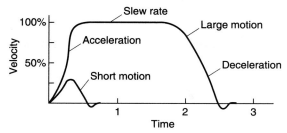

FIGURE 11   It is common for robots to be offered with abbreviated specifications that list the slew rates and the repeatability of each articulation. What is really needed is the total amount of time required to go from position to position and net accuracy of all articulations in consort. Shown here are two typical velocity traces for a short-arm motion and a large-arm motion. It is evident that the slew rate is no measure of elapsed time in making a motion, particularly a short motion in which the slew rate may not be attained at all.

Resolution and Repeatability

Repeatability is affected by resolution and component inaccuracy. Both short- and long-term repeatability exist in a robotic system. Long-term repeatability is of concern in robot systems that must perform tasks over a several-month period. During long-term repetitive use, components wear and age to the extent that repeatability must be checked periodically. Short-term repeatability is influenced most by temperature changes within the control and the environment as well as by transient conditions between shutdown and start-up of the system. These factors frequently are grouped under the umbrella term, drift. The accuracy of a robotic system can range from several hundredths of an inch for a simple robot to several thousandths of an inch for a robot doing precision assembly or handling small parts. In the case of a robot used in testing printed-circuit boards and other electronic manufacturing operations, the need for precision is paramount. Repeatability claims for standard or stock robots are listed in Table 1.

TABLE 1   Manufacturers' Claimed Repeatability of Randomly Selected Contemporary Robots

Load capacity		Claimed repeatability (±)	
lb	kg	inches	mm
5	2.2	0.004	0.1
14	6	0.004	0.1
22	10	0.008	0.2
35	16	0.001	0.03
66	30	0.002	0.05
110	50	0.020	0.5
132	60	0.020	0.5
150	68	0.020	0.5
176	80	0.020	0.5
200	90	0.010	0.25
264	120	0.04	1.0

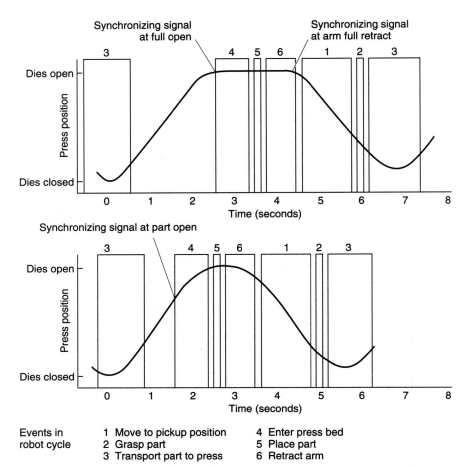

Events in	1 Move to pickup position	4 Enter press bed
robot cycle	2 Grasp part	5 Place part
	3 Transport part to press	6 Retract arm

FIGURE 12   For some operations, program time is critical, such as when a robot is serving heavy, expensive capital equipment. If the production rate is paced by the robot rather than the equipment, the project would not seem viable because of loss of throughput. Optimizing such a program may involve a range of techniques. A typical application might be press-to-press transfer of sheet-metal parts. A line of presses runs at a gross production rate of up to 700 parts per hour. At this rate a robot must make a complete transfer and return for the next pickup in 5.16 seconds. With presses on center-to-center distances of 6 meters, this is a demanding transfer speed. To meet this rate, a robot was modified by increasing the capacity of both hydraulic supply and servo valves. Acceleration and deceleration times were reduced at some sacrifice in damping and accuracy. This was compensated by providing the die nests with leads or strike bars. Finally, interlocks were refined so that the robot could make approaches and departures during the rise and fall of the moving platens of the press. The curves given here show how the time can be shortened by tight interlocks that do not wait for press-cycle operation. For safety, this approach cannot be used with human operators.

Compliance

The compliance of a manipulator is indicated by its displacement relative to a fixed frame in response to a force (torque) exerted on it. The force may be a reaction force (torque) that arises when the manipulator pushes (twists) the tool against an object, or it may be the result of the object pushing (twisting) the tool. High compliance means that the tool moves a lot in response to a small force, and the manipulator is then said to be spongy or springy. If it moves very little, the compliance is low and the manipulator is said to be stiff.

Compliance is a complex quantity to measure. In practice, a manipulator may be defined as a nonlinear, anisotropic tensor quantity that varies with time and with the manipulator's posture and motion. It is a tensor because a force in one direction can result in displacements in other directions and even rotation. A torque can result in rotation about any axis and displacement in any direction. A 6-by-6 matrix is a convenient representation for a compliance tensor. Time can affect compliance through changes in temperature and hence the viscosity of hydraulic fluid. Compliance will often be found to be a function of the frequency of the applied force or torque. A manipulator, for example, may be very compliant at frequencies around 2 Hz, but very stiff in response to slower disturbances. Compliance may exhibit hysteresis. For example, the servos in one design of hydraulic manipulator turn off when the arm stops moving. In this condition all the servo valves are closed, and the compliance has a value that is determined by the volume of incompressible hydraulic fluid trapped in the hydraulic hoses and the elasticity of the hoses. However, if an outside force on the tool should move any of the joints more than a given distance from the position at which they are supposed to remain, then the servos on all joints will turn on again. The compliance then changes to a completely different value (presumably stiffer in some sense).

Both electric and hydraulic manipulators have complex compliance properties. In an electric manipulator the motors generally connect to the joints through a mechanical coupling. The sticking and sliding friction in such a coupling and in the motor itself can cause strange effects on the compliance measured at the tool tip. In particular, some of these couplings are not very back-drivable. For example, if one pushes on the nut of a lead screw (backdrive), the lead screw will not turn unless the screw's pitch is very coarse and ball bearings are used between the threads to reduce friction. But one can turn the screw easily, and the nut will move.

For manipulators that (1) operate open-loop in the sense that (2) they go blindly to a given point in space, (3) without any regard to the actual position in the environment, or (4) without regard to any reaction forces (feedback) that those objects exert on the arm (or tool)—then less compliance than that of the surrounding objects would be an advantage. High-frequency oscillations can be filtered out without degrading the overall response. Such filtering actually requires no special effort inasmuch as the combinations of servo valves and actuators commonly used have relatively low bandwidths.

Tactile sensors, which measure forces and moments exerted on the tool, can allow the manipulator to track or locate objects. Even in such cases, however, oscillations may arise in the force-feedback control loop if the compliance at the point of sensing is too low (stiff). Examination of a particular servo design is needed to predict reliably whether it will provide the kind of compliance needed for a specific task. There is no substitute for an actual test with the real tool on the manipulator.

END-EFFECTORS (GRIPPERS)

The device that is fastened to the free end of a manipulator is known as an end-effector or gripper. The usual function of the device is to grasp an object or a tool and then hold it while the manipulator moves, thereby moving the object, and finally releasing the object. Many end-effectors are widely used and available from stock. Custom-designed end-effectors, however, are not uncommon where standard designs do not satisfy specific application needs. Normally the end-effector is not included in the price of a robot per se. Hence their cost often can create substantial additional expense. Mechanical clamps (grippers) are commonly used. Other forms include vacuum-operated holders and both permanent and electromagnetic holders (Fig. 13).

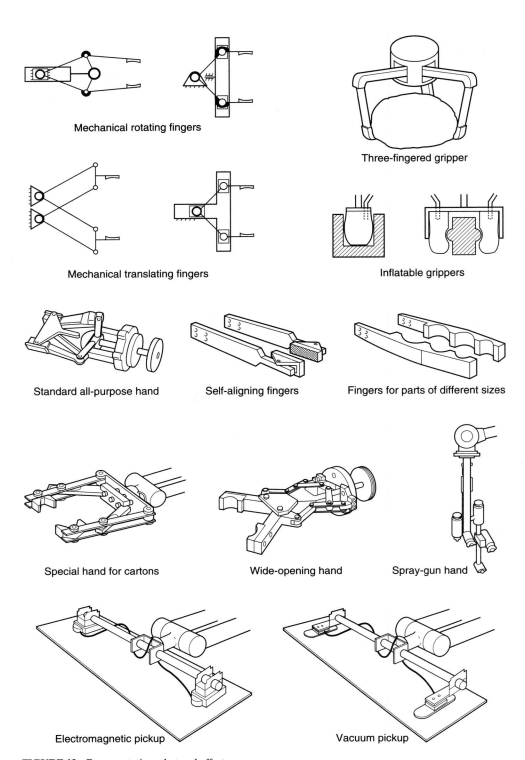

Mechanical rotating fingers

Three-fingered gripper

Mechanical translating fingers

Inflatable grippers

Standard all-purpose hand

Self-aligning fingers

Fingers for parts of different sizes

Special hand for cartons

Wide-opening hand

Spray-gun hand

Electromagnetic pickup

Vacuum pickup

FIGURE 13 Representative robot end-effectors.

WORKPLACE CONFIGURATIONS

There are four basic situations pertaining to the flow of work and the location of the robot:

1. Work may be arranged around the robot (Fig. 14).
2. Work is brought to the robot (Fig. 15).
3. Work may travel past the robot.
4. The robot travels to the work.

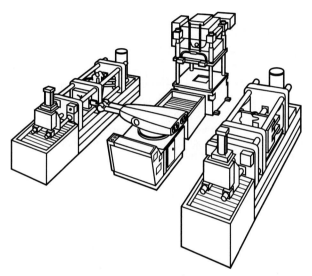

FIGURE 14 Die-casting installations to unload, quench, and dispose of part. In this installation, quite exemplary of earlier robot installations, the work is arranged around the robot.

Work Cells

A robotic work cell may be defined as a cluster of two or more robots and several machine tools or transfer lines that are interconnected in such a way that they work in unison. All of the necessary accessory equipment is embraced within the work cell and, together, establishes a particular work environment. The cell level is one step higher than the station level in the hierarchy of control and command. Keeping very close supervision over statistical quality control has become a paramount consideration in instrumenting the cell-level concept (Fig. 16).

ROBOT PROGRAMMING AND CONTROL

The two general classifications of robots from a control standpoint are (1) nonservo robots and (2) servo-controlled robots. For noncritical, simple applications, nonservo robots may suffice, particularly where low cost is a major consideration. Most of the early designs were in this category. However, the wide acceptance of the servo-controlled robot, as technological advancements made them possible, are evident from the inspection of Table 2.

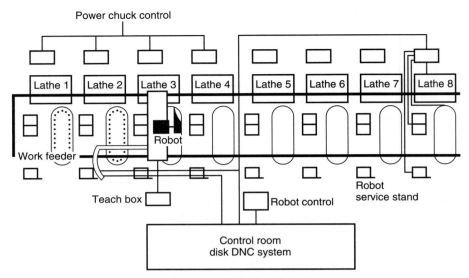

FIGURE 15   Overhead robot system, where the robot travels to the work. In this system an overhead robot system allows one robot to serve eight numerically controlled lathes.

Of historical interest, when robots and other automation techniques were largely associated with replicating the skills of human operators, the detailed steps and operations of the human operator were carefully studied and recorded. Initially this was the main source of robot programming information. Out of these studies the early "playback" concept was developed. In this method, the robot was "taught" by manually recording all of the movements that robot had to take to accomplish a given task. Obviously, at that time this method represented a "shortcut" because the path of the robot did not have to be measured or described in complex mathematical terms. Since those earlier years, of course, the techniques of mathematical simulation and appropriate motion algorithms have been developed. Much research along these lines has been conducted over the past decade or so by a combination effort made by robot designers and manufacturers, by large robot-using firms, and by academic institutions. For example, a pioneer in the field has been the Robotics Institute, Carnegie Mellon University, Pittsburgh, Pennsylvania, which initially developed a program known as VAST (versatile robot-arm dynamic simulation tool). As has become an accepted practice pertaining to instrumentation and control in the process industries, many robots have developed a high dependence on manufacturers and consultants for robot programming and software systems. In cases where applications from one user to the next may differ only in minor detail, packaged computer controls and software are now available and may be used with few, if any, alterations.

For example, large numbers of robot designs have been refined over several generations. Thus special controllers, visual operating displays, and software programs are available from robot suppliers. Standard applications that fit these criteria include those used in general materials handling, palletizing, arc welding, and, more recently, laser cutting, welding, etching, and surface-hardening applications.

Customized software for palletizing, for example, provides quick setup, easy modification of existing applications, and automatic calculation of all robot paths, eliminating the process of position teaching. In connection with robot welding applications, touch-sensing systems adaptively locate weld joints, and a through-arc seam-tracking system offers further enhancement by allowing the robot to compensate for weld-joint deviations and to correct the robot's path in real time.

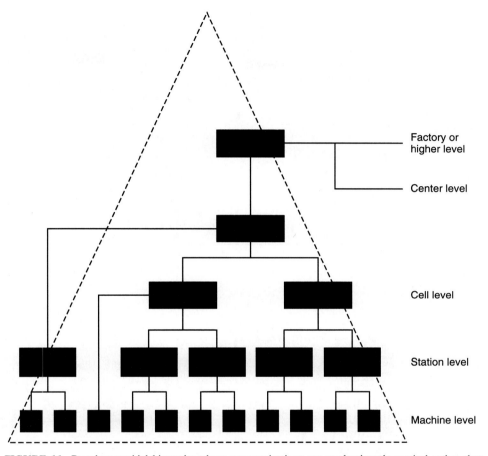

Factory or
higher level

Center level

Cell level

Station level

Machine level

FIGURE 16 Pseudopyramidal hierarchy where communications are predominantly vertical rather than horizontal.

For parts- and piece-dispensing operations, a software package minimizes the amount of code that is needed. For more complex operations, vision systems equip robots with advanced gray-scale vision capabilities. This is covered in more detail elsewhere in this handbook. Software also has been developed for painting applications and adds many teaching, editing, and programming capabilities. In these cases the robot can be controlled by a specially developed electronics package, or by using a teach pendant. A representative grouping of contemporary industrial robots is given in Fig. 17.

TABLE 2   Nonservo versus Servo-Controlled Robots

Characteristic	Nonservo robot	Servo-controlled robot
Flexibility	Limited in terms of program capacity and positioning capability. Arms can travel at only one speed and can stop only at end points of their axes.	Maximum flexibility provided by ability to program axes of manipulator to any position within limits of travel. Can vary speed at any point within envelope. Ability to move heavy loads in a controlled fashion.
Speed	Relatively high.	Relatively slow.
Repeatability	Approximately ±0.5 mm.	±0.1 to 0.5 mm and better, depending on design and application.
Cost	Comparatively low.	Comparatively high.
Complexity	Simple operation, programming, and maintenance.	Permits storage and execution of more than one program, with random selection of programs from memory via externally generated signals. Subrouting and branching capabilities may be available, permitting robot to take alternative actions within a program when commanded.

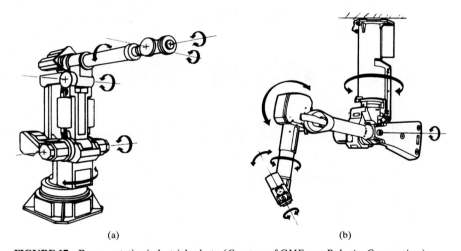

(a)   (b)

FIGURE 17   Representative industrial robots. (*Courtesy of GMFanuc Robotics Corporation.*)
(*a*) Spot welding, heavy part or tool handling, parts transfer, palletizing, material removal. Payload 120 kg (264 lb). Six axes of motion; floor- or wall-mounted; repeatability ±0.5 mm (0.02 inch); base rotation 300°; vertical travel 2731 mm (107.5 inches); reach 2413 mm (95 inches). (*S-420*)
(*b*) Arc welding of large parts on conveyors and fixtures. Payload 5 kg (11 lb). Six axes of motion; overhead-mounted; repeatability ±0.1 mm (0.004 inch); base rotation 300°; reach 1309 mm (51.5 inch). (*ArcMate OH*)

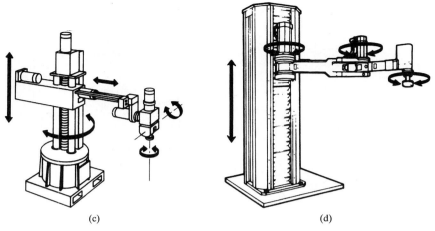

(c)   (d)

FIGURE 17 (*Continued*)

(*c*) Material handling, machine loading, palletizing, mechanical assembly in severe environments. Payload 50 kg (110 lb). Three to five axes of motion; floor-mounted; repeatability ±0.5 mm (0.02 inch); base rotation 300°; vertical travel 550 or 1300 mm (21.6 or 51.2 inches); horizontal travel 500 to 1100 mm (19.7 to 43.3 inches). (*M-100*)

(*d*) Palletizing and machine loading. Payload 50 kg (110 lb). Four to five axes of motion; floor-mounted; repeatability ±0.5 mm (0.02 inch); vertical travel 1850 mm (72.8 inches); radius reach 1930 mm (76 inches); access to two or more conveyors. (*M-400*)

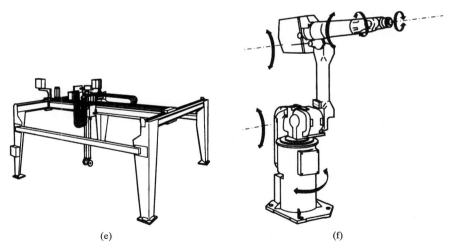

(e)   (f)

FIGURE 17 (*Continued*)

(*e*) Gantry robot for medium- to heavy-payload machine load and unload uses. Also palletizing, mechanical assembly, parts transfer. Cartesian coordinates. Area (shown) or linear configurations. Payload 50 kg (110 lb). Repeatability ±0.5 mm (0.02 inch); very large work envelope; two to four axes of motion for linear design; three to five for area design. (*G-500*)

(*f*) Multipurpose material handling; light-payload applications. Payload 10 kg (22 lb). Six axes of motion; floor-, ceiling-, or wall-mounted; repeatability 0.2 mm (0.008 inch); base rotation 300°; front reach 1529 mm (60.2 inches). (S-10)

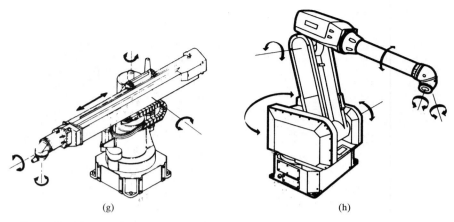

(g)   (h)

FIGURE 17   (*Continued*)

(*g*) Laser robot for integration with a laser generator. For precision-path laser processing—welding, cutting, heat treating, and cladding. Payload 5 kg (11 lb). Five axes of motion; floor-mounted; antibacklash drive; repeatability ±0.05 mm (0.002 inch); base rotation 200°; vertical travel 1968.5 mm (77.4 inches); horizontal travel 3964 mm (156 inches). Complete robotic laser cells available. (*L-100*)

(*h*) Industrial and automotive paint finishing of stationary or moving parts. Payload 7 kg (15.5 lb). Six or seven axes of motion; floor- or rail-mounted; repeatability ±0.5 mm (0.02 inch); maximum reach 2613 mm (103 inches), large work envelope. Robot also used for dispensing and applying antichip sealers and underbody deadeners. (P-155)

CURRENT-TO-PRESSURE TRANSDUCERS FOR CONTROL-VALVE ACTUATION

by Len Auer[1]

Current-to-pressure transducers (I/Ps) are used primarily in process control to change a 4- to 20-mA electronic signal from a computer controller into a 3- to 15-psi (21-103 kPa) pneumatic signal. The output signal from the I/P is then used to fill a diaphragm or piston actuator, which, in turn, modulates a valve. An effective I/P must provide air to the receiver quickly, accurately, and in sufficient quantity. The I/P device also must be able to exhaust air quickly when the signal decreases, consume a minimum amount of supply air for operation, and be easy to repair. In most industrial applications the I/P also must be sufficiently rugged to withstand difficult environmental conditions, including vibration, dirty supply air, temperature extremes, and corrosive conditions.

Since the mid-1960s I/Ps have used the traditional flapper-nozzle design concept with relatively few alterations. However, in the late 1980s several I/P manufacturers introduced new technologies in an attempt to overcome some of the difficulties encounted with the traditional design. These technologies vary in approach, and application success often depends on the integration of several design features within the same I/P.

---

[1] Product Marketing Manager, Rosemount Inc., Eden Prairie, Minnesota.

TRADITIONAL FLAPPER-NOZZLE DESIGN

The traditional flapper-nozzle design is shown in Fig. 1. The input current (4 to 20 mA) is applied to a coil-armature arrangement that acts on a beam. The beam ("flapper") positions itself against a nozzle that has air flowing through it. The gap between flapper and nozzle determines the back-pressure, also called pilot pressure, that builds up in the nozzle. Other variations of the flapper-nozzle concept are shown in Fig. 2. A bellows sometimes is connected to the nozzle area to balance the forces on the armature-flapper. The pilot pressure usually is channeled to a pneumatic booster or relay. This booster translates the low pilot pressure into a higher output pressure and capacity, which typically is 3 to 15 psi (20 to 100 kPa) and 4 ft^3/min (0.11 m^3/min), respectively.

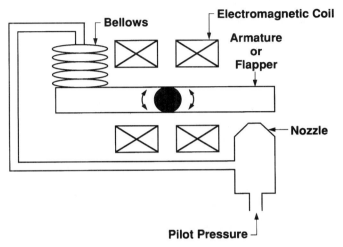

FIGURE 1   Traditional flapper-nozzle design. (*Rosemount Inc.*)

I/Ps that use the flapper-nozzle principle alone may have difficulty with some of the environmental factors faced in an application. Such I/Ps essentially are mechanical in nature and do not use electronic feedback sensors. The flapper is susceptible to vibration and traditionally has forced users to mount the I/P separately on a pipe or rack. This requires additional tubing to carry the I/P output signal to the valve. Output tubing installation costs offset any benefits from mounting the I/Ps together in a common location. The dead time and lag time introduced into the loop by longer output signal tubing can have a significant impact on loop performance. In addition to vibration, traditional I/Ps also can be adversely affected by fluctuations in air supply, downstream tubing leaks, temperature changes, and aging of the magnetic coil within the I/P. Periodic calibration checks are required in order to maintain the output of the I/P within the desired range.

Dirty supply air can be a major cause of I/P downtime. While mechanical I/Ps do not have electronic feedback to compensate for partial plugging, the nozzle opening traditionally has been designed to be at least 0.015 inch (0.4 mm) in diameter. This has reduced the likelihood of the nozzles plugging and is a strong point for those I/Ps that have maintained the larger nozzle diameters.

INTRODUCTION OF NEW I/P CONCEPTS

Since the late 1980s several new technologies have been introduced within the I/P. These new concepts have changed the nature of the pilot stage and incorporated sensor-based electron-

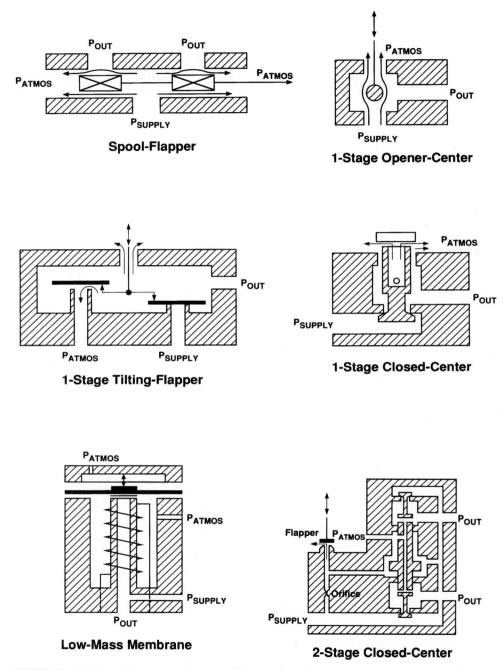

FIGURE 2   Variations of flapper-nozzle concept. (*Rosemount Inc.*)

ics. There are some inherent tradeoffs when combining different versions of these new technologies, and they have met with mixed results in the field. Two concepts are described here.

Piezoceramic Bender-Nozzle.   This device, another variation of the flapper concept, is shown in Fig. 3. The unit does not use the coil to move the flapper, but instead, the flapper itself is made of layers of different materials which are laminated together. These different materials flex or bend when a voltage is applied across them. The 4- to 20-mA input signal to the I/P must be converted to a voltage in the range of 20 to 30 volts dc.

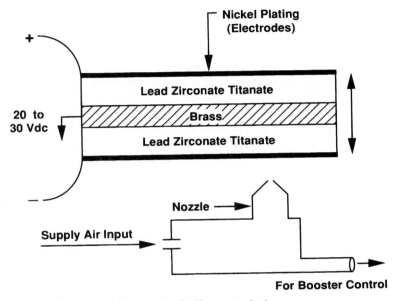

FIGURE 3   Piezoceramic bender-nozzle. (*Rosemount Inc.*)

This design tends to be more stable in vibration than the typical flapper armature, particularly when combined with an electronic feedback loop. Several drawbacks also have become evident from field applications. The bender does not have a very good "memory" and will tend to locate in a different position for the same input signal. This creep can be cumulative and eventually will exceed the adjustment range of the calibration mechanism. An electronic feedback sensor can be combined with the piezoceramic bender to compensate for the creep temporarily, but the feedback circuit typically uses much of the power available from the input signal. This leaves little power to energize the bender. The bender cannot balance against the force of the nozzle air, unless the nozzle is kept relatively small. Thus larger nozzles must be traded for improved bender control. Plugging of small nozzles or orifices typically is the leading cause of I/P field failure.

Deflector Bar Design.   Shown in Fig. 4, this pilot stage concept also was introduced in the late 1980s. This design uses an electromagnetic coil similar to the traditional flappers. The deflector bar, however, replaces the flapper as the main moving part in the assembly. The flapper no longer is used to block the airflow coming out of the nozzle. The deflector bar design is based on the Coanda effect, which may be defined as the tendency of an airstream to attach itself to a surface with which it makes oblique contact.

The actual hardware consists of two opposed 0.015-inch (0.4-mm)-diameter nozzles, fixed on the same centerline, spaced about 0.15 inch (0.4 mm) apart. One nozzle provides a high-velocity airstream from the air supply, the second nozzle recaptures the airstream and converts its kinetic energy to a pressure (potential energy).

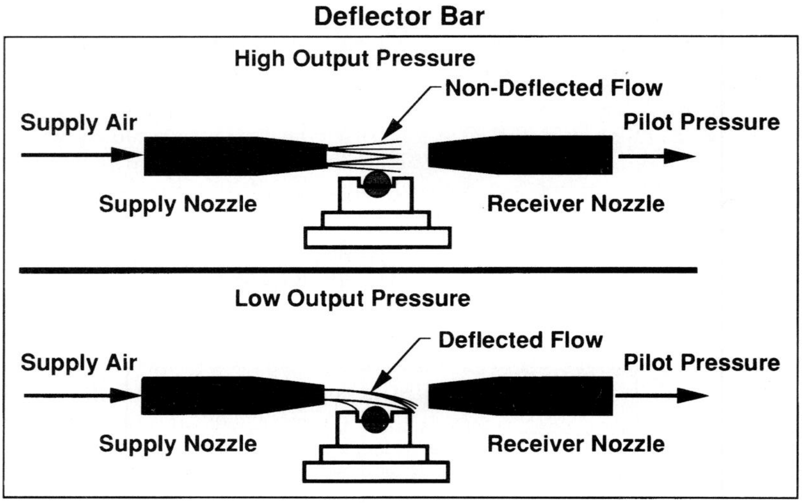

Deflector Bar

FIGURE 4   Deflector bar design. (*Rosemount Inc.*)

To vary the pilot stage output pressure, a 0.019-inch (0.5-mm)-diameter solid deflector bar, which is positioned crosswise and midway between the nozzles, is caused to move into the airstream. The stream attaches to the surface of the bar and follows its curvature for some distance before separating. This deflection of the airstream away from the receiver nozzle results in a decrease in the pilot output pressure. To increase the output pressure again, the bar is simply pulled out of the airstream.

The deflector bar is low mass, which adds vibration stability, and only travels about 0.003 inch (0.08 mm) to produce a full-scale output change. This small travel requirement, coupled with the fact that the bar movement does not directly oppose the airflow, allows for a low-power magnetic actuator coil. The nozzle diameter is not limited by force-balance versus power tradeoffs, as is the case with flapper designs. The nozzle diameter is relatively large at 0.016 inch (0.15 mm), and is only limited by the desirable range for air consumption.

When combined with an electronic feedback sensor, this type of pilot stage is virtually unaffected by vibration and provides quick response to input changes. Lag time, the rate of change to reach a new output pressure, actually can be reduced by the quick response, as compared with other pilot stage designs.

ELECTRONIC FEEDBACK

Several I/Ps introduced since the late 1980s have incorporated pressure sensors and electronic feedback control. This feedback, shown in Fig. 5, detects the actual output pressure and is completely internal to the I/P. There are several types of sensors used, the most common of which is a solid-state silicon strain-gauge type. The electronics in the I/P contain an error-correction circuit that continuously compares the output sensor reading with the input signal. The electronics then adjust the current to the pilot stage to make any needed corrections to the output of the I/P.

Electronic pressure sensor feedback is the foundation on which the installed advantages of recent I/Ps are based. These advantages include vibration immunity, calibration stability, repeatability, quick dynamic response, and reduced downtime. Performance of the electronic

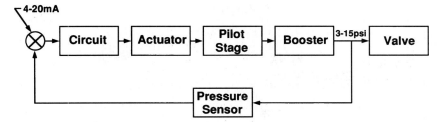

FIGURE 5   Electronic feedback control as applied to I/P device. (*Rosemount Inc.*)

feedback circuits is relatively consistent across the I/Ps available as of the early 1990s. However, many of the advantages just given are contingent on integration of the electronic feedback with the optimum pilot stage design. If tradeoffs are made in order to incorporate electronics into the I/P, reliability can be adversely affected. In addition, dynamic response may vary considerably, as shown in Fig. 6. In particular this can be evident when the same I/P design is used to fill a wide range of output end volumes. Proper balance of damping and responsiveness is critical to the operation of the I/P into the full range of typical output volumes. The importance of the role played by the I/P in the performance of a loop cannot be overemphasized.

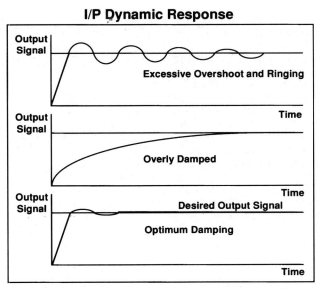

FIGURE 6   Possible variations in dynamic response of I/P devices. (*Rosemount Inc.*)

INDEX

ABOUT THE EDITOR IN CHIEF

Douglas M. Considine, P.E., is an engineering consultant to numerous corporations and a prolific editor and writer of technical books and encyclopedias. Previously, he was director of advanced control engineering at Hughes Aircraft Co. Among his many books are *Handbook of Applied Instrumentation* (1964), *Encyclopedia of Instrumentation and Control* (1971), *Energy Technology Handbook* (1977), *Van Nostrand's Scientific Encyclopedia* (8th Edition, 1993), and *Encyclopedia of Chemistry* (5th Edition, 1993).

Mr. Considine is a Fellow of both the Instrument Society of America and the American Association for the Advancement of Science, as well as a senior member of the American Institute of Chemical Engineers.